BIG IDEAS MATH®
Modeling Real Life
Grade 6 Advanced

Ron Larson
Laurie Boswell

Erie, Pennsylvania
BigIdeasLearning.com

Big Ideas Learning, LLC
1762 Norcross Road
Erie, PA 16510-3838
USA

For product information and customer support, contact Big Ideas Learning at 1-877-552-7766 or visit us at BigIdeasLearning.com.

Cover Image
Valdis Torms, Bogahan/iStock/Getty Images Plus

Copyright © 2019 by Big Ideas Learning, LLC. All rights reserved.

No part of this work may be reproduced or transmitted in any form or by any means, electronic or mechanical, including, but not limited to, photocopying and recording, or by any information storage or retrieval system, without prior written permission of Big Ideas Learning, LLC, unless such copying is expressly permitted by copyright law. Address inquiries to Permissions, Big Ideas Learning, LLC, 1762 Norcross Road, Erie, PA 16510.

Big Ideas Learning and Big Ideas Math are registered trademarks of Larson Texts, Inc.

Printed in the U.S.A.

ISBN 13: 978-1-64245-063-7

2 3 4 5 6 7 8 9 10—22 21 20 19 18

About the Authors

Ron Larson

Ron Larson, Ph.D., is well known as the lead author of a comprehensive program for mathematics that spans school mathematics and college courses. He holds the distinction of Professor Emeritus from Penn State Erie, The Behrend College, where he taught for nearly 40 years. He received his Ph.D. in mathematics from the University of Colorado. Dr. Larson's numerous professional activities keep him actively involved in the mathematics education community and allow him to fully understand the needs of students, teachers, supervisors, and administrators.

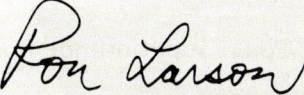

Laurie Boswell

Laurie Boswell, Ed.D., is the former Head of School at Riverside School in Lyndonville, Vermont. In addition to textbook authoring, she provides mathematics consulting and embedded coaching sessions. Dr. Boswell received her Ed.D. from the University of Vermont in 2010. She is a recipient of the Presidential Award for Excellence in Mathematics Teaching and is a Tandy Technology Scholar. Laurie has taught math to students at all levels, elementary through college. In addition, Laurie has served on the NCTM Board of Directors and as a Regional Director for NCSM. Along with Ron, Laurie has co-authored numerous math programs and has become a popular national speaker.

Dr. Ron Larson and Dr. Laurie Boswell began writing together in 1992. Since that time, they have authored over four dozen textbooks. This successful collaboration allows for one voice from Kindergarten through Algebra 2.

Contributors, Reviewers,

Big Ideas Learning would like to express our gratitude to the mathematics education and instruction experts who served as our advisory panel, contributing specialists, and reviewers during the writing of *Big Ideas Math: Modeling Real Life*. Their input was an invaluable asset during the development of this program.

Contributing Specialists and Reviewers

- **Sophie Murphy**, Ph.D. Candidate, Melbourne School of Education, Melbourne, Australia
  Learning Targets and Success Criteria Specialist and Visible Learning Reviewer

- **Linda Hall**, Mathematics Educational Consultant, Edmond, OK
  Advisory Panel and Teaching Edition Contributor

- **Michael McDowell**, Ed.D., Superintendent, Ross, CA
  Project-Based Learning Specialist

- **Kelly Byrne**, Math Supervisor and Coordinator of Data Analysis, Downingtown, PA
  Advisory Panel and Content Reviewer

- **Jean Carwin**, Math Specialist/TOSA, Snohomish, WA
  Advisory Panel and Content Reviewer

- **Nancy Siddens**, Independent Language Teaching Consultant, Las Cruces, NM
  English Language Learner Specialist

- **Nancy Thiele**, Mathematics Consultant, Mesa, AZ
  Teaching Edition Contributor

- **Kristen Karbon**, Curriculum and Assessment Coordinator, Troy, MI
  Advisory Panel and Content Reviewer

- **Kery Obradovich**, K–8 Math/Science Coordinator, Northbrook, IL
  Advisory Panel and Content Reviewer

- **Jennifer Rollins**, Math Curriculum Content Specialist, Golden, CO
  Advisory Panel

- **Becky Walker**, Ph.D., School Improvement Services Director, Green Bay, WI
  Advisory Panel

- **Anthony Smith**, Ph.D., Associate Professor, Associate Dean, University of Washington Bothell, Seattle, WA
  Reading/Writing Reviewer

- **Nicole Dimich Vagle**, Educator, Author, and Consultant, Hopkins, MN
  Assessment Reviewer

- **Jill Kalb**, Secondary Math Content Specialist, Arvada, CO
  Content Reviewer

- **Janet Graham**, District Math Specialist, Manassas, VA
  Response to Intervention and Differentiated Instruction Reviewer

- **Sharon Huber**, Director of Elementary Mathematics, Chesapeake, VA
  Universal Design for Learning Reviewer

Student Reviewers

- Jackson Currier
- Mason Currier
- Taylor DeLuca
- Ajalae Evans
- Malik Goodwine
- Majesty Hamilton
- Reilly Koch
- Kyla Kramer
- Matthew Lindemuth
- Greer Lippert
- Zane Lippert
- Jeffrey Lobaugh
- Riley Moran
- Zoe Morin
- Deke Patton
- Brooke Smith
- Dylan Throop
- Jenna Urso
- Madison Whitford
- Jenna Wigham

and Research

Research

Ron Larson and Laurie Boswell used the latest in educational research, along with the body of knowledge collected from expert mathematics instructors, to develop the *Modeling Real Life* series. The pedagogical approach used in this program follows the best practices outlined in the most prominent and widely accepted educational research, including:

- *Visible Learning*
  John Hattie © 2009

- *Visible Learning for Teachers*
  John Hattie © 2012

- *Visible Learning for Mathematics*
  John Hattie © 2017

- *Principles to Actions: Ensuring Mathematical Success for All*
  NCTM © 2014

- *Adding It Up: Helping Children Learn Mathematics*
  National Research Council © 2001

- *Mathematical Mindsets: Unleashing Students' Potential through Creative Math, Inspiring Messages and Innovative Teaching*
  Jo Boaler © 2015

- *What Works in Schools: Translating Research into Action*
  Robert Marzano © 2003

- *Classroom Instruction That Works: Research-Based Strategies for Increasing Student Achievement*
  Marzano, Pickering, and Pollock © 2001

- *Principles and Standards for School Mathematics*
  NCTM © 2000

- *Rigorous PBL by Design: Three Shifts for Developing Confident and Competent Learners*
  Michael McDowell © 2017

- *Universal Design for Learning Guidelines*
  CAST © 2011

- *Rigor/Relevance Framework®*
  International Center for Leadership in Education

- *Understanding by Design*
  Grant Wiggins and Jay McTighe © 2005

- Achieve, ACT, and The College Board

- *Elementary and Middle School Mathematics: Teaching Developmentally*
  John A. Van de Walle and Karen S. Karp © 2015

- *Evaluating the Quality of Learning: The SOLO Taxonomy*
  John B. Biggs & Kevin F. Collis © 1982

- *Unlocking Formative Assessment: Practical Strategies for Enhancing Students' Learning in the Primary and Intermediate Classroom*
  Shirley Clarke, Helen Timperley, and John Hattie © 2004

- *Formative Assessment in the Secondary Classroom*
  Shirley Clarke © 2005

- *Improving Student Achievement: A Practical Guide to Assessment for Learning*
  Toni Glasson © 2009

Mathematical Processes and Proficiencies

Big Ideas Math: Modeling Real Life reinforces the Process Standards from NCTM and the Five Strands of Mathematical Proficiency endorsed by the National Research Council. With *Big Ideas Math*, students get the practice they need to become well-rounded, mathematically proficient learners.

Problem Solving/Strategic Competence
- *Modeling Real Life Examples* use problem-solving strategies, such as drawing a diagram, making a table, and solving a simpler problem. They also use a formal problem-solving plan: understand the problem, make a plan, and solve and check.
- Real-life problems are provided to help students learn to apply the mathematics that they are learning to everyday life.
- Real-life problems help students use the structure of mathematics to break down and solve more difficult problems.

Reasoning and Proof/Adaptive Reasoning
- *Explorations* allow students to investigate math and make conjectures.
- Questions ask students to explain and justify their reasoning.
- Questions encourage students to formulate consistent and appropriate reasoning.

Communication
- Cooperative learning opportunities support precise communication.
- Exercises, such as *You Be The Teacher*; *Different Words, Same Question*; and *Which One Doesn't Belong?*, provide students the opportunity to critique the reasoning of others.
- *Self-Assessment for Concepts & Skills* and *Self-Assessment for Problem Solving* features allow students to demonstrate their understanding of the lesson up to that point.
- *ELL Support* notes provide insights into how to support English learners.

Connections
- Prior knowledge is continually brought back and tied in with current learning.
- Real-life problems incorporate other disciplines to help students see that math is used across content areas.

Representations/Productive Disposition
- Real-life problems are translated into diagrams, tables, equations, and graphs to help students analyze relations and to draw conclusions.
- Visual problem-solving models help students create a coherent representation of the problem.
- Multiple representations are presented to help students move from concrete to representative and into abstract thinking.
- *Learning Targets* and *Success Criteria* at the start of each chapter and section help students understand what they are going to learn.

Conceptual Understanding
- *Explorations* allow students to investigate math to understand the reasoning behind the rules.
- *Self-Assessment for Concepts & Skills* and *Self-Assessment for Problem Solving* features allow students to demonstrate their understanding of the lesson up to that point.

Procedural Fluency
- *Key Ideas* present procedures needed to solve mathematical problems. Skill exercises are provided to continually practice these procedures.
- Prior knowledge is continually brought back and tied in with current learning.

Meeting Proficiency and Major Topics

Meeting Proficiency

As standards shift to prepare students for college and careers, the importance of focus, coherence, and rigor continues to grow.

FOCUS   *Big Ideas Math: Modeling Real Life* emphasizes a narrower and deeper curriculum, ensuring students spend their time on the major topics of each grade.

COHERENCE   The program was developed around coherent progressions from Kindergarten through eighth grade, guaranteeing students develop and progress their foundational skills through the grades while maintaining a strong focus on the major topics.

RIGOR   *Big Ideas Math: Modeling Real Life* uses a balance of procedural fluency, conceptual understanding, and real-life applications. Students develop conceptual understanding in every *Exploration*, continue that development in the *Lessons* while gaining procedural fluency during the *Concepts and Skills Examples*, and then tie it all together with *Modeling Real Life Examples*. Every set of *Exercises* reflects this balance, giving students the rigorous practice they need to be college- and career-ready.

Major Topics in Grade 6 Advanced

Ratios and Proportional Relationships
- Understand ratio concepts and use ratio reasoning to solve problems.
- Analyze proportional relationships and use them to solve real-world and mathematical problems.

The Number System
- Apply and extend previous understandings of multiplication and division to divide fractions by fractions.
- Apply and extend previous understandings of numbers to the system of rational numbers.
- Apply and extend previous understandings of operations with fractions to add, subtract, multiply, and divide rational numbers.

Expressions and Equations
- Apply and extend previous understandings of arithmetic to algebraic expressions.
- Reason about and solve one-variable equations and inequalities.
- Represent and analyze quantitative relationships between dependent and independent variables.
- Use properties of operations to generate equivalent expressions.

Use the color-coded Table of Contents to determine where the major topics, supporting topics, and additional topics occur throughout the curriculum.

- 🟩 Major Topic
- 🟦 Supporting Topic
- 🟨 Additional Topic

1 Numerical Expressions and Factors

	STEAM Video/Performance Task	1
	Getting Ready for Chapter 1	2
Section 1.1	**Powers and Exponents**	
	Exploration	3
	Lesson	4
Section 1.2	**Order of Operations**	
	Exploration	9
	Lesson	10
Section 1.3	**Prime Factorization**	
	Exploration	15
	Lesson	16
Section 1.4	**Greatest Common Factor**	
	Exploration	21
	Lesson	22
Section 1.5	**Least Common Multiple**	
	Exploration	27
	Lesson	28
	Connecting Concepts	33
	Chapter Review	34
	Practice Test	38
	Cumulative Practice	39

■ Major Topic
■ Supporting Topic
■ Additional Topic

Fractions and Decimals

	STEAM Video/Performance Task	43
	Getting Ready for Chapter 2	44
Section 2.1	**Multiplying Fractions**	
	Exploration	45
	Lesson	46
Section 2.2	**Dividing Fractions**	
	Exploration	53
	Lesson	54
Section 2.3	**Dividing Mixed Numbers**	
	Exploration	61
	Lesson	62
Section 2.4	**Adding and Subtracting Decimals**	
	Exploration	67
	Lesson	68
Section 2.5	**Multiplying Decimals**	
	Exploration	73
	Lesson	74
Section 2.6	**Dividing Whole Numbers**	
	Exploration	81
	Lesson	82
Section 2.7	**Dividing Decimals**	
	Exploration	87
	Lesson	88
	Connecting Concepts	95
	Chapter Review	96
	Practice Test	100
	Cumulative Practice	101

3 Ratios and Rates

	STEAM Video/Performance Task	105
	Getting Ready for Chapter 3	106
Section 3.1	**Ratios**	
	Exploration	107
	Lesson	108
Section 3.2	**Using Tape Diagrams**	
	Exploration	115
	Lesson	116
Section 3.3	**Using Ratio Tables**	
	Exploration	121
	Lesson	122
Section 3.4	**Graphing Ratio Relationships**	
	Exploration	129
	Lesson	130
Section 3.5	**Rates and Unit Rates**	
	Exploration	135
	Lesson	136
Section 3.6	**Converting Measures**	
	Exploration	141
	Lesson	142
	Connecting Concepts	149
	Chapter Review	150
	Practice Test	156
	Cumulative Practice	157

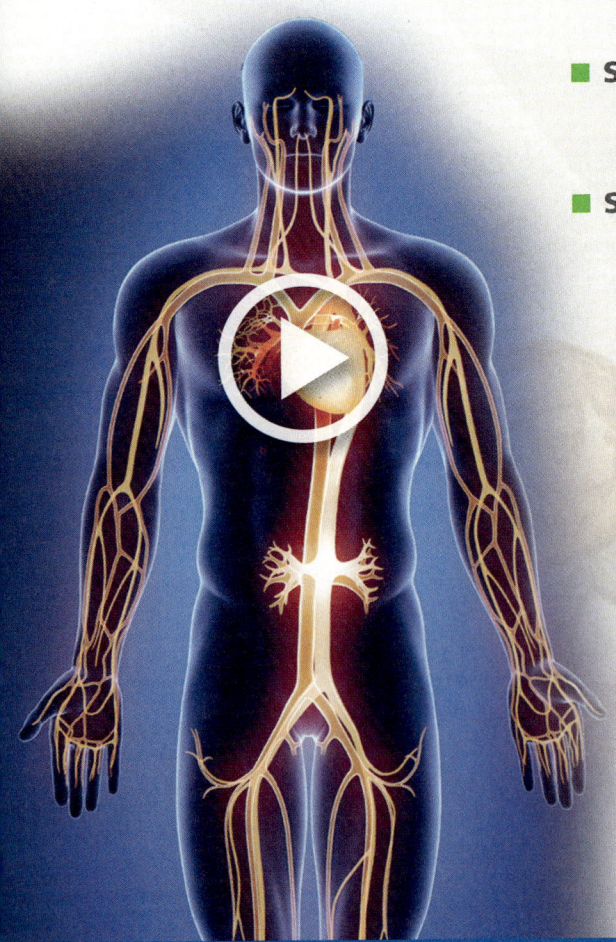

Percents

	STEAM Video/Performance Task	161
	Getting Ready for Chapter 4	162
Section 4.1	**Percents and Fractions**	
	Exploration	163
	Lesson	164
Section 4.2	**Percents and Decimals**	
	Exploration	169
	Lesson	170
Section 4.3	**Comparing and Ordering Fractions, Decimals, and Percents**	
	Exploration	175
	Lesson	176
Section 4.4	**Solving Percent Problems**	
	Exploration	181
	Lesson	182
	Connecting Concepts	189
	Chapter Review	190
	Practice Test	194
	Cumulative Practice	195

5 Algebraic Expressions and Properties

	STEAM Video/Performance Task	199
	Getting Ready for Chapter 5	200
Section 5.1	**Algebraic Expressions**	
	Exploration	201
	Lesson	202
Section 5.2	**Writing Expressions**	
	Exploration	209
	Lesson	210
Section 5.3	**Properties of Addition and Multiplication**	
	Exploration	215
	Lesson	216
Section 5.4	**The Distributive Property**	
	Exploration	221
	Lesson	222
Section 5.5	**Factoring Expressions**	
	Exploration	227
	Lesson	228
	Connecting Concepts	233
	Chapter Review	234
	Practice Test	238
	Cumulative Practice	239

Equations

	STEAM Video/Performance Task	243
	Getting Ready for Chapter 6	244
■ Section 6.1	**Writing Equations in One Variable**	
	Exploration	245
	Lesson	246
■ Section 6.2	**Solving Equations Using Addition or Subtraction**	
	Exploration	251
	Lesson	252
■ Section 6.3	**Solving Equations Using Multiplication or Division**	
	Exploration	259
	Lesson	260
■ Section 6.4	**Writing Equations in Two Variables**	
	Exploration	265
	Lesson	266
	Connecting Concepts	273
	Chapter Review	274
	Practice Test	278
	Cumulative Practice	279

7 Area, Surface Area, and Volume

STEAM Video/Performance Task 283
Getting Ready for Chapter 7 284

- **Section 7.1** **Areas of Parallelograms**
  - Exploration 285
  - Lesson 286
- **Section 7.2** **Areas of Triangles**
  - Exploration 291
  - Lesson 292
- **Section 7.3** **Areas of Trapezoids and Kites**
  - Exploration 297
  - Lesson 298
- **Section 7.4** **Three-Dimensional Figures**
  - Exploration 305
  - Lesson 306
- **Section 7.5** **Surface Areas of Prisms**
  - Exploration 311
  - Lesson 312
- **Section 7.6** **Surface Areas of Pyramids**
  - Exploration 319
  - Lesson 320
- **Section 7.7** **Volumes of Rectangular Prisms**
  - Exploration 325
  - Lesson 326

Connecting Concepts 331
Chapter Review 332
Practice Test 338
Cumulative Practice 339

Integers, Number Lines, and the Coordinate Plane

	STEAM Video/Performance Task	343
	Getting Ready for Chapter 8	344
■ Section 8.1	**Integers**	
	Exploration	345
	Lesson	346
■ Section 8.2	**Comparing and Ordering Integers**	
	Exploration	351
	Lesson	352
■ Section 8.3	**Rational Numbers**	
	Exploration	357
	Lesson	358
■ Section 8.4	**Absolute Value**	
	Exploration	363
	Lesson	364
■ Section 8.5	**The Coordinate Plane**	
	Exploration	369
	Lesson	370
■ Section 8.6	**Polygons in the Coordinate Plane**	
	Exploration	377
	Lesson	378
■ Section 8.7	**Writing and Graphing Inequalities**	
	Exploration	383
	Lesson	384
■ Section 8.8	**Solving Inequalities**	
	Exploration	391
	Lesson	392
	Connecting Concepts	399
	Chapter Review	400
	Practice Test	406
	Cumulative Practice	407

9 Statistical Measures

	STEAM Video/Performance Task	411
	Getting Ready for Chapter 9	412
Section 9.1	**Introduction to Statistics**	
	Exploration	413
	Lesson	414
Section 9.2	**Mean**	
	Exploration	419
	Lesson	420
Section 9.3	**Measures of Center**	
	Exploration	425
	Lesson	426
Section 9.4	**Measures of Variation**	
	Exploration	433
	Lesson	434
Section 9.5	**Mean Absolute Deviation**	
	Exploration	439
	Lesson	440
	Connecting Concepts	445
	Chapter Review	446
	Practice Test	450
	Cumulative Practice	451

Data Displays

	STEAM Video/Performance Task	455
	Getting Ready for Chapter 10	456
Section 10.1	**Stem-and-Leaf Plots**	
	Exploration	457
	Lesson	458
Section 10.2	**Histograms**	
	Exploration	463
	Lesson	464
Section 10.3	**Shapes of Distributions**	
	Exploration	471
	Lesson	472
Section 10.4	**Choosing Appropriate Measures**	
	Exploration	477
	Lesson	478
Section 10.5	**Box-and-Whisker Plots**	
	Exploration	483
	Lesson	484
	Connecting Concepts	491
	Chapter Review	492
	Practice Test	496
	Cumulative Practice	497

xvii

Adding and Subtracting Rational Numbers

	STEAM Video/Performance Task	501
	Getting Ready for Chapter A	502
■ Section A.1	**Rational Numbers**	
	Exploration	503
	Lesson	504
■ Section A.2	**Adding Integers**	
	Exploration	509
	Lesson	510
■ Section A.3	**Adding Rational Numbers**	
	Exploration	517
	Lesson	518
■ Section A.4	**Subtracting Integers**	
	Exploration	523
	Lesson	524
■ Section A.5	**Subtracting Rational Numbers**	
	Exploration	529
	Lesson	530
	Connecting Concepts	537
	Chapter Review	538
	Practice Test	542
	Cumulative Practice	543

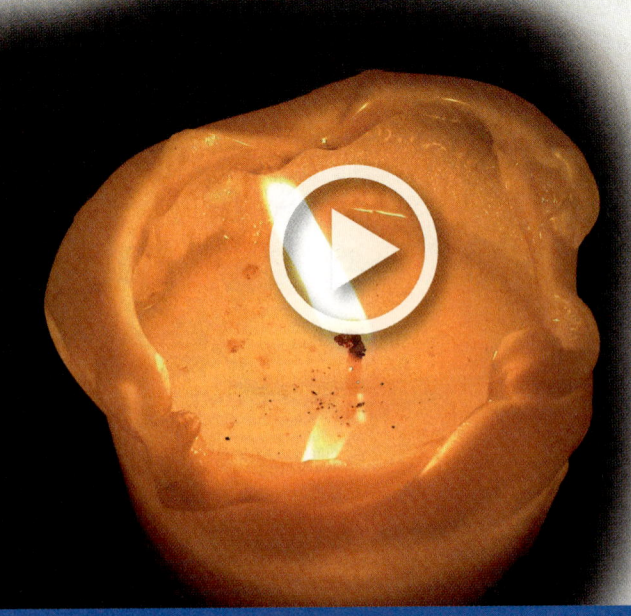

Multiplying and Dividing Rational Numbers

	STEAM Video/Performance Task	547
	Getting Ready for Chapter B	548
■ Section B.1	**Multiplying Integers**	
	Exploration	549
	Lesson	550
■ Section B.2	**Dividing Integers**	
	Exploration	555
	Lesson	556
■ Section B.3	**Converting Between Fractions and Decimals**	
	Exploration	561
	Lesson	562
■ Section B.4	**Multiplying Rational Numbers**	
	Exploration	567
	Lesson	568
■ Section B.5	**Dividing Rational Numbers**	
	Exploration	573
	Lesson	574
	Connecting Concepts	579
	Chapter Review	580
	Practice Test	584
	Cumulative Practice	585

xix

C Expressions

	STEAM Video/Performance Task	589
	Getting Ready for Chapter C	590
Section C.1	**Algebraic Expressions**	
	Exploration	591
	Lesson	592
Section C.2	**Adding and Subtracting Linear Expressions**	
	Exploration	597
	Lesson	598
Section C.3	**The Distributive Property**	
	Exploration	603
	Lesson	604
Section C.4	**Factoring Expressions**	
	Exploration	609
	Lesson	610
	Connecting Concepts	615
	Chapter Review	616
	Practice Test	620
	Cumulative Practice	621

Ratios and Proportions

	STEAM Video/Performance Task	625
	Getting Ready for Chapter D	626
■ Section D.1	**Ratios and Ratio Tables**	
	Exploration	627
	Lesson	628
■ Section D.2	**Rates and Unit Rates**	
	Exploration	633
	Lesson	634
■ Section D.3	**Identifying Proportional Relationships**	
	Exploration	639
	Lesson	640
■ Section D.4	**Writing and Solving Proportions**	
	Exploration	647
	Lesson	648
■ Section D.5	**Graphs of Proportional Relationships**	
	Exploration	655
	Lesson	656
■ Section D.6	**Scale Drawings**	
	Exploration	661
	Lesson	662
	Connecting Concepts	667
	Chapter Review	668
	Practice Test	672
	Cumulative Practice	673

E Percents

	STEAM Video/Performance Task	677
	Getting Ready for Chapter E	678
■ Section E.1	**Fractions, Decimals, and Percents**	
	Exploration	679
	Lesson	680
■ Section E.2	**The Percent Proportion**	
	Exploration	685
	Lesson	686
■ Section E.3	**The Percent Equation**	
	Exploration	691
	Lesson	692
■ Section E.4	**Percents of Increase and Decrease**	
	Exploration	697
	Lesson	698
■ Section E.5	**Discounts and Markups**	
	Exploration	703
	Lesson	704
■ Section E.6	**Simple Interest**	
	Exploration	709
	Lesson	710
	Connecting Concepts	715
	Chapter Review	716
	Practice Test	720
	Cumulative Practice	721

Selected Answers A1
English-Spanish Glossary A37
Index A47
Mathematics Reference Sheet B1

How to Use Your Math Book

▶ Get ready for the chapter by watching the **STEAM Video**, completing the **Chapter Exploration**, and recording your thoughts on **Vocabulary**.

▶ Read the **Learning Target** and **Success Criteria** for each section. Work with a partner to complete the **EXPLORATIONS**. Discuss the **Math Practice** question with your partner.

▶ Find the **Key Vocabulary** words highlighted in yellow. Read their definitions. Study the concepts in each **Key Idea**. If you forget a definition, you can look it up online in the **Multi-Language Glossary** at *BigIdeasMath.com*.

▶ During the **Lessons**, study each **EXAMPLE** and then complete the *Try It* exercises. Pay special attention to the push-pin notes and other helpful tips, such as **Common Errors**, **Remember**, and **Reading**.

▶ Use the **Self-Assessment** *for Concepts & Skills* and **Self-Assessment** *for Problem Solving* to assess your understanding of the success criteria from the lesson.

▶ The **Practice** is broken into two parts: *Review & Refresh* and *Concepts, Skills, & Problem Solving*. The *Review & Refresh* reviews prior skills and prepares you for the next lesson. The *Concepts, Skills, & Problem Solving* exercises are for the current lesson and are color-coded red for concepts and skills and blue for real-life problem solving.

▶ Use your *Problem-Solving Strategies* to complete the **Connecting Concepts** lesson, where you will practice previously learned skills with current concepts.

▶ Use the **Chapter Review** to study for your test, where you will *Review Vocabulary*, use *Graphic Organizers*, and complete the *Chapter Self-Assessment*. Remember to assess your understanding of each learning target! You can also take a **Practice Test** on the concepts from the chapter.

▶ Use the **Cumulative Practice** to prepare for high-stakes tests, where you will complete standardized test questions from throughout the course.

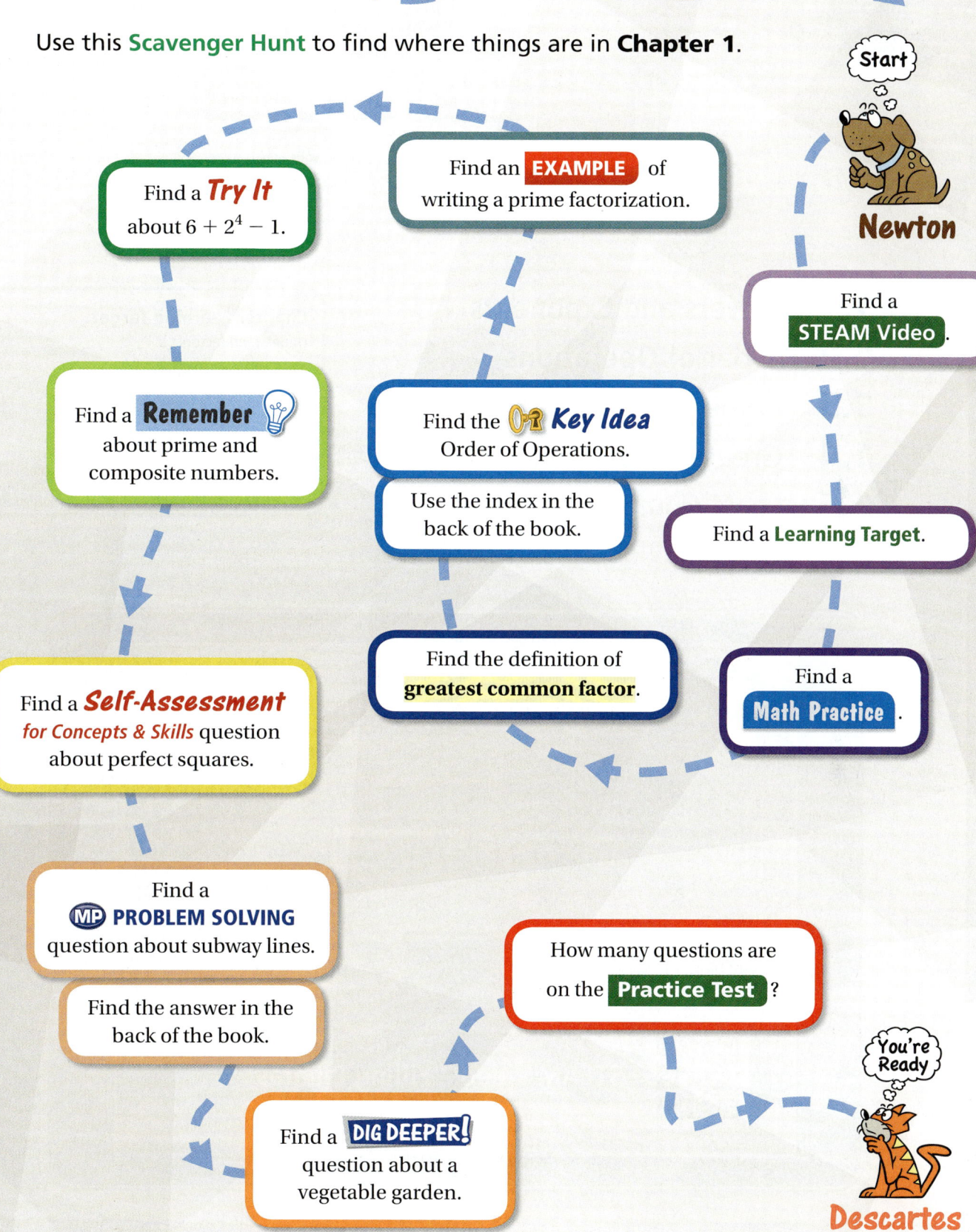

1 Numerical Expressions and Factors

- **1.1** Powers and Exponents
- **1.2** Order of Operations
- **1.3** Prime Factorization
- **1.4** Greatest Common Factor
- **1.5** Least Common Multiple

Chapter Learning Target:
Understand factors.

Chapter Success Criteria:
- I can identify factors of a number.
- I can explain order of operations.
- I can solve a problem using factors.
- I can model different types of multiples of numbers.

STEAM Video: "Filling Piñatas"

STEAM Video

Filling Piñatas

Common factors can be used to make identical groups of objects. Can you think of any situations in which you would want to separate objects into equal groups? Are there any common factors that may be more useful than others? Can you think of any other ways to use common factors?

Watch the STEAM Video "Filling Piñatas." Then answer the following questions. The table below shows the numbers of party favors that Alex and Enid use to make piñatas.

Party Favor	Taffies	Key Chains	Kazoos	Bubbles	Mints
Number	50	12	16	24	100

1. When finding the number of identical piñatas that can be made, why is it helpful for Alex and Enid to list the factors of each number given in the table?

2. You want to create 6 identical piñatas. How can you change the numbers of party favors in the table to make this happen? Can you do this without changing the total number of party favors?

Performance Task

Setting the Table

After completing this chapter, you will be able to use the concepts you learned to answer the questions in the *STEAM Video Performance Task*. You will be asked to plan a fundraising event with the items below.

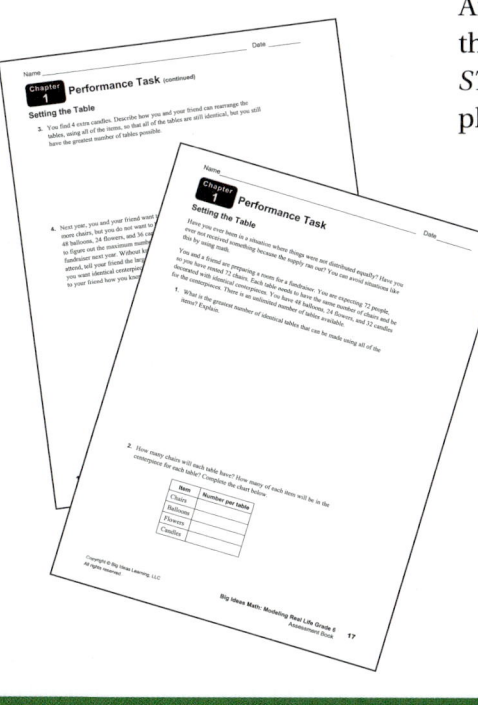

72 chairs

48 balloons

24 flowers

32 candles

You will find the greatest number of identical tables that can be prepared, and what will be in each centerpiece. When making arrangements for a party, should a party planner always use the greatest number of identical tables possible? Explain why or why not.

Getting Ready for Chapter 1

Chapter Exploration

Work with a partner. In Exercises 1 and 2, use the table.

	2	3	4	5	6	7	8	9	10
11	12	13	14	15	16	17	18	19	20
21	22	23	24	25	26	27	28	29	30
31	32	33	34	35	36	37	38	39	40
41	42	43	44	45	46	47	48	49	50
51	52	53	54	55	56	57	58	59	60
61	62	63	64	65	66	67	68	69	70
71	72	73	74	75	76	77	78	79	80
81	82	83	84	85	86	87	88	89	90
91	92	93	94	95	96	97	98	99	100

Eratosthenes
(c. 276–c. 194 B.C.)

This table is called the *Sieve of Eratosthenes*. Eratosthenes was a Greek mathematician who was the chief librarian at the Library of Alexandria in Egypt. He was the first person to calculate the circumference of Earth.

1. Cross out the multiples of 2 that are greater than 2. Do the same for 3, 5, and 7.

2. The numbers that are *not* crossed out are called *prime numbers*. The numbers that are crossed out are called *composite numbers*. In your own words, describe the characteristics of prime numbers and composite numbers.

3. **MODELING REAL LIFE** Work with a partner. Cicadas are insects that live underground and emerge from the ground after x or $x + 4$ years. Is it possible that both x and $x + 4$ are prime? Give some examples.

Vocabulary

The following vocabulary terms are defined in this chapter. Think about what each term might mean and record your thoughts.

exponent
numerical expression
order of operations

common factors
greatest common factor

common multiples
least common multiple

1.1 Powers and Exponents

Learning Target: Write and evaluate expressions involving exponents.

Success Criteria:
- I can write products of repeated factors as powers.
- I can evaluate powers.

EXPLORATION 1

Writing Expressions Using Exponents

Work with a partner. Copy and complete the table.

Repeated Factors	Using an Exponent	Value
a. 10×10		
b. 4×4		
c. 6×6		
d. $10 \times 10 \times 10$		
e. $100 \times 100 \times 100$		
f. $3 \times 3 \times 3 \times 3$		
g. $4 \times 4 \times 4 \times 4 \times 4$		
h. $2 \times 2 \times 2 \times 2 \times 2 \times 2$		

Math Practice

Repeat Calculations

What patterns do you notice in the expressions? How does this help you write exponents?

i. In your own words, describe what the two numbers in the expression 3^5 mean.

EXPLORATION 2

Using a Calculator to Find a Pattern

Work with a partner. Copy the diagram. Use a calculator to find each value. Write one digit of the value in each box. Describe the pattern in the digits of the values.

a. 11^1

b. 11^2

c. 11^3

d. 11^4

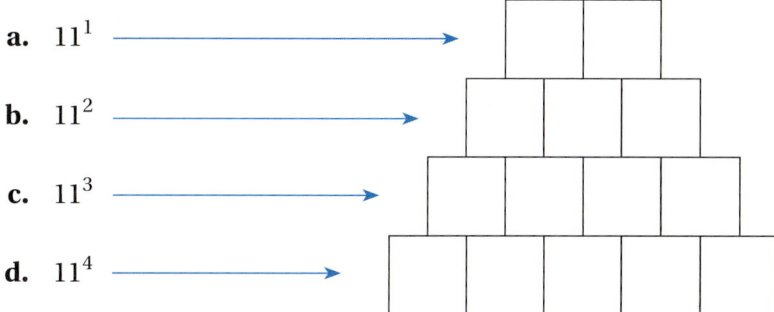

Section 1.1 Powers and Exponents 3

1.1 Lesson

Key Vocabulary
power, p. 4
base, p. 4
exponent, p. 4
perfect square, p. 5

A **power** is a product of repeated factors. The **base** of a power is the repeated factor. The **exponent** of a power indicates the number of times the base is used as a factor.

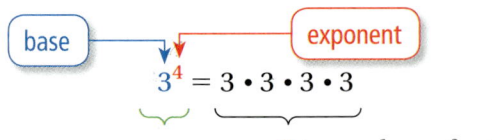

$3^4 = 3 \cdot 3 \cdot 3 \cdot 3$

power   3 is used as a factor 4 times.

Remember
You can use the dot symbol • to indicate multiplication. For example, the product of 3 and 5 can be expressed as 3×5 or $3 \cdot 5$.

Power	Words
3^2	Three *squared*, or three to the second
3^3	Three *cubed*, or three to the third
3^4	Three to the fourth
3^5	Three to the fifth

EXAMPLE 1   Writing Expressions as Powers

Write each product as a power.

a. $7 \cdot 7 \cdot 7 \cdot 7 \cdot 7$

Because 7 is used as a factor 5 times, its exponent is 5.

▶ So, $7 \cdot 7 \cdot 7 \cdot 7 \cdot 7 = 7^5$.

b. $12 \times 12 \times 12$

Because 12 is used as a factor 3 times, its exponent is 3.

▶ So, $12 \times 12 \times 12 = 12^3$.

c. $100 \times 100 \times 100 \times 100 \times 100 \times 100$

Because 100 is used as a factor 6 times, its exponent is 6.

▶ So, $100 \times 100 \times 100 \times 100 \times 100 \times 100 = 100^6$.

Try It   Write the product as a power.

1. $2 \times 2 \times 2$
2. $6 \cdot 6 \cdot 6 \cdot 6 \cdot 6 \cdot 6$
3. $15 \times 15 \times 15 \times 15$
4. $20 \cdot 20 \cdot 20 \cdot 20 \cdot 20 \cdot 20 \cdot 20$

EXAMPLE 2 — Finding Values of Powers

Find the value of each power.

a. 7^2

$7^2 = 7 \cdot 7$ Write as repeated multiplication.
$= 49$ Simplify.

b. 5^3

$5^3 = 5 \cdot 5 \cdot 5$
$= 125$

Try It Find the value of the power.

5. 6^3 6. 9^2 7. 3^4 8. 18^2

The square of a whole number is a **perfect square**.

EXAMPLE 3 — Identifying Perfect Squares

Determine whether each number is a perfect square.

a. 64

 Because $8^2 = 64$, 64 is a perfect square.

b. 20

 No whole number squared equals 20. So, 20 is not a perfect square.

Try It Determine whether the number is a perfect square.

9. 25 10. 2 11. 99 12. 36

Self-Assessment for Concepts & Skills

Solve each exercise. Then rate your understanding of the success criteria in your journal.

FINDING VALUES OF POWERS Find the value of the power.

13. 8^2 14. 3^5 15. 11^3

16. **VOCABULARY** How are exponents and powers different?

17. **VOCABULARY** Is 10 a perfect square? Is 100 a perfect square? Explain.

18. **WHICH ONE DOESN'T BELONG?** Which one does *not* belong with the other three? Explain your reasoning.

$2^4 = 2 \times 2 \times 2 \times 2$ $3^2 = 3 \times 3$

$3 + 3 + 3 + 3 = 3 \times 4$ $5 \cdot 5 \cdot 5 = 5^3$

> **Remember**
> The *area* of a figure is the amount of surface it covers. Area is measured in square units.

You can use powers to find areas of squares. The area of a square is equal to its side length squared.

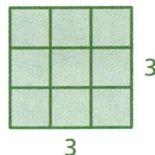

Area = 3^2 = 9 square units

EXAMPLE 4 Modeling Real Life

A life-size MONOPOLY® game board is a square with a side length of 11 yards. What is the area of the game board?

Use a verbal model to solve the problem.

$$\boxed{\text{Area of game board}} = \left(\boxed{\text{Side length}}\right)^2$$

$\phantom{\text{Area of game board}} = 11^2$ Substitute 11 for side length.

$\phantom{\text{Area of game board}} = 121$ Multiply.

▶ The area of the game board is 121 square yards.

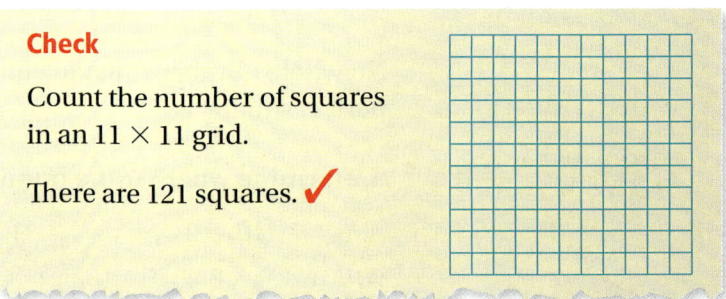

Check

Count the number of squares in an 11 × 11 grid.

There are 121 squares. ✓

Self-Assessment for Problem Solving

Solve each exercise. Then rate your understanding of the success criteria in your journal.

19. A square solar panel has an area of 16 square feet. Write the area as a power. Then find the side lengths of the panel.

20. The four-square court shown is a square made up of four identical smaller squares. What is the area of the court?

21. **DIG DEEPER!** Each face of a number cube is a square with a side length of 16 millimeters. What is the total area of all of the faces of the number cube?

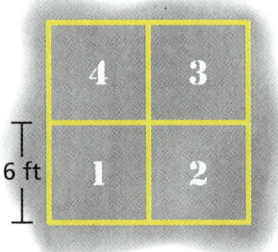

6 ft

1.1 Practice

Go to *BigIdeasMath.com* to get HELP with solving the exercises.

▶ Review & Refresh

Multiply.

1. 150×2
2. 175×8
3. 123×3
4. 151×9

Write the sentence as a numerical expression.

5. Add 5 and 8, then multiply by 4.
6. Subtract 7 from 11, then divide by 2.

Round the number to the indicated place value.

7. 4.03785 to the tenths
8. 12.89503 to the hundredths

Complete the sentence.

9. $\frac{1}{10}$ of 30 is ____.
10. $\frac{4}{5}$ of 25 is ____.

▶ Concepts, Skills, & Problem Solving

WRITING EXPRESSIONS USING EXPONENTS Copy and complete the table. (See Exploration 1, p. 3.)

Repeated Factors	Using an Exponent	Value
11. 8×8		
12. $4 \times 4 \times 4$		
13. $9 \times 9 \times 9 \times 9$		
14. $12 \times 12 \times 12 \times 12$		

WRITING EXPRESSIONS AS POWERS Write the product as a power.

15. 9×9
16. 13×13
17. $15 \times 15 \times 15$
18. $2 \cdot 2 \cdot 2 \cdot 2 \cdot 2$
19. $14 \times 14 \times 14$
20. $8 \cdot 8 \cdot 8 \cdot 8$
21. $11 \times 11 \times 11 \times 11 \times 11$
22. $7 \cdot 7 \cdot 7 \cdot 7 \cdot 7 \cdot 7$
23. $16 \cdot 16 \cdot 16 \cdot 16$
24. $43 \times 43 \times 43 \times 43 \times 43$
25. $167 \cdot 167 \cdot 167$
26. $245 \cdot 245 \cdot 245 \cdot 245$

FINDING VALUES OF POWERS Find the value of the power.

27. 5^2
28. 4^3
29. 6^2
30. 1^7
31. 0^3
32. 8^4
33. 2^4
34. 12^2
35. 7^3
36. 5^4
37. 2^5
38. 14^2

USING TOOLS Use a calculator to find the value of the power.

39. 7^6
40. 4^8
41. 12^4
42. 17^5

Section 1.1 Powers and Exponents 7

43. YOU BE THE TEACHER Your friend finds the value of 8^3. Is your friend correct? Explain your reasoning.

$$8^3 = 8 \cdot 3 = 24$$

IDENTIFYING PERFECT SQUARES Determine whether the number is a perfect square.

44. 8 **45.** 4 **46.** 81 **47.** 44

48. 49 **49.** 125 **50.** 150 **51.** 144

52. MODELING REAL LIFE On each square centimeter of a person's skin, there are about 39^2 bacteria. How many bacteria does this expression represent?

53. MP REPEATED REASONING The smallest figurine in a gift shop is 2 inches tall. The height of each figurine is twice the height of the previous figurine. What is the height of the tallest figurine?

54. MODELING REAL LIFE A square painting measures 2 meters on each side. What is the area of the painting in square centimeters?

55. MP NUMBER SENSE Write three powers that have values greater than 120 and less than 130.

56. DIG DEEPER! A landscaper has 125 tiles to build a square patio. The patio must have an area of at least 80 square feet.

   a. What are the possible arrangements for the patio?

   b. How many tiles are not used in each arrangement?

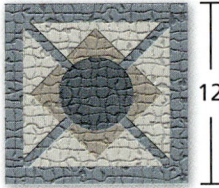

12 in.

12 in.

57. MP PATTERNS Copy and complete the table. Describe what happens to the value of the power as the exponent decreases. Use this pattern to find the value of 4^0.

Power	4^6	4^5	4^4	4^3	4^2	4^1
Value	4096	1024				

58. MP REPEATED REASONING How many blocks do you need to add to Square 6 to get Square 7? to Square 9 to get Square 10? to Square 19 to get Square 20? Explain.

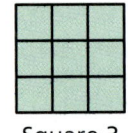
Square 3

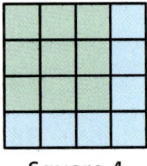
Square 4

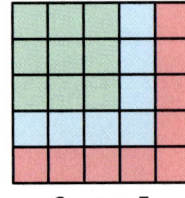
Square 5

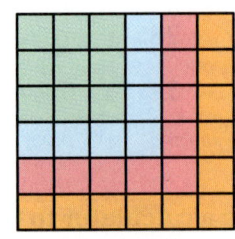
Square 6

8   Chapter 1   Numerical Expressions and Factors

1.2 Order of Operations

Learning Target: Write and evaluate numerical expressions using the order of operations.

Success Criteria:
- I can explain why there is a need for a standard order of operations.
- I can evaluate numerical expressions involving several operations, exponents, and grouping symbols.
- I can write numerical expressions involving exponents to represent a real-life problem.

EXPLORATION 1
Comparing Different Orders

Work with a partner. Find the value of each expression by using different orders of operations. Are your answers the same?

a. Add, then multiply.
$3 + 2 \times 2$

Multiply, then add.
$3 + 2 \times 2$

b. Subtract, then multiply.
$18 - 3 \cdot 3$

Multiply, then subtract.
$18 - 3 \cdot 3$

c. Multiply, then subtract.
$8 \times 8 - 2$

Subtract, then multiply.
$8 \times 8 - 2$

d. Multiply, then add.
$6 \cdot 6 + 2$

Add, then multiply.
$6 \cdot 6 + 2$

EXPLORATION 2
Determining Order of Operations

Work with a partner.

a. Scientific calculators use a standard order of operations when evaluating expressions. Why is a standard order of operations needed?

b. Use a scientific calculator to evaluate each expression in Exploration 1. Enter each expression exactly as written. For each expression, which order of operations is correct?

c. What order of operations should be used to evaluate $3 + 2^2$, $18 - 3^2$, $8^2 - 2$, and $6^2 + 2$?

d. Do $18 \div 3 \cdot 3$ and $18 \div 3^2$ have the same value? Justify your answer.

e. How does evaluating powers fit into the order of operations?

Math Practice

Use Technology to Explore

How does a scientific calculator help you explore order of operations?

Section 1.2 Order of Operations

1.2 Lesson

Key Vocabulary 🔊
numerical expression, p. 10
evaluate, p. 10
order of operations, p. 10

A **numerical expression** is an expression that contains numbers and operations. To **evaluate**, or find the value of, a numerical expression, use a set of rules called the **order of operations**.

 **Key Idea**

Order of Operations
1. Perform operations in grouping symbols.
2. Evaluate numbers with exponents.
3. Multiply and divide from left to right.
4. Add and subtract from left to right.

EXAMPLE 1 **Using Order of Operations**

a. Evaluate $12 - 2 \times 4$.

$12 - 2 \times 4 = 12 - 8$ Multiply 2 and 4.
$ = 4$ Subtract 8 from 12.

b. Evaluate $60 \div [(4 + 2) \times 5]$.

$60 \div [(4 + 2) \times 5] = 60 \div [6 \times 5]$ Perform operation in parentheses.
$ = 60 \div 30$ Perform operation in brackets.
$ = 2$ Divide 60 by 30.

Try It Evaluate the expression.

1. $7 \cdot 5 + 3$
2. $(28 - 20) \div 4$
3. $[6 + (15 - 10)] \times 5$

EXAMPLE 2 **Using Order of Operations with Exponents**

Evaluate $30 \div (7 + 2^3) \times 6$.

$30 \div (7 + 2^3) \times 6 = 30 \div (7 + 8) \times 6$ Evaluate power in parentheses.
$ = 30 \div 15 \times 6$ Perform operation in parentheses.
$ = 2 \times 6$ Divide 30 by 15.
$ = 12$ Multiply 2 and 6.

Remember to multiply and divide from left to right. In Example 2, you should divide before multiplying because the division symbol comes first when reading from left to right.

Try It Evaluate the expression.

4. $6 + 2^4 - 1$
5. $4 \cdot 3^2 + 18 - 9$
6. $16 + (5^2 - 7) \div 3$

The symbols × and • are used to indicate multiplication. You can also use parentheses to indicate multiplication. For example, 3(2 + 7) is the same as 3 × (2 + 7).

EXAMPLE 3 Using Order of Operations

Remember
You can interpret a fraction as division of the numerator by the denominator.
$\frac{a}{b} = a \div b$

a. Evaluate $9 + \frac{8-2}{3}$.

$9 + \frac{8-2}{3} = 9 + (8 - 2) \div 3$ Rewrite fraction as division.
$= 9 + 6 \div 3$ Perform operation in parentheses.
$= 9 + 2$ Divide 6 by 3.
$= 11$ Add 9 and 2.

b. Evaluate $10 - 8(13 + 7) \div 4^2$.

$10 - 8(13 + 7) \div 4^2 = 10 - 8(20) \div 4^2$ Perform operation in parentheses.
$= 10 - 8(20) \div 16$ Evaluate 4^2.
$= 10 - 160 \div 16$ Multiply 8 and 20.
$= 10 - 10$ Divide 160 by 16.
$= 0$ Subtract 10 from 10.

Try It Evaluate the expression.

7. $50 + 6(12 \div 4) - 8^2$ 8. $5^2 - \frac{1}{5}(10 - 5)$ 9. $\frac{8(2+5)}{7}$

Self-Assessment for Concepts & Skills

Solve each exercise. Then rate your understanding of the success criteria in your journal.

USING ORDER OF OPERATIONS Evaluate the expression.

10. $7 + 2 \cdot 4$ 11. $8 \div 4 \times 2$ 12. $3(5 + 1) \div 3^2$

13. **WRITING** Why does $12 - 8 \div 2 = 8$, but $(12 - 8) \div 2 = 2$?

14. **MP REASONING** Describe the steps in evaluating the expression $8 \div (6 - 4) + 3^2$.

15. **WHICH ONE DOESN'T BELONG?** Which expression does *not* belong with the other three? Explain your reasoning.

$5^2 - 8 \times 2$ $5^2 - (8 \times 2)$ $5^2 - 2 \times 8$ $(5^2 - 8) \times 2$

Section 1.2 Order of Operations

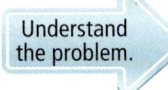

EXAMPLE 4 **Modeling Real Life**

The diagram shows landing zones for skydivers. Zone 1 is for experts. The remaining space is divided in half and designated as Zones 2 and 3 for tandem divers. What is the area of Zone 2?

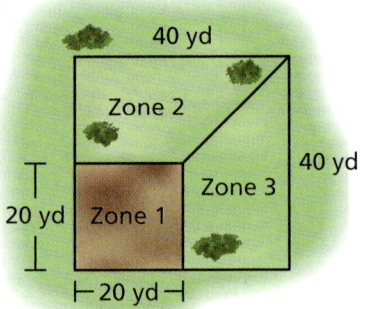

You are given the dimensions of landing zones and that the areas of Zones 2 and 3 are equal. You are asked to find the area of Zone 2.

Use a verbal model to write an expression. Subtract the area of Zone 1 from the total area to find the combined area of Zones 2 and 3. Then multiply the combined area by one-half.

Verbal Model One-half (Total area − Area of Zone 1)

Expression $\frac{1}{2}$ (40^2 − 20^2)

$$\frac{1}{2}(40^2 - 20^2) = \frac{1}{2}(1600 - 400)$$ Evaluate powers in parentheses.

$$= \frac{1}{2}(1200)$$ Perform operation in parentheses.

$$= 600$$ Multiply $\frac{1}{2}$ and 1200.

Check Verify that the areas of the three zones have a sum equal to the total area.

$400 + 600 + 600 \stackrel{?}{=} 1600$

$1600 = 1600$ ✓

▶ The area of Zone 2 is 600 square yards.

Self-Assessment for Problem Solving

Solve each exercise. Then rate your understanding of the success criteria in your journal.

16. A square plot of land has side lengths of 40 meters. An archaeologist divides the land into 64 equal parts. What is the area of each part?

17. A glass block window is made of two different-sized glass squares. The window has side lengths of 40 inches. The large glass squares have side lengths of 10 inches. Find the total area of the small glass squares.

18. **DIG DEEPER!** A square vegetable garden has side lengths of 12 feet. You plant flowers in the center portion as shown. You divide the remaining space into 4 equal sections and plant tomatoes, onions, zucchini, and peppers. What is the area of the onion section?

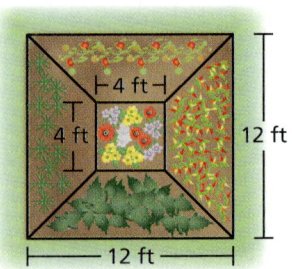

1.2 Practice

Go to BigIdeasMath.com to get HELP with solving the exercises.

Review & Refresh

Write the product as a power.

1. $11 \times 11 \times 11 \times 11$
2. $13 \times 13 \times 13 \times 13 \times 13$

Find the missing dimension of the rectangular prism.

3.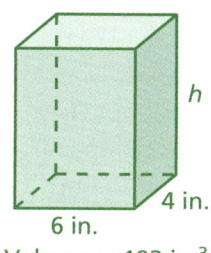
   6 in., 4 in., h
   Volume = 192 in.³

4.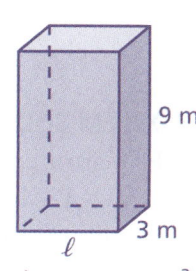
   9 m, 3 m, ℓ
   Volume = 135 m³

Tell whether the number is prime or composite.

5. 9
6. 11
7. 23

Concepts, Skills, & Problem Solving

COMPARING DIFFERENT ORDERS Find the value of the expression by using different orders of operations. Are your answers the same? *(See Exploration 1, p. 9.)*

8. Add, then multiply.   Multiply, then add.
   $4 + 6 \times 6$ $4 + 6 \times 6$

9. Subtract, then multiply.   Multiply, then subtract.
   $5 \times 5 - 3$ $5 \times 5 - 3$

USING ORDER OF OPERATIONS Evaluate the expression.

10. $5 + 18 \div 6$
11. $(11 - 3) \div 2 + 1$
12. $45 \div 9 \times 12$
13. $6^2 - 3 \cdot 4$
14. $42 \div (15 - 2^3)$
15. $4^2 \cdot 2 + 8 \cdot 7$
16. $(5^2 - 2) \times 1^5 + 4$
17. $4 + 2 \times 3^2 - 9$
18. $8 \div 2 \times 3 + 4^2 \div 4$
19. $3^2 + 12 \div (6 - 3) \times 8$
20. $(10 + 4) \div (26 - 19)$
21. $(5^2 - 4) \cdot 2 - 18$
22. $2 \times [(16 - 8) \times 2]$
23. $12 + 8 \times 3^3 - 24$
24. $6^2 \div [(2 + 4) \times 2^3]$

YOU BE THE TEACHER Your friend evaluates the expression. Is your friend correct? Explain your reasoning.

25. $9 + 3 \times 3^2 = 12 \times 9$
    $= 108$

26. $19 - 6 + 12 = 19 - 18$
    $= 1$

27. **MP PROBLEM SOLVING** You need to read 20 poems in 5 days for an English project. Each poem is 2 pages long. Evaluate the expression $20 \times 2 \div 5$ to find how many pages you need to read each day.

Section 1.2 Order of Operations 13

USING ORDER OF OPERATIONS Evaluate the expression.

28. $12 - 2(7 - 4)$

29. $4(3 + 5) - 3(6 - 2)$

30. $6 + \frac{1}{4}(12 - 8)$

31. $9^2 - 8(6 + 2)$

32. $4(3 - 1)^3 + 7(6) - 5^2$

33. $8\left[\left(1\frac{1}{6} + \frac{5}{6}\right) \div 4\right]$

34. $7^2 - 2\left(\frac{11}{8} - \frac{3}{8}\right)$

35. $8(7.3 + 3.7 - 8) \div 2$

36. $2^4(5.2 - 3.2) \div 4$

37. $\frac{6^2(3 + 5)}{4}$

38. $\frac{12^2 - 4(6) + 1}{11^2}$

39. $\frac{26 \div 2 + 5}{3^2 - 3}$

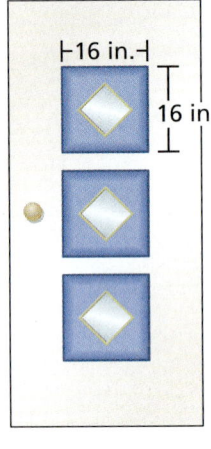

40. **MP PROBLEM SOLVING** Before a show, there are 8 people in a theater. Five groups of 4 people enter, and then three groups of 2 people leave. Evaluate the expression $8 + 5(4) - 3(2)$ to find how many people are in the theater.

41. **MODELING REAL LIFE** The front door of a house is painted white and blue. Each window is a square with a side length of 7 inches. What is the area of the door that is painted blue?

42. **MP PROBLEM SOLVING** You buy 6 notebooks, 10 folders, 1 pack of pencils, and 1 lunch box for school. After using a $10 gift card, how much do you owe? Explain how you solved the problem.

43. **OPEN-ENDED** Use all four operations and at least one exponent to write an expression that has a value of 100.

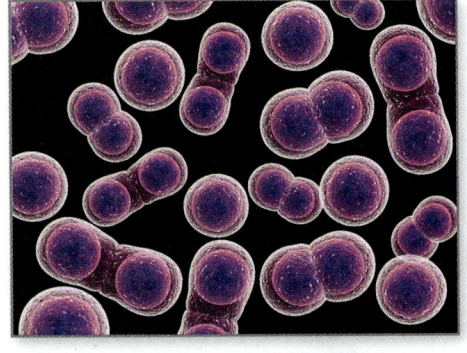

44. **MP REPEATED REASONING** A Petri dish contains 35 cells. Every day, each cell in the Petri dish divides into 2 cells in a process called *mitosis*. How many cells are there after 14 days? Justify your answer.

45. **MP REASONING** Two groups collect litter along the side of a road. It takes each group 5 minutes to clean up a 200-yard section. How long does it take both groups working together to clean up 2 *miles*? Explain how you solved the problem.

46. **MP NUMBER SENSE** Copy each statement. Insert $+$, $-$, $\times$, or $\div$ symbols to make each statement true.

   a. $27 \quad 3 \quad 5 \quad 2 = 19$

   b. $9^2 \quad 11 \quad 8 \quad 4 \quad 1 = 60$

   c. $5 \quad 6 \quad 15 \quad 9 = 24$

   d. $14 \quad 2 \quad 7 \quad 3 \quad 9 = 10$

1.3 Prime Factorization

Learning Target: Write a number as a product of prime factors and represent the product using exponents.

Success Criteria:
- I can find factor pairs of a number.
- I can explain the meanings of prime and composite numbers.
- I can create a factor tree to find the prime factors of a number.
- I can write the prime factorization of a number.

EXPLORATION 1

Rewriting Numbers as Products of Factors

Work with a partner. Two students use *factor trees* to write 108 as a product of factors, as shown below.

Student A

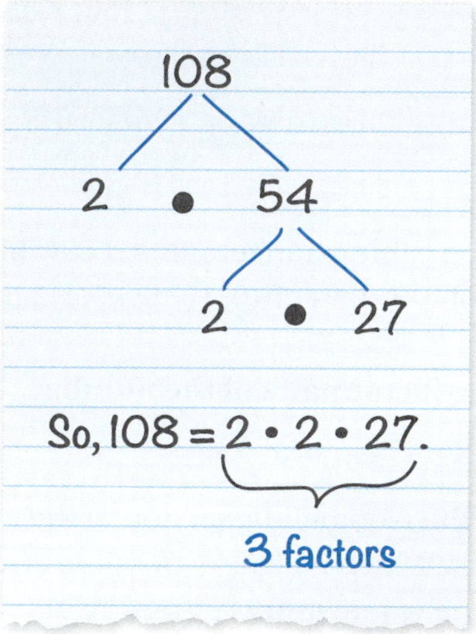

```
       108
      /   \
     2     54
          /  \
         2    27
```

So, 108 = 2 • 2 • 27.

3 factors

Student B

```
       108
      /   \
     3     36
          /  \
         3    12
              / \
             2   6
```

So, 108 = 3 • 3 • 2 • 6.

4 factors

a. Without using 1 as a factor, can you write 108 as a product with more factors than each student used? Justify your answer.

Math Practice

Interpret Results
How do you know your answer makes sense?

b. Use factor trees to write 80, 162, and 300 as products of as many factors as possible. Do not use 1 as a factor.

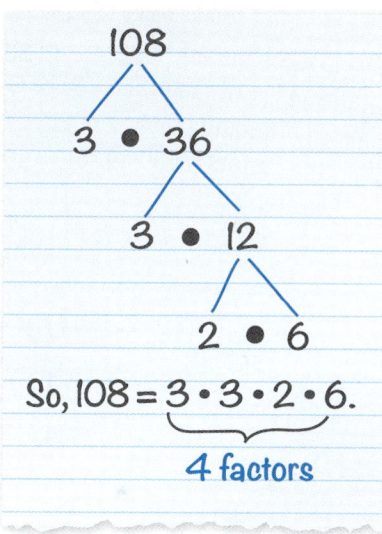

c. Compare your results in parts (a) and (b) with other groups. For each number, identify the product with the greatest number of factors. What do these factors have in common?

Section 1.3 Prime Factorization 15

1.3 Lesson

Because 2 is a factor of 10 and 2 • 5 = 10, 5 is also a factor of 10. The pair 2, 5 is called a **factor pair** of 10.

EXAMPLE 1   Finding Factor Pairs

Key Vocabulary
factor pair, p. 16
prime factorization, p. 16
factor tree, p. 16

The brass section of a marching band has 30 members. The band director arranges the brass section in rows. Each row has the same number of members. How many possible arrangements are there?

Use the factor pairs of 30 to find the number of arrangements.

30 = 1 • 30	There could be 1 row of 30 or 30 rows of 1.
30 = 2 • 15	There could be 2 rows of 15 or 15 rows of 2.
30 = 3 • 10	There could be 3 rows of 10 or 10 rows of 3.
30 = 5 • 6	There could be 5 rows of 6 or 6 rows of 5.
30 = 6 • 5	The factors 5 and 6 are already listed.

When making an organized list of factor pairs, stop finding pairs when the factors begin to repeat.

There are 8 possible arrangements: 1 row of 30, 30 rows of 1, 2 rows of 15, 15 rows of 2, 3 rows of 10, 10 rows of 3, 5 rows of 6, or 6 rows of 5.

Try It List the factor pairs of the number.

1. 18
2. 24
3. 51

4. **WHAT IF?** The woodwinds section of the marching band has 38 members. Which has more possible arrangements, the brass section or the woodwinds section? Explain.

Key Idea

Prime Factorization

The **prime factorization** of a composite number is the number written as a product of its prime factors.

You can use factor pairs and a **factor tree** to help find the prime factorization of a number. The factor tree is complete when only prime factors appear in the product. A factor tree for 60 is shown.

Remember
A *prime number* is a whole number greater than 1 with exactly two factors, 1 and itself. A *composite number* is a whole number greater than 1 with factors in addition to 1 and itself.

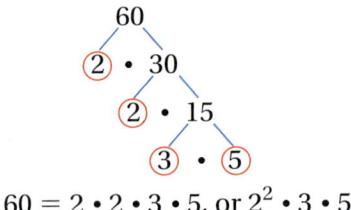

$60 = 2 \cdot 2 \cdot 3 \cdot 5$, or $2^2 \cdot 3 \cdot 5$

EXAMPLE 2 Writing a Prime Factorization

Write the prime factorization of 48.

Choose any factor pair of 48 to begin the factor tree.

Tree 1

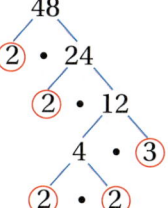

$48 = 2 \cdot 2 \cdot 3 \cdot 2 \cdot 2$

Find a factor pair and draw "branches."

Circle the prime factors as you find them.

Find factors until each branch ends at a prime factor.

Tree 2

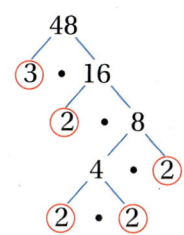

$48 = 3 \cdot 2 \cdot 2 \cdot 2 \cdot 2$

> Notice that beginning with different factor pairs results in the same prime factorization. Every composite number has only one prime factorization.

▶ The prime factorization of 48 is $2 \cdot 2 \cdot 2 \cdot 2 \cdot 3$, or $2^4 \cdot 3$.

Try It Write the prime factorization of the number.

5. 20   **6.** 88   **7.** 90   **8.** 462

Self-Assessment for Concepts & Skills

Solve each exercise. Then rate your understanding of the success criteria in your journal.

WRITING A PRIME FACTORIZATION Write the prime factorization of the number.

9. 14   **10.** 86   **11.** 40   **12.** 516

13. WRITING Explain the difference between prime numbers and composite numbers.

14. MP STRUCTURE Your friend lists the following factor pairs and concludes that there are 6 factor pairs of 12. Explain why your friend is incorrect.

| 1, 12 | 2, 6 | 3, 4 |
| 12, 1 | 6, 2 | 4, 3 |

15. WHICH ONE DOESN'T BELONG? Which factor pair does *not* belong with the other three? Explain your reasoning.

| 2, 28 | 4, 14 | 6, 9 | 7, 8 |

Section 1.3 Prime Factorization

EXAMPLE 3 Using a Prime Factorization

What is the greatest perfect square that is a factor of 1575?

Because 1575 has many factors, it is not efficient to list all of its factors and check for perfect squares. Use a factor tree to write the prime factorization of 1575. Then analyze the prime factors to find perfect square factors.

```
        1575
       /    \
      25  •  63
     /\     /\
    5• 5  7 • 9
                /\
               3• 3
```

$$1575 = 3 \cdot 3 \cdot 5 \cdot 5 \cdot 7$$

The prime factorization shows that 1575 has three factors other than 1 that are perfect squares.

$$3 \cdot 3 = 9$$

$$5 \cdot 5 = 25$$

$$(3 \cdot 5) \cdot (3 \cdot 5) = 15 \cdot 15 = 225$$

▶ So, the greatest perfect square that is a factor of 1575 is 225.

Self-Assessment for Problem Solving

Solve each exercise. Then rate your understanding of the success criteria in your journal.

16. A group of 20 friends plays a card game. The game can be played with 2 or more teams of equal size. Each team must have at least 2 members. List the possible numbers and sizes of teams.

17. You arrange 150 chairs in rows for a school play. You want each row to have the same number of chairs. How many possible arrangements are there? Are all of the possible arrangements appropriate for the play? Explain.

18. What is the least perfect square that is a factor of 4536? What is the greatest perfect square that is a factor of 4536?

19. **DIG DEEPER!** The prime factorization of a number is $2^4 \times 3^4 \times 5^4 \times 7^2$. Is the number a perfect square? Explain your reasoning.

1.3 Practice

Go to **BigIdeasMath.com** to get HELP with solving the exercises.

▶ Review & Refresh

Evaluate the expression.

1. $2 + 4^2(5 - 3)$
2. $2^3 + 4 \times 3^2$
3. $9 \times 5 - 2^4\left(\dfrac{5}{2} - \dfrac{1}{2}\right)$

Plot the points in a coordinate plane. Draw a line segment connecting the points.

4. $(1, 1)$ and $(4, 3)$
5. $(2, 3)$ and $(5, 9)$
6. $(2, 5)$ and $(4, 8)$

Use the Distributive Property to find the quotient. Justify your answer.

7. $408 \div 4$
8. $628 \div 2$
9. $969 \div 3$

Classify the triangle in as many ways as possible.

10.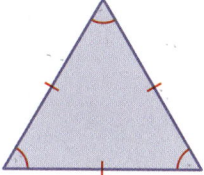
11.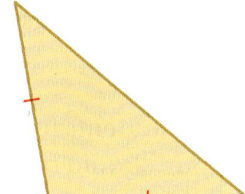
12.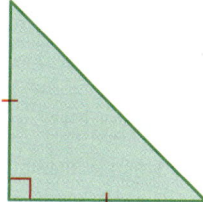

▶ Concepts, Skills, & Problem Solving

REWRITING A NUMBER Write the number as a product of as many factors as possible. *(See Exploration 1, p. 15.)*

13. 60
14. 63
15. 120
16. 150

FINDING FACTOR PAIRS List the factor pairs of the number.

17. 15
18. 22
19. 34
20. 39
21. 45
22. 54
23. 59
24. 61
25. 100
26. 58
27. 25
28. 76
29. 52
30. 88
31. 71
32. 91

WRITING A PRIME FACTORIZATION Write the prime factorization of the number.

33. 16
34. 25
35. 30
36. 26
37. 84
38. 54
39. 65
40. 77
41. 46
42. 39
43. 99
44. 24
45. 315
46. 490
47. 140
48. 640

USING A PRIME FACTORIZATION Find the number represented by the prime factorization.

49. $2^2 \cdot 3^2 \cdot 5$
50. $3^2 \cdot 5^2 \cdot 7$
51. $2^3 \cdot 11^2 \cdot 13$

Section 1.3 Prime Factorization 19

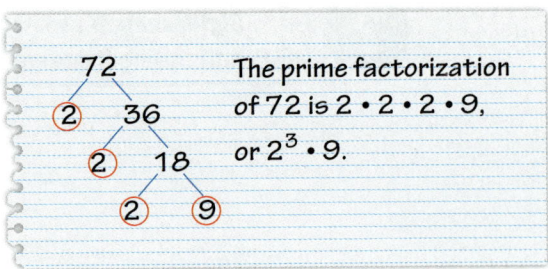

The prime factorization of 72 is 2 • 2 • 2 • 9, or $2^3 • 9$.

52. **YOU BE THE TEACHER** Your friend finds the prime factorization of 72. Is your friend correct? Explain your reasoning.

USING A PRIME FACTORIZATION Find the greatest perfect square that is a factor of the number.

53. 250
54. 275
55. 392
56. 338
57. 244
58. 650
59. 756
60. 1290
61. 2205
62. 1890
63. 495
64. 4725

65. **VOCABULARY** A botanist separates plants into equal groups of 5 for an experiment. Is the total number of plants in the experiment *prime* or *composite*? Explain.

66. **MP REASONING** A teacher divides 36 students into equal groups for a scavenger hunt. Each group should have at least 4 students but no more than 8 students. What are the possible group sizes?

67. **CRITICAL THINKING** Is 2 the only even prime number? Explain.

68. **MP LOGIC** One table at a bake sale has 75 cookies. Another table has 60 cupcakes. Which table allows for more rectangular arrangements? Explain.

69. **PERFECT NUMBERS** A *perfect number* is a number that equals the sum of its factors, not including itself. For example, the factors of 28 are 1, 2, 4, 7, 14, and 28. Because 1 + 2 + 4 + 7 + 14 = 28, 28 is a perfect number. What are the perfect numbers between 1 and 27?

70. **MP REPEATED REASONING** Choose any two perfect squares and find their product. Then multiply your answer by another perfect square. Continue this process. Are any of the products perfect squares? What can you conclude?

71. **MP PROBLEM SOLVING** The stage manager of a school play creates a rectangular stage that has whole number dimensions and an area of 42 square yards. String lights will outline the stage. What is the least number of yards of string lights needed to enclose the stage?

72. **DIG DEEPER!** Consider the rectangular prism shown. Using only whole number dimensions, how many different prisms are possible? Explain.

Rectangular Prism

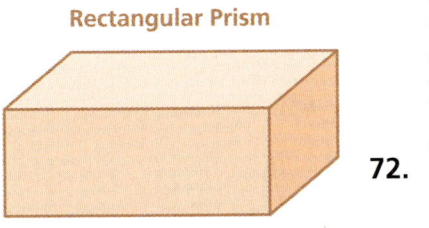

Volume = 40 cubic inches

1.4 Greatest Common Factor

Learning Target: Find the greatest common factor of two numbers.

Success Criteria:
- I can explain the meaning of factors of a number.
- I can use lists of factors to identify the greatest common factor of numbers.
- I can use prime factors to identify the greatest common factor of numbers.

A **Venn diagram** uses circles to describe relationships between two or more sets. The Venn diagram shows the factors of 12 and 15. Numbers that are factors of both 12 and 15 are represented by the overlap of the two circles.

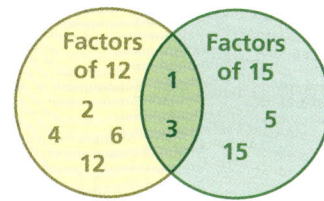

EXPLORATION 1
Identifying Common Factors

Work with a partner. In parts (a)–(d), create a Venn diagram that represents the factors of each number and identify any *common factors*.

a. 36 and 48
b. 16 and 56
c. 30 and 75
d. 54 and 90

e. Look at the Venn diagrams in parts (a)–(d). Explain how to identify the *greatest common factor* of each pair of numbers. Then circle it in each diagram.

EXPLORATION 2
Using Prime Factors

Work with a partner.

a. Each Venn diagram represents the prime factorizations of two numbers. Identify each pair of numbers. Explain your reasoning.

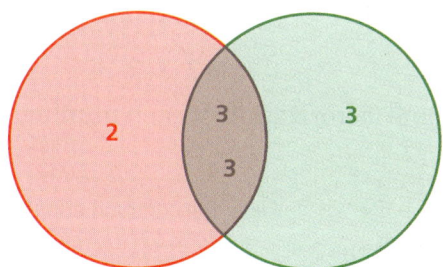
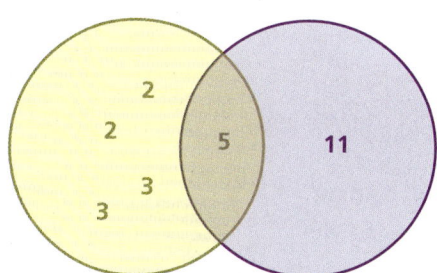

Math Practice

Interpret a Solution
What does the diagram representing the prime factorizations mean?

b. Create a Venn diagram that represents the prime factorizations of 36 and 48.

c. Repeat part (b) for the remaining number pairs in Exploration 1.

d. **MP STRUCTURE** Make a conjecture about the relationship between the greatest common factors you found in Exploration 1 and the numbers in the overlaps of the Venn diagrams you just created.

1.4 Lesson

Factors that are shared by two or more numbers are called **common factors**. The greatest of the common factors is called the **greatest common factor** (GCF). One way to find the GCF of two or more numbers is by listing factors.

EXAMPLE 1 **Finding the GCF Using Lists of Factors**

Find the GCF of 24 and 40.

List the factors of each number.

Factors of 24: ①, ②, 3, ④, 6, ⑧, 12, 24   Circle the common factors.

Factors of 40: ①, ②, ④, 5, ⑧, 10, 20, 40

The common factors of 24 and 40 are 1, 2, 4, and 8. The greatest of these common factors is 8.

▸ So, the GCF of 24 and 40 is 8.

Key Vocabulary
Venn diagram, *p. 21*
common factors, *p. 22*
greatest common factor, *p. 22*

Try It Find the GCF of the numbers using lists of factors.

1. 8, 36
2. 18, 72
3. 14, 28, 49

Another way to find the GCF of two or more numbers is by using prime factors. The GCF is the product of the common prime factors of the numbers.

EXAMPLE 2 **Finding the GCF Using Prime Factorizations**

Find the GCF of 12 and 56.

Make a factor tree for each number.

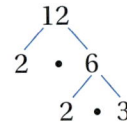 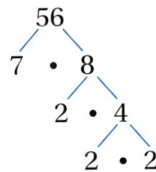

Write the prime factorization of each number.

$12 = ②\cdot ②\cdot 3$
$56 = ②\cdot ②\cdot 2 \cdot 7$

Circle the common prime factors.

$2 \cdot 2 = 4$ Find the product of the common prime factors.

▸ So, the GCF of 12 and 56 is 4.

> Examples 1 and 2 show two different methods for finding the GCF. After solving with one method, you can use the other method to check your answer.

Try It Find the GCF of the numbers using prime factorizations.

4. 20, 45
5. 32, 90
6. 45, 75, 120

EXAMPLE 3 Finding Two Numbers with a Given GCF

Which pair of numbers has a GCF of 15?

A. 10, 15 **B.** 30, 60 **C.** 21, 45 **D.** 45, 75

The number 15 cannot be a factor of the lesser number 10. So, you can eliminate Choice A.

The number 15 cannot be a factor of a number that does not have a 0 or 5 in the ones place. So, you can eliminate Choice C.

List the factors for Choices B and D. Then identify the GCF for each.

Choice B: Factors of 30: ①, ②, ③, ⑤, ⑥, ⑩, ⑮, ㉚

Factors of 60: ①, ②, ③, 4, ⑤, ⑥, ⑩, 12, ⑮, 20, ㉚, 60

The GCF of 30 and 60 is 30.

Choice D: Factors of 45: ①, ③, ⑤, 9, ⑮, 45

Factors of 75: ①, ③, ⑤, ⑮, 25, 75

The GCF of 45 and 75 is 15.

 The correct answer is **D**.

Try It

7. Write a pair of numbers whose greatest common factor is 10.

Self-Assessment for Concepts & Skills

Solve each exercise. Then rate your understanding of the success criteria in your journal.

FINDING THE GCF Find the GCF of the numbers.

8. 16, 40   **9.** 35, 63   **10.** 18, 72, 144

11. MULTIPLE CHOICE Which number is *not* a factor of 10? Explain.

A. 1   **B.** 2   **C.** 4   **D.** 5

12. DIFFERENT WORDS, SAME QUESTION Which is different? Find "both" answers.

> What is the greatest common factor of 24 and 32?

> What is the greatest common divisor of 24 and 32?

> What is the greatest common prime factor of 24 and 32?

> What is the product of the common prime factors of 24 and 32?

Section 1.4 Greatest Common Factor

EXAMPLE 4 Modeling Real Life

You are filling piñatas for your friend's birthday party. The list shows the gifts you are putting into the piñatas. You want identical groups of gifts in each piñata with no gifts left over. What is the greatest number of piñatas you can make?

* 18 kazoos
* 24 mints
* 42 lollipops

The GCF of the numbers of gifts represents the greatest number of identical groups of gifts you can make with no gifts left over. So, to find the number of piñatas, find the GCF.

Write the prime factorization of each number.

$18 = \boxed{2} \cdot \boxed{3} \cdot 3$
$24 = \boxed{2} \cdot \boxed{3} \cdot 2 \cdot 2$ Circle the common prime factors.
$42 = \boxed{2} \cdot \boxed{3} \cdot 7$

$2 \cdot 3 = 6$ Find the product of the common prime factors.

The GCF of 18, 24, and 42 is 6.

 So, you can make at most 6 piñatas.

> **Check** Verify that 6 identical piñatas will use all of the gifts.
> 18 kazoos ÷ 6 piñatas = 3 kazoos per piñata
> 24 mints ÷ 6 piñatas = 4 mints per piñata
> 42 lollipops ÷ 6 piñatas = 7 lollipops per piñata ✓

Self-Assessment for Problem Solving

Solve each exercise. Then rate your understanding of the success criteria in your journal.

13. You use 30 sandwiches and 42 granola bars to make identical picnic baskets. You make the greatest number of picnic baskets with no food left over. How many sandwiches and how many granola bars are in each basket?

14. You fill bags with cookies to give to your friends. You bake 45 chocolate chip cookies, 30 peanut butter cookies, and 15 oatmeal cookies. You want identical groups of cookies in each bag with no cookies left over. What is the greatest number of bags you can make?

1.4 Practice

 Go to *BigIdeasMath.com* to get HELP with solving the exercises.

Review & Refresh

List the factor pairs of the number.

1. 20
2. 16
3. 56
4. 87

Tell whether the statement is *always*, *sometimes*, or *never* true.

5. A rectangle is a rhombus.
6. A rhombus is a square.
7. A square is a rectangle.
8. A trapezoid is a parallelogram.

Concepts, Skills, & Problem Solving

USING A VENN DIAGRAM Use a Venn diagram to find the greatest common factor of the numbers. (See Exploration 1, p. 21.)

9. 12, 30
10. 32, 54
11. 24, 108

FINDING THE GCF Find the GCF of the numbers using lists of factors.

12. 6, 15
13. 14, 84
14. 45, 76
15. 39, 65
16. 51, 85
17. 40, 63
18. 12, 48
19. 24, 52
20. 30, 58

FINDING THE GCF Find the GCF of the numbers using prime factorizations.

21. 45, 60
22. 27, 63
23. 36, 81
24. 72, 84
25. 61, 73
26. 38, 95
27. 60, 75
28. 42, 60
29. 42, 63
30. 24, 96
31. 189, 200
32. 90, 108

OPEN-ENDED Write a pair of numbers with the indicated GCF.

33. 5
34. 12
35. 37

36. **MODELING REAL LIFE** A teacher is making identical activity packets using 92 crayons and 23 sheets of paper. What is the greatest number of packets the teacher can make with no items left over?

37. **MODELING REAL LIFE** You are making balloon arrangements for a birthday party. There are 16 white balloons and 24 red balloons. Each arrangement must be identical. What is the greatest number of arrangements you can make using every balloon?

YOU BE THE TEACHER Your friend finds the GCF of the two numbers. Is your friend correct? Explain your reasoning.

38.
$42 = 2 \cdot 3 \cdot 7$
$154 = 2 \cdot 7 \cdot 11$
The GCF is 7.

39.
$36 = 2^2 \cdot 3^2$
$60 = 2^2 \cdot 3 \cdot 5$
The GCF is $2^2 \cdot 3 = 12$.

FINDING THE GCF Find the GCF of the numbers.

40. 35, 56, 63
41. 30, 60, 78
42. 42, 70, 84
43. 40, 55, 72
44. 18, 54, 90
45. 16, 48, 88
46. 52, 78, 104
47. 96, 120, 156
48. 280, 300, 380

49. **OPEN-ENDED** Write three numbers that have a GCF of 16. What method did you use to find your answer?

CRITICAL THINKING Tell whether the statement is *always*, *sometimes*, or *never* true. Explain your reasoning.

50. The GCF of two even numbers is 2.
51. The GCF of two prime numbers is 1.

52. When one number is a multiple of another, the GCF of the numbers is the greater of the numbers.

53. **MP PROBLEM SOLVING** A science museum makes gift bags for students using 168 magnets, 48 robot figurines, and 24 packs of freeze-dried ice cream. What is the greatest number of gift bags that can be made using all of the items? How many of each item are in each gift bag?

54. **VENN DIAGRAM** Consider the numbers 252, 270, and 300.

   a. Create a Venn diagram using the prime factors of the numbers.

   b. Use the Venn diagram to find the GCF of 252, 270, and 300.

   c. What is the GCF of 252 and 270? 252 and 300? 270 and 300? Explain how you found your answers.

55. **MP REASONING** You are making fruit baskets using 54 apples, 36 oranges, and 73 bananas.

   a. Explain why you cannot make identical fruit baskets without leftover fruit.

   b. What is the greatest number of identical fruit baskets you can make with the least amount of fruit left over? Explain how you found your answer.

56. **DIG DEEPER!** Two rectangular, adjacent rooms share a wall. One-foot-by-one-foot tiles cover the floor of each room. Describe how the greatest possible length of the adjoining wall is related to the total number of tiles in each room. Draw a diagram that represents one possibility.

1.5 Least Common Multiple

Learning Target: Find the least common multiple of two numbers.

Success Criteria:
- I can explain the meaning of multiples of a number.
- I can use lists of multiples to identify the least common multiple of numbers.
- I can use prime factors to identify the least common multiple of numbers.

EXPLORATION 1
Identifying Common Multiples

Work with a partner. In parts (a)–(d), create a Venn diagram that represents the first several multiples of each number and identify any *common multiples*.

a. 8 and 12

b. 4 and 14

c. 10 and 15

d. 20 and 35

e. Look at the Venn diagrams in parts (a)–(d). Explain how to identify the *least common multiple* of each pair of numbers. Then circle it in each diagram.

EXPLORATION 2
Using Prime Factors

Work with a partner.

Math Practice

Analyze Conjectures

How can you test your conjecture in part (c)?

a. Create a Venn diagram that represents the prime factorizations of 8 and 12.

b. Repeat part (a) for the remaining number pairs in Exploration 1.

c. **MP STRUCTURE** Make a conjecture about the relationship between the least common multiples you found in Exploration 1 and the numbers in the Venn diagrams you just created.

d. The Venn diagram shows the prime factors of two numbers.

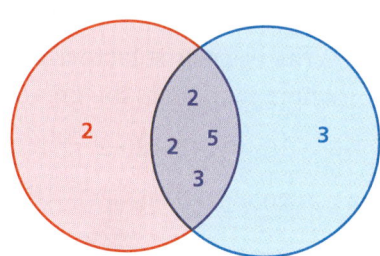

Use the diagram to complete the following tasks.

- Identify the two numbers.
- Find the greatest common factor.
- Find the least common multiple.

Section 1.5   Least Common Multiple   27

1.5 Lesson

Key Vocabulary
common multiples, p. 28
least common multiple, p. 28

Multiples that are shared by two or more numbers are called **common multiples**. The least of the common multiples is called the **least common multiple** (LCM). You can find the LCM of two or more numbers by listing multiples or using prime factors.

EXAMPLE 1 — Finding the LCM Using Lists of Multiples

Find the LCM of 4 and 6.

List the multiples of each number.

Multiples of 4: 4, 8, ⓐ, 16, 20, ㉔, 28, 32, ㊱, . . .   Circle the common multiples.

Multiples of 6: 6, ⓐ, 18, ㉔, 30, ㊱, . . .

Some common multiples of 4 and 6 are 12, 24, and 36. The least of these common multiples is 12.

▸ So, the LCM of 4 and 6 is 12.

Try It Find the LCM of the numbers using lists of multiples.

1. 3, 8
2. 9, 12
3. 6, 10

EXAMPLE 2 — Finding the LCM Using Prime Factorizations

Find the LCM of 16 and 20.

Make a factor tree for each number.

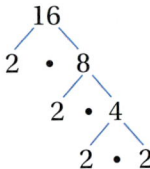 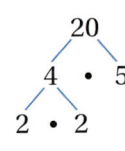

Write the prime factorization of each number. Circle each different factor where it appears the greater number of times.

16 = ②·②·②·②   2 appears more often here, so circle all 2s.
20 = 2 · 2 · ⑤   5 appears once. Do not circle the 2s again.
2 · 2 · 2 · 2 · 5 = 80   Find the product of the circled factors.

▸ So, the LCM of 16 and 20 is 80.

Try It Find the LCM of the numbers using prime factorizations.

4. 14, 18
5. 28, 36
6. 24, 90

EXAMPLE 3 **Finding the LCM of Three Numbers**

Find the LCM of 4, 15, and 18.

Write the prime factorization of each number. Circle each different factor where it appears the greatest number of times.

4 = ②•②	2 appears most often here, so circle both 2s.
15 = 3 •⑤	5 appears here only, so circle 5.
18 = 2 •③•③	3 appears most often here, so circle both 3s.
2 • 2 • 5 • 3 • 3 = 180	Find the product of the circled factors.

 So, the LCM of 4, 15, and 18 is 180.

Try It

Find the LCM of the numbers.

7. 2, 5, 8   **8.** 6, 10, 12

9. Write three numbers that have a least common multiple of 100.

Self-Assessment for Concepts & Skills

Solve each exercise. Then rate your understanding of the success criteria in your journal.

FINDING THE LCM Find the LCM of the numbers.

10. 6, 9   **11.** 30, 40   **12.** 5, 11

13. **MP REASONING** Write two numbers such that 18 and 30 are multiples of the numbers. Justify your answer.

14. **MP REASONING** You need to find the LCM of 13 and 14. Would you rather list their multiples or use their prime factorizations? Explain.

15. **MP CHOOSE TOOLS** A student writes the prime factorizations of 8 and 12 in a table as shown. She claims she can use the table to find the greatest common factor and the least common multiple of 8 and 12. How is this possible?

8 =	2	2	2	
12 =	2	2		3

16. **CRITICAL THINKING** How can you use least common multiples to add or subtract fractions with different denominators?

EXAMPLE 4   Modeling Real Life

One firefly flashes every 8 seconds. Another firefly flashes every 10 seconds. Both fireflies just flashed. After how many seconds will both fireflies flash at the same time again?

Understand the problem.

You are given the numbers of seconds between flashes for two different fireflies. You are asked when the fireflies will flash at the same time again.

Make a plan.

The LCM of the numbers of seconds between flashes represents the number of seconds it will take for both fireflies to flash at the same time again. So, find the LCM of 8 and 10 by listing the multiples of each number.

Solve and check.

Multiples of 8: 8, 16, 24, 32, (40), . . .

Multiples of 10: 10, 20, 30, (40), 50, . . .

The LCM of 8 and 10 is 40.

▶ So, both fireflies will flash at the same time again after 40 seconds.

> **Another Method** Find the LCM using prime factorizations.
>
> $8 = 2 \cdot 2 \cdot 2$ $10 = 2 \cdot 5$
>
> So, the LCM is $2 \cdot 2 \cdot 2 \cdot 5 = 40$. ✓

Self-Assessment for Problem Solving

Solve each exercise. Then rate your understanding of the success criteria in your journal.

17. A geyser erupts every fourth day. Another geyser erupts every sixth day. Today both geysers erupted. In how many days will both geysers erupt on the same day again?

18. A water park has two large buckets that slowly fill with water. One bucket dumps water every 12 minutes. The other bucket dumps water every 10 minutes. Five minutes ago, both buckets dumped water. When will both buckets dump water at the same time again?

19. **DIG DEEPER!** You purchase disposable plates, cups, and forks for a cookout. Plates are sold in packages of 24, cups in packages of 32, and forks in packages of 48. What are the least numbers of packages you should buy in order to have the same number of plates, cups, and forks?

1.5 Practice

Review & Refresh

Find the GCF of the numbers.

1. 18, 42
2. 72, 96
3. 38, 76, 114

Divide.

4. 900 ÷ 6
5. 1944 ÷ 9
6. 672 ÷ 12

Write an ordered pair that corresponds to the point.

7. Point A
8. Point B
9. Point C
10. Point D

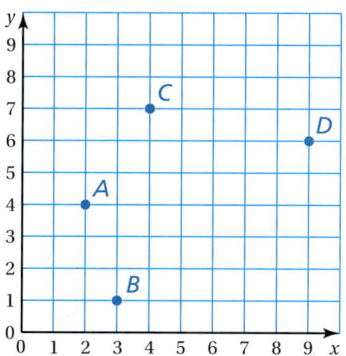

Concepts, Skills, & Problem Solving

USING A VENN DIAGRAM Use a Venn diagram to find the least common multiple of the numbers. (See Exploration 1, p. 27.)

11. 3, 7
12. 6, 8
13. 4, 5

FINDING THE LCM Find the LCM of the numbers using lists of multiples.

14. 1, 5
15. 2, 6
16. 2, 3
17. 2, 9
18. 3, 4
19. 8, 9
20. 5, 8
21. 11, 12
22. 12, 18

FINDING THE LCM Find the LCM of the numbers using prime factorizations.

23. 7, 12
24. 5, 9
25. 4, 11
26. 9, 10
27. 12, 27
28. 18, 45
29. 22, 33
30. 36, 60
31. 35, 50

32. **YOU BE THE TEACHER** Your friend finds the LCM of 6 and 9. Is your friend correct? Explain your reasoning.

    > 6 × 9 = 54
    > The LCM of 6 and 9 is 54.

33. **MODELING REAL LIFE** You have diving lessons every fifth day and swimming lessons every third day. Today you have both lessons. In how many days will you have both lessons on the same day again?

34. **MP REASONING** Which model represents an LCM that is different from the other three? Explain your reasoning.

A.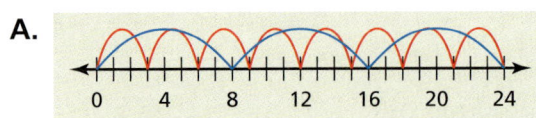

B.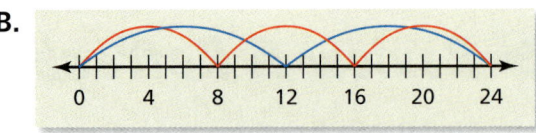

C.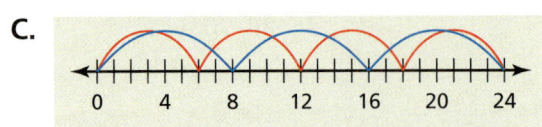

D.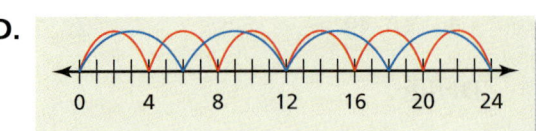

FINDING THE LCM Find the LCM of the numbers.

35. 2, 3, 7

36. 3, 5, 11

37. 4, 9, 12

38. 6, 8, 15

39. 7, 18, 21

40. 9, 10, 28

41. **MP PROBLEM SOLVING** At Union Station, you notice that three subway lines just arrived at the same time. How long must you wait until all three lines arrive at Union Station at the same time again?

Subway Line	Arrival Time
A	Every 10 min
B	Every 12 min
C	Every 15 min

42. **DIG DEEPER!** A radio station gives away $15 to every 15th caller, $25 to every 25th caller, and a free concert ticket to every 100th caller. When will the station first give away *all* three prizes to one caller? When this happens, how much money and how many tickets are given away?

43. **MP LOGIC** You and a friend are running on treadmills. You run 0.5 mile every 3 minutes, and your friend runs 2 miles every 14 minutes. You both start and stop running at the same time and run a whole number of miles. What are the least possible numbers of miles you and your friend can run?

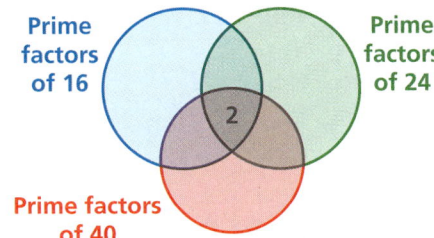

44. **VENN DIAGRAM** Refer to the Venn diagram.

   a. Copy and complete the Venn diagram.

   b. What is the LCM of 16, 24, and 40?

   c. What is the LCM of 16 and 40? 24 and 40? 16 and 24? Explain how you found your answers.

CRITICAL THINKING Tell whether the statement is *always*, *sometimes*, or *never* true. Explain your reasoning.

45. The LCM of two different prime numbers is their product.

46. The LCM of a set of numbers is equal to one of the numbers in the set.

47. The GCF of two different numbers is the LCM of the numbers.

Connecting Concepts

Problem-Solving Strategies

Using an appropriate strategy will help you make sense of problems as you study the mathematics in this course. You can use the following strategies to solve problems that you encounter.

- Use a verbal model.
- Draw a diagram.
- Write an equation.
- Solve a simpler problem.
- Sketch a graph or number line.
- Make a table.
- Make a list.
- Break the problem into parts.

Using the Problem-Solving Plan

1. A sports team gives away shirts at the stadium. There are 60 large shirts, 1.6 times as many small shirts as large shirts, and 1.5 times as many medium shirts as small shirts. The team wants to divide the shirts into identical groups to be distributed throughout the stadium. What is the greatest number of groups that can be formed using every shirt?

   Understand the problem. You know the number of large shirts and two relationships among the numbers of small, medium, and large shirts. You are asked to find the greatest number of identical groups that can be formed using every shirt.

   Make a plan. Break the problem into parts. First use multiplication to find the number of each size shirt. Then find the GCF of these numbers.

   Solve and check. Use the plan to solve the problem. Then check your solution.

2. An escape artist fills the tank shown with water. Find the number of cubic feet of water needed to fill the tank. Then find the number of cubic yards of water that are needed to fill the tank. Justify your answer.

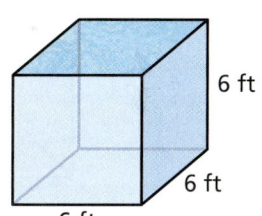

Performance Task

Setting the Table

At the beginning of this chapter, you watched a STEAM video called "Filling Piñatas." You are now ready to complete the performance task for this video, available at *BigIdeasMath.com*. Be sure to use the problem-solving plan as you work through the performance task.

Connecting Concepts 33

Chapter Review

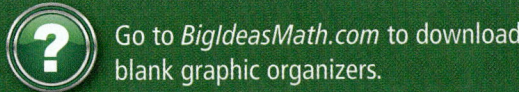

Review Vocabulary

Write the definition and give an example of each vocabulary term.

power, *p. 4*
base, *p. 4*
exponent, *p. 4*
perfect square, *p. 5*
numerical expression, *p. 10*
evaluate, *p. 10*
order of operations, *p. 10*
factor pair, *p. 16*
prime factorization, *p. 16*
factor tree, *p. 16*
Venn diagram, *p. 21*
common factors, *p. 22*
greatest common factor, *p. 22*
common mutliples, *p. 28*
least common multiple, *p. 28*

Graphic Organizers

You can use an **Information Frame** to organize and remember concepts. Here is an example of an Information Frame for the vocabulary term *power*.

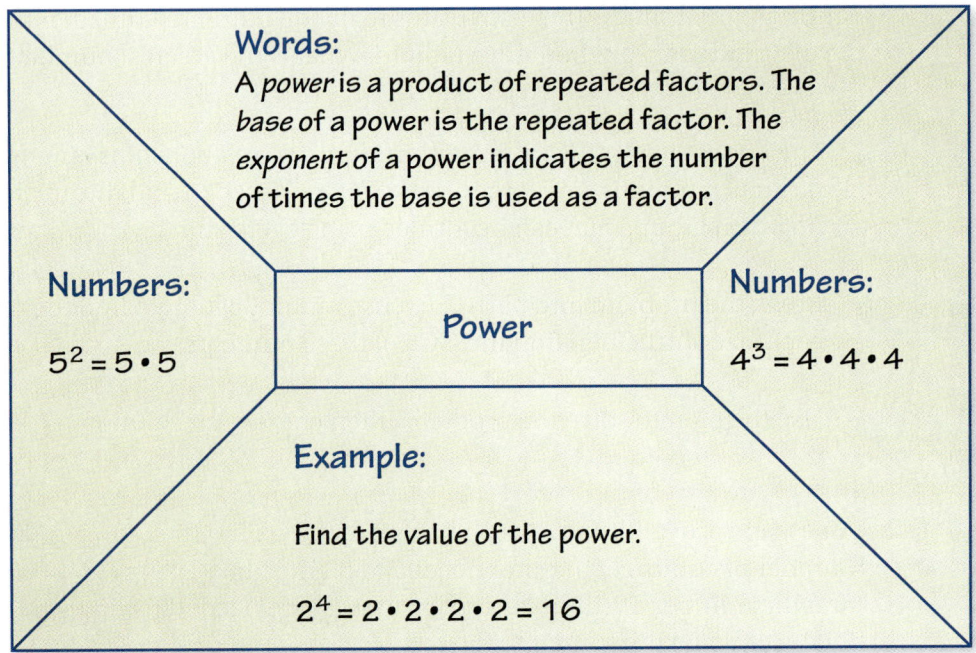

Choose and complete a graphic organizer to help you study the concept.

1. perfect square
2. numerical expression
3. order of operations
4. prime factorization
5. greatest common factor (GCF)
6. least common multiple (LCM)

34 Chapter 1   Numerical Expressions and Factors

Chapter Self-Assessment

As you complete the exercises, use the scale below to rate your understanding of the success criteria in your journal.

1	2	3	4
I do not understand.	I can do it with help.	I can do it on my own.	I can teach someone else.

1.1 Powers and Exponents (pp. 3–8)

Learning Target: Write and evaluate expressions involving exponents.

Write the product as a power.

1. $3 \times 3 \times 3 \times 3 \times 3 \times 3$
2. $5 \times 5 \times 5$
3. $17 \cdot 17 \cdot 17 \cdot 17 \cdot 17$

Find the value of the power.

4. 3^3
5. 2^6
6. 4^4

7. Write a power that has a value greater than 2^3 and less than 3^3.

8. Without evaluating, determine whether 2^5 or 4^2 is greater. Explain.

9. The bases on a softball field are square. What is the area of each base?

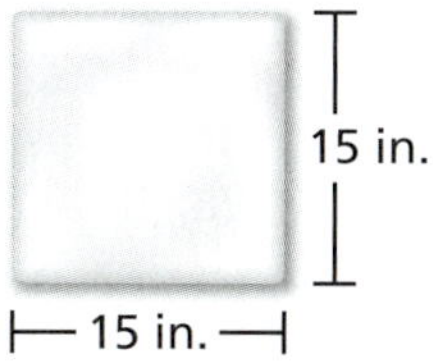

15 in. × 15 in.

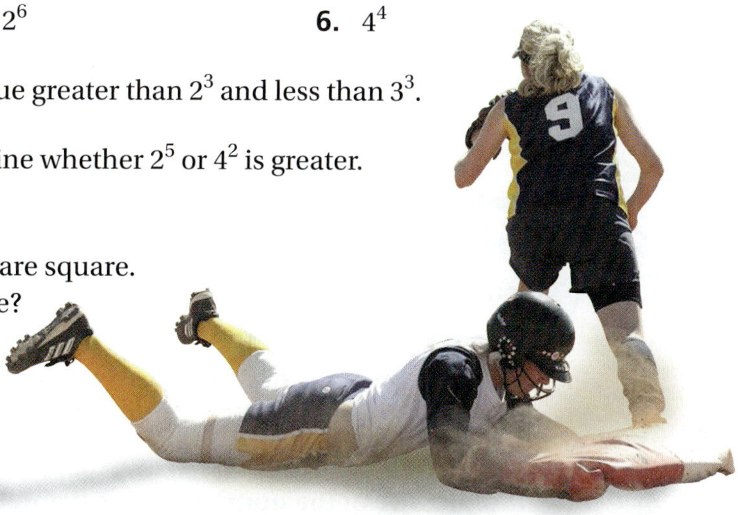

1.2 Order of Operations (pp. 9–14)

Learning Target: Write and evaluate numerical expressions using the order of operations.

Evaluate the expression.

10. $3 \times 6 - 12 \div 6$
11. $30 \div (14 - 2^2) \times 5$
12. $\dfrac{5(2.3 + 3.7)}{2}$
13. $4^3 - \dfrac{1}{2}(7^2 + 5)$
14. $20 \times (3^2 - 4) \div 50$
15. $5 + 3(4^2 - 2) \div 6$

16. Use grouping symbols and at least one exponent to write a numerical expression that has a value of 80.

1.3 Prime Factorization (pp. 15–20)

Learning Target: Write a number as a product of prime factors and represent the product using exponents.

List the factor pairs of the number.

17. 28 **18.** 44 **19.** 96

20. There are 36 graduated cylinders to put away on a shelf after science class. The shelf can fit a maximum of 20 cylinders across and 4 cylinders deep. The teacher wants each row to have the same number of cylinders. List the possible arrangements of the graduated cylinders on the shelf.

Write the prime factorization of the number.

21. 42 **22.** 50 **23.** 66

1.4 Greatest Common Factor (pp. 21–26)

Learning Target: Find the greatest common factor of two numbers.

Find the GCF of the numbers using lists of factors.

24. 27, 45 **25.** 30, 48 **26.** 28, 48

Find the GCF of the numbers using prime factorizations.

27. 24, 80 **28.** 52, 68 **29.** 32, 56

30. Write a pair of numbers that have a GCF of 20.

31. What is the greatest number of friends you can invite to an arcade using the coupon such that the tokens and slices of pizza are equally split between you and your friends with none left over? How many slices of pizza and tokens will each person receive?

1.5 Least Common Multiple (pp. 27–32)

Learning Target: Find the least common multiple of two numbers.

Find the LCM of the numbers using lists of multiples.

32. 4, 14　　　　**33.** 6, 20　　　　**34.** 12, 28

Find the LCM of the numbers using prime factorizations.

35. 6, 45　　　　**36.** 10, 12　　　　**37.** 18, 27

38. Find the LCM of 8, 12, and 18.

39. Write a pair of numbers that have an LCM of 84.

40. Write three numbers that have an LCM of 45.

41. You water your roses every sixth day and your hydrangeas every fifth day. Today you water both plants. In how many days will you water both plants on the same day again?

42. Hamburgers are sold in packages of 20, while buns are sold in packages of 12. What are the least numbers of packages you should buy in order to have the same number of hamburgers and buns?

43. A science museum is giving away a magnetic liquid kit to every 50th guest and a plasma ball to every 35th guest until someone receives both prizes.

　a. Which numbered guest will receive both a magnetic liquid kit and a plasma ball?

　b. How many people will receive a plasma ball?

Practice Test

1. Find the value of 2^3.

2. Evaluate $\dfrac{5 + 4(12 - 2)}{3^2}$.

3. Write 264 • 264 • 264 as a power.

4. List the factor pairs of 66.

5. Write the prime factorization of 56.

Find the GCF of the numbers.

6. 24, 54
7. 16, 32, 72
8. 52, 65

Find the LCM of the numbers.

9. 9, 24
10. 26, 39
11. 6, 12, 14

12. You have 16 yellow beads, 20 red beads, and 24 orange beads to make identical bracelets. What is the greatest number of bracelets that you can make using all of the beads?

13. A bag contains equal numbers of green marbles and blue marbles. You can divide all of the green marbles into groups of 12 and all the blue marbles into groups of 16. What is the least number of each color of marble that can be in the bag?

14. The ages of the members of a family are 65, 58, 27, 25, 5, and 2 years old. What is the total admission price for the family to visit the zoo?

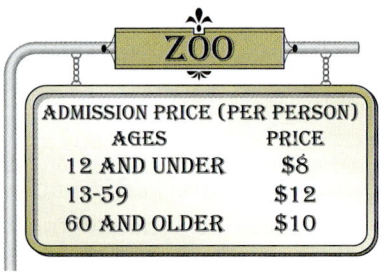

15. A competition awards prizes for fourth, third, second, and first place. The fourth place winner receives $5. Each place above that receives a prize that is five times the amount of the previous prize. How much prize money is awarded?

16. You buy tealight candles and mints as party favors for a baby shower. The tealight candles come in packs of 12 for $3.50. The mints come in packs of 50 for $6.25. What is the least amount of money you can spend to buy the same number of candles and mints?

1 Cumulative Practice

1. What is the value of 8×135?

2. Which number is equivalent to the expression below?

   $3 \cdot 2^3 - 8 \div 4$

   A. 0
   B. 4
   C. 22
   D. 214

3. The top of an end table is a square with a side length of 16 inches. What is the area of the tabletop?

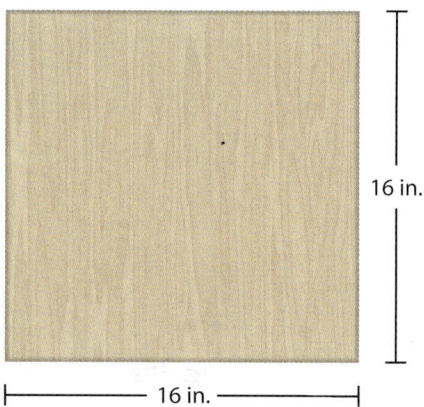

   F. 16 in.^{2}
   G. 32 in.^{2}
   H. 64 in.^{2}
   I. 256 in.^{2}

4. You are filling baskets using 18 green eggs, 36 red eggs, and 54 blue eggs. What is the greatest number of baskets that you can fill so that the baskets are identical and there are no eggs left over?

   A. 3
   B. 6
   C. 9
   D. 18

5. What is the value of $2^3 \cdot 3^2 \cdot 5$?

6. You hang the two strands of decorative lights shown below.

   Strand 1: changes between red and blue every 15 seconds

   Strand 2: changes between green and gold every 18 seconds

   Both strands just changed color. After how many seconds will the strands change color at the same time again?

   F. 3 seconds   **G.** 30 seconds

   H. 90 seconds   **I.** 270 seconds

7. Point *P* is plotted in the coordinate plane below.

   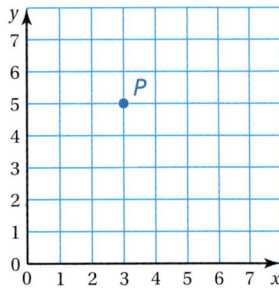

   What are the coordinates of Point *P*?

   A. (5, 3)   **B.** (4, 3)

   C. (3, 5)   **D.** (3, 4)

8. What is the prime factorization of 1100?

   F. $2 \times 5 \times 11$ **G.** $2^2 \times 5^2 \times 11$

   H. $4 \times 5^2 \times 11$ **I.** $2^2 \times 5 \times 55$

9. What is the least common multiple of 3, 8, and 10?

　A. 24　　　　　　　　**B.** 30

　C. 80　　　　　　　　**D.** 120

10. What is the area of the shaded region of the figure below?

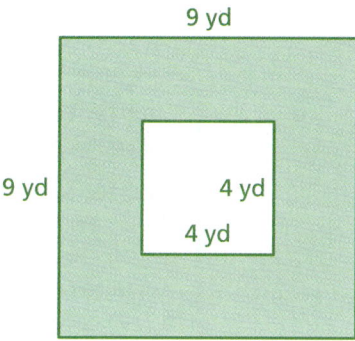

　F. 16 yd²　　　　　　**G.** 65 yd²

　H. 81 yd²　　　　　　**I.** 97 yd²

11. Which expression represents a prime factorization?

　A. $4 \times 4 \times 7$　　　　　**B.** $2^2 \times 21 \times 23$

　C. $3^4 \times 5 \times 7$　　　　　**D.** $5 \times 5 \times 9 \times 11$

12. Find the greatest common factor for each pair of numbers.

　　10 and 15　　10 and 21　　15 and 21

What can you conclude about the greatest common factor of 10, 15, and 21? Explain your reasoning.

Cumulative Practice　41

2 Fractions and Decimals

- **2.1** Multiplying Fractions
- **2.2** Dividing Fractions
- **2.3** Dividing Mixed Numbers
- **2.4** Adding and Subtracting Decimals
- **2.5** Multiplying Decimals
- **2.6** Dividing Whole Numbers
- **2.7** Dividing Decimals

Chapter Learning Target:
Understand fractions and decimals.

Chapter Success Criteria:
- ■ I can identify a fraction and a decimal.
- ■ I can add, subtract, multiply, and divide fractions and decimals.
- ■ I can evaluate expressions involving fractions and decimals using the order of operations.
- ■ I can solve a problem using fractions and decimals.

STEAM Video: "Space is Big"

STEAM Video

Space is Big

An astronomical unit (AU) is the average distance between Earth and the Sun, about 93 million miles. Why do astronomers use astronomical units to measure distances in space? In what different ways can you compare the distances between objects and the locations of objects using the four mathematical operations?

Watch the STEAM Video "Space is Big." Then answer the following questions.

1. You know the distances between the Sun and each planet. How can you find the minimum and maximum distances between two planets as they rotate around the Sun?

2. The table shows the distances of three celestial bodies from Earth. It takes about three days to travel from Earth to the Moon. How can you estimate the amount of time it would take to travel from Earth to the Sun or to Venus?

Celestial body	Sun	Moon	Venus
Distance from Earth (AU)	1	0.00256	0.277

Performance Task

Space Explorers

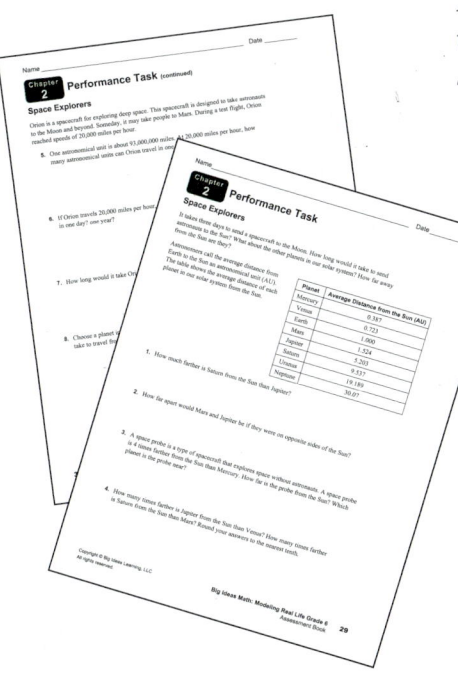

After completing this chapter, you will be able to use the concepts you learned to answer the questions in the *STEAM Video Performance Task*.

You will use a table that shows the average distances between the Sun and each planet in our solar system to find several distances in space. Then you will use the speed of the Orion spacecraft to answer questions about time and distance.

Is it realistic for a manned spacecraft to travel to each planet in our solar system? Explain why or why not.

43

Getting Ready for Chapter 2

Chapter Exploration

Work with a partner. The area model represents the multiplication of two fractions. Copy and complete the statement.

1.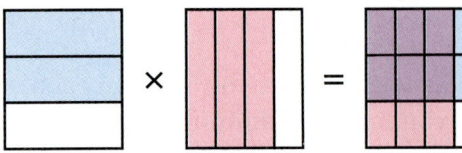

   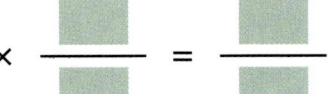

2.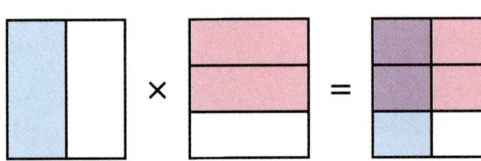

   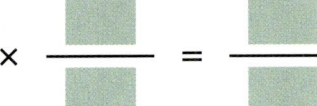

3.

   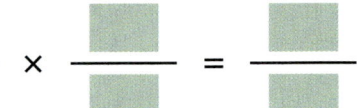

4.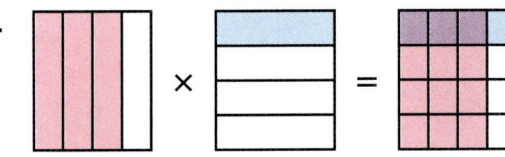

   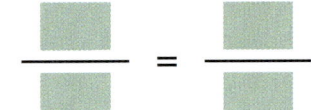

Work with a partner. Use an area model to find the product.

5. $\dfrac{1}{2} \times \dfrac{1}{3}$

6. $\dfrac{4}{5} \times \dfrac{1}{4}$

7. $\dfrac{1}{6} \times \dfrac{3}{4}$

8. $\dfrac{3}{5} \times \dfrac{1}{4}$

9. **MODELING REAL LIFE** You have a recipe that serves 6 people. The recipe uses three-fourths of a cup of milk.

   a. How can you use the recipe to serve *more* people? How much milk would you need? Give 2 examples.

   b. How can you use the recipe to serve *fewer* people? How much milk would you need? Give 2 examples.

Vocabulary

The following vocabulary terms are defined in this chapter. Think about what each term might mean and record your thoughts.

reciprocals   multiplicative inverses

2.1 Multiplying Fractions

Learning Target: Find products involving fractions and mixed numbers.

Success Criteria:
- I can draw a model to explain fraction multiplication.
- I can multiply fractions.
- I can find products involving mixed numbers.
- I can interpret products involving fractions and mixed numbers to solve real-life problems.

EXPLORATION 1

Using Models to Solve a Problem

Work with a partner. A bottle of water is $\frac{1}{2}$ full. You drink $\frac{2}{3}$ of the water.

Use one of the models to find the portion of the bottle of water that you drink. Explain your steps.

- number line

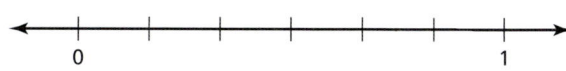

- area model

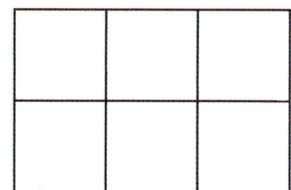

- tape diagram

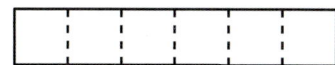

EXPLORATION 2

Solving a Problem Involving Fractions

Math Practice

Find General Methods
How can you use your answer to find a method for multiplying fractions?

Work with a partner. A park has a playground that is $\frac{3}{4}$ of its width and $\frac{4}{5}$ of its length.

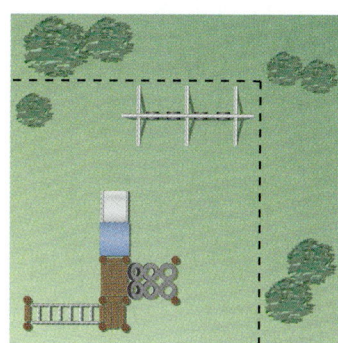

a. Use a model to find the portion of the park that is covered by the playground. Explain your steps.

b. How can you find the solution of part (a) without using a model?

Section 2.1 Multiplying Fractions 45

2.1 Lesson

EXAMPLE 1 **Multiplying Fractions**

Think: What is one-fourth of one-half?

Find $\dfrac{1}{4} \times \dfrac{1}{2}$.

Use a model to find $\dfrac{1}{4}$ of $\dfrac{1}{2}$.

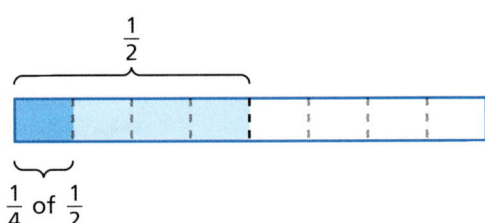

▶ So, the product is $\dfrac{1}{8}$.

Try It Multiply.

1. $\dfrac{1}{3} \times \dfrac{1}{5}$ 2. $\dfrac{2}{3} \times \dfrac{3}{4}$ 3. $\dfrac{1}{2} \cdot \dfrac{5}{6}$

🔑 Key Idea

Multiplying Fractions

Words Multiply the numerators and multiply the denominators.

Numbers $\dfrac{3}{7} \times \dfrac{1}{2} = \dfrac{3 \times 1}{7 \times 2} = \dfrac{3}{14}$

Algebra $\dfrac{a}{b} \cdot \dfrac{c}{d} = \dfrac{a \cdot c}{b \cdot d}$, where $b, d \neq 0$

EXAMPLE 2 **Multiplying Fractions**

Find $\dfrac{8}{9} \cdot \dfrac{3}{4}$.   **Estimate** $1 \cdot \dfrac{3}{4} = \dfrac{3}{4}$

Remember

To simplify a fraction, write an equivalent fraction whose numerator and denominator have no common factors other than 1.

$\dfrac{24}{36} = \dfrac{24 \div 12}{36 \div 12} = \dfrac{2}{3}$

$\dfrac{8}{9} \cdot \dfrac{3}{4} = \dfrac{8 \cdot 3}{9 \cdot 4}$ Multiply the numerators and the denominators.

$= \dfrac{24}{36}$, or $\dfrac{2}{3}$ Simplify.

The symbol $\approx$ means *is approximately equal to.*

▶ The product is $\dfrac{2}{3}$.   **Reasonable?** $\dfrac{2}{3} \approx \dfrac{3}{4}$ ✓

Try It Multiply. Write the answer in simplest form.

4. $\dfrac{3}{7} \times \dfrac{2}{3}$ 5. $\dfrac{4}{9} \cdot \dfrac{3}{10}$ 6. $\dfrac{6}{5} \cdot \dfrac{5}{8}$

46   Chapter 2   Fractions and Decimals

EXAMPLE 3 Solving a Problem Involving Fractions

You have $\frac{2}{3}$ of a bag of flour. You use $\frac{3}{4}$ of the flour to make empanada dough. How much of the entire bag do you use to make the dough?

You use $\frac{3}{4}$ of $\frac{2}{3}$ of the bag. To find $\frac{3}{4}$ of $\frac{2}{3}$, multiply.

$$\frac{3}{4} \times \frac{2}{3} = \frac{3 \times 2}{4 \times 3} \quad \text{Multiply the numerators and the denominators.}$$

$$= \frac{6}{12}, \text{ or } \frac{1}{2} \quad \text{Simplify.}$$

▶ So, you use $\frac{1}{2}$ of the entire bag.

Try It

7. **WHAT IF?** You use $\frac{1}{4}$ of the flour to make the dough. How much of the entire bag do you use to make the dough?

🔑 Key Idea

Multiplying Mixed Numbers

Write each mixed number as an improper fraction. Then multiply as you would with fractions.

EXAMPLE 4 Multiplying a Fraction and a Mixed Number

Find $\frac{1}{2} \times 2\frac{3}{4}$.   Estimate $\frac{1}{2} \times 3 = 1\frac{1}{2}$

Another Method
Use a model.

So, $\frac{1}{2} \times 2\frac{3}{4} = 1 + \frac{3}{8} = 1\frac{3}{8}$. ✓

$$\frac{1}{2} \times 2\frac{3}{4} = \frac{1}{2} \times \frac{11}{4} \quad \text{Write } 2\frac{3}{4} \text{ as the improper fraction } \frac{11}{4}.$$

$$= \frac{1 \times 11}{2 \times 4} \quad \text{Multiply the numerators and the denominators.}$$

$$= \frac{11}{8}, \text{ or } 1\frac{3}{8} \quad \text{Simplify.}$$

▶ The product is $1\frac{3}{8}$.   **Reasonable?** $1\frac{3}{8} \approx 1\frac{1}{2}$ ✓

Try It Multiply. Write the answer in simplest form.

8. $\frac{1}{3} \times 1\frac{1}{6}$

9. $3\frac{1}{2} \times \frac{4}{9}$

10. $4\frac{2}{3} \cdot \frac{3}{4}$

EXAMPLE 5 Multiplying Mixed Numbers

Find $1\frac{4}{5} \times 3\frac{2}{3}$.　　　　　Estimate $2 \times 4 = 8$

$1\frac{4}{5} \times 3\frac{2}{3} = \frac{9}{5} \times \frac{11}{3}$　　Write $1\frac{4}{5}$ and $3\frac{2}{3}$ as improper fractions.

$= \frac{9 \times 11}{5 \times 3}$　　Multiply the numerators and the denominators.

$= \frac{99}{15}$, or $6\frac{3}{5}$　　Simplify.

▶ The product is $6\frac{3}{5}$.　　Reasonable? $6\frac{3}{5} \approx 8$ ✓

Try It Multiply. Write the answer in simplest form.

11. $1\frac{7}{8} \cdot 2\frac{2}{5}$

12. $5\frac{5}{7} \times 2\frac{1}{10}$

13. $2\frac{1}{3} \cdot 7\frac{2}{3}$

Self-Assessment for Concepts & Skills

Solve each exercise. Then rate your understanding of the success criteria in your journal.

MULTIPLYING FRACTIONS AND MIXED NUMBERS Multiply. Write the answer in simplest form.

14. $\frac{1}{8} \times \frac{1}{6}$

15. $\frac{3}{8} \cdot \frac{2}{3}$

16. $2\frac{1}{6} \cdot 4\frac{2}{5}$

17. **MP REASONING** What is the missing denominator?

$$\frac{3}{7} \times \frac{1}{\boxed{}} = \frac{3}{28}$$

18. **USING TOOLS** Write a multiplication problem involving fractions that is represented by the model. Explain your reasoning.

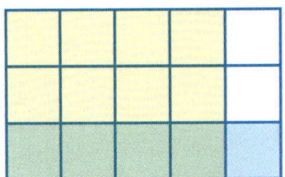

19. **USING TOOLS** Use the number line to find $\frac{3}{4} \times \frac{1}{2}$. Explain your reasoning.

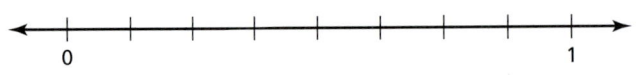

EXAMPLE 6 Modeling Real Life

A city is resurfacing a basketball court. Find the area of the court.

Understand the problem. You are given the dimensions of a basketball court. You are asked to find the area of the court.

Make a plan. Use the formula for the area of a rectangle. Find the product of the length and the width of the court.

Solve and check.

$$A = \ell w \quad \text{Write the formula.}$$

$$= 21\tfrac{1}{3} \cdot 13\tfrac{1}{2} \quad \text{Substitute for } \ell \text{ and } w.$$

$$= \tfrac{64}{3} \cdot \tfrac{27}{2} \quad \text{Write } 21\tfrac{1}{3} \text{ and } 13\tfrac{1}{2} \text{ as improper fractions.}$$

$$= \tfrac{64 \cdot 27}{3 \cdot 2} \quad \text{Multiply the numerators and the denominators.}$$

$$= \tfrac{1728}{6}, \text{ or } 288 \quad \text{Simplify.}$$

So, the area of the court is 288 square meters.

Check Reasonableness
Find an underestimate and an overestimate.

Underestimate:
13 • 21 = 273

Overestimate:
14 • 22 = 308

The answer is reasonable because 273 < 288 < 308. ✓

(Court dimensions: $21\tfrac{1}{3}$ m by $13\tfrac{1}{2}$ m)

Self-Assessment for Problem Solving

Solve each exercise. Then rate your understanding of the success criteria in your journal.

20. You spend $\tfrac{5}{12}$ of a day at an amusement park. You spend $\tfrac{2}{5}$ of that time riding waterslides. How many hours do you spend riding waterslides? Draw a model to show why your answer makes sense.

21. A venue is preparing for a concert on the floor shown. The width of the red carpet is $\tfrac{1}{6}$ of the width of the floor. What is the area of the red carpet?

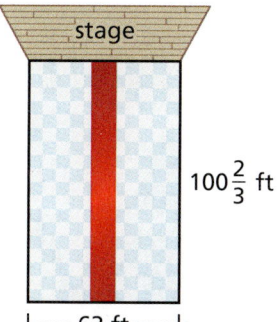

(Floor: $100\tfrac{2}{3}$ ft by 63 ft, with stage)

22. You travel $9\tfrac{3}{8}$ miles from your house to a shopping mall. You travel $\tfrac{2}{3}$ of that distance on an interstate. The only road construction you encounter is on the first $\tfrac{2}{5}$ of the interstate. On how many miles of your trip do you encounter construction?

Section 2.1 Multiplying Fractions

2.1 Practice

Go to **BigIdeasMath.com** to get HELP with solving the exercises.

Review & Refresh

Find the LCM of the numbers.

1. 8, 10
2. 5, 7
3. 2, 5, 7
4. 6, 7, 10

Divide. Use a diagram to justify your answer.

5. $6 \div \frac{1}{2}$
6. $\frac{1}{4} \div 8$
7. $4 \div \frac{1}{3}$
8. $\frac{1}{5} \div 4$

Write the product as a power.

9. $10 \times 10 \times 10$
10. $5 \times 5 \times 5 \times 5$

11. How many inches are in $5\frac{1}{2}$ yards?

    A. $15\frac{1}{2}$
    B. $16\frac{1}{2}$
    C. 66
    D. 198

Concepts, Skills, & Problem Solving

CHOOSE TOOLS A bottle of water is $\frac{2}{3}$ full. You drink the given portion of the water. Use a model to find the portion of the bottle of water that you drink. (See Exploration 1, p. 45.)

12. $\frac{1}{2}$
13. $\frac{1}{4}$
14. $\frac{3}{4}$

MULTIPLYING FRACTIONS Multiply. Write the answer in simplest form.

15. $\frac{1}{7} \times \frac{2}{3}$
16. $\frac{5}{8} \cdot \frac{1}{2}$
17. $\frac{1}{4} \times \frac{2}{5}$
18. $\frac{3}{7} \times \frac{1}{4}$

19. $\frac{2}{3} \times \frac{4}{7}$
20. $\frac{5}{7} \times \frac{7}{8}$
21. $\frac{3}{8} \cdot \frac{1}{9}$
22. $\frac{5}{6} \cdot \frac{2}{5}$

23. $\frac{5}{12} \times 10$
24. $6 \cdot \frac{7}{8}$
25. $\frac{3}{4} \times \frac{8}{15}$
26. $\frac{4}{9} \times \frac{4}{5}$

27. $\frac{3}{7} \cdot \frac{3}{7}$
28. $\frac{5}{6} \times \frac{2}{9}$
29. $\frac{13}{18} \times \frac{6}{7}$
30. $\frac{7}{9} \cdot \frac{21}{10}$

31. **MODELING REAL LIFE** In an aquarium, $\frac{2}{5}$ of the fish are surgeonfish. Of these, $\frac{3}{4}$ are yellow tangs. What portion of all fish in the aquarium are yellow tangs?

32. **MODELING REAL LIFE** You exercise for $\frac{3}{4}$ of an hour. You jump rope for $\frac{1}{3}$ of that time. What portion of the hour do you spend jumping rope?

50 Chapter 2 Fractions and Decimals

MP REASONING Without finding the product, copy and complete the statement using <, >, or =. Explain your reasoning.

33. $\frac{4}{7}$ ☐ $\frac{9}{10} \cdot \frac{4}{7}$

34. $\frac{5}{8} \times \frac{22}{15}$ ☐ $\frac{5}{8}$

35. $\frac{5}{6}$ ☐ $\frac{5}{6} \times \frac{7}{7}$

MULTIPLYING FRACTIONS AND MIXED NUMBERS Multiply. Write the answer in simplest form.

36. $1\frac{1}{3} \cdot \frac{2}{3}$

37. $6\frac{2}{3} \times \frac{3}{10}$

38. $2\frac{1}{2} \cdot \frac{4}{5}$

39. $\frac{3}{5} \cdot 3\frac{1}{3}$

40. $7\frac{1}{2} \times \frac{2}{3}$

41. $\frac{5}{9} \times 3\frac{3}{5}$

42. $\frac{3}{4} \cdot 1\frac{1}{3}$

43. $3\frac{3}{4} \times \frac{2}{5}$

44. $4\frac{3}{8} \cdot \frac{4}{5}$

45. $\frac{3}{7} \times 2\frac{5}{6}$

46. $1\frac{3}{10} \times 18$

47. $15 \cdot 2\frac{4}{9}$

48. $1\frac{1}{6} \times 6\frac{3}{4}$

49. $2\frac{5}{12} \cdot 2\frac{2}{3}$

50. $5\frac{5}{7} \cdot 3\frac{1}{8}$

51. $2\frac{4}{5} \times 4\frac{1}{16}$

YOU BE THE TEACHER Your friend finds the product. Is your friend correct? Explain your reasoning.

52.
$$4 \times 3\frac{7}{10} = 12\frac{7}{10}$$

53.
$$2\frac{1}{2} \times 7\frac{4}{5} = (2 \times 7) + \left(\frac{1}{2} \times \frac{4}{5}\right)$$
$$= 14 + \frac{2}{5}$$
$$= 14\frac{2}{5}$$

54. **MODELING REAL LIFE** A vitamin C tablet contains $\frac{1}{4}$ of a gram of vitamin C. You take $1\frac{1}{2}$ tablets every day. How many grams of vitamin C do you take every day?

55. **MP PROBLEM SOLVING** You make a banner for a football rally.

   a. What is the area of the banner?

   b. You add a $\frac{1}{4}$-foot border on each side. What is the area of the new banner?

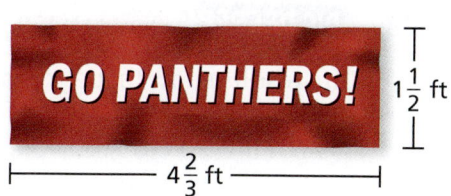

MULTIPLYING FRACTIONS AND MIXED NUMBERS Multiply. Write the answer in simplest form.

56. $\frac{1}{2} \times \frac{3}{5} \times \frac{4}{9}$

57. $\frac{4}{7} \cdot 4\frac{3}{8} \cdot \frac{5}{6}$

58. $1\frac{1}{15} \times 5\frac{2}{5} \times 4\frac{7}{12}$

59. $\left(\frac{3}{5}\right)^3$

60. $\left(\frac{4}{5}\right)^2 \times \left(\frac{3}{4}\right)^2$

61. $\left(\frac{5}{6}\right)^2 \cdot \left(1\frac{1}{10}\right)^2$

62. **OPEN-ENDED** Find a fraction that, when multiplied by $\frac{1}{2}$, is less than $\frac{1}{4}$.

Section 2.1   Multiplying Fractions   51

63. **MP LOGIC** You are in a bike race. When you get to the first checkpoint, you are $\frac{2}{5}$ of the distance to the second checkpoint. When you get to the second checkpoint, you are $\frac{1}{4}$ of the distance to the finish. What is the distance from the start to the first checkpoint?

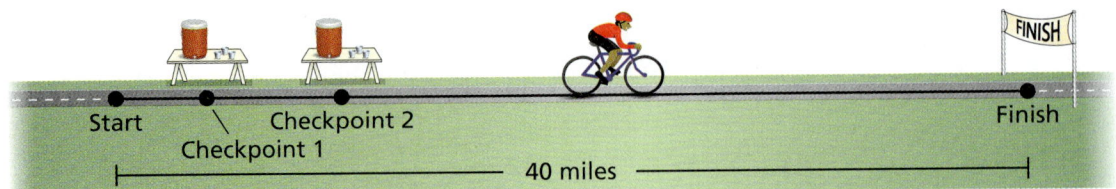

64. **MP NUMBER SENSE** Is the product of two positive mixed numbers ever less than 1? Explain.

65. **MP REASONING** You plan to add a fountain to your garden.

   a. Draw a diagram of the fountain in the garden. Label the dimensions.

   b. Describe two methods for finding the area of the garden that surrounds the fountain.

   c. Find the area. Which method did you use, and why?

66. **MP PROBLEM SOLVING** The cooking time for a ham is $\frac{2}{5}$ of an hour for each pound. What time should you start cooking a ham that weighs $12\frac{3}{4}$ pounds so that it is done at 4:45 P.M.?

67. **MP PRECISION** Complete the Four Square for $\frac{7}{8} \times \frac{1}{3}$.

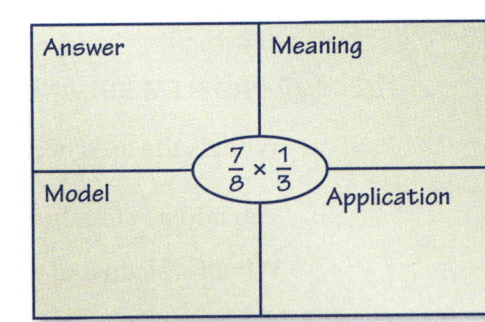

68. **DIG DEEPER!** You ask 150 people about their pets. The results show that $\frac{9}{25}$ of the people own a dog. Of the people who own a dog, $\frac{1}{6}$ of them also own a cat.

   a. What portion of the people own a dog and a cat?

   b. How many people own a dog but not a cat? Explain.

69. **MP NUMBER SENSE** Use each of the numbers from 1 to 9 exactly once to create three mixed numbers with the greatest possible product. Then use each of the numbers exactly once to create three mixed numbers with the least possible product. Find each product. Explain your reasoning. The fraction portion of each mixed number should be proper.

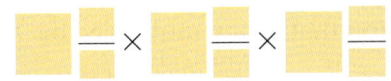

2.2 Dividing Fractions

Learning Target: Compute quotients of fractions and solve problems involving division by fractions.

Success Criteria:
- I can draw a model to explain division of fractions.
- I can find reciprocals of numbers.
- I can divide fractions by fractions.
- I can divide fractions and whole numbers.

EXPLORATION 1
Dividing by Fractions

It may help to create a context for each question in Exploration 1. In part (a), suppose you want to cut 4 yards of string into pieces that are $\frac{2}{3}$ of a yard. How many pieces can you cut?

Work with a partner. Answer each question using a model.

a. How many two-thirds are in four?

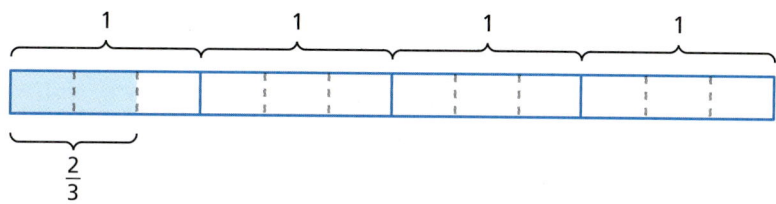

b. How many three-fourths are in three?

c. How many two-fifths are in four-fifths?

d. How many two-thirds are in three?

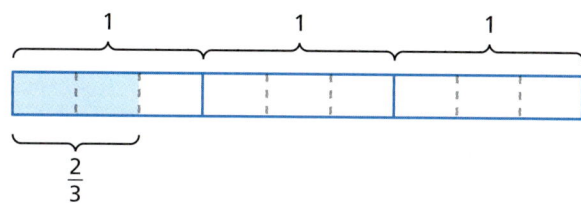

e. How many one-thirds are in five-sixths?

EXPLORATION 2
Finding a Pattern

Math Practice

Look for Structure
Can the pattern you found be applied to division by a whole number? Why or why not?

Work with a partner. The table shows the division expressions from Exploration 1. Complete each multiplication expression so that it has the same value as the division expression above it. What can you conclude about dividing by fractions?

Division Expression	$4 \div \frac{2}{3}$	$3 \div \frac{3}{4}$	$\frac{4}{5} \div \frac{2}{5}$	$3 \div \frac{2}{3}$	$\frac{5}{6} \div \frac{1}{3}$
Multiplication Expression	$4 \times ?$	$3 \times ?$	$\frac{4}{5} \times ?$	$3 \times ?$	$\frac{5}{6} \times ?$

Section 2.2   Dividing Fractions   53

2.2 Lesson

Key Vocabulary
reciprocals, *p. 54*
multiplicative inverses, *p. 54*

Two numbers whose product is 1 are **reciprocals**, or **multiplicative inverses**. To write the reciprocal of a number, first write the number as a fraction. Then invert the fraction. So, the reciprocal of a fraction $\frac{a}{b}$ is $\frac{b}{a}$, where $a \neq 0$ and $b \neq 0$.

The Meaning of a Word ▶ Invert

When you **invert** a glass, you turn it over.

EXAMPLE 1 Writing Reciprocals

When any number is multiplied by 0, the product is 0. So, the number 0 does not have a reciprocal.

	Original Number	Fraction	Reciprocal	Check
a.	$\frac{3}{5}$	$\frac{3}{5}$	$\frac{5}{3}$	$\frac{3}{5} \times \frac{5}{3} = 1$
b.	$\frac{9}{5}$	$\frac{9}{5}$	$\frac{5}{9}$	$\frac{9}{5} \times \frac{5}{9} = 1$
c.	2	$\frac{2}{1}$	$\frac{1}{2}$	$\frac{2}{1} \times \frac{1}{2} = 1$

Try It Write the reciprocal of the number.

1. $\frac{3}{4}$ 2. 5   3. $\frac{7}{2}$ 4. $\frac{4}{9}$

 Key Idea

Dividing Fractions

Words To divide a number by a fraction, multiply the number by the reciprocal of the fraction.

Numbers $\frac{1}{5} \div \frac{3}{4} = \frac{1}{5} \times \frac{4}{3} = \frac{1 \times 4}{5 \times 3}$

Algebra $\frac{a}{b} \div \frac{c}{d} = \frac{a}{b} \cdot \frac{d}{c} = \frac{a \cdot d}{b \cdot c}$, where b, c, and $d \neq 0$

Chapter 2   Fractions and Decimals   Multi-Language Glossary at *BigIdeasMath.com*

EXAMPLE 2 Dividing a Fraction by a Fraction

a. Find $\dfrac{3}{4} \div \dfrac{5}{12}$.

$\dfrac{3}{4} \div \dfrac{5}{12} = \dfrac{3}{4} \cdot \dfrac{12}{5}$ Multiply by the reciprocal of $\dfrac{5}{12}$, which is $\dfrac{12}{5}$.

$= \dfrac{3 \cdot 12}{4 \cdot 5}$ Multiply fractions.

$= \dfrac{36}{20}$, or $1\dfrac{4}{5}$ Simplify.

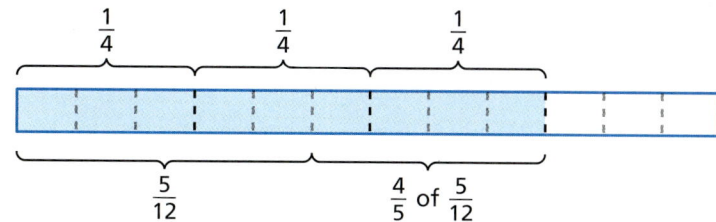

Think: How many five-twelfths are in three-fourths?

b. Find $\dfrac{1}{6} \div \dfrac{2}{3}$.

$\dfrac{1}{6} \div \dfrac{2}{3} = \dfrac{1}{6} \times \dfrac{3}{2}$ Multiply by the reciprocal of $\dfrac{2}{3}$, which is $\dfrac{3}{2}$.

$= \dfrac{1 \times 3}{6 \times 2}$ Multiply fractions.

$= \dfrac{3}{12}$, or $\dfrac{1}{4}$ Simplify.

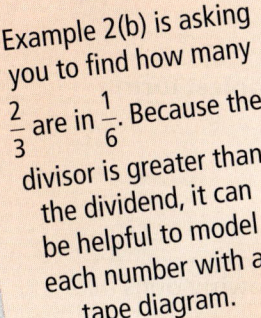

Example 2(b) is asking you to find how many $\dfrac{2}{3}$ are in $\dfrac{1}{6}$. Because the divisor is greater than the dividend, it can be helpful to model each number with a tape diagram.

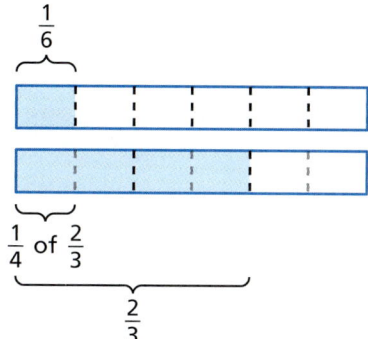

Try It Divide. Write the answer in simplest form. Use a model to justify your answer.

5. $\dfrac{1}{2} \div \dfrac{1}{8}$

6. $\dfrac{2}{5} \div \dfrac{3}{10}$

7. $\dfrac{3}{8} \div \dfrac{3}{4}$

8. $\dfrac{2}{7} \div \dfrac{9}{14}$

EXAMPLE 3 Dividing a Fraction by a Whole Number

Find $\frac{4}{5} \div 2$.

$\frac{4}{5} \div 2 = \frac{4}{5} \div \frac{2}{1}$ Write 2 as an improper fraction.

$= \frac{4}{5} \times \frac{1}{2}$ Multiply by the reciprocal of $\frac{2}{1}$, which is $\frac{1}{2}$.

$= \frac{4 \times 1}{5 \times 2}$ Multiply fractions.

$= \frac{4}{10}$, or $\frac{2}{5}$ Simplify.

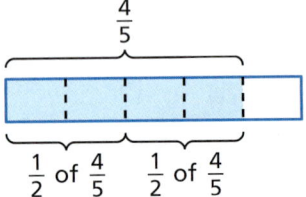

Remember
When dividing unit fractions by whole numbers, remember that dividing by a number n is equivalent to multiplying by $\frac{1}{n}$.

Try It Divide. Write the answer in simplest form.

9. $\frac{1}{3} \div 3$ 10. $\frac{2}{3} \div 10$ 11. $\frac{5}{8} \div 4$ 12. $\frac{6}{7} \div 4$

Self-Assessment for Concepts & Skills

Solve each exercise. Then rate your understanding of the success criteria in your journal.

DIVIDING FRACTIONS Divide. Write the answer in simplest form. Draw a model to justify your answer.

13. $\frac{2}{3} \div \frac{5}{6}$ 14. $\frac{6}{7} \div 3$

15. **WHICH ONE DOESN'T BELONG?** Which of the following does *not* belong with the other three? Explain your reasoning.

$\frac{2}{3} \div \frac{4}{5}$ $\frac{3}{2} \cdot \frac{4}{5}$ $\frac{5}{4} \times \frac{2}{3}$ $\frac{5}{4} \div \frac{3}{2}$

MATCHING Match the expression with its value.

16. $\frac{2}{5} \div \frac{8}{15}$ 17. $\frac{8}{15} \div \frac{2}{5}$ 18. $\frac{2}{15} \div \frac{8}{5}$ 19. $\frac{8}{5} \div \frac{2}{15}$

A. $\frac{1}{12}$ B. $\frac{3}{4}$ C. 12   D. $1\frac{1}{3}$

56 Chapter 2 Fractions and Decimals

EXAMPLE 4 Modeling Real Life

A piece of wood is 4 feet long. How many $\frac{3}{4}$-foot pieces can you cut from the piece of wood? How much wood is left over?

Divide the length of the entire piece of wood by the length of a smaller piece. So, divide 4 by $\frac{3}{4}$. The quotient can be used to answer both questions.

$$4 \div \frac{3}{4} = 4 \times \frac{4}{3} \qquad \text{Multiply by the reciprocal of } \frac{3}{4}, \text{ which is } \frac{4}{3}.$$

$$= \frac{4 \times 4}{3} \qquad \text{Multiply.}$$

$$= \frac{16}{3}, \text{ or } 5\frac{1}{3} \qquad \text{Simplify.}$$

▸ You can cut five $\frac{3}{4}$-foot pieces from the piece of wood. The remaining piece is $\frac{1}{3}$ of a $\frac{3}{4}$-foot section. So, $\frac{1}{3} \cdot \frac{3}{4} = \frac{3}{12} = \frac{1}{4}$ foot of wood is left over.

Another Method Use a diagram.

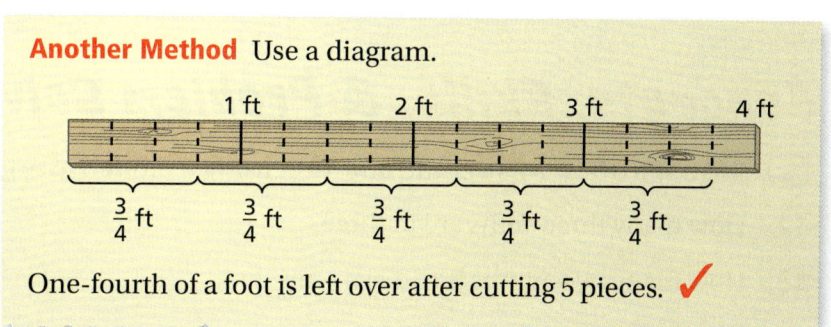

One-fourth of a foot is left over after cutting 5 pieces. ✓

 **Self-Assessment** for Problem Solving

Solve each exercise. Then rate your understanding of the success criteria in your journal.

20. You have 5 cups of rice to make *bibimbap*, a popular Korean meal. The recipe calls for $\frac{4}{5}$ cup of rice per serving. How many full servings of bibimbap can you make? How much rice is left over?

21. A band earns $\frac{2}{3}$ of their profit from selling concert tickets and $\frac{1}{5}$ of their profit from selling merchandise. The band earns a profit of $1500 from selling concert tickets. How much profit does the band earn from selling merchandise?

2.2 Practice

> Go to **BigIdeasMath.com** to get HELP with solving the exercises.

Review & Refresh

Multiply. Write the answer in simplest form.

1. $\dfrac{7}{10} \cdot \dfrac{3}{4}$
2. $\dfrac{5}{6} \times 2\dfrac{1}{3}$
3. $\dfrac{4}{9} \times \dfrac{3}{8}$
4. $2\dfrac{2}{5} \cdot 6\dfrac{2}{3}$

Match the expression with its value.

5. $3 + 2 \times 4^2$
6. $(3 + 2) \times 4^2$
7. $2 + 3 \times 4^2$
8. $4^2 + 2 \times 3$

   A. 22 **B.** 35 **C.** 50 **D.** 80

Find the area of the rectangle.

9. 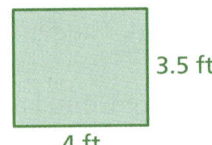 3.5 ft by 4 ft

10. 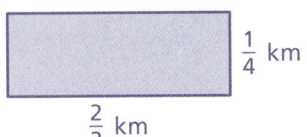 $\dfrac{1}{4}$ km by $\dfrac{2}{3}$ km

11. 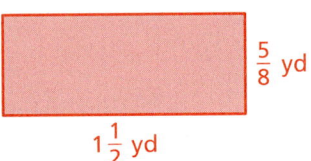 $\dfrac{5}{8}$ yd by $1\dfrac{1}{2}$ yd

Concepts, Skills, & Problem Solving

CHOOSE TOOLS Answer the question using a model. *(See Exploration 1, p. 53.)*

12. How many three-fifths are in three?

13. How many two-ninths are in eight-ninths?

14. How many three-fourths are in seven-eighths?

WRITING RECIPROCALS Write the reciprocal of the number.

15. 8
16. $\dfrac{6}{7}$
17. $\dfrac{2}{5}$
18. $\dfrac{11}{8}$

DIVIDING FRACTIONS Divide. Write the answer in simplest form.

19. $\dfrac{1}{3} \div \dfrac{1}{2}$
20. $\dfrac{1}{8} \div \dfrac{1}{4}$
21. $\dfrac{2}{7} \div 2$
22. $\dfrac{6}{5} \div 3$

23. $\dfrac{2}{3} \div \dfrac{4}{9}$
24. $\dfrac{5}{6} \div \dfrac{2}{7}$
25. $12 \div \dfrac{3}{4}$
26. $8 \div \dfrac{2}{5}$

27. $\dfrac{3}{7} \div 6$
28. $\dfrac{12}{25} \div 4$
29. $\dfrac{2}{9} \div \dfrac{2}{3}$
30. $\dfrac{8}{15} \div \dfrac{4}{5}$

31. $\dfrac{1}{3} \div \dfrac{1}{9}$
32. $\dfrac{7}{10} \div \dfrac{3}{8}$
33. $\dfrac{14}{27} \div 7$
34. $\dfrac{5}{8} \div 15$

35. $\dfrac{27}{32} \div \dfrac{7}{8}$
36. $\dfrac{4}{15} \div \dfrac{10}{13}$
37. $9 \div \dfrac{4}{9}$
38. $10 \div \dfrac{5}{12}$

58 Chapter 2 Fractions and Decimals

YOU BE THE TEACHER Your friend finds the quotient. Is your friend correct? Explain your reasoning.

39.
$$\frac{4}{7} \div \frac{13}{28} = \frac{4}{7} \times \frac{28}{13}$$
$$= \frac{4 \times 28}{7 \times 13}$$
$$= \frac{112}{91}, \text{ or } 1\frac{3}{13}$$

40.
$$\frac{2}{5} \div \frac{8}{9} = \frac{5}{2} \times \frac{8}{9}$$
$$= \frac{5 \times 8}{2 \times 9}$$
$$= \frac{40}{18}, \text{ or } 2\frac{2}{9}$$

41. **REASONING** You have $\frac{3}{5}$ of an apple pie. You divide the remaining pie into 5 equal slices. What portion of the original pie is each slice?

42. **PROBLEM SOLVING** How many times longer is the baby alligator than the baby gecko?

OPEN-ENDED Write a real-life problem for the expression. Then solve the problem.

43. $\frac{5}{6} \div 4$ 44. $\frac{2}{5} \div \frac{3}{8}$ 45. $10 \div \frac{2}{3}$ 46. $\frac{2}{7} \div \frac{4}{9}$

NUMBER SENSE Copy and complete the statement.

47. $\frac{5}{12} \times \boxed{} = 1$ 48. $3 \times \boxed{} = 1$ 49. $7 \div \boxed{} = 56$

REASONING Without finding the quotient, copy and complete the statement using <, >, or =. Explain your reasoning.

50. $\frac{7}{9} \div 5 \;\boxed{}\; \frac{7}{9}$ 51. $\frac{3}{7} \div 1 \;\boxed{}\; \frac{3}{7}$ 52. $8 \div \frac{3}{4} \;\boxed{}\; 8$ 53. $\frac{5}{6} \div \frac{7}{8} \;\boxed{}\; \frac{5}{6}$

ORDER OF OPERATIONS Evaluate the expression. Write the answer in simplest form.

54. $\frac{1}{6} \div 6 \div 6$ 55. $\frac{7}{12} \div 14 \div 6$ 56. $\frac{3}{5} \div \frac{4}{7} \div \frac{9}{10}$

57. $4 \div \frac{8}{9} - \frac{1}{2}$ 58. $\frac{3}{4} + \frac{5}{6} \div \frac{2}{3}$ 59. $\frac{7}{8} - \frac{3}{8} \div 9$

60. $\frac{9}{16} \div \frac{3}{4} \cdot \frac{2}{13}$ 61. $\frac{3}{14} \cdot \frac{2}{5} \div \frac{6}{7}$ 62. $\frac{10}{27} \cdot \left(\frac{3}{8} \div \frac{5}{24} \right)$

63. **NUMBER SENSE** When is the reciprocal of a fraction a whole number? Explain.

64. **MODELING REAL LIFE** You use $\frac{1}{8}$ of your battery for every $\frac{2}{5}$ of an hour that you video chat. You use $\frac{3}{4}$ of your battery video chatting. How long did you video chat?

65. **MP PROBLEM SOLVING** The table shows the portions of a family budget that are spent on several expenses.

    a. How many times more is the expense for housing than for automobiles?

    b. How many times more is the expense for food than for recreation?

    c. The expense for automobile fuel is $\frac{1}{60}$ of the total expenses. What portion of the automobile expense is spent on fuel?

Expense	Portion of Budget
Housing	$\frac{2}{5}$
Food	$\frac{4}{9}$
Automobiles	$\frac{1}{15}$
Recreation	$\frac{1}{40}$

66. **CRITICAL THINKING** A bottle of juice is $\frac{2}{3}$ full. The bottle contains $\frac{4}{5}$ of a cup of juice.

    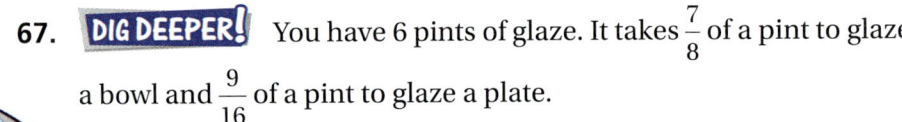

    a. Write a division expression that represents the capacity of the bottle.

    b. Write a related multiplication expression that represents the capacity of the bottle.

    c. Explain how you can use the diagram to verify the expression in part (b).

    d. Find the capacity of the bottle.

67. **DIG DEEPER!** You have 6 pints of glaze. It takes $\frac{7}{8}$ of a pint to glaze a bowl and $\frac{9}{16}$ of a pint to glaze a plate.

    a. How many bowls can you completely glaze? How many plates can you completely glaze?

    b. You want to glaze 5 bowls, and then use the rest for plates. How many plates can you completely glaze? How much glaze will be left over?

    c. How many of each object can you completely glaze so that there is no glaze left over? Explain how you found your answer.

68. **MP REASONING** A water tank is $\frac{1}{8}$ full. The tank is $\frac{3}{4}$ full when 42 gallons of water are added to the tank.

    a. How much water can the tank hold?

    b. How much water was originally in the tank?

    c. How much water is in the tank when it is $\frac{1}{2}$ full?

2.3 Dividing Mixed Numbers

Learning Target: Compute quotients with mixed numbers and solve problems involving division with mixed numbers.

Success Criteria:
- I can draw a model to explain division of mixed numbers.
- I can write a mixed number as an improper fraction.
- I can divide with mixed numbers.
- I can evaluate expressions involving mixed numbers using the order of operations.

EXPLORATION 1

Dividing Mixed Numbers

Work with a partner. Write a real-life problem that represents each division expression described. Then solve each problem using a model. Check your answers.

Math Practice

Make Sense of Quantities
What values do the parts of the model represent?

a. How many three-fourths are in four and one-half?

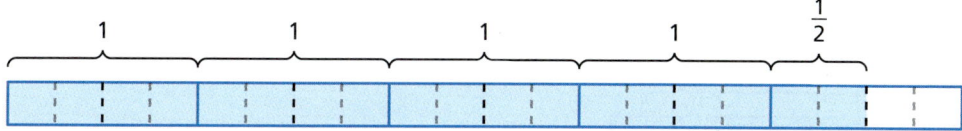

b. How many three-eighths are in two and one-fourth?

c. How many one and one-halves are in six?

d. How many seven-sixths are in three and one-third?

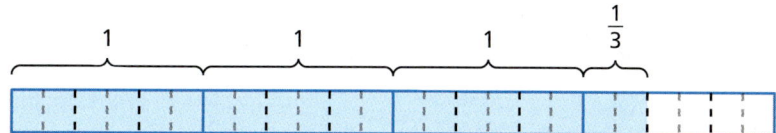

e. How many one and one-fifths are in five?

f. How many three and one-halves are in two and one-half?

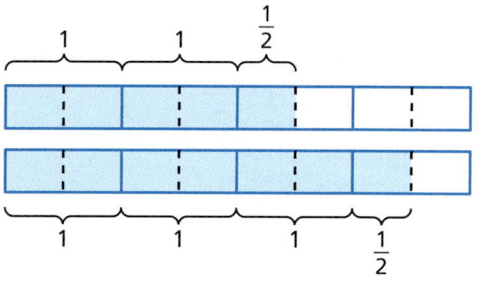

g. How many four and one-halves are in one and one-half?

2.3 Lesson

Key Idea

Dividing Mixed Numbers
Write each mixed number as an improper fraction. Then divide as you would with proper fractions.

EXAMPLE 1 **Dividing with Mixed Numbers**

a. Find $2\frac{2}{3} \div \frac{2}{3}$.

Think: How many two-thirds are in two and two-thirds?

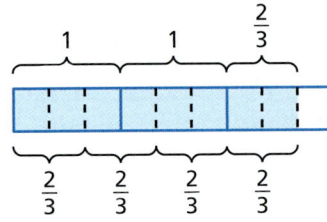

$2\frac{2}{3} \div \frac{2}{3} = \frac{8}{3} \div \frac{2}{3}$ Write $2\frac{2}{3}$ as the improper fraction $\frac{8}{3}$.

$= \frac{8}{3} \times \frac{3}{2}$ Multiply by the reciprocal of $\frac{2}{3}$, which is $\frac{3}{2}$.

$= \frac{8 \times 3}{3 \times 2}$ Multiply fractions.

$= \frac{24}{6}$, or 4 Simplify.

▶ So, the quotient is 4.

b. Find $3\frac{5}{6} \div 1\frac{2}{3}$.

Estimate $4 \div 2 = 2$

$3\frac{5}{6} \div 1\frac{2}{3} = \frac{23}{6} \div \frac{5}{3}$ Write each mixed number as an improper fraction.

$= \frac{23}{6} \times \frac{3}{5}$ Multiply by the reciprocal of $\frac{5}{3}$, which is $\frac{3}{5}$.

$= \frac{23 \times 3}{6 \times 5}$ Multiply fractions.

$= \frac{69}{30}$, or $2\frac{3}{10}$ Simplify.

▶ So, the quotient is $2\frac{3}{10}$. **Reasonable?** $2\frac{3}{10} \approx 2$ ✓

Try It Divide. Write the answer in simplest form.

1. $3\frac{2}{3} \div \frac{1}{3}$
2. $1\frac{3}{7} \div \frac{2}{3}$
3. $2\frac{1}{6} \div \frac{3}{4}$
4. $6\frac{1}{2} \div 2$

5. $10\frac{2}{3} \div 2\frac{2}{3}$
6. $8\frac{1}{4} \div 1\frac{1}{2}$
7. $3 \div 1\frac{3}{4}$
8. $\frac{3}{4} \div 2\frac{1}{2}$

62 Chapter 2 Fractions and Decimals

EXAMPLE 2   Using Order of Operations

Evaluate $5\frac{1}{4} \div 1\frac{1}{8} - \frac{2}{3}$.

$5\frac{1}{4} \div 1\frac{1}{8} - \frac{2}{3} = \frac{21}{4} \div \frac{9}{8} - \frac{2}{3}$ Write each mixed number as an improper fraction.

$= \frac{21}{4} \times \frac{8}{9} - \frac{2}{3}$ Multiply by the reciprocal of $\frac{9}{8}$, which is $\frac{8}{9}$.

$= \frac{21 \times 8}{4 \times 9} - \frac{2}{3}$ Multiply fractions.

$= \frac{168}{36} - \frac{2}{3}$ Multiply.

$= \frac{14}{3} - \frac{2}{3}$ Simplify.

$= \frac{12}{3}$, or 4   Subtract.

> **Remember**
> Be sure to check your answers whenever possible. In Example 2, you can use estimation to check that your answer is reasonable.
> $5\frac{1}{4} \div 1\frac{1}{8} - \frac{2}{3}$
> $\approx 5 \div 1 - 1$
> $= 5 - 1$
> $= 4$ ✓

Try It Evaluate the expression. Write the answer in simplest form.

9. $1\frac{1}{2} \div \frac{1}{6} - \frac{7}{8}$

10. $3\frac{1}{3} \div \frac{5}{6} + \frac{8}{9}$

11. $\frac{2}{5} + 2\frac{4}{5} \div 2$

12. $\frac{2}{3} - 1\frac{4}{7} \div 4\frac{5}{7}$

Self-Assessment for Concepts & Skills

Solve each exercise. Then rate your understanding of the success criteria in your journal.

EVALUATING EXPRESSIONS Evaluate the expression. Write the answer in simplest form.

13. $4\frac{4}{7} \div \frac{4}{7}$

14. $\frac{1}{2} \div 5\frac{1}{4}$

15. $\frac{3}{4} + 6\frac{2}{5} \div 1\frac{3}{5}$

16. **MP NUMBER SENSE** Is $2\frac{1}{2} \div 1\frac{1}{4}$ the same as $1\frac{1}{4} \div 2\frac{1}{2}$? Use models to justify your answer.

17. **DIFFERENT WORDS, SAME QUESTION** Which is different? Find "both" answers.

What is $5\frac{1}{2}$ divided by $\frac{1}{8}$?	What is the quotient of $5\frac{1}{2}$ and $\frac{1}{8}$?
What is $5\frac{1}{2}$ times 8?	What is $\frac{1}{8}$ of $5\frac{1}{2}$?

Section 2.3   Dividing Mixed Numbers

EXAMPLE 3 Modeling Real Life

One serving of tortilla soup is $1\frac{2}{3}$ cups. A restaurant cook makes 50 cups of soup. Is there enough to serve 35 people? Explain.

Divide the total amount of soup by the amount per serving to find the number of available servings. So, divide 50 by $1\frac{2}{3}$.

$50 \div 1\frac{2}{3} = 50 \div \frac{5}{3}$ Write $1\frac{2}{3}$ as the improper fraction $\frac{5}{3}$.

$\phantom{50 \div 1\frac{2}{3}} = 50 \cdot \frac{3}{5}$ Multiply by the reciprocal of $\frac{5}{3}$, which is $\frac{3}{5}$.

$\phantom{50 \div 1\frac{2}{3}} = \frac{50 \cdot 3}{5}$ Multiply.

$\phantom{50 \div 1\frac{2}{3}} = \frac{150}{5}$, or 30   Simplify.

▸ No. Because 30 is less than 35, there is not enough soup to serve 35 people.

> **Another Method** Multiply the amount in one serving, $1\frac{2}{3}$ cups, by 35 to see how much soup is needed for 35 people.
>
> $$1\frac{2}{3} \cdot 35 = \frac{5}{3} \cdot 35 = \frac{175}{3} = 58\frac{1}{3}$$
>
> Because $58\frac{1}{3}$ is greater than 50, there is not enough soup to serve 35 people. ✓

Self-Assessment for Problem Solving

Solve each exercise. Then rate your understanding of the success criteria in your journal.

18. A watercooler contains 160 cups of water. During practice, each person on a team fills a water bottle with $3\frac{1}{3}$ cups of water from the cooler. Is there enough water for all 45 people on the team to fill their water bottles? Explain.

19. A cyclist is $7\frac{3}{4}$ kilometers from the finish line of a race. The cyclist rides at a rate of $25\frac{5}{6}$ kilometers per hour. How many minutes will it take the cyclist to finish the race?

64   Chapter 2   Fractions and Decimals

2.3 Practice

> Go to *BigIdeasMath.com* to get HELP with solving the exercises.

▶ Review & Refresh

Divide. Write the answer in simplest form.

1. $\dfrac{1}{8} \div \dfrac{1}{7}$
2. $\dfrac{7}{9} \div \dfrac{2}{3}$
3. $\dfrac{5}{6} \div 10$
4. $12 \div \dfrac{3}{8}$

Find the LCM of the numbers.

5. 8, 14
6. 9, 11, 12
7. 12, 27, 30

Find the volume of the rectangular prism.

8.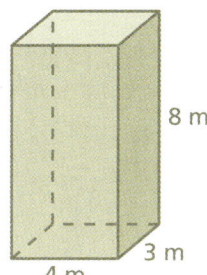
9.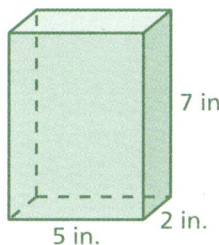
10.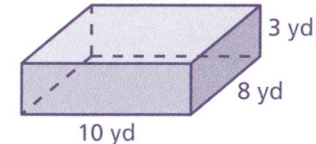

11. Which number is *not* a prime factor of 286?

   A. 2   B. 7   C. 11   D. 13

▶ Concepts, Skills, & Problem Solving

MP CHOOSE TOOLS Write a real-life problem that represents the division expression described. Then solve the problem using a model. Check your answer algebraically. *(See Exploration 1, p. 61.)*

12. How many two-thirds are in three and one-third?
13. How many one and one-sixths are in five and five-sixths?
14. How many two and one-halves are in eight and three-fourths?

DIVIDING WITH MIXED NUMBERS Divide. Write the answer in simplest form.

15. $2\dfrac{1}{4} \div \dfrac{3}{4}$
16. $3\dfrac{4}{5} \div \dfrac{2}{5}$
17. $8\dfrac{1}{8} \div \dfrac{5}{6}$
18. $7\dfrac{5}{9} \div \dfrac{4}{7}$

19. $7\dfrac{1}{2} \div 1\dfrac{9}{10}$
20. $3\dfrac{3}{4} \div 2\dfrac{1}{12}$
21. $7\dfrac{1}{5} \div 8$
22. $8\dfrac{4}{7} \div 15$

23. $8\dfrac{1}{3} \div \dfrac{2}{3}$
24. $9\dfrac{1}{6} \div \dfrac{5}{6}$
25. $13 \div 10\dfrac{5}{6}$
26. $12 \div 5\dfrac{9}{11}$

27. $\dfrac{7}{8} \div 3\dfrac{1}{16}$
28. $\dfrac{4}{9} \div 1\dfrac{7}{15}$
29. $4\dfrac{5}{16} \div 3\dfrac{3}{8}$
30. $6\dfrac{2}{9} \div 5\dfrac{5}{6}$

31. **YOU BE THE TEACHER** Your friend finds the quotient of $3\frac{1}{2}$ and $1\frac{2}{3}$. Is your friend correct? Explain your reasoning.

$$3\frac{1}{2} \div 1\frac{2}{3} = 3\frac{1}{2} \times 1\frac{3}{2} = \frac{7}{2} \times \frac{5}{2} = \frac{7 \times 5}{2 \times 2} = \frac{35}{4}, \text{ or } 8\frac{3}{4}$$

32. **MP PROBLEM SOLVING** A platinum nugget weighs $3\frac{1}{2}$ ounces. How many $\frac{1}{4}$-ounce pieces can be cut from the nugget?

ORDER OF OPERATIONS Evaluate the expression. Write the answer in simplest form.

33. $3 \div 1\frac{1}{5} + \frac{1}{2}$

34. $4\frac{2}{3} - 1\frac{1}{3} \div 2$

35. $\frac{2}{5} + 2\frac{1}{6} \div \frac{5}{6}$

36. $5\frac{5}{6} \div 3\frac{3}{4} - \frac{2}{9}$

37. $6\frac{1}{2} - \frac{7}{8} \div 5\frac{11}{16}$

38. $9\frac{1}{6} \div 5 + 3\frac{1}{3}$

39. $3\frac{3}{5} + 4\frac{4}{15} \div \frac{4}{9}$

40. $\frac{3}{5} \times \frac{7}{12} \div 2\frac{7}{10}$

41. $4\frac{3}{8} \div \frac{3}{4} \cdot \frac{4}{7}$

42. $1\frac{9}{11} \times 4\frac{7}{12} \div \frac{2}{3}$

43. $3\frac{4}{15} \div \left(8 \cdot 6\frac{3}{10}\right)$

44. $2\frac{5}{14} \div \left(2\frac{5}{8} \times 1\frac{3}{7}\right)$

45. **MP LOGIC** Your friend uses the model shown to state that $2\frac{1}{2} \div 1\frac{1}{6} = 2\frac{1}{6}$. Is your friend correct? Justify your answer using the model.

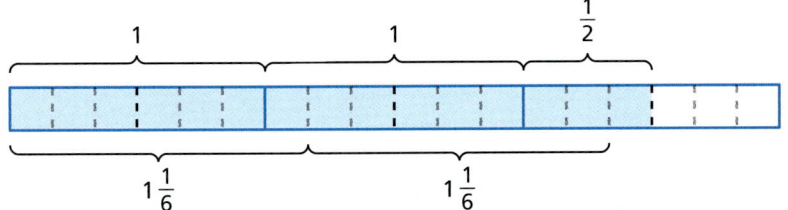

46. **MODELING REAL LIFE** A bag contains 42 cups of dog food. Your dog eats $2\frac{1}{3}$ cups of dog food each day. Is there enough food to last 3 weeks? Explain.

47. **DIG DEEPER!** You have 12 cups of granola and $8\frac{1}{2}$ cups of peanuts to make trail mix. What is the greatest number of full batches of trail mix you can make? Explain how you found your answer.

Trail Mix
$2\frac{3}{4}$ cups granola
$1\frac{1}{3}$ cups peanuts

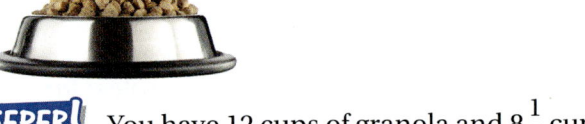

48. **MP REASONING** At a track and field meet, the longest shot-put throw by a boy is 25 feet 8 inches. The longest shot-put throw by a girl is 19 feet 3 inches. How many times greater is the longest shot-put throw by the boy than by the girl?

2.4 Adding and Subtracting Decimals

Learning Target: Add and subtract decimals and solve problems involving addition and subtraction of decimals.

Success Criteria:
- I can explain why it is necessary to line up the decimal points when adding and subtracting decimals.
- I can add decimals.
- I can subtract decimals.
- I can evaluate expressions involving addition and subtraction of decimals.

EXPLORATION 1
Using Number Lines

Work with a partner. Use each number line to find $A + B$ and $B - A$. Explain how you know you are correct.

a.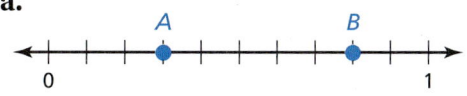

b.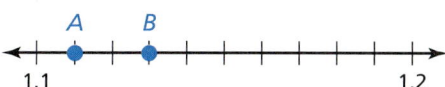

c.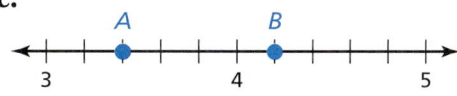

d.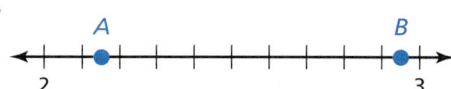

EXPLORATION 2
Extending the Place Value Chart

Work with a partner. Explain how you can use the place value chart below to add and subtract decimals beyond hundredths. Then find each sum or difference.

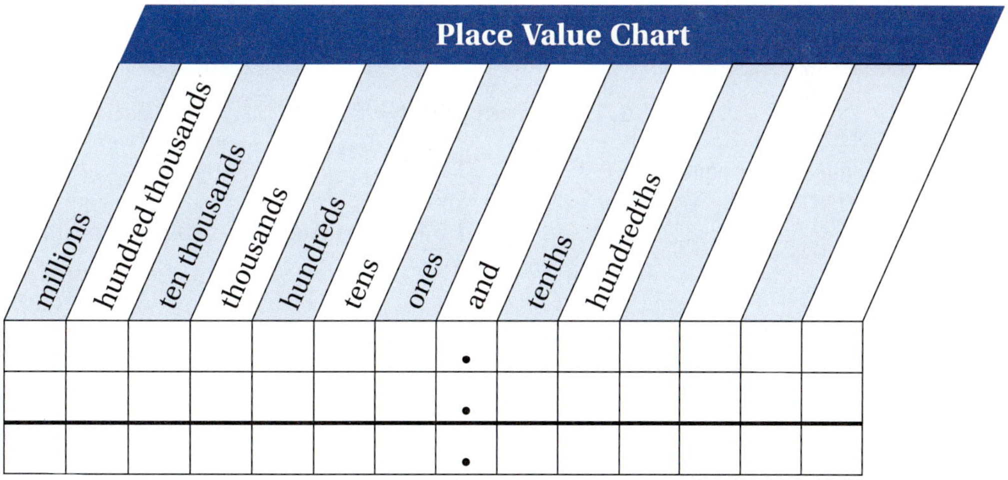

Place Value Chart

(columns: millions, hundred thousands, ten thousands, thousands, hundreds, tens, ones, and, tenths, hundredths)

Math Practice

Recognize Usefulness of Tools
How does the place value chart help you perform the operations? How can you perform the operations without the chart?

a. $16.05 + 2.945$

b. $7.421 + 8.058$

c. $38.72 - 8.618$

d. $64.968 - 51.167$

e. $225.1 + 85.0465$

f. $1107.20592 - 102.3056$

2.4 Lesson

Key Idea

Adding and Subtracting Decimals

To add or subtract decimals, write the numbers vertically and line up the decimal points. Then bring down the decimal point and add or subtract as you would with whole numbers.

EXAMPLE 1 Adding Decimals

a. Add $8.13 + 2.76$.   Estimate $8 + 3 = 11$

Line up the decimal points.

$$\begin{array}{r} 8.13 \\ +\ 2.76 \\ \hline 10.89 \end{array}$$

Add as you would with whole numbers.

Reasonable? $10.89 \approx 11$ ✓

b. Add $1.459 + 23.7$.

$$\begin{array}{r} 1 \\ 1.459 \\ +\ 23.700 \\ \hline 25.159 \end{array}$$

Insert zeros so that both numbers have the same number of decimal places.

Be sure to add or subtract only digits that have the same place value.

Try It Add.

1. $4.206 + 10.85$ 2. $15.5 + 8.229$ 3. $78.41 + 90.99$

EXAMPLE 2 Subtracting Decimals

a. Subtract $5.508 - 3.174$.   Estimate $6 - 3 = 3$

Line up the decimal points.

$$\begin{array}{r} \overset{4\ 10}{5.\cancel{5}\cancel{0}8} \\ -\ 3.174 \\ \hline 2.334 \end{array}$$

Subtract as you would with whole numbers.

Reasonable? $2.334 \approx 3$ ✓

b. Subtract $21.9 - 1.605$.

$$\begin{array}{r} 21.900 \\ -\ 1.605 \\ \hline 20.295 \end{array}$$

Insert zeros so that both numbers have the same number of decimal places.

Try It Subtract.

4. $6.34 - 5.33$ 5. $27.9 - 0.905$ 6. $18.626 - 13.88$

EXAMPLE 3 Adding and Subtracting Decimals

A gymnast's score for a routine is calculated by finding the sum of a difficulty score and an execution score, minus any penalties. The execution score starts at 10 and is reduced by deductions. The gymnast's score can be represented by

$$6.9 + (10 - 1.534) - 0.3.$$

What is the gymnast's score?

Evaluate the expression to solve the problem.

$6.9 + (10 - 1.534) - 0.3$

$= 6.9 + 8.466 - 0.3$ Subtract inside parentheses.

$= 15.366 - 0.3$ Add.

$= 15.066$ Subtract.

Difficulty score	6.9
Execution score deductions	1.534
Penalties	0.3

 So, the gymnast's score is 15.066.

Try It

7. **WHAT IF?** The execution score is adjusted and has 1.467 in deductions. What is the gymnast's score?

Self-Assessment for Concepts & Skills

Solve each exercise. Then rate your understanding of the success criteria in your journal.

ADDING AND SUBTRACTING DECIMALS Evaluate the expression.

8. $23.557 - 17.601 + 5.216$

9. $16.5263 + 12.404 - 11.73$

10. **CHOOSE TOOLS** Why is it helpful to estimate the answer before adding or subtracting decimals?

11. **WRITING** When adding or subtracting decimals, how can you be sure to add or subtract only digits that have the same place value?

12. **OPEN-ENDED** Describe two real-life examples of when you would need to add and subtract decimals.

13. **STRUCTURE** You add $3.841 + 29.999$ as shown. Describe a method for adding the numbers using mental math. Which method do you prefer? Explain.

```
  1 1 1 1
    3.841
 + 29.999
   33.840
```

14. **OPEN-ENDED** Write three decimals that have a sum of 27.905.

Section 2.4 Adding and Subtracting Decimals

EXAMPLE 4 Modeling Real Life

The Lincoln Memorial Reflecting Pool is approximately rectangular. Its width is 50.9 meters, and its length is 618.44 meters. You walk the perimeter of the pool. About how many meters do you walk?

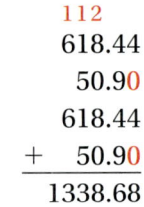

You are given the dimensions of a rectangular pool. You are asked to find the number of meters you walk around the perimeter of the pool.

Draw a diagram and label the dimensions. Write the side lengths vertically and line up the decimal points to find the sum.

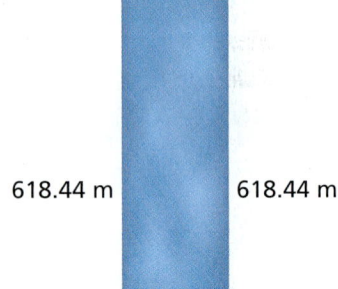

```
  1 1 2
  618.44
   50.90
  618.44
+  50.90
 1338.68
```

So, you walk about 1339 meters.

Check Reasonableness Estimate the distance by finding the sum of the side lengths rounded to the nearest ten.

620 + 50 + 620 + 50 = 1340 meters

The answer is reasonable because 1340 ≈ 1338.68.

Self-Assessment for Problem Solving

Solve each exercise. Then rate your understanding of the success criteria in your journal.

15. A field hockey field is rectangular. Its width is 54.88 meters, and its perimeter is 289.76 meters. Find the length of the field.

16. **DIG DEEPER!** You mix 23.385 grams of sugar and 12.873 grams of baking soda in a glass container for an experiment. You place the container on a scale to find that the total mass is 104.2 grams. What is the mass of the container?

17. One molecule of water is made of two hydrogen atoms and one oxygen atom. The masses (in atomic mass units) for one atom of hydrogen and oxygen are shown. What is the mass (in atomic mass units) of one molecule of water?

70   Chapter 2   Fractions and Decimals

2.4 Practice

> Go to *BigIdeasMath.com* to get HELP with solving the exercises.

▶ Review & Refresh

Divide. Write the answer in simplest form.

1. $3\frac{1}{4} \div \frac{3}{4}$
2. $4\frac{1}{6} \div 5$
3. $6\frac{2}{3} \div 3\frac{1}{5}$

Find the GCF of the numbers.

4. 16, 28, 40
5. 39, 54, 63
6. 24, 72, 132

Find the value of the power.

7. 1^{12}
8. 2^4
9. 3^6
10. 5^4

Classify the quadrilateral.

11.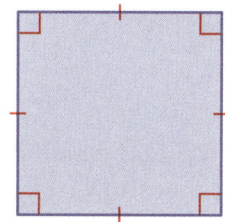
12.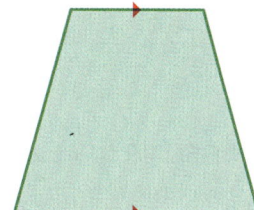
13.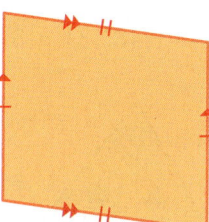

▶ Concepts, Skills, & Problem Solving

USING TOOLS Use a place value chart to find the sum or difference. (See Exploration 2, p. 67.)

14. $4.63 + 8.547$
15. $3.6257 - 2.98$
16. $14.065 + 13.8542$

ADDING DECIMALS Add.

17. $7.82 + 3.209$
18. $3.7 + 2.774$
19. $12.829 + 10.07$
20. $20.35 + 13.748$
21. $11.212 + 7.36$
22. $14.91 + 2.095$
23. $31.994 + 8.006$
24. $3.946 + 6.052$
25. $41.226 + 102.774$
26. $122.781 + 19.228$
27. $17.440 + 12.497$
28. $15.255 + 19.058$

SUBTRACTING DECIMALS Subtract.

29. $4.58 - 3.12$
30. $8.629 - 5.309$
31. $6.98 - 2.614$
32. $15.131 - 11.57$
33. $13.5 - 10.856$
34. $25.82 - 22.936$
35. $17.651 - 12.04$
36. $19.255 - 6.194$
37. $56.217 - 35.8$
38. $62.486 - 18.549$
39. $152.883 - 35.6247$
40. $129.343 - 125.0372$

Section 2.4 Adding and Subtracting Decimals

YOU BE THE TEACHER Your friend finds the sum or difference. Is your friend correct? Explain your reasoning.

41.
```
  6.058
+ 3.95
 10.008
```

42.
```
  9.5
− 7.18
  2.48
```

43. **MP PROBLEM SOLVING** Vehicles must weigh no more than 10.75 tons to cross a bridge. A truck weighs 11.638 tons. By how many tons does the truck exceed the weight limit?

ADDING AND SUBTRACTING DECIMALS Evaluate the expression.

44. $6.105 + 10.4 + 3.075$

45. $22.6 - 12.286 - 3.542$

46. $15.35 + 7.604 - 12.954$

47. $16.5 - 13.45 + 7.293$

48. $25.92 - 18.478 + 8.164$

49. $23.45 + 17.75 - 19.618$

50. $14.549 - (8.131 + 3.7024)$

51. $41.563 - (18.65 + 15.9214) + 9.6$

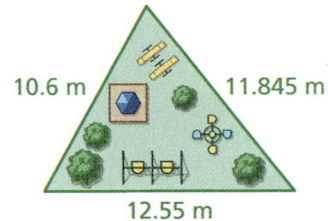

52. **MODELING REAL LIFE** A day-care center is building a new outdoor play area. The diagram shows the dimensions in meters. How much fencing is needed to enclose the play area?

53. **MP PROBLEM SOLVING** On a fantasy football team, a tight end scores 11.15 points and a running back scores 11.75 points. A wide receiver scores 1.05 points less than the running back. How many total points do the three players score?

MODELING REAL LIFE An astronomical unit (AU) is the average distance between Earth and the Sun. In Exercises 54–57, use the table that shows the average distance of each planet in our solar system from the Sun.

54. How much farther is Jupiter from the Sun than Mercury?

55. How much farther is Neptune from the Sun than Mars?

56. Estimate the greatest distance between Earth and Uranus.

57. Estimate the greatest distance between Venus and Saturn.

58. **MP STRUCTURE** When is the sum of two decimals equal to a whole number? When is the difference of two decimals equal to a whole number? Explain.

Planet	Average Distance from the Sun (AU)
Mercury	0.387
Venus	0.723
Earth	1.000
Mars	1.524
Jupiter	5.203
Saturn	9.537
Uranus	19.189
Neptune	30.07

2.5 Multiplying Decimals

Learning Target: Multiply decimals and solve problems involving multiplication of decimals.

Success Criteria:
- I can multiply decimals by whole numbers.
- I can multiply decimals by decimals.
- I can evaluate expressions involving multiplication of decimals.

EXPLORATION 1

Multiplying Decimals

Work with a partner.

a. Write the multiplication expression represented by each area model. Then find the product. Explain how you found your answer.

i.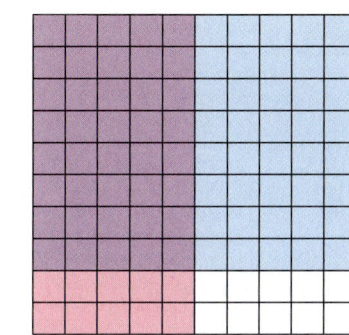

ii.

iii.

iv.

Math Practice

View as Components
How can you use an area model to find the product?

b. How can you find the products in part (a) without using a model? How do you know where to place the decimal points in the answers?

c. Find the product of 0.55 and 0.45. Explain how you found your answer.

Section 2.5 Multiplying Decimals 73

2.5 Lesson

 **Key Idea**

Multiplying Decimals by Whole Numbers

Words Multiply as you would with whole numbers. Then count the number of decimal places in the decimal factor. The product has the same number of decimal places.

Numbers

$$\begin{array}{r} 13.91 \\ \times 7 \\ \hline 97.37 \end{array}$$ ← 2 decimal places

$$\begin{array}{r} 6.218 \\ \times 4 \\ \hline 24.872 \end{array}$$ ← 3 decimal places

EXAMPLE 1 Multiplying Decimals and Whole Numbers

a. Find 6×3.91.   **Estimate** $6 \times 4 = 24$

$$\begin{array}{r} \overset{5}{3.91} \\ \times 6 \\ \hline 23.46 \end{array}$$

3.91 ← 2 decimal places
23.46 ← Count 2 decimal places from right to left.

▸ So, $6 \times 3.91 = 23.46$.   **Reasonable?** $23.46 \approx 24$ ✓

b. Find 3×0.016.   **Estimate** $3 \times 0 = 0$

$$\begin{array}{r} \overset{1}{0.016} \\ \times 3 \\ \hline 0.048 \end{array}$$

0.016 ← 3 decimal places
0.048 ← To have 3 decimal places, insert zeros to the left of 48.

▸ So, $3 \times 0.016 = 0.048$.   **Reasonable?** $0.048 \approx 0$ ✓

c. Find $5.1024 \cdot 12$.   **Estimate** $5 \cdot 12 = 60$

$$\begin{array}{r} 5.1024 \\ \times 12 \\ \hline 1\,0\,2\,0\,4\,8 \\ 5\,1\,0\,2\,4 \\ \hline 61.2288 \end{array}$$

5.1024 ← 4 decimal places
61.2288 ← Count 4 decimal places from right to left.

▸ So, $5.1024 \cdot 12 = 61.2288$.   **Reasonable?** $61.2288 \approx 60$ ✓

Try It Multiply. Use estimation to check your answer.

1. 12.3×8 **2.** 5×14.51 **3.** $20 \cdot 0.008$ **4.** $2.3275 \cdot 90$

The rule for multiplying two decimals is similar to the rule for multiplying a decimal by a whole number.

Key Idea

Multiplying Decimals by Decimals

Words Multiply as you would with whole numbers. Then add the number of decimal places in the factors. The sum is the number of decimal places in the product.

Numbers
```
  4.716   ←    3 decimal places
× 0.2     ←  + 1 decimal place
─────
  0.9432  ←    4 decimal places
```

EXAMPLE 2 Multiplying Decimals

a. Multiply 4.8 × 7.2.   **Estimate** 5 × 7 = 35

```
    4.8   ←   1 decimal place
×   7.2   ← + 1 decimal place
─────
     96
    336
─────
  34.56   ←   2 decimal places
```

So, 4.8 × 7.2 = 34.56.   **Reasonable?** 34.56 ≈ 35 ✓

b. Multiply 3.1 × 0.005.   **Estimate** 3 × 0 = 0

```
     3.1   ←   1 decimal place
× 0.005   ← + 3 decimal places
──────
  0.0155   ←   4 decimal places
```

So, 3.1 × 0.005 = 0.0155.   **Reasonable?** 0.0155 ≈ 0 ✓

c. Multiply 4.25 • 1.75.   **Estimate** 4 × 2 = 8

```
    4.25   ←   2 decimal places
×   1.75   ← + 2 decimal places
──────
    2125
    2975
    425
──────
  7.4375   ←   4 decimal places
```

So, 4.25 • 1.75 = 7.4375.   **Reasonable?** 7.4375 ≈ 8 ✓

Try It Multiply. Use estimation to check your answer.

5. 8.1 × 5.6   **6.** 2.7 × 9.04

7. 6.32 × 0.09   **8.** 1.785 × 0.2

Section 2.5 Multiplying Decimals 75

EXAMPLE 3 Evaluating an Expression

What is the value of 2.44(4.5 − 3.175)?

   A. 3.233   **B.** 3.599   **C.** 7.805   **D.** 32.33

Step 1: Evaluate the expression in parentheses first.

```
       9
   4 10 10
  4.5 0̸ 0̸
 − 3.1 7 5
  1.3 2 5
```

So, 2.44(4.5 − 3.175) = 2.44(1.325).

Step 2: Multiply the result from Step 1 by 2.44.

```
    1.3 2 5    ← 3 decimal places
  ×   2.4 4    ← + 2 decimal places
    5 3 0 0
    5 3 0 0
    2 6 5 0
   3.2 3 3 0 0  ← 5 decimal places
```

 The correct answer is **A**.

Try It Evaluate the expression.

9. 12.67 + 8.2 • 1.9   **10.** 6.4(1.8 • 7.5)

Self-Assessment for Concepts & Skills

Solve each exercise. Then rate your understanding of the success criteria in your journal.

EVALUATING AN EXPRESSION Evaluate the expression.

11. 8 × 11.215   **12.** 9.42 • 6.83   **13.** 0.15(4.3 − 2.417)

14. **MP NUMBER SENSE** If you know 12 × 24 = 288, how can you find 0.12 × 0.24?

15. **MP NUMBER SENSE** Is the product 1.23 × 8 greater than or less than 8? Explain.

16. **MP REASONING** Copy the problem and place the decimal point in the product.

$$\begin{array}{r} 1.78 \\ \times\ 4.9 \\ \hline 8722 \end{array}$$

76   Chapter 2   Fractions and Decimals

EXAMPLE 4 Modeling Real Life

Zinc: $2.41 per ounce

A science teacher buys 3.25 ounces of zinc for an experiment. The teacher pays with a $10 bill. How much change does the teacher receive?

Find the cost of 3.25 ounces of zinc at $2.41 per ounce. Then subtract that amount from $10.

Step 1: Multiply 2.41 by 3.25 to find the cost of the zinc.

```
    2.41    ←   2 decimal places
  × 3.25    ←  + 2 decimal places
   1205
    482
   723
  7.8325    ←   4 decimal places
```

The cost of 3.25 ounces of zinc is $7.83.

Step 2: Subtract the cost of the zinc from the amount of money the teacher uses to buy the zinc.

$$10.00 - 7.83 = 2.17$$

So, the teacher receives $2.17 in change.

Self-Assessment for Problem Solving

Solve each exercise. Then rate your understanding of the success criteria in your journal.

17. You earn $9.15 per hour painting a fence. It takes 6.75 hours to paint the fence. Did you earn enough money to buy the jersey shown? If so, how much money do you have left? If not, how much money do you need to earn?

18. A sand volleyball court is a rectangle that has a length of 52.5 feet and a width that is half of the length. In case of rain, the court is covered with a tarp. How many square feet of tarp are needed to cover the court?

19. **DIG DEEPER!** You buy 4 cases of bottled water and 5 bottles of fruit punch for a birthday party. Each case of bottled water costs $2.75, and each bottle of fruit punch costs $1.35. You hand the cashier a $20 bill. How much change will you receive?

2.5 Practice

Go to **BigIdeasMath.com** to get HELP with solving the exercises.

Review & Refresh

Add or subtract.

1. $12.29 - 6.15$
2. $4.6 + 11.81$
3. $9.34 + 17.009$
4. $18.247 - 16.262$

Divide.

5. $78 \div 3$
6. $65 \div 13$
7. $57 \div 19$
8. $84 \div 12$

9. What is $4\frac{1}{3} \times \frac{4}{5}$?

   A. $2\frac{1}{8}$
   B. $3\frac{7}{15}$
   C. $4\frac{4}{15}$
   D. $5\frac{5}{12}$

Evaluate the expression.

10. $4 + 6^2 \div 2$
11. $(35 + 9) \div 4 - 3^2$
12. $8^2 \div [(14 - 12) \times 2^3]$

Concepts, Skills, & Problem Solving

USING TOOLS Use an area model to find the product. *(See Exploration 1, p. 73.)*

13. 2.1×1.5
14. 0.6×0.4
15. 0.7×0.3
16. 2.7×2.3

MULTIPLYING DECIMALS AND WHOLE NUMBERS Multiply. Use estimation to check your answer.

17. 4.8×7
18. 6.3×5
19. 7.19×16
20. 0.87×21

21. 1.95×11
22. 5.89×5
23. 3.472×4
24. 8.188×12

25. 100×0.024
26. 19×0.004
27. 3.27×14
28. $46 \cdot 5.448$

29. 50×12.21
30. $104 \cdot 4.786$
31. 0.0038×9
32. 10×0.0093

YOU BE THE TEACHER Your friend finds the product. Is your friend correct? Explain your reasoning.

33.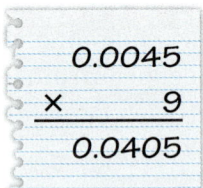
    $$0.0045 \times 9 = 0.0405$$

34.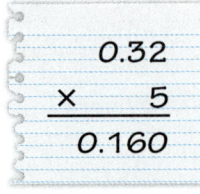
    $$0.32 \times 5 = 0.160$$

35. **MODELING REAL LIFE** The weight of an object on the Moon is about 0.167 of its weight on Earth. How much does a 180-pound astronaut weigh on the Moon?

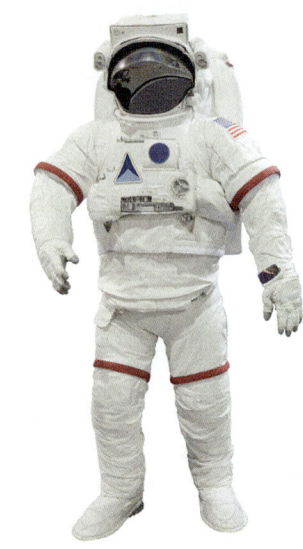

78   Chapter 2   Fractions and Decimals

MULTIPLYING DECIMALS Multiply.

36. 0.7 × 0.2

37. 0.08 × 0.3

38. 0.007 × 0.03

39. 0.0008 × 0.09

40. 0.004 × 0.9

41. 0.06 × 0.5

42. 0.0008 × 0.004

43. 0.0002 × 0.06

44. 12.4 × 0.2

45. 18.6 • 5.9

46. 7.91 × 0.72

47. 1.16 × 3.35

48. 6.478 × 18.21

49. 1.9 × 7.216

50. 0.0021 × 18.2

51. 6.109 • 8.4

52. YOU BE THE TEACHER Your friend finds the product of 4.9 and 3.8. Is your friend correct? Explain your reasoning.

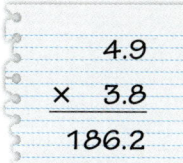

53. MP PROBLEM SOLVING A Chinese restaurant offers buffet takeout for $4.99 per pound. How much does your takeout meal cost?

54. MP PROBLEM SOLVING On a tour of an old gold mine, you find a nugget containing 0.82 ounce of gold. Gold is worth $1323.80 per ounce. How much is your nugget worth?

55. MP PRECISION One meter is approximately 3.28 feet. Find the height of each building in feet.

Continent	Tallest Building	Height (meters)
Africa	Carlton Centre	223
Asia	Burj Khalifa	828
Australia	Q1	323
Europe	Federation Tower	374
North America	One World Trade Center	541
South America	Gran Torre Santiago	300

56. MP REASONING Show how to evaluate (7.12 × 8.22) × 100 without multiplying two decimals.

EVALUATING AN EXPRESSION Evaluate the expression.

57. 2.4 × 16 + 7

58. 6.85 × 2 × 10

59. 1.047 × 5 − 0.88

60. 4.32(3.7 + 1.65)

61. $23.98 - 1.7^2$ • 7.6

62. 12 • 5.16 + 10.064

63. 0.9(8.2 • 20.35)

64. $7.5^2(6.084 - 5.44)$

65. 0.629[81 ÷ (10 × 2.7)]

66. **MP REASONING** Without multiplying, how many decimal places does 3.4^2 have? 3.4^3? 3.4^4? Explain your reasoning.

67. **MODELING REAL LIFE** You buy 2.6 pounds of apples and 1.475 pounds of peaches. You hand the cashier a $20 bill. How much change will you receive?

MP PATTERNS Describe the pattern. Find the next three numbers.

68. 1, 0.6, 0.36, 0.216, . . .

69. 15, 1.5, 0.15, 0.015, . . .

70. 0.04, 0.02, 0.01, 0.005, . . .

71. 5, 7.5, 11.25, 16.875, . . .

72. **DIG DEEPER!** You are preparing for a trip to Canada. At the time of your trip, each U.S. dollar is worth 1.293 Canadian dollars and each Canadian dollar is worth 0.773 U.S. dollar.

   a. You exchange 150 U.S. dollars for Canadian dollars. How many Canadian dollars do you receive?

   b. You spend 120 Canadian dollars on the trip. Then you exchange the remaining Canadian dollars for U.S. dollars. How many U.S. dollars do you receive?

73. **OPEN-ENDED** You and four friends have dinner at a restaurant.

   a. Draw a restaurant menu that has main items, desserts, and beverages, with their prices.

   b. Write a guest check that shows what each of you ate. Find the subtotal.

   c. Multiply by 0.07 to find the tax. Then find the total.

   d. Round the total to the nearest whole number. Multiply by 0.20 to estimate a tip. Including the tip, how much did the dinner cost?

74. **GEOMETRY** A rectangular painting has an area of 9.52 square feet.

   a. Draw three different ways in which this can happen.

   b. The cost of a frame depends on the perimeter of the painting. Which of your drawings from part (a) is the least expensive to frame? Explain your reasoning.

   c. The thin, black framing costs $1 per foot. The fancy framing costs $5 per foot. Will the fancy framing cost five times as much as the black framing? Explain why or why not.

   d. Suppose the cost of a frame depends on the outside perimeter of the frame. Does this change your answer to part (c)? Explain why or why not.

2.6 Dividing Whole Numbers

Learning Target: Divide whole numbers and solve problems involving division of whole numbers.

Success Criteria:
- I can use long division to divide whole numbers.
- I can write a remainder as a fraction.
- I can interpret quotients in real-life problems.

EXPLORATION 1

Using a Double Bar Graph

Work with a partner. The double bar graph shows the history of a citywide cleanup day.

a. Make five conclusions from the graph.

b. Compare the results of the city cleanup day in 2016 to the results in 2014.

c. What is the average combined amount of trash and recyclables collected each year over the four-year period?

d. Make a prediction about the amount of trash collected in a future year.

Math Practice

Calculate Accurately

How can you extend what you know about long division to divide any pair of multi-digit whole numbers accurately?

Section 2.6 Dividing Whole Numbers

2.6 Lesson

You have used long division to divide whole numbers. When the *divisor* divides evenly into the *dividend*, the *quotient* is a whole number.

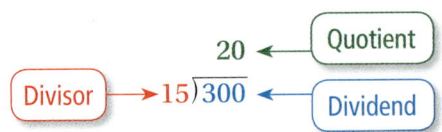

When the divisor does not divide evenly into the dividend, you obtain a remainder. When this occurs, you can write the quotient as a mixed number.

EXAMPLE 1 **Dividing Whole Numbers**

a. Find $672 \div 8$.

$$
\begin{array}{r}
84 \\
8\overline{)672} \\
-64\downarrow \\
\hline
32 \\
-32 \\
\hline
0
\end{array}
$$

There are eight groups of 8 in 67.

There are four groups of 8 in 32.

There is no remainder.

▶ So, $672 \div 8 = 84$.

Check Find the product of the quotient and the divisor.

$$
\begin{array}{r}
84 \quad \text{quotient} \\
\times 8 \quad \text{divisor} \\
\hline
672 \quad \text{dividend} \checkmark
\end{array}
$$

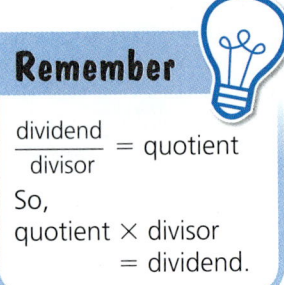

Remember

$\dfrac{\text{dividend}}{\text{divisor}} = \text{quotient}$

So,
quotient × divisor
= dividend.

b. Find the quotient of 9216 and 150.

$$
\begin{array}{r}
61 \text{ R}66 \\
150\overline{)9216} \\
-900\downarrow \\
\hline
216 \\
-150 \\
\hline
66
\end{array}
$$

There are six groups of 150 in 921.

There is one group of 150 in 216.

The remainder is 66.

▶ The quotient is $61\dfrac{66}{150}$, or $61\dfrac{11}{25}$.

Try It Divide. Use estimation to check your answer.

1. $234 \div 9$
2. $\dfrac{6096}{30}$
3. $45{,}691 \div 28$

4. Find the quotient of 9920 and 320.

82   Chapter 2   Fractions and Decimals

EXAMPLE 2 **Solving a Problem Using Division**

You make 12 equal payments for a go-kart. You pay a total of $1380. How much is each payment?

You want to find the number of groups of 12 in $1380. So, find the quotient of 1380 and 12 using long division.

```
      115
   ┌──────
12 │ 1380     There is one group of 12 in 13.
   − 12↓
   ─────
       18     There is one group of 12 in 18.
     − 12↓
     ─────
       60     There are five groups of 12 in 60.
     − 60
     ─────
        0     There is no remainder.
```

The quotient of 1380 and 12 is 115.

▶ So, each payment is $115.

Check Find the product of the quotient and the divisor.

```
    115    quotient
  ×  12    divisor
  ─────
    230
    115
  ─────
   1380    dividend ✓
```

Try It

5. **WHAT IF?** You make 30 equal payments for the go-kart. How much is each payment?

Self-Assessment for Concepts & Skills

Solve each exercise. Then rate your understanding of the success criteria in your journal.

DIVIDING WHOLE NUMBERS Divide. Use estimation to check your answer.

6. 876 ÷ 12   7. 3024 ÷ 7   8. 1043 ÷ 22

9. **VOCABULARY** Use the division problem shown to tell whether the number is the divisor, dividend, or quotient.

   a. 884   b. 26   c. 34

10. **MP NUMBER SENSE** Without calculating, decide which is greater: 3999 ÷ 129 or 3834 ÷ 142. Explain.

11. **MP REASONING** In a division problem, can the remainder be greater than the divisor? Explain.

Section 2.6 Dividing Whole Numbers

EXAMPLE 3 Modeling Real Life

A 301-foot-high swing at an amusement park can take 64 people on each ride. A total of 10,250 people ride the swing today. All the rides are full except for the last ride. How many rides are given? How many people are on the last ride?

To find the number of rides given, first you need to find the number of groups of 64 people in 10,250 people. So, find the quotient of 10,250 and 64.

```
       160 R10
   64)10,250
      − 64
        385      There are six groups of 64 in 385.
      − 384
          10     There are no groups of 64 in 10.
        −  0
          10     The remainder is 10.
```

There is one group of 64 in 102.

Do not stop here. You must write a 0 in the ones place of the quotient.

The quotient is $160\frac{10}{64}$. This means there are 160 full rides, with 10 people remaining.

 So, 161 rides are given, with 10 people on the last ride.

Check Find the product of the quotient and the divisor.

$$64 \cdot 160\frac{10}{64} = 64\left(160 + \frac{10}{64}\right) = 64(160) + 64\left(\frac{10}{64}\right) = 10{,}240 + 10 = 10{,}250 \checkmark$$

Self-Assessment for Problem Solving

Solve each exercise. Then rate your understanding of the success criteria in your journal.

12. In a movie's opening weekend, 879,575 tickets are sold in 755 theaters. The average cost of a ticket is $9.50. What is the average amount of money earned by each theater?

13. A boat can carry 582 passengers to the base of a waterfall. A total of 13,105 people ride the boat today. All the rides are full except for the first ride. How many rides are given? How many people are on the first ride?

14. **DIG DEEPER!** A new year begins at 12:00 A.M. on January 1. What is the date and time 12,345 minutes after the start of a new year?

2.6 Practice

Go to *BigIdeasMath.com* to get HELP with solving the exercises.

Review & Refresh

Multiply.

1. 8×3.79
2. 12.1×2.42
3. 6.43×0.28
4. $9.526 \cdot 6.61$

List the factor pairs of the number.

5. 26
6. 72
7. 50
8. 98

Match the expression with its value.

9. $\dfrac{6}{7} \div \dfrac{3}{5}$
10. $\dfrac{3}{7} \div \dfrac{6}{5}$
11. $\dfrac{6}{5} \div \dfrac{3}{7}$
12. $\dfrac{3}{5} \div \dfrac{6}{7}$

A. $\dfrac{7}{10}$
B. $1\dfrac{3}{7}$
C. $\dfrac{5}{14}$
D. $2\dfrac{4}{5}$

Concepts, Skills, & Problem Solving

OPERATIONS WITH WHOLE NUMBERS The bar graph shows the attendance at a food festival. Use the graph to answer the question.
(See Exploration 1, p. 81.)

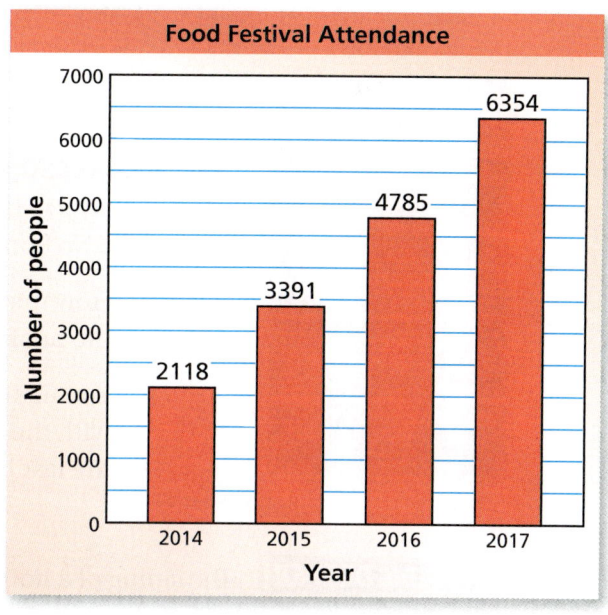

13. What is the total attendance at the food festival from 2014 to 2017?

14. How many times more people attended the food festival in 2017 than in 2014?

15. What is the average attendance at the festival each year over the four-year period?

16. The festival projects that the attendance for 2018 will be twice the attendance in 2016. What is the projected attendance for 2018?

DIVIDING WHOLE NUMBERS Divide. Use estimation to check your answer.

17. $837 \div 27$
18. $1088 \div 34$
19. $903 \div 72$
20. $6409 \div 61$
21. $\dfrac{5986}{82}$
22. $6200 \div 163$
23. $6255 \div 118$
24. $\dfrac{588}{84}$
25. $7440 \div 124$
26. $26{,}862 \div 407$
27. $8241 \div 173$
28. $\dfrac{33{,}505}{160}$

Section 2.6 Dividing Whole Numbers 85

29. **MODELING REAL LIFE** A pharmacist divides 364 pills into prescription bottles. Each bottle contains 28 pills. How many bottles does the pharmacist fill?

YOU BE THE TEACHER Your friend finds the quotient. Is your friend correct? Explain your reasoning.

30.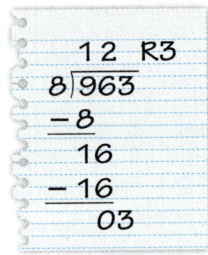

31.
```
      19
  12)1308
   -12
    108
   -108
      0
```

GEOMETRY Find the perimeter of the rectangle.

32. Area = 35 in.²
    7 in.

33.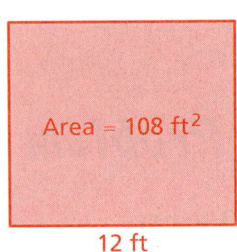
    Area = 108 ft²
    12 ft

34.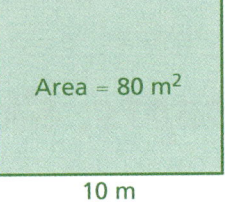
    Area = 80 m²
    10 m

35. **REASONING** You borrow bookcases like the one shown to display 943 books at a book sale. You plan to put 22 books on each shelf. No books will be on top of the bookcases.

    a. How many bookcases must you borrow to display all the books?

    b. You fill the shelves of each bookcase in order, starting with the top shelf. How many books are on each shelf of the last bookcase?

36. **DIG DEEPER!** The siding of a house is 2250 square feet. The siding needs two coats of paint.

    a. What is the minimum cost of the paint needed to complete the job?

    b. How much paint is left over when you spend the minimum amount?

Can Size	Cost	Coverage
1 quart	$18	80 square feet
1 gallon	$29	320 square feet

37. **CRITICAL THINKING** Use the digits 3, 4, 6, and 9 to complete the division problem. Use each digit once.

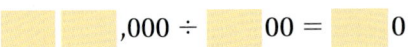
☐☐,000 ÷ ☐00 = ☐0

2.7 Dividing Decimals

Learning Target: Divide decimals and solve problems involving division of decimals.

Success Criteria:
- I can divide decimals by whole numbers.
- I can divide decimals by decimals.
- I can divide whole numbers by decimals.

EXPLORATION 1

Dividing Decimals

Work with a partner.

a. Write two division expressions represented by each area model. Then find the quotients. Explain how you found your answer.

i.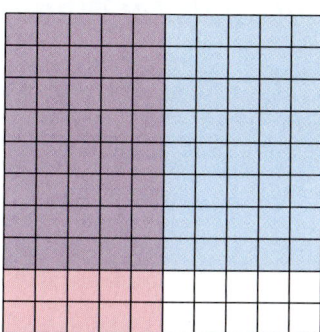

ii.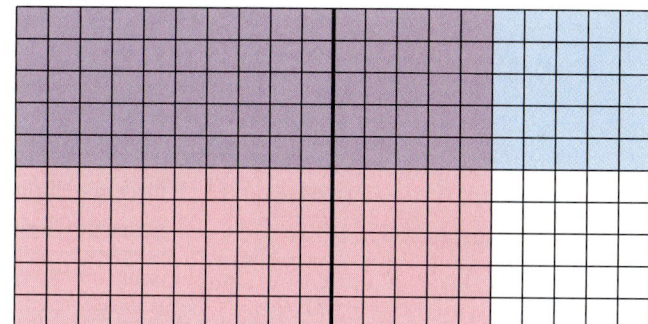

iii.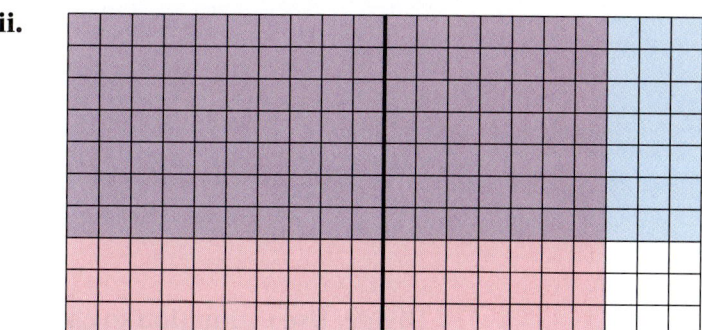

b. Use a calculator to find $119 \div 17$, $11.9 \div 1.7$, $1.19 \div 0.17$, and $0.119 \div 0.017$. What do you notice? Explain how you can use long division to divide any pair of multi-digit decimals.

Math Practice

Construct Arguments

Why do the quotients in part (b) have the relationship you observed?

2.7 Lesson

🔑 Key Idea

Dividing Decimals by Whole Numbers

Words Place the decimal point in the quotient above the decimal point in the dividend. Then divide as you would with whole numbers. Continue until there is no remainder.

Numbers

1.83
$4\overline{)7.32}$

Place the decimal point in the quotient above the decimal point in the dividend.

EXAMPLE 1 Dividing Decimals by Whole Numbers

a. Find $7.6 \div 4$. **Estimate** $8 \div 4 = 2$

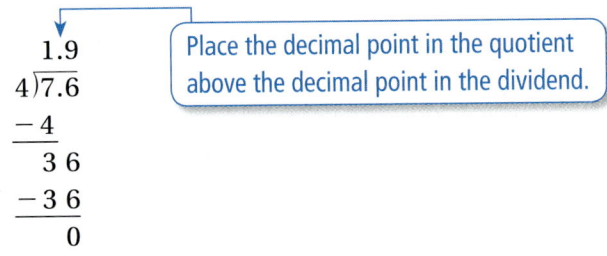

Place the decimal point in the quotient above the decimal point in the dividend.

▶ So, $7.6 \div 4 = 1.9$. **Reasonable?** $1.9 \approx 2$ ✓

b. Find $4.374 \div 12$.

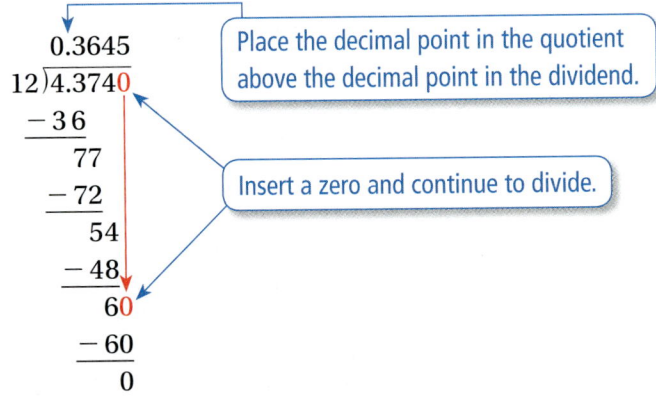

Place the decimal point in the quotient above the decimal point in the dividend.

Insert a zero and continue to divide.

▶ So, $4.374 \div 12 = 0.3645$. **Check** $0.3645 \times 12 = 4.374$ ✓

Try It Divide. Use estimation to check your answer.

1. $36.4 \div 2$
2. $22.2 \div 6$
3. $59.64 \div 7$
4. $3.12 \div 16$
5. $6.224 \div 4$
6. $43.407 \div 14$

 **Key Idea**

Dividing Decimals by Decimals

Words Multiply the divisor *and* the dividend by a power of 10 to make the divisor a whole number. Then place the decimal point in the quotient above the decimal point in the dividend and divide as you would with whole numbers. Continue until there is no remainder.

Numbers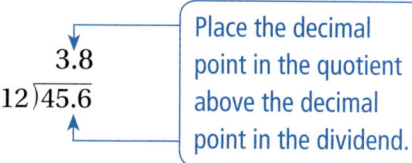

Multiply each number by 10.

Place the decimal point in the quotient above the decimal point in the dividend.

EXAMPLE 2 **Dividing Decimals**

Multiplying the divisor and the dividend by a power of 10 does not change the quotient.

For example:
18.2 ÷ 1.4 = 13
182 ÷ 14 = 13
1820 ÷ 140 = 13

a. Find 18.2 ÷ 1.4.

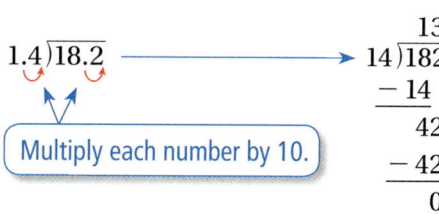

Multiply each number by 10.

Place the decimal point in the quotient above the decimal point in the dividend.

▶ So, 18.2 ÷ 1.4 = 13.

Check 13 × 1.4 = 18.2 ✓

b. Find 0.273 ÷ 0.39.

$$0.39\overline{)0.273} \longrightarrow 39\overline{)27.3}$$
$$\phantom{0.39\overline{)0.273}} \underline{-273}$$
$$\phantom{0.39\overline{)0.273}} 0$$

quotient: 0.7

Multiply each number by 100.

▶ So, 0.273 ÷ 0.39 = 0.7.

Check 0.7 × 0.39 = 0.273 ✓

Try It Divide. Check your answer.

7. $1.2\overline{)9.6}$

8. $3.4\overline{)57.8}$

9. 21.643 ÷ 2.3

10. 0.459 ÷ 0.51

Section 2.7 Dividing Decimals 89

EXAMPLE 3 Inserting Zeros in the Dividend and the Quotient

a. Find 2.45 ÷ 0.007.

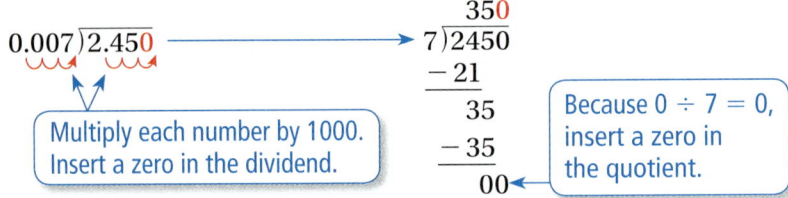

Remember to check your answer by multiplying the quotient by the divisor.

So, 2.45 ÷ 0.007 = 350.

b. Find 32 ÷ 1.25.

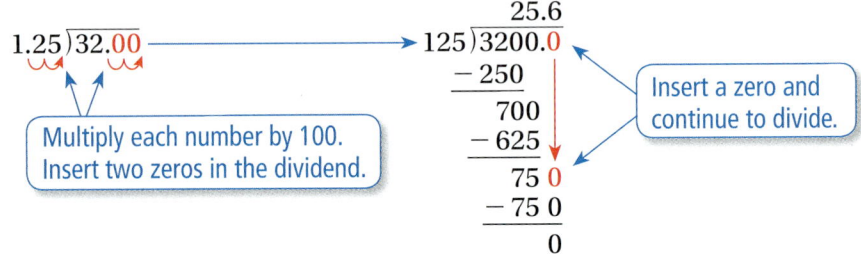

So, 32 ÷ 1.25 = 25.6.

Try It Divide. Check your answer.

11. 3.8 ÷ 0.16
12. 15.6 ÷ 0.78
13. 7.2 ÷ 0.048
14. 42 ÷ 3.75

Self-Assessment for Concepts & Skills

Solve each exercise. Then rate your understanding of the success criteria in your journal.

DIVIDING DECIMALS Divide. Check your answer.

15. 37.7 ÷ 13
16. 33 ÷ 4.4
17. 2.16 ÷ 0.009

18. **MP NUMBER SENSE** Fix the one that is not correct.

$$\frac{6.1}{4\overline{)24.4}} \qquad \frac{61}{4\overline{)244}} \qquad \frac{6.1}{4\overline{)2.44}}$$

19. **MP NUMBER SENSE** Rewrite $2.16\overline{)18.5}$ so that the divisor is a whole number.

20. **MP STRUCTURE** Write 1.8 ÷ 6 as a multiplication problem with a missing factor. Explain your reasoning.

EXAMPLE 4 Modeling Real Life

How many times more cell phone subscribers were there in 2015 than in 1990? Round to the nearest whole number.

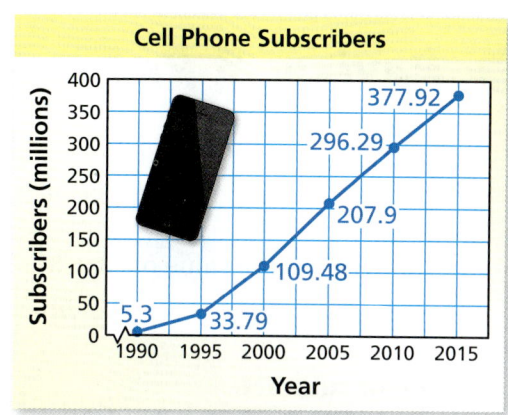

Divide the number of subscribers in 2015 by the number of subscribers in 1990.

From the graph, there were 377.92 million subscribers in 2015 and 5.3 million in 1990. So, divide 377.92 by 5.3.

$$5.3\overline{)377.92} \longrightarrow 53\overline{)3779.2}$$

Rounds to 71.

$$\begin{array}{r} 71.3 \\ 53\overline{)3779.2} \\ -371 \\ \hline 69 \\ -53 \\ \hline 16\,2 \\ -15\,9 \\ \hline 3 \end{array}$$

Check Reasonableness
Find an underestimate and an overestimate.
Underestimate:
$370 \div 6 = 61\frac{2}{3}$

Overestimate:
$380 \div 5 = 76$

The answer is reasonable because
$61\frac{2}{3} < 71 < 76.$ ✓

So, there were about 71 times more subscribers in 2015 than in 1990.

Self-Assessment for Problem Solving

Solve each exercise. Then rate your understanding of the success criteria in your journal.

21. A magazine subscription costs $29.88 for 12 issues or $15.24 for 6 issues. Which subscription costs more per issue? How much more?

22. The track of a roller coaster is 1.265 miles long. The ride lasts for 2.3 minutes. What is the average speed of the roller coaster in miles per hour?

23. **DIG DEEPER!** The table shows the number of visitors to a website each year for 4 years. Does the number of visitors increase more from Year 1 to Year 2 or from Year 3 to Year 4? How many times greater is the increase?

Year	Visitors (millions)
1	2.4
2	32.22
3	88.4
4	102.6

Section 2.7 Dividing Decimals

2.7 Practice

Go to *BigIdeasMath.com* to get HELP with solving the exercises.

▶ Review & Refresh

Divide.

1. $84 \div 14$
2. $391 \div 23$
3. $1458 \div 54$
4. $\dfrac{68{,}134}{163}$

5. What is the value of $18 + 3^2 \div [3 \times (8 - 5)]$?

   A. 3   **B.** 19   **C.** 27   **D.** 49

Add or subtract.

6. $7.635 - 5.046$
7. $12.177 + 3.09$
8. $14.008 - 9.433$

▶ Concepts, Skills, & Problem Solving

DIVIDING DECIMALS Write two division expressions represented by the area model. Then find the quotients. Explain how you found your answer. (See Exploration 1, p. 87.)

9.

10.

DIVIDING DECIMALS BY WHOLE NUMBERS Divide. Use estimation to check your answer.

11. $6\overline{)25.2}$
12. $5\overline{)33.5}$
13. $7\overline{)3.5}$
14. $8\overline{)10.4}$

15. $38.79 \div 9$
16. $37.72 \div 4$
17. $43.4 \div 7$
18. $22.505 \div 7$

19. $44.64 \div 8$
20. $0.294 \div 3$
21. $3.6 \div 24$
22. $52.014 \div 20$

YOU BE THE TEACHER Your friend finds the quotient. Is your friend correct? Explain your reasoning.

23.

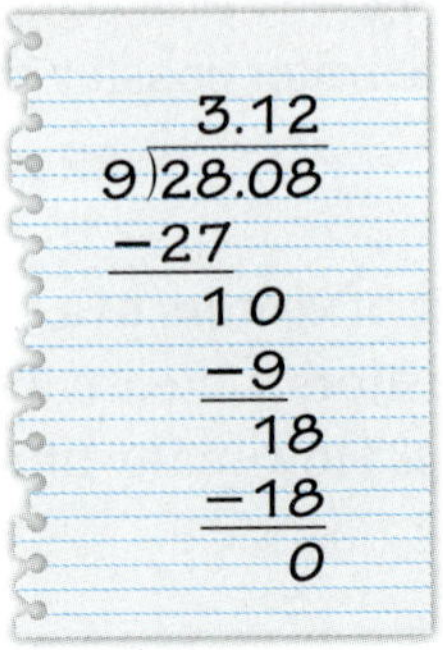

24.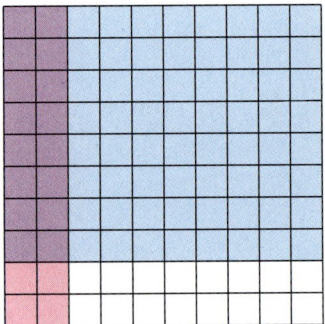

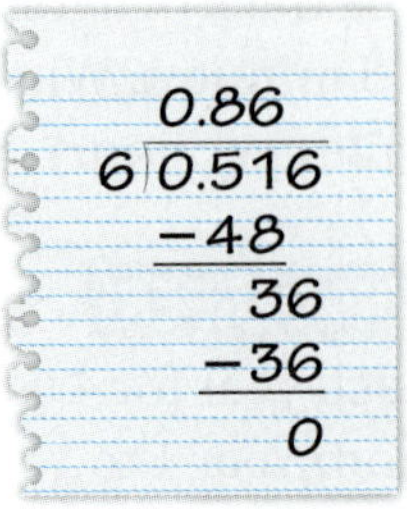

92   Chapter 2   Fractions and Decimals

25. **MP PROBLEM SOLVING** You buy the same pair of pants in 3 different colors for $89.85. How much does each pair of pants cost?

26. **MP REASONING** Which pack of fruit punch is the best buy? Explain.

DIVIDING DECIMALS Divide. Check your answer.

27. 2.1)25.2   28. 3.8)34.2   29. 36.47 ÷ 0.7   30. 0.984 ÷ 12.3
31. 6.64 ÷ 8.3   32. 83.266 ÷ 13.43   33. 0.09)1.053   34. 35.903 ÷ 16.1
35. 0.996 ÷ 0.12   36. 4.63)12.501   37. 0.005)0.00115   38. 56.7175 ÷ 4.63
39. 4.23 ÷ 0.012   40. 0.52 ÷ 0.0013   41. 95.04 ÷ 0.0132   42. 32.2 ÷ 0.07
43. 1.37)54.8   44. 44.2 ÷ 3.25   45. 4.04)50.5   46. 250 ÷ 0.008
47. 11.16 ÷ 0.062   48. 12.5)835   49. 597.6 ÷ 12.45   50. 0.435)118.32

DIVIDING DECIMALS Divide. Round to the nearest hundredth.

51. 80.89 ÷ 8.425   52. 0.8 ÷ 0.6   53. 38.9 ÷ 6.44   54. 11.6 ÷ 0.95

55. **YOU BE THE TEACHER** Your friend rewrites the problem. Is your friend correct? Explain your reasoning.

ORDER OF OPERATIONS Evaluate the expression.

56. 7.68 + 3.18 ÷ 12   57. 10.56 ÷ 3 − 1.9   58. 19.6 ÷ 7 × 9
59. 5.5 × 16.56 ÷ 9   60. 35.25 ÷ 5 ÷ 3   61. 13.41 × (5.4 ÷ 9)
62. 6.2 • (5.16 ÷ 6.45)   63. 132.06 ÷ (4^2 + 2.6) 64. 4.8[23.9841 ÷ (1.16 + 1.27)]

65. **MODELING REAL LIFE** A person's running stride is about 1.14 times the person's height. Your friend's stride is 5.472 feet. How tall is your friend?

66. **MP PROBLEM SOLVING** You have 3.4 gigabytes available on your tablet. A song is about 0.004 gigabyte. How many songs can you download onto your tablet?

MP REASONING Without finding the quotient, copy and complete the statement using <, >, or =. Explain your reasoning.

67. 6.66 ÷ 0.74 ▢ 66.6 ÷ 7.4   68. 32.2 ÷ 0.7 ▢ 3.22 ÷ 7
69. 160.72 ÷ 16.4 ▢ 160.72 ÷ 1.64   70. 75.6 ÷ 63 ▢ 7.56 ÷ 0.63

71. **DIG DEEPER!** The table shows the top three times in a swimming event at the Summer Olympics. The event consists of a team of four women swimming 100 meters each.

Women's 4 × 100 Freestyle Relay		
Medal	Country	Time (seconds)
Gold	Australia	210.65
Silver	United States	211.89
Bronze	Canada	212.89

    a. Suppose the times of all four swimmers on each team were the same. For each team, how much time does it take a swimmer to swim 100 meters?

    b. Suppose each U.S. swimmer completed 100 meters a quarter second faster. Would the U.S. team have won the gold medal? Explain your reasoning.

72. **MP PROBLEM SOLVING** To approximate the number of bees in a hive, multiply the number of bees that leave the hive in one minute by 3 and divide by 0.014. You count 25 bees leaving a hive in one minute. How many bees are in the hive?

73. **MP PROBLEM SOLVING** You are saving money to buy a new bicycle that costs $155.75. You have $30 and plan to save $5 each week. Your aunt decides to give you an additional $10 each week.

    a. How many weeks will you have to save until you have enough money to buy the bicycle?

    b. How many more weeks would you have to save to buy a new bicycle that costs $203.89? Explain how you found your answer.

74. **MP PRECISION** A store sells applesauce in two sizes.

    Applesauce
    3.9-ounce bowl $0.52
    24-ounce jar $2.63

    a. How many bowls of applesauce fit in a jar? Round your answer to the nearest hundredth.

    b. Explain two ways to find the better buy.

    c. Which is the better buy?

75. **GEOMETRY** The large rectangle's dimensions are three times the dimensions of the small rectangle.

    23.1 ft
    49.2 ft

    a. How many times greater is the perimeter of the large rectangle than the perimeter of the small rectangle?

    b. How many times greater is the area of the large rectangle than the area of the small rectangle?

    c. Are the answers to parts (a) and (b) the same? Explain why or why not.

    d. What happens in parts (a) and (b) if the dimensions of the large rectangle are two times the dimensions of the small rectangle?

2 Connecting Concepts

Using the Problem-Solving Plan

1. You change the water jug on the watercooler. How many glasses can be completely filled before you need to change the water jug again?

   Understand the problem. You know the capacities of the water jug and the glass. You are asked to determine how many glasses the water jug can fill.

   Make a plan. First, use what you know about converting measures to find the number of fluid ounces in 5 gallons. Then divide this amount by the capacity of the glass to find the number of glasses that can be filled.

   Solve and check. Use the plan to solve the problem. Then check your solution.

2. Two ferries just departed from their docks at the same time. Ferry A departs from its dock every 1.2 hours. Ferry B departs from its dock every 1.8 hours. How long will it be until both ferries depart from their docks at the same time again?

3. You want to paint the ceiling of your bedroom. The ceiling has two square skylights as shown. Each skylight has a side length of $1\frac{7}{8}$ feet. How many square feet will you paint? Justify your answer.

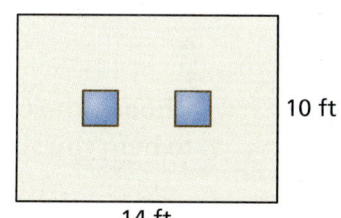

Performance Task

Space Explorers

At the beginning of this chapter, you watched a STEAM video called "Space is Big." You are now ready to complete the performance task for this video, available at **BigIdeasMath.com**. Be sure to use the problem-solving plan as you work through the performance task.

Chapter Review

Review Vocabulary

Write the definition and give an example of each vocabulary term.

reciprocals, *p. 54*

multiplicative inverses, *p. 54*

Graphic Organizers

You can use a **Summary Triangle** to explain a concept. Here is an example of a Summary Triangle for *dividing fractions*.

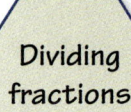

Dividing fractions

Procedure: To divide a number by a fraction, multiply the number by the reciprocal of the fraction.

Algebra:
$\dfrac{a}{b} \div \dfrac{c}{d} = \dfrac{a}{b} \cdot \dfrac{d}{c} = \dfrac{a \cdot d}{b \cdot c}$, where b, c, and $d \neq 0$

Example: $\dfrac{1}{5} \div \dfrac{3}{4} = \dfrac{1}{5} \times \dfrac{4}{3} = \dfrac{1 \times 4}{5 \times 3} = \dfrac{4}{15}$

Choose and complete a graphic organizer to help you study the concept.

1. multiplying fractions
2. multiplying mixed numbers
3. reciprocals
4. dividing mixed numbers
5. adding and subtracting decimals
6. multiplying decimals by decimals
7. dividing whole numbers
8. dividing decimals by decimals

"I finished my **Summary Triangle** about characteristics of igloos. I built one to use as my dog house. I'm calling it a dog-gloo."

Chapter Self-Assessment

As you complete the exercises, use the scale below to rate your understanding of the success criteria in your journal.

1	2	3	4
I do not understand.	I can do it with help.	I can do it on my own.	I can teach someone else.

2.1 Multiplying Fractions (pp. 45–52)

Learning Target: Find products involving fractions and mixed numbers.

Multiply. Write the answer in simplest form.

1. $\dfrac{2}{9} \times \dfrac{8}{11}$
2. $\dfrac{3}{10} \cdot \dfrac{4}{5}$
3. $2\dfrac{3}{10} \times 5\dfrac{1}{3}$
4. $\dfrac{2}{7} \times 4\dfrac{4}{9}$

5. Write two fractions whose product is $\dfrac{21}{32}$.

6. A costume designer needs to make 12 costumes for the school play. Each costume requires $2\dfrac{2}{3}$ yards of fabric. How many yards of fabric does the costume designer need to make all the costumes?

7. You spend $\dfrac{4}{5}$ of an hour on your homework. You spend $\dfrac{1}{2}$ of that time working on your science homework. How many minutes do you spend working on science homework?

2.2 Dividing Fractions (pp. 53–60)

Learning Target: Compute quotients of fractions and solve problems involving division by fractions.

Divide. Write the answer in simplest form.

8. $\dfrac{3}{4} \div \dfrac{5}{6}$
9. $\dfrac{2}{5} \div 8$
10. $5 \div \dfrac{1}{3}$
11. $\dfrac{8}{9} \div \dfrac{3}{10}$

12. A box contains 10 cups of pancake mix. You use $\dfrac{2}{3}$ cup each time you make pancakes. How many times can you make pancakes?

13. Write two fractions whose quotient is $\dfrac{28}{45}$.

Chapter Review

2.3 Dividing Mixed Numbers (pp. 61–66)

Learning Target: Compute quotients with mixed numbers and solve problems involving division with mixed numbers.

Divide. Write the answer in simplest form.

14. $1\frac{2}{5} \div \frac{4}{7}$

15. $5\frac{5}{8} \div 3$

16. $5 \div 2\frac{6}{7}$

17. $4\frac{1}{8} \div 2\frac{1}{4}$

18. Evaluate $5\frac{5}{7} \div 1\frac{3}{5} \cdot 4\frac{2}{3}$. Write the answer in simplest form.

19. You have $23\frac{1}{2}$ pounds of blueberries to store in freezer bags. Each bag holds $3\frac{3}{4}$ pounds of blueberries. What is the minimum number of freezer bags needed to store all the blueberries?

20. A squirrel feeder holds $4\frac{1}{2}$ cups of seeds. Another squirrel feeder holds $6\frac{7}{8}$ cups of seeds. One scoop of seeds is $1\frac{5}{8}$ cups. How many scoops of seeds do you need to fill both squirrel feeders?

2.4 Adding and Subtracting Decimals (pp. 67–72)

Learning Target: Add and subtract decimals and solve problems involving addition and subtraction of decimals.

Add.

21. $3.78 + 8.94$

22. $19.89 + 4.372$

23. $24.916 + 17.385$

Subtract.

24. $7.638 - 2.365$

25. $14.21 - 4.103$

26. $5.467 - 2.736$

27. Write three decimals that have a sum of 10.806.

28. To make fuel for the main engines of a space shuttle, 102,619.377 kilograms of liquid hydrogen and 616,496.4409 kilograms of liquid oxygen are mixed together in the external tank. How much fuel is stored in the external tank?

2.5 Multiplying Decimals (pp. 73–80)

Learning Target: Multiply decimals and solve problems involving multiplication of decimals.

Multiply. Use estimation to check your answer.

29. 26.174×79 **30.** 9.475×8.03 **31.** 0.051×0.244

32. Evaluate $3.76(2.43 + 9.8)$.

33. Hair grows about 1.27 centimeters each month. How much does hair grow in 4 months?

2.6 Dividing Whole Numbers (pp. 81–86)

Learning Target: Divide whole numbers and solve problems involving division of whole numbers.

Divide. Use estimation to check your answer.

34. $7296 \div 38$ **35.** $5081 \div 203$ **36.** $\dfrac{17{,}264}{128}$

37. Your local varsity basketball team offers bus transportation for a playoff game. Each bus holds 56 people. A total of 328 people sign up. All buses are full except for the last bus. How many buses are used? How many people are in the last bus?

38. You have 600 elastic bands to make railroad bracelets. How many complete bracelets can you make?

> **Railroad Bracelet Supplies**
> - loom
> - hook
> - clip
> - 28 elastic bands for outer rails
> - 10 elastic bands for inner track

2.7 Dividing Decimals (pp. 87–94)

Learning Target: Divide decimals and solve problems involving division of decimals.

Divide. Check your answer.

39. $0.498 \div 6$ **40.** $8.9 \div 0.356$ **41.** $21.85 \div 3.8$

42. Evaluate $\dfrac{14.075 + 24.67}{3.15}$.

43. Your beginning balance on your lunch account is $42. You buy lunch for $1.80 every day and sometimes buy a snack for $0.85. After 20 days, you have a balance of $0.05. How many snacks did you buy?

Chapter Review   **99**

2 Practice Test

Evaluate the expression. Write the answer in simplest form.

1. $5.138 + 2.624$
2. $\dfrac{5}{6} \div \dfrac{10}{21}$
3. $0.25 \overline{)5.46}$
4. $\dfrac{9}{16} \times \dfrac{2}{3}$
5. $8\dfrac{3}{4} \div 2\dfrac{7}{8}$
6. 4.87×7.23
7. $1875 \div 125$
8. $10 \div \dfrac{2}{5}$
9. $57.82 \div 0.784$
10. $5.316 - 1.942$
11. 6.729×8.3
12. $\dfrac{13{,}376}{248}$

13. On a road trip, you notice that the gas tank is $\dfrac{1}{4}$ full. The gas tank can hold 18 gallons, and the vehicle averages 22 miles per gallon. Will you make it to your destination 110 miles away before you run out of gas? Explain.

14. For a diving event, the highest and the lowest of seven scores are discarded. Next, the total of the remaining scores is multiplied by the degree of difficulty of the dive. That value is then multiplied by 0.6 to determine the final score. Find the final score for the dive.

15. You are cutting as many $20\dfrac{1}{2}$-inch pieces from the board to make ladder steps for a tree fort. How many steps can you make? How much wood is left over?

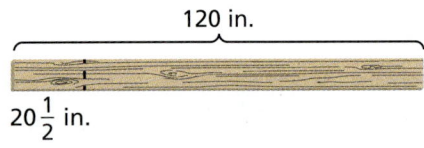

16. You spend $2\dfrac{1}{2}$ hours online. You spend $\dfrac{1}{5}$ of that time writing a blog. How long do you spend writing your blog?

17. You and a friend take pictures at a motocross event. Your camera can take 24 pictures in 3.75 seconds. Your friend's camera can take 36 pictures in 4.5 seconds. Evaluate the expression $(36 \div 4.5) \div (24 \div 3.75)$ to find how many times faster your friend's camera is than your camera.

2 Cumulative Practice

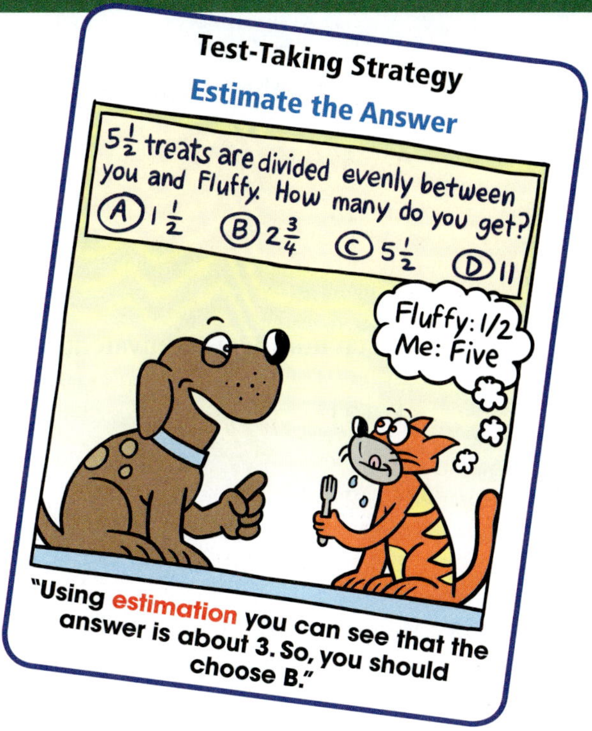

1. Which number is equivalent to the expression below?

    $6 \times 8 - 2 \times 3^2$

    A. 12
    B. 30
    C. 324
    D. 414

2. What is the greatest common factor of 48 and 120?

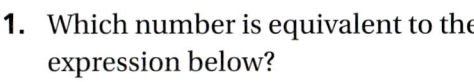

Test-Taking Strategy
Estimate the Answer

"Using estimation you can see that the answer is about 3. So, you should choose B."

3. Which number is equivalent to $5.139 - 2.64$?

    F. 2.499
    G. 2.599
    H. 3.519
    I. 3.599

4. Which number is equivalent to $\frac{4}{9} \div \frac{5}{7}$?

    A. $\frac{20}{63}$
    B. $\frac{28}{45}$
    C. $\frac{45}{28}$
    D. $\frac{63}{20}$

5. You buy orange and black streamers for a party. The orange streamers are 9 feet long, and the black streamers are 12 feet long. What are the least numbers of streamers you should buy in order for the total length of the orange streamers to be the same as the total length of the black streamers?

    F. 36 orange streamers and 36 black streamers
    G. 12 orange streamers and 9 black streamers
    H. 3 orange streamers and 4 black streamers
    I. 4 orange streamers and 3 black streamers

Cumulative Practice 101

6. Which number is a prime factor of 572?

    A. 4
    B. 7
    C. 13
    D. 22

7. Which number is equivalent to 7059 ÷ 301?

    F. 23
    G. $23\frac{136}{7059}$
    H. $23\frac{136}{301}$
    I. 136

8. A square wall tile has side lengths of 4 inches. You use 360 of the tiles. What is the area of the wall covered by the tiles?

    A. 16 in.^{2}
    B. 360 in.^{2}
    C. 1440 in.^{2}
    D. 5760 in.^{2}

9. Which expression is equivalent to a perfect square?

    F. $3 + 2^2 \times 7$
    G. $34 + 18 \div 3^2$
    H. $(80 + 4) \div 4$
    I. $3^2 + 6 \times 5 \div 3$

10. What is the missing denominator in the expression below?

    $$\frac{4}{8} \div \frac{2}{} = \frac{3}{4}$$

    A. 1
    B. 2
    C. 3
    D. 8

11. What is 4.56 × 0.7?

12. The area of the large rectangle is how many times the area of the small rectangle?

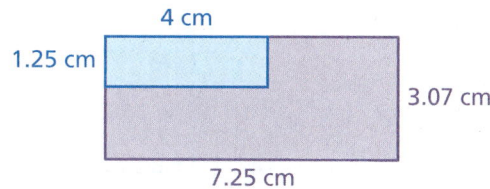

- **F.** 4.4515
- **G.** 5.915
- **H.** 17.2575
- **I.** 111.2875

13. Which expression is equivalent to $5 \times 5 \times 5 \times 5$?

- **A.** 5×4
- **B.** 4^5
- **C.** 5^4
- **D.** 5^5

14. A walkway is built using identical concrete blocks.

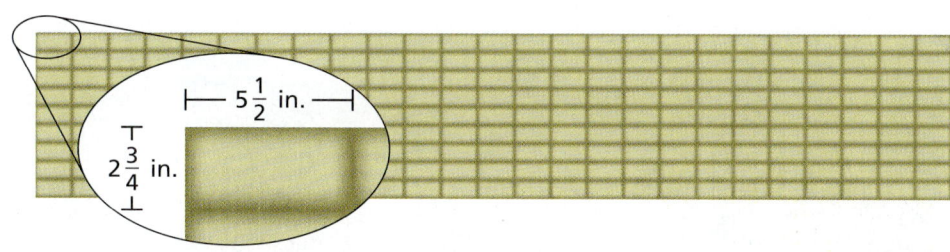

Part A How much longer, in inches, is the length of the walkway than the width of the walkway? Show your work and explain your reasoning.

Part B How many times longer is the length of the walkway than the width of the walkway? Show your work and explain your reasoning.

15. A meteoroid moving at a constant speed travels $6\frac{7}{8}$ miles in 30 seconds. How far does the meteoroid travel in 1 second?

- **F.** $\frac{1}{5}$ mile
- **G.** $\frac{11}{48}$ mile
- **H.** $2\frac{7}{24}$ miles
- **I.** $206\frac{1}{4}$ miles

3 Ratios and Rates

- **3.1** Ratios
- **3.2** Using Tape Diagrams
- **3.3** Using Ratio Tables
- **3.4** Graphing Ratio Relationships
- **3.5** Rates and Unit Rates
- **3.6** Converting Measures

Chapter Learning Target:
Understand ratios.

Chapter Success Criteria:
- ■ I can write and interpret ratios.
- ■ I can name ratios equivalent to a given ratio.
- ■ I can solve a problem using ratios.
- ■ I can convert units of measure using ratio reasoning.

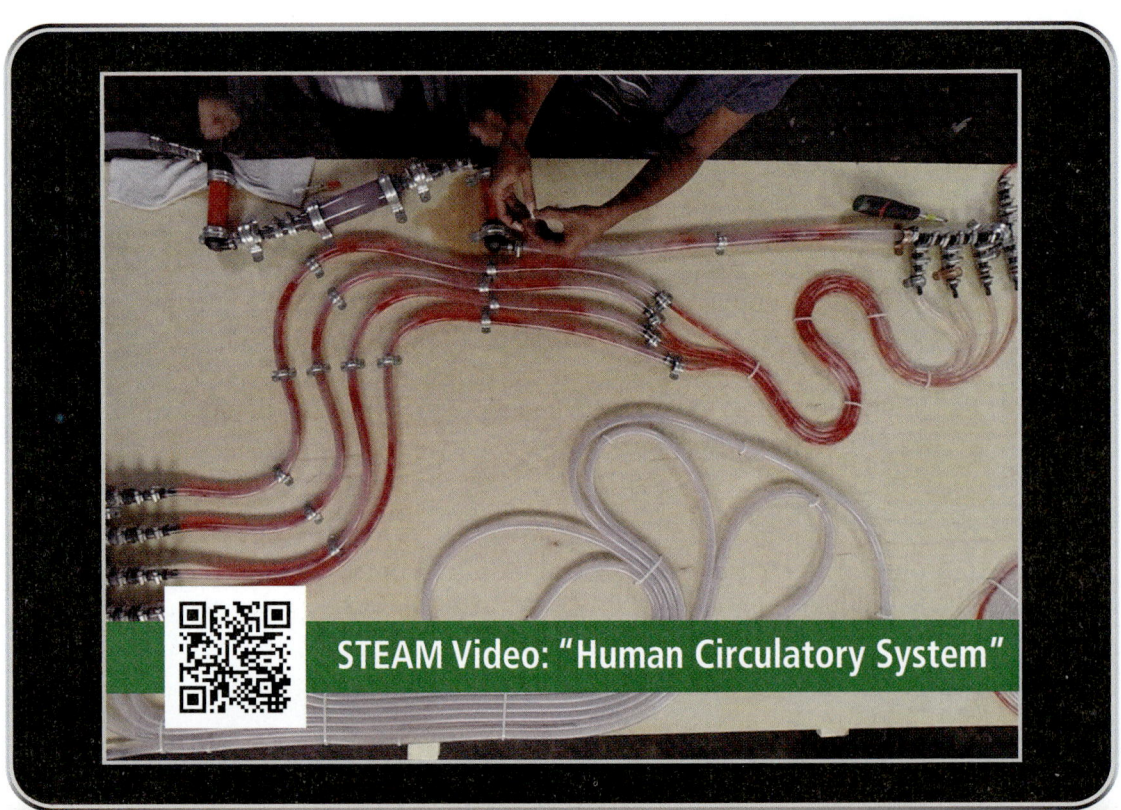

STEAM Video: "Human Circulatory System"

STEAM Video

Human Circulatory System

Watch the STEAM Video "Human Circulatory System." Then answer the following questions.

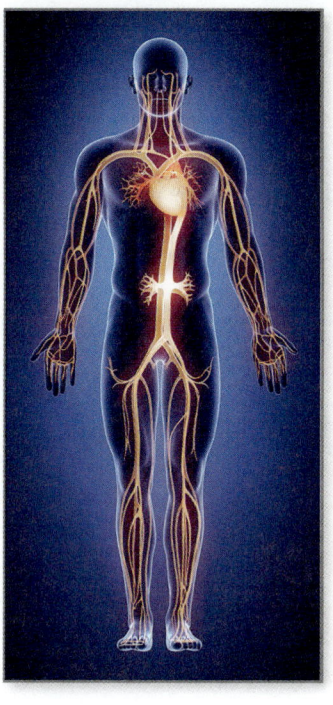

1. Enid says the heart pumps about 5 liters of blood each minute. How can you find the amount of blood the heart pumps for any given number of minutes?

2. Explain how you can estimate the amount of blood your heart pumps in one heart beat.

3. The table shows the amounts of blood contained in several different types of blood vessels. How can you make meaningful comparisons of the amounts?

Vessel	Volume
Aorta and large arteries	300 mL
Small arteries	0.4 L
Small veins	2.43 qt
Large veins	0.24 gal

Performance Task

Oops! Unit Conversion Mistakes

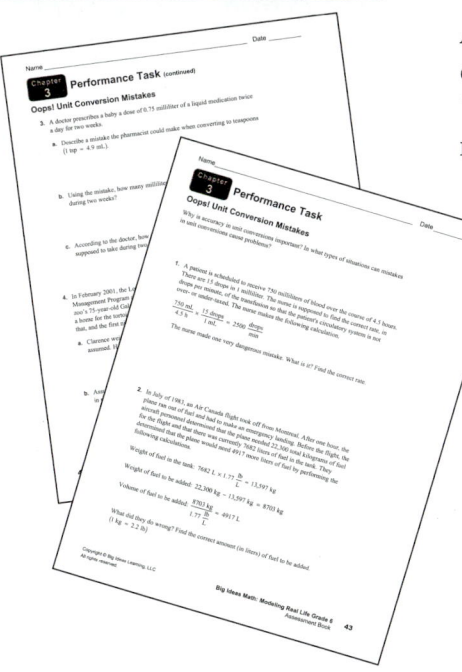

After completing this chapter, you will be able to use the concepts you learned to answer the questions in the *STEAM Video Performance Task*. You will be shown unit conversion mistakes in the following real-life situations.

- **Blood transfusion**
- **Airplane fuel**
- **Baby medication**
- **Zoo enclosure**

In each situation, you will analyze and correct the mistake in the unit conversion. How accurate must conversions be in real-life situations?

Getting Ready for Chapter

Chapter Exploration

Work with a partner. What portion of the rectangle is red? How did you write your answer?

1. 2. 3.

4. 5.  6.

7.  8.  9.

10. Work with a partner. In Exercises 1–9, which of the rectangles have the same portion of red tiles? Explain your reasoning.

Work with a partner. Use square color tiles to build two different-sized rectangles that represent the description.

11. Five-sixths of the tiles are blue.

12. Three-fourths of the tiles are yellow.

13. Four-fifths of the tiles are green.

14. Five-sevenths of the tiles are red.

15. **MODELING REAL LIFE** Work with a partner. The soccer committee has 8 girls and 6 boys. The tennis committee has 9 girls and 8 boys. A friend tells you that the tennis committee has a greater portion of girls than the soccer committee. Is your friend correct? Explain. If not, how many boys could you add to the soccer committee so that your friend is correct?

Vocabulary

The following vocabulary terms are defined in this chapter. Think about what each term might mean and record your thoughts.

ratio	rate	equivalent rates
equivalent ratios	unit rate	

3.1 Ratios

Learning Target: Understand the concepts of ratios and equivalent ratios.

Success Criteria:
- I can write and interpret ratios using appropriate notation and language.
- I can recognize multiplicative relationships in ratios.
- I can describe how to determine whether ratios are equivalent.
- I can name ratios equivalent to a given ratio.

A **ratio** is a comparison of two quantities. Consider two quantities a and b. The ratio $a : b$ indicates that there are a units of the first quantity for every b units of the second quantity.

EXPLORATION 1

Writing Ratios

Work with a partner. A science class has two times as many girls as it has boys.

a. Discuss possible numbers of boys and girls in the science class.

b. What comparisons can you make between your class and the science class? Can you determine which class has more girls? more boys? Explain your reasoning.

c. Write three ratios that you observe in your classroom. Describe what each ratio represents.

Math Practice

Use a Table
How can you use a table to represent the relationship between the numbers of girls and boys?

EXPLORATION 2

Using Ratios in a Recipe

Work with a partner. The ratio of iced tea to lemonade in a recipe is 3 : 1. You begin by combining 3 cups of iced tea with 1 cup of lemonade.

Iced Tea   Lemonade

a. You add 1 cup of iced tea and 1 cup of lemonade to the mixture. Does this change the taste of the mixture?

b. Describe how you can make larger amounts without changing the taste.

Section 3.1   Ratios   107

3.1 Lesson

Key Vocabulary
ratio, *p. 108*
value of a ratio, *p. 109*
equivalent ratios, *p. 109*

 Key Idea

Ratio

Words A **ratio** is a comparison of two quantities. Ratios can be part-to-part, part-to-whole, or whole-to-part comparisons. Ratios may or may not include units.

Examples 2 cats *to* 6 dogs
1 cat *for every* 3 dogs
3 dogs *per* 1 cat
3 dogs *for each* cat
3 dogs *out of every* 4 pets
2 cats *out of* 8 pets

Algebra The ratio of *a* to *b* can be written as *a* : *b*.

Reading
Phrases that indicate ratios include *for each*, *for every*, and *per*.

EXAMPLE 1 Writing Ratios

You have the coins shown.

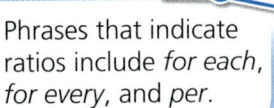

a. **Write the ratio of pennies to quarters.**

6 pennies → 6 to 7 ← 7 quarters

▸ So, the ratio of pennies to quarters is 6 to 7, or 6 : 7.

b. **Write the ratio of quarters to dimes.**

7 quarters → 7 to 3 ← 3 dimes

▸ So, the ratio of quarters to dimes is 7 to 3, or 7 : 3.

c. **Write the ratio of dimes to the total number of coins.**

3 dimes → 3 to 16 ← 16 coins

▸ So, the ratio of dimes to the total number of coins is 3 to 16, or 3 : 16.

Try It Write the indicated ratio using the coins in Example 1.

1. dimes to pennies

2. quarters to the total number of coins

The number $\frac{a}{b}$ associated with the ratio $a:b$ is called the **value of the ratio**. It describes the multiplicative relationship between the quantities in a ratio.

EXAMPLE 2 Writing and Interpreting Ratios

a. **The ratio of rubies to diamonds on a ring is 1 : 2. Find and interpret the value of the ratio.**

The value of the ratio 1 : 2 is $\frac{1}{2}$. So, the multiplicative relationship is $\frac{1}{2}$.

▸ The number of rubies is $\frac{1}{2}$ times the number of diamonds.

b. **On another ring, the number of rubies is 4 times the number of diamonds. Write the ratio of rubies to diamonds.**

Because the number of rubies is 4 times the number of diamonds, there are 4 rubies per diamond.

▸ So, the ratio of rubies to diamonds is 4 : 1.

Try It

3. An elephant sanctuary contains adult and baby elephants. The ratio of adult elephants to baby elephants is 5 : 1. Find and interpret the value of the ratio.

Key Idea

Equivalent Ratios

Words Two ratios that describe the same relationship are **equivalent ratios**. Two ratios are equivalent when you can multiply each quantity in one ratio by the same positive number to obtain the other ratio. The values of equivalent ratios are equivalent.

Example 1 : 3 and 2 : 6 are equivalent.

$$\times 2 \begin{pmatrix} 1:3 \\ 2:6 \end{pmatrix} \times 2 \qquad \frac{1}{3} = \frac{1 \times 2}{3 \times 2} = \frac{2}{6}$$

Algebra Two ratios $a:b$ and $c:d$ are equivalent when there exists a positive number n such that $a \times n = c$ and $b \times n = d$.

Section 3.1 Ratios 109

EXAMPLE 3 **Determining Whether Ratios Are Equivalent**

Determine whether the ratios are equivalent.

a. 4 : 3 and 20 : 15

You can multiply each number in the first ratio by 5 to obtain the numbers in the second ratio.

Also, the values of the ratios are equivalent.

×5 ⤻ 4 : 3 ⤺ ×5
20 : 15

$$\frac{4}{3} = \frac{4 \times 5}{3 \times 5} = \frac{20}{15}$$

▸ So, the ratios are equivalent.

b. 5 : 7 and 10 : 21

You need to multiply each number in the ratio by different amounts to obtain the numbers in the second ratio.

Also, the values of the ratios are not equivalent.

×2 ⤻ 5 : 7 ⤺ ×3
10 : 21

$$\frac{5}{7} \neq \frac{10}{21}$$

▸ So, the ratios are not equivalent.

Try It Determine whether the ratios are equivalent.

4. 1 : 1 and 6 : 6
5. 1 : 2 and 3 : 4
6. 8 : 3 and 6 : 16

Self-Assessment for Concepts & Skills

Solve each exercise. Then rate your understanding of the success criteria in your journal.

WRITING AND INTERPRETING RATIOS Write the ratio. Then find and interpret the value of the ratio.

7. sharks to dolphins
8. dolphins : animals

IDENTIFYING EQUIVALENT RATIOS Determine whether the ratios are equivalent. Explain your reasoning.

9. 2 : 3 and 24 : 36
10. 5 : 7 and 20 : 28
11. 3 : 10 and 9 : 25

12. **DIFFERENT WORDS, SAME QUESTION** Which is different? Find "both" answers.

What kind of relationship is 2 peaches to 5 pears?	What kind of relationship is 2 pears out of every 5 fruit?
What kind of relationship is 2 pears per 5 peaches?	What kind of relationship is 2 peaches for every 5 pears?

110 Chapter 3 Ratios and Rates

EXAMPLE 4 Modeling Real Life

Meg is speedwalking at a pace of 5 meters every 2 seconds. Sean's pace is 10 meters every 5 seconds. Are they speedwalking at the same pace? If not, who is faster?

Write a ratio to represent each person's pace.

    Meg's pace: 5 : 2

    Sean's pace: 10 : 5

To decide whether Meg and Sean are speedwalking at the same pace, determine whether the ratios are equivalent.

$$\times 2 \overset{5 : 2}{\underset{10 : 5}{\curvearrowright}} \times 2.5 \quad \text{Not equivalent}$$

So, they are not speedwalking at the same pace. To decide who is faster, use an equivalent ratio to compare Meg's pace to Sean's pace.

Meg's pace: $\quad \times 2 \overset{5 : 2}{\underset{10 : 4}{\curvearrowright}} \times 2$

Meg walks 10 meters every 4 seconds. It takes Sean 5 seconds to walk the same distance.

▶ So, Meg is speedwalking faster than Sean.

Self-Assessment for Problem Solving

Solve each exercise. Then rate your understanding of the success criteria in your journal.

13. The ratio of wolves to cougars in a forest is 5 : 3. Find and interpret the value of the ratio.

14. You are kayaking at a pace of 63 feet every 12 seconds. Your friend's pace is 21 feet every 3 seconds. Are you and your friend kayaking at the same pace? If not, who is faster?

15. **DIG DEEPER!** The ratio of Jet Ski rentals to boat rentals at a store is 7 : 2. If the number of boat rentals doubles and the number of Jet Ski rentals stays the same, then the number of boat rentals is how many times the number of Jet Ski rentals?

3.1 Practice

Go to *BigIdeasMath.com* to get HELP with solving the exercises.

Review & Refresh

Divide. Check your answer.

1. $15.4 \div 2.2$
2. $56.07 \div 8.9$
3. $8.43 \overline{)12.645}$
4. $11.6 \overline{)51.62}$

Find the value of the power.

5. 8^2
6. 1^6
7. 3^4
8. 2^6

The Venn diagram shows the prime factors of two numbers. Identify the numbers. Then find the GCF and the LCM of the two numbers.

9.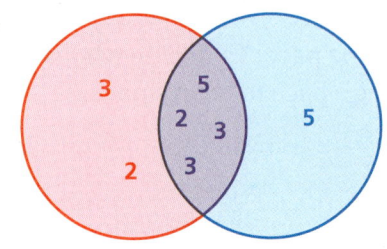
10.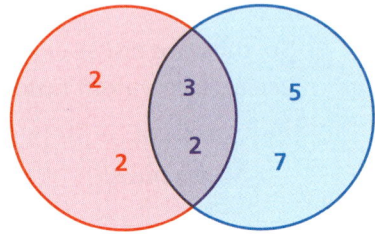

Concepts, Skills, & Problem Solving

USING RATIOS You mix the amounts of iced tea and lemonade shown. Describe how you can make larger amounts without changing the taste. *(See Exploration 2, p. 107.)*

11.
12.

WRITING RATIOS Write the ratio.

13. frogs to turtles

14. basketballs to soccer balls

15. calculators : pencils

16. shirts : pants

112   Chapter 3   Ratios and Rates

17. **MODELING REAL LIFE** Twelve of the 28 students in a class own a dog. What is the ratio of students who own a dog to students who do not?

18. **LOGIC** Name two things that you would like to have in a ratio of 5 : 1 but not in a ratio of 1 : 5. Explain your reasoning.

OPEN-ENDED Describe a real-life relationship that can be represented by the ratio.

19. 1 out of every 7
20. 5 to 26
21. 2 per 5
22. 7 : 1

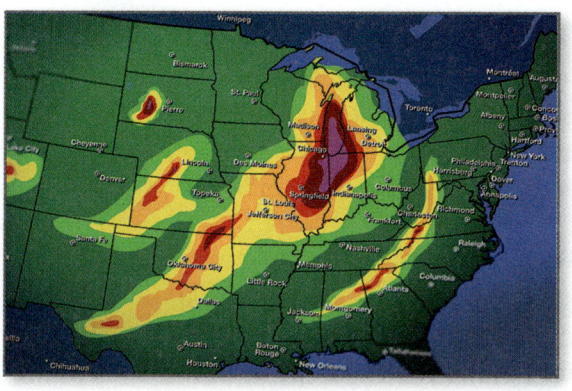

23. **MODELING REAL LIFE** During a given month, the ratio of sunny days to rainy days is 4 : 1.

   a. Find and interpret the value of the ratio.

   b. In another month, the number of sunny days is 5 times the number of rainy days. Write the ratio of sunny days to rainy days.

IDENTIFYING EQUIVALENT RATIOS Determine whether the ratios are equivalent.

24. 2 : 3 and 4 : 9
25. 3 : 8 and 9 : 24
26. 1 : 4 and 2 : 6
27. 5 : 3 and 15 : 12
28. 6 : 10 and 12 : 20
29. 2 : 3 and 4 : 5
30. 28 : 32 and 7 : 8
31. 24 : 100 and 6 : 25
32. 85 : 210 and 340 : 735

WRITING EQUIVALENT RATIOS Write a ratio that is equivalent to the given ratio. Justify your answer.

33. 3 : 1
34. 7 : 2
35. 6 : 6
36. 0 : 8

WRITING EQUIVALENT RATIOS Fill in the blank so that the ratios are equivalent.

37. 3 : 9 and 6 : ▢
38. 2 : 6 and 8 : ▢
39. ▢ : 6 and 7 : 2

40. **YOU BE THE TEACHER** Your friend says that the two ratios are equivalent. Is your friend correct? Explain your reasoning.

> +4 ⟶ 4 : 8 ⟵ +4
> 8 : 12
> Because you can add 4 to each number in the first ratio to obtain the numbers in the second ratio, the ratios are equivalent.

41. **OPEN-ENDED** A *non-Newtonian* liquid demonstrates properties of both a solid and a liquid. A recipe for a non-Newtonian liquid calls for 1 cup of water and 2 cups of cornstarch. Find two possible combinations of water and cornstarch that you can use to make a larger batch. Justify your answer.

42. **PROBLEM SOLVING** You are downloading songs to your tablet. The ratio of pop songs to rock songs is 5 : 4. You download 40 pop songs. How many rock songs do you download?

43. **MP PROBLEM SOLVING** In the contiguous United States, the ratio of states that border an ocean to states that do not border an ocean is 7 : 9. How many of the states border an ocean?

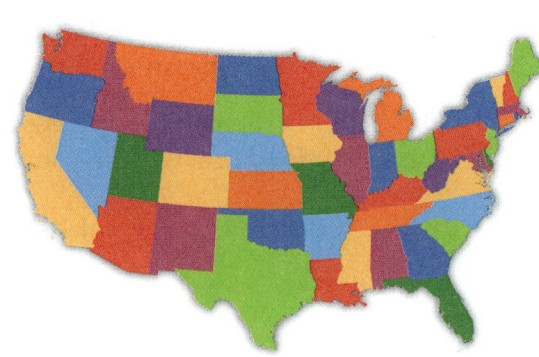

44. **MP REASONING** The value of a ratio is $\frac{4}{3}$. The second quantity in the ratio is how many times the first quantity in the ratio? Explain your reasoning.

45. **MODELING REAL LIFE** A train moving at a constant speed travels 3 miles every 5 minutes. A car moving at a constant speed travels 12 miles every 20 minutes. Are the vehicles traveling at the same speed? If not, which is faster?

46. **CRITICAL THINKING** To win a relay race, you must swim 200 yards before your opponent swims 190 yards. You swim at a pace of 50 yards every 40 seconds. Your opponent swims at a pace of 10 yards every 8.5 seconds. Who wins the race? Justify your answer.

47. **DIG DEEPER!** There are 3 boys for every 2 girls in a dance competition. Does it make sense for there to be a total of 9 people in the competition? Explain.

48. **GEOMETRY** Use the blue and green rectangles.

    a. Find the ratio of the length of the blue rectangle to the length of the green rectangle. Repeat this for width, perimeter, and area.

    b. Compare your ratios in part (a).

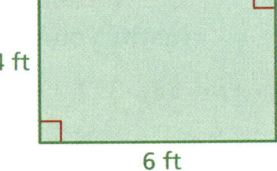

49. **MP STRUCTURE** The ratio of the side lengths of a triangle is 2 : 3 : 4. The shortest side is 15 inches. What is the perimeter of the triangle? Explain.

50. **MP PROBLEM SOLVING** A restaurant sells tokens that customers use to play games while waiting for their orders.

TOKENS	
1 Token	$0.50
10 Tokens	$5.00
25 Tokens	$10.00
50 Tokens	$25.00
90 Tokens	$40.00

    a. Which option is the best deal? Justify your answer.

    b. What suggestions, if any, would you give to the restaurant about how it could modify the prices of tokens?

51. **DIG DEEPER!** There are 12 boys and 10 girls in your gym class. If 6 boys joined the class, how many girls would need to join for the ratio of boys to girls to remain the same? Justify your answer.

3.2 Using Tape Diagrams

Learning Target: Use tape diagrams to model and solve ratio problems.

Success Criteria:
- I can interpret tape diagrams that represent ratio relationships.
- I can draw tape diagrams to model ratio relationships.
- I can find the value of one part of a tape diagram.
- I can use tape diagrams to solve ratio problems.

You can use a visual model, called a *tape diagram*, to represent the relationship between two quantities in a ratio.

EXPLORATION 1

Using a Tape Diagram

Work with a partner. The tape diagram models the lengths of two snowboarding trails.

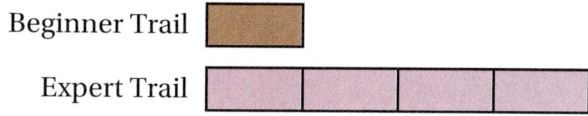

Math Practice

Make Sense of Quantities

How does the tape diagram help you make sense of the quantities and the relationship between the quantities?

a. What can you determine from the tape diagram?

b. Choose a length for one of the trails. What conclusions can you make from the tape diagram? Explain your reasoning.

c. Suppose you know the combined length of the trails or the difference in the lengths of the trails. Explain how you can use that information to find the lengths of the two trails. Provide an example with your explanation.

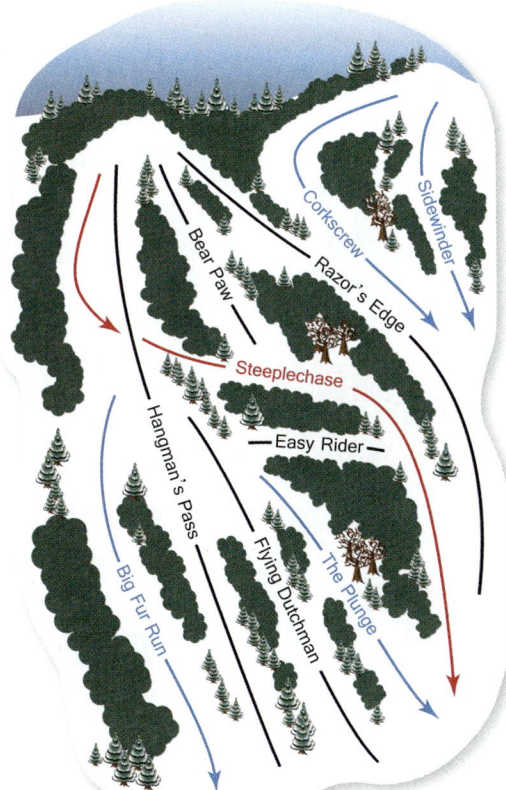

Section 3.2 Using Tape Diagrams 115

3.2 Lesson

You can use tape diagrams to represent ratios and solve ratio problems.

EXAMPLE 1 Interpreting a Tape Diagram

The tape diagram represents the ratio of blue monsters to green monsters you caught in a game. You caught 10 green monsters. How many blue monsters did you catch?

Reading
The tape diagram shows that the ratio of blue monsters to green monsters is 3 : 1.

Blue [| |]

Green [ ]

The 1 part for green represents 10 monsters. So, the 3 parts for blue represents $3 \times 10 = 30$ monsters.

▸ You caught 30 blue monsters.

Try It

1. The tape diagram represents the ratio of gifts received to gifts given. You received 4 gifts. How many gifts did you give?

   Received [ ]

   Given [| | |]

EXAMPLE 2 Drawing a Tape Diagram

There are 3 bones in a cat for every 4 bones in a dog. The cat has 240 bones. How many bones does the dog have?

The ratio of bones in the cat to bones in the dog is 3 : 4. Represent the ratio using a tape diagram.

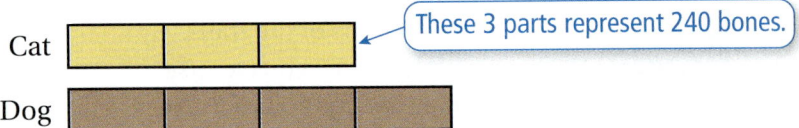

Cat [| |]   These 3 parts represent 240 bones.

Dog [| | |]

One part represents $240 \div 3 = 80$ bones. So, 4 parts represent $4 \times 80 = 320$ bones.

▸ The dog has 320 bones.

Try It

2. There are 8 bones in a large snake for every 3 bones in a small snake. The small snake has 150 bones. How many bones does the large snake have?

EXAMPLE 3 **Using a Tape Diagram to Solve a Ratio Problem**

The ratio of your monthly allowance to your friend's monthly allowance is 5 : 3. The monthly allowances total $40. How much is each allowance?

Represent the ratio 5 : 3 using a tape diagram.

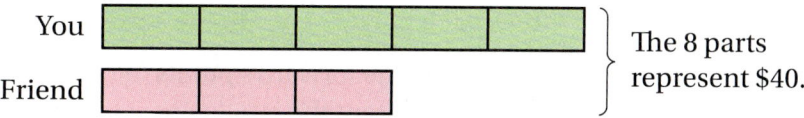

The 8 parts represent $40.

1 part represents $40 ÷ 8 = $5.
5 parts represent 5 × $5 = $25.
3 parts represent 3 × $5 = $15.

▸ So, your allowance is $25, and your friend's allowance is $15.

Check Verify that the ratio of allowances is 5 : 3.

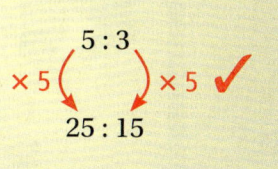

Try It

3. **WHAT IF?** Repeat Example 3 when the ratio of your monthly allowance to your friend's monthly allowance is 2 to 3.

Self-Assessment for Concepts & Skills

Solve each exercise. Then rate your understanding of the success criteria in your journal.

4. **MP STRUCTURE** What ratio is represented by the tape diagram? Can you use the tape diagram to model the ratio 6 : 9? Can you use the tape diagram to model the ratio 8 : 16? Explain your reasoning.

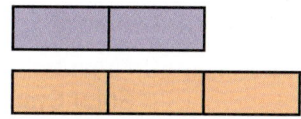

5. **MP REASONING** You are given a tape diagram and the total value of the parts. How can you find the value of 1 part?

6. **DRAWING A TAPE DIAGRAM** Describe two ways that you can represent the ratio 12 : 4 using a tape diagram.

USING A TAPE DIAGRAM You are given the number of tickets in a bag and the ratio of winning tickets to losing tickets. How many of each kind of ticket are in the bag?

7. 35 tickets; 1 to 4

8. 80 tickets; 2 : 8

Section 3.2 Using Tape Diagrams 117

EXAMPLE 4 — Modeling Real Life

In a seven-game basketball series, a team's power forward scores 8 points for every 5 points the center scores. The forward scores 60 more points than the center in the series. How many points does each player score in the series?

The ratio of the forward's points to the center's points is 8 : 5. Represent the ratio using a tape diagram.

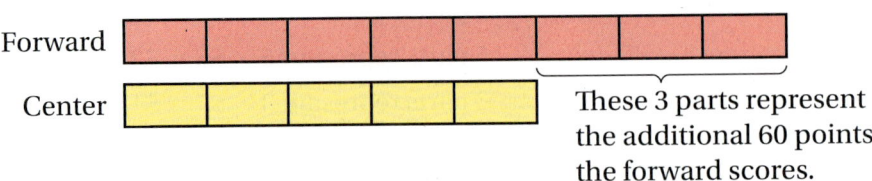

1 part represents 60 ÷ 3 = 20 points.
8 parts represent 8 × 20 = 160 points.
5 parts represent 5 × 20 = 100 points.

▸ So, the forward scores 160 points in the series and the center scores 100 points in the series.

Self-Assessment for Problem Solving

Solve each exercise. Then rate your understanding of the success criteria in your journal.

9. The tape diagram represents the ratio of the numbers of planets in two different solar systems. There are 8 planets in Solar System B. How many planets are in Solar System A?

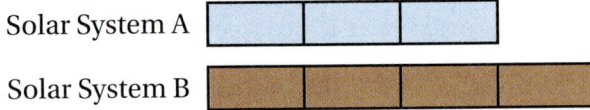

10. You and your friend play an arcade game. You score 5 points for every 9 points that your friend scores. You score 320 points less than your friend. How many points do you each score?

11. **DIG DEEPER!** Your team wins 18 medals at a track meet. The medals are gold, silver, and bronze in a ratio of 2 : 2 : 5. How many of each medal were won by your team?

118   Chapter 3   Ratios and Rates

3.2 Practice

Review & Refresh

Determine whether the ratios are equivalent.

1. 11 : 4 and 22 : 8
2. 12 : 18 and 2 : 3
3. 56 : 81 and 7 : 9
4. 2 : 12 and 6 : 24

Multiply. Write the answer in simplest form.

5. $\dfrac{7}{10} \cdot \dfrac{5}{7}$
6. $2\dfrac{1}{3} \cdot \dfrac{3}{4}$
7. $5\dfrac{3}{8} \cdot 2\dfrac{1}{2}$

8. Melissa earns $7.40 per hour working at a grocery store. She works 14.25 hours this week. How much does she earn?

   A. $83.13   **B.** $105.45   **C.** $156.75   **D.** $1054.50

Concepts, Skills, & Problem Solving

USING A TAPE DIAGRAM Use the tape diagram in Exploration 1 to answer the question. *(See Exploration 1, p. 115.)*

9. The beginner trail is 200 meters long. How long is the expert trail?
10. The expert trail is 1200 meters long. How long is the beginner trail?
11. The combined length of the trails is 2000 meters. How long is each trail?
12. The expert trail is 750 meters longer than the beginner trail. How long is each trail?

INTERPRETING A TAPE DIAGRAM The tape diagram represents the ratio of the time you spend tutoring to the time your friend spends tutoring. You tutor for 3 hours. How many hours does your friend spend tutoring?

13.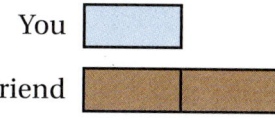

14.

DRAWING A TAPE DIAGRAM A bag contains red marbles and blue marbles. You are given the number of red marbles in the bag and the ratio of red marbles to blue marbles. Find the number of blue marbles in the bag.

15. 10 red marbles; 5 to 1
16. 3 red marbles; 3 : 7
17. 12 red marbles; 4 : 3
18. 6 red marbles; 2 for every 5
19. 18 red marbles; 6 to 9
20. 12 red marbles; 3 : 4

USING A TAPE DIAGRAM A bowl contains blueberries and strawberries. You are given the total number of berries in the bowl and the ratio of blueberries to strawberries. How many of each berry are in the bowl?

21. 16 berries; 3 : 1

22. 10 berries; 2 for every 3

23. 12 berries; 1 to 2

24. 20 berries; 4 : 1

25. 48 berries; 9 to 3

26. 46 berries; 11 for every 12

Clove Bulb

27. **MP PROBLEM SOLVING** You separate bulbs of garlic into two groups: one for planting and one for cooking. The tape diagram represents the ratio of bulbs for planting to bulbs for cooking. You use 6 bulbs for cooking. Each bulb has 8 cloves. How many cloves of garlic will you plant?

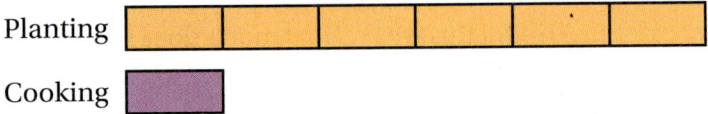

28. **MODELING REAL LIFE** Methane gas contains carbon atoms and hydrogen atoms in the ratio of 1 : 4. A sample of methane gas contains 92 hydrogen atoms. How many carbon atoms are in the sample? How many total atoms are in the sample?

CH_4 Methane

29. **MODELING REAL LIFE** There are 8 more girls than boys in a school play. The ratio of boys to girls is 5 : 7. How many boys and how many girls are in the play?

30. **DIG DEEPER!** A baseball team sells tickets for two games. The ratio of sold tickets to unsold tickets for the first game was 7 : 3. For the second game, the ratio was 13 : 2. There were 240 unsold tickets for the second game. How many tickets were sold for the first game?

31. **MP PROBLEM SOLVING** You have $150 in a savings account and you have some cash. The tape diagram represents the ratio of the amounts of money. You want to have twice the amount of money in your savings account as you have in cash. How much of your cash should you deposit into your savings account?

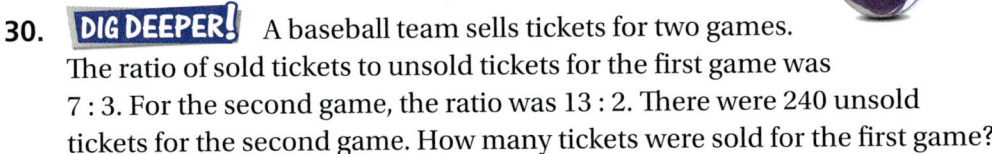

32. **DIG DEEPER!** A fish tank contains tetras, guppies, and minnows. The ratio of tetras to guppies is 4 : 2. The ratio of minnows to guppies is 1 : 3. There are 60 fish in the tank. How many more tetras are there than minnows? Justify your answer.

3.3 Using Ratio Tables

Learning Target: Use ratio tables to represent equivalent ratios and solve ratio problems.

Success Criteria:
- I can use various operations to create tables of equivalent ratios.
- I can use ratio tables to solve ratio problems.
- I can use ratio tables to compare ratios.

EXPLORATION 1
Making a Table of Equivalent Ratios

Work with a partner. You buy milk that contains 180 calories per 2 cups.

a. You measure 2 cups of the milk for a recipe and pour it into a pitcher. You repeat this four more times. Make a table to show the numbers of calories and cups in the pitcher as you add the milk.

b. Describe any relationships you see in your table.

c. Describe ways that you can find equivalent ratios using different operations.

Math Practice

Compare Arguments
Compare your explanations in part (c) with another group. If they are different, are they both correct?

EXPLORATION 2
Creating a Double Number Line

Work with a partner.

a. Represent the ratio in Exploration 1 by labeling the increments on the *double number line* below. Can you label the increments in more than one way?

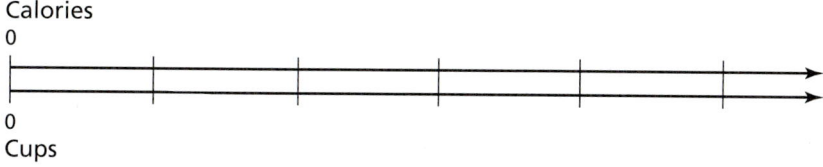

b. How can you use the double number line to find the number of calories in 3 cups of milk? 3.5 cups of milk?

Section 3.3 Using Ratio Tables

3.3 Lesson

Key Vocabulary
ratio table, p. 122

You can find and organize equivalent ratios in a **ratio table**. You can generate a ratio table by using repeated addition or multiplication.

EXAMPLE 1 Completing Ratio Tables

Find the missing values in each ratio table. Then write the equivalent ratios.

a.
Triangles	1	2		4
Sides	3		9	

b.
Frogs	4	20		
Toads	6		90	180

a. Because the original ratio is 1 triangle to 3 sides, you can repeatedly add 1 to the first row and repeatedly add 3 to the second row.

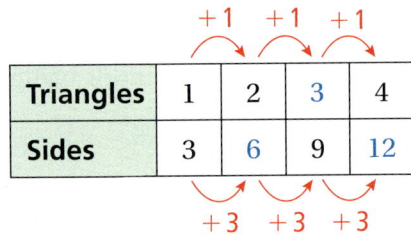

▸ The equivalent ratios are 1 : 3, 2 : 6, 3 : 9, and 4 : 12.

b. You can use multiplication to find the missing values.

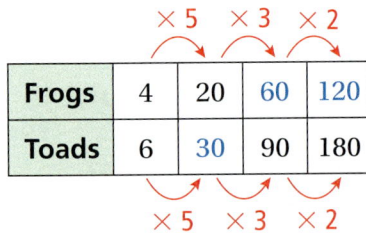

▸ The equivalent ratios are 4 : 6, 20 : 30, 60 : 90, and 120 : 180.

Try It Find the missing values in the ratio table. Then write the equivalent ratios.

1.
Hands	4		12	16
People	2	4		

2.
Miles	4		24	96
Hours	3	6		

You can also generate a ratio table by using subtraction or division. In summary, you can find equivalent ratios by:

- adding or subtracting quantities in equivalent ratios.
- multiplying or dividing each quantity in a ratio by the same number.

EXAMPLE 2 Completing Ratio Tables

Find the missing values in each ratio table. Then write the equivalent ratios.

a.
Dollars	1			8
Cents	100	300	900	

b.
Meters	3		2	
Minutes	1	2		$\frac{5}{3}$

> In Example 2(a), notice that you obtain the fourth column by subtracting the values in the first column from the values in the third column.
> $9 - 1 = 8$
> $900 - 100 = 800$

a. You can use a combination of operations to find the missing values.

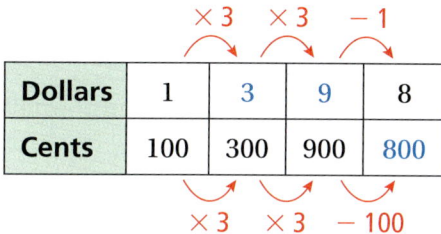

The equivalent ratios are 1 : 100, 3 : 300, 9 : 900, and 8 : 800.

b. You can use a combination of operations to find the missing values.

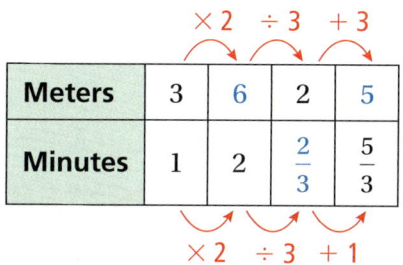

The equivalent ratios are 3 : 1, 6 : 2, 2 : $\frac{2}{3}$, and 5 : $\frac{5}{3}$.

Try It Find the missing values in the ratio table. Then write the equivalent ratios.

3.
Flowers	1	2	4	
Petals	5			15

4.
Students	24	12		36
Teachers	2		4	

Section 3.3 Using Ratio Tables 123

EXAMPLE 3 Solving a Ratio Problem

A nutrition label shows that there are 75 milligrams of sodium in every 12 crackers. You eat 30 crackers. How much sodium do you consume?

Method 1: Use a double number line. Increment the number lines using the original ratio of 75 to 12.

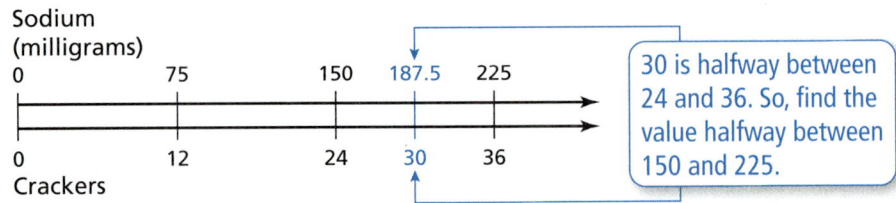

30 is halfway between 24 and 36. So, find the value halfway between 150 and 225.

▸ So, you consume 187.5 milligrams of sodium in 30 crackers.

Method 2: Use a ratio table. The ratio of milligrams of sodium to crackers is 75 to 12. Find an equivalent ratio with 30 crackers.

÷ 2   × 5

Sodium (milligrams)	75	37.5	187.5
Crackers	12	6	30

÷ 2   × 5

▸ So, you consume 187.5 milligrams of sodium in 30 crackers.

Try It

5. WHAT IF? You eat 21 crackers. How much sodium do you consume?

Self-Assessment for Concepts & Skills

Solve each exercise. Then rate your understanding of the success criteria in your journal.

COMPLETING A RATIO TABLE Find the missing values in the ratio table. Then write the equivalent ratios.

6.
Fruit	2		6
Vegetables	6	12	

7.
Gnats	2	14		5
Flies	8		28	

8. WRITING Explain how creating a ratio table using repeated addition is similar to creating a ratio table using multiplication.

EXAMPLE 4 Modeling Real Life

You and your teacher make colored frosting. You add 3 drops of red food coloring for every 1 drop of blue food coloring. Your teacher adds 5 drops of red for every 3 drops of blue. Whose frosting is redder?

Understand the problem. You are given the numbers of drops of food coloring that you and your teacher use to make frosting. You are asked to determine whose frosting is redder.

Make a plan. Use ratio tables to compare the frostings. Find ratios in which the number of drops of red, the number of drops of blue, or the total number of drops is the same. Then compare the quantities to determine which is redder.

Solve and check. Create ratio tables for 3 : 1 and 5 : 3 using repeated addition. Include a column for the total number of drops in each frosting.

Your Frosting		
Drops of Red	Drops of Blue	Total Drops
3	1	4
6	2	8
9	3	12
12	4	16
15	5	20

Your Teacher's Frosting		
Drops of Red	Drops of Blue	Total Drops
5	3	8
10	6	16
15	9	24
20	12	32
25	15	40

Look Back
The tables show that when both frostings have a total of 16 drops, your frosting has 2 more drops of red and 2 fewer drops of blue. So, your frosting is redder.

When both frostings have 3 drops of blue, your frosting has $9 - 5 = 4$ more drops of red than your teacher's frosting.

▶ So, your frosting is redder than your teacher's frosting.

 ## Self-Assessment for Problem Solving

Solve each exercise. Then rate your understanding of the success criteria in your journal.

9. You mix 7 tablespoons of vinegar for every 4 tablespoons of baking soda to produce a chemical reaction. You use 15 tablespoons of baking soda. How much vinegar do you use?

10. You make a carbonated beverage by adding 7 ounces of soda water for every 3 ounces of regular water. Your friend uses 11 ounces of soda water for every 4 ounces of regular water. Whose beverage is more carbonated?

Section 3.3 Using Ratio Tables 125

3.3 Practice

 Go to *BigIdeasMath.com* to get HELP with solving the exercises.

▶ Review & Refresh

A bag contains green tokens and black tokens. You are given the number of green tokens in the bag and the ratio of green tokens to black tokens. Find the number of black tokens in the bag.

1. 8 green tokens; 4 for every 1
2. 6 green tokens; 2 : 7
3. 24 green tokens; 8 to 5
4. 36 green tokens; 3 for every 4

Find the GCF of the numbers.

5. 8, 16
6. 48, 80
7. 15, 45, 100

Evaluate the expression.

8. $35 - 2 \times 4^2$
9. $12 \div (1 + 3^3 - 2^4)$
10. $8^2 \div [(11 - 3) \cdot 2]$

Find the perimeter of the rectangle.

11.
Area = 48 yd²
8 yd

12.
Area = 132 mm²
12 mm

▶ Concepts, Skills, & Problem Solving

USING A RATIO TABLE Use a ratio table to find the number of calories in the indicated number of cups of milk from Exploration 1. Explain your method. (See Exploration 1, p. 121.)

13. 16 cups
14. 18 cups
15. 5.5 cups

COMPLETING RATIO TABLES Find the missing value(s) in the ratio table. Then write the equivalent ratios.

16.
Boys	1	
Girls	5	10

17.
Burgers	3		9
Hot Dogs	5	10	

18.
People	6		18
Benches	3	12	

19.
Adults	2	1		18
Children	14		21	

20.
Pies	5		$\frac{10}{3}$	
Cakes	3	12		5

21.
Plums	14	42		
Grapes	7		3	24

126 Chapter 3 Ratios and Rates

22. YOU BE THE TEACHER Your friend creates a ratio table for the ratio 5 : 3. Is your friend correct? Explain your reasoning.

A	5	25	125
B	3	9	27

COMPLETING RATIO TABLES Complete the ratio table to solve the problem.

23. For every 3 tickets you sell, your friend sells 4 tickets. You sell a total of 12 tickets. How many tickets does your friend sell?

You	3		12
Friend	4		

24. A store sells 2 printers for every 5 computers. The store sells 40 computers. How many printers does the store sell?

Printers	2		8	
Computers	5	10		40

25. First and second place in a contest use a ratio to share a cash prize. When first place pays $100, second place pays $60. How much does first place pay when second place pays $36?

First	100		
Second	60		36

26. A grade has 81 girls and 72 boys. The grade is split into groups that have the same ratio of girls to boys as the whole grade. How many girls are in a group that has 16 boys?

Girls	81		
Boys	72		16

USING A DOUBLE NUMBER LINE Find the missing quantity in the double number line.

27. Pounds: 0, 460, ?
    Pallets: 0, 4, 16

28. Trees: 0, 700, ?
    Acres: 0, 14, 21

29. **MP PROBLEM SOLVING** A company sets sales goals for employees each month.

   a. At her current pace, how many items will Kristina sell in 28 days? Is she on track to meet the goal? Explain.

   b. At his current pace, how many dollars worth of product will Jim sell in 28 days? Is he on track to meet the goal? Explain.

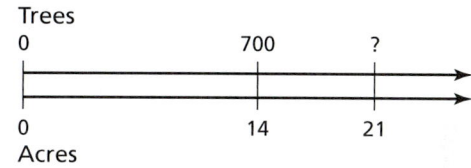

Monthly Goals
Sell 400 items in 28 days.
Sell $16,800 worth of items in 28 days.

Weekly Report
Date: February 7, 8:37 P.M.

Employee	Dollars	Items	Transactions
Jim	$4100.00	105	30
Kristina	$4250.00	98	35

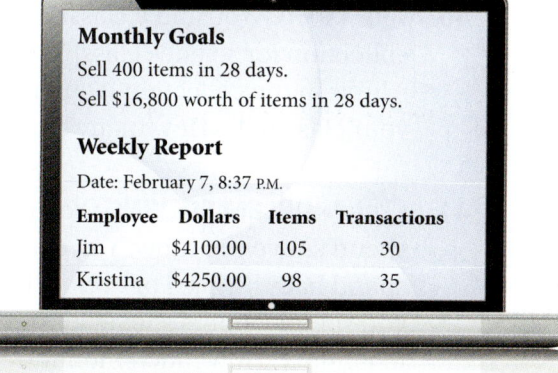

30. **MODELING REAL LIFE** A gold alloy contains 15 milligrams of gold for every 4 milligrams of copper. A jeweler uses 48 milligrams of copper to make the alloy. How much gold does the jeweler use to make the alloy?

31. **MODELING REAL LIFE** You make candles by adding 2 fluid ounces of scented oil for every 22 fluid ounces of wax. Your friend makes candles by adding 3 fluid ounces of the same scented oil for every 37 fluid ounces of wax. Whose candles are more fragrant? Explain your reasoning.

32. **MODELING REAL LIFE** A mint milk shake contains 1.25 fluid ounces of milk for every 4 ounces of ice cream. A strawberry milk shake contains 1.75 fluid ounces of milk for every 5 ounces of ice cream. Which milk shake is thicker? Explain.

CRITICAL THINKING Two whole numbers A and B satisfy the following conditions. Find A and B.

33. $A + B = 30$
    $A : B$ is equivalent to $2 : 3$.

34. $A + B = 44$
    $A : B$ is equivalent to $4 : 7$.

35. $A - B = 18$
    $A : B$ is equivalent to $11 : 5$.

36. $A - B = 25$
    $A : B$ is equivalent to $13 : 8$.

37. **MODELING REAL LIFE** A nutrition label shows that there are 161 calories in 28 grams of dry roasted cashews. You eat 9 cashews totaling 12 grams.

    a. Do you think it is possible to find the number of calories you consume? Explain your reasoning.

    b. How many cashews are in one serving?

38. **REASONING** The ratio of three numbers is $4 : 5 : 3$. The sum of the numbers is 54. What are the three numbers?

39. **CRITICAL THINKING** Seven out of every 8 students surveyed own a bike. The difference between the number of students who own a bike and those who do not is 72. How many students were surveyed?

40. **LOGIC** You and a classmate have a bug collection for science class. You find 5 out of every 9 bugs in the collection. You find 4 more bugs than your classmate. How many bugs are in the collection?

41. **PROBLEM SOLVING** You earn $72 for every 8 hours you spend shoveling snow. You earn $60 for every 5 hours you spend babysitting. For every 3 hours you spend babysitting, you spend 2 hours shoveling snow. You babysit for 15 hours in January. How much money do you earn in January?

42. **DIG DEEPER!** You and a friend each have a collection of tokens. Initially, for every 8 tokens you had, your friend had 3. After you give half of your tokens to your friend, your friend now has 18 more tokens than you. Initially, how many more tokens did you have than your friend?

3.4 Graphing Ratio Relationships

Learning Target: Represent ratio relationships in a coordinate plane.

Success Criteria:
- I can create and plot ordered pairs from a ratio relationship.
- I can create graphs to solve ratio problems.
- I can create graphs to compare ratios.

EXPLORATION 1
Using a Coordinate Plane

Work with a partner. An airplane travels 300 miles per hour.

a. Represent the relationship between distance and time in a coordinate plane. Explain your choice for labeling and scaling the axes.

b. Write a question that can be answered using the graph. Exchange your question with another group. Answer their question and discuss the solution with the other group.

Math Practice

Label Axes
Could you have placed each quantity on the other axis? Explain why or why not.

EXPLORATION 2
Identifying Relationships in Graphs

Work with a partner. Use the graphs to make a ratio table. Explain how the blue, red, and green arrows correspond to the ratio table.

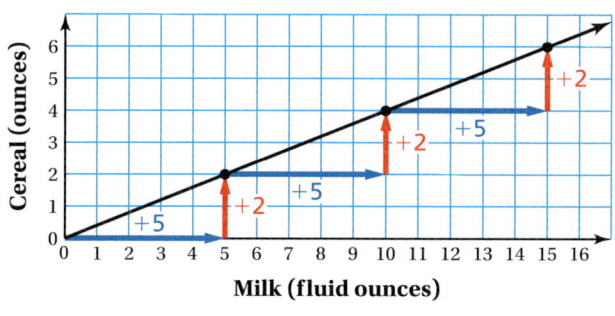

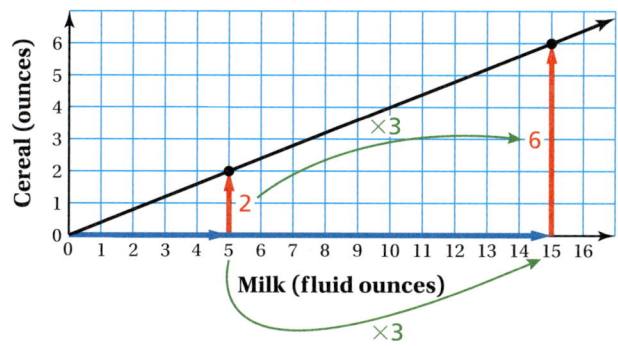

Section 3.4 Graphing Ratio Relationships **129**

3.4 Lesson

For a ratio of two quantities, you can use equivalent ratios to create ordered pairs of the form (first quantity, second quantity). You can plot these ordered pairs in a coordinate plane and draw a line, starting at (0, 0), through the points.

EXAMPLE 1   Graphing Ratio Relationships

Represent each ratio relationship using a graph.

a.

Flour (cups)	4	8	12
Water (cups)	2	4	6

Remember

When plotting an ordered pair (x, y), the number x corresponds to the horizontal axis and the number y corresponds to the vertical axis.

The ordered pairs (flour, water) are (4, 2), (8, 4), and (12, 6).

Plot the ordered pairs. Starting at (0, 0), draw a line through the points.

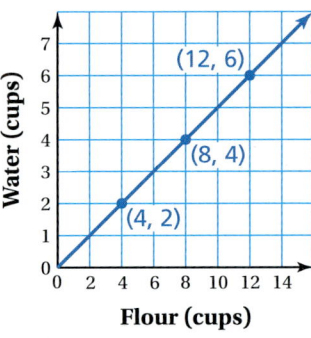

b.

Time (hours)	1	2	3
Earnings (dollars)	10	20	30

Relationships involving time are often graphed with time on the horizontal axis.

The ordered pairs (time, earnings) are (1, 10), (2, 20), and (3, 30).

Plot the ordered pairs. Starting at (0, 0), draw a line through the points.

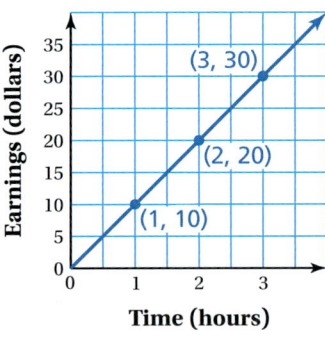

Try It Represent the ratio relationship using a graph.

1.

Time (minutes)	Number of Words
1	50
2	100
3	150

2.

Number of 6th Graders	Number of 7th Graders
5	4
10	8
15	12

EXAMPLE 2 Using a Graph to Solve a Ratio Problem

You buy dark chocolate cashews for $12.50 per pound.

a. **Represent the ratio relationship using a graph.**

Create a ratio table.

Cashews (pounds)	1	2	3
Cost (dollars)	12.5	25	37.5

The ordered pairs (cashews, cost) are (1, 12.5), (2, 25), and (3, 37.5). Plot the ordered pairs and draw a line, starting at (0, 0), through the points.

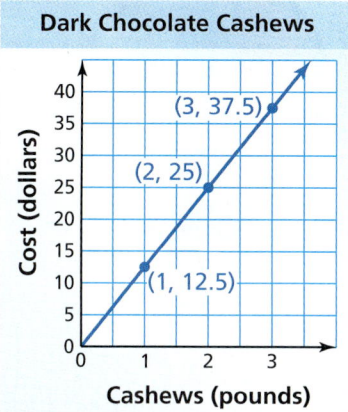

Relationships involving cost are often graphed with cost on the vertical axis.

b. **How much does 2.5 pounds of dark chocolate cashews cost?**

Using the graph, you can see that the cost of 2.5 pounds is halfway between $25 and $37.50.

So, 2.5 pounds of dark chocolate cashews cost $31.25.

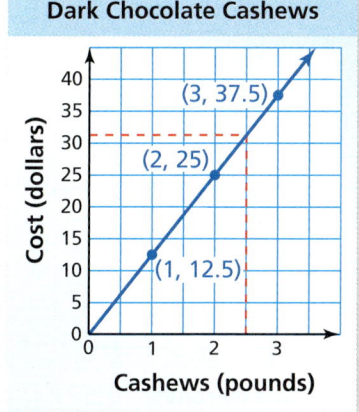

Another Method Use a double number line to find the cost.

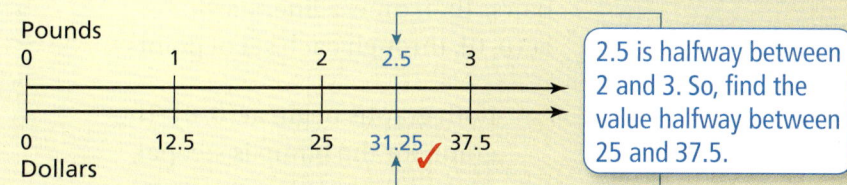

2.5 is halfway between 2 and 3. So, find the value halfway between 25 and 37.5.

Try It

3. **WHAT IF?** Repeat Example 2 when the cost of the dark chocolate cashews is $15 per pound.

Self-Assessment for Concepts & Skills

Solve each exercise. Then rate your understanding of the success criteria in your journal.

Rain (inches)	Snow (inches)
3	5
6	10
9	15

4. **GRAPHING A RATIO RELATIONSHIP** Represent the ratio relationship using a graph.

5. **CRITICAL THINKING** Use what you know about equivalent ratios to explain why the graph of a ratio relationship passes through (0, 0).

6. **WHICH ONE DOESN'T BELONG?** Which ordered pair does *not* belong with the other three? Explain your reasoning.

(4, 1)   (8, 2)   (12, 3)   (24, 4)

EXAMPLE 3 Modeling Real Life

A hot-air balloon rises 9 meters every 3 seconds. A blimp rises 7 meters every 2 seconds. Graph each ratio relationship in the same coordinate plane. Which rises faster?

Create ratio tables for each rising object. Then plot the ordered pairs (time, height) from the table and use the graph to determine which rises faster.

Balloon	
Time (seconds)	Height (meters)
3	9
6	18
9	27

Blimp	
Time (seconds)	Height (meters)
2	7
4	14
6	21

Balloon: (3, 9), (6, 18), (9, 27)

Blimp: (2, 7), (4, 14), (6, 21)

Plot and label each set of ordered pairs. Then draw a line, starting at (0, 0), through each set of points.

▶ Both graphs begin at (0, 0). The graph for the blimp is steeper, so the blimp rises faster than the hot-air balloon.

Check From the ratio tables, you can see that every 6 seconds, the balloon rises 18 meters and the blimp rises 21 meters. So, the blimp rises faster. ✓

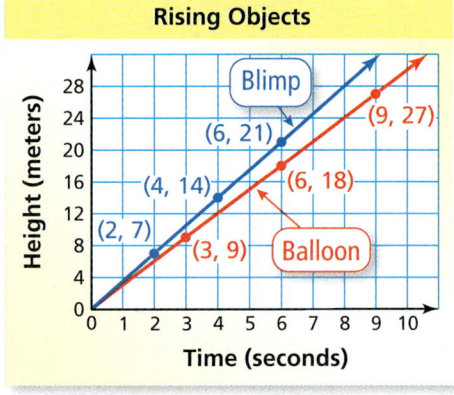

Self-Assessment for Problem Solving

Solve each exercise. Then rate your understanding of the success criteria in your journal.

7. You are skateboarding at a pace of 30 meters every 5 seconds. Your friend is in-line skating at a pace of 9 meters every 2 seconds. Graph each ratio relationship in the same coordinate plane. Who is faster?

8. You buy 2.5 pounds of pumpkin seeds and 2.5 pounds of sunflower seeds. Use a graph to find your total cost. Then use the graph to determine how much more you pay for pumpkin seeds than for sunflower seeds.

3.4 Practice

Go to *BigIdeasMath.com* to get HELP with solving the exercises.

Review & Refresh

Find the missing values in the ratio table. Then write the equivalent ratios.

1.
Chickens	8		24
Eggs		6	12

2.
Fish	6	3		15
Snails		2	4	

Write the name of the decimal number.

3. 7.1   4. 3.54   5. 13.6   6. 8.132

Write two equivalent ratios that describe the relationship.

7. baseballs to gloves

8. ladybugs to bees

Concepts, Skills, & Problem Solving

USING A COORDINATE PLANE Represent the relationship between distance and time in a coordinate plane. (See Exploration 1, p. 129.)

9. A train travels 45 miles per hour.

10. A motorcycle travels 70 kilometers per hour.

11. A snail travels 80 centimeters per minute.

12. A whale travels 800 yards per minute.

GRAPHING RATIO RELATIONSHIPS Represent the ratio relationship using a graph.

13.
Height (inches)	20	40	60
Weight (pounds)	30	60	90

14.
Students	9	18	27
Computers	4	8	12

15.
Ribbon (inches)	1	2	3
String (inches)	3	6	9

16.
Water (gallons)	30	60	90
Soda (gallons)	5	10	15

17.
Cherries	5	10	15
Limes	8	16	24

18.
Jog (miles)	2	4	6
Sprint (meters)	400	800	1200

19. **MODELING REAL LIFE** A radio station collects donations for a new broadcast tower. The cost to construct the tower is $25.50 per inch.

    a. Represent the ratio relationship using a graph.

    b. How much does it cost to fund 4.5 inches of the construction?

20. **MODELING REAL LIFE** Your school organizes a clothing drive as a fundraiser for a class trip. The school earns $100 for every 400 pounds of donated clothing.

    a. Represent the ratio relationship using a graph.

    b. How much money does your school earn for donating 2200 pounds of clothing?

21. **NUMBER SENSE** Just by looking at the graph, determine who earns a greater hourly wage. Explain.

22. **MODELING REAL LIFE** An airplane traveling from Chicago to Los Angeles travels 15 miles every 2 minutes. On the return trip, the plane travels 25 miles every 3 minutes. Graph each ratio relationship in the same coordinate plane. Does the plane fly faster when traveling to Los Angeles or to Chicago?

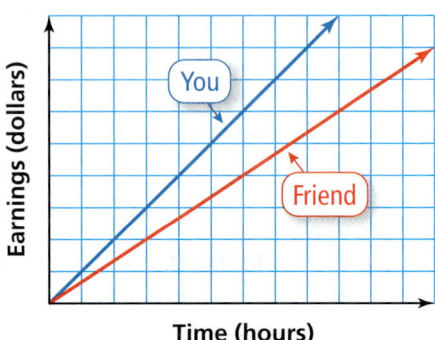

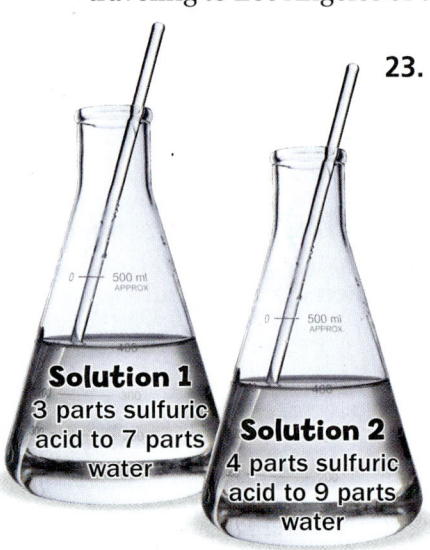

23. **MODELING REAL LIFE** Your freezer produces 8 ice cubes every 2 hours. Your friend's freezer produces 24 ice cubes every 5 hours. Graph each ratio relationship in the same coordinate plane. Whose freezer produces ice faster?

24. **CHOOSE TOOLS** A chemist prepares two acid solutions.

    a. Use a ratio table to determine which solution is more acidic.

    b. Use a graph to determine which solution is more acidic.

    c. Which method do you prefer? Explain.

25. **DIG DEEPER!** A company offers a nut mixture with 7 peanuts for every 3 almonds. The company changes the mixture to have 9 peanuts for every 5 almonds, but the number of nuts per container does not change.

    a. How many nuts are in the smallest possible container?

    b. Graph each ratio relationship. What can you conclude?

    c. Almonds cost more than peanuts. Should the company change the price of the mixture? Explain your reasoning.

26. **STRUCTURE** The point (p, q) is on the graph of values from a ratio table. What are two additional points on the graph?

3.5 Rates and Unit Rates

Learning Target: Understand the concept of a unit rate and solve rate problems.

Success Criteria:
- I can find unit rates.
- I can use unit rates to solve rate problems.
- I can use unit rates to compare rates.

EXPLORATION 1

Using a Diagram

Work with a partner. The diagram shows a story problem.

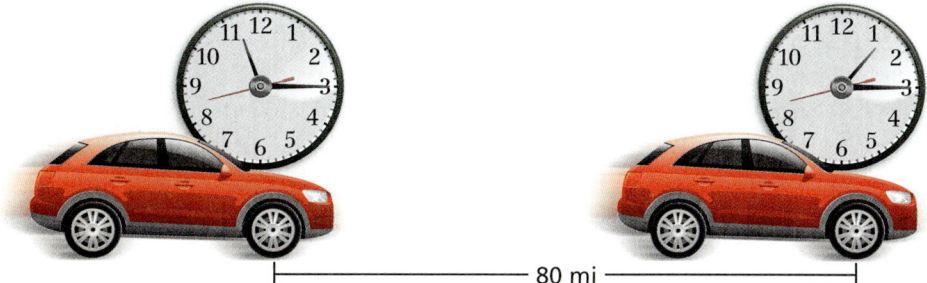

Math Practice

Specify Units
What are the units for the speed in part (c)? Why is it important to keep track of units in ratio problems?

a. What information can you obtain from the diagram?

b. Assuming that the car travels at a constant speed, how far does the car travel in 3.25 hours? Explain your method.

c. Draw a speedometer that shows the speed of the car. How can you use the speedometer to answer part (b)?

EXPLORATION 2

Using Equivalent Ratios

Work with a partner. Count the number of times you can clap your hands in 12 seconds. Have your partner record your results. Then switch roles with your partner and repeat the process.

a. Using your results and your partner's results, write ratios that represent the numbers of claps for every 12 seconds.

b. Explain how you can use the ratios in part (a) to find the numbers of times you and your partner can clap your hands in 2 minutes, in 2.5 minutes, and in 3 minutes.

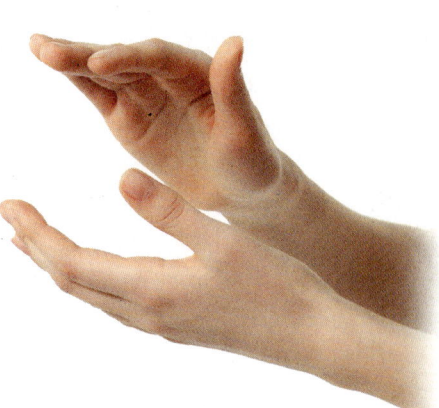

3.5 Lesson

Key Vocabulary
rate, *p. 136*
unit rate, *p. 136*
equivalent rates, *p. 136*

A **rate** is a ratio of two quantities using different units. You solved various ratio problems in the previous sections that involved rates. Now you will use *unit rates* to solve rate problems.

 **Key Idea**

Unit Rate

Words A **unit rate** compares a quantity to one unit of another quantity. **Equivalent rates** have the same unit rate.

Numbers You pay $27 for 3 pizzas.

Rate: $27 : 3 pizzas

Unit rate: $9 : 1 pizza

Algebra Rate: a units : b units   Unit rate: $\frac{a}{b}$ units : 1 unit

---

EXAMPLE 1 **Finding Unit Rates**

You make fruit juice by adding 4 pints of water for every 2 cups of concentrate.

a. Find the unit rate.

The ratio of pints of water to cups of concentrate is 4 : 2. Divide each quantity by 2 to find the unit rate.

Water (pints)	4	2
Concentrate (cups)	2	1

÷ 2

 The unit rate is 2 pints of water per cup of concentrate.

b. Write the ratio of cups of concentrate to pints of water. Then find the unit rate.

The ratio of cups of concentrate to pints of water is 2 : 4. Divide each quantity by 4 to find the unit rate.

Concentrate (cups)	2	$\frac{1}{2}$
Water (pints)	4	1

÷ 4

 The unit rate is $\frac{1}{2}$ cup of concentrate per pint of water.

For a rate $a : b$, the associated rate $b : a$ has a unit rate of $\frac{b}{a} : 1$. This can be useful when solving rate problems.

Try It

1. **WHAT IF?** Repeat Example 1 when you add 4 pints of water for every 3 cups of concentrate.

136 Chapter 3 Ratios and Rates   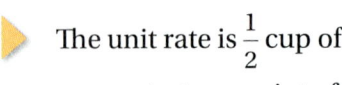 Multi-Language Glossary at *BigIdeasMath.com*

EXAMPLE 2 Using a Unit Rate to Solve a Rate Problem

A piece of space junk travels 5 miles every 6 seconds.

a. **How far does the space junk travel in 30 seconds?**

The ratio of miles to seconds is 5 : 6. Divide by 6 to find the unit rate in miles per second. Then multiply each quantity by 30 to find the distance traveled in 30 seconds.

▶ The space junk travels 25 miles in 30 seconds.

Distance (miles)	5	5/6	25
Time (seconds)	6	1	30

÷ 6   × 30

b. **How many seconds does it take the space junk to travel 2 miles?**

The ratio of seconds to miles is 6 : 5. Divide by 5 to find the unit rate in seconds per mile. Then multiply each quantity by 2 to find the time to travel 2 miles.

▶ It takes $\frac{12}{5} = 2\frac{2}{5}$ seconds for the space junk to travel 2 miles.

Time (seconds)	6	6/5	12/5
Distance (miles)	5	1	2

÷ 5   × 2

In Example 2, notice that you can use one step in each ratio table. Multiply by
$\frac{1}{6} \times 30 = 5$ *in part (a)*
and $\frac{1}{5} \times 2 = \frac{2}{5}$
in part (b).

Try It

2. **WHAT IF?** Repeat Example 2 when the space junk travels 3 miles every 5 seconds.

Self-Assessment for Concepts & Skills

Solve each exercise. Then rate your understanding of the success criteria in your journal.

FINDING UNIT RATES Write a unit rate for the situation.

3. 5 revolutions in 50 seconds
4. 1400 words for every 4 pages

5. **WHICH ONE DOESN'T BELONG?** Which rate does *not* belong with the other three? Explain your reasoning.

| 8 pounds for every 2 feet | 12 pounds per 3 feet |
| 20 pounds per 4 feet | 24 pounds for every 6 feet |

Section 3.5 Rates and Unit Rates

EXAMPLE 3   Modeling Real Life

You buy 2 pounds of salmon filets at Store A for $21.50. Your friend buys 3 pounds of salmon filets at Store B for $33.75. How much less would you spend for 5 pounds of salmon filets at the store with the better deal?

Use ratio tables to find the cost of 5 pounds of salmon filets at each store.

Store A	
Cost (dollars)	Salmon (pounds)
21.50	2
10.75	1
53.75	5

Store B	
Cost (dollars)	Salmon (pounds)
33.75	3
11.25	1
56.25	5

Find the unit rate at each store.
Find the cost of 5 pounds at each store.

▶ So, you would spend $56.25 − $53.75 = $2.50 less for 5 pounds of salmon filets at Store A.

Check The cost at Store A is $11.25 − $10.75 = $0.50 cheaper per pound than the cost at Store B. Because you are buying 5 pounds, you will pay 5 × $0.50 = $2.50 less at Store A. ✓

Self-Assessment for Problem Solving

Solve each exercise. Then rate your understanding of the success criteria in your journal.

6. You buy 10 pounds of bird seed at Store A for $11.50. Your friend buys 15 pounds of bird seed at Store B for $19.50. How much less would you spend by buying 20 pounds of bird seed at the store with the better deal?

7. A person hikes 4 miles in 2.5 hours. Find the unit rate in miles per hour. Then find the unit rate in hours per mile. How is each unit rate useful in a real-life situation?

8. **DIG DEEPER!** You buy 11 bagels with a $20 bill. How much change do you receive? How many more bagels could you buy?

3.5 Practice

> Go to **BigIdeasMath.com** to get HELP with solving the exercises.

▶ Review & Refresh

Represent the ratio relationship using a graph.

1.
Push-Ups	5	10	15
Sit-Ups	10	20	30

2.
Texts Sent	4	8	12
Texts Received	3	6	9

3.
Seeds	15	30	45
Plants	12	24	36

4.
Run (minutes)	6	12	18
Walk (minutes)	2	4	6

Divide. Write the answer in simplest form.

5. $\frac{1}{5} \div \frac{3}{10}$

6. $\frac{3}{8} \div 6$

7. $3\frac{1}{6} \div 2$

8. $5\frac{1}{3} \div 2\frac{2}{3}$

Add or subtract.

9. $6.94 + 12.301$

10. $8.753 - 7.71$

11. $14.532 - 6.613$

12. The winner in an election for class president received $\frac{3}{4}$ of the 240 votes. How many votes did the winner receive?

   A. 60   **B.** 150   **C.** 180   **D.** 320

▶ Concepts, Skills, & Problem Solving

USING EQUIVALENT RATIOS Use the ratio in Exploration 2 to estimate the number of times you can clap your hands in the given amount of time. *(See Exploration 2, p. 135.)*

13. 0.5 minute

14. 1.75 minutes

15. 2.25 minutes

FINDING UNIT RATES Write a unit rate for the situation.

16. 24 animals in 2 square miles

17. $100 for every 5 guests

18. $28 saved in 4 weeks

19. 18 necklaces made in 3 hours

20. 270 miles in 6 hours

21. 228 students in 12 classes

22. 2520 kilobytes in 18 seconds

23. 880 calories in 8 servings

24. 1080 miles on 15 gallons

25. $12.50 for 5 ounces

USING UNIT RATES Find the missing values in the ratio table.

26.
Inches	2		
Years	3	1	8

27.
Gallons	30	1	17
Seconds	5		

Section 3.5 Rates and Unit Rates 139

28. **MODELING REAL LIFE** Lightning strikes Earth 1000 times in 10 seconds.

    a. How many times does lightning strike in 12 seconds?

    b. How many seconds does it take for lightning to strike 7250 times?

29. **MODELING REAL LIFE** You earn $35 for washing 7 cars.

    a. How much do you earn for washing 4 cars?

    b. You earn $45. How many cars did you wash?

COMPARING RATES Decide whether the rates are equivalent.

30. 24 laps in 6 minutes
    72 laps in 18 minutes

31. 126 points for every 3 games
    210 points for every 5 games

32. 15 breaths for every 36 seconds
    90 breaths for every 3 minutes

33. $16 for 4 pounds
    $1 for 4 ounces

34. **MODELING REAL LIFE** An office printer prints 25 photos in 12.5 minutes. A home printer prints 15 photos in 6 minutes. Which printer is faster? How many more photos can you print in 12 minutes using the faster printer?

35. **MODELING REAL LIFE** You jog 2 kilometers in 12 minutes. Your friend jogs 3 kilometers in 16.5 minutes. Who jogs faster? How much sooner will the faster jogger finish a five-kilometer race?

36. **PROBLEM SOLVING** A softball team has a budget of $200 for visors. The athletic director pays $90 for 12 sun visors. Is there enough money in the budget to purchase 15 more sun visors? Explain your reasoning.

37. **DIG DEEPER!** The table shows the amounts of food collected by two homerooms. Homeroom A collects 21 additional items of food. How many more items does Homeroom B need to collect to have more items per student?

	Homeroom A	Homeroom B
Students	24	16
Canned Food	30	22
Dry Food	42	24

38. **REASONING** A runner completed a 26.2-mile marathon in 210 minutes.

    a. Estimate the unit rate, in miles per minute.

    b. Estimate the unit rate, in minutes per mile.

    c. Another runner says, "I averaged 10-minute miles in the marathon." Is this runner talking about the unit rate described in part (a) or in part (b)? Explain your reasoning.

39. **DIG DEEPER!** You can complete one-half of a job in an hour. Your friend can complete one-third of the same job in an hour. How long will it take to complete the job if you work together?

3.6 Converting Measures

Learning Target: Use ratio reasoning to convert units of measure.

Success Criteria:
- I can write conversion facts as unit rates.
- I can convert units of measure using ratio tables.
- I can convert units of measure using conversion factors.
- I can convert rates using conversion factors.

EXPLORATION 1
Estimating Unit Conversions

Work with a partner. You are given 4 one-liter containers and a one-gallon container.

a. A full one-gallon container can be used to fill the one-liter containers, as shown below. Write a unit rate that estimates the number of liters per gallon.

b. A full one-liter container can be used to partially fill the one-gallon container, as shown below. Write a unit rate that estimates the number of gallons per liter.

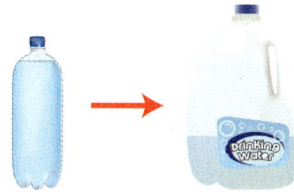

c. Estimate the number of liters in 5.5 gallons and the number of gallons in 12 liters. What method(s) did you use? What other methods could you have used?

EXPLORATION 2
Converting Units in a Rate

Math Practice

Recognize Usefulness of Tools
When would rulers be a useful tool for converting between centimeters and inches? When would they not be useful?

Work with a partner. The rate that a caterpillar moves is given in inches per minute. Using the rulers below, how can you convert the rate to centimeters per second? Justify your answer.

Section 3.6 Converting Measures 141

3.6 Lesson

Key Vocabulary
U.S. customary system, p. 142
metric system, p. 142
conversion factor, p. 143
unit analysis, p. 143

The **U.S. customary system** is a system of measurement that contains units for length, capacity, and weight. The **metric system** is a decimal system of measurement, based on powers of 10, that contains units for length, capacity, and mass.

You can use unit rates and ratio tables to convert measures within the same system and between systems.

EXAMPLE 1 Converting Measures within the Same System

Convert 36 quarts to gallons.

Because 1 gallon = 4 quarts, there are 4 quarts per gallon, and $\frac{1}{4}$ gallon per quart. You can use either of these unit rates to find an equivalent rate with 36 quarts.

For a list of conversion facts, see the Mathematics Reference Sheet in the back of this book.

Method 1: Create a ratio table using the unit rate 4 quarts per gallon. Multiply each quantity by 9 to find the number of gallons in 36 quarts.

Quarts	4	36
Gallons	1	9

× 9

 So, 36 quarts is 9 gallons.

Method 2: Create a ratio table using the unit rate $\frac{1}{4}$ gallon per quart. Multiply each quantity by 36 to find the number of gallons in 36 quarts.

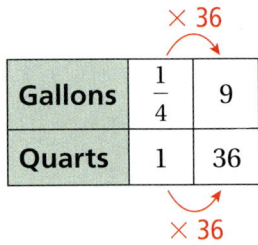

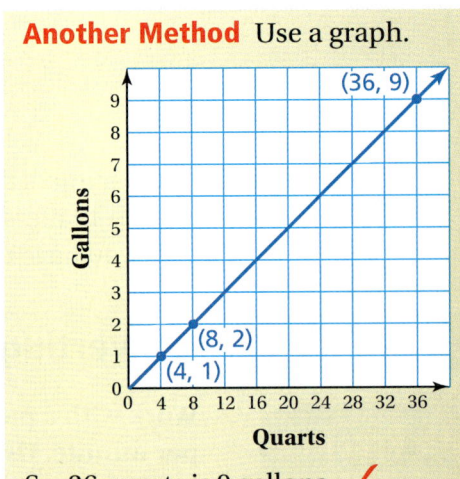

Another Method Use a graph.

So, 36 quarts is 9 gallons. ✓

So, 36 quarts is 9 gallons.

Try It

1. Convert 48 feet to yards.

EXAMPLE 2 Converting Measures Between Systems

Convert 10 meters to feet.

Because 1 meter ≈ 3.28 feet, there are about 3.28 feet per meter. Because 1 foot ≈ 0.3 meter, there is about 0.3 meter per foot. You can use either of these unit rates to find an equivalent rate with 10 meters.

Method 1: Create a ratio table using the unit rate 3.28 feet per meter.

× 10

Feet	3.28	32.8
Meters	1	10

× 10

So, 10 meters is about 32.8 feet.

Method 2: Create a ratio table using the unit rate 0.3 meter per foot.

÷ 0.3   × 10

Meters	0.3	1	10
Feet	1	$\frac{1}{0.3}$	$\frac{10}{0.3}$

÷ 0.3   × 10

So, 10 meters is about $\frac{10}{0.3}$ ≈ 33.33 feet.

> In Example 2, it is more efficient to use the unit rate in Method 1 because there is 1 unit of the quantity you are converting. The answers are slightly different due to differences in the rounding used with conversion facts, but both answers are acceptable.

Try It

2. Convert 7 miles to kilometers. Round to the nearest hundredth if necessary.

Another way to convert units of measure is to multiply by one or more *conversion factors*.

Key Idea

Conversion Factor

A **conversion factor** is a rate in which the two quantities are equal. When using conversion factors, write rates using fraction notation.

	Relationship	Conversion Factors
Example	1 ft = 12 in.	1 ft per 12 in., or $\frac{1 \text{ ft}}{12 \text{ in.}}$
		12 in. per 1 ft, or $\frac{12 \text{ in.}}{1 \text{ ft}}$

> Because the quantities in a conversion factor are equal, conversion factors are equal to 1. So, you can multiply a quantity by a conversion factor and not change its value.

You can use **unit analysis** to decide which conversion factor will produce the appropriate units. You can "cross out" a unit that appears in both a numerator and a denominator of a product.

Section 3.6 Converting Measures

EXAMPLE 3 Using Conversion Factors

a. Convert 4 pounds to kilograms.

Use a conversion factor.

$$4 \text{ lb} \approx 4 \text{ lb} \times \frac{0.45 \text{ kg}}{1 \text{ lb}} = 1.8 \text{ kg}$$

(1 lb ≈ 0.45 kg)

▸ So, 4 pounds is about 1.8 kilograms.

Another Method Use a double number line.

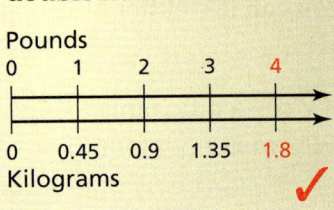

b. Convert 5 yards per second to yards per minute.

Write 5 yards per second as a fraction. Then use a conversion factor.

$$5 \text{ yards per second} = \frac{5 \text{ yd}}{1 \text{ sec}} \times \frac{60 \text{ sec}}{1 \text{ min}} = \frac{300 \text{ yd}}{1 \text{ min}}$$

(1 min = 60 sec)

▸ So, 5 yards per second is 300 yards per minute.

Try It

3. Convert 20 quarts to liters. Round to the nearest hundredth if necessary.

4. Convert 60 kilometers per hour to miles per hour. Round to the nearest hundredth if necessary.

Self-Assessment for Concepts & Skills

Solve each exercise. Then rate your understanding of the success criteria in your journal.

5. **DIFFERENT WORDS, SAME QUESTION** Which is different? Find "both" answers.

Convert 5 inches to centimeters.	Find the number of inches in 5 centimeters.
How many centimeters are in 5 inches?	Five inches equals how many centimeters?

CONVERTING MEASURES Copy and complete the statement. Round to the nearest hundredth if necessary.

6. $\dfrac{12 \text{ m}}{\text{min}} \approx \dfrac{\boxed{} \text{ ft}}{\text{min}}$

7. $\dfrac{12 \text{ ft}}{\text{sec}} = \dfrac{\boxed{} \text{ yd}}{\text{min}}$

144 Chapter 3 Ratios and Rates

EXAMPLE 4 Modeling Real Life

A runner's goal is to complete a mile in 4 minutes or less. The runner's speed is 20 feet per second. Does the runner meet the goal? If not, how much faster (in feet per second) must the runner be to meet the goal?

To meet the goal, the runner must complete 1 mile in 4 minutes or less. The minimum speed required is 1 mile per 4 minutes.

To compare this speed to the runner's speed of 20 feet per second, convert the minimum speed of 1 mile per 4 minutes to feet per second.

$$\boxed{1 \text{ min} = 60 \text{ sec}} \qquad \boxed{1 \text{ mi} = 5280 \text{ ft}}$$

$$\frac{1 \text{ mi}}{4 \text{ min}} = \frac{1 \text{ mi}}{4 \text{ min}} \times \frac{1 \text{ min}}{60 \text{ sec}} \times \frac{5280 \text{ ft}}{1 \text{ mi}} = \frac{22 \text{ ft}}{1 \text{ sec}}$$

So, the runner did not meet the goal because a speed of 20 feet per second is below the minimum speed of 22 feet per second. The runner must be $22 - 20 = 2$ feet per second faster to meet the goal.

Check Verify that the distances traveled in 4 minutes at the runner's speed and in 4 minutes at the additional speed have a sum of 1 mile.

At runner's speed

$$4 \text{ min} \times \frac{60 \text{ sec}}{1 \text{ min}} \times \frac{20 \text{ ft}}{1 \text{ sec}} = 4800 \text{ ft}$$

At additional speed

$$4 \text{ min} \times \frac{60 \text{ sec}}{1 \text{ min}} \times \frac{2 \text{ ft}}{1 \text{ sec}} = 480 \text{ ft}$$

The sum of the distances is $4800 + 480 = 5280$ feet, or 1 mile. ✓

Self-Assessment for Problem Solving

Solve each exercise. Then rate your understanding of the success criteria in your journal.

8. Will all of the water from a full two-liter bottle fit into a two-quart pitcher? Explain.

9. **DIG DEEPER!** The speed of light is about 300,000 kilometers per second. The Sun is about 93 million miles from Earth. How many minutes does it take for sunlight to reach Earth?

10. A race car driver's goal is to complete a 1000-kilometer auto race in 4 hours or less. The driver's average speed is 4200 meters per minute. Does the driver meet the goal? If not, how much faster (in meters per minute) must the driver be to meet the goal?

Section 3.6 Converting Measures

3.6 Practice

> Go to *BigIdeasMath.com* to get HELP with solving the exercises.

Review & Refresh

Write a unit rate for the situation.

1. 102 beats per 2 minutes
2. 60 shirts for every 5 clothing racks
3. $100 donated for every 5 volunteers
4. 30 milliliters every 4 hours
5. What is the LCM of 6, 12, and 18?
   - **A.** 6
   - **B.** 18
   - **C.** 36
   - **D.** 72

Write the prime factorization of the number.

6. 56
7. 74
8. 63
9. 132

Write the product as a power.

10. 6×6
11. $18 \times 18 \times 18 \times 18$
12. $12 \times 12 \times 12 \times 12 \times 12$

Concepts, Skills, & Problem Solving

COMPARING MEASURES Answer the question. Explain your answer.
(See Explorations 1 & 2, p. 141.)

13. Which juice container is larger: 2 L or 1 gal?

14. Which is longer: 1 in. or 2 cm?

CONVERTING MEASURES Copy and complete the statement.

15. 3 pt = ___ c
16. 1500 mL = ___ L
17. 40 oz = ___ lb
18. 5 ft = ___ in.
19. 6 gal = ___ qt
20. 48 cm = ___ mm
21. 500 cm = ___ m
22. 6000 g = ___ kg
23. 32 fl oz = ___ c

CONVERTING MEASURES Copy and complete the statement. Round to the nearest hundredth if necessary.

24. 12 L ≈ ___ qt
25. 14 m ≈ ___ ft
26. 4 ft ≈ ___ m
27. 64 lb ≈ ___ kg
28. 0.3 km ≈ ___ mi
29. 75.2 in. ≈ ___ cm
30. 17 kg ≈ ___ lb
31. 15 cm ≈ ___ in.
32. 9 mi ≈ ___ km

33. **GRAPHING RELATIONSHIPS** Represent the relationship between each pair of units in a coordinate plane.
    - **a.** feet and yards
    - **b.** pounds and kilograms

34. **MODELING REAL LIFE** Earth travels 30 kilometers each second as it revolves around the Sun. How many miles does Earth travel in 1 second?

35. **MODELING REAL LIFE** The Mackinac Bridge in Michigan is the third-longest suspension bridge in the United States.

   a. How high above the water is the roadway in meters?

   b. The bridge has a length of 26,372 feet. What is the length in kilometers?

USING CONVERSION FACTORS Copy and complete the statement. Round to the nearest hundredth if necessary.

36. 12 cu ft ≈ ___ gal

37. 6 qt ≈ ___ L

38. 5 L ≈ ___ gal

39. $\dfrac{13 \text{ km}}{\text{h}} \approx \dfrac{\underline{} \text{ mi}}{\text{h}}$

40. $\dfrac{22 \text{ L}}{\text{min}} = \dfrac{\underline{} \text{ L}}{\text{h}}$

41. $\dfrac{63 \text{ mi}}{\text{h}} = \dfrac{\underline{} \text{ mi}}{\text{sec}}$

42. **YOU BE THE TEACHER** Your friend converts 8 liters to quarts. Is your friend correct? Explain your reasoning.

$$8 \text{ L} \approx 8 \cancel{\text{L}} \cdot \dfrac{1.06 \text{ qt}}{1 \cancel{\text{L}}}$$
$$= 8.48 \text{ qt}$$

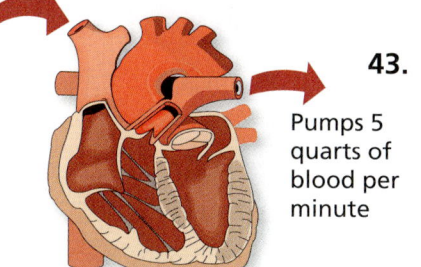

Pumps 5 quarts of blood per minute

43. **MODELING REAL LIFE** The diagram shows the number of quarts of blood the human heart pumps per minute.

   a. How many quarts of blood does the human heart pump per hour?

   b. How many liters of blood does the human heart pump per minute?

44. **MP PROBLEM SOLVING** After washing dishes, water drips from the faucet. The graph shows the number of cups of water that drip from the faucet over time. How many gallons of water drip from the faucet in 24 hours?

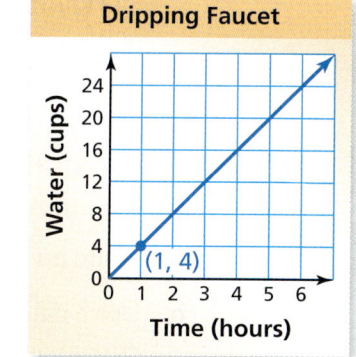

COMPARING MEASURES Copy and complete the statement using < or >.

45. 30 oz ___ 8 kg

46. 6 ft ___ 300 cm

47. 3 gal ___ 6 L

48. 10 in. ___ 200 mm

49. 5 lb ___ 1200 g

50. 1500 m ___ 3000 ft

USING DERIVED UNITS Copy and complete the statement. Round to the nearest hundredth if necessary.

51. $\dfrac{3 \text{ km}}{\text{min}} \approx \dfrac{\boxed{} \text{ mi}}{\text{h}}$

52. $\dfrac{17 \text{ gal}}{\text{h}} = \dfrac{\boxed{} \text{ qt}}{\text{min}}$

53. $\dfrac{600 \text{ cm}}{\text{min}} \approx \dfrac{\boxed{} \text{ in.}}{\text{sec}}$

54. **MODELING REAL LIFE** You are riding on a zip line. Your speed is 15 miles per hour. What is your speed in feet per second?

55. **MP PROBLEM SOLVING** Thunder is the sound caused by lightning. You hear thunder 5 seconds after a lightning strike. The speed of sound is about 1225 kilometers per hour. About how many miles away was the lightning?

56. **MP PROBLEM SOLVING** Boston, Massachusetts, and Buffalo, New York, are hit by snowstorms that last 3 days. Boston accumulates snow at a rate of 1.5 feet every 36 hours. Buffalo accumulates snow at a rate of 0.01 inch every minute. Which city accumulates more snow in 3 days? How much more snow?

57. **DIG DEEPER!** You travel 4000 feet every minute on a snowmobile.

   a. The evening speed limit for snowmobiles in your state is 55 miles per hour. Is your speed less than or equal to the speed limit? Justify your answer.

   b. What is your pace in minutes per mile?

   c. You are 22 miles from your house at 6:00 P.M. If you continue to travel at this speed, do you reach your house in time for dinner at 6:30 P.M.?

58. **MP REASONING** The table shows the flying speeds of several birds.

   a. Which bird is the fastest? Which is the slowest?

   b. The peregrine falcon has a dive speed of 322 kilometers per hour. Is the dive speed of the peregrine falcon faster than the flying speed of any of the birds? Explain.

Bird	Speed
Spine-tailed swift	2843.2 m/min
Spur-winged goose	129.1 ft/sec
Eider duck	31.3 m/sec
Mallard	65 mi/h

59. **MP STRUCTURE** Consider the conversion facts 1 inch = 2.54 centimeters and 1 centimeter ≈ 0.39 inch.

   a. Write an expression for the exact number of inches in 1 centimeter.

   b. Use a calculator to evaluate your expression in part (a). Explain why measurement conversions may be slightly different when converting between metric units and U.S. customary units using the conversion facts in the back of the book.

60. **DIG DEEPER!** One liter of paint covers 100 square feet. How many gallons of paint does it take to cover a room whose walls have an area of 800 square meters?

3 Connecting Concepts

▶ Using the Problem-Solving Plan

1. You mix water, glue, and borax in the ratio of 3 : 1 : 2 to make slime. How many gallons of each ingredient should you use to make 0.75 gallon of slime?

   Understand the problem. You know the ratio of the ingredients in the slime and that you are making 0.75 gallon of slime. You are asked to find the number of gallons of each ingredient needed to make 0.75 gallon of slime.

   Make a plan. Represent the ratio 3 : 1 : 2 using a tape diagram. Because there are 6 parts that represent 0.75 gallon, divide 0.75 by 6 to find the value of one part of the tape diagram. Then use the value of one part to find the number of gallons of each ingredient you should use.

   Solve and check. Use the plan to solve the problem. Then check your solution.

2. You buy yogurt cups and frozen fruit bars for a party. Yogurt cups are sold in packages of six. The ratio of the number of yogurt cups in a package to the number of frozen fruit bars in a package is 3 : 2. What are the least numbers of packages you should buy in order to have the same numbers of yogurt cups and frozen fruit bars?

3. The greatest common factor of two whole numbers is 9. The ratio of the greater number to the lesser number is 6 : 5. What are the two numbers? Justify your answer.

Performance Task

Oops! Unit Conversion Mistakes

At the beginning of this chapter, you watched a STEAM Video called "Human Circulatory System." You are now ready to complete the performance task related to this video, available at *BigIdeasMath.com*. Be sure to use the problem-solving plan as you work through the performance task.

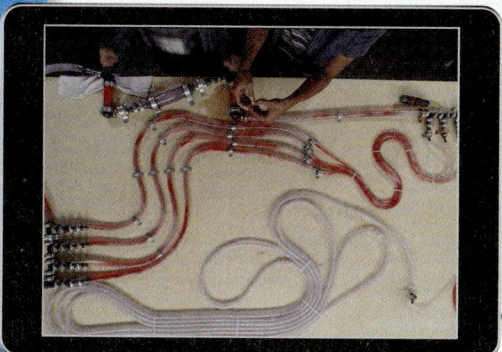

Connecting Concepts 149

Chapter Review

Go to *BigIdeasMath.com* to download blank graphic organizers.

▶ Review Vocabulary

Write the definition and give an example of each vocabulary term.

ratio, *p. 108*
value of a ratio, *p. 109*
equivalent ratios, *p. 109*
ratio table, *p. 122*
rate, *p. 136*

unit rate, *p. 136*
equivalent rates, *p. 136*
U.S. customary system, *p. 142*

metric system, *p. 142*
conversion factor, *p. 143*
unit analysis, *p. 143*

▶ Graphic Organizers

You can use a **Definition and Example Chart** to organize information about a concept. Here is an example of a Definition and Example Chart for the vocabulary term *ratio*.

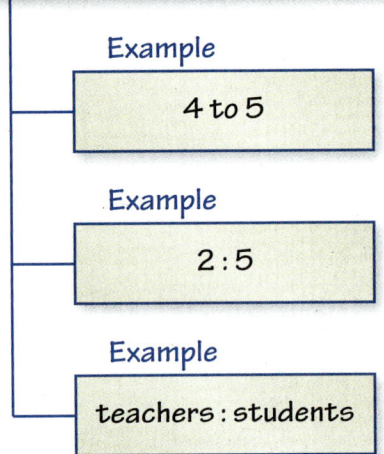

Choose and complete a graphic organizer to help you study the concept.

1. value of a ratio
2. equivalent ratios
3. tape diagram
4. ratio table
5. rate
6. unit rate
7. conversion factor

"My math teacher taught us how to make a Definition and Example Chart."

150 Chapter 3 Ratios and Rates

Chapter Self-Assessment

As you complete the exercises, use the scale below to rate your understanding of the success criteria in your journal.

1	2	3	4
I do not understand.	I can do it with help.	I can do it on my own.	I can teach someone else.

3.1 Ratios (pp. 107–114)

Learning Target: Understand the concepts of ratios and equivalent ratios.

Write the ratio.

1. butterflies : caterpillars

2. saxophones : trumpets

3. The ratio of hydrogen atoms to nitrogen atoms in a container is 2 : 3.

   a. Find and interpret the value of the ratio.

   b. In another container, the number of hydrogen atoms is 3 times the number of nitrogen atoms. Write the ratio of hydrogen atoms to nitrogen atoms.

Determine whether the ratios are equivalent.

4. 5 : 2 and 30 : 12

5. 4 : 3 and 8 : 7

6. 6 : 4 and 18 : 6

7. 18 : 12 and 3 : 2

8. Write two equivalent ratios that have values of $\frac{5}{7}$.

9. During a chess match, there are 12 pieces left on the board. The ratio of white pieces to black pieces is 2 : 1. How many white pieces are on the board?

10. You run at a pace of 2 miles every 17 minutes. Your friend runs at a pace of 3 miles every 24 minutes. Are you and your friend running at the same pace? If not, who is running faster?

Chapter Review 151

3.2 Using Tape Diagrams (pp. 115–120)

Learning Target: Use tape diagrams to model and solve ratio problems.

The tape diagram represents the ratio of the time you spend reading to the time your friend spends reading. You read for 8 hours. How many hours does your friend spend reading?

11. You / Friend

12. You / Friend

13. The tape diagram represents the ratio of customers to guides on a mountain climbing trip. There are 6 guides on the trip. How many customers are on the trip?

   Customers / Guides

A container has peppermint gum and spearmint gum. You are given the number of pieces of peppermint gum in the container and the ratio of peppermint gum to spearmint gum. Find the number of pieces of spearmint gum in the container.

14. 24 peppermint; 8 to 5

15. 18 peppermint; 2 : 3

16. 32 peppermint; 8 to 7

17. 40 peppermint; 5 : 2

A theater sells adult tickets and student tickets. You are given the total number of tickets sold and the ratio of adult tickets sold to student tickets sold. How many of each type of ticket are sold?

18. 120 tickets; 6 to 4

19. 165 tickets; 8 to 7

20. 210 tickets; 16 : 5

21. 248 tickets; 5 : 3

22. You perform 7 sit-ups for every 2 pull-ups as part of an exercise routine. You perform 25 more sit-ups than pull-ups. How many sit-ups and how many pull-ups do you perform?

3.3 Using Ratio Tables (pp. 121–128)

Learning Target: Use ratio tables to represent equivalent ratios and solve ratio problems.

Find the missing values in the ratio table. Then write the equivalent ratios.

23.
Levers	6		18
Pulleys	3	6	

24.
Nails	5	10	
Screws	2		6

25.
Cars	3	6	
Trucks	4		24

26.
Customers	8	4	
Servings	12		30

Find the missing quantity in the double number line.

27.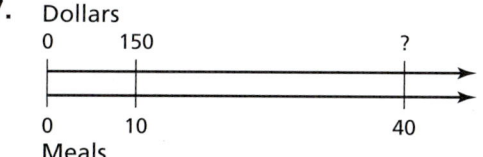

28.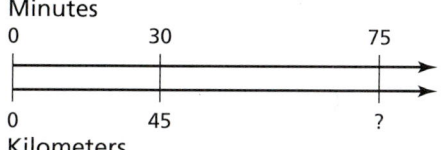

29. Use all four operations to complete the ratio table. Justify your answer.

Pumpkins Grown	4			
Seeds Planted	12			

30. A song has 12 beats every 5 seconds. How many beats are there in 30 seconds?

31. On New Year's Eve, the Times Square ball is lowered 47 feet every 20 seconds. How long does it take for the ball to be lowered 141 feet?

32. Welder A charges $300 for every 4 hours of labor. Welder B charges $240 for every 3 hours of labor. Which welder offers a better deal?

33. You make lemonade by adding 11 cups of water for every 3 cups of lemon juice. Your friend makes lemonade by adding 9 cups of water for every 2 cups of lemon juice. Whose lemonade is more watered down?

3.4 Graphing Ratio Relationships (pp. 129–134)

Learning Target: Represent ratio relationships in a coordinate plane.

Represent the ratio relationship using a graph.

34.
Time (years)	1	2	3
Penguins	6	12	18

35.
Televisions	12	24	36
Houses	4	8	12

36. You buy magnesium sulfate for $1.50 per pound.

    a. Represent the ratio relationship using a graph.

    b. How much does 3.5 pounds of magnesium sulfate cost?

37. A 5-ounce can of tuna costs $0.90. A 12-ounce can of tuna costs $2.40. Graph each ratio relationship in the same coordinate plane. Which is the better buy?

3.5 Rates and Unit Rates (pp. 135–140)

Learning Target: Understand the concept of a unit rate and solve rate problems.

Write a unit rate for the situation.

38. 12 stunts in 4 movies

39. 3600 stitches in 3 minutes

40. $18 for 6 pounds

41. 240 people in 5 buses

42. A train travels 120 miles in 3 hours. Write two unit rates that describe the relationship between the number of miles and the number of hours the train travels.

43. Mercury orbits the Sun 3 times in 264 days.

    a. How many times does Mercury orbit the Sun in 440 days?

    b. How many days does it take Mercury to orbit the Sun 8 times?

44. A cyclist travels 4 miles in 20 minutes. At this rate, how many miles does the cyclist travel in 30 minutes?

Decide whether the rates are equivalent.

45. 18 keystrokes in 3 seconds
    48 keystrokes in 16 seconds

46. 210 miles in 3 hours
    780 miles in 12 hours

47. You and a friend are picking up trash on a beach. You fill 2 bags with trash in 28 minutes. Your friend fills 3 bags with trash in 48 minutes. Who fills bags with trash faster? How much sooner will the faster person fill 7 bags with trash?

3.6 Converting Measures (pp. 141–148)

Learning Target: Use ratio reasoning to convert units of measure.

Copy and complete the statement. Round to the nearest hundredth if necessary.

48. 2.5 c = ____ fl oz
49. 12 ft = ____ yd
50. 3500 mg = ____ g
51. 3 L ≈ ____ qt
52. 9.2 in. ≈ ____ cm
53. 15 lb ≈ ____ kg
54. $\dfrac{\$2}{\min} = \dfrac{\$\_\_\_\_}{\text{h}}$
55. $\dfrac{13 \text{ gal}}{\text{h}} = \dfrac{\_\_\_\_ \text{ qt}}{\text{h}}$
56. $\dfrac{8 \text{ ft}}{\text{h}} \approx \dfrac{\_\_\_\_ \text{ m}}{\min}$

57. Explain how to use conversion factors to find the number of fluid ounces in any given number of quarts of a liquid.

58. Water flows through a pipe at a rate of 10 gallons per minute. How many gallons of water flow through the pipe in an hour?

59. Germany suggests a speed limit of 130 kilometers per hour on highways. Is the speed shown greater than the suggested limit?

60. The distance between two stars increases at a rate of 3 centimeters per month. What is the rate in inches per year?

Chapter Review 155

3 Practice Test

1. Write the ratio of scooters to bikes.

2. Determine whether the ratios 8 : 7 and 15 : 14 are equivalent.

Find the missing values in the ratio table. Then write the equivalent ratios.

3.
Lemons	4		36
Limes	2	6	

4.
Rabbits	2	12	
Hamsters	9		45

5. Represent the ratio relationship using a graph.

Cranberry Juice (cups)	2	4	6
Grape Juice (cups)	3	6	9

6. You travel 224 miles in 4 hours. Find the unit rate.

Copy and complete the statement. Round to the nearest hundredth if necessary.

7. 6 cm ≈ ▢ in.

8. 30 L ≈ ▢ gal

9. $\dfrac{10 \text{ gal}}{\text{h}} = \dfrac{▢ \text{ gal}}{\text{wk}}$

10. $\dfrac{4 \text{ ft}}{\text{sec}} \approx \dfrac{▢ \text{ m}}{\text{sec}}$

11. During a baseball season, Team A scores 9 runs for every 7 runs that Team B scores. The total number of runs scored by both teams is 1440. How many runs does each team score?

12. At a movie theater, the ratio of filled seats to empty seats is 6 : 5. There are 120 empty seats. How many seats are filled?

13. You and your friend mix water and citric acid. You add 3 cups of citric acid for every 16 cups of water. Your friend adds 2 cups of citric acid for every 12 cups of water. Whose mixture is more acidic?

14. Determine which windsurfer is faster. Explain your reasoning.

Speed: 720 feet per minute

Speed: 5 meters per second

15. In a rectangle, the ratio of the length to the width is 5 : 2. The length of the rectangle is 13.875 feet greater than the width. What are the perimeter and the area of the rectangle?

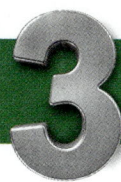

3 Cumulative Practice

Test-Taking Strategy

Solve Problem before Looking at Choices

"Solve the problem before looking at the choices. You know one-fourth of 20 is 5, so the answer is 5 miles."

1. Which number is equivalent to $\frac{2}{9} \div \frac{4}{5}$?

A. $\frac{8}{45}$ **B.** $\frac{5}{18}$

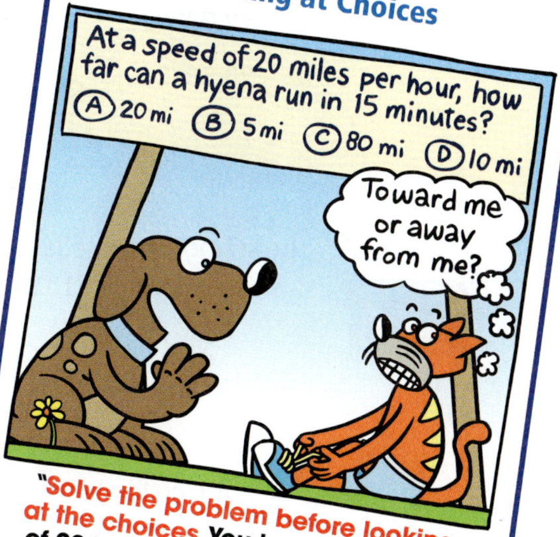

C. $\frac{7}{13}$ **D.** $3\frac{3}{5}$

2. Your speed while waterskiing is 22 miles per hour. How fast are you traveling in kilometers per hour? Round your answer to the nearest hundredth.

3. Which number is equivalent to the expression below?

$$2 \cdot 4^2 + 3(6 \div 2)$$

F. 25 **G.** 41

H. 73 **I.** 105

4. The tape diagram models the ratio of red beads to green beads in a bracelet. The bracelet uses 12 red beads. How many green beads are in the bracelet?

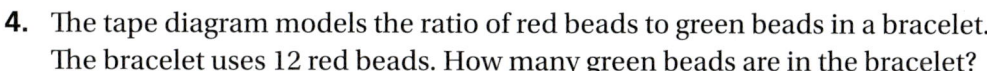

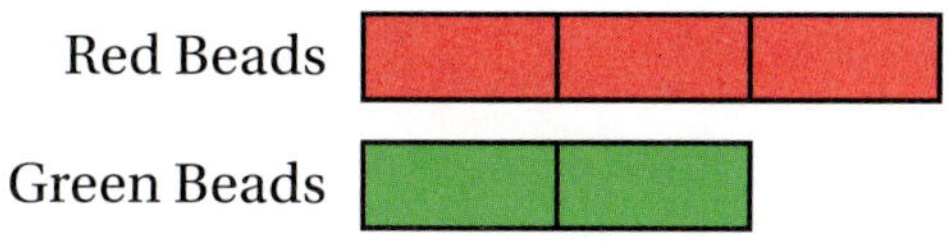

A. 4 green beads **B.** 8 green beads

C. 12 green beads **D.** 20 green beads

5. What is the least common multiple of 8, 12, and 20?

F. 24 **G.** 40

H. 60 **I.** 120

6. Which number is equivalent to 2.34 × 1.08 × 5.6?

　A. 12.787632　　　　**B.** 14.15232

　C. 23.5872　　　　　**D.** 14,152.32

7. The school store sells 4 pencils for $0.50. At this rate, what is the cost (in dollars) of 10 pencils?

8. A factor tree for 14,700 is shown. Which factor of 14,700 is *not* a perfect square?

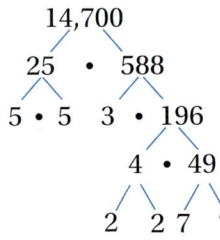

　F. 25　　　　　　　**G.** 49

　H. 196　　　　　　 **I.** 588

9. Which of the following is a ratio of frogs to snakes?

　A. 4 : 8　　　　　　**B.** 8 : 12

　C. 8 : 4　　　　　　**D.** 4 : 12

158　Chapter 3　Ratios and Rates

10. Which expression is equivalent to 3^5?

    F. $3 \times 3 \times 3 \times 3 \times 3$
    G. 3×5
    H. $5 \times 5 \times 5$
    I. $3 + 3 + 3 + 3 + 3$

11. Which is the correct order of operations when evaluating $5 + 4 \times 2^3$?

    k. Add 5 and 4.
    l. Add 5 and 32.
    m. Evaluate 2^3.
    n. Multiply 4 and 2.
    p. Multiply 4 and 8.
    q. Add 5 and 512.
    r. Multiply 9 and 8.
    s. Evaluate 8^3.
    t. Multiply 9 and 2.
    u. Evaluate 18^3.

    A. k, t, u
    B. n, s, q
    C. m, k, r
    D. m, p, l

12. The ratio of scrambled eggs to hard-boiled eggs served at a restaurant is 6 : 2.

    Part A Make a ratio table showing three possible combinations of scrambled eggs and hard-boiled eggs.

    Part B Represent the ratio relationship using a graph.

    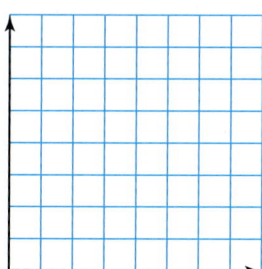

    Part C Use the graph to find the number of hard-boiled eggs served when the restaurant serves 15 scrambled eggs.

4 Percents

- **4.1** Percents and Fractions
- **4.2** Percents and Decimals
- **4.3** Comparing and Ordering Fractions, Decimals, and Percents
- **4.4** Solving Percent Problems

Chapter Learning Target:
Understand percents.

Chapter Success Criteria:
- I can write fractions and decimals as percents.
- I can write percents as fractions and as decimals.
- I can order fractions, decimals, and percents.
- I can solve percent problems.

STEAM Video: "Chargaff's Rules"

STEAM Video

Chargaff's Rules

DNA is a molecule made up of four *nucleotide bases* called adenine (A), thymine (T), cytosine (C), and guanine (G). DNA contains the genetic information for a living organism. What can you learn about an organism from its DNA?

Watch the STEAM Video "Chargaff's Rules." Then answer the following questions.

1. Veronica says that the DNA of most mammals contains about 60 percent A and T nucleotides and 40 percent C and G nucleotides. What do you think this means?

2. Use your answer in Question 1 to determine which of the following DNA samples is most likely to belong to a mammal. Explain your reasoning.

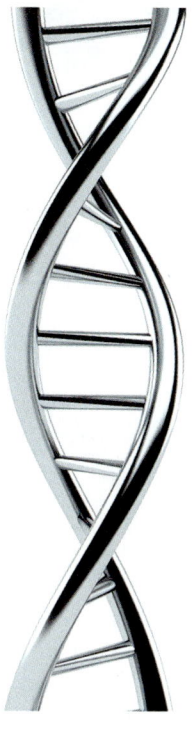

	Sample 1	Sample 2	Sample 3
A and T Nucleotides	38	61	60
C and G Nucleotides	62	39	100

Performance Task

Genetic Ancestry

After completing this chapter, you will be able to use the concepts you learned to answer the questions in the *STEAM Video Performance Task*. You will be given the results of your friend's ancestry test.

Native American: $\frac{1}{50}$

African: $\frac{30}{40}$

Northern European: 0.10

Southwest Asian: $\frac{3}{75}$

Mediterranean: 0.06

You will be asked to compare the portions of your friend's ancestry from different regions of the world. How can you order the portions of your friend's ancestry from least to greatest.

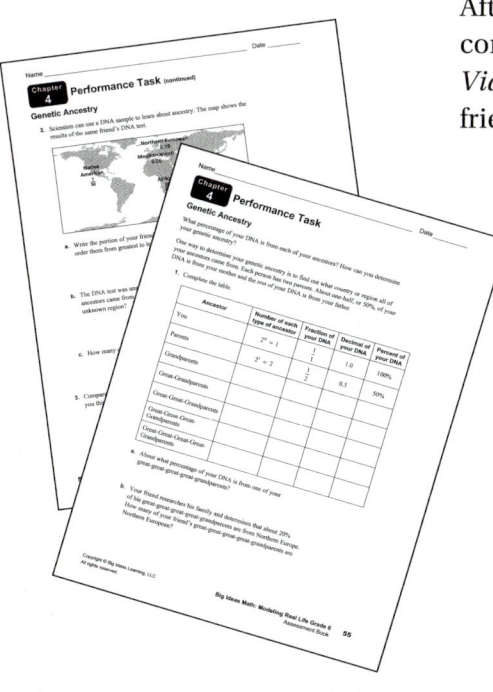

Getting Ready for Chapter 4

Chapter Exploration

1. **THE MEANING OF A WORD** Work with a partner. Match the "cent" word with its definition.

    a. centipede **b.** centimeter **c.** centennial **d.** centiliter
    e. centuple **f.** century **g.** centenarian **h.** centigram

    J. 100-year anniversary **K.** one-hundredth of a gram
    L. multiply by 100 **M.** bug with many legs
    N. one-hundredth of a liter **O.** one-hundredth of a meter
    P. 100 years **Q.** person who is at least 100 years old

2. **THE MEANING OF A WORD** Work with a partner. Describe a situation where you have seen the word *per* used and explain its use. In your own words, what do you think the word percent means? Use the word *percent* in a sentence.

Work with a partner. Represent the shaded portion of the square as (a) a fraction whose denominator is 100, (b) a fraction in simplest form, (c) a decimal, and (d) a percent.

3.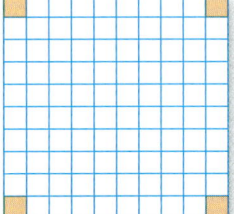

4.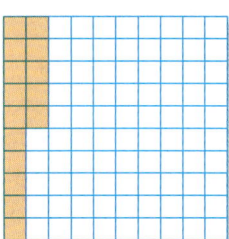

5.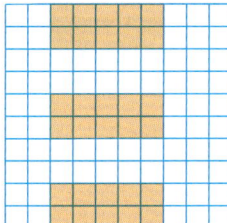

6.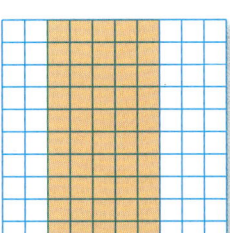

7.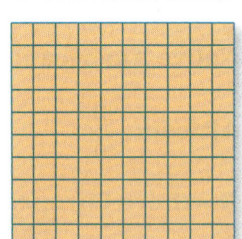

8.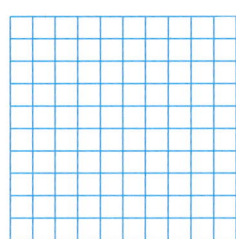

Vocabulary

The following terms are used in this chapter. Think about what each term might mean and record your thoughts.

percent   fraction   decimal

4.1 Percents and Fractions

Learning Target: Write percents as fractions and fractions as percents.

Success Criteria:
- I can draw models to represent fractions and percents.
- I can write percents as fractions.
- I can write equivalent fractions with denominators of 100.
- I can write fractions as percents.

The Meaning of a Word ▶ Percent

A century is 100 years.

A cent is one hundredth of a dollar.

In Mexico, a centavo is one hundredth of a peso.

Cent means *one hundred*, so **percent** means *per one hundred*. The symbol for percent is %.

EXPLORATION 1   Interpreting Models

Work with a partner. Write a percent, a fraction, and a ratio shown by each model. How are percents, fractions, and ratios related?

Math Practice

Use Definitions
How does the meaning of *percent* help you determine how to write the percent of each diagram that is shaded?

a.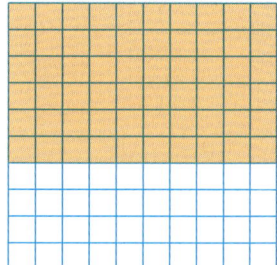

b.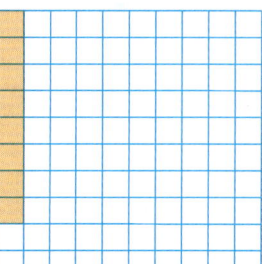

c.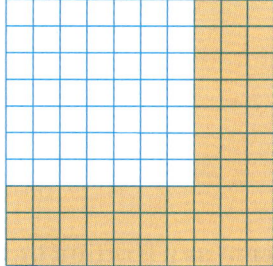

d.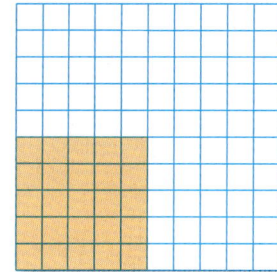

4.1 Lesson

Key Vocabulary 🔊
percent, p. 164

Key Idea

Writing Percents as Fractions

Words A **percent** is the value of a part-to-whole ratio where the whole is 100. So, you can write a percent as a fraction with a denominator of 100. The symbol % is used to denote a percent.

Numbers $60\% = 60 \text{ out of } 100 = \dfrac{60}{100}$

(part / per one hundred (whole) / part / per / one hundred (whole))

Algebra $n\% = \dfrac{n}{100}$

EXAMPLE 1 Writing Percents as Fractions

Fractions and percents that are equivalent represent the same number using different notations.

a. Write 35% as a fraction in simplest form.

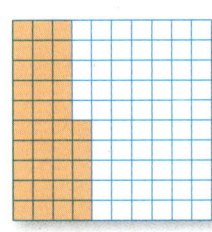

$35\% = \dfrac{35}{100}$ Write as a fraction with a denominator of 100.

$= \dfrac{7}{20}$ Simplify.

▸ So, $35\% = \dfrac{7}{20}$.

b. Write 100% as a fraction in simplest form.

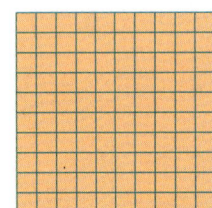

$100\% = \dfrac{100}{100}$ Write as a fraction with a denominator of 100.

$= 1$ Simplify.

▸ So, $100\% = 1$.

c. Write 174% as a mixed number in simplest form.

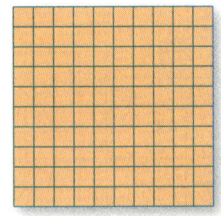 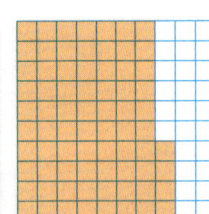

$174\% = \dfrac{174}{100}$ Write as a fraction with a denominator of 100.

$= \dfrac{87}{50}, \text{ or } 1\dfrac{37}{50}$ Simplify.

▸ So, $174\% = 1\dfrac{37}{50}$.

Try It Write the percent as a fraction or mixed number in simplest form.

1. 5%   2. 168%   3. 36%   4. 83%

 **Key Idea**

Writing Fractions as Percents

Words Write an equivalent fraction with a denominator of 100. Then write the numerator with the percent symbol.

Numbers $\dfrac{1}{4} = \dfrac{1 \times 25}{4 \times 25} = \dfrac{25}{100} = 25\%$

EXAMPLE 2 **Writing Fractions as Percents**

a. Write $\dfrac{3}{50}$ as a percent.

$\dfrac{3}{50} = \dfrac{3 \times 2}{50 \times 2} = \dfrac{6}{100} = 6\%$

Because 50 × 2 = 100, multiply the numerator and denominator by 2. Write the numerator with a percent symbol.

b. Write $\dfrac{9}{1000}$ as a percent.

$\dfrac{9}{1000} = \dfrac{9 \div 10}{1000 \div 10} = \dfrac{0.9}{100} = 0.9\%$

Because 1000 ÷ 10 = 100, divide the numerator and denominator by 10. Write the numerator with a percent symbol.

Try It Write the fraction or mixed number as a percent.

5. $\dfrac{31}{50}$ 6. $\dfrac{19}{20}$ 7. $\dfrac{1}{200}$ 8. $1\dfrac{1}{2}$

 **Self-Assessment** for Concepts & Skills

Solve each exercise. Then rate your understanding of the success criteria in your journal.

9. **WRITING A FRACTION** Write 40% as a fraction in simplest form. Draw a model that represents this fraction.

10. **WRITING A PERCENT** Write $\dfrac{9}{20}$ as a percent.

11. **WHICH ONE DOESN'T BELONG?** Which number does *not* belong with the other three? Explain your reasoning.

$\dfrac{10}{100}$ 10% $\dfrac{1}{10}$ 0.01

12. **OPEN-ENDED** Write three different fractions that are less than 40%.

13. **MP NUMBER SENSE** Can $1\dfrac{1}{4}$ be written as a percent? Explain.

EXAMPLE 3 Modeling Real Life

Midwestern United States

A drought affects 9 out of 12 midwestern states. What percent of the midwestern states are affected by the drought?

Write a fraction that represents the portion of midwestern states affected by the drought. Then convert the fraction to a percent.

$$\frac{\text{number of states affected}}{\text{total number of midwestern states}} = \frac{9}{12}$$

The denominator of 12 is not a factor of 100. But 4 is a factor of both 12 and 100. So, first write an equivalent fraction whose denominator is 4. Then convert that fraction to a percent.

$\frac{9}{12} = \frac{3}{4}$ Divide the numerator and denominator by 3.

$= \frac{75}{100}$ Multiply the numerator and denominator by 25.

$= 75\%$ Write the numerator with a percent symbol.

▸ So, 75% of the midwestern states are affected by the drought.

Check Reasonableness Notice that $\frac{9}{12} = \frac{9 \times 8}{12 \times 8} = \frac{72}{96}$. Because $\frac{72}{96} > \frac{72}{100}$, you know that $\frac{9}{12} > 72\%$. The answer is reasonable because $75\% > 72\%$. ✓

Self-Assessment for Problem Solving

Solve each exercise. Then rate your understanding of the success criteria in your journal.

14. You and a friend fill balloons with water. Two out of every 25 balloons pop while they are being filled. What percent of the balloons do *not* pop while they are being filled?

15. During the month of April, it rains 2 days for every 3 days that it does not rain. What percent of the days in April does it rain?

16. **DIG DEEPER!** There are 100 students in a band. Forty percent are 12 years old, $\frac{1}{2}$ are 13 years old, and the rest are 14 years old. Write the portion of the band for each age as a fraction and a percent.

4.1 Practice

Review & Refresh

Copy and complete the statement. Round to the nearest hundredth if necessary.

1. 3 km = ▢ m
2. 7 gal = ▢ c
3. 54 in. = ▢ yd
4. 4 mi ≈ ▢ km
5. 6.5 L ≈ ▢ gal
6. 45 lb ≈ ▢ kg

Divide.

7. 120 ÷ 12
8. 208 ÷ 4
9. 195 ÷ 15
10. 1428 ÷ 34
11. 8528 ÷ 164
12. 295 ÷ 8

Use the table to write the ratio.

13. birch to willow
14. trees : oak
15. oak : willow
16. birch to trees

Tree	Oak	Birch	Willow
Number	4	3	1

Concepts, Skills, & Problem Solving

USING TOOLS Use a 10-by-10 grid to model the percent. *(See Exploration 1, p. 163.)*

17. 10%
18. 55%
19. 45%

WRITING PERCENTS AS FRACTIONS Write the percent as a fraction or mixed number in simplest form.

20. 45%
21. 90%
22. 15%
23. 7%
24. 34%
25. 79%
26. 77.5%
27. 188%
28. 8%
29. 224%
30. 0.25%
31. 0.4%

32. **YOU BE THE TEACHER** Your friend writes 225% as a fraction. Is your friend correct? Explain your reasoning.

$$225\% = \frac{225}{100} = \frac{9}{4}$$

WRITING FRACTIONS AS PERCENTS Write the fraction or mixed number as a percent.

33. $\frac{1}{10}$
34. $\frac{1}{5}$
35. $\frac{11}{20}$
36. $\frac{1}{400}$
37. $\frac{2}{25}$
38. $\frac{27}{50}$
39. $\frac{3}{250}$
40. $\frac{18}{25}$
41. $1\frac{17}{20}$
42. $2\frac{41}{50}$
43. $3\frac{1}{200}$
44. $4\frac{7}{500}$

45. **YOU BE THE TEACHER** Your friend writes $\frac{14}{25}$ as a percent. Is your friend correct? Explain your reasoning.

$$\frac{14}{25} = \frac{14 \times 4}{25 \times 4} = \frac{56}{100} = 0.56\%$$

46. **MODELING REAL LIFE** During a 10-year period, 6 out of 30 Major League Baseball teams won the World Series. What percent of Major League Baseball teams won the World Series during the 10-year period?

47. **MODELING REAL LIFE** A doctor conducts an experiment to test new treatments for a medical condition. Of the 16 volunteers in the experiment, 4 do not receive any treatment. What percent of the volunteers do not receive any treatment?

48. **MP LOGIC** Of the students in your class, 12% are left-handed and the rest are right-handed. What fraction of the students are left-handed? Are there more right-handed or left-handed students? Explain.

49. **MP NUMBER SENSE** You have 125% of the tickets required for a prize. What fraction of the required tickets do you have? Do you need more tickets for the prize? Explain.

FINDING PERCENTS Find the percent.

50. 3 is what percent of 8?

51. 13 is what percent of 16?

52. 9 is what percent of 16?

53. 33 is what percent of 40?

54. **MODELING REAL LIFE** A survey asked students to choose their favorite social media website.

    a. What fraction of the students chose Website A?

    b. What percent of the students chose Website C?

Social Media Website	Number of Students
Website A	35
Website B	13
Website C	22
Website D	10

55. **DIG DEEPER!** The percent of the total area of the United States that is in each of four states is shown.

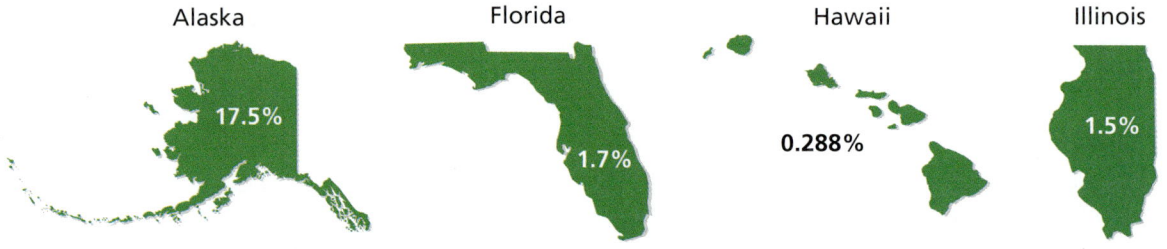

a. Write the percents as fractions in simplest form.

b. Compared to the map of Florida, is the map of Alaska the correct size? Explain your reasoning.

c. **RESEARCH** Which of the 50 states are larger than Illinois?

56. **CRITICAL THINKING** A school fundraiser raised 120% of last year's goal and 25% of this year's goal. Did the fundraiser raise more money this year? Explain your reasoning.

57. **CRITICAL THINKING** How can you use a 10-by-10 grid to model $\frac{1}{2}$%?

58. **MP REASONING** Write $\frac{1}{12}$ as a percent. Explain how you found your answer.

4.2 Percents and Decimals

Learning Target: Write percents as decimals and decimals as percents.

Success Criteria:
- I can draw models to represent decimals.
- I can explain why the decimal point moves when multiplying and dividing by 100.
- I can write percents as decimals.
- I can write decimals as percents.

EXPLORATION 1

Interpreting Models

Work with a partner. Write a percent and a decimal shown by each model. How are percents and decimals related?

a.

b.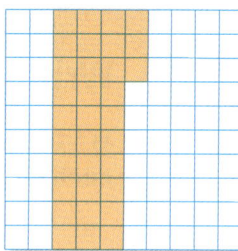

c.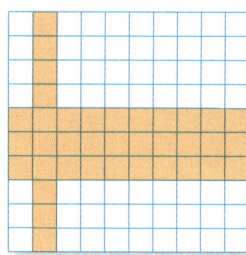

d.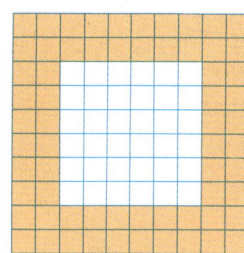

e.

f.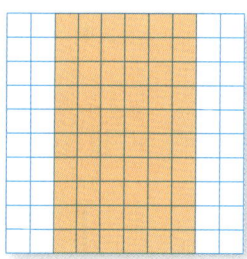

Math Practice

Look for Patterns

How does the decimal point move when you rewrite a percent as a decimal and a decimal as a percent? Explain why this occurs.

g.

Section 4.2   Percents and Decimals   169

4.2 Lesson

 Key Idea

Writing Percents as Decimals

Words Remove the percent symbol. Then divide by 100, which moves the decimal point two places to the left.

Numbers 23% = 23.%̶ = 0.23

Model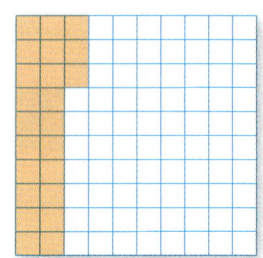

> Notice that 23 *percent* is equivalent to 23 *hundredths*.

EXAMPLE 1 **Writing Percents as Decimals**

Write (a) 52% and (b) 7% as decimals. Use a model to represent each decimal.

a. 52% = 52.%̶ = 0.52

b. 7% = 07.%̶ = 0.07

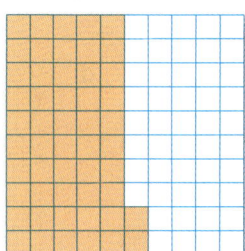 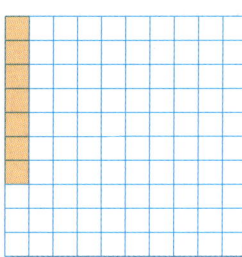

> When writing a percent as a decimal, you may need to include one or more zeros so that the digits are in the correct place-value positions.

Try It Write the percent as a decimal. Use a model to represent the decimal.

1. 24% 2. 3% 3. 107% 4. 92.5%

 Key Idea

Writing Decimals as Percents

Words Multiply by 100, which moves the decimal point two places to the right. Then add a percent symbol.

Numbers 0.36 = 0.36 = 36%

Model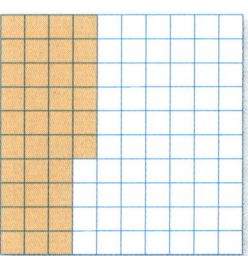

170   Chapter 4   Percents

EXAMPLE 2 Writing Decimals as Percents

a. Write 0.47 as a percent.

$0.47 = 0.47 = 47\%$

b. Write 0.663 as a percent.

$0.663 = 0.663 = 66.3\%$

c. Write 1.8 as a percent.

$1.8 = 1.80 = 180\%$

d. Write 0.009 as a percent.

$0.009 = 0.009 = 0.9\%$

Try It Write the decimal as a percent.

5. 0.94   6. 1.2   7. 0.316   8. 0.005

EXAMPLE 3 Writing a Fraction as a Percent and a Decimal

On a math test, you earn 92 out of a possible 100 points. Which of the following is *not* another way of expressing 92 out of 100?

A. $\dfrac{23}{25}$ B. 92%   C. $\dfrac{17}{20}$ D. 0.92

Write "92 out of 100" as a fraction, a decimal, and a percent.

$$92 \text{ out of } 100 = \dfrac{92}{100} \begin{cases} = 92\% & \text{Eliminate Choice B.} \\ = \dfrac{23}{25} & \text{Eliminate Choice A.} \\ = 0.92 & \text{Eliminate Choice D.} \end{cases}$$

▶ So, the correct answer is **C**.

Try It

9. **WHAT IF?** You earn 90 out of a possible 100 points on the test. Write "90 out of 100" as a fraction, a decimal, and a percent.

Self-Assessment for Concepts & Skills

Solve each exercise. Then rate your understanding of the success criteria in your journal.

WRITING PERCENTS AS DECIMALS Write the percent as a decimal. Use a model to represent the decimal.

10. 32%   11. 54.5%   12. 108%

WRITING DECIMALS AS PERCENTS Write the decimal as a percent.

13. 0.71   14. 0.052   15. 9.66

16. **WRITING** Explain why the decimal point moves left when dividing a number by 100.

Section 4.2   Percents and Decimals   171

EXAMPLE 4 Modeling Real Life

The figure shows the portions of ultraviolet (UV) rays reflected by four different surfaces. How many times more UV rays are reflected by water than by sea foam?

The diagram shows that sea foam reflects 25% of UV rays and water reflects $\frac{21}{25}$ of UV rays. First, write 25% and $\frac{21}{25}$ as decimals.

Sea foam: $25\% = 25.\% = 0.25$ **Water:** $\frac{21}{25} = \frac{84}{100} = 0.84$

Next, divide 0.84 by 0.25.

$0.25 \overline{)0.84} \longrightarrow 25 \overline{)84.00}$

```
        3.36
  25 )84.00
     -75
       9 0
      -7 5
       1 50
      -1 50
          0
```

Another Method First, write 25% as a fraction:
$25\% = \frac{25}{100} = \frac{1}{4}$.
Then divide $\frac{21}{25}$ by $\frac{1}{4}$.

$\frac{21}{25} \div \frac{1}{4} = \frac{21}{25} \cdot 4$

$= \frac{84}{25}$

$= 3\frac{9}{25}$, or 3.36 ✓

So, water reflects 3.36 times more UV rays than sea foam.

Self-Assessment for Problem Solving

Solve each exercise. Then rate your understanding of the success criteria in your journal.

17. Write the amount of occupied space on the computer as a percent.

Volume	Capacity	Free Space
(C:)	150 GB	132 GB

18. *Salinity* is a measure of the salt content of a body of water. One researcher measures the salinity of the Indian Ocean as 3.2%. Another researcher measures the salinity of the Dead Sea as 34%. A bucket of water from the Indian Ocean contains 56 grams of salt. How much salt is contained in the same amount of water from the Dead Sea? Justify your answer.

4.2 Practice

▶ Review & Refresh

Write the fraction or mixed number as a percent.

1. $\dfrac{7}{50}$
2. $\dfrac{2}{5}$
3. $\dfrac{1}{250}$
4. $5\dfrac{2}{25}$

Represent the ratio relationship using a graph.

5.
Bags of Gravel	6	12	18
Cost (dollars)	15	30	45

6.
Teachers	1	2	3
Students	12	24	36

Multiply. Write the answer in simplest form.

7. $\dfrac{3}{7} \times \dfrac{2}{9}$
8. $\dfrac{5}{12} \times \dfrac{1}{5}$
9. $4\dfrac{2}{3} \times \dfrac{9}{10}$
10. $2\dfrac{1}{6} \times 3\dfrac{1}{2}$

▶ Concepts, Skills, & Problem Solving

INTERPRETING MODELS Write the percent and the decimal shown by the model. *(See Exploration 1, p. 169.)*

11.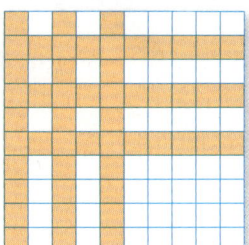

12.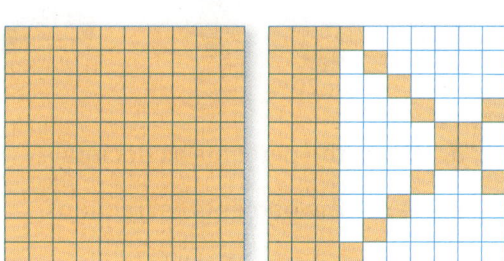

WRITING PERCENTS AS DECIMALS Write the percent as a decimal.

13. 78%
14. 55%
15. 18.5%
16. 57.4%
17. 33%
18. 9%
19. 47.63%
20. 91.25%
21. 166%
22. 217%
23. 0.06%
24. 0.034%

WRITING DECIMALS AS PERCENTS Write the decimal as a percent.

25. 0.74
26. 0.52
27. 0.89
28. 0.768
29. 0.99
30. 0.49
31. 0.487
32. 0.128
33. 3.68
34. 5.12
35. 0.0371
36. 0.0046

37. **YOU BE THE TEACHER** Your friend writes 0.86 as a percent. Is your friend correct? Explain your reasoning.

   0.86 = 00.86 = 0.0086%

Section 4.2 Percents and Decimals 173

MATCHING Match the decimal with its equivalent percent.

38. 0.42   **39.** 4.02   **40.** 0.042   **41.** 0.0402

    A. 4.02%   **B.** 42%   **C.** 4.2%   **D.** 402%

42. MODELING REAL LIFE About 80% of the precipitation that enters Crater Lake falls directly on the surface of the lake. Write this percent as a decimal.

43. MODELING REAL LIFE About 0.34 of the length of a cat is its tail. Write this decimal as a percent.

44. OPEN-ENDED Write three different decimals that are between 10% and 20%.

WRITING PERCENTS AS FRACTIONS AND DECIMALS Write the percent as a fraction in simplest form and as a decimal.

45. 36%   **46.** 23.5%   **47.** 16.24%

48. DIG DEEPER! The percents of students who travel to school by car, bus, and bicycle are shown for a school of 825 students.

Car: 20%   School bus: 48%   Bicycle: 8%

   a. Write the percents as decimals.

   b. Write the percents as fractions.

   c. What percent of students use another method to travel to school?

   d. RESEARCH Make a bar graph that represents how the students in your class travel to school.

49. LOGIC A running back was the MVP (most valuable player) in 0.14 of the first 50 Super Bowls.

   a. What percent of the MVPs were running backs?

   b. What fraction of the MVPs were *not* running backs?

50. CHOOSING A METHOD Students in a class were asked to tell their favorite color.

   a. What percent said red, blue, or yellow?

   b. How many times more students said red than yellow?

   c. Use two methods to find the percent of students who said green. Which method do you prefer? Explain.

4.3 Comparing and Ordering Fractions, Decimals, and Percents

Learning Target: Compare and order fractions, decimals, and percents.

Success Criteria:
- I can rewrite a group of fractions, decimals, and percents using the same representation.
- I can explain how to compare fractions, decimals, and percents.
- I can order fractions, decimals, and percents from least to greatest.

EXPLORATION 1

Using a Number Line to Order Numbers

Work with three partners. Create a number line on the floor. Have your group stand on the number line to represent the four numbers in each list. Use the results to order each list of numbers from least to greatest. How did you know where to stand?

Math Practice

Find Entry Points
What strategies can you use to determine where to stand?

a.
| 0.25 | 0.9% | 40% | 0.5 |

b.
| 0% | $\frac{3}{4}$ | 30% | $\frac{1}{20}$ |

c.
| 100% | 0.125 | 75% | $\frac{3}{10}$ |

d.
| 12.5% | 1.02 | $\frac{1}{100}$ | 25% |

e.
| 0.3 | $\frac{1}{8}$ | 4% | 0.75 |

f.
| $\frac{51}{50}$ | 105% | 1.5 | $\frac{9}{10}$ |

Section 4.3 Comparing and Ordering Fractions, Decimals, and Percents 175

4.3 Lesson

When comparing and ordering fractions, decimals, and percents, write the numbers as all fractions, all decimals, or all percents.

EXAMPLE 1 **Comparing Fractions, Decimals, and Percents**

a. Which is greater, $\frac{3}{20}$ or 16%?

Write $\frac{3}{20}$ as a percent: $\frac{3}{20} = \frac{3 \times 5}{20 \times 5} = \frac{15}{100} = 15\%$.

▸ 15% is less than 16%. So, 16% is the greater number.

b. Which is greater, 79% or 0.08?

Write 79% as a decimal: $79\% = 79.\% = 0.79$.

▸ 0.79 is greater than 0.08. So, 79% is the greater number.

It is usually easier to order decimals or percents than to order fractions.

Try It Tell which number is greater.

1. 25%, $\frac{7}{25}$
2. 0.49, 94%

EXAMPLE 2 **Ordering Fractions, Decimals, and Percents**

Order 3%, $\frac{1}{50}$, 0.005, 0.8%, and 0.025 from least to greatest.

One way to order the numbers is to first write each number as a percent.

$\frac{1}{50} = \frac{1 \times 2}{50 \times 2} = \frac{2}{100} = 2\%$ $0.005 = 0.005 = 0.5\%$ $0.025 = 0.025 = 2.5\%$

Graph the percents on a number line.

▸ So, the order from least to greatest is 0.005, 0.8%, $\frac{1}{50}$, 0.025, and 3%.

Remember
To order numbers from least to greatest, write them as they appear on a number line from left to right.

Try It Order the numbers from least to greatest.

3. $\frac{3}{10}$, 15%, 0.2, $\frac{3}{8}$, 0.09
4. 100%, 0.95, 1.2, $\frac{5}{4}$, 110%

EXAMPLE 3 **Comparing Numbers to Solve a Real-Life Problem**

You, your sister, and a friend each take the same number of shots at a soccer goal. You make 72% of your shots, your sister makes $\frac{19}{25}$ of her shots, and your friend makes 0.67 of his shots. Who made the fewest shots?

To determine who made the fewest shots, find who made the least portion of their shots. One way to do this is to first write each number as a decimal.

You: $72\% = 72.\% = 0.72$ **Sister:** $\frac{19}{25} = \frac{19 \times 4}{25 \times 4} = \frac{76}{100} = 0.76$

Graph the decimals on a number line.

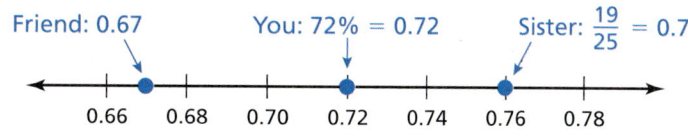

▶ 0.67 is the least number. So, your friend made the fewest shots.

Try It

5. **WHAT IF?** Your friend makes $\frac{3}{4}$ of his shots. Did your friend make more shots than you? your sister?

Self-Assessment for Concepts & Skills

Solve each exercise. Then rate your understanding of the success criteria in your journal.

Fraction	Decimal	Percent
$\frac{18}{25}$	0.72	
$\frac{17}{20}$		85%
$\frac{13}{50}$		
	0.62	
		45%

6. **NUMBER SENSE** Copy and complete the table.

7. **NUMBER SENSE** How would you decide whether $\frac{3}{5}$ or 59% is greater? Explain.

COMPARING NUMBERS Tell which number is greater.

8. 33%, 0.34   9. 0.85, $\frac{4}{5}$ 10. $\frac{9}{50}$, 17%

ORDERING NUMBERS Order the numbers from least to greatest.

11. 12%, 0.1, $\frac{4}{25}$ 12. 1.35, 125%, $\frac{6}{5}$, 1.5, 130%

Section 4.3   Comparing and Ordering Fractions, Decimals, and Percents

EXAMPLE 4 Modeling Real Life

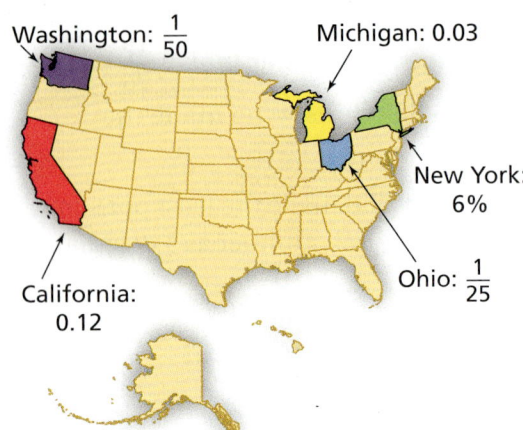

The map shows the portions of the U.S. population that live in five states. List the five states in order by population from least to greatest.

Begin by writing each portion as a fraction, a decimal, and a percent.

State	Fraction	Decimal	Percent
Michigan	$\frac{3}{100}$	0.03	3%
New York	$\frac{6}{100}$	0.06	6%
Washington	$\frac{1}{50}$	0.02	2%
California	$\frac{12}{100}$	0.12	12%
Ohio	$\frac{1}{25}$	0.04	4%

Graph the percent for each state on a number line.

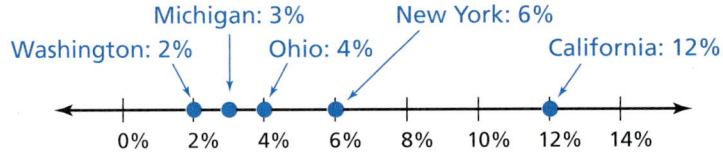

▸ The states in order by population from least to greatest are Washington, Michigan, Ohio, New York, and California.

Self-Assessment for Problem Solving

Solve each exercise. Then rate your understanding of the success criteria in your journal.

Ring	Portion that is Copper
Red gold	25%
Pink gold	0.2
Rose gold	$\frac{9}{40}$

13. The table shows the portions of copper in three rings. Which ring has the highest portion of copper?

14. **DIG DEEPER!** The table shows the results of five teams competing in a scavenger hunt. List the five teams in order by the portion of items collected from least to greatest. What is the minimum number of items in the scavenger hunt? Explain.

Team	1	2	3	4	5
Portion Collected	$\frac{3}{4}$	0.8	77.5%	0.825	$\frac{13}{20}$

178   Chapter 4   Percents

4.3 Practice

▶ Review & Refresh

Write the percent as a decimal.

1. 12%
2. 98.37%
3. 0.046%

Find the missing value(s) in the ratio table. Then write the equivalent ratios.

4.
Cashews	5	
Almonds	4	36

5.
Cost (dollars)	2		13.5
Time (minutes)	50	75	

Multiply.

6. 5.3 × 6
7. 18.1 × 9
8. 12.43 × 1.51
9. 0.852 × 6.7

Find the GCF of the numbers.

10. 15, 36
11. 51, 85
12. 88, 112

▶ Concepts, Skills, & Problem Solving

USING A NUMBER LINE Use a number line to order the numbers from least to greatest. (See Exploration 1, p. 175.)

13. 80%, 0.65, $\frac{3}{10}$, 40%
14. 0.27, 20%, $\frac{1}{25}$, 9%
15. $\frac{1}{8}$, 0.25, 15%, $\frac{27}{100}$

COMPARING NUMBERS Tell which number is greater.

16. 0.9, 95%
17. 20%, 0.02
18. $\frac{37}{50}$, 37%
19. 50%, $\frac{13}{25}$
20. 0.086, 86%
21. 76%, 0.67
22. 60%, $\frac{5}{8}$
23. 0.12, 1.2%
24. 17%, $\frac{4}{25}$
25. 140%, 0.14
26. $\frac{3}{8}$, 30%
27. 80%, $\frac{7}{10}$

ORDERING NUMBERS Order the numbers from least to greatest.

28. 38%, $\frac{8}{25}$, 0.41
29. 68%, 0.63, $\frac{13}{20}$
30. $\frac{43}{50}$, 0.91, $\frac{7}{8}$, 84%
31. 0.15%, $\frac{3}{20}$, 0.015
32. 2.62, $2\frac{2}{5}$, 26.8%, 2.26, 271%
33. $\frac{87}{200}$, 0.44, 43.7%, $\frac{21}{50}$

34. **NUMBER SENSE** You answer 21 out of 25 questions correctly on a test. Do you reach your goal of answering at least 80% of the questions correctly?

Section 4.3 Comparing and Ordering Fractions, Decimals, and Percents 179

35. MODELING REAL LIFE The table shows the approximate portions of the world population that live in four countries. Order the countries by population from least to greatest.

Country	Brazil 	India 	Russia 	United States 
Portion of World Population	2.8%	$\dfrac{7}{40}$	$\dfrac{1}{50}$	0.044

MP PRECISION Order the numbers from least to greatest.

36. 66.1%, 0.66, $\dfrac{133}{200}$, 0.667

37. $\dfrac{111}{500}$, 21%, 0.211, $\dfrac{11}{50}$

MATCHING Tell which letter shows the graph of the number.

38. $\dfrac{2}{5}$ **39.** 45.2%   **40.** 0.435   **41.** $\dfrac{89}{200}$

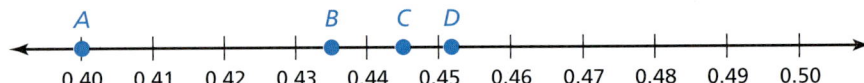

42. MP PRECISION The Tour de France is a bicycle road race. The whole race is made up of 21 small races called *stages*. The table shows how several stages compare to the whole Tour de France in a recent year. Order the stages by distance from shortest to longest.

Stage	1	7	8	17	21
Portion of Total Distance	$\dfrac{11}{200}$	0.044	$\dfrac{6}{125}$	0.06	4%

43. MP PRECISION The table shows the portions of the day that several animals sleep.

a. Order the animals by sleep time from least to greatest.

b. Estimate the portion of the day that you sleep.

c. Where do you fit on the ordered list?

Animal	Portion of Day Sleeping
Dolphin	0.433
Lion	56.3%
Rabbit	$\dfrac{19}{40}$
Squirrel	$\dfrac{31}{50}$
Tiger	65.8%

44. MP NUMBER SENSE Tell what whole number you can substitute for *a* in each list so the numbers are ordered from least to greatest. If there is none, explain why.

a. $\dfrac{1}{a}, \dfrac{a}{20}, 28\%$ b. $\dfrac{3}{a}, \dfrac{a}{5}, 75\%$

180   Chapter 4   Percents

4.4 Solving Percent Problems

Learning Target: Find a percent of a quantity and solve percent problems.

Success Criteria:
- I can represent percents of numbers using an equation, a ratio table, or a model.
- I can find percents of numbers.
- I can find the whole given a part and the percent.

EXPLORATION 1

Using Percent Models

Work with a partner.

a. Find the missing values. What does the model represent?

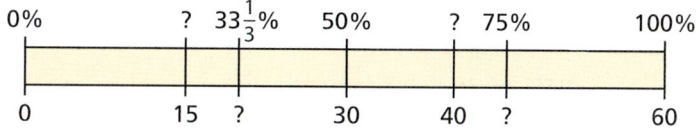

b. Label at least three percents and their corresponding numbers on the model below. How do you know you are correct?

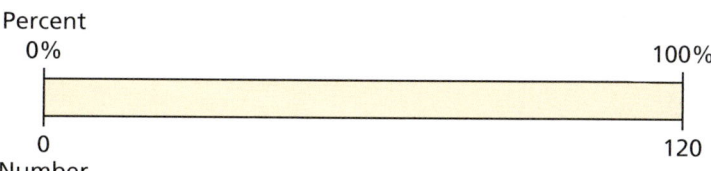

EXPLORATION 2

Solving a Percent Problem

Work with a partner. You purchase a national parks annual pass for 75% of the full price of the pass.

a. Suppose you know the full price or the discounted price. How can you find the other price? Compare your answers with other students in your class.

b. Suppose the full price of the pass is $80. How can you use a percent model to find the purchase price?

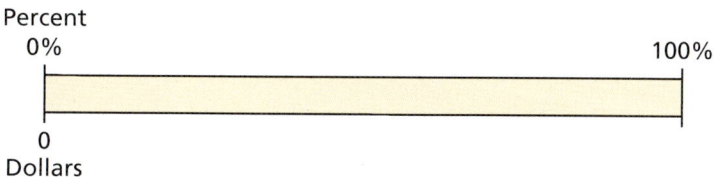

Math Practice

Consider Simpler Forms

Suppose you know 1% and 10% of the full price. How can you use these values to find the purchase price?

Section 4.4 Solving Percent Problems 181

4.4 Lesson

 **Key Idea**

Finding the Percent of a Number

Words Write the percent as a fraction or decimal. Then multiply by the whole. The percent times the whole equals the part.

Numbers 20% of 60 is 12.

$0.2 \times 60 = 12$

$\frac{1}{5} \times 60 = 12$

Model

Percent: 0%, 20%, 40%, 60%, 80%, 100%
Number: 0, 12, 24, 36, 48, 60

EXAMPLE 1 — Finding the Percent of a Number

25% of 40 is what number?

$25\% \text{ of } 40 = \frac{1}{4} \times 40$ Write the percent as a fraction and multiply.

$= \frac{40}{4} = 10$ Simplify.

So, 25% of 40 is 10.

Percent: 0%, 25%, 50%, 75%, 100%
Number: 0, 10, 20, 30, 40

You can use mental math to check your answer in Example 1.

10% of 40 = 4
5% of 40 = 2
So, 25% of 40 is
4 + 4 + 2 = 10.

Try It Find the percent of the number.

1. 90% of 20
2. 75% of 32

Because $n\%$ means n per 100, you can also solve percent problems using part-to-whole ratios.

EXAMPLE 2 — Finding the Percent of a Number Using a Ratio Table

60% of 150 is what number?

Use a ratio table to find the part. Let one row be the *part*, and let the other row be the *whole*. Find an equivalent ratio with 150 as the whole.

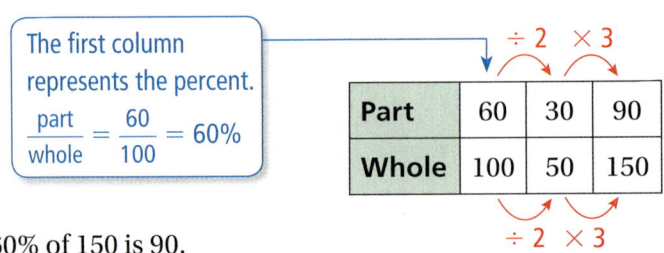

The first column represents the percent.
$\frac{\text{part}}{\text{whole}} = \frac{60}{100} = 60\%$

Part	60	30	90
Whole	100	50	150

÷2   ×3

So, 60% of 150 is 90.

Try It Find the percent of the number.

3. 10% of 110
4. 30% of 75

182   Chapter 4   Percents

You can use a related division equation to find the whole given the part and the percent.

 **Key Idea**

Finding the Whole

Write the percent as a fraction or decimal. Then divide the part by the fraction or decimal.

Words The part divided by the percent equals the whole.

Numbers 20% of 60 is 12.

$$\frac{1}{5} \times 60 = 12 \longrightarrow 12 \div \frac{1}{5} = 60$$

Multiplication equation   Related division equation

EXAMPLE 3   Finding the Whole

50% of what number is 48?

The question can be represented by the multiplication equation $50\% \times \square = 48$. By definition, $\square = 48 \div 50\%$.

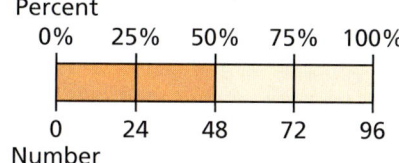

$$48 \div 50\% = 48 \div \frac{1}{2}$$ Write the percent as a fraction.

$$= 96$$ Multiply 48 by the reciprocal of $\frac{1}{2}$, which is 2.

 So, 50% of 96 is 48.

Try It   Find the whole.

5. 15% of what number is 9?   **6.** 5% of what number is 10?

EXAMPLE 4   Finding the Whole Using a Ratio Table

120% of what number is 72?

Use a ratio table to find the whole. Find an equivalent ratio with 72 as the part.

The first column represents the percent.
$\frac{\text{part}}{\text{whole}} = \frac{120}{100} = 120\%$

÷ 20   × 12

Part	120	6	72
Whole	100	5	60

÷ 20   × 12

 So, 120% of 60 is 72.

Try It   Find the whole.

7. 62% of what number is 31?   **8.** 125% of what number is 50?

Section 4.4   Solving Percent Problems   183

EXAMPLE 5 Solving a Percent Problem

The width of a rectangular room is 80% of its length. What is the area of the room?

Find 80% of 15 feet to determine the width of the room.

80% of 15 = 0.8 × 15
= 12

The width is 12 feet.

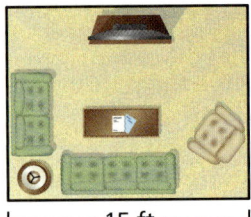

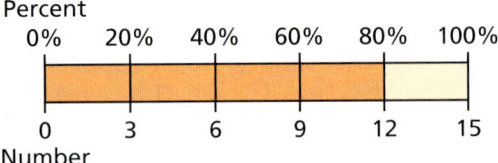

Use the formula for the area of a rectangle.

Area = ℓw
= 15(12)
= 180

So, the area of the room is 180 square feet.

Try It

9. The width of a rectangular stage is 55% of its length. The stage is 120 feet long. What is the area of the stage?

Self-Assessment for Concepts & Skills

Solve each exercise. Then rate your understanding of the success criteria in your journal.

10. **DIFFERENT WORDS, SAME QUESTION** Which is different? Find "both" answers.

    What is twenty percent of 30?   What is one-fifth of 30?

    Twenty percent of what number is 30?   What is two-tenths of 30?

11. **FINDING THE PERCENT OF A NUMBER** Find 12% of 75.

12. **FINDING THE WHOLE** 35% of what number is 21?

13. **MP NUMBER SENSE** If 52 is 130% of a number, is the number greater than or less than 52? Explain.

14. **MP STRUCTURE** How can you find 10% of any number without multiplying or dividing? Explain your reasoning.

EXAMPLE 6   Modeling Real Life

You win an online auction for concert tickets. Your winning bid is 60% of your maximum bid. How much more were you willing to pay for the tickets than you actually paid?

A. $72   **B.** $80

C. $120   **D.** $200

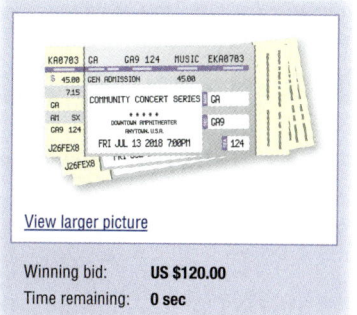

Understand the problem. You are given the winning bid and the percent of your maximum bid represented by the winning bid. You are asked to find how much more you were willing to pay for the tickets than you actually paid.

Make a plan. Your maximum bid is the whole and your winning bid is the part. Create a model using the fact that 60% of the whole is $120 to find the maximum bid. Then subtract the winning bid from the maximum bid to determine how much more you were willing to pay.

Solve and check.

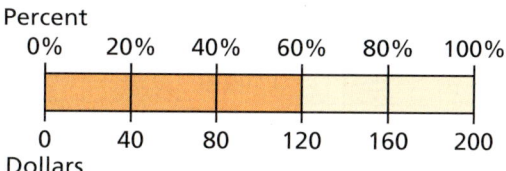

Your maximum bid is $200, and your winning bid is $120. So, you were willing to pay $200 − $120 = $80 more for the tickets.

> The correct answer is **B**.

Look Back
Verify that the additional $80 you were willing to pay is 100% − 60% = 40% of the maximum bid.

The model shows that 40% of the maximum bid is $80. ✓

Self-Assessment for Problem Solving

Solve each exercise. Then rate your understanding of the success criteria in your journal.

15. You raise $420 during a fundraising event. The amount of money that you raise is 120% of your goal. How much more did you raise than your goal?

16. A shirt is on sale for 60% of the original price. The original price is $28 more than the sale price. What was the original price?

17. You have a meal at a restaurant. The sales tax is 8%. You leave a tip for the waitress that is 20% of the pretax price. You spend a total of $23.04. What is the pretax price of the meal?

4.4 Practice

Go to *BigIdeasMath.com* to get HELP with solving the exercises.

▶ Review & Refresh

Order the numbers from least to greatest.

1. $\frac{1}{8}$, 35%, 0.33
2. 0.3, $\frac{9}{25}$, 0.35, 33%
3. $\frac{13}{50}$, 22%, 0.28, $\frac{1}{5}$, 0.41

Write the percent as a fraction or mixed number in simplest form.

4. 65%
5. 0.45%
6. 110%

Divide.

7. $19.2 \div 1.6$
8. $0.61 \overline{)0.244}$
9. $0.9 \overline{)0.558}$
10. $4.65 \div 0.003$

The tape diagram represents the ratio of the time you spend online to the time your friend spends online. You are online for 30 minutes. How many minutes does your friend spend online?

11. You / Friend
12. You / Friend

▶ Concepts, Skills, & Problem Solving

USING TOOLS An annual pass to a park costs $120. Use a percent model to find the given percent of the full price of the annual pass. *(See Exploration 2, p. 181.)*

13. 25%
14. 50%
15. 200%

FINDING THE PERCENT OF A NUMBER Find the percent of the number. Explain your method.

16. 20% of 60
17. 10% of 40
18. 50% of 70
19. 30% of 30
20. 10% of 90
21. 15% of 20
22. 25% of 50
23. 5% of 60
24. 30% of 70
25. 75% of 48
26. 45% of 45
27. 92% of 19
28. 40% of 60
29. 38% of 22
30. 70% of 20
31. 87% of 55
32. 140% of 60
33. 120% of 33
34. 175% of 54
35. 250% of 146

36. **MODELING REAL LIFE** The tail of the spider monkey is 64% of the length shown. What is the length of the tail?

37. **MP PROBLEM SOLVING** A family pays $45 each month for cable television. The cost increases 7%.
    a. How many dollars is the increase?
    b. What is the new monthly cost?

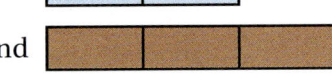

186 Chapter 4 Percents

38. YOU BE THE TEACHER Your friend finds 40% of 75. Is your friend correct? Explain your reasoning.

$$40\% \text{ of } 75 = \frac{2}{5} \times 75 = 30$$

FINDING THE WHOLE Find the whole. Explain your method.

39. 10% of what number is 14?

40. 20% of what number is 18?

41. 25% of what number is 21?

42. 75% of what number is 27?

43. 15% of what number is 12?

44. 85% of what number is 17?

45. 140% of what number is 35?

46. 160% of what number is 32?

47. 125% of what number is 25?

48. 175% of what number is 42?

49. YOU BE THE TEACHER Your friend answers the question "20% of what number is 5?" Is your friend correct? Explain your reasoning.

$$5 \div 20\% = \frac{5}{20} = \frac{1}{4}$$

50. PROBLEM SOLVING You have a coupon for a restaurant. You save $3 on a meal. What was the original cost of the meal?

51. PROBLEM SOLVING The results of a survey are shown at the right. In the survey, 12 students said that they would like to learn French. How many of the students surveyed would like to learn Spanish?

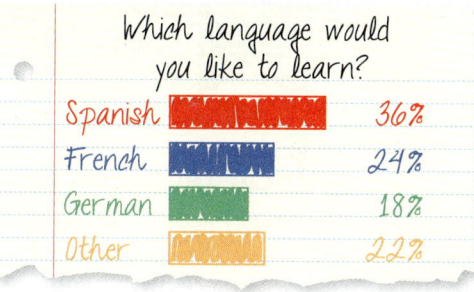

52. MODELING REAL LIFE A sixth grader weighs 90 pounds, which is 120% of what he weighed in fourth grade. How much did he weigh in fourth grade?

53. LOGIC In an asteroid field, 75% of the asteroids are *carbonaceous* asteroids. There are 375,000 carbonaceous asteroids in the asteroid field. How many asteroids are *not* carbonaceous?

54. DIG DEEPER! A bottle contains 20 fluid ounces of lotion and sells for $5.80. The 20-fluid-ounce bottle contains 125% of the lotion in the next smallest size, which sells for $5.12. Which is the better buy? Explain.

COMPARING PERCENTS Copy and complete the statement using <, >, or =.

55. 80% of 60 ◯ 60% of 80

56. 20% of 30 ◯ 30% of 40

57. 120% of 5 ◯ 0.8% of 250

58. 85% of 40 ◯ 25% of 136

59. TIME How many minutes is 40% of 2 hours?

60. LENGTH How many inches is 78% of 3 feet?

61. GEOMETRY The width of the rectangle is 75% of its length.

   a. What is the area of the rectangle?

   b. The length of the rectangle is doubled. What percent of the length is the width now? Explain your reasoning.

24 in.

62. PRECISION To pass inspection, a new basketball should bounce between 68% and 75% of the starting height. A new ball is dropped from 6 feet and bounces back 4 feet 1 inch. Does the ball pass inspection? Explain.

63. REASONING You know that 15% of a number n is 12. How can you use this to find 30% of n? 45% of n? Explain.

64. REASONING You have a coupon for 10% off the sale price of a surfboard. Which is the better buy? Explain your reasoning.

- 40% off the regular price
- 30% off the regular price and then 10% off the sale price

65. CRITICAL THINKING Consider two different numbers x and y. Is x% of y the same as y% of x? Justify your answer.

66. GEOMETRY Square $ABCD$ and Square $EFGH$ both have side lengths of 8 inches. The squares overlap and form Rectangle $ABGH$, which has a length of 10 inches. What percent of Rectangle $ABGH$ is shaded purple?

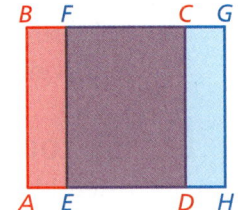

67. NUMBER SENSE On three 150-point geography tests, you earned grades of 88%, 94%, and 90%. The final test is worth 250 points. What *percent* do you need on the final to earn 93% of the total points on all tests?

Connecting Concepts

Using the Problem-Solving Plan

1. During a football game, a total of 63 points are scored by the two teams. Team A scores 80% of the number of points that Team B scores. What is the final score of the game?

   Understand the problem. You know that Team A's score is 80% of Team B's score, and the total points scored is 63. You are asked to find the final score of the football game.

   Make a plan. Because 80% means 80 per 100, write the relationship between the scores of the teams as a ratio. Then represent the situation using a tape diagram. Use the total points to find the value of each part of the tape diagram and the final score of the game.

   Solve and check. Use the plan to solve the problem. Then check your solution.

2. A pen at a pet store contains male and female guinea pigs. The ratio of female guinea pigs to male guinea pigs is 7 to 3. Find the percent of guinea pigs in the pen that are male. Justify your answer.

3. You multiply two numbers. The first number, 21, is 6.25% of the product. What is the second number? Justify your answer.

4. You have a bag containing dollar coins, dimes, and pennies. The bag contains 40 coins. The number of dollar coins is 20% of the total number of coins. The number of pennies is $\frac{7}{9}$ of the number of dimes. How much money is in the bag?

Performance Task

Genetic Ancestry

At the beginning of this chapter, you watched a STEAM video called "Chargaff's Rules." You are now ready to complete the performance task for this video, available at *BigIdeasMath.com*. Be sure to use the problem-solving plan as you work through the performance task.

4 Chapter Review

Go to *BigIdeasMath.com* to download blank graphic organizers.

Review Vocabulary

Write the definition and give an example of the vocabulary term.

percent, *p. 164*

Graphic Organizers

You can use a **Four Square** to organize information about a concept. Each of the four squares can be a category, such as definition, vocabulary, example, non-example, words, algebra, table, numbers, visual, graph, or equation. Here is an example of a Four Square for *finding the percent of a number*.

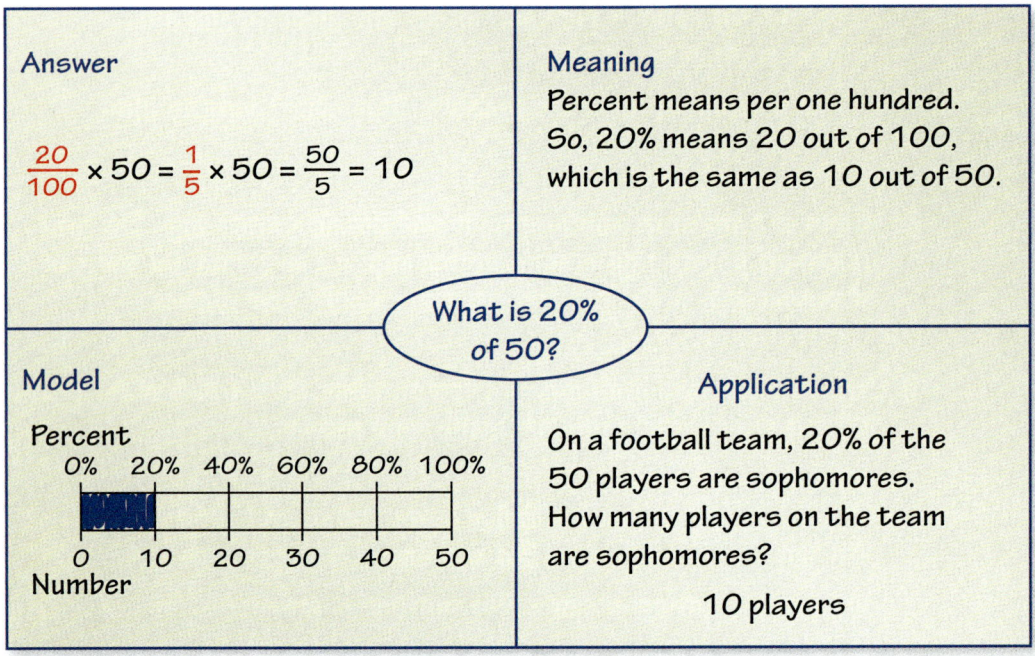

Choose and complete a graphic organizer to help you study the concept.

1. percent
2. writing percents as fractions
3. writing fractions as percents
4. writing percents as decimals
5. writing decimals as percents
6. finding the whole

"Sorry, but I have limited space in my **Four Square**. I needed pet names with only three letters."

▶ Chapter Self-Assessment

As you complete the exercises, use the scale below to rate your understanding of the success criteria in your journal.

1	2	3	4
I do not understand.	I can do it with help.	I can do it on my own.	I can teach someone else.

4.1 Percents and Fractions (pp. 163–168)

Learning Target: Write percents as fractions and fractions as percents.

Write the percent as a fraction or mixed number in simplest form.

1. 12%
2. 88%
3. 0.8%
4. 127%
5. 2.5%
6. 18%

Write the fraction or mixed number as a percent.

7. $\dfrac{3}{5}$
8. $1\dfrac{18}{25}$
9. $1\dfrac{21}{50}$
10. $\dfrac{14}{35}$
11. $5\dfrac{7}{25}$
12. $\dfrac{7}{400}$

13. Write a fraction in simplest form that is greater than 43% and less than 47%.

14. Write a percent that is greater than $3\dfrac{3}{4}$ and less than $3\dfrac{4}{5}$.

15. Your computer displays the progress of a downloading video. What fraction of the video is downloaded?

16. You complete 40% of your homework problems before dinner. What fraction of the problems did you complete before dinner?

17. There are nine different colonies of bacteria on the Petri dish. What percent of the bacteria on the Petri dish is from Colony 3?

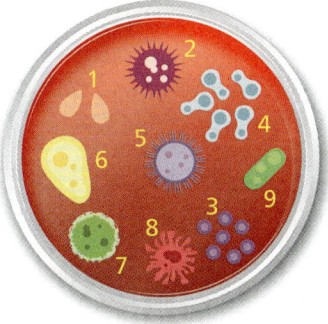

4.2 Percent and Decimals (pp. 169–174)

Learning Target: Write percents as decimals and decimals as percents.

Write the percent as a decimal.

18. 76%
19. 6%
20. 17%
21. 0.8%
22. 0.016%
23. 334%

Write the decimal as a percent.

24. 0.15
25. 0.77
26. 0.56
27. 1.06
28. 1.24
29. 0.097

30. Write a decimal that is greater than 0.62% and less than 0.64%.

31. Write a percent that is greater than 0.026 and less than 0.028.

32. On a fishing trip, 38% of the fish that you catch are perch. Write this percent as a decimal.

4.3 Comparing and Ordering Fractions, Decimals, and Percents (pp. 175–180)

Learning Target: Compare and order fractions, decimals, and percents.

Tell which number is greater.

33. $\frac{1}{2}$, 52%
34. $\frac{12}{5}$, 245%
35. 0.46, 43%
36. 0.023, 22%

Order the numbers from least to greatest.

37. $\frac{9}{4}$, 220%, 2.15, 218%
38. 0.88, $\frac{7}{8}$, 92%, $\frac{9}{10}$, 0.89

39. Write a percent that is greater than $\frac{13}{25}$ and less than 0.54.

40. The table shows the portions of students in your grade who participate in five activities. List the activities in order by number of students from least to greatest.

Activity	Band	Chorus	Debate	Gymnastics	Theater
Portion of Students	0.14	$\frac{3}{25}$	$\frac{1}{20}$	11%	0.08

4.4 Solving Percent Problems (pp. 181–188)

Learning Target: Find a percent of a quantity and solve percent problems.

Find the percent of the number. Explain your method.

41. 60% of 80
42. 80% of 55
43. 150% of 48
44. 42% of 150
45. 112% of 75
46. 45% of 42

Find the whole. Explain your method.

47. 70% of what number is 35?
48. 28% of what number is 21?
49. 56% of what number is 84?
50. 20% of what number is 96?
51. 140% of what number is 56?
52. 175% of what number is 112?

53. Each cell of a dog contains 78 chromosomes. Exactly 50% of the chromosomes are inherited from the father. How many chromosomes in each cell of the dog are inherited from the father?

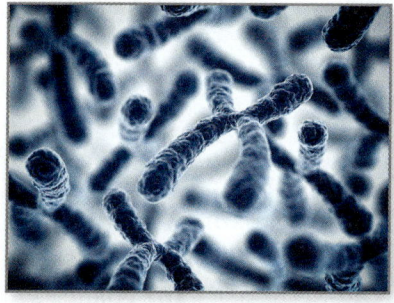

54. You went to the mall with $80. You spent 25% of your money on a pair of shorts and 65% of the remainder on sandals. How much did you spend on the sandals?

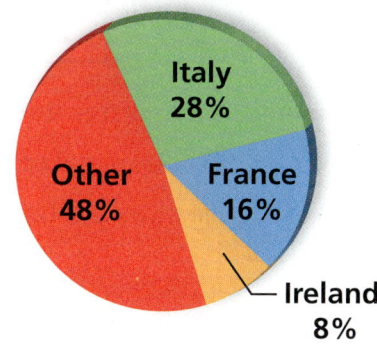

55. The results of a survey are shown at the left. In the survey, 7 students said that they would most like to visit Italy. How many of the students surveyed would most like to visit Ireland?

56. You answer 24 questions on a 100-point test correctly and earn a 96%.

   a. All of the questions are worth the same number of points. How many questions are on the test? How many points is each question worth?

   b. Your friend earns a grade of 76% on the same test. How many questions did your friend answer correctly?

4 Practice Test

Write the fraction or mixed number as a percent.

1. $\frac{21}{25}$
2. $\frac{17}{20}$
3. $1\frac{2}{5}$

Write the decimal as a percent.

4. 0.42
5. 7.88
6. 0.5854

Write the percent as a fraction in simplest form and as a decimal.

7. 0.96%
8. 65%
9. 25.7%

Tell which number is greater.

10. $\frac{16}{25}$, 65%
11. 56%, 5.6

Order the numbers from least to greatest.

12. 85%, $\frac{7}{10}$, 0.74, $\frac{4}{5}$
13. 130%, 1.32, $\frac{6}{5}$, $\frac{5}{4}$, 1.28

14. 80% of 90 is what number?
15. 120% of 75 is what number?

16. 40% of what number is 34?
17. 130% of what number is 52?

18. A goalie's saves (•) and goals scored against (×) are shown. What percent of shots did the goalie save? Explain.

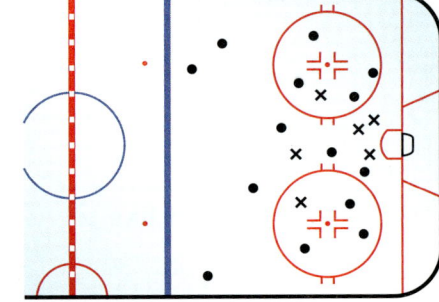

19. About 62% of the human body is composed of water. Write this percent as a fraction in simplest form.

20. You, your cousin, and a friend each take the same number of free throws at a basketball hoop. You make $\frac{17}{20}$ of your free throws, your cousin makes 0.8 of her free throws, and your friend makes 87.5% of his free throws. Who made the most free throws?

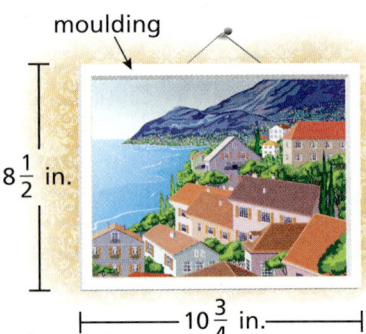

moulding

$8\frac{1}{2}$ in.

$10\frac{3}{4}$ in.

21. In a class of 20 students, 40% are boys. Twenty-five percent of the boys and 50% of the girls wear glasses. How many students in the class wear glasses?

22. Eighty percent of the picture frame is glass. What is the area of the moulding?

4 Cumulative Practice

1. How many pints are in 8 quarts?

   A. 2 pints   B. 4 pints

   C. 16 pints   D. 32 pints

Test-Taking Strategy
Answer Easy Questions First

"Answer the easy questions first. Then try the hard ones."

2. Which fraction is *not* equivalent to 25%?

   F. $\frac{1}{4}$ G. $\frac{2}{5}$

   H. $\frac{5}{20}$ I. $\frac{25}{100}$

3. What is the missing value in the ratio table?

Pairs of Shoes	7		56
Pairs of Boots	2	8	16

4. Your friend was finding the percent of a number in the box below.

   25% of 24 is what number?

   25% of 24 = 24 ÷ $\frac{1}{4}$

   = 96

   What should your friend do to correct the error?

   A. Divide 24 by 25.   B. Divide $\frac{1}{4}$ by 24.

   C. Multiply 24 by 25.   D. Multiply 24 by $\frac{1}{4}$.

5. Which percent is equivalent to $\frac{4}{5}$?

   F. 20%   G. 45%

   H. 80%   I. 125%

6. Which pair of numbers does *not* have a least common multiple less than 100?

   A. 10, 15
   B. 12, 16
   C. 16, 18
   D. 18, 24

7. You are comparing the costs of buying bottles of water at the supermarket. Which of the following has the least cost per liter?

   F. 6 one-liter bottles for $1.80

   G. 1 two-liter bottle for $0.65

   H. 8 half-liter bottles for $1.50

   I. 12 half-liter bottles for $1.98

8. What is 75% of 36?

9. Which number is equivalent to $\frac{5}{12} \times \frac{4}{9}$?

   A. $\frac{5}{27}$
   B. $\frac{3}{7}$
   C. $\frac{15}{16}$
   D. $\frac{5}{3}$

10. Which list of numbers is in order from least to greatest?

    F. $0.8, \frac{5}{8}, 70\%, 0.09$
    G. $\frac{5}{8}, 70\%, 0.8, 0.09$
    H. $0.09, \frac{5}{8}, 0.8, 70\%$
    I. $0.09, \frac{5}{8}, 70\%, 0.8$

11. Which number is equivalent to 1.32 ÷ 0.006?

    A. 2.2   **B.** 22

    C. 220   **D.** 2200

12. Which ratio is equivalent to 4 : 14?

    F. 2 : 12   **G.** 10 : 35

    H. 18 : 28   **I.** 8 : 18

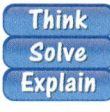

13. For a party, you make a gelatin dessert in a rectangular pan and cut the dessert into equal-sized pieces, as shown below.

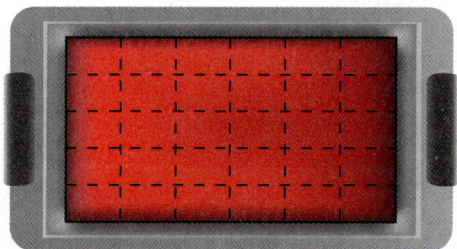

The dessert consists of 5 layers of equal height. Each layer is a different flavor, as shown below by a side view of the pan.

Your guests eat $\frac{3}{5}$ of the pieces of the dessert.

Part A   Write the amount of cherry gelatin that your guests eat as a fraction of the total dessert. Justify your answer.

Part B   Write the amount of cherry gelatin that your guests eat as a percent of the total dessert. Justify your answer.

5 Algebraic Expressions and Properties

- **5.1** Algebraic Expressions
- **5.2** Writing Expressions
- **5.3** Properties of Addition and Multiplication
- **5.4** The Distributive Property
- **5.5** Factoring Expressions

Chapter Learning Target:
Understand algebraic expressions.

Chapter Success Criteria:
- I can identify parts of an algebraic expression.
- I can write algebraic expressions.
- I can solve a problem using algebraic expressions.
- I can interpret algebraic expressions in real-life problems.

STEAM Video: "Shadow Drawings"

STEAM Video

Shadow Drawings

Expressions can be used to represent the growth of living things over time. Can you think of any other real-life situations in which you would want to use an expression to represent a changing quantity?

Watch the STEAM Video "Shadow Drawings." Then answer the following questions.

1. Tory traces the shadow of a plant each week on the same day of the week and at the same time of day. Why does she need to be so careful about the timing of the drawing?

2. The table shows the height of the plant each week for the first three weeks. About how tall was the plant after 1.5 weeks? Explain your reasoning.

Week	1	2	3
Height (inches)	7	14	22

3. Predict the height of the plant when Tory makes her next three weekly drawings.

Performance Task

Describing Change

After completing this chapter, you will be able to use the concepts you learned to answer the questions in the *STEAM Video Performance Task*. You will be given data sets for the following real-life situations.

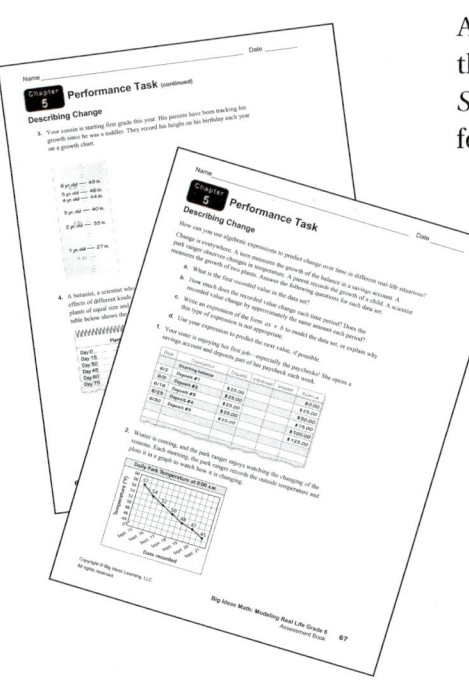

- **Savings account**
- **Temperature**
- **Human growth**
- **Plant growth**

You will be asked to use given data to write expressions and make predictions. Do the expressions provide accurate predictions far into the future?

199

Getting Ready for Chapter 5

Chapter Exploration

1. Work with a partner.

   a. You babysit for 3 hours. You receive $24. What is your hourly wage?

   - Write the problem. Underline the important numbers and units you need to solve the problem.
   - Read the problem carefully a second time. Circle the key phrase for the question.

     > You babysit for 3 hours. You receive $24.
     > What is your hourly wage?

   - Write each important number or phrase, with its units, on a piece of paper. Write $+, -, \times, \div,$ and $=$ on five other pieces of paper.

   - Arrange the pieces of paper to answer the question, "What is your hourly wage?"
   - Evaluate the expression that represents the hourly wage.

     hourly wage = ☐ ÷ ☐   Write.

     = ☐   Evaluate.

   ▶ So, your hourly wage is $ ☐ per hour.

   b. How can you use your hourly wage to find how much you will receive for any number of hours worked?

Vocabulary

The following vocabulary terms are defined in this chapter. Think about what each term might mean and record your thoughts.

algebraic expression   equivalent expressions
variable   factoring an expression
constant

5.1 Algebraic Expressions

Learning Target: Evaluate algebraic expressions given values of their variables.

Success Criteria:
- I can identify parts of an algebraic expression.
- I can evaluate algebraic expressions with one or more variables.
- I can evaluate algebraic expressions with one or more operations.

EXPLORATION 1

Evaluating Expressions

Math Practice

Make Sense of Quantities
What are the units in the problem? How does this help you write an expression?

Work with a partner. Identify any missing information that is needed to answer each question. Then choose a reasonable quantity and write an expression for each problem. After you have written the expression, evaluate it using mental math or some other method.

a. You receive $24 for washing cars. How much do you earn per hour?

b. You buy 5 silicone baking molds at a craft store. How much do you spend?

c. You are running in a mud race. How much farther do you have to go after running 2000 feet?

d. A rattlesnake is 25 centimeters long when it hatches. The snake grows at a rate of about 1.6 centimeters per month for several months. What is the length of the rattlesnake?

5.1 Lesson

Key Vocabulary
algebraic expression, p. 202
variable, p. 202
term, p. 202
coefficient, p. 202
constant, p. 202

An **algebraic expression** is an expression that may contain numbers, operations, and one or more *variables*. A **variable** is a symbol that represents one or more numbers. Each number or variable by itself, or product of numbers and variables in an algebraic expression, is called a **term**.

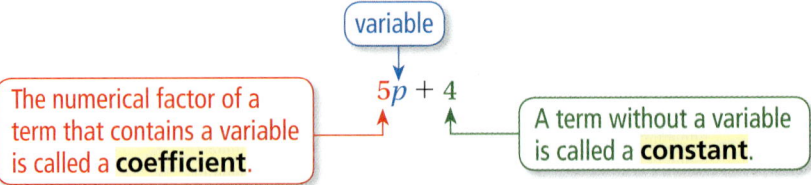

The numerical factor of a term that contains a variable is called a **coefficient**.

A term without a variable is called a **constant**.

EXAMPLE 1 Identifying Parts of an Algebraic Expression

Identify the terms, coefficients, and constants in each expression.

a. $5x + 13$

Terms: $5x$, 13
Coefficient: 5
Constant: 13

b. $2z^2 + y + 3$

Terms: $2z^2$, $1y$, 3
Coefficients: 2, 1
Constant: 3

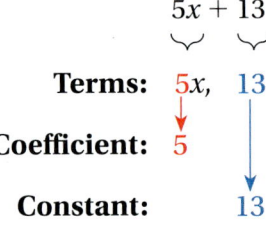

A variable by itself has a coefficient of 1. So, the term *y* in Example 1(b) has a coefficient of 1.

Try It Identify the terms, coefficients, and constants in the expression.

1. $12 + 10c$
2. $15 + 3w + \dfrac{1}{2}$
3. $z^2 + 9z$

EXAMPLE 2 Writing Algebraic Expressions Using Exponents

Write each expression using exponents.

a. $d \cdot d \cdot d \cdot d$

Because *d* is used as a factor 4 times, its exponent is 4.

▶ So, $d \cdot d \cdot d \cdot d = d^4$.

b. $1.5 \cdot h \cdot h \cdot h$

Because *h* is used as a factor 3 times, its exponent is 3.

▶ So, $1.5 \cdot h \cdot h \cdot h = 1.5h^3$.

Try It Write the expression using exponents.

4. $j \cdot j \cdot j \cdot j \cdot j$
5. $9 \cdot k \cdot k \cdot k \cdot k$

202 Chapter 5 Algebraic Expressions and Properties

To evaluate an algebraic expression, substitute a number for each variable. Then use the order of operations to find the value of the numerical expression.

EXAMPLE 3 Evaluating Algebraic Expressions

a. **Evaluate $k + 10$ when $k = 25$.**

$k + 10 = 25 + 10$ Substitute 25 for k.

$= 35$ Add 25 and 10.

b. **Evaluate $4 \cdot n$ when $n = 12$.**

$4 \cdot n = 4 \cdot 12$ Substitute 12 for n.

$= 48$ Multiply 4 and 12.

> You can write the product of 4 and n in several ways.
> $4 \cdot n$
> $4n$
> $4(n)$

Try It

6. Evaluate $24 + c$ when $c = 9$.

7. Evaluate $d - 17$ when $d = 30$.

8. Evaluate $18 \div q$ when $q = \dfrac{1}{2}$.

EXAMPLE 4 Evaluating an Expression with Two Variables

a. **Evaluate $n - m$ when $m = 12$ and $n = 30$.**

$n - m = 30 - 12$ Substitute 30 for n and 12 for m.

$= 18$ Subtract 12 from 30.

b. **Evaluate $a \div b$ when $a = 16$ and $b = \dfrac{2}{3}$.**

$a \div b = 16 \div \dfrac{2}{3}$ Substitute 16 for a and $\dfrac{2}{3}$ for b.

$= 16 \cdot \dfrac{3}{2}$ Multiply by the reciprocal of $\dfrac{2}{3}$, which is $\dfrac{3}{2}$.

$= 24$ Multiply.

Try It Evaluate the expression when $p = 24$ and $q = 8$.

9. $p \div q$

10. $q + p$

11. $p - q$

12. $p \cdot q$

Section 5.1 Algebraic Expressions

EXAMPLE 5 **Evaluating Expressions with Two Operations**

a. Evaluate $3x - 14$ when $x = 5$.

$3x - 14 = 3(5) - 14$ — Substitute 5 for x.
$= 15 - 14$ — Using order of operations, multiply 3 and 5.
$= 1$ — Subtract 14 from 15.

b. Evaluate $n^2 + 8.5$ when $n = 2$.

$n^2 + 8.5 = 2^2 + 8.5$ — Substitute 2 for n.
$= 4 + 8.5$ — Using order of operations, evaluate 2^2.
$= 12.5$ — Add 4 and 8.5.

Try It Evaluate the expression when $y = 6$.

13. $5y + 1$
14. $30 - 24 \div y$
15. $y^2 - 7$
16. $1.5 + y^2$

Self-Assessment for Concepts & Skills

Solve each exercise. Then rate your understanding of the success criteria in your journal.

17. **WHICH ONE DOESN'T BELONG?** Which expression does *not* belong with the other three? Explain your reasoning.

    $2x + 1$ $5w \cdot c$ $3(4) + 5$ $2y \cdot z$

18. **ALGEBRAIC EXPRESSIONS** Identify the terms, coefficients, and constants in the expression $9h + 1$.

EVALUATING EXPRESSIONS Evaluate the expression when $m = 8$.

19. $m - 7$
20. $5m + 4$

21. **MP NUMBER SENSE** Does the value of the expression $20 - x$ *increase*, *decrease*, or *stay the same* as x increases? Explain.

22. **OPEN-ENDED** Write an algebraic expression using more than one operation. When you evaluate the expression, how do you know which operation to perform first?

23. **MP STRUCTURE** Is the expression $8.2 \div m \cdot m \cdot m \cdot m$ the same as the expression $8.2 \div m^4$? Explain your reasoning.

204 Chapter 5 Algebraic Expressions and Properties

EXAMPLE 6 Modeling Real Life

You are saving to buy a meteorite fragment for $125. You begin with $45 and you save $3 each week. The expression $45 + 3w$ gives the amount of money you save after w weeks. Can you buy the meteorite after 20 weeks?

Understand the problem. You are given an expression that represents your savings after w weeks. You are asked whether you have enough money to buy a $125 meteorite after 20 weeks.

Make a plan. To find the amount of money you save after 20 weeks, evaluate the expression when $w = 20$. Then compare the value of the expression to the price of the meteorite.

Solve and check.

$45 + 3w = 45 + 3(20)$ Substitute 20 for w.

$ = 45 + 60$ Multiply 3 and 20.

$ = 105$ Add 45 and 60.

▸ You cannot buy the $125 meteorite after 20 weeks because you only have $105.

> **Another Method** You start with $45, so you need to save another $125 - 45 = \$80$. At $3 per week, it will take you $\dfrac{80}{3} \approx 27$ weeks of saving.
>
> $45 + 3(27) = 45 + 81 = \$126$ ✓

Self-Assessment for Problem Solving

Solve each exercise. Then rate your understanding of the success criteria in your journal.

24. The expression $12.25m + 29.99$ gives the cost (in dollars) of a gym membership for m months. You have $180 to spend on a membership. Can you buy a one-year membership?

25. **DIG DEEPER!** The expression $p - 15$ gives the amount (in dollars) you pay after using the coupon when the original amount of a purchase is p dollars. The expression $30 + 6n$ gives the amount of money (in dollars) you save after n weeks. A jacket costs $78. Can you buy the jacket after 6 weeks? Explain.

Coupon

Good for $15 off any purchase of $75 or more

5.1 Practice

Go to **BigIdeasMath.com** to get HELP with solving the exercises.

Review & Refresh

You ask 40 students which of three items from the cafeteria they like the best. You record the results on the piece of paper shown.

1. What percent of students answered salad?
2. How many students answered pizza?
3. What percent of students answered pasta?

Find the missing quantity in the double number line.

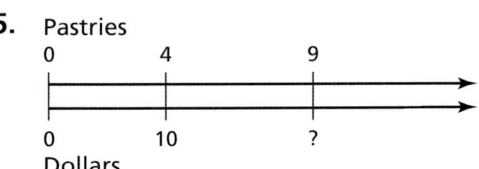

Divide. Write the answer in simplest form.

6. $1\dfrac{3}{8} \div \dfrac{3}{4}$
7. $2\dfrac{7}{9} \div 2$
8. $4 \div 4\dfrac{2}{5}$
9. $3\dfrac{2}{3} \div 1\dfrac{2}{7}$

Concepts, Skills, & Problem Solving

EVALUATING EXPRESSIONS Write and evaluate an expression for the problem. *(See Exploration 1, p. 201.)*

10. The scores on your first two history tests are 82 and 95. By how many points did you improve on your second test?

11. You buy a hat for $12 and give the cashier a $20 bill. How much change do you receive?

12. You receive $8 for raking leaves for 2 hours. What is your hourly wage?

13. Music lessons cost $20 per week. How much do 6 weeks of lessons cost?

ALGEBRAIC EXPRESSIONS Identify the terms, coefficients, and constants in the expression.

14. $7h + 3$
15. $g + 12 + 9g$
16. $5c^2 + 7d$
17. $2m^2 + 15 + 2p^2$
18. $6 + n^2 + \dfrac{1}{2}d$
19. $8x + \dfrac{x^2}{3}$

> Terms: 2, x^2, y
> Coefficient: 2
> Constant: none

20. **YOU BE THE TEACHER** Your friend finds the terms, coefficients, and constants in the algebraic expression $2x^2y$. Is your friend correct? Explain your reasoning.

206   Chapter 5   Algebraic Expressions and Properties

21. **PERIMETER** You can use the expression $2\ell + 2w$ to find the perimeter of a rectangle, where ℓ is the length and w is the width.

   a. Identify the terms, coefficients, and constants in the expression.

   b. Interpret the coefficients of the terms.

USING EXPONENTS Write the expression using exponents.

22. $b \cdot b \cdot b$
23. $g \cdot g \cdot g \cdot g \cdot g$
24. $8 \cdot w \cdot w \cdot w \cdot w$
25. $5.2 \cdot y \cdot y \cdot y$
26. $a \cdot a \cdot c \cdot c$
27. $2.1 \cdot x \cdot z \cdot z \cdot z \cdot z$

28. **YOU BE THE TEACHER** Your friend writes the product using exponents. Is your friend correct? Explain your reasoning.

$3 \cdot n \cdot n \cdot n \cdot n = 3n^4$

29. **AREA** Write an expression using exponents that represents the area of the square.

30. **REASONING** Suppose the man in the St. Ives poem has x wives, each wife has x sacks, each sack has x cats, and each cat has x kits. Write an expression using exponents that represents the total number of kits, cats, sacks, and wives.

As I was going to St. Ives
I met a man with seven wives
Each wife had seven sacks
Each sack had seven cats
Each cat had seven kits
Kits, cats, sacks, wives
How many were going to St. Ives?

EVALUATING EXPRESSIONS Evaluate the expression when $a = 3$, $b = 2$, and $c = 12$.

31. $6 + a$
32. $b \cdot 5$
33. $c - 1$
34. $27 \div a$
35. $12 - b$
36. $c + 5$
37. $2a$
38. $c \div 6$
39. $a + b$
40. $c + a$
41. $c - a$
42. $a - b$
43. $\dfrac{c}{a}$
44. $\dfrac{c}{b}$
45. $b \cdot c$
46. $c(a)$

47. **PROBLEM SOLVING** You earn $15n$ dollars for mowing n lawns. How much do you earn for mowing 1 lawn? 7 lawns?

EVALUATING EXPRESSIONS Copy and complete the table.

48.

x	3	6	9
x · 8			

49.

x	2	4	8
64 ÷ x			

50. **MODELING REAL LIFE** Due to gravity, an object falls $16t^2$ feet in t seconds. You drop a rock from a bridge that is 75 feet above the water. Will the rock hit the water in 2 seconds? Explain.

EVALUATING EXPRESSIONS Evaluate the expression when $a = 10$, $b = 9$, and $c = 4$.

51. $2a + 3$

52. $4c - 7.8$

53. $\dfrac{a}{4} + \dfrac{1}{3}$

54. $\dfrac{24}{b} + 8$

55. $c^2 + 6$

56. $a^2 - 18$

57. $a + 9c$

58. $bc + 12.3$

59. $3a + 2b - 6c$

60. **YOU BE THE TEACHER** Your friend evaluates the expression when $m = 8$. Is your friend correct? Explain your reasoning.

$5m + 3 = 5 \cdot 8 + 3$
$ = 5 \cdot 11$
$ = 55$

61. **MP PROBLEM SOLVING** After m months, the height of a plant is $(10 + 3m)$ millimeters. How tall is the plant after 8 months? 3 years?

62. **MP STRUCTURE** You use a video streaming service to rent x new releases and y standard rentals. Which expression tells you how much money you will need?

$3x + 4y$ $4x + 3y$ $7(x + y)$

Standard Rentals $3

New Releases $4

 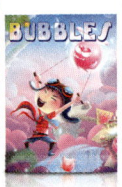 

63. **OPEN-ENDED** You float 2000 feet along a lazy river water ride. The ride takes less than 10 minutes. Give two examples of possible times and speeds.

64. **DIG DEEPER!** The expression $20a + 13c$ is the cost (in dollars) for a adults and c students to enter a science center.

  a. How much does it cost for an adult? a student? Explain your reasoning.

  b. Find the total cost for 4 adults and 24 students.

  c. You find the cost for a group. Then the numbers of adults and students in the group both double. Does the cost double? Explain your answer using an example.

  d. In part (b), the number of adults is cut in half, but the number of students doubles. Is the cost the same? Explain your answer.

65. **MP REASONING** The volume of the cube (in cubic inches) is equal to four times the area of one of its faces (in square inches). What is the volume of the cube?

x in.

5.2 Writing Expressions

Learning Target: Write algebraic expressions and solve problems involving algebraic expressions.

Success Criteria:
- I can write numerical expressions.
- I can write algebraic expressions.
- I can write and evaluate algebraic expressions that represent real-life problems.

EXPLORATION 1

Writing Expressions

Work with a partner. You use a $20 bill to buy lunch at a café. You order a sandwich from the menu board shown.

Math Practice

Use Expressions
How do the numerical expressions help you generalize the situation and write an algebraic expression?

a. Complete the table. In the last column, write a numerical expression for the amount of change you receive.

Sandwich	Price (dollars)	Change Received (dollars)
Reuben		
BLT		
Egg salad		
Roast beef		

b. **MP REPEATED REASONING** Write an algebraic expression that represents the amount of change you receive when you order any sandwich from the menu board.

c. The expression $20 - 4.65s$ represents the amount of change one customer receives after ordering from the menu board. Explain what each part of the expression represents. Do you know what the customer ordered? Explain your reasoning.

Section 5.2 Writing Expressions 209

5.2 Lesson

Some words can imply math operations.

Operation	Addition	Subtraction	Multiplication	Division
Key Words and Phrases	added to plus sum of more than increased by total of and	subtracted from minus difference of less than decreased by fewer than take away	multiplied by times product of twice	divided by quotient of

EXAMPLE 1 Writing Numerical Expressions

Write each phrase as an expression.

a. 8 fewer than 21

$21 - 8$ *The phrase fewer than means subtraction.*

b. the product of 30 and 9

30×9, or $30 \cdot 9$ *The phrase product of means multiplication.*

Try It Write the phrase as an expression.

1. the sum of 18 and 35
2. 6 times 50

EXAMPLE 2 Writing Algebraic Expressions

Write each phrase as an expression.

a. 14 more than a number x

$x + 14$ *The phrase more than means addition.*

b. a number y minus 75

$y - 75$ *The word minus means subtraction.*

c. the quotient of 3 and a number z

$3 \div z$, or $\dfrac{3}{z}$ *The phrase quotient of means division.*

Common Error

When writing expressions involving subtraction or division, order is important. For example, the quotient of a number x and 2 means $x \div 2$, not $2 \div x$.

Try It Write the phrase as an expression.

3. 25 less than a number b
4. a number x divided by 4
5. the total of a number t and 11
6. 100 decreased by a number k

EXAMPLE 3 Writing an Algebraic Expression

The length of Interstate 90 from the West Coast to the East Coast is 153.5 miles more than 2 times the length of Interstate 15 from southern California to northern Montana. Let m be the length of Interstate 15. Which expression can you use to represent the length of Interstate 90?

- **A.** $2m + 153.5$
- **B.** $2m - 153.5$
- **C.** $153.5 - 2m$
- **D.** $153.5m + 2$

> Variables can be lowercase or uppercase. Make sure you consistently use the same case for a variable when solving a problem.

The word *times* means *multiplication*. So, multiply 2 and m.

The phrase *more than* means *addition*. So, add $2m$ and 153.5.

$$2m + 153.5$$

▶ The correct answer is **A**.

Try It

7. Your friend has 5 more than twice as many game tokens as you. Let t be the number of game tokens you have. Write an expression for the number of game tokens your friend has.

Self-Assessment for Concepts & Skills

Solve each exercise. Then rate your understanding of the success criteria in your journal.

WRITING EXPRESSIONS Write the phrase as an expression.

8. the sum of 7 and 11

9. 5 subtracted from 9

10. DIFFERENT WORDS, SAME QUESTION Which is different? Write "both" expressions.

- 12 more than x
- x increased by 12
- x take away 12
- the sum of x and 12

11. **MP PRECISION** Your friend says that the phrases below have the same meaning. Is your friend correct? Explain your reasoning.

- the difference of a number x and 12
- the difference of 12 and a number x

Section 5.2 Writing Expressions 211

EXAMPLE 4 Modeling Real Life

You plant a cypress tree that is 10 inches tall. Each year, its height increases by 15 inches. Write an expression that represents the height (in inches) after t years. What is the height after 9 years?

Make a table showing the height of the tree each year for the first several years. Use the results to write an expression and evaluate the expression when $t = 9$.

The height is *increasing*, so *add* 15 each year, as shown in the table.

Year, t	Height (inches)
0	10
1	$10 + 15(1) = 25$
2	$10 + 15(2) = 40$
3	$10 + 15(3) = 55$
4	$10 + 15(4) = 70$

When t is 0, the height is 10 inches.

You can see that an expression is $10 + 15t$.

Sometimes, as in Example 3, a variable represents a single value. Other times, as in Example 4, a variable can represent more than one value.

Evaluate $10 + 15t$ when $t = 9$.

$$10 + 15t = 10 + 15(9) = 145$$

▸ So, the height (in inches) after t years is $10 + 15t$. After 9 years, the height of the tree is 145 inches.

Self-Assessment for Problem Solving

Solve each exercise. Then rate your understanding of the success criteria in your journal.

12. A company rents paddleboards by charging a rental fee plus an hourly rate. Write an expression that represents the cost (in dollars) of renting a paddleboard for h hours. How much does an eight-hour rental cost?

13. **DIG DEEPER!** A county fair charges an entry fee of $7 and $0.75 for each ride token. You have $15. Write an expression that represents the amount (in dollars) you have left after entering the fair and purchasing n tokens. How many tokens can you purchase? How much money do you have left after purchasing 6 tokens?

5.2 Practice

Go to *BigIdeasMath.com* to get HELP with solving the exercises.

▶ Review & Refresh

Identify the terms, coefficients, and constants in the expression.

1. $4f + 8$
2. $\dfrac{4}{5} + 3s + 2$
3. $9h^2 + \dfrac{8}{9}p + 1$

Copy and complete the statement.

4. $\dfrac{2\text{ c}}{\text{min}} = \dfrac{\boxed{}\text{ gal}}{\text{h}}$
5. $\dfrac{12\text{ m}}{\text{sec}} \approx \dfrac{\boxed{}\text{ ft}}{\text{min}}$
6. $\dfrac{3\text{ lb}}{\text{sec}} \approx \dfrac{\boxed{}\text{ kg}}{\text{h}}$

Divide. Write the answer in simplest form.

7. $\dfrac{1}{2} \div \dfrac{5}{8}$
8. $\dfrac{1}{3} \div \dfrac{3}{4}$
9. $\dfrac{2}{5} \div 3$
10. $3 \div \dfrac{6}{7}$

▶ Concepts, Skills, & Problem Solving

MP STRUCTURE The expression represents the amount of change you receive after buying *n* sandwiches. Explain what each part of the expression represents. *(See Exploration 1, p. 209.)*

11. $10 - 5.25n$
12. $20 - 4.95n$
13. $100 - 6.75n$

WRITING EXPRESSIONS Write the phrase as an expression.

14. 5 less than 8
15. the product of 3 and 12
16. 28 divided by 7
17. the total of 6 and 10
18. 3 fewer than 18
19. 17 added to 15
20. 13 subtracted from a number x
21. 5 times a number d
22. the quotient of 18 and a number a
23. the difference of a number s and 6
24. 7 increased by a number w
25. a number t cubed

YOU BE THE TEACHER Your friend writes the phrase as an expression. Is your friend correct? Explain your reasoning.

26.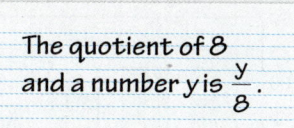
The quotient of 8 and a number y is $\dfrac{y}{8}$.

27.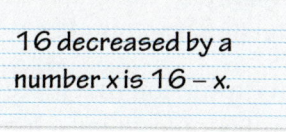
16 decreased by a number x is $16 - x$.

28. **MP NUMBER SENSE** Five friends share the cost of a dinner equally.

   a. Write an expression that represents the cost (in dollars) per person.

   b. Make up a reasonable total cost and test your expression.

29. **MODELING REAL LIFE** A biologist analyzes 15 bacteria samples each day.

   a. Copy and complete the table.

   b. Write an expression that represents the total number of samples analyzed after n days.

Days	1	2	3	4	5
Total Samples					

Section 5.2 Writing Expressions

30. **MP PROBLEM SOLVING** To rent a moving truck for the day, it costs $33 plus $1 for each mile driven.

   a. Write an expression that represents the cost (in dollars) to rent the truck.

   b. You drive the truck 300 miles. How much do you pay?

WRITING PHRASES Give two ways to write the expression as a phrase.

31. $n + 6$
32. $4w$
33. $15 - b$
34. $14 - 3z$

EVALUATING EXPRESSIONS Write the phrase as an expression. Then evaluate the expression when $x = 5$ and $y = 20$.

35. 3 less than the quotient of a number y and 4

36. the sum of a number x and 4, all divided by 3

37. 6 more than the product of 8 and a number x

38. the quotient of 40 and the difference of a number y and 16

39. **MODELING REAL LIFE** It costs $3 to bowl a game and $2 for shoe rental.

   a. Write an expression that represents the total cost (in dollars) of g games.

   b. Use your expression to find the total cost of 8 games.

40. **MODELING REAL LIFE** Florida has 8 less than 5 times the number of counties in Arizona. Georgia has 25 more than twice the number of counties in Florida.

   a. Write an expression that represents the number of counties in Florida.

   b. Write an expression that represents the number of counties in Georgia.

   c. Arizona has 15 counties. How many do Florida and Georgia have?

41. **MP PATTERNS** There are 140 people in a singing competition. The graph shows the results for the first five rounds.

   a. Write an expression that represents the number of people after each round.

   b. Assuming this pattern continues, how many people compete in the ninth round? Explain your reasoning.

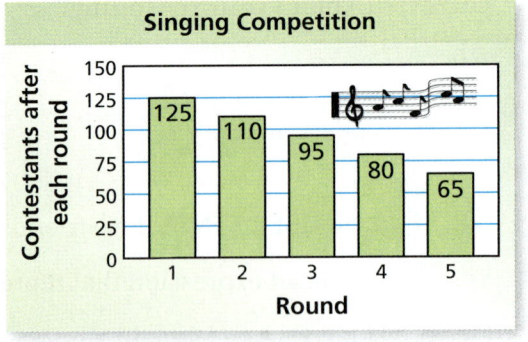

42. **MP NUMBER SENSE** The difference between two numbers is 8. The lesser number is a. Write an expression that represents the greater number.

43. **MP NUMBER SENSE** One number is four times another. The greater number is x. Write an expression that represents the lesser number.

5.3 Properties of Addition and Multiplication

Learning Target: Identify equivalent expressions and apply properties to generate equivalent expressions.

Success Criteria:
- I can explain the meaning of equivalent expressions.
- I can use properties of addition to generate equivalent expressions.
- I can use properties of multiplication to generate equivalent expressions.

EXPLORATION 1

Identifying Equivalent Expressions

Work with a partner.

a. Choose four values for a variable x. Then evaluate each expression for each value of x. Are any of the expressions *equivalent*? Explain your reasoning.

x			
$4 + x + 4$			

x			
$16x$			

x			
$4 \cdot (x \cdot 4)$			

x			
$x + 4 + 4$			

x			
$x + 8$			

x			
$(4 \cdot x) \cdot 4$			

b. You have used the following properties in a previous course. Use the examples to explain the meaning of each property.

Commutative Property of Addition: $3 + 5 = 5 + 3$

Commutative Property of Multiplication: $9 \cdot 3 = 3 \cdot 9$

Associative Property of Addition: $8 + (3 + 1) = (8 + 3) + 1$

Associative Property of Multiplication: $12 \cdot (6 \cdot 2) = (12 \cdot 6) \cdot 2$

Are these properties true for algebraic expressions? Explain your reasoning.

> **Math Practice**
>
> **Use Counterexamples**
> Use a counterexample to show that the Commutative Property is not true for division.

Section 5.3 Properties of Addition and Multiplication 215

5.3 Lesson

Key Vocabulary
equivalent expressions, p. 216

Expressions that result in the same number for any value of each variable are **equivalent expressions**. You can use the Commutative and Associative Properties to write equivalent expressions.

Key Ideas

Commutative Properties

Words Changing the order of addends or factors does not change the sum or product.

Numbers $5 + 8 = 8 + 5$ **Algebra** $a + b = b + a$
$5 \cdot 8 = 8 \cdot 5$ $a \cdot b = b \cdot a$

Associative Properties

Words Changing the grouping of addends or factors does not change the sum or product.

Numbers $(7 + 4) + 2 = 7 + (4 + 2)$
$(7 \cdot 4) \cdot 2 = 7 \cdot (4 \cdot 2)$

Algebra $(a + b) + c = a + (b + c)$
$(a \cdot b) \cdot c = a \cdot (b \cdot c)$

EXAMPLE 1 Using Properties to Write Equivalent Expressions

a. Simplify the expression $7 + (12 + x)$.

$7 + (12 + x) = (7 + 12) + x$ Associative Property of Addition
$= 19 + x$ Add 7 and 12.

b. Simplify the expression $(6.1 + x) + 8.4$.

$(6.1 + x) + 8.4 = (x + 6.1) + 8.4$ Commutative Property of Addition
$= x + (6.1 + 8.4)$ Associative Property of Addition
$= x + 14.5$ Add 6.1 and 8.4.

c. Simplify the expression $5(11y)$.

$5(11y) = (5 \cdot 11)y$ Associative Property of Multiplication
$= 55y$ Multiply 5 and 11.

One way to check whether expressions are equivalent is to evaluate each expression for any value of the variable. In Example 1(a), use $x = 2$.

$7 + (12 + x) \stackrel{?}{=} 19 + x$
$7 + (12 + 2) \stackrel{?}{=} 19 + 2$
$21 = 21$ ✓

Try It Simplify the expression. Explain each step.

1. $10 + (a + 9)$
2. $\left(c + \dfrac{2}{3}\right) + \dfrac{1}{2}$
3. $5(4n)$

216 Chapter 5 Algebraic Expressions and Properties   Multi-Language Glossary at BigIdeasMath.com

 **Key Ideas**

Addition Property of Zero

Words The sum of any number and 0 is that number.

Numbers $7 + 0 = 7$ **Algebra** $a + 0 = a$

Multiplication Properties of Zero and One

Words The product of any number and 0 is 0.

The product of any number and 1 is that number.

Numbers $9 \cdot 0 = 0$ **Algebra** $a \cdot 0 = 0$
$4 \cdot 1 = 4$ $a \cdot 1 = a$

EXAMPLE 2 **Using Properties to Write Equivalent Expressions**

a. Simplify the expression $9 \cdot 0 \cdot p$.

$9 \cdot 0 \cdot p = (9 \cdot 0) \cdot p$ Associative Property of Multiplication
$= 0 \cdot p$ Multiplication Property of Zero
$= 0$ Multiplication Property of Zero

b. Simplify the expression $4.5 \cdot r \cdot 1$.

$4.5 \cdot r \cdot 1 = 4.5 \cdot (r \cdot 1)$ Associative Property of Multiplication
$= 4.5 \cdot r$ Multiplication Property of One
$= 4.5r$ Rewrite.

Try It Simplify the expression. Explain each step.

4. $12 \cdot b \cdot 0$ **5.** $1 \cdot m \cdot 24$ **6.** $(t + 15) + 0$

 **Self-Assessment** for Concepts & Skills

Solve each exercise. Then rate your understanding of the success criteria in your journal.

USING PROPERTIES Simplify the expression. Explain each step.

7. $(7 + c) + 4$ **8.** $4(b \cdot 6)$ **9.** $0 \cdot b \cdot 9$

10. WRITING Explain what it means for expressions to be equivalent. Then give an example of equivalent expressions.

11. OPEN-ENDED Write an algebraic expression that can be simplified using the Associative Property of Multiplication and the Multiplication Property of One.

Section 5.3   Properties of Addition and Multiplication

EXAMPLE 3 Modeling Real Life

You and six friends play on a basketball team. A sponsor paid $100 for the league fee, x dollars for each player's T-shirt, and $68.25 for basketballs. Write an expression that represents the total amount (in dollars) the sponsor paid. Then find the total amount paid when each T-shirt costs $14.50.

Use a verbal model to write an expression that represents the sum of the league fee, the cost of the T-shirts, and the cost of the basketballs. Then evaluate the expression when $x = 14.5$.

Common Error
You **and** six friends are on the team, so use 7, not 6, to represent the number of T-shirts.

League fee (dollars)	+	Number of T-shirts	·	Cost per T-shirt (dollars)	+	Cost of basketballs (dollars)
$100		7		x		$68.25

$100 + 7x + 68.25 = 7x + 100 + 68.25$ Commutative Property of Addition
$\qquad\qquad\qquad\quad = 7x + (100 + 68.25)$ Associative Property of Addition
$\qquad\qquad\qquad\quad = 7x + 168.25$ Add 100 and 68.25.

Evaluate $7x + 168.25$ when $x = 14.5$.

$7x + 168.25 = 7(14.5) + 168.25 = 101.5 + 168.25 = 269.75$

▸ An expression that represents the total amount (in dollars) is $7x + 168.25$. When each T-shirt costs $14.50, the sponsor pays $269.75.

Self-Assessment for Problem Solving

Solve each exercise. Then rate your understanding of the success criteria in your journal.

12. You and five friends form a team for an outdoor adventure race. Your team needs to raise money to pay for $130 of travel fees, x dollars for each team member's entry fee, and $85.50 for food. Use an algebraic expression to find the total amount your team needs to raise when the entry fee is $25.50 per person.

13. You have $50 and a $15 gift card to spend online. You purchase a pair of headphones for $34.99 and 8 songs for x dollars each. Use an algebraic expression to find the amount you have left when each song costs $1.10.

5.3 Practice

Go to BigIdeasMath.com to get HELP with solving the exercises.

Review & Refresh

Write the phrase as an expression.

1. 10 added to a number p
2. the product of 6 and a number m
3. the quotient of a number b and 15
4. 7 fewer than a number s

Write the prime factorization of the number.

5. 36
6. 144
7. 147
8. 205

Evaluate the expression.

9. $8.092 + 3.5$
10. $16.78 - 12.237$
11. $9.17 + 1.83 + 2.641$
12. $8.43 - 6.218 + 4.2$

Represent the ratio relationship using a graph.

13.

Oil (teaspoons)	8	16	24
Flour (cups)	1	2	3

14.

Atoms	4	8	12
Protons	64	128	192

Concepts, Skills, & Problem Solving

MATCHING Match the expression with an equivalent expression. (See Exploration 1, p. 215.)

15. $3 + 3 + y$
16. $(y \cdot y) \cdot 3$
17. $3 \cdot 1 \cdot y$
18. $(3 + 0) + (y + y)$

 A. $y \cdot 3$
 B. $y + 3 + 3$
 C. $y(3 \cdot y)$
 D. $(3 + y) + y$

IDENTIFYING PROPERTIES Tell which property the statement illustrates.

19. $5 \cdot p = p \cdot 5$
20. $2 + (12 + r) = (2 + 12) + r$
21. $4 \cdot (x \cdot 10) = (4 \cdot x) \cdot 10$
22. $x + 7.5 = 7.5 + x$
23. $(c + 2) + 0 = c + 2$
24. $a \cdot 1 = a$

25. **YOU BE THE TEACHER** Your friend states the property that the statement illustrates. Is your friend correct? Explain your reasoning.

 $(7 + x) + 3 = (x + 7) + 3$
 Associative Property of Addition

USING PROPERTIES Simplify the expression. Explain each step.

26. $6 + (5 + x)$
27. $(14 + y) + 3$
28. $6(2b)$
29. $7(9w)$
30. $3.2 + (x + 5.1)$
31. $(0 + a) + 8$
32. $9 \cdot c \cdot 4$
33. $(18.6 \cdot d) \cdot 1$
34. $\left(3k + 4\dfrac{1}{5}\right) + 8\dfrac{3}{5}$
35. $(2.4 + 4n) + 9$
36. $(3s) \cdot 8$
37. $z \cdot 0 \cdot 12$

38. GEOMETRY The expression $12 + x + 4$ represents the perimeter of a triangle. Simplify the expression.

39. MP PRECISION A case of scout cookies has 10 cartons. A carton has 12 boxes. The amount you earn on a whole case is $10(12x)$ dollars.

   a. What does x represent?

   b. Simplify the expression.

40. MODELING REAL LIFE A government estimates the cost to design new radar technology over a period of m months. The government estimates $840,000 for equipment, $15,000 for software, and $40,000 per month for wages. Use an algebraic expression to find the total cost the government estimates when the project takes 16 months to complete.

WRITING EXPRESSIONS Write the phrase as an expression. Then simplify the expression.

41. 7 plus the sum of a number x and 5

42. the product of 8 and a number y, multiplied by 9

USING PROPERTIES Copy and complete the statement using the specified property.

	Property	Statement
43.	Associative Property of Multiplication	$7(2y) =$
44.	Commutative Property of Multiplication	$13.2 \cdot (x \cdot 1) =$
45.	Associative Property of Addition	$17 + (6 + 2x) =$
46.	Addition Property of Zero	$2 + (c + 0) =$
47.	Multiplication Property of One	$1 \cdot w \cdot 16 =$

48. GEOMETRY Five identical triangles form the trapezoid shown.

   a. What is the perimeter of the trapezoid?

   b. How can you use some or all of the triangles to form a new trapezoid with a perimeter of $3x + 14$? Explain your reasoning.

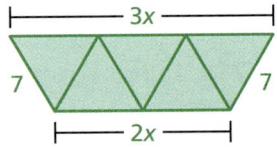

49. DIG DEEPER! You and a friend sell hats at a fair booth. You sell 16 hats on the first shift and 21 hats on the third shift. Your friend sells x hats on the second shift.

   a. The expression $37(14) + 10x$ represents the amount (in dollars) that you both earn. How can you tell that your friend is selling the hats for a lower price?

   b. You earn more money than your friend. What can you say about the value of x?

5.4 The Distributive Property

Learning Target: Apply the Distributive Property to generate equivalent expressions.

Success Criteria:
- I can explain how to apply the Distributive Property.
- I can use the Distributive Property to simplify algebraic expressions.
- I can use the Distributive Property to combine like terms.

EXPLORATION 1

Using Models to Simplify Expressions

Work with a partner.

a. Use the models to simplify the expressions. Explain your reasoning.

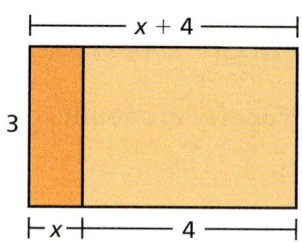

$3(x + 4) = $ ☐

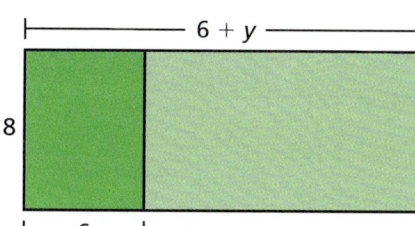

$8(6 + y) = $ ☐

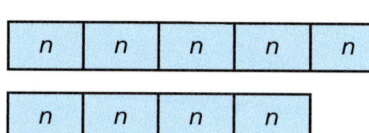

$5n + 4n = $ ☐

$5n - 4n = $ ☐

b. In part (a), check that the original expressions are equivalent to the simplified expressions.

c. You used the Distributive Property in a previous course. Use the example to explain the meaning of the property.

Distributive Property: $6(20 + 3) = 6(20) + 6(3)$

Is this property true for algebraic expressions? Explain your reasoning.

Math Practice

Find Entry Points

How can the Distributive Property be used to find the product of 9 and 32?

5.4 Lesson

Key Vocabulary
like terms, p. 223

 **Key Idea**

Distributive Property

Words To multiply a sum or difference by a number, multiply each term in the sum or difference by the number outside the parentheses. Then simplify.

Numbers $3(7 + 2) = 3 \times 7 + 3 \times 2$ **Algebra** $a(b + c) = ab + ac$

$3(7 - 2) = 3 \times 7 - 3 \times 2$ $a(b - c) = ab - ac$

EXAMPLE 1   Simplifying Algebraic Expressions

Use the Distributive Property to simplify each expression.

a. $4(n + 5)$

$4(n + 5) = 4(n) + 4(5)$ — Distributive Property

$= 4n + 20$ — Multiply.

b. $12(2y - 3)$

$12(2y - 3) = 12(2y) - 12(3)$ — Distributive Property

$= 24y - 36$ — Multiply.

c. $\frac{1}{2}(6y - 2z)$

$\frac{1}{2}(6y - 2z) = \frac{1}{2}(6y) - \frac{1}{2}(2z)$ — Distributive Property

$= 3y - z$ — Multiply.

d. $9(6 + x + 2)$

$9(6 + x + 2) = 9(6) + 9(x) + 9(2)$ — Distributive Property

$= 54 + 9x + 18$ — Multiply.

$= 9x + 54 + 18$ — Commutative Property of Addition

$= 9x + (54 + 18)$ — Associative Property of Addition

$= 9x + 72$ — Add 54 and 18.

You can use the Distributive Property when there are more than two terms in the sum or difference.

Try It Use the Distributive Property to simplify the expression.

1. $7(a + 2)$

2. $3(d - 11)$

3. $12\left(a + \frac{2}{3}b\right)$

4. $7(2 + 6 - 4d)$

In an algebraic expression, **like terms** are terms that have the same variables raised to the same exponents. Constant terms are also like terms.

like terms — $5x + 2x$ like terms — $2 + 19$

$$5x + 2x + 2 + 19$$

You can use the Distributive Property to *combine* like terms.

EXAMPLE 2 Combining Like Terms

Simplify each expression.

a. $3x + 9 + 2x - 5$

$3x + 9 + 2x - 5 = 3x + 2x + 9 - 5$	Commutative Property of Addition
$= (3 + 2)x + 9 - 5$	Distributive Property
$= 5x + 4$	Simplify.

b. $y + y + y$

$y + y + y = 1y + 1y + 1y$	Multiplication Property of One
$= (1 + 1 + 1)y$	Distributive Property
$= 3y$	Add coefficients.

c. $7z + 2(z - 5y)$

$7z + 2(z - 5y) = 7z + 2(z) - 2(5y)$	Distributive Property
$= 7z + 2z - 10y$	Multiply.
$= (7 + 2)z - 10y$	Distributive Property
$= 9z - 10y$	Add coefficients.

When you combine like terms, you are using the Distributive Property. You are applying the rules $ab + ac = a(b + c)$ and $ab - ac = a(b - c)$.

Try It Simplify the expression.

5. $8 + 3z - z$

6. $3(b + 5) + b + 2$

Self-Assessment for Concepts & Skills

Solve each exercise. Then rate your understanding of the success criteria in your journal.

7. **WRITING** One meaning of the word *distribute* is *to give something to each member of a group*. How can this help you remember the Distributive Property?

SIMPLIFYING EXPRESSIONS Use the Distributive Property to simplify the expression.

8. $3(x + 10)$

9. $15(4n - 2)$

10. $2w + 4 + 13w + 1$

EXAMPLE 3 Modeling Real Life

José is x years old. His brother, Felipe, is 2 years older than José. Their aunt, Maria, is three times as old as Felipe. Write and simplify an expression that represents Maria's age in years.

Use a table to organize the given information and write an expression that represents each person's age in years.

Name	Description	Expression
José	He is x years old.	x
Felipe	He is 2 years *older* than José. So, *add* 2 to x.	$x + 2$
Maria	She is three *times* as old as Felipe. So, *multiply* 3 and $(x + 2)$.	$3(x + 2)$

Look Back
If José is 10 years old, then Felipe is $10 + 2 = 12$ years old and Maria is $3(12) = 36$ years old. So, you should obtain 36 when you evaluate $3x + 6$ for $x = 10$.

$3x + 6 = 3(10) + 6$
$ = 36$

Simplify the expression that represents Maria's age.

$3(x + 2) = 3(x) + 3(2)$ Distributive Property
$ = 3x + 6$ Multiply.

▶ Maria's age in years is represented by the expression $3x + 6$.

Self-Assessment for Problem Solving

Solve each exercise. Then rate your understanding of the success criteria in your journal.

11. You purchase a remote-controlled drone for d dollars. Your friend purchases a drone that costs $35 more than your drone. Your brother purchases a drone that costs three times as much as your friend's drone. Write and simplify an expression that represents the cost (in dollars) of your brother's drone.

12. Write and simplify an expression that represents the total cost (in dollars) of buying the items shown for each member of a baseball team.

Pants: $10   and   Belt: $x

13. **DIG DEEPER!** One molecule of caffeine contains x oxygen atoms, twice as many nitrogen atoms as oxygen atoms, 4 more carbon atoms than nitrogen atoms, and 1.25 times as many hydrogen atoms as carbon atoms. Write and simplify an expression that represents the number of hydrogen atoms in one molecule of caffeine.

5.4 Practice

> Go to BigIdeasMath.com to get HELP with solving the exercises.

Review & Refresh

Simplify the expression. Explain each step.

1. $(s + 4) + 8$
2. $(12 + x) + 2$
3. $3(4n)$

You are given the difference of the numbers of boys and girls in a class and the ratio of boys to girls. How many boys and how many girls are in the class?

4. 3 more boys; 5 for every 4
5. 8 more girls; 3 for every 5
6. 4 more girls; 9 for every 13
7. 6 more boys; 7 for every 4

Divide.

8. $301 \div 7$
9. $1722 \div 14$
10. $629 \div 12$
11. $8068 \div 31$

Concepts, Skills, & Problem Solving

USING MODELS Use the model to simplify the expression. Explain your reasoning. *(See Exploration 1, p. 221.)*

12. $5(z + 6) = $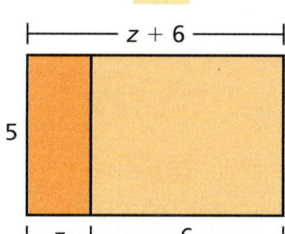

13. $4s + 2s = $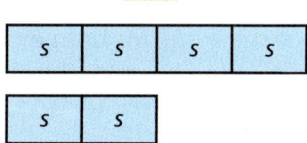

SIMPLIFYING EXPRESSIONS Use the Distributive Property to simplify the expression.

14. $3(x + 4)$
15. $10(b - 6)$
16. $6(s - 9)$
17. $7(8 + y)$
18. $8(12 + a)$
19. $9(2n + 1)$
20. $12(6 - k)$
21. $18(5 - 3w)$
22. $9(3 + c + 4)$
23. $\frac{1}{4}(8 + x + 4)$
24. $8(5g + 5 - 2)$
25. $6(10 + z + 3)$
26. $4(x + y)$
27. $25(x - y)$
28. $7(p + q + 9)$
29. $\frac{1}{2}(2n + 4 + 6m)$

MATCHING Match the expression with an equivalent expression.

30. $6(n + 4)$
31. $2(3n + 9)$
32. $6(n + 2)$
33. $3(2n + 3)$

   A. $3(2n + 6)$
   B. $6n + 9$
   C. $3(2n + 8)$
   D. $6n + 12$

34. **MP STRUCTURE** Each day, you run on a treadmill for r minutes and lift weights for 15 minutes. Which expressions can you use to find how many minutes of exercise you do in 5 days? Explain your reasoning.

   $5(r + 15)$ $5r + 5 \cdot 15$ $5r + 15$ $r(5 + 15)$

35. **MODELING REAL LIFE** A cheetah can run 103 feet per second. A zebra can run x feet per second. Write and simplify an expression that represents how many feet farther the cheetah can run in 10 seconds.

COMBINING LIKE TERMS Simplify the expression.

36. $6(x + 4) + 1$
37. $5 + 8(3 + x)$
38. $x + 3 + 5x$
39. $7y + 6 - 1 + 12y$
40. $4d + 9 - d - 8$
41. $n + 3(n - 1)$
42. $2v + 8v - 5v$
43. $5(z + 4) + 5(2 - z)$
44. $2.7(w - 5.2)$
45. $\frac{2}{3}y + \frac{1}{6}y + y$
46. $\frac{3}{4}\left(z + \frac{2}{5}\right) + 2z$
47. $7(x + y) - 7x$
48. $4x + 9y + 3(x + y)$

49. **YOU BE THE TEACHER** Your friend simplifies the expression. Is your friend correct? Explain your reasoning.

$8x - 2x + 5x = 8x - 7x$
$= (8 - 7)x$
$= x$

50. **MP REASONING** Evaluate each expression by (1) using the Distributive Property and (2) evaluating inside the parentheses first. Which method do you prefer? Is your preference the same for both expressions? Explain your reasoning.

   a. $2(3.22 - 0.12)$
   b. $12\left(\frac{1}{2} + \frac{2}{3}\right)$

51. **DIG DEEPER!** An art club sells 42 large candles and 56 small candles.

   a. Write and simplify an expression that represents the profit.
   b. A large candle costs $5, and a small candle costs $3. What is the club's profit?

Profit = Price − Cost

52. **MP REASONING** Find the difference between the perimeters of the rectangle and the hexagon. Interpret your answer.

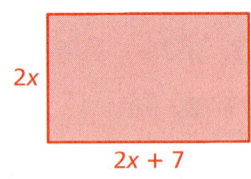
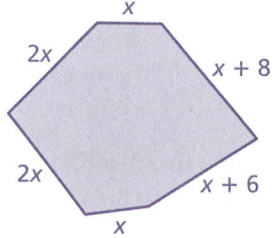

53. **PUZZLE** Add one set of parentheses to the expression $7 \cdot x + 3 + 8 \cdot x + 3 \cdot x + 8 - 9$ so that it is equivalent to $2(9x + 10)$.

5.5 Factoring Expressions

Learning Target: Factor numerical and algebraic expressions.

Success Criteria:
- I can use the Distributive Property to factor numerical expressions.
- I can identify the greatest common factor of terms including variables.
- I can use the Distributive Property to factor algebraic expressions.
- I can interpret factored expressions in real-life problems.

EXPLORATION 1 Finding Dimensions

Work with a partner.

a. The models show the area (in square units) of each part of a rectangle. Use the models to find missing values that complete the expressions. Explain your reasoning.

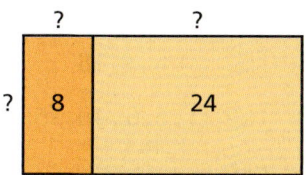

$8 + 24 = \boxed{}(\boxed{} + \boxed{})$

Math Practice

Evaluate Results

Do your answers in the first two models seem reasonable? How can you check your answers?

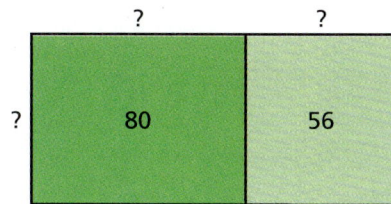

$80 + 56 = \boxed{}(\boxed{} + \boxed{})$

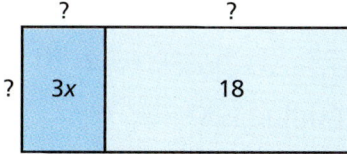

$3x + 18 = \boxed{}(\boxed{} + \boxed{})$

b. In part (a), check that the original expressions are equivalent to the expressions you wrote. Explain your reasoning.

c. Explain how you can use the Distributive Property to rewrite a sum of two whole numbers with a common factor.

5.5 Lesson

Key Vocabulary 🔊
factoring an expression, p. 228

Key Idea

Factoring an Expression

Words Writing a numerical expression or algebraic expression as a product of factors is called **factoring the expression**. You can use the Distributive Property to factor expressions.

Numbers $3 \cdot 7 + 3 \cdot 2 = 3(7 + 2)$ **Algebra** $ab + ac = a(b + c)$
$3 \cdot 7 - 3 \cdot 2 = 3(7 - 2)$ $ab - ac = a(b - c)$

EXAMPLE 1 Factoring Numerical Expressions

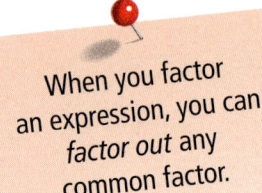

When you factor an expression, you can *factor out* any common factor.

a. **Factor 18 + 30 using the GCF.**

One way to find the GCF of 18 and 30 is to list their factors.

Factors of 18: ①,②,③,⑥, 9, 18
Factors of 30: ①,②,③, 5,⑥, 10, 15, 30   Circle the common factors.

The GCF of 18 and 30 is 6.

Write each term of the expression as a product of the GCF and the remaining factor. Then use the Distributive Property to factor the expression.

$18 + 30 = 6(3) + 6(5)$ Rewrite using GCF.
$ = 6(3 + 5)$ Distributive Property

b. **Factor 20 − 12 using the GCF.**

One way to find the GCF of 20 and 12 is to list their factors.

Factors of 20: ①,②,④, 5, 10, 20
Factors of 12: ①,②, 3,④, 6, 12   Circle the common factors.

The GCF of 20 and 12 is 4.

Write each term of the expression as a product of the GCF and the remaining factor. Then use the Distributive Property to factor the expression.

$20 - 12 = 4(5) - 4(3)$ Rewrite using GCF.
$ = 4(5 - 3)$ Distributive Property

Try It Factor the expression using the GCF.

1. $9 + 15$ 2. $60 + 45$ 3. $30 - 20$

EXAMPLE 2 Factoring Algebraic Expressions

a. **Factor $3x + 42$ using the GCF.**

You can find the GCF of $3x$ and 42 by writing their prime factorizations.

$3x = ③ \cdot x$
$42 = 2 \cdot ③ \cdot 7$ Circle the common prime factor.

The GCF of $3x$ and 42 is 3. Use the GCF to factor the expression.

$3x + 42 = 3(x) + 3(14)$ Rewrite using GCF.
$\qquad\quad = 3(x + 14)$ Distributive Property

b. **Factor $63z - 27y$ using the GCF.**

You can find the GCF of $63z$ and $27y$ by writing their prime factorizations.

$63z = ③ \cdot ③ \cdot 7 \cdot z$
$27y = ③ \cdot ③ \cdot 3 \cdot y$ Circle the common prime factors.

The GCF of $63z$ and $27y$ is $3 \cdot 3 = 9$. Use the GCF to factor the expression.

$63z - 27y = 9(7z) - 9(3y)$ Rewrite using GCF.
$\qquad\qquad\; = 9(7z - 3y)$ Distributive Property

Try It Factor the expression using the GCF.

4. $7x + 49$ 5. $8y - 44$ 6. $25a + 10b$

Self-Assessment for Concepts & Skills

Solve each exercise. Then rate your understanding of the success criteria in your journal.

FACTORING EXPRESSIONS Factor the expression using the GCF.

7. $16 + 24$ 8. $49 - 28$ 9. $8y + 14$

10. **WHICH ONE DOESN'T BELONG?** Which expression does *not* belong with the other three? Explain your reasoning.

   $3(8n + 12)$ $4(6n + 9)$ $6(4n + 3)$ $12(2n + 3)$

11. **MP REASONING** Use what you know about factoring to explain how you can factor the expression $18x + 30y + 9z$. Then factor the expression.

12. **CRITICAL THINKING** Identify the GCF of the terms $(x \cdot x)$ and $(4 \cdot x)$. Explain your reasoning. Then use the GCF to factor the expression $x^2 + 4x$.

Section 5.5 Factoring Expressions

EXAMPLE 3 Modeling Real Life

You receive a discount on each book you buy for your electronic reader. The original price of each book is x dollars. You buy 5 books for a total of $(5x - 15)$ dollars. Factor the expression. What can you conclude about the discount?

To factor $5x - 15$, you can find the GCF of $5x$ and 15 by writing their prime factorizations.

$$5x = \boxed{5} \cdot x$$
$$15 = \boxed{5} \cdot 3$$

Circle the common prime factor.

So, the GCF of $5x$ and 15 is 5. Use the GCF to factor the expression.

$5x - 15 = 5(x) - 5(3)$ Rewrite using GCF.
$ = 5(x - 3)$ Distributive Property

The factor 5 represents the number of books purchased. The factor $(x - 3)$ represents the discounted price of each book. This factor is a difference of two terms, showing that the original price, $\$x$, of each book is decreased by $\$3$.

▶ So, the factored expression shows a $\$3$ discount for every book you buy. The original expression shows a total savings of $\$15$.

Check Suppose that the original price of each book is $\$6$. Verify that each expression has the same value when $x = 6$.

$5x - 15 = 5(6) - 15 = 15$
$5(x - 3) = 5(6 - 3) = 15$ ✓

Self-Assessment for Problem Solving

Solve each exercise. Then rate your understanding of the success criteria in your journal.

13. A youth club receives a discount on each pizza purchased for a party. The original price of each pizza is x dollars. The club leader purchases 8 pizzas for a total of $(8x - 32)$ dollars. Factor the expression. What can you conclude about the discount?

14. Three crates of food are packed on a shuttle departing for the Moon. Each crate weighs x pounds. On the Moon, the combined weight of the crates is $(3x - 81)$ pounds. What can you conclude about the weight of each crate on the Moon?

5.5 Practice

Review & Refresh

Use the Distributive Property to simplify the expression.

1. $2(n + 8)$
2. $3(4 + m)$
3. $7(b - 3)$
4. $10(4 - w)$

Write the phrase as an expression.

5. 5 plus a number p
6. 18 less than a number r
7. 11 times a number d
8. a number c divided by 25

Decide whether the rates are equivalent.

9. 84 feet in 12 seconds
   217 feet in 31 seconds
10. 12 cups of soda for every 54 cups of juice
    8 cups of soda for every 36 cups of juice

Match the decimal with its equivalent percent.

11. 0.36
12. 3.6
13. 0.0036
14. 0.036

A. 0.36%
B. 360%
C. 36%
D. 3.6%

Concepts, Skills, & Problem Solving

FINDING DIMENSIONS The model shows the area (in square units) of each part of a rectangle. Use the model to find missing values that complete the expression. Explain your reasoning. (See Exploration 1, p. 227.)

15. $12 + 16 = \boxed{}(\boxed{} + \boxed{})$

16. $48 + 32 = \boxed{}(\boxed{} + \boxed{})$

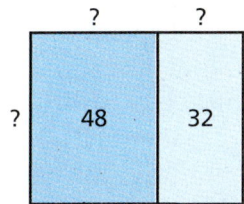

FACTORING NUMERICAL EXPRESSIONS Factor the expression using the GCF.

17. $7 + 14$
18. $12 + 42$
19. $22 + 11$
20. $70 + 95$
21. $60 - 36$
22. $100 - 80$
23. $84 + 28$
24. $48 + 80$
25. $19 + 95$
26. $44 - 11$
27. $18 - 12$
28. $48 + 16$
29. $98 - 70$
30. $58 + 28$
31. $72 - 39$
32. $69 + 84$

33. **MP REASONING** The whole numbers a and b are divisible by c, where b is greater than a. Is $a + b$ divisible by c? Is $b - a$ divisible by c? Explain your reasoning.

34. **MULTIPLE CHOICE** Which expression is *not* equivalent to $81x + 54$?

A. $27(3x + 2)$
B. $3(27x + 18)$
C. $9(9x + 6)$
D. $6(13x + 9)$

Section 5.5 Factoring Expressions 231

FACTORING ALGEBRAIC EXPRESSIONS Factor the expression using the GCF.

35. $2x + 10$
36. $15x + 6$
37. $26x - 13$
38. $50x - 60$
39. $36x + 9$
40. $14x - 98$
41. $18p + 26$
42. $16m + 40$
43. $24 + 72n$
44. $50 + 65h$
45. $76d - 24$
46. $27 - 45c$
47. $18t + 38x$
48. $90y + 65z$
49. $10x - 25y$
50. $24y + 88x$

51. OPEN-ENDED Use the Distributive Property to write two expressions that are equivalent to $8x + 16$.

MATCHING Match the expression with an equivalent expression.

52. $8x + 16y$
53. $4x + 8y$
54. $16x + 8y$
55. $8x + 4y$

A. $4(2x + y)$
B. $2(4y + 2x)$
C. $4(2x + 4y)$
D. $8(y + 2x)$

56. YOU BE THE TEACHER Your friend factors the expression $24x + 56$. Is your friend correct? Explain your reasoning.

$24x + 56 = 8(3x) + 8(7)$
$= (8 + 8) \cdot (3x + 7)$
$= 16(3x + 7)$

57. MODELING REAL LIFE You sell soup mixes for a fundraiser. For each soup mix you sell, the company that makes the soup receives x dollars, and you receive the remaining amount. You sell 16 soup mixes for a total of $(16x + 96)$ dollars. How much money do you receive for each soup mix that you sell?

58. MP PROBLEM SOLVING A clothing store is having a sale on holiday socks. Each pair of socks costs x dollars. You leave the store with 6 pairs of socks and spend a total of $(6x - 14)$ dollars. You pay with $40. How much change do you receive? Explain your reasoning.

59. MP STRUCTURE You buy 37 concert tickets for $8 each, and then sell all 37 tickets for $11 each. The work below shows two ways you can determine your profit. Describe each solution method. Which do you prefer? Explain your reasoning.

Profit = $37(11) - 37(8)$
= $407 - 296$
= $111

Profit = $37(11) - 37(8)$
= $37(11 - 8)$
= $37(3)$
= $111

60. MP NUMBER SENSE The prime factorizations of two numbers are shown, where a and b represent prime numbers. Write the sum of the two numbers as an expression of the form $14(\square + \square)$. Explain your reasoning.

Number 1: $2 \cdot 11 \cdot 5 \cdot a$
Number 2: $7 \cdot b \cdot 3 \cdot 3$

5 Connecting Concepts

▶ Using the Problem-Solving Plan

1. A store sells 18 pairs of the wireless earbuds shown. Customers saved a total of $882 on the earbuds. Find the original price of the earbuds.

   35% off wireless earbuds

   Understand the problem. You know the percent discount on a pair of wireless earbuds, the number of pairs of earbuds sold, and the total amount of money that customers saved. You are asked to find the original price of the earbuds.

   Make a plan. First, write an expression that represents the total amount of money that customers pay for the earbuds. Then factor the expression to find the discount (in dollars) on each pair of earbuds. Finally, solve a percent problem to find the original price.

   Solve and check. Use the plan to solve the problem. Then check your solution.

2. All of the weight plates in a gym are labeled in kilograms. You want to convert the weights to pounds. Write an expression to find the number of pounds in z kilograms. Then find the weight in pounds of a plate that weighs 20.4 kilograms.

3. You buy apple chips and banana chips in the ratio of 2 : 7.

   a. How many ounces of banana chips do you buy when you buy n ounces of apple chips? Explain.

   b. You buy 12 ounces of apple chips. How many ounces of banana chips do you buy?

Performance Task

Describing Change

At the beginning of this chapter, you watched a STEAM video called "Shadow Drawings." You are now ready to complete the performance task related to this video, available at *BigIdeasMath.com*. Be sure to use the problem-solving plan as you work through the performance task.

Chapter 5 Chapter Review

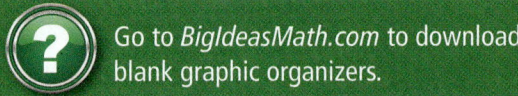

Go to BigIdeasMath.com to download blank graphic organizers.

Review Vocabulary

Write the definition and give an example of each vocabulary term.

algebraic expression, *p. 202*
variable, *p. 202*
term, *p. 202*
coefficient, *p. 202*
constant, *p. 202*
equivalent expressions, *p. 216*
like terms, *p. 223*
factoring an expression, *p. 228*

Graphic Organizers

You can use an **Example and Non-Example Chart** to list examples and non-examples of a concept. Here is an Example and Non-Example Chart for the *Commutative Property of Addition*.

Commutative Property of Addition

Examples	Non-Examples
$a + b = b + a$	$a \cdot b = b \cdot a$
$2.1 + 9 = 9 + 2.1$	$(7 + 4) + 2 = 7 + (4 + 2)$
$17 + 34 = 34 + 17$	$b \cdot 0 = 0$
$(6 + x) + 8 = (x + 6) + 8$	$46 \cdot 1 = 46$
$(3 + y) + 1 = 1 + (3 + y)$	$2(12 + x) = 2(12) + 2(x)$

Choose and complete a graphic organizer to help you study the concept.

1. algebraic expressions
2. variable
3. Commutative Property of Multiplication
4. Associative Property of Addition
5. Associative Property of Multiplication
6. Addition Property of Zero
7. Multiplication Property of Zero
8. Multiplication Property of One
9. Distributive Property

"I finished my **Example and Non-Example Chart** about things we need on the moon."

Chapter Self-Assessment

As you complete the exercises, use the scale below to rate your understanding of the success criteria in your journal.

1	2	3	4
I do not understand.	I can do it with help.	I can do it on my own.	I can teach someone else.

5.1 Algebraic Expressions (pp. 201–208)

Learning Target: Evaluate algebraic expressions given values of their variables.

Identify the terms, coefficients, and constants in the expression.

1. $9x + 2 + 8y$
2. $3x^2 + x + 7$
3. $1 + \frac{q}{4} + 7q$

Evaluate the expression when $x = 20$, $y = 4$, and $z = 7$.

4. $x \div 5$
5. $12 - z$
6. $4y$
7. $y + x$
8. $x \cdot z$
9. $x - y$
10. $3z + 8$
11. $8y - x$
12. $\frac{x^2}{y}$

13. The amount earned (in dollars) for recycling p pounds of copper is $2p$. How much do you earn for recycling 28 pounds of copper?

14. While playing a video game, you score p game points and b triple bonus points. An expression for your score is $p + 3b$. What is your score when you earn 245 game points and 20 triple bonus points?

15. Tickets for a baseball game cost a dollars for adults and c dollars for children. The expression $2a + 3c$ represents the cost (in dollars) for a family to go to the game. What is the cost for the family when an adult ticket is $17 and a child ticket is $12?

16. Add one set of parentheses to the expression $2x^2 + 4 - 5$ so that the value of the expression is 75 when $x = 6$.

Chapter Review 235

5.2 Writing Expressions (pp. 209–214)

Learning Target: Write algebraic expressions and solve problems involving algebraic expressions.

Write the phrase as an expression.

17. 9 fewer than 23

18. 6 more than the quotient of 15 and 3

19. the product of a number d and 32

20. a number t decreased by 17

21. Your basketball team scored 4 fewer than twice as many points as the other team.

   a. Write an expression that represents the number of points your team scored.

   b. The other team scored 24 points. How many points did your team score?

22. The boiling temperature (in degrees Celsius) of platinum is 199 more than four times the boiling temperature (in degrees Celsius) of zinc.

   a. Write an expression that represents the boiling temperature (in degrees Celsius) of platinum.

   b. The boiling temperature of zinc is 907 degrees Celsius. What is the boiling temperature of platinum?

23. Write an algebraic expression with two variables, x and y, that has a value of 50 when $x = 3$ and $y = 5$.

5.3 Properties of Addition and Multiplication (pp. 215–220)

Learning Target: Identify equivalent expressions and apply properties to generate equivalent expressions.

Simplify the expression. Explain each step.

24. $10 + (2 + y)$

25. $(21 + b) + 1$

26. $3(7x)$

27. $1(3.2w)$

28. $5.3 + (w + 1.2)$

29. $(0 + t) + 9$

30. The expression $7 + 3x + 4$ represents the perimeter of the triangle. Simplify the expression.

31. Write an algebraic expression that can be simplified using the Associative Property of Addition.

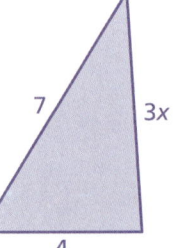

5.4 The Distributive Property (pp. 221–226)

Learning Target: Apply the Distributive Property to generate equivalent expressions.

Use the Distributive Property to simplify the expression.

32. $2(x + 12)$ **33.** $11(4b - 3)$ **34.** $8(s - 1)$ **35.** $6(6 + y)$

Simplify the expression.

36. $5(n + 3) + 4n$ **37.** $t + 2 + 6t$ **38.** $3z + 14 + 5z - 9$

39. A family of three goes to a salon. Each person gets a haircut and highlights. The cost of each haircut is $15, and the cost per person for highlights is x dollars. Write and simplify an expression that represents the total cost (in dollars) for the family at the salon.

40. Each day, you take vocal lessons for v minutes and trumpet lessons for 30 minutes. Write and simplify an expression to find how many minutes of lessons you take in 4 days.

5.5 Factoring Expressions (pp. 227–232)

Learning Target: Factor numerical and algebraic expressions.

Factor the expression using the GCF.

41. $42 - 12$ **42.** $15 + 35$ **43.** $36x - 28$

44. $24 + 64x$ **45.** $60 - 150x$ **46.** $16x + 56y$

47. A soccer team receives a discount on each jersey purchased. The original price of each jersey is x dollars. The team buys 18 jerseys for a total of $(18x - 36)$ dollars. What can you conclude about the discount?

48. You sell apple cider for a fundraiser. For each gallon of cider you sell, the company that makes the cider receives x dollars, and you receive the remaining amount. You sell 15 gallons of cider for $(15x + 45)$ dollars. How much money do you receive for each gallon of cider that you sell?

5 Practice Test

1. Identify the terms, coefficients, and constants of $\frac{q}{3} + 6 + 9q$.

2. Evaluate $4b - a$ when $a = 12$ and $b = 7$.

Write the phrase as an expression.

3. 25 more than 50
4. 6 less than the quotient of 32 and a number y

Simplify the expression. Explain each step.

5. $3.1 + (8.6 + m)$
6. $\left(\frac{2}{3} \cdot t\right) \cdot 1\frac{1}{2}$
7. $4(x + 8)$
8. $4t + 7 + 2t - 2$

Factor the expression using the GCF.

9. $18 + 24$
10. $15x + 20$
11. $32x - 40y$

12. Playing time is added at the end of a soccer game to make up for stoppages. An expression for the length (in minutes) of a 90-minute soccer game with x minutes of stoppage time is $90 + x$. How long is a game with 4 minutes of stoppage time?

13. The expression $15 \cdot x \cdot 6$ represents the volume of a rectangular prism with a length of 15, a width of x, and a height of 6. Simplify the expression.

14. The Coiling Dragon Cliff Skywalk in China is 128 feet longer than the length x (in feet) of the Tianmen Skywalk in China. The world's longest glass-bottom bridge, located in China's Zhangjiajie National Park, is about 4.3 times longer than the Coiling Dragon Cliff Skywalk. Write and simplify an expression that represents the length (in feet) of the world's longest glass-bottom bridge.

15. A youth group is making and selling sandwiches to raise money. The cost to make each sandwich is h dollars. The group sells 150 sandwiches for a total of $(150h + 450)$ dollars. How much profit does the group earn for each sandwich sold?

16. You make party favors for an event. You tie 9 inches of ribbon around each party favor. Write an expression for the number of inches of ribbon needed for n party favors. The ribbon costs $3 for each *yard*. Write an expression for the total cost (in dollars) of the ribbon.

5. Cumulative Practice

1. The student council is organizing a school fair. Council members are making signs to show the prices for admission and for each game a person can play.

   SCHOOL FAIR
   Admission $2.00
   Price per game $0.25

   Let x represent the number of games. Which expression can you use to determine the total amount (in dollars) a person pays for admission and playing x games?

   A. 2.25
   B. $2.25x$
   C. $2 + 0.25x$
   D. $2x + 0.25$

Test-Taking Strategy
After Answering Easy Questions, Relax

"After answering easy questions, relax and try the harder ones. For this, $3(4) = 12$. So, it is B."

2. Which ratio relationship is represented in the graph?

   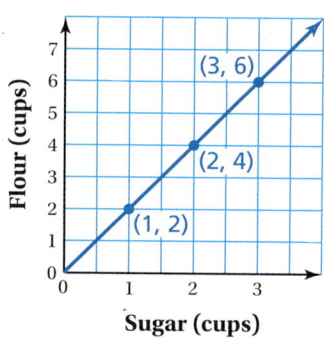

   F. 2 cups of flour for every $\frac{1}{2}$ cup of sugar
   G. 6 cups of flour for every 3 cups of sugar
   H. 1 cup of flour for every 4 cups of sugar
   I. $\frac{1}{2}$ cup of flour for every 1 cup of sugar

3. At a used bookstore, you can purchase two types of books.

   You can use the expression $3h + 2p$ to find the total cost (in dollars) for h hardcover books and p paperback books. What is the total cost (in dollars) for 6 hardcover books and 4 paperback books?

4. Your friend divided two decimal numbers. Her work is shown in the box below. What should your friend change in order to divide the two decimal numbers correctly?

$$0.07\overline{)14.56} \rightarrow 7\overline{)14.56}^{\,2.08}$$

   A. Rewrite the problem as $0.07\overline{)0.1456}$.
   B. Rewrite the problem as $0.07\overline{)1456}$.
   C. Rewrite the problem as $7\overline{)0.1456}$.
   D. Rewrite the problem as $7\overline{)1456}$.

5. What is the value of $4.391 + 5.954$?

   F. 9.12145
   G. 9.245
   H. 9.345
   I. 10.345

6. The circle graph shows the eye color of students in a sixth-grade class. Nine students in the class have brown eyes. How many students are in the class?

   A. 4 students
   B. 18 students
   C. 20 students
   D. 405 students

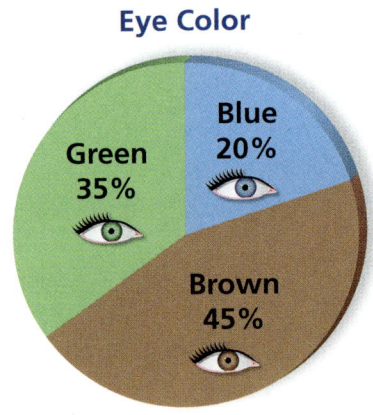

Eye Color
Blue 20%
Green 35%
Brown 45%

7. Properties of Addition and Multiplication are used to simplify an expression.

$$36 \cdot 23 + 33 \cdot 64 = 36 \cdot 23 + 64 \cdot 33$$
$$= 36 \cdot 23 + 64 \cdot (23 + 10)$$
$$= 36 \cdot 23 + 64 \cdot 23 + 64 \cdot 10$$
$$= x \cdot 23 + 64 \cdot 10$$

What number belongs in place of the x?

8. What is the prime factorization of 1350?

  F. 10 • 135
  G. 2 • 3 • 5
  H. 6 • 225
  I. $2 \cdot 3^3 \cdot 5^2$

9. A horse gallops at a speed of 44 feet per second. What is the speed of the horse in miles per hour?

  A. $\frac{1}{2}$ mile per hour
  B. 30 miles per hour
  C. $64\frac{8}{15}$ miles per hour
  D. 158,400 miles per hour

10. Which equation correctly demonstrates the Distributive Property?

  F. $a(b + c) = ab + c$
  G. $a(b + c) = ab + ac$
  H. $a + (b + c) = (a + b) + (a + c)$
  I. $a + (b + c) = (a + b) \cdot (a + c)$

11. Which number is equivalent to $2\frac{4}{5} \cdot 1\frac{2}{7}$?

  A. $2\frac{8}{45}$
  B. $2\frac{8}{35}$
  C. $3\frac{3}{5}$
  D. $4\frac{3}{35}$

12. Which pair of numbers does *not* have a least common multiple of 24?

  F. 2, 12
  G. 3, 8
  H. 6, 8
  I. 12, 24

13. Use the Properties of Multiplication to simplify the expression in an efficient way. Show your work and explain how you used the Properties of Multiplication.

$$(25 \times 18) \times 4$$

14. Which number is *not* a perfect square?

  A. 64
  B. 81
  C. 96
  D. 100

Cumulative Practice   241

6 Equations

6.1 Writing Equations in One Variable

6.2 Solving Equations Using Addition or Subtraction

6.3 Solving Equations Using Multiplication or Division

6.4 Writing Equations in Two Variables

Chapter Learning Target:
Understand equations.

Chapter Success Criteria:
- I can identify key words and phrases.
- I can write word sentences as equations.
- I can solve equations using properties of equality.
- I can model different types of equations to solve real-life problems.

STEAM Video: "Rock Climbing"

STEAM Video

Rock Climbing

Equations can be used to solve many different kinds of problems in real life, such as estimating the amount of time it will take to climb a rock wall. Can you think of any other real-life situations where equations are useful?

In rock climbing, a *pitch* is a section of a climbing route between two anchor points. Watch the STEAM Video "Rock Climbing." Then answer the following questions.

1. How can you use pitches to estimate the amount of time it will take to climb a rock wall?

2. Are there any other methods you could use to estimate the amount of time it will take to climb a rock wall? Explain.

3. You know two of the three pieces of information below. Explain how you can find the missing piece of information.

   Average climbing speed

   Height of rock wall

   Time to complete climb

Performance Task

Planning the Climb

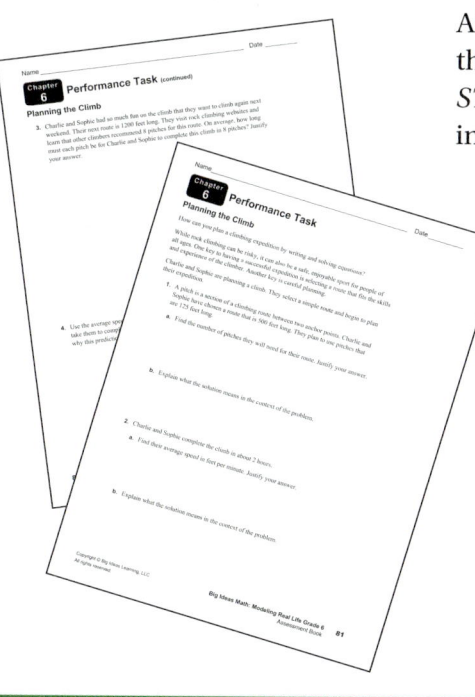

After completing this chapter, you will be able to use the concepts you learned to answer the questions in the *STEAM Video Performance Task*. You will be given information about two rock-climbing routes.

Route 1: 500 feet, 125 feet per pitch

Route 2: 1200 feet, 8 pitches

You will find the average speed of the climbers on Route 1 and the amount of time it takes to complete Route 2. Will the average speed of the climbers on Route 1 provide accurate predictions for the amount of time it takes to climb other routes? Explain why or why not.

243

Getting Ready for Chapter 6

Chapter Exploration

Work with a partner. Every equation that has an unknown variable can be written as a question. Write a question that represents the equation. Then answer the question.

	Equation	Question	Answer to Question
1.	$x + 3 = 7$		
2.	$5 - x = 2$		
3.	$3x = 12$		
4.	$x \div 5 = 3$		
5.	$20 \div x = 4$		

Work with a partner. Write an equation that represents the question. Then answer the question.

	Question	Equation	Answer to Question
6.	What number can be added to 7 to get 12?		
7.	What number can be subtracted from 11 to get 3?		
8.	What number can be multiplied by 10 to get 30?		
9.	What number can be divided by 7 to get 3?		
10.	**MODELING REAL LIFE** Your friend says that he will be 21 years old in 7 years. How old is he now?		

Vocabulary

The following vocabulary terms are defined in this chapter. Think about what each term might mean and record your thoughts.

equation
inverse operations
equation in two variables
independent variable
dependent variable

6.1 Writing Equations in One Variable

Learning Target: Write equations in one variable and write equations that represent real-life problems.

Success Criteria:
- I can identify key words and phrases that indicate equality.
- I can write word sentences as equations.
- I can create equations to represent real-life problems.

EXPLORATION 1

Writing Equations

Work with a partner. Customers order sandwiches at a café from the menu board shown.

a. The equation $6.75x = 20.25$ represents the purchase of one customer from the menu board. What does the equation tell you about the purchase? What cannot be determined from the equation?

b. The four customers in the table buy multiple sandwiches of the same type. For each customer, write an equation that represents the situation. Then determine how many sandwiches each customer buys. Explain your reasoning.

Math Practice

Analyze Givens
What information do you need to solve the problem?

	Sandwich	Amount Used for Payment	Change Received
Customer A	Reuben	$20	$0.65
Customer B	Chicken salad	$10	$0.10
Customer C	BLT	$30	$9.00
Customer D	Egg salad	$50	$26.75

Section 6.1 Writing Equations in One Variable 245

6.1 Lesson

Key Vocabulary
equation, p. 246

An **equation** is a mathematical sentence that uses an equal sign, =, to show that two expressions are equal.

Expressions	Equations
$4 + 8$	$4 + 8 = 12$
$x + 8$	$x + 8 = 12$

To write a word sentence as an equation, look for key words or phrases such as *is*, *the same as*, or *equals* to determine where to place the equal sign.

EXAMPLE 1 Writing Equations

Write each word sentence as an equation.

a. The sum of a number n and 7 is 15.

$$\underbrace{\text{The sum of a number } n \text{ and } 7}_{n + 7} \; \underbrace{\text{is}}_{=} \; \underbrace{15}_{15}$$

▸ An equation is $n + 7 = 15$.   *Sum of means addition.*

b. A number y decreased by 4 is 3.

$$\underbrace{\text{A number } y \text{ decreased by } 4}_{y - 4} \; \underbrace{\text{is}}_{=} \; \underbrace{3}_{3}$$

▸ An equation is $y - 4 = 3$.   *Decreased by means subtraction.*

c. 48 equals 12 times a number p.

$$\underbrace{48}_{48} \; \underbrace{\text{equals}}_{=} \; \underbrace{12 \text{ times a number } p}_{12p}$$

▸ An equation is $48 = 12p$.   *Times means multiplication.*

Try It Write the word sentence as an equation.

1. 9 less than a number b equals 2.
2. The product of a number g and 5 is 30.
3. A number k increased by 10 is the same as 24.
4. The quotient of a number q and 4 is 12.
5. $2\frac{1}{2}$ is the same as the sum of a number w and $\frac{1}{2}$.

EXAMPLE 2 Writing an Equation

Ten servers decorate 25 tables for a wedding. Each table is decorated as shown. Let c be the total number of yellow and purple candles. Which equation can you use to find c?

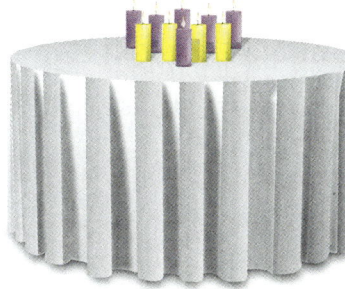

A. $c = 25 + (4 \times 6)$ **B.** $c = 25(4 + 6)$

C. $c = 10(25 + 4 + 6)$ **D.** $c = 10(4 + 6)$

Use a verbal model to write an equation.

Verbal Model | Total number of candles | = | Number of tables | $\cdot$ | Number of candles on each table |

Variable Let c be the total number of candles.

Equation $c \ = \ 25 \ \cdot \ (4 + 6)$

 The correct answer is **B**.

Try It

6. **WHAT IF?** Each server decorates one table. Which equation can you use to find c?

Self-Assessment for Concepts & Skills

Solve each exercise. Then rate your understanding of the success criteria in your journal.

7. **VOCABULARY** How are expressions and equations different?

8. **DIFFERENT WORDS, SAME QUESTION** Which is different? Write "both" equations.

   4 less than a number n is 8.

   A number n is 4 less than 8.

   A number n minus 4 equals 8.

   4 subtracted from a number n is 8.

9. **OPEN-ENDED** Write a word sentence for the equation $28 - n = 5$.

10. **WRITING** You purchase x items for $4 each. Explain how the variable in the expression $4x$ and the variable in the equation $4x = 20$ are similar. Explain how they are different.

Section 6.1 Writing Equations in One Variable **247**

EXAMPLE 3 Modeling Real Life

After two rounds, 24 students are eliminated from a spelling bee. There are 96 students remaining. Find the number of students who started the spelling bee.

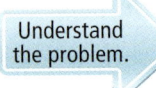
Understand the problem.

You are given the numbers of students who have and have not been eliminated from a spelling bee. You are asked for the number of students who started the spelling bee.

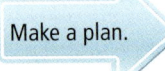
Make a plan.

Write and solve an equation relating the number of students who started, the number eliminated, and the number remaining. Use a verbal model.

Solve and check.

Verbal Model   Number of students who started − Number of students eliminated = Number of students remaining

Reading
The word *eliminated* indicates *subtraction*.

Variable   Let s be the number of students who started.

Equation $s - 24 = 96$

An equation is $s - 24 = 96$. Because you can think of the equation as saying, "*What number minus 24 is 96?*," you know that s is 24 more than 96.

▸ So, $96 + 24 = 120$ students started the spelling bee.

Check Verify that $120 - 24 = 96$.
$$120 - 24 = 120 - 20 - 4$$
$$= 100 - 4$$
$$= 96 \checkmark$$

Self-Assessment for Problem Solving

Solve each exercise. Then rate your understanding of the success criteria in your journal.

11. After four rounds, 74 teams are eliminated from a robotics competition. There are 18 teams remaining. Write and solve an equation to find the number of teams that started the competition.

12. The mass of the blue copper sulfate crystal is two-thirds the mass of the red fluorite crystal. Write an equation you can use to find the mass (in grams) of the blue copper sulfate crystal.

30 grams

m grams

13. **DIG DEEPER!** You print photographs from a vacation. Find the number of photographs you can print for $3.60.

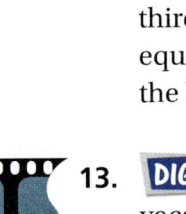

6.1 Practice

Review & Refresh

Factor the expression using the GCF.

1. $6 + 27$
2. $9w + 72$
3. $42 + 24n$
4. $18h + 30k$

5. Which number is *not* equal to 225%?

   A. $2\frac{1}{4}$
   B. $\frac{9}{4}$
   C. $\frac{50}{40}$
   D. $\frac{45}{20}$

Evaluate the expression when $a = 7$.

6. $6 + a$
7. $a - 4$
8. $4a$
9. $\frac{35}{a}$

Find the perimeter of the rectangle.

10.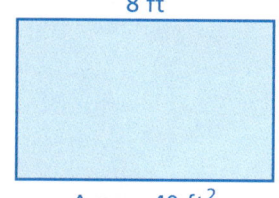
    8 ft
    Area = 40 ft²

11.
    13 cm
    Area = 52 cm²

12.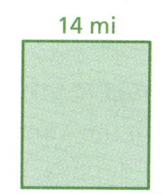
    14 mi
    Area = 224 mi²

Concepts, Skills, & Problem Solving

WRITING EQUATIONS A roast beef sandwich costs $6.75. A customer buys multiple roast beef sandwiches. Write an equation that represents the situation. Then determine how many sandwiches the customer buys. *(See Exploration 1, p. 245.)*

13.
Amount Used for Payment	$50
Change Received	$16.25

14.
Amount Used for Payment	$80
Change Received	$19.25

WRITING EQUATIONS Write the word sentence as an equation.

15. A number y decreased by 9 is 8.
16. The sum of a number x and 4 equals 12.
17. 9 times a number b is 36.
18. A number w divided by 5 equals 6.
19. 54 equals 9 more than a number t.
20. 5 is one-fourth of a number c.
21. 9.5 less than a number n equals 27.
22. $11\frac{3}{4}$ is the quotient of a number y and $6\frac{1}{4}$.

23. **YOU BE THE TEACHER** Your friend writes the word sentence as an equation. Is your friend correct? Explain your reasoning.

    > 5 less than a number n is 12.
    > $n - 5 = 12$

24. **MODELING REAL LIFE** Students and faculty raise $6042 for band uniforms. The faculty raised $1780. Write an equation you can use to find the amount a (in dollars) the students raised.

25. **MODELING REAL LIFE** You hit a golf ball 90 yards. It travels three-fourths of the distance to the hole. Write an equation you can use to find the distance *d* (in yards) from the tee to the hole.

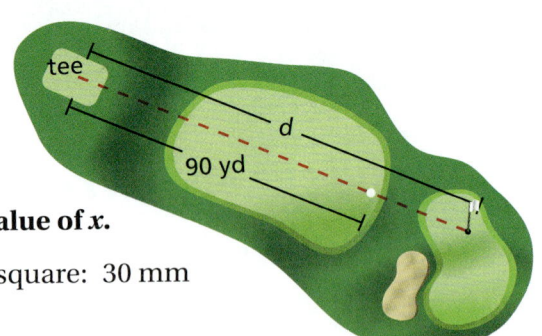

GEOMETRY Write an equation you can use to find the value of *x*.

26. Perimeter of triangle: 16 in.

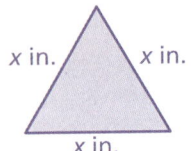

27. Perimeter of square: 30 mm

28. **MODELING REAL LIFE** You sell instruments at a Caribbean music festival. You earn $326 by selling 12 sets of maracas, 6 sets of claves, and *x* djembe drums. Find the number of djembe drums you sold.

29. **MP PROBLEM SOLVING** Neil Armstrong set foot on the Moon 109.4 hours after *Apollo 11* departed from the Kennedy Space Center. *Apollo 11* landed on the Moon about 6.6 hours before Armstrong's first step. How many hours did it take for *Apollo 11* to reach the Moon?

30. **MP LOGIC** You buy a basket of 24 strawberries. You eat them as you walk to the beach. It takes the same amount of time to walk each block. When you are halfway there, half of the berries are gone. After walking 3 more blocks, you still have 5 blocks to go. You reach the beach 28 minutes after you began. One-sixth of your strawberries are left.

   a. Is there enough information to find the time it takes to walk each block? Explain.

   b. Is there enough information to find how many strawberries you ate while walking the last block? Explain.

31. **DIG DEEPER!** Find a sales receipt from a store that shows the total price of the items and the total amount paid including sales tax.

   a. Write an equation you can use to find the sales tax rate *r*.

   b. Can you use *r* to find the *percent* for the sales tax? Explain.

32. **GEOMETRY** A square is cut from a rectangle. The side length of the square is half of the unknown width *w*. The area of the shaded region is 84 square inches. Write an equation you can use to find the width (in inches).

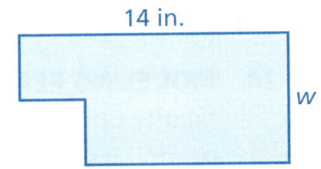

250 Chapter 6 Equations

6.2 Solving Equations Using Addition or Subtraction

Learning Target: Write and solve equations using addition or subtraction.

Success Criteria:
- I can determine whether a value is a solution of an equation.
- I can apply the Addition and Subtraction Properties of Equality to generate equivalent equations.
- I can solve equations using addition or subtraction.
- I can create equations involving addition or subtraction to solve real-life problems.

EXPLORATION 1

Solving an Equation Using a Tape Diagram

Work with a partner. A student solves an equation using the tape diagrams below.

Step 1: | 12 |
        | x | 4 |

Step 2: | 8 | 4 |
        | x | 4 |

Step 3: | 8 |
        | x |

a. What equation did the student solve? What is the solution?

b. Explain how the tape diagrams in Steps 2 and 3 relate to the equation and its solution.

EXPLORATION 2

Solving an Equation Using a Model

Work with a partner.

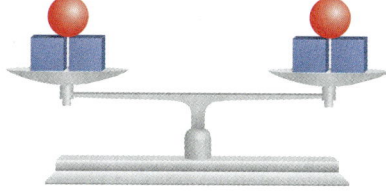

When two sides of a scale weigh the same, the scale will balance.

Math Practice

Analyze Relationships
How can you use the relationship between addition and subtraction to solve $x + 2 = 7$?

a. How are the two sides of an equation similar to a balanced scale?

b. When you add weight to one side of a balanced scale, what can you do to balance the scale? What if you subtract weight from one side of a balanced scale? How does this relate to solving an equation?

c. Use a model to solve $x + 2 = 7$. Describe how you can solve the equation algebraically.

Section 6.2 Solving Equations Using Addition or Subtraction 251

6.2 Lesson

Key Vocabulary
solution, p. 252
inverse operations, p. 253

Equations may be true for some values and false for others. A **solution** of an equation is a value that makes the equation true.

Value of x	x + 3 = 7	Are both sides equal?
3	$3 + 3 \stackrel{?}{=} 7$ $6 \neq 7$ ✗	no
4	$4 + 3 \stackrel{?}{=} 7$ $7 = 7$ ✓	yes
5	$5 + 3 \stackrel{?}{=} 7$ $8 \neq 7$ ✗	no

Reading
The symbol $\neq$ means *is not equal to*.

So, the value $x = 4$ is a solution of the equation $x + 3 = 7$.

EXAMPLE 1 Checking Solutions

Tell whether the given value is a solution of the equation.

a. $p + 10 = 38$; $p = 18$

$18 + 10 \stackrel{?}{=} 38$ Substitute 18 for *p*.

$28 \neq 38$ ✗   Sides are *not* equal.

▸ So, $p = 18$ is *not* a solution.

b. $4y = 56$; $y = 14$

$4(14) \stackrel{?}{=} 56$ Substitute 14 for *y*.

$56 = 56$ ✓   Sides are equal.

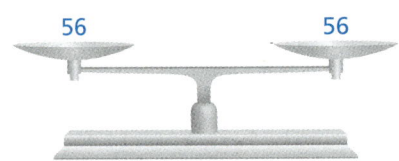

▸ So, $y = 14$ is a solution.

Try It Tell whether the given value is a solution of the equation.

1. $a + 6 = 17$; $a = 9$
2. $9 - g = 5$; $g = 3$
3. $35 = 7n$; $n = 5$
4. $\dfrac{q}{2} = 28$; $q = 14$

252   Chapter 6   Equations   Multi-Language Glossary at BigIdeasMath.com

You can use *inverse operations* to solve equations. **Inverse operations** "undo" each other. Addition and subtraction are inverse operations.

 **Key Ideas**

> You can add or subtract any number on each side of an equation to generate an equivalent equation. For example, $x + 4 = 5$ and $x + 2 = 3$ are equivalent equations.
> $$\begin{array}{r} x + 4 = 5 \\ -2 = -2 \\ \hline x + 2 = 3 \end{array}$$

Addition Property of Equality

Words When you add the same number to each side of an equation, the two sides remain equal.

Numbers
$$\begin{array}{r} 8 = 8 \\ +5 +5 \\ \hline 13 = 13 \end{array}$$

Algebra
$$\begin{array}{r} x - 4 = 5 \\ +4 +4 \\ \hline x = 9 \end{array}$$

Subtraction Property of Equality

Words When you subtract the same number from each side of an equation, the two sides remain equal.

Numbers
$$\begin{array}{r} 8 = 8 \\ -5 -5 \\ \hline 3 = 3 \end{array}$$

Algebra
$$\begin{array}{r} x + 4 = 5 \\ -4 -4 \\ \hline x = 1 \end{array}$$

EXAMPLE 2 Solving Equations Using Addition

a. Solve $x - 2 = 6$.

$x - 2 = 6$		Write the equation.
$+2 +2$		Addition Property of Equality
$x = 8$		Simplify.

Undo the subtraction.

 The solution is $x = 8$.

Check
$$x - 2 = 6$$
$$8 - 2 \stackrel{?}{=} 6$$
$$6 = 6 \checkmark$$

b. Solve $18 = x - 7$.

$18 = x - 7$	Write the equation.
$+7 +7$	Addition Property of Equality
$25 = x$	Simplify.

 The solution is $x = 25$.

Check Verify that $25 - 7 = 18$.

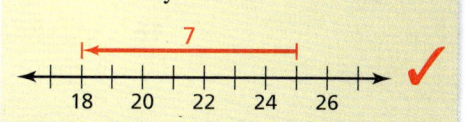

Try It Solve the equation. Check your solution.

5. $k - 3 = 1$
6. $n - 10 = 4$
7. $15 = r - 6$

EXAMPLE 3 Solving Equations Using Subtraction

a. Solve $x + 2 = 9$.

	$x + 2 =$	9	Write the equation.
Undo the addition. →	-2	-2	Subtraction Property of Equality
	$x =$	7	Simplify.

 The solution is $x = 7$.

Check
$x + 2 = 9$
$7 + 2 \stackrel{?}{=} 9$
$9 = 9$ ✓

b. Solve $26 = 11 + x$.

Another Method

26	
11	x
11	15	✓

$26 =$	$11 + x$	Write the equation.
-11	-11	Subtraction Property of Equality
$15 =$	x	Simplify.

 The solution is $x = 15$.

Try It Solve the equation. Check your solution.

8. $s + 8 = 17$
9. $9 = y + 6$
10. $13 + m = 20$

Self-Assessment for Concepts & Skills

Solve each exercise. Then rate your understanding of the success criteria in your journal.

CHECKING SOLUTIONS Tell whether the given value is a solution of the equation.

11. $n + 8 = 42; n = 36$
12. $g - 9 = 24; g = 35$

SOLVING EQUATIONS Solve the equation. Check your solution.

13. $x - 8 = 12$
14. $b + 14 = 33$

15. **WRITING** When solving $x + 5 = 16$, why do you subtract 5 from the left side of the equation? Why do you subtract 5 from the right side of the equation?

16. **MP REASONING** Do the equations have the same solution? Explain your reasoning.
$$x - 8 = 6$$
$$x - 6 = 8$$

17. **MP STRUCTURE** Just by looking at the equation $x + 6 + 2x = 2x + 6 + 4$, find the value of x. Explain your reasoning.

254 Chapter 6 Equations

EXAMPLE 4 Modeling Real Life

Your aunt gives you $50 to help buy the reflector telescope shown. After you buy the telescope, you have $25.50 left. How much money did you have before your aunt gave you $50?

Use a verbal model to write an equation that represents the situation. Then solve the equation.

Verbal Model Starting amount (dollars) + Aunt's money (dollars) − Telescope price (dollars) = Amount left (dollars)

Variable Equation Let s be the starting amount.

$s + 50 - 275.95 = 25.50$	Write the equation.
$s + 50 - 275.95 + 275.95 = 25.50 + 275.95$	Addition Property of Equality
$s + 50 = 301.45$	Simplify.
$s + 50 - 50 = 301.45 - 50$	Subtraction Property of Equality
$s = 251.45$	Simplify.

> Addition and subtraction preserve units. For example, when you add dollars to dollars, you get dollars.

▸ You had $251.45 before your aunt gave you money.

Another Method Solve the problem arithmetically by working backwards from $25.50.

$$25.50 + 275.95 - 50 = \$251.45 \checkmark$$

Self-Assessment for Problem Solving

Solve each exercise. Then rate your understanding of the success criteria in your journal.

24 in.

18. An emperor penguin is 45 inches tall. It is 24 inches taller than a rockhopper penguin. Write and solve an equation to find the height (in inches) of a rockhopper penguin. Is your answer reasonable? Explain.

19. **DIG DEEPER!** You get in an elevator and go up 2 floors and down 8 floors before exiting. Then you get back in the elevator and go up 4 floors before exiting on the 12th floor. On what floors did you enter the elevator?

Section 6.2 Solving Equations Using Addition or Subtraction

6.2 Practice

Go to **BigIdeasMath.com** to get HELP with solving the exercises.

Review & Refresh

Write the word sentence as an equation.

1. The sum of a number x and 9 is 15.
2. 12 less than a number m equals 20.
3. The product of a number d and 7 is 63.
4. 18 divided by a number s equals 3.

Divide. Write the answer in simplest form.

5. $\frac{1}{2} \div \frac{1}{4}$
6. $12 \div \frac{3}{8}$
7. $8 \div \frac{4}{5}$
8. $\frac{7}{9} \div \frac{3}{2}$

9. Which ratio is *not* equivalent to 72 : 18?

    A. 36 : 9   **B.** 18 : 6   **C.** 4 : 1   **D.** 288 : 72

Evaluate the expression.

10. $(2 + 5^2) \div 3$
11. $6 + 2^3 \cdot 3 - 5$
12. $4 \cdot [3 + 3(20 - 4^2 - 2)]$

13. Find the missing values in the ratio table. Then write the equivalent ratios.

Snakes	2		24
Mice	5	20	

Concepts, Skills, & Problem Solving

CHOOSE TOOLS Use a model to solve the equation. *(See Explorations 1 and 2, p. 251.)*

14. $n + 7 = 9$
15. $t + 4 = 5$
16. $c + 2 = 8$

CHECKING SOLUTIONS Tell whether the given value is a solution of the equation.

17. $x + 42 = 85;\ x = 43$
18. $8b = 48;\ b = 6$
19. $19 - g = 7;\ g = 15$
20. $\frac{m}{4} = 16;\ m = 4$
21. $w + 23 = 41;\ w = 28$
22. $s - 68 = 11;\ s = 79$

SOLVING EQUATIONS Solve the equation. Check your solution.

23. $y - 7 = 3$
24. $z - 3 = 13$
25. $8 = r - 14$
26. $p + 5 = 8$
27. $k + 6 = 18$
28. $64 = h + 30$
29. $f - 27 = 19$
30. $25 = q + 14$
31. $\frac{3}{4} = j - \frac{1}{2}$
32. $x + \frac{2}{3} = \frac{9}{10}$
33. $1.2 = m - 2.5$
34. $a + 5.5 = 17.3$

YOU BE THE TEACHER Your friend solves the equation. Is your friend correct? Explain your reasoning.

35.
```
x + 7 = 13
  + 7   + 7
  x = 20
```

36.
```
34 = y − 12
− 12   + 12
22 = y
```

37. **MODELING REAL LIFE** The main span of the Sunshine Skyway Bridge is 366 meters long. The bridge's main span is 30 meters shorter than the main span of the Dames Point Bridge. Write and solve an equation to find the length (in meters) of the main span of the Dames Point Bridge.

Dames Point Bridge
main span

38. **PROBLEM SOLVING** A park has 22 elm trees. Elm leaf beetles have been attacking the trees. After removing several of the diseased trees, there are 13 healthy elm trees left. Write and solve an equation to find the number of elm trees that were removed.

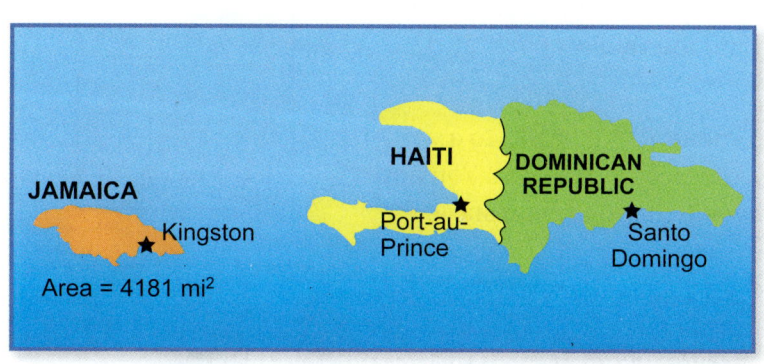
JAMAICA — Kingston — Area = 4181 mi²
HAITI — Port-au-Prince
DOMINICAN REPUBLIC — Santo Domingo

39. **PROBLEM SOLVING** The area of Jamaica is 6460 square miles less than the area of Haiti. Find the area (in square miles) of Haiti.

40. **REASONING** The solution of the equation $x + 3 = 12$ is shown. Explain each step. Use a property, if possible.

$x + 3 = 12$ Write the equation.
$x + 3 - 3 = 12 - 3$
$x + 0 = 9$
$x = 9$

WRITING EQUATIONS Write the word sentence as an equation. Then solve the equation.

41. 13 subtracted from a number w is 15.
42. A number k increased by 7 is 34.
43. 9 is the difference of a number n and 7.
44. 93 is the sum of a number g and 58.
45. 11 more than a number k equals 29.
46. A number p decreased by 19 is 6.
47. 46 is the total of 18 and a number d.
48. 84 is 99 fewer than a number c.

Section 6.2   Solving Equations Using Addition or Subtraction

SOLVING EQUATIONS Solve the equation. Check your solution.

49. $b + 7 + 12 = 30$ **50.** $y + 4 - 1 = 18$ **51.** $m + 18 + 23 = 71$

52. $v - 7 = 9 + 12$ **53.** $5 + 44 = 2 + r$ **54.** $22 + 15 = d - 17$

GEOMETRY Solve for x.

55. Perimeter = 48 ft   **56.** Perimeter = 132 in.   **57.** Perimeter = 93 ft

Right triangle with legs x and 12 ft, hypotenuse 20 ft.

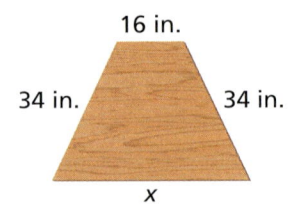

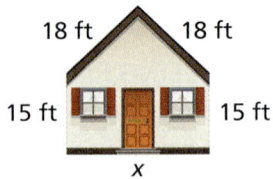

58. SIMPLIFYING AND SOLVING Compare and contrast the two problems.

> Simplify the expression $2(x + 3) - 4$.
>
> $2(x + 3) - 4 = 2x + 6 - 4$
>
> $\qquad\qquad\quad\; = 2x + 2$

> Solve the equation $x + 3 = 4$.
>
> $\begin{aligned} x + 3 &= 4 \\ \underline{-\,3} &\underline{-\,3} \\ x &= 1 \end{aligned}$

59. PUZZLE In a *magic square*, the sum of the numbers in each row, column, and diagonal is the same. Find the values of a, b, and c. Justify your answers.

a	37	16
19	25	b
34	c	28

60. **MP REASONING** On Saturday, you spend $33, give $15 to a friend, and receive $20 for mowing your neighbor's lawn. You have $21 left. Use two methods to find how much money you started with that day.

61. DIG DEEPER! You have $15.

> Bumper Cars: $1.75
> Super Pendulum: $1.25 more than Ferris Wheel
> Giant Slide: $0.50 less than Bumper Cars
> Ferris Wheel: $1.50 more than Giant Slide

  a. How much money do you have left if you ride each ride once?

  b. Do you have enough money to ride each ride twice? Explain.

62. CRITICAL THINKING Consider the equation $15 - y = 8$. Explain how you can solve the equation using the Addition and Subtraction Properties of Equality.

6.3 Solving Equations Using Multiplication or Division

Learning Target: Write and solve equations using multiplication or division.

Success Criteria:
- I can apply the Multiplication and Division Properties of Equality to generate equivalent equations.
- I can solve equations using multiplication or division.
- I can create equations involving multiplication or division to solve real-life problems.

EXPLORATION 1

Solving an Equation Using a Tape Diagram

Work with a partner. A student solves an equation using the tape diagrams below.

Step 1: | 20 |
        | 4x |

Step 2: | 5 | 5 | 5 | 5 |
        | x | x | x | x |

Step 3: | 5 |
        | x |

Math Practice

Maintain Oversight

How do these explorations help you solve equations of the form $ax = b$?

a. What equation did the student solve? What is the solution?

b. Explain how the tape diagrams in Steps 2 and 3 relate to the equation and its solution.

EXPLORATION 2

Solving an Equation Using a Model

Work with a partner. Three robots go out to lunch. They decide to split the $12 bill evenly. The scale represents the number of robots and the price of the meal.

a. How much does each robot pay?

b. When you triple the weight on one side of a balanced scale, what can you do to balance the scale? What if you divide the weight on one side of a balanced scale in half? How does this relate to solving an equation?

c. Use a model to solve $5x = 15$. Describe how you can solve the equation algebraically.

Section 6.3 Solving Equations Using Multiplication or Division 259

6.3 Lesson

Remember

Inverse operations "undo" each other. Multiplication and division are inverse operations.

Key Ideas

Multiplication Property of Equality

Words When you multiply each side of an equation by the same nonzero number, the two sides remain equal.

Numbers
$$\frac{8}{4} = 2$$
$$\frac{8}{4} \cdot 4 = 2 \cdot 4$$
$$8 = 8$$

Algebra
$$\frac{x}{4} = 2$$
$$\frac{x}{4} \cdot 4 = 2 \cdot 4$$
$$x = 8$$

Multiplicative Inverse Property

Words The product of a nonzero number n and its reciprocal, $\frac{1}{n}$, is 1.

Numbers $5 \cdot \frac{1}{5} = 1$

Algebra $n \cdot \frac{1}{n} = \frac{1}{n} \cdot n = 1, n \neq 0$

EXAMPLE 1 **Solving Equations Using Multiplication**

a. Solve $\dfrac{w}{4} = 12$.

$\dfrac{w}{4} = 12$ Write the equation.

Undo the division. → $\dfrac{w}{4} \cdot 4 = 12 \cdot 4$ Multiplication Property of Equality

$w = 48$ Simplify.

 The solution is $w = 48$.

Check
$\dfrac{w}{4} = 12$
$\dfrac{48}{4} \stackrel{?}{=} 12$
$12 = 12$ ✓

b. Solve $\dfrac{2}{7}x = 6$.

$\dfrac{2}{7}x = 6$ Write the equation.

Use the Multiplicative Inverse Property. → $\dfrac{7}{2} \cdot \left(\dfrac{2}{7}x\right) = \dfrac{7}{2} \cdot 6$ Multiplication Property of Equality

$x = 21$ Simplify.

 The solution is $x = 21$.

Try It Solve the equation. Check your solution.

1. $\dfrac{a}{8} = 6$

2. $14 = \dfrac{2y}{5}$

3. $3z \div 2 = 9$

260   Chapter 6   Equations

> You can multiply or divide each side of an equation by any number to generate an equivalent equation. For example, $4x = 32$ and $2x = 16$ are equivalent equations.
>
> $4x = 32$
> $\dfrac{4x}{2} = \dfrac{32}{2}$
> $2x = 16$

 **Key Idea**

Division Property of Equality

Words When you divide each side of an equation by the same nonzero number, the two sides remain equal.

Numbers $\quad 8 \cdot 4 = 32 \quad\quad$ **Algebra** $\quad 4x = 32$

$$8 \cdot 4 \div 4 = 32 \div 4 \quad\quad\quad \dfrac{4x}{4} = \dfrac{32}{4}$$

$$8 = 8 \quad\quad\quad\quad\quad\quad x = 8$$

EXAMPLE 2 Solving an Equation Using Division

Solve $65 = 5b$.

$65 = 5b \quad$ Write the equation.

Undo the multiplication. $\rightarrow \dfrac{65}{5} = \dfrac{5b}{5} \quad$ Division Property of Equality

$13 = b \quad$ Simplify.

▶ The solution is $b = 13$.

Check
$65 = 5b$
$65 \stackrel{?}{=} 5(13)$
$65 = 65$ ✓

Try It Solve the equation. Check your solution.

4. $p \cdot 3 = 18$ **5.** $12q = 60$ **6.** $81 = 9r$

 ## Self-Assessment for Concepts & Skills

Solve each exercise. Then rate your understanding of the success criteria in your journal.

SOLVING EQUATIONS Solve the equation. Check your solution.

7. $6 = \dfrac{2y}{3}$ **8.** $8s = 56$

9. WHICH ONE DOESN'T BELONG? Which equation does *not* belong with the other three? Explain your reasoning.

$\dfrac{1}{4}x = 27 \quad\quad 3x = 36 \quad\quad \dfrac{3}{4}x = 9 \quad\quad 4x = 48$

MP STRUCTURE Just by looking at the equation, find the value of x. Explain your reasoning.

10. $5x + 3x = 5x + 18$ **11.** $8x + \dfrac{x}{2} = 8x + 6$

Section 6.3   Solving Equations Using Multiplication or Division

EXAMPLE 3 Modeling Real Life

The area of a rectangular LED "sky screen" in Beijing, China, is 7500 square meters. The width of the sky screen is 30 meters. What is the length of the sky screen?

Use the given information to draw a diagram. Then substitute for the area A and the width w in the formula for the area of a rectangle, $A = \ell w$. Solve for the length ℓ.

$A = 7500$ m²   30 m

$$A = \ell w \quad \text{Use the formula for area of a rectangle.}$$
$$7500 = 30\ell \quad \text{Substitute 7500 for } A \text{ and 30 for } w.$$
$$\frac{7500}{30} = \frac{30\ell}{30} \quad \text{Division Property of Equality}$$
$$250 = \ell \quad \text{Simplify.}$$

▶ So, the sky screen is 250 meters long.

 ## Self-Assessment for Problem Solving

Solve each exercise. Then rate your understanding of the success criteria in your journal.

Area = 1625 mm²

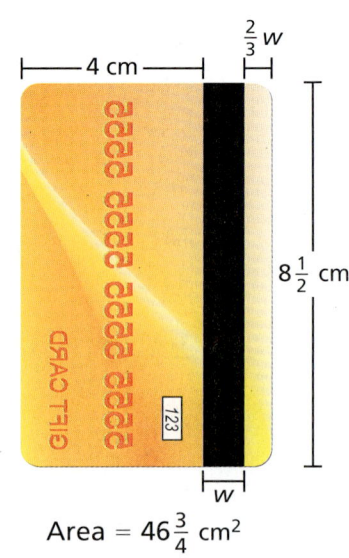

Area = $46\frac{3}{4}$ cm²

12. The area of the screen of the smart watch is shown. What are possible dimensions for the length and the width of the screen? Justify your answer.

13. A rock climber climbs at a rate of 720 feet per hour. Write and solve an equation to find the number of minutes it takes for the rock climber to climb 288 feet.

14. **DIG DEEPER!** A gift card stores data using a black, magnetic stripe on the back of the card. Find the width w of the stripe.

6.3 Practice

Go to *BigIdeasMath.com* to get HELP with solving the exercises.

Review & Refresh

Solve the equation. Check your solution.

1. $y - 5 = 6$
2. $m + 7 = 8$
3. $\frac{7}{8} = \frac{1}{4} + p$

4. What is the value of a^3 when $a = 4$?
   - **A.** 12
   - **B.** 43
   - **C.** 64
   - **D.** 81

Multiply. Write the answer in simplest form.

5. $\frac{1}{5} \cdot \frac{2}{9}$
6. $\frac{5}{12} \times \frac{4}{7}$
7. $2\frac{1}{3} \cdot \frac{3}{10}$
8. $1\frac{3}{4} \times 2\frac{2}{3}$

Multiply.

9. 0.4×0.9
10. 0.78×0.5
11. 2.63×4.31
12. 1.115×3.28

Concepts, Skills, & Problem Solving

MP CHOOSE TOOLS Use a model to solve the equation. *(See Explorations 1 and 2, p. 259.)*

13. $8x = 8$
14. $9 = 3y$
15. $2z = 14$

SOLVING EQUATIONS Solve the equation. Check your solution.

16. $\frac{s}{10} = 7$
17. $6 = \frac{t}{5}$
18. $5x \div 6 = 20$
19. $24 = \frac{3}{4}r$

20. $3a = 12$
21. $5 \cdot z = 35$
22. $40 = 4y$
23. $42 = 7k$

24. $7x = 105$
25. $75 = 6 \cdot w$
26. $13 = d \div 6$
27. $9 = v \div 5$

28. $\frac{5d}{9} = 10$
29. $\frac{3}{5} = 4m$
30. $136 = 17b$
31. $\frac{2}{3} = \frac{1}{4}k$

32. $\frac{2c}{15} = 8.8$
33. $7b \div 12 = 4.2$
34. $12.5 \cdot n = 32$
35. $3.4m = 20.4$

36. **YOU BE THE TEACHER** Your friend solves the equation $x \div 4 = 28$. Is your friend correct? Explain your reasoning.

$$x \div 4 = 28$$
$$\frac{x \div 4}{4} = \frac{28}{4}$$
$$x = 7$$

37. **ANOTHER WAY** Show how you can solve the equation $3x = 9$ by multiplying each side by the reciprocal of 3.

38. **MODELING REAL LIFE** Forty-five basketball players participate in a three-on-three tournament. Write and solve an equation to find the number of three-person teams in the tournament.

39. **MODELING REAL LIFE** A theater has 1200 seats. Each row has 20 seats. Write and solve an equation to find the number of rows in the theater.

GEOMETRY Solve for *x*. Check your answer.

40. Area = 45 square units

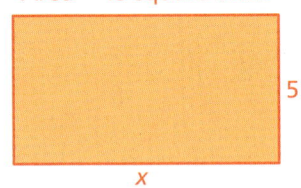

41. Area = 176 square units

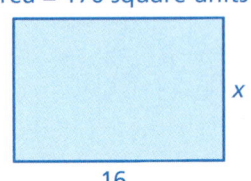

42. **MP LOGIC** On a test, you earn 92% of the possible points by correctly answering 6 five-point questions and 8 two-point questions. How many points *p* is the test worth?

43. **MODELING REAL LIFE** You use index cards to play a homemade game. The object is to be the first to get rid of all your cards. How many cards are in your friend's stack?

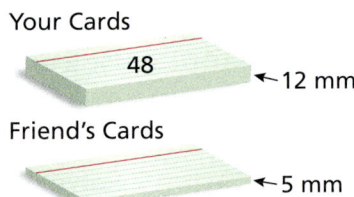

44. **DIG DEEPER!** A slush drink machine fills 1440 cups in 24 hours.

   a. Find the number *c* of cups each symbol represents.

   b. To lower costs, you replace the cups with paper cones that hold 20% less. Find the number *n* of paper cones that the machine can fill in 24 hours.

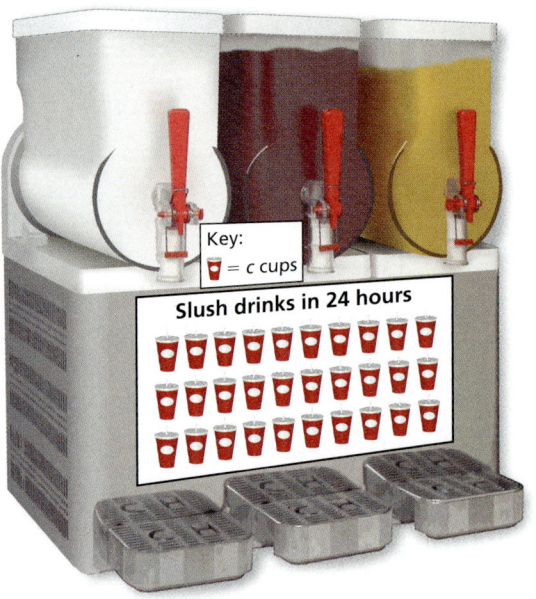

45. **MP NUMBER SENSE** The area of the picture is 100 square inches. The length is 4 times the width. Find the length and width of the picture.

6.4 Writing Equations in Two Variables

Learning Target: Write equations in two variables and analyze the relationship between the two quantities.

Success Criteria:
- I can determine whether an ordered pair is a solution of an equation in two variables.
- I can distinguish between independent and dependent variables.
- I can write and graph an equation in two variables.
- I can create equations in two variables to solve real-life problems.

EXPLORATION 1

Writing Equations in Two Variables

Work with a partner. In Section 3.4 Exploration 1, you used a ratio table to create a graph for an airplane traveling 300 miles per hour. Below is one possible ratio table and graph.

Time (hours)	1	2	3	4
Distance (miles)	300	600	900	1200

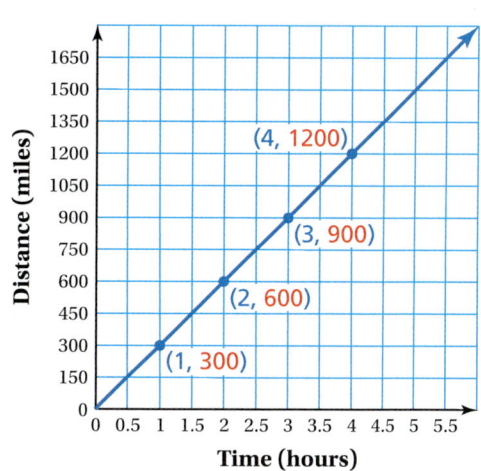

Math Practice

Look for Patterns

How can you use the patterns in the table to help you write an equation?

a. Describe the relationship between the two quantities. Which quantity *depends* on the other quantity?

b. Use variables to write an equation that represents the relationship between the time and the distance. What can you do with this equation? Provide an example.

c. Suppose the airplane is 1500 miles away from its destination. Write an equation that represents the relationship between time and distance from the destination. How can you represent this relationship using a graph?

Section 6.4 Writing Equations in Two Variables 265

6.4 Lesson

An **equation in two variables** represents two quantities that change in relationship to one another. A **solution of an equation in two variables** is an ordered pair that makes the equation true.

Key Vocabulary
equation in two variables, p. 266
solution of an equation in two variables, p. 266
independent variable, p. 266
dependent variable, p. 266

EXAMPLE 1 Identifying Solutions of Equations in Two Variables

Tell whether the ordered pair is a solution of the equation.

a. $y = 2x$; (3, 6)

$6 \stackrel{?}{=} 2(3)$ Substitute.

$6 = 6$ ✓   Compare.

▸ So, (3, 6) is a solution.

b. $y = 4x - 3$; (4, 12)

$12 \stackrel{?}{=} 4(4) - 3$

$12 \neq 13$ ✗

▸ So, (4, 12) is *not* a solution.

Try It Tell whether the ordered pair is a solution of the equation.

1. $y = 7x$; (2, 21)
2. $y = 5x + 1$; (3, 16)

Equations in two variables have an *independent variable* and a *dependent variable*. The variable representing the quantity that can change freely is the **independent variable**. The other variable is called the **dependent variable** because its value *depends* on the independent variable.

EXAMPLE 2 Using an Equation in Two Variables

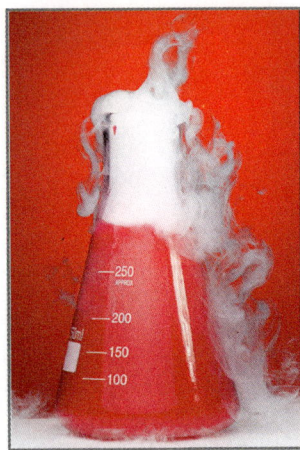

The equation $y = 64 - 8x$ represents the amount y (in fluid ounces) of chemical remaining in a flask after you pour x cups. Identify the independent and dependent variables. How much of the chemical remains in the flask after you pour 5 cups?

Because the amount y of fluid ounces remaining depends on the number x of cups you pour, y is the dependent variable and x is the independent variable.

Use the equation to find the value of y when $x = 5$.

$y = 64 - 8x$ Write the equation.

$ = 64 - 8(5)$ Substitute 5 for x.

$ = 24$ Simplify.

▸ There are 24 fluid ounces remaining.

Try It

3. The equation $y = 10x + 25$ represents the amount y (in dollars) in your savings account after x weeks. Identify the independent and dependent variables. How much is in your savings account after 8 weeks?

266   Chapter 6   Equations

 **Key Idea**

Tables, Graphs, and Equations

You can use tables and graphs to represent equations in two variables. The independent variable is graphed on the horizontal axis, and the dependent variable is graphed on the vertical axis. The table and graph below represent the equation $y = x + 2$.

Independent Variable, x	Dependent Variable, y	Ordered Pair, (x, y)
1	3	(1, 3)
2	4	(2, 4)
3	5	(3, 5)

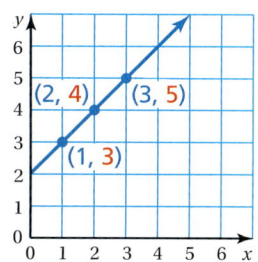

When you draw a line through the points, you graph all the solutions of the equation.

EXAMPLE 3 Graphing an Equation in Two Variables

Graph $y = 2x + 1$.

To graph the equation, first make a table.

Independent Variable, x	$y = 2x + 1$	Dependent Variable, y	Ordered Pair, (x, y)
0	$y = 2(0) + 1$	1	(0, 1)
1	$y = 2(1) + 1$	3	(1, 3)
2	$y = 2(2) + 1$	5	(2, 5)
3	$y = 2(3) + 1$	7	(3, 7)

Then plot the ordered pairs and draw a line through the points.

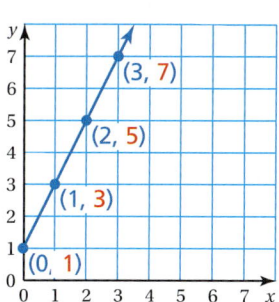

Remember

In a coordinate plane, the horizontal axis is often called the *x*-axis. The vertical axis is often called the *y*-axis. In real-life problems, other variables can be used.

Try It Graph the equation.

4. $y = 3x$
5. $y = 4x + 1$
6. $y = \dfrac{1}{2}x + 2$

Section 6.4 Writing Equations in Two Variables

EXAMPLE 4 Writing and Graphing an Equation in Two Variables

An athlete burns 200 calories weight lifting. The athlete then works out on an elliptical trainer and burns 10 calories for every minute. Write and graph an equation that represents the total number of calories burned during the workout.

Use a verbal model to write an equation.

Verbal Model Total number of calories burned = Calories burned weight lifting + Calories burned per minute · Number of minutes

Variables Let c be the total number of calories burned, and let m be the number of minutes on the elliptical trainer.

Equation $c = 200 + 10 \cdot m$

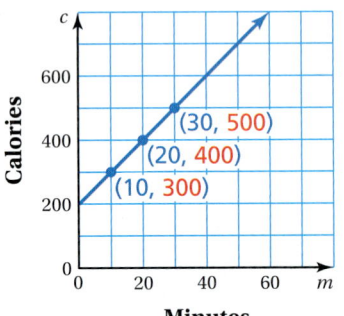

To graph the equation, first notice that the total number of calories burned depends on the number of minutes. So, create a table and plot the ordered pairs with minutes m on the horizontal axis and calories c on the vertical axis. Then draw a line through the points.

Minutes, m	Calories, c
10	300
20	400
30	500

Try It

7. It costs $25 to rent a kayak plus $8 for each hour. Write and graph an equation that represents the total cost (in dollars) of renting the kayak.

Self-Assessment for Concepts & Skills

Solve each exercise. Then rate your understanding of the success criteria in your journal.

8. **WRITING** Describe the difference between independent variables and dependent variables.

IDENTIFYING SOLUTIONS Tell whether the ordered pair is a solution of the equation.

9. $y = 3x + 8$; (4, 20)

10. $y = 6x - 14$; (7, 29)

11. **MP PRECISION** Explain how to graph an equation in two variables.

12. **WHICH ONE DOESN'T BELONG?** Which one does *not* belong with the other three? Explain your reasoning.

| $y = 12x + 25$ | $c = 10t - 5$ | $a = 7b + 11$ | $n = 4n - 6$ |

You can model many rate problems by using the *distance formula*, $d = rt$, where d is the distance traveled, r is the speed, and t is the time.

EXAMPLE 5 Modeling Real Life

A train averages 40 miles per hour between two cities. Write and graph an equation that represents the relationship between the time and the distance traveled. How long does it take the train to travel 220 miles?

The rate r is 40 miles per hour. Using the distance formula, an equation that represents the relationship between time and distance traveled is $d = 40t$.

Make a table and graph the equation.

Time (hours), t	1	2	3	4
Distance (miles), d	40	80	120	160

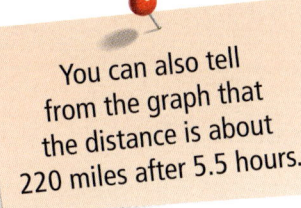

Use the equation to find the value of t when $d = 220$.

$d = 40t$ Write the equation.
$220 = 40t$ Substitute 220 for d.
$5.5 = t$ Divide each side by 40.

 The train travels 220 miles in 5.5 hours.

Another Method Use a ratio table.

	÷2	×11
Time (hours), t	1	0.5	5.5 ✓
Distance (miles), d	40	20	220
	÷2	×11

You can also tell from the graph that the distance is about 220 miles after 5.5 hours.

Self-Assessment for Problem Solving

Solve each exercise. Then rate your understanding of the success criteria in your journal.

13. A sky lantern rises at an average speed of 8 feet per second. Write and graph an equation that represents the relationship between the time and the distance risen. How long does it take the lantern to rise 100 feet?

14. You and a friend start biking in opposite directions from the same point. You travel 108 feet every 8 seconds. Your friend travels 63 feet every 6 seconds. How far apart are you and your friend after 15 minutes?

Section 6.4 Writing Equations in Two Variables

6.4 Practice

Go to *BigIdeasMath.com* to get HELP with solving the exercises.

Review & Refresh

Solve the equation.

1. $4x = 36$
2. $\dfrac{x}{8} = 5$
3. $\dfrac{4x}{3} = 8$
4. $\dfrac{2}{5}x = 6$

Divide. Write the answer in simplest form.

5. $3\dfrac{1}{2} \div \dfrac{4}{5}$
6. $7 \div 5\dfrac{1}{4}$
7. $\dfrac{3}{11} \div 1\dfrac{1}{8}$
8. $7\dfrac{1}{2} \div 1\dfrac{1}{3}$

9. Find the area of the carpet tile. Then find the area covered by 120 carpet tiles.

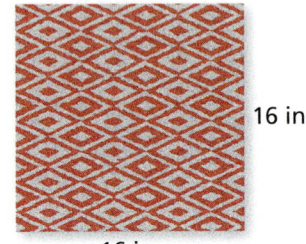

16 in.

16 in.

Copy and complete the statement. Round to the nearest hundredth if necessary.

10. 8 m = ▢ cm
11. 88 oz = ▢ lb
12. 3 c ≈ ▢ mL
13. 15 km ≈ ▢ mi

Divide.

14. $6 \overline{)34.8}$
15. $4 \overline{)12.8}$
16. $45.92 \div 2.8$
17. $39.525 \div 4.25$

Concepts, Skills, & Problem Solving

WRITING EQUATIONS Use variables to write an equation that represents the relationship between the time and the distance. *(See Exploration 1, p. 265.)*

18. An eagle flies 40 miles per hour.
19. A person runs 175 yards per minute.

IDENTIFYING SOLUTIONS Tell whether the ordered pair is a solution of the equation.

20. $y = 4x$; $(0, 4)$
21. $y = 3x$; $(2, 6)$
22. $y = 5x - 10$; $(3, 5)$
23. $y = x + 7$; $(1, 6)$
24. $y = x + 4$; $(2, 4)$
25. $y = x - 5$; $(6, 11)$
26. $y = 6x + 1$; $(2, 13)$
27. $y = 7x + 2$; $(2, 0)$
28. $y = 2x - 3$; $(4, 5)$
29. $y = 3x - 3$; $(1, 0)$
30. $7 = y - 5x$; $(4, 28)$
31. $y + 3 = 6x$; $(3, 15)$

32. **YOU BE THE TEACHER** Your friend determines whether $(5, 1)$ is a solution of $y = 3x + 2$. Is your friend correct? Explain your reasoning.

> $y = 3x + 2$; $(5, 1)$
> $1 \stackrel{?}{=} 3(5) + 2$
> $1 \neq 17$
> So, $(5, 1)$ is not a solution.

270 Chapter 6 Equations

IDENTIFYING VARIABLES Identify the independent and dependent variables.

33. The equation $A = 25w$ represents the area A (in square feet) of a rectangular dance floor with a width of w feet.

34. The equation $c = 0.09s$ represents the amount c (in dollars) of commission a salesperson receives for making a sale of s dollars.

35. The equation $t = 12p + 12$ represents the total cost t (in dollars) of a meal with a tip of p percent (in decimal form).

36. The equation $h = 60 - 4m$ represents the height h (in inches) of the water in a tank m minutes after it starts to drain.

OPEN-ENDED Complete the table by describing possible independent or dependent variables.

	Independent Variable	Dependent Variable
37.	The number of hours you study for a test	
38.	The speed you are pedaling a bike	
39.		Your monthly cell phone bill
40.		The amount of money you earn

GRAPHING EQUATIONS Graph the equation.

41. $y = 2x$
42. $y = 5x$
43. $y = 6x$
44. $y = x + 2$
45. $y = x + 0.5$
46. $y = x + 4$
47. $y = x + 10$
48. $y = 3x + 2$
49. $y = 2x + 4$
50. $y = \frac{2}{3}x + 8$
51. $y = \frac{1}{4}x + 6$
52. $y = 2.5x + 12$

53. **MODELING REAL LIFE** A cheese pizza costs $5. Additional toppings cost $1.50 each. Write and graph an equation that represents the total cost (in dollars) of a pizza.

54. **MODELING REAL LIFE** It costs $35 for a membership at a wholesale store. The monthly fee is $15. Write and graph an equation that represents the total cost (in dollars) of a membership.

55. **MP PROBLEM SOLVING** The maximum size of a text message is 160 characters. A space counts as one character.

   a. Write an equation that represents the number of remaining (unused) characters in a text message as you type.

   b. Identify the independent and dependent variables.

   c. How many characters remain in the message shown?

Section 6.4 Writing Equations in Two Variables

56. **MP CHOOSE TOOLS** A car averages 60 miles per hour on a road trip. Use a graph to represent the relationship between the time and the distance traveled.

MP PRECISION Write and graph an equation that represents the relationship between the time and the distance traveled.

57.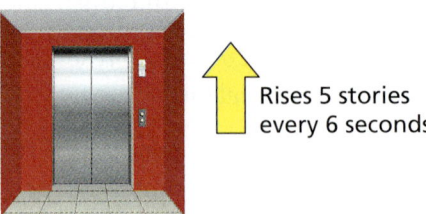
Moves 2 meters every 3 hours

58. Rises 5 stories every 6 seconds

59.  60.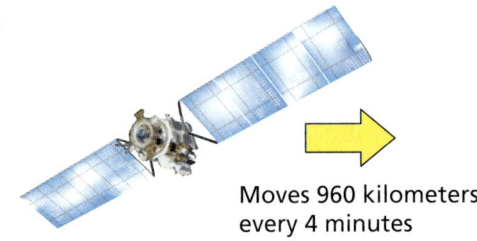
Moves 660 feet every 10 seconds

Moves 960 kilometers every 4 minutes

IDENTIFYING SOLUTIONS Fill in the blank so that the ordered pair is a solution of the equation.

61. $y = 8x + 3$; (1, ____)   62. $y = 12x + 2$; (____ , 14)   63. $y = 9x + 4$; (____ , 22)

12 in.

16 in.

64. **DIG DEEPER!** Can the dependent variable cause a change in the independent variable? Explain.

65. **OPEN-ENDED** Write an equation that has (3, 4) as a solution.

66. **MODELING REAL LIFE** You walk 5 city blocks in 12 minutes. How many city blocks can you walk in 2 hours?

67. **GEOMETRY** How fast should the ant walk to go around the rectangle in 4 minutes?

68. **MODELING REAL LIFE** To estimate how far you are from lightning (in miles), count the number of seconds between a lightning flash and the thunder that follows. Then divide the number of seconds by 5. Use two different methods to find the number of seconds between a lightning flash and the thunder that follows when a storm is 2.4 miles away.

69. **MP REASONING** The graph represents the cost c (in dollars) of buying n tickets to a baseball game.

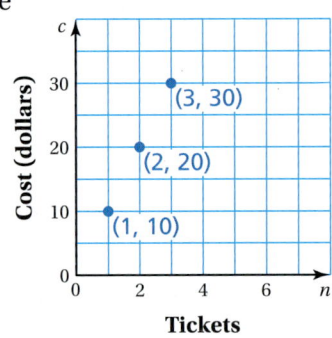

a. Should the points be connected with a line to show all the solutions? Explain your reasoning.

b. Write an equation that represents the graph.

272   Chapter 6   Equations

6 Connecting Concepts

▶ Using the Problem-Solving Plan

1. A tornado forms 12.25 miles from a weather station. It travels away from the station at an average speed of 440 yards per minute. How far from the station is the tornado after 30 minutes?

   Understand the problem. You know the initial distance between the tornado and the station, and the average speed the tornado is traveling away from the station. You are asked to determine how far the tornado is from the station after 30 minutes.

   Make a plan. First, convert the average speed to miles per minute. Then write an equation that represents the distance d (in miles) between the tornado and the station after t minutes. Use the equation to find the value of d when $t = 30$.

   Solve and check. Use the plan to solve the problem. Then check your solution.

2. You buy 96 cans of soup to donate to a food bank. The store manager discounts the cost of each case for a total discount of $40. Use an equation in two variables to find the discount for each case of soup. What is the total cost when each can of soup originally costs $1.20?

   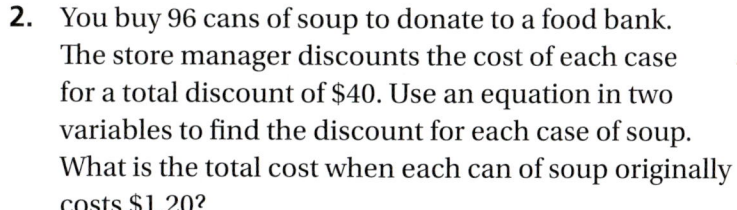
   1 case = 12 cans

3. The diagram shows the initial amount raised by an organization for cancer research. A business agrees to donate $2 for every $5 donated by the community during an additional fundraising event. Write an equation that represents the total amount raised (in dollars). How much money does the community need to donate for the organization to reach its fundraising goal?

Performance Task

Planning the Climb

At the beginning of this chapter, you watched a STEAM video called "Rock Climbing." You are now ready to complete the performance task related to this video, available at *BigIdeasMath.com*. Be sure to use the problem-solving plan as you work through the performance task.

Connecting Concepts 273

6 Chapter Review

Go to *BigIdeasMath.com* to download blank graphic organizers.

Review Vocabulary

Write the definition and give an example of each vocabulary term.

equation, *p. 246*
solution, *p. 252*
inverse operations, *p. 253*
equation in two variables, *p. 266*
solution of an equation in two variables, *p. 266*
independent variable, *p. 266*
dependent variable, *p. 266*

Graphic Organizers

You can use an **Example and Non-Example Chart** to list examples and non-examples of a concept. Here is an Example and Non-Example Chart for the vocabulary term *equation*.

Equation

Examples	Non-Examples
$x = 5$	5
$2a = 16$	$2a$
$x + 4 = 19$	$x + 4$
$5 = x + 3$	$x + 3$
$12 - 7 = 5$	$12 - 7$
$\frac{3}{4}y = 6$	$\frac{3}{4}$

Choose and complete a graphic organizer to help you study the concept.

1. inverse operations

2. solving equations using addition or subtraction

3. solving equations using multiplication or division

4. equations in two variables

5. independent variables

6. dependent variables

"I need a good non-example of a cool animal for my **Example and Non-Example Chart**."

Chapter Self-Assessment

As you complete the exercises, use the scale below to rate your understanding of the success criteria in your journal.

1	2	3	4
I do not understand.	I can do it with help.	I can do it on my own.	I can teach someone else.

6.1 Writing Equations in One Variable (pp. 245–250)

Learning Target: Write equations in one variable and write equations that represent real-life problems.

Write the word sentence as an equation.

1. The product of a number m and 2 is 8.

2. 6 less than a number t is 7.

3. A number m increased by 5 is 7.

4. 8 is the quotient of a number g and 3.

5. The height of the 50-milliliter beaker is one-third the height of the 2000-milliliter beaker. Write an equation you can use to find the height (in centimeters) of the 2000-milliliter beaker.

6. There are 16 teams in a basketball tournament. After two rounds, 12 teams are eliminated. Write and solve an equation to find the number of teams remaining after two rounds.

7. Write an equation that has a solution of $x = 8$.

8. Write a word sentence for the equation $y + 3 = 5$.

6.2 Solving Equations Using Addition or Subtraction (pp. 251–258)

Learning Target: Write and solve equations using addition or subtraction.

9. Tell whether $x = 7$ is a solution of $x + 9 = 16$.

Solve the equation. Check your solution.

10. $x - 1 = 8$
11. $m + 7 = 11$
12. $21 = p - 12$

Write the word sentence as an equation. Then solve the equation.

13. 5 more than a number x is 9.
14. 82 is the difference of a number b and 24.

15. A stuntman is running on the roof of a train. His combined speed is the sum of the speed of the train and his running speed. The combined speed is 73 miles per hour, and his running speed is 15 miles per hour. Find the speed of the train.

16. Before swallowing a large rodent, a python weighs 152 pounds. After swallowing the rodent, the python weighs 164 pounds. Find the weight of the rodent.

6.3 Solving Equations Using Multiplication or Division (pp. 259–264)

Learning Target: Write and solve equations using multiplication or division.

Solve the equation. Check your solution.

17. $6 \cdot q = 54$
18. $k \div 3 = 21$
19. $\dfrac{5}{7}a = 25$

20. The weight of an object on the Moon is about 16.5% of its weight on Earth. The weight of an astronaut on the Moon is 24.75 pounds. How much does the astronaut weigh on Earth?

21. Write an equation that can be solved using multiplication and has a solution of $x = 12$.

22. At a farmers' market, you buy 4 pounds of tomatoes and 2 pounds of sweet potatoes. You spend 80% of the money in your wallet. How much money is in your wallet before you pay?

6.4 Writing Equations in Two Variables (pp. 265–272)

Learning Target: Write equations in two variables and analyze the relationship between the two quantities.

Tell whether the ordered pair is a solution of the equation.

23. $y = 3x + 1$; (2, 7)

24. $y = 7x - 4$; (4, 22)

25. The equation $E = 360m$ represents the kinetic energy E (in joules) of a roller-coaster car with a mass of m kilograms. Identify the independent and dependent variables.

Graph the equation.

26. $y = x + 1$

27. $y = 7x$

28. $y = 4x + 3$

29. $y = \dfrac{1}{2}x + 5$

30. A taxi ride costs $3 plus $2.50 per mile. Write and graph an equation that represents the total cost (in dollars) of a taxi ride. What is the total cost of a five-mile taxi ride?

31. Write and graph an equation that represents the total cost (in dollars) of renting the bounce house. How much does it cost to rent the bounce house for 6 hours?

32. A car averages 50 miles per hour on a trip. Write and graph an equation that represents the relationship between the time and the distance traveled. How long does it take the car to travel 525 miles?

Chapter Review

6 Practice Test

1. Write "7 times a number s is 84" as an equation.

Solve the equation. Check your solution.

2. $15 = 7 + b$

3. $v - 6 = 16$

4. $5x = 70$

5. $\dfrac{6m}{7} = 30$

6. Tell whether (3, 27) is a solution of $y = 9x$.

7. Tell whether (8, 36) is a solution of $y = 4x + 2$.

8. The drawbridge shown consists of two identical sections that open to allow boats to pass. Write an equation you can use to find the length s (in feet) of each section of the drawbridge.

366 feet

9. Each ticket to a school dance is $4. The total amount collected in ticket sales is $332. Find the number of students attending the dance.

10. A soccer team sells T-shirts for a fundraiser. The company that makes the T-shirts charges $10 per shirt plus a $20 shipping fee per order.

    a. Write and graph an equation that represents the total cost (in dollars) of ordering the shirts.

    b. Choose an ordered pair that lies on your graph in part (a). Interpret it in the context of the problem.

11. You hand in 2 homework pages to your teacher. Your teacher now has 32 homework pages to grade. Find the number of homework pages that your teacher originally had to grade.

12. Write an equation that represents the total cost (in dollars) of the meal shown with a tip that is a percent of the check total. What is the total cost of the meal when the tip is 15%?

6 Cumulative Practice

1. You buy roses at a flower shop for $3 each. How many roses can you buy with $27?

    A. 9 **B.** 10
    C. 24 **D.** 81

Test-Taking Strategy
Work Backwards

"You like taking x catnaps each day, where $3x=24$. How many is that?
Ⓐ 6 Ⓑ 7 Ⓒ 8 Ⓓ 9"

"**Work backwards** by trying 6, 7, 8, and 9. You will see that $3(8)=24$. So, C is correct."

2. You are making identical fruit baskets using 16 apples, 24 pears, and 32 bananas. What is the greatest number of baskets you can make using all of the fruit?

    F. 2 **G.** 4
    H. 8 **I.** 16

3. Which equation represents the word sentence?

    > The sum of 18 and 5 is equal to 9 less than a number y.

    A. $18 - 5 = 9 - y$ **B.** $18 + 5 = 9 - y$
    C. $18 + 5 = y - 9$ **D.** $18 - 5 = y - 9$

4. The tape diagram shows the ratio of tickets sold by you and your friend. How many more tickets did you sell than your friend?

    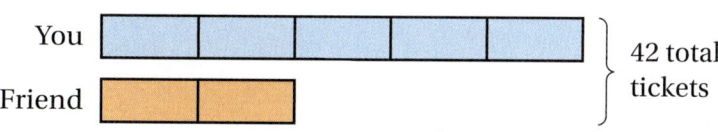

    F. 6 **G.** 12
    H. 18 **I.** 30

5. What is the value of x that makes the equation true?

    $$59 + x = 112$$

6. The steps your friend took to divide two mixed numbers are shown.

$$3\frac{3}{5} \div 1\frac{1}{2} = \frac{18}{5} \times \frac{3}{2}$$
$$= \frac{27}{5}$$
$$= 5\frac{2}{5}$$

What should your friend change in order to divide the two mixed numbers correctly?

A. Find a common denominator of 5 and 2.

B. Multiply by the reciprocal of $\frac{18}{5}$.

C. Multiply by the reciprocal of $\frac{3}{2}$.

D. Rename $3\frac{3}{5}$ as $2\frac{8}{5}$.

7. A company ordering parts receives a charge of $25 for shipping and handling plus $20 per part. Which equation represents the cost c (in dollars) of ordering p parts?

F. $c = 25 + 20p$

G. $c = 20 + 25p$

H. $p = 25 + 20c$

I. $p = 20 + 25c$

8. Which property is illustrated by the statement?

$$5(a + 6) = 5(a) + 5(6)$$

A. Associative Property of Multiplication

B. Commutative Property of Multiplication

C. Commutative Property of Addition

D. Distributive Property

9. What is the value of the expression?

$$46.8 \div 0.156$$

10. In the mural below, the squares that are painted red are marked with the letter R.

What percent of the mural is painted red?

F. 24%   G. 25%

H. 48%   I. 50%

11. Which expression is equivalent to $28x + 70$?

A. $14(2x + 5)$ B. $14(5x + 2)$

C. $2(14x + 5)$ D. $14(7x)$

12. What is the first step in evaluating the expression?

$$3 \cdot (5 + 2)^2 \div 7$$

F. Multiply 3 and 5.   G. Add 5 and 2.

H. Evaluate 5^2.   I. Evaluate 2^2.

13. Jeff wants to save $4000 to buy a used car. He has already saved $850. He plans to save an additional $150 each week.

Part A Write and solve an equation to represent the number of weeks remaining until he can afford the car.

Jeff saves $150 per week by saving $\frac{3}{4}$ of what he earns at his job each week. He works 20 hours per week.

Part B Write an equation to represent the amount per hour that Jeff must earn to save $150 per week. Explain your reasoning.

Part C What is the amount per hour that Jeff must earn? Show your work and explain your reasoning.

Cumulative Practice   281

7 Area, Surface Area, and Volume

- **7.1** Areas of Parallelograms
- **7.2** Areas of Triangles
- **7.3** Areas of Trapezoids and Kites
- **7.4** Three-Dimensional Figures
- **7.5** Surface Areas of Prisms
- **7.6** Surface Areas of Pyramids
- **7.7** Volumes of Rectangular Prisms

Chapter Learning Target:
Understand measurement.

Chapter Success Criteria:
- ■ I can explain how to find areas of figures.
- ■ I can explain how to find surface areas and volumes of solids.
- ■ I can describe and draw three-dimensional figures.
- ■ I can apply units of measurement to solve real-life problems.

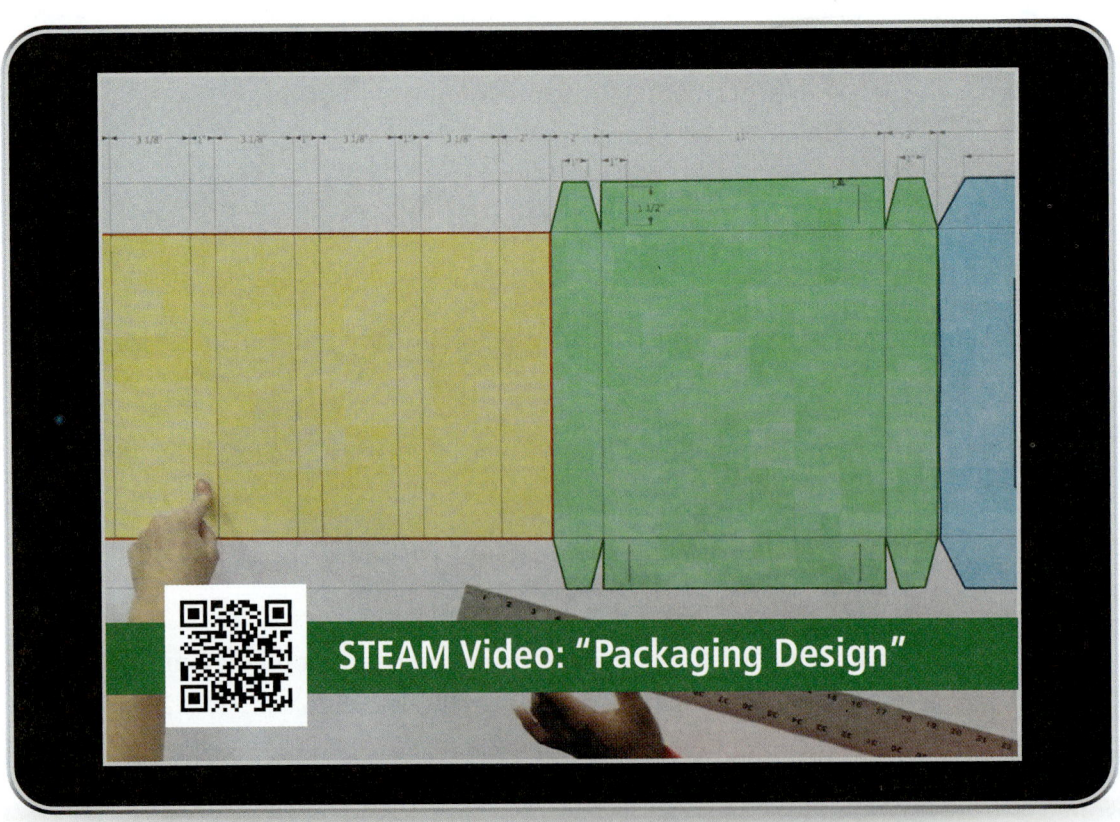

STEAM Video: "Packaging Design"

STEAM Video

Packaging Design

Surface area can be used to determine amounts of materials needed to create objects. Describe another situation in which you need to find the surface area of an object.

Watch the STEAM Video "Packaging Design." Then answer the following questions. Alex is cutting a design out of paper and folding it to form a box.

1. Tory says that the length of the design is 47 inches and the width is 16 inches. Show several ways that you can arrange three of the designs on a roll of paper that is 48 inches wide. You can make the length of the paper as long as is needed.

2. Tory says that the cut-out design has an area of 619 square inches. What is the least possible amount of paper that is wasted when you cut out three of the designs?

Performance Task

Maximizing the Volumes of Boxes

After completing this chapter, you will be able to use the concepts you learned to answer the questions in the *STEAM Video Performance Task*. You will be given the dimensions of a small box and two larger boxes.

Small box: $4\frac{1}{2}$ in. × $4\frac{1}{2}$ in. × 8 in.

Large box 1: $24\frac{1}{4}$ in. × 18 in. × 24 in.

Large box 2: $20\frac{1}{2}$ in. × $18\frac{1}{2}$ in. × 24 in.

You will be asked to determine how many small boxes can be placed in each larger box. When a company is deciding on packaging for a product, why should the surface area of the packaging also be considered?

Getting Ready for Chapter

Chapter Exploration

The formulas for the areas of polygons can be derived from one area formula, the area of a rectangle.

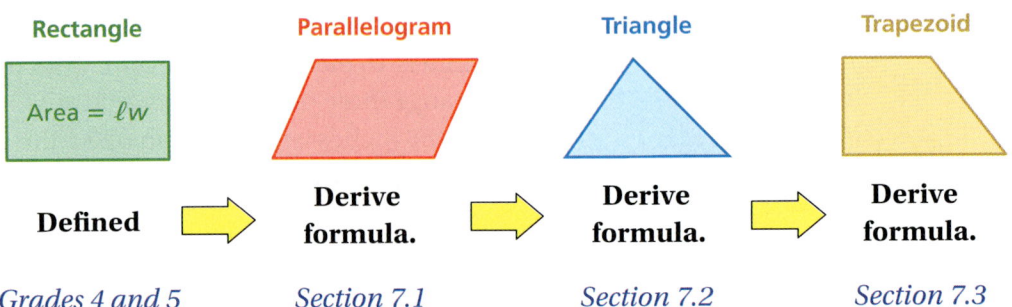

Work with a partner. Find (a) the dimensions of the figures and (b) the areas of the figures. What do you notice?

1.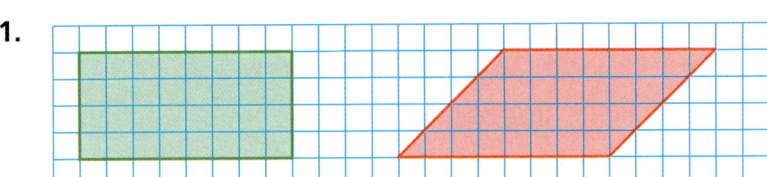

2.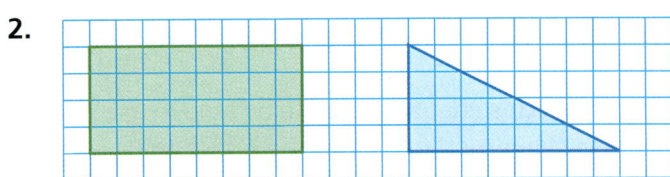

3.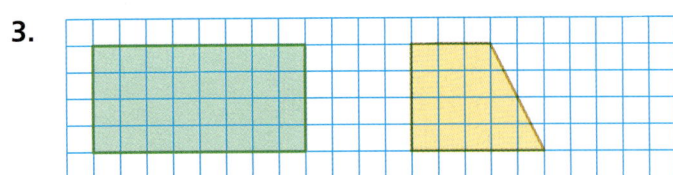

Vocabulary

The following vocabulary terms are defined in this chapter. Think about what each term might mean and record your thoughts.

prism   surface area
pyramid   volume

7.1 Areas of Parallelograms

Learning Target: Find areas and missing dimensions of parallelograms.

Success Criteria:
- I can explain how the area of a rectangle is used to find the area of a parallelogram.
- I can use the base and the height of a parallelogram to find its area.
- I can use the area of a parallelogram and one of its dimensions to find the other dimension.

A **polygon** is a closed figure in a plane that is made up of three or more line segments that intersect only at their endpoints. Several examples of polygons are parallelograms, rhombuses, triangles, trapezoids, and kites.

The formula for the area of a parallelogram can be derived from the definition of the area of a rectangle. Recall that the area of a rectangle is the product of its length ℓ and its width w. The process you use to derive this and other area formulas in this chapter is called *deductive reasoning*.

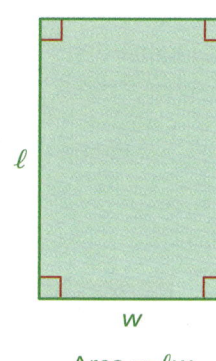

Area = ℓw

EXPLORATION 1

Deriving the Area Formula of a Parallelogram

Work with a partner.

a. Draw *any* rectangle on a piece of centimeter grid paper. Cut the rectangle into two pieces that can be arranged to form a parallelogram. What do you notice about the areas of the rectangle and the parallelogram?

b. Copy the parallelogram below on a piece of centimeter grid paper. Cut the parallelogram and rearrange the pieces to find its area.

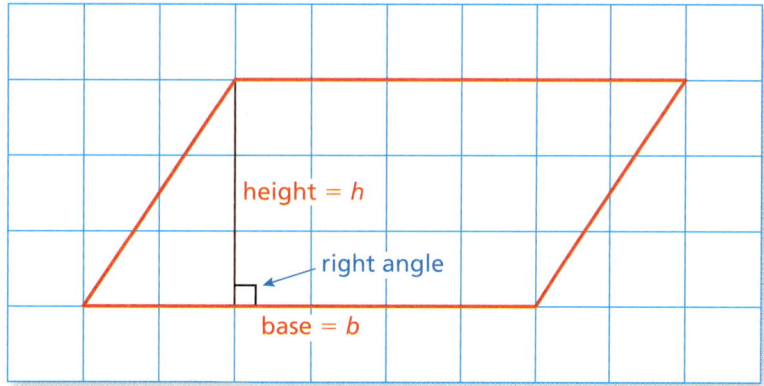

Math Practice

Justify Conclusions

How does decomposing the parallelogram into other figures help you justify your formula?

c. Draw *any* parallelogram on a piece of centimeter grid paper and find its area. Does the area change when you use a different side as the base? Explain your reasoning.

d. Use your results to write a formula for the area A of a parallelogram.

7.1 Lesson

Key Vocabulary
polygon, *p. 285*

The *area* of a polygon is the amount of surface it covers. You can find the area of a parallelogram in much the same way as you can find the area of a rectangle.

Key Idea

Area of a Parallelogram

Words The area A of a parallelogram is the product of its base b and its height h.

Algebra $A = bh$

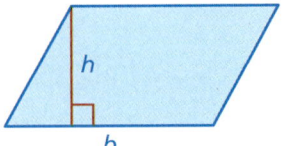

> The base of a parallelogram does not have to be horizontal. Any side of a parallelogram can be the base.

EXAMPLE 1 Finding Areas of Parallelograms

Find the area of each parallelogram.

a.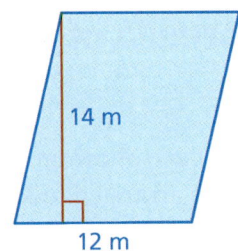

b.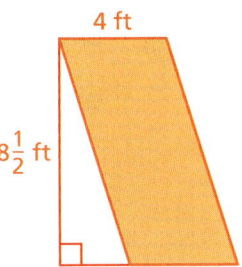

Remember
Area is measured in square units.

$A = bh$ Write formula. $A = bh$

$= 12(14)$ Substitute values. $= 4\left(8\frac{1}{2}\right)$

$= 168$ Multiply. $= 34$

▸ The area of the parallelogram is 168 square meters.

▸ The area of the parallelogram is 34 square feet.

Try It Find the area of the parallelogram.

1.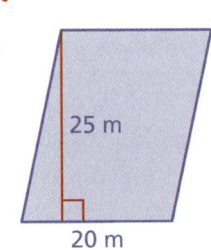

2.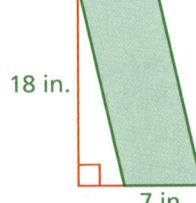

3.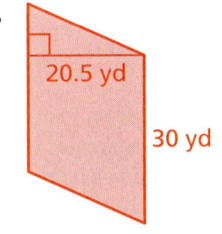

286   Chapter 7   Area, Surface Area, and Volume   Multi-Language Glossary at *BigIdeasMath.com*

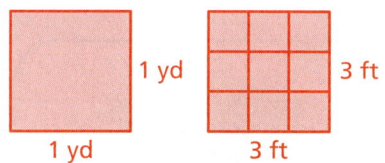

When finding areas, you may need to convert square units. The diagrams at the left show that there are 9 square feet per square yard.

$$1 \text{ yd}^2 = (1 \text{ yd})(1 \text{ yd}) = (3 \text{ ft})(3 \text{ ft}) = 9 \text{ ft}^2$$

You can use a similar procedure to convert other square units.

EXAMPLE 2 Finding the Area of a Parallelogram

Find the area of the parallelogram in square inches.

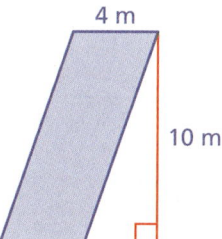

$A = bh$ Write formula.
$= 2.5(2)$ Substitute values.
$= 5$ Multiply.

The area of the parallelogram is 5 square feet.
To convert the area to square inches, use a conversion factor.
Notice that $1 \text{ ft}^2 = (1 \text{ ft})(1 \text{ ft}) = (12 \text{ in.})(12 \text{ in.}) = 144 \text{ in.}^2$.

$$5 \text{ ft}^2 = 5 \,\cancel{\text{ft}^2} \times \frac{144 \text{ in.}^2}{1 \,\cancel{\text{ft}^2}} = 720 \text{ in.}^2$$

▸ The area of the parallelogram is 5 square feet, or 720 square inches.

Try It

4. Find the area of the parallelogram in square centimeters.

Self-Assessment for Concepts & Skills

Solve each exercise. Then rate your understanding of the success criteria in your journal.

5. **WRITING** Explain how to use the area of a rectangle to find the area of a parallelogram.

FINDING AREA Find the area of the parallelogram.

6.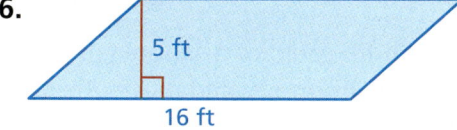

7.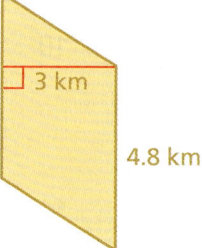

8. **MP REASONING** Draw a parallelogram that has an area of 24 square inches.

EXAMPLE 3 Modeling Real Life

The area of the parallelogram-shaped forest bordered by roads is 99,000 square yards. What is the length of the deer trail?

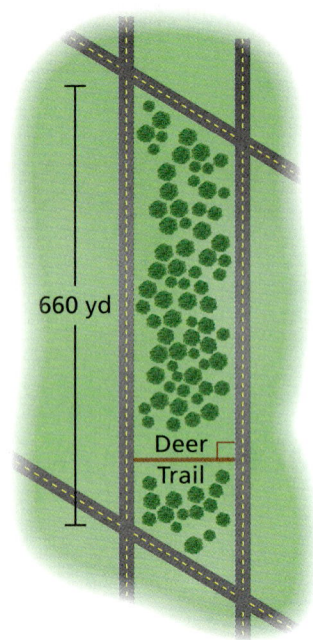

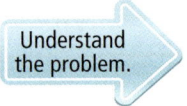 Understand the problem.

You are given the area of a parallelogram-shaped forest, and the map shows the length of one of its bases. You are asked to find the length of the deer trail, which represents the height of the parallelogram.

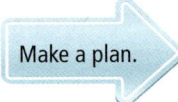 Make a plan.

Use the formula for the area of a parallelogram. Substitute for the area and the base, then solve for the height.

 Solve and check.

$A = bh$ Write formula for area of a parallelogram.

$99,000 = 660h$ Substitute 99,000 for A and 660 for b.

$\dfrac{99,000}{660} = \dfrac{660h}{660}$ Division Property of Equality

$150 = h$ Simplify.

Check
$A = bh$
$99,000 \stackrel{?}{=} 660(150)$
$99,000 = 99,000$ ✓

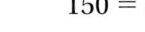 So, the deer trail is 150 yards long.

Self-Assessment for Problem Solving

Solve each exercise. Then rate your understanding of the success criteria in your journal.

9. The side of an office building in Hamburg, Germany, is in the shape of a parallelogram. The area of the side of the building is about 2150 square meters. What is the length x of the portion of the building that extends over the river?

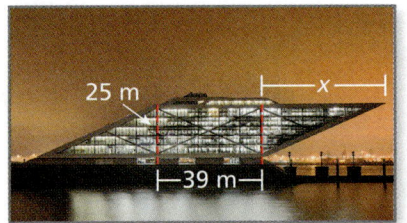

10. You make a photo prop for a school fair. You cut a 10-inch square out of a parallelogram-shaped piece of wood. What is the area of the photo prop?

11. **DIG DEEPER!** A galaxy contains a parallelogram-shaped dust field. The dust field has a base of 150 miles. The height is 14% of the base. What is the area of the dust field?

7.1 Practice

▶ Review & Refresh

Graph the equation.

1. $y = 4x$ **2.** $y = x + 3$ **3.** $y = 2x + 5$

Represent the ratio relationship using a graph.

4.

Length (inches)	4	8	12
Width (inches)	3	6	9

5.

Red Blood Cells	600	1200	1800
White Blood Cells	1	2	3

Write the prime factorization of the number.

6. 55 **7.** 60 **8.** 150 **9.** 126

Add or subtract.

10. $2.36 + 15.71$ **11.** $9.035 - 6.144$ **12.** $28.351 - 19.3518$

▶ Concepts, Skills, & Problem Solving

USING TOOLS Rearrange the parallelogram as a rectangle. Then find the area. (See Exploration 1, p. 285.)

13. 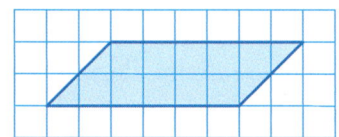 **14.** 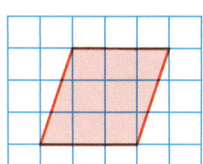 **15.**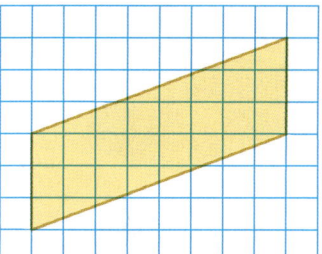

FINDING AREA Find the area of the parallelogram.

16. 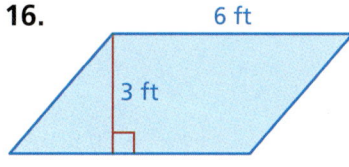 **17.** 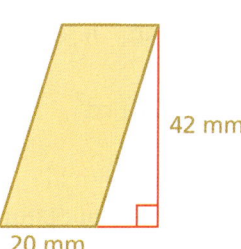 **18.**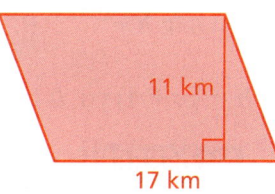

19. 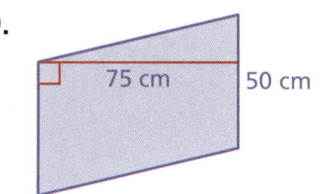 **20.** 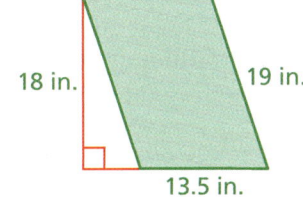 **21.**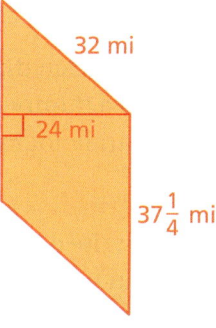

Section 7.1 Areas of Parallelograms **289**

22. **YOU BE THE TEACHER** Your friend finds the area of the parallelogram. Is your friend correct? Explain your reasoning.

23. **MODELING REAL LIFE** A ceramic tile in the shape of a parallelogram has a base of 4 inches and a height of 1.5 inches. What is the area of the tile?

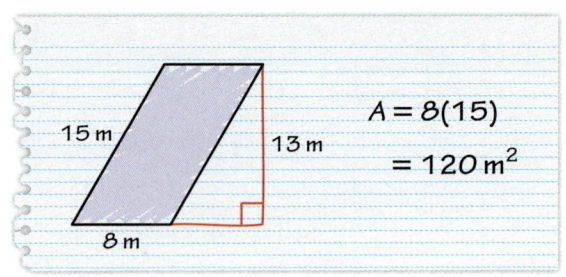

FINDING AREA Find the area of the parallelogram. Round to the nearest hundredth if necessary.

24. Area = ⬚ cm²   25. Area ≈ ⬚ mi²   26. Area ≈ ⬚ ft²

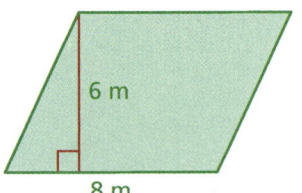 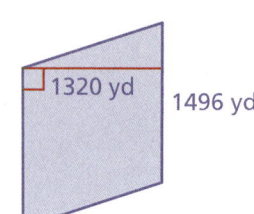 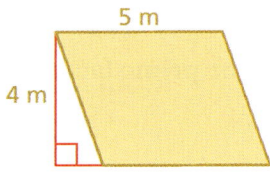

27. **OPEN-ENDED** Your deck has an area of 128 square feet. After adding a section, the area will be $(s^2 + 128)$ square feet. Draw a diagram of how this can happen.

28. **MODELING REAL LIFE** You use the parallelogram-shaped sponge to create the T-shirt design. The area of the design is 66 square inches. How many times do you use the sponge to create the design? Draw a diagram to support your answer.

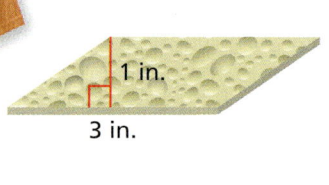

FINDING A MISSING DIMENSION Find the missing dimension of the parallelogram described.

29. $b = 6$ ft, $h =$ ⬚ ft, $A = 54$ ft²   30. $b =$ ⬚ cm, $h = 2.5$ cm, $A = 16$ cm²

31. $b = (x + 4)$ yd, $h =$ ⬚ yd, $A = (5x + 20)$ yd²

32. **DIG DEEPER!** The staircase has three identical parallelogram-shaped panels. The horizontal distance between each panel is 4.25 inches. The area of each panel is 287 square inches. What is the value of x?

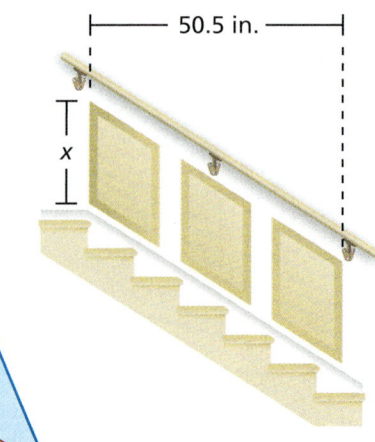

33. **MP LOGIC** Each dimension of a parallelogram is multiplied by a positive number n. Write an expression for the area of the new parallelogram.

34. **CRITICAL THINKING** Rearrange the rhombus shown to write a formula for the area of a rhombus in terms of its diagonals.

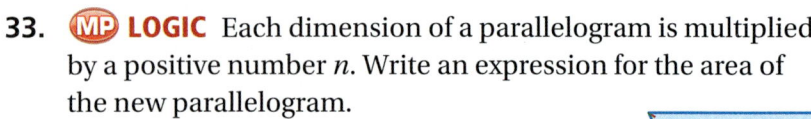

290   Chapter 7   Area, Surface Area, and Volume

7.2 Areas of Triangles

Learning Target: Find areas and missing dimensions of triangles, and find areas of composite figures.

Success Criteria:
- I can explain how the area of a parallelogram is used to find the area of a triangle.
- I can use the base and the height of a triangle to find its area.
- I can use the area of a triangle and one of its dimensions to find the other dimension.
- I can use decomposition to find the area of a figure.

EXPLORATION 1

Deriving the Area Formula of a Triangle

Work with a partner.

a. Draw *any* parallelogram on a piece of centimeter grid paper. Cut the parallelogram into two identical triangles. How can you use the area of the parallelogram to find the area of each triangle?

b. Copy the triangle below on a piece of centimeter grid paper. Find the area of the triangle. Explain how you found the area.

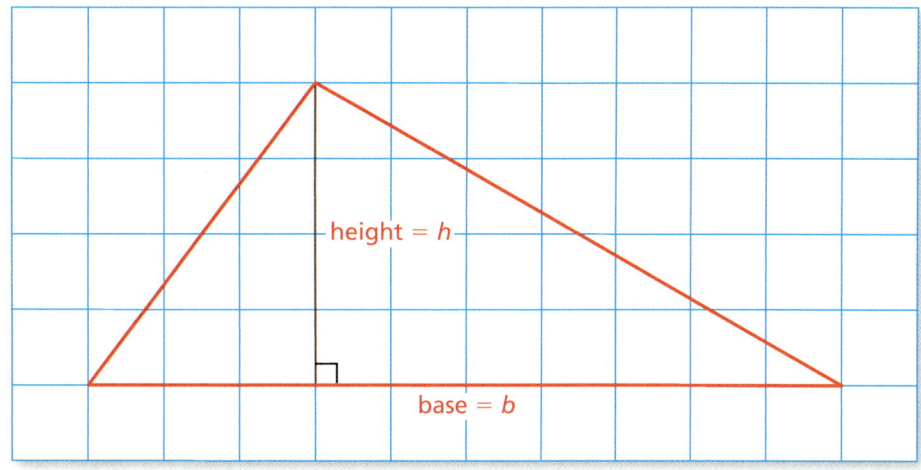

c. Draw *any* acute triangle on a piece of centimeter grid paper and find its area. Repeat this process for a right triangle and an obtuse triangle.

Math Practice

Calculate Accurately

If you use the base and the height to calculate the area in part (d), how can you estimate the dimensions so that your calculations are accurate?

d. Do the areas change in part (c) when you use different sides as the base? Explain your reasoning.

e. Use your results to write a formula for the area A of a triangle. Use the formula to find the area of the triangle shown.

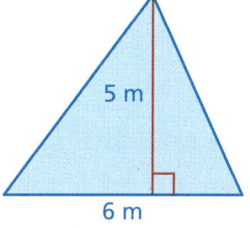

7.2 Lesson

Key Vocabulary
composite figure, p. 294

Key Idea

Area of a Triangle

Words The area A of a triangle is one-half the product of its base b and its height h.

Algebra $A = \dfrac{1}{2}bh$

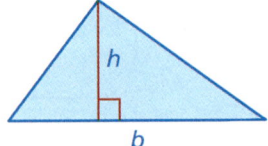

EXAMPLE 1 Finding Areas of Triangles

Find the area of each triangle.

a.

b.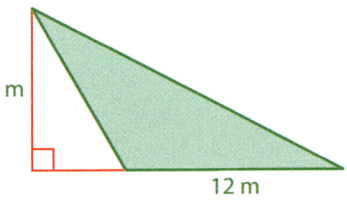

> The base of a triangle does not have to be horizontal. Any side of a triangle can be the base. In Example 1(a), you could have used 8 inches as the base and 5 inches as the height.

$A = \dfrac{1}{2}bh$ Write formula. $A = \dfrac{1}{2}bh$

$= \dfrac{1}{2}(5)(8)$ Substitute values. $= \dfrac{1}{2}(12)(9)$

$= \dfrac{1}{2}(40)$ Multiply. $= 6(9)$

$= 20$ Multiply. $= 54$

▸ The area of the triangle is 20 square inches.

▸ The area of the triangle is 54 square meters.

Try It Find the area of the triangle.

1.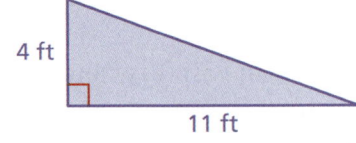

2.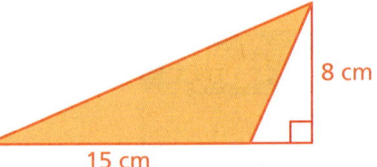

3.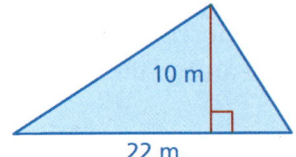

4.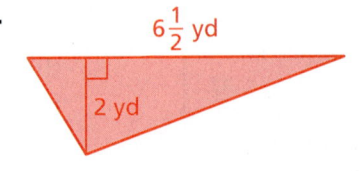

Chapter 7   Area, Surface Area, and Volume   Multi-Language Glossary at BigIdeasMath.com

EXAMPLE 2 Finding a Missing Dimension

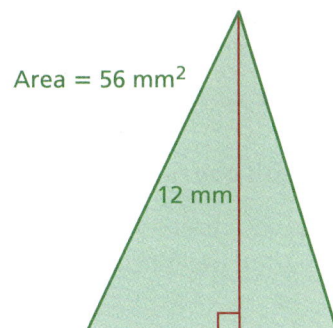

Area = 56 mm²
12 mm

Find the base of the triangle.

Use the formula for the area of a triangle. Substitute for the area and the height, then solve for the base.

$A = \dfrac{1}{2}bh$ Write formula for area of a triangle.

$56 = \dfrac{1}{2}b(12)$ Substitute 56 for A and 12 for h.

$56 = 6b$ Simplify.

$\dfrac{56}{6} = \dfrac{6b}{6}$ Division Property of Equality

$9\dfrac{1}{3} = b$ Simplify.

▶ So, the base is $9\dfrac{1}{3}$ millimeters.

Try It Find the missing dimension of the triangle.

5.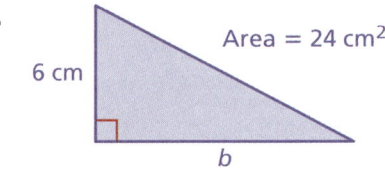
Area = 24 cm²
6 cm
b

6.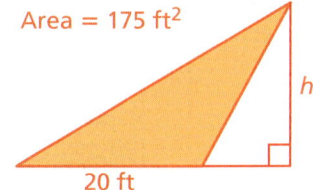
Area = 175 ft²
h
20 ft

Self-Assessment for Concepts & Skills

Solve each exercise. Then rate your understanding of the success criteria in your journal.

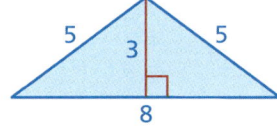
5, 3, 5, 8

7. FINDING AREA Find the area of the triangle at the left.

8. WRITING Explain how to use the area of a parallelogram to find the area of a triangle.

FINDING A MISSING DIMENSION Find the missing dimension of the triangle.

9. Area = 30 mm²
6 mm
b

10.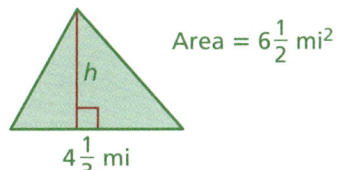
Area = $6\dfrac{1}{2}$ mi²
h
$4\dfrac{1}{3}$ mi

Section 7.2 Areas of Triangles

A **composite figure** is made up of triangles, squares, rectangles, and other two-dimensional figures. To find the area of a composite figure, separate it into figures with areas you know how to find. This is called *decomposition*.

EXAMPLE 3 Modeling Real Life

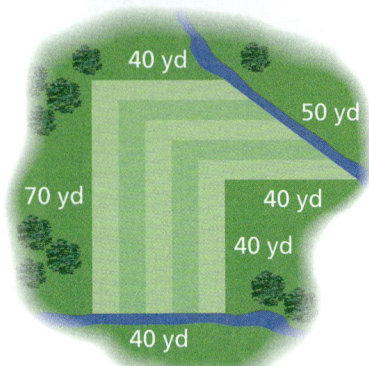

Find the area of the fairway between two streams on a golf course.

There are several ways to separate the fairway into figures with areas you can find using formulas. It appears that one way is to separate the fairway into a rectangle and a right triangle.

Identify each shape and find any missing dimensions. Then find the area of each shape.

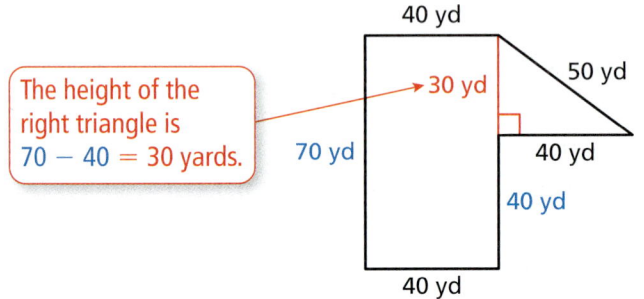

The height of the right triangle is $70 - 40 = 30$ yards.

Another Method It appears that you can separate the fairway into a parallelogram, a triangle, and a square.

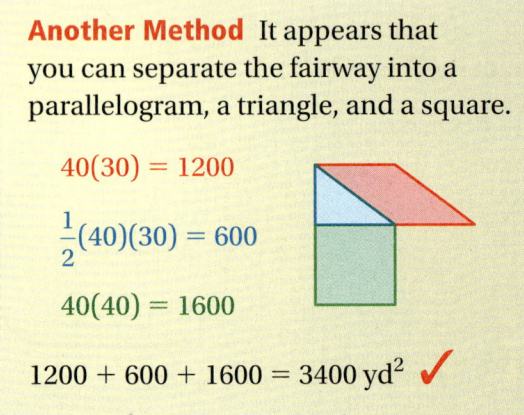

$40(30) = 1200$

$\frac{1}{2}(40)(30) = 600$

$40(40) = 1600$

$1200 + 600 + 1600 = 3400 \text{ yd}^2$ ✓

Area of Rectangle

$A = \ell w$

$= 70(40)$

$= 2800$

Area of Right Triangle

$A = \frac{1}{2}bh$

$= \frac{1}{2}(40)(30)$

$= 600$

▶ So, the area of the fairway is $2800 + 600 = 3400$ square yards.

Self-Assessment for Problem Solving

Solve each exercise. Then rate your understanding of the success criteria in your journal.

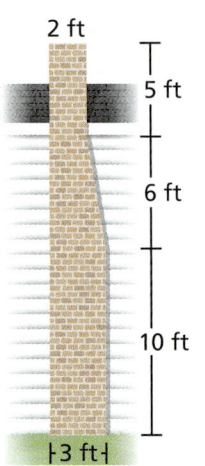

11. A wildlife conservation group buys the 9 square miles of land shown. What is the distance from Point *A* to Point *B*?

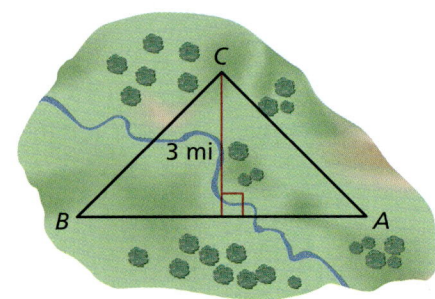

12. **DIG DEEPER!** Find the area of the side of the chimney. Explain how you found the area.

7.2 Practice

▶ Review & Refresh

Find the area of the parallelogram.

1.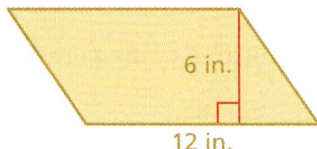
2.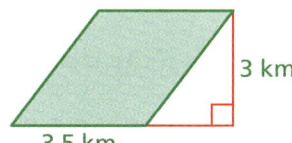
3.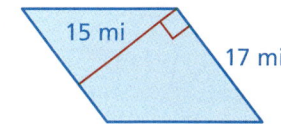

Tell which property the statement illustrates.

4. $n \cdot 1 = n$
5. $4 \cdot m = m \cdot 4$
6. $(x + 2) + 5 = x + (2 + 5)$

7. What is the first step when using order of operations?

   A. Multiply and divide from left to right. B. Add and subtract from left to right.

   C. Perform operations in grouping symbols. D. Evaluate numbers with exponents.

▶ Concepts, Skills, & Problem Solving

USING TOOLS Find the area of the triangle by forming a parallelogram. (See Exploration 1, p. 291.)

8.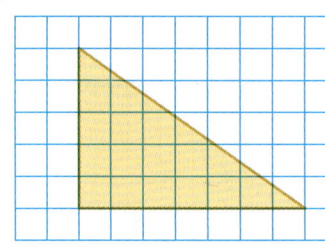
9.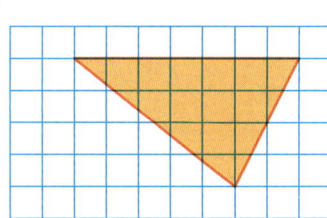
10.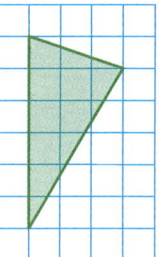

FINDING AREA Find the area of the triangle.

11.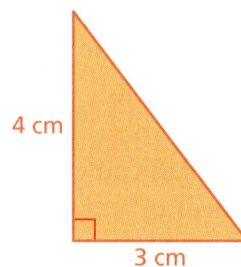
12.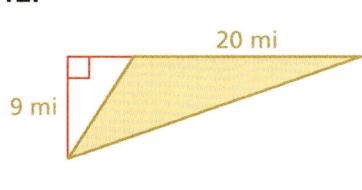
13.
14.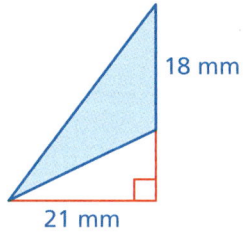
15.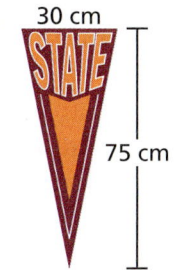
16.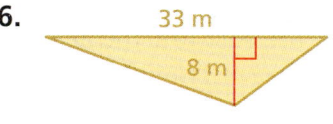

17. **YOU BE THE TEACHER** Your friend finds the area of the triangle. Is your friend correct? Explain your reasoning.

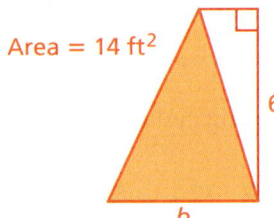

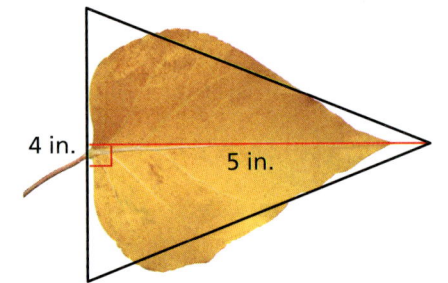

18. **MODELING REAL LIFE** Estimate the area of the cottonwood leaf.

19. **MODELING REAL LIFE** A shelf has the shape of a triangle. The base of the shelf is 36 centimeters, and the height is 18 centimeters. Find the area of the shelf in square inches.

FINDING A MISSING DIMENSION Find the missing dimension of the triangle.

20. Area = 28 m²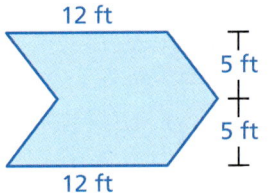
21. Area = 14 ft²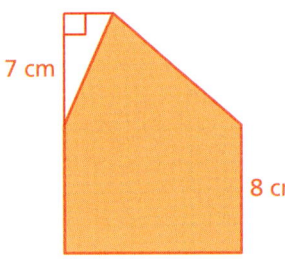
22. 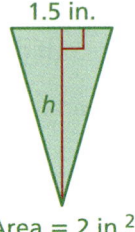 Area = 2 in.²

COMPOSITE FIGURES Find the area of the figure.

23.
24.
25.

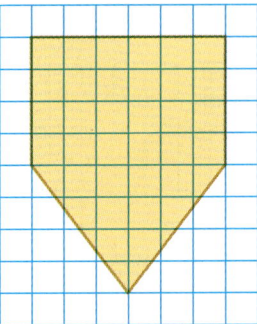

26. **WRITING** You know the height and the perimeter of an equilateral triangle. Explain how to find the area of the triangle. Draw a diagram to support your reasoning.

27. **CRITICAL THINKING** The total area of the polygon is 176 square feet. What is the value of x?

28. **MP REASONING** The base and the height of Triangle A are one-half the base and the height of Triangle B. How many times greater is the area of Triangle B?

29. **MP STRUCTURE** Use what you know about finding areas of triangles to write a formula for the area of a rhombus in terms of its diagonals. Compare the formula with your answer to Section 7.1 Exercise 34.

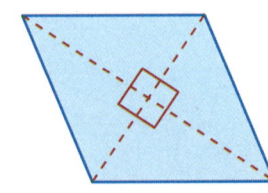

296 Chapter 7 Area, Surface Area, and Volume

7.3 Areas of Trapezoids and Kites

Learning Target: Find areas of trapezoids, kites, and composite figures.

Success Criteria:
- I can explain how the area of a parallelogram is used to find the area of a trapezoid.
- I can decompose trapezoids and kites into smaller shapes.
- I can use decomposition to find the area of a figure.
- I can use the bases and the height of a trapezoid to find its area.

EXPLORATION 1

Deriving the Area Formula of a Trapezoid

Work with a partner.

a. Draw *any* parallelogram on a piece of centimeter grid paper. Cut the parallelogram into two identical trapezoids. How can you use the area of the parallelogram to find the area of each trapezoid?

b. Copy the trapezoid below on a piece of centimeter grid paper. Find the area of the trapezoid. Explain how you found the area.

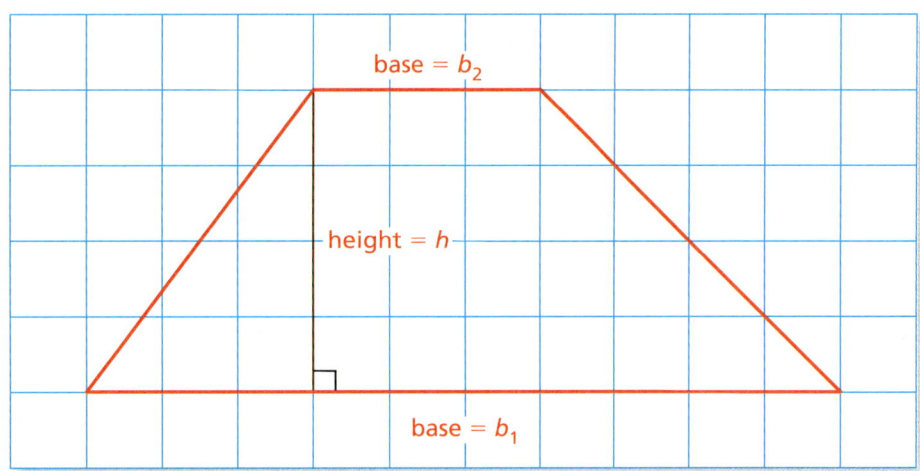

c. Draw *any* trapezoid on a piece of centimeter grid paper and find its area.

d. Use your results to write a formula for the area A of a trapezoid. Use the formula to find the area of the trapezoid shown.

Math Practice

Make a Plan
How can you use the diagram below to justify the formula you wrote in part (d)?

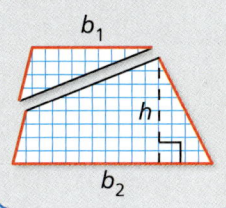

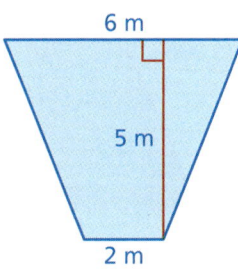

Section 7.3 Areas of Trapezoids and Kites 297

7.3 Lesson

Key Vocabulary
kite, *p. 298*

You can use decomposition to find areas of trapezoids and *kites*. A **kite** is a quadrilateral that has two pairs of adjacent sides with the same length and opposite sides with different lengths.

EXAMPLE 1 — Finding Areas of Trapezoids and Kites

Find the area of each figure.

a.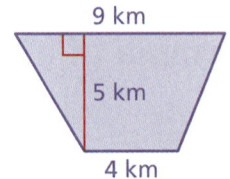

b.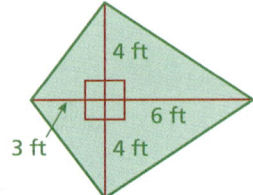

Decompose the trapezoid into a triangle and a rectangle. Find the sum of the areas of the figures.

Decompose the kite into two triangles. Find the sum of the areas of the triangles.

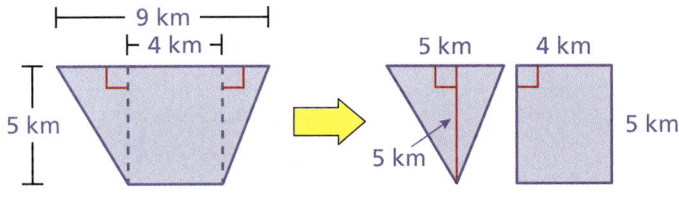

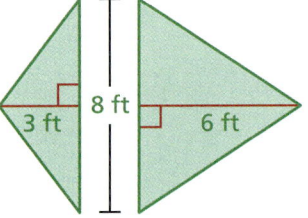

$A = \dfrac{1}{2}(5)(5) + 5(4)$

$= 12\dfrac{1}{2} + 20$

$= 32\dfrac{1}{2}$

$A = \dfrac{1}{2}(8)(3) + \dfrac{1}{2}(8)(6)$

$= 12 + 24$

$= 36$

▶ The area of the trapezoid is $32\dfrac{1}{2}$ square kilometers.

▶ The area of the kite is 36 square feet.

Try It — Find the area of the figure.

1.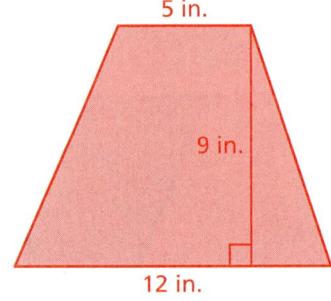

2.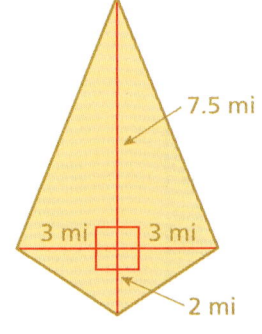

In Example 1(a), you could have used a copy of the trapezoid to form a parallelogram. As you may have discovered in the exploration, this leads to the following formula for the area of a trapezoid.

Key Idea

Area of a Trapezoid

Words The area A of a trapezoid is one-half the product of its height h and the sum of its bases b_1 and b_2.

Algebra $A = \dfrac{1}{2}h(b_1 + b_2)$

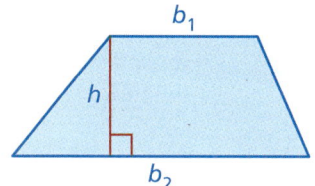

EXAMPLE 2 Finding Areas of Trapezoids

Find the area of each trapezoid.

a.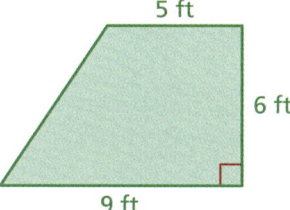

b.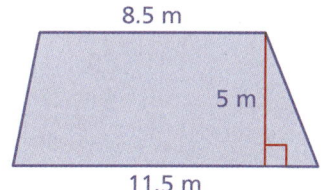

$A = \dfrac{1}{2}h(b_1 + b_2)$	Write formula.	$A = \dfrac{1}{2}h(b_1 + b_2)$
$= \dfrac{1}{2}(6)(5 + 9)$	Substitute.	$= \dfrac{1}{2}(5)(8.5 + 11.5)$
$= \dfrac{1}{2}(6)(14)$	Add.	$= \dfrac{1}{2}(5)(20)$
$= 42$	Multiply.	$= 50$

▶ The area of the trapezoid is 42 square feet.

▶ The area of the trapezoid is 50 square meters.

Try It Find the area of the trapezoid.

3.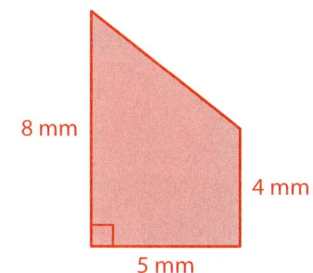

4.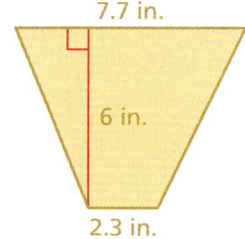

Section 7.3   Areas of Trapezoids and Kites   299

EXAMPLE 3 Finding the Area of a Composite Figure

Find the area of the figure.

You can separate the figure into a rectangle and a trapezoid. Identify the height of the trapezoid. Then find the area of each shape.

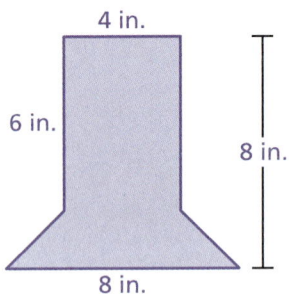

> There is often more than one way to separate composite figures. In Example 3, you can separate the figure into one rectangle and two triangles.

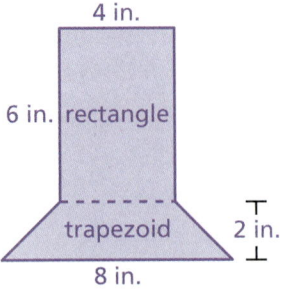

Area of Rectangle

$A = \ell w$
$= 6(4)$
$= 24$

Area of Trapezoid

$A = \dfrac{1}{2}h(b_1 + b_2)$
$= \dfrac{1}{2}(2)(4 + 8)$
$= 12$

▸ So, the area of the figure is $24 + 12 = 36$ square inches.

Try It Find the area of the figure.

5.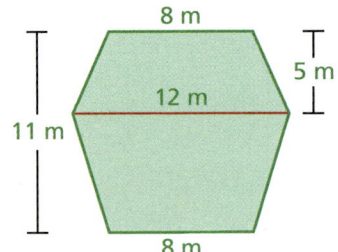

6.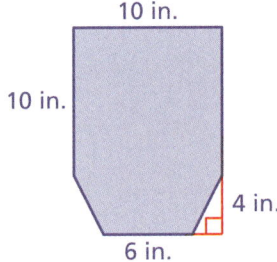

Self-Assessment for Concepts & Skills

Solve each exercise. Then rate your understanding of the success criteria in your journal.

7. **WRITING** Explain how to use the area of a parallelogram to find the area of a trapezoid.

8. **MP REASONING** What measures do you need to find the area of a kite?

FINDING AREA Find the area of the figure.

9.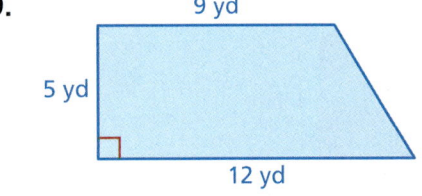

10.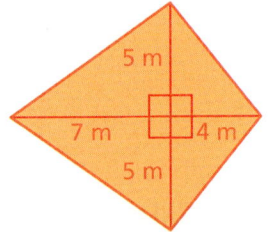

EXAMPLE 4 Modeling Real Life

You can use a trapezoid to approximate the shape of Scott County, Virginia. The population is about 22,100. About how many people are there per square mile?

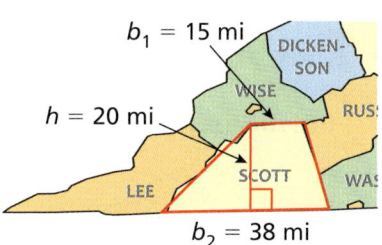

Understand the problem.

You are given the population and the dimensions of a county shaped like a trapezoid. You are asked to find the number of people per square mile.

Make a plan.

Use the formula for the area of a trapezoid to find the area of Scott County. Then divide the population by the area to find the number of people per square mile.

Solve and check.

$$A = \frac{1}{2}h(b_1 + b_2)$$ Write formula for area of a trapezoid.

$$= \frac{1}{2}(20)(15 + 38)$$ Substitute 20 for h, 15 for b_1, and 38 for b_2.

$$= \frac{1}{2}(20)(53)$$ Add.

$$= 530$$ Multiply.

The area of Scott County is about 530 square miles.

So, there are about $\frac{22{,}100 \text{ people}}{530 \text{ mi}^2} \approx 42$ people per square mile.

Check Reasonableness
Round the population to 20,000 and the area to 500 square miles to obtain an estimate that is simpler to calculate.

$$20{,}000 \div 500 = 40$$

The answer is reasonable because $40 \approx 42$. ✓

 Self-Assessment for Problem Solving

Solve each exercise. Then rate your understanding of the success criteria in your journal.

11. **DIG DEEPER!** An archaeologist estimates that the manuscript shown was originally a rectangle with a length of 20 inches. Estimate the area of the fragment that is missing.

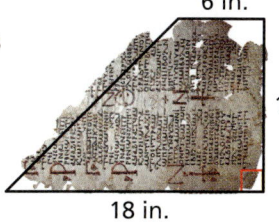

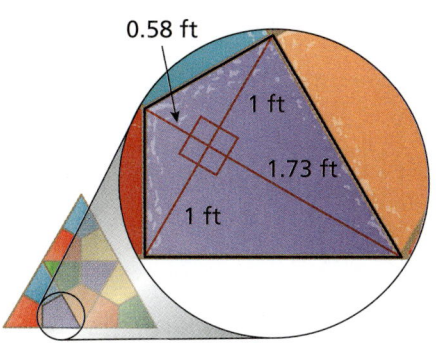

12. The stained-glass window is made of identical kite-shaped glass panes. The approximate dimensions of one pane are shown. The glass used to make the window costs $12.50 per square foot. Find the total cost of the glass used to make the window.

7.3 Practice

▶ Review & Refresh

Find the area of the triangle.

1.
7 in.
18 in.

2.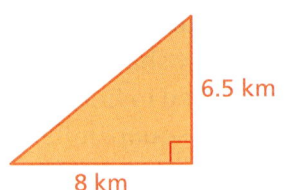
6.5 km
8 km

3.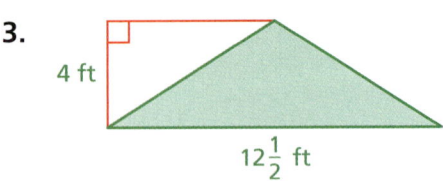
4 ft
$12\frac{1}{2}$ ft

Classify the quadrilateral.

4.

5.

6.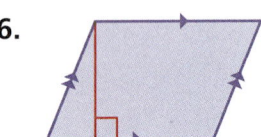

7. On a normal day, 12 airplanes arrive at an airport every 15 minutes. Which rate does *not* represent this situation?

   A. 24 airplanes every 30 minutes
   B. 4 airplanes every 5 minutes
   C. 6 airplanes every 5 minutes
   D. 48 airplanes each hour

▶ Concepts, Skills, & Problem Solving

USING TOOLS Find the area of the trapezoid by forming a parallelogram. (See Exploration 1, p. 297.)

8.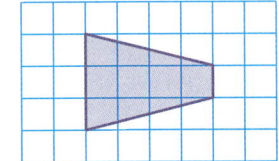

9.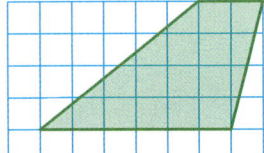

10.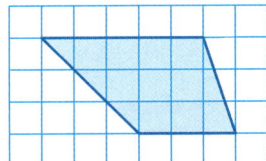

FINDING AREA Use decomposition to find the area of the figure.

11.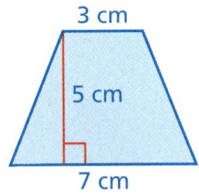
3 cm
5 cm
7 cm

12.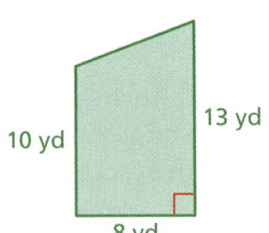
10 yd
13 yd
8 yd

13.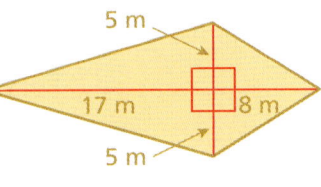
5 m
17 m
8 m
5 m

14.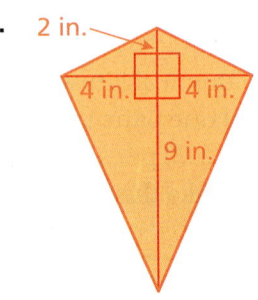
2 in.
4 in. 4 in.
9 in.

15.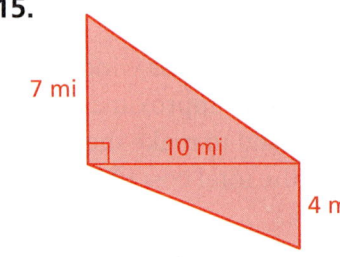
7 mi
10 mi
4 mi

16.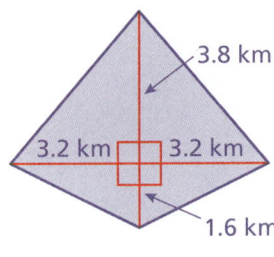
3.8 km
3.2 km 3.2 km
1.6 km

302 Chapter 7 Area, Surface Area, and Volume

FINDING AREA Find the area of the trapezoid.

17.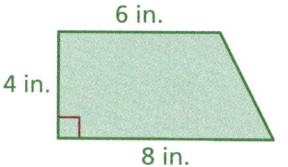

18.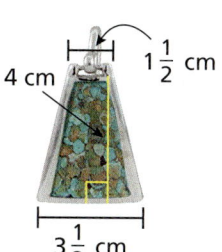

19.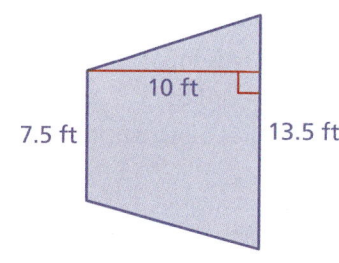

20. **YOU BE THE TEACHER** Your friend finds the area of the trapezoid. Is your friend correct? Explain your reasoning.

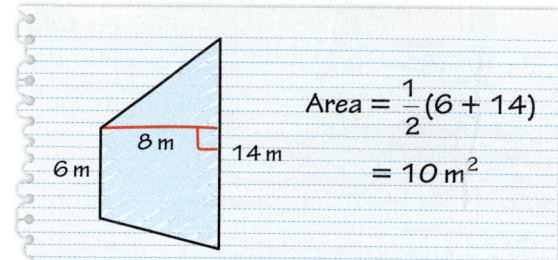

21. **MODELING REAL LIFE** Light shines through a window. What is the area of the trapezoid-shaped region created by the light?

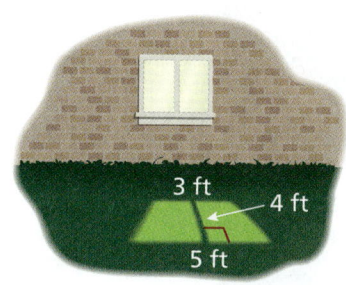

COMPOSITE FIGURES Find the area of the figure.

22.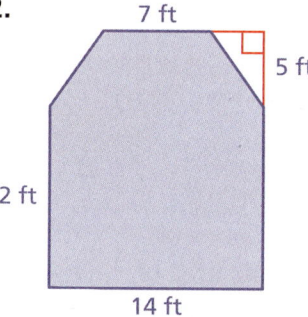

23.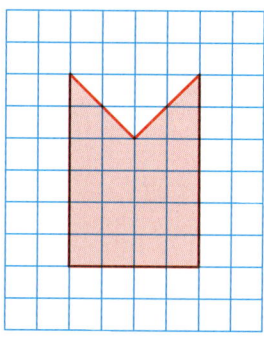

24.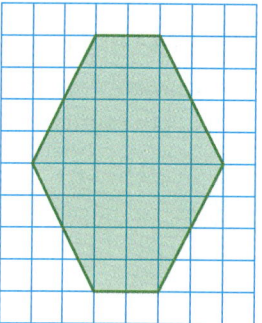

FINDING A MISSING DIMENSION Find the height of the trapezoid.

25. Area = 180 km^2

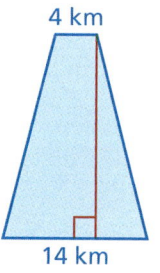

26. Area = 600 yd^2

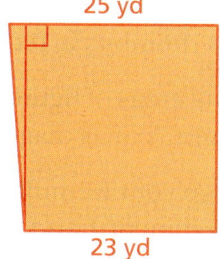

27. Area = $21\frac{1}{5}$ mm^2

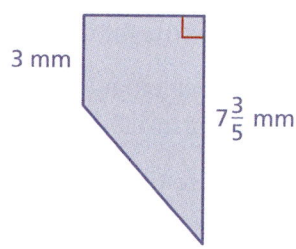

Section 7.3 Areas of Trapezoids and Kites

FINDING AREA Find the area (in square feet) of a trapezoid with height h and bases b_1 and b_2.

28. $h = 6$ in.
    $b_1 = 9$ in.
    $b_2 = 12$ in.

29. $h = 12$ yd
    $b_1 = 5$ yd
    $b_2 = 7$ yd

30. $h = 6$ m
    $b_1 = 3$ m
    $b_2 = 8$ m

31. **OPEN-ENDED** The area of the trapezoidal student election sign is 5 square feet. Find two possible values for each base length.

32. **MP REASONING** How many times greater is the area of the floor covered by the larger speaker than by the smaller speaker?

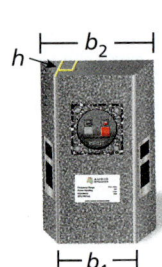

33. **MP REASONING** The rectangle and the trapezoid have the same area. What is the length ℓ of the rectangle?

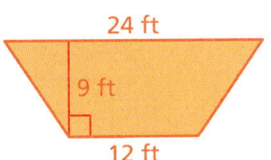

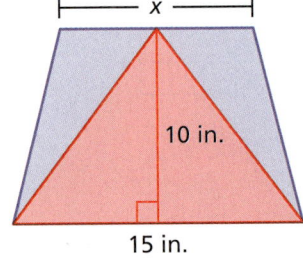

34. **CRITICAL THINKING** In the figure shown, the area of the trapezoid is less than twice the area of the triangle. Find the possible values of x. Can the trapezoid have the same area as the triangle? Explain your reasoning.

35. **MP STRUCTURE** In Section 7.1 Exercise 34 and Section 7.2 Exercise 29, you wrote a formula for the area of a rhombus in terms of its diagonals.

   a. Use what you know about finding areas of figures to write a formula for the area of a kite in terms of its diagonals.

   b. Are there any similarities between your formula in part (a) and the formula you found in Sections 7.1 and 7.2? Explain why or why not.

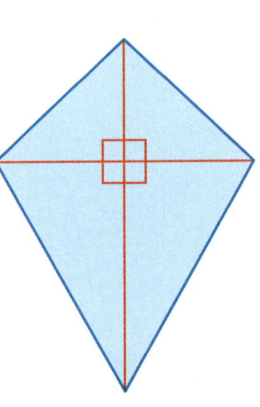

7.4 Three-Dimensional Figures

Learning Target: Describe and draw three-dimensional figures.

Success Criteria:
- I can find the numbers of faces, edges, and vertices of a three-dimensional figure.
- I can draw prisms and pyramids.
- I can draw the front, side, and top views of a three-dimensional figure.

EXPLORATION 1

Exploring Faces, Edges, and Vertices

Work with a partner. Use the rectangular prism shown.

a. Prisms have *faces*, *edges*, and *vertices*. What does each of these terms mean?

b. What does it mean for lines or planes to be parallel or perpendicular in three dimensions? Use drawings to identify one pair of each of the following.

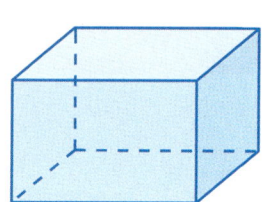

- parallel faces
- perpendicular faces
- parallel edges
- perpendicular edges
- edge parallel to a face
- edge perpendicular to a face

Math Practice

View as Components

What are the different parts of a solid? How can you use these parts to help you draw a solid?

EXPLORATION 2

Drawing Views of a Solid

Work with a partner. Draw the front, side, and top views of each stack of cubes. Then find the number of cubes in the stack. An example is shown at the left.

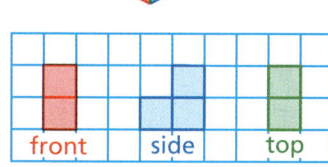

Number of cubes: 3

a.

b.

c.

d.

Section 7.4 Three-Dimensional Figures 305

7.4 Lesson

Key Vocabulary
solid, *p. 306*
polyhedron, *p. 306*
face, *p. 306*
edge, *p. 306*
vertex, *p. 306*
prism, *p. 306*
pyramid, *p. 306*

A **solid** is a three-dimensional figure that encloses a space. A **polyhedron** is a solid whose *faces* are all polygons.

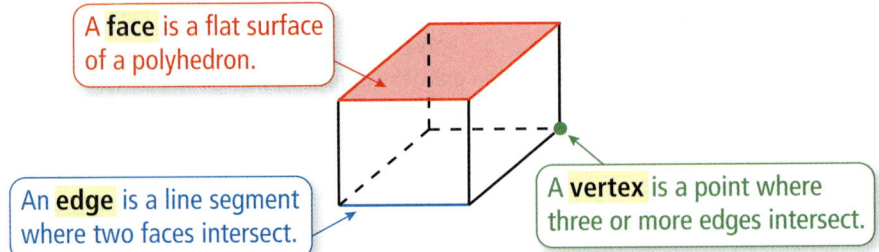

A **face** is a flat surface of a polyhedron.

An **edge** is a line segment where two faces intersect.

A **vertex** is a point where three or more edges intersect.

EXAMPLE 1 Finding the Numbers of Faces, Edges, and Vertices

Find the numbers of faces, edges, and vertices of the solid.

The solid has 1 face on the bottom, 1 face on the top, and 4 faces on the sides.

The faces intersect at 12 different line segments.

The edges intersect at 8 different points.

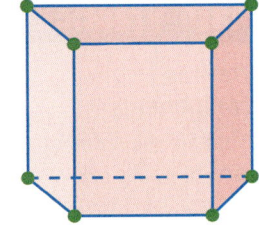

▸ So, the solid has 6 faces, 12 edges, and 8 vertices.

Try It

1. Find the numbers of faces, edges, and vertices of the solid.

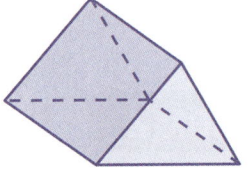

 Key Ideas

Prisms

A **prism** is a polyhedron that has two parallel, identical *bases*. The *lateral faces* are parallelograms.

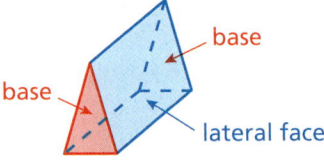

base
base
lateral face

Triangular Prism

Pyramids

A **pyramid** is a polyhedron that has one base. The lateral faces are triangles.

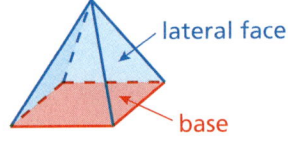

lateral face
base

Rectangular Pyramid

The shape of the base tells the name of the prism or the pyramid.

306 Chapter 7 Area, Surface Area, and Volume

EXAMPLE 2 Drawing Solids

a. Draw a rectangular prism.

Step 1:
Draw identical rectangular bases.

Step 2:
Connect corresponding vertices.

Step 3:
Change any *hidden* lines to dashed lines.

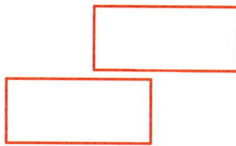 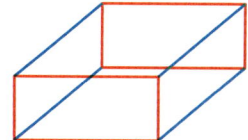 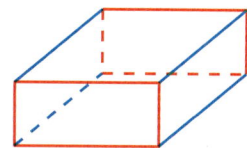

b. Draw a triangular pyramid.

Step 1:
Draw a triangular base and a point.

Step 2:
Connect the vertices of the triangle to the point.

Step 3:
Change any *hidden* lines to dashed lines.

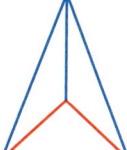 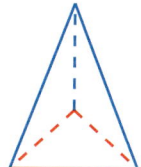

Try It Draw the solid.

2. square prism

3. pentagonal pyramid

Self-Assessment for Concepts & Skills

Solve each exercise. Then rate your understanding of the success criteria in your journal.

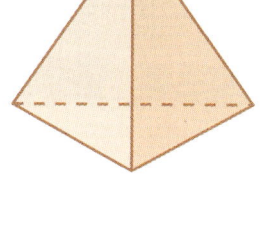

4. **FACES, EDGES, AND VERTICES** Find the numbers of faces, edges, and vertices of the solid at the left.

5. **DRAWING A SOLID** Draw an octagonal prism.

6. **WHICH ONE DOESN'T BELONG?** Which figure does *not* belong with the other three? Explain your reasoning.

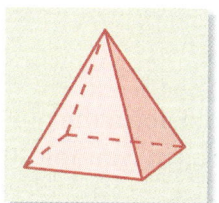 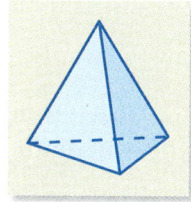 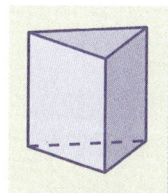 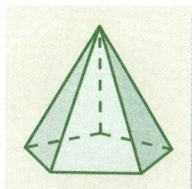

Section 7.4 Three-Dimensional Figures

EXAMPLE 3 Modeling Real Life

a. **Find the numbers of faces, edges, and vertices of the table-cut diamond.**

The diamond has 1 face on the top and 8 faces on the sides.

The faces intersect at 16 different line segments.

The edges intersect at 9 different points.

▸ So, the diamond has 9 faces, 16 edges, and 9 vertices.

b. **Draw the front, side, and top views of the diamond.**

Front view   Side view   Top view

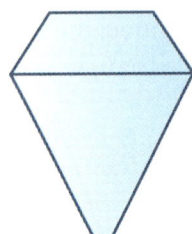 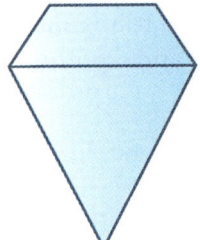 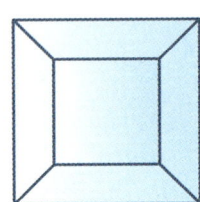

Self-Assessment for Problem Solving

Solve each exercise. Then rate your understanding of the success criteria in your journal.

7. The Flatiron Building in New York City is in the shape of a triangular prism. Draw a sketch of the building.

8. The Pyramid of the Niches is in El Tajín, an archaeological site in Veracruz, Mexico. Draw the front, side, and top views of the pyramid. Explain.

9. Use the point-cut diamond shown.

   a. Find the numbers of faces, edges, and vertices of the diamond.

   b. Draw the front, side, and top views of the diamond.

   c. How can a jeweler transform the point-cut diamond into a table-cut diamond as in Example 3?

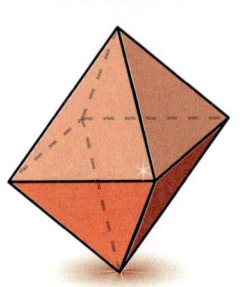

308   Chapter 7   Area, Surface Area, and Volume

7.4 Practice

 Go to BigIdeasMath.com to get HELP with solving the exercises.

▶ Review & Refresh

Find the area of the figure.

1. 6 ft (top), 3 ft (height), 4 ft (bottom)

2. 5 km (top), 3 km (height), 7 km (bottom)

3. 6 m, 9 m, 9 m, 6 m (rhombus)

Find the LCM of the numbers.

4. 8, 12
5. 15, 25
6. 32, 44
7. 3, 7, 10

A bucket contains stones and seashells. You are given the number of seashells in the bucket and the ratio of stones to seashells. Find the number of stones in the bucket.

8. 18 seashells; 2 to 1
9. 30 seashells; 4 : 3
10. 40 seashells; 7 : 4

▶ Concepts, Skills, & Problem Solving

DRAWING VIEWS OF A SOLID Draw the front, side, and top views of the stack of cubes. Then find the number of cubes in the stack. (See Exploration 2, p. 305.)

11.
12.
13.

FACES, EDGES, AND VERTICES Find the numbers of faces, edges, and vertices of the solid.

14.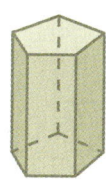
15.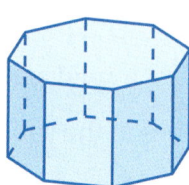
16.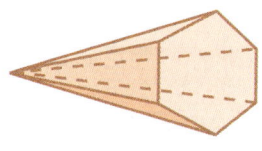

DRAWING SOLIDS Draw the solid.

17. triangular prism
18. pentagonal prism
19. rectangular pyramid
20. hexagonal pyramid

21. **MODELING REAL LIFE** The Pyramid of Cestius in Rome, Italy, is in the shape of a square pyramid. Draw a sketch of the pyramid.

22. **RESEARCH** Use the Internet to find a picture of the Washington Monument. Describe its shape.

DRAWING VIEWS OF A SOLID Draw the front, side, and top views of the solid.

23.

24.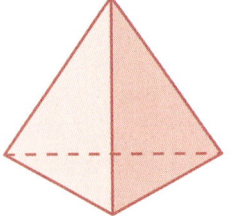

25.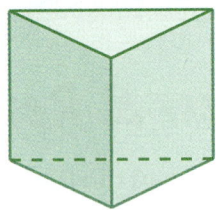

26.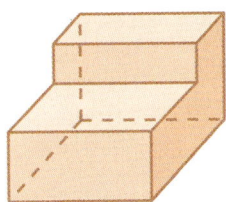

27.

28.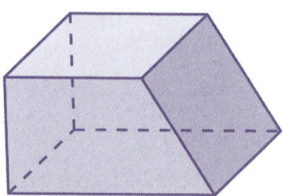

DRAWING SOLIDS Draw a solid with the following front, side, and top views.

29.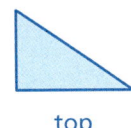
front   side   top

30.
front   side   top

31. **MODELING REAL LIFE** Design and draw a house. Name the different solids that you can use to make a model of the house.

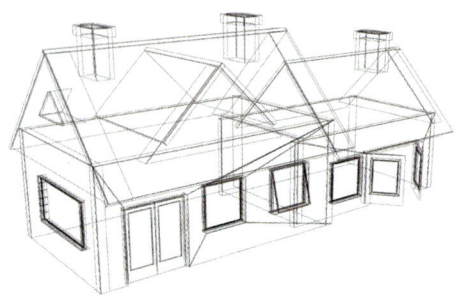 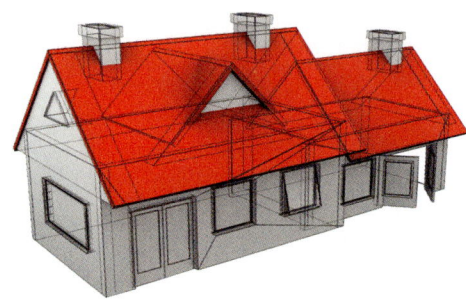

32. **DIG DEEPER!** Two of the three views of a solid are shown.

    a. What is the greatest number of cubes in the solid?
    b. What is the least number of cubes in the solid?
    c. Draw the front views of both solids in parts (a) and (b).

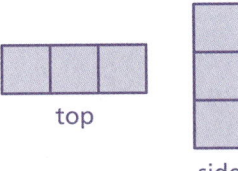

33. **OPEN-ENDED** Draw two different solids with five faces.

    a. Write the numbers of vertices and edges for each solid.
    b. Explain how knowing the numbers of edges and vertices helps you draw a three-dimensional figure.

34. **CRITICAL THINKING** The base of a pyramid has n sides. Find the numbers of faces, edges, and vertices of the pyramid. Explain your reasoning.

7.5 Surface Areas of Prisms

Learning Target: Represent prisms using nets and use nets to find surface areas of prisms.

Success Criteria:
- I can draw nets to represent prisms.
- I can use nets to find surface areas of prisms.
- I can use a formula to find the surface area of a cube.
- I can apply surface areas of prisms to solve real-life problems.

EXPLORATION 1

Using Grid Paper to Construct a Solid

Work with a partner. Copy the figure shown below onto grid paper.

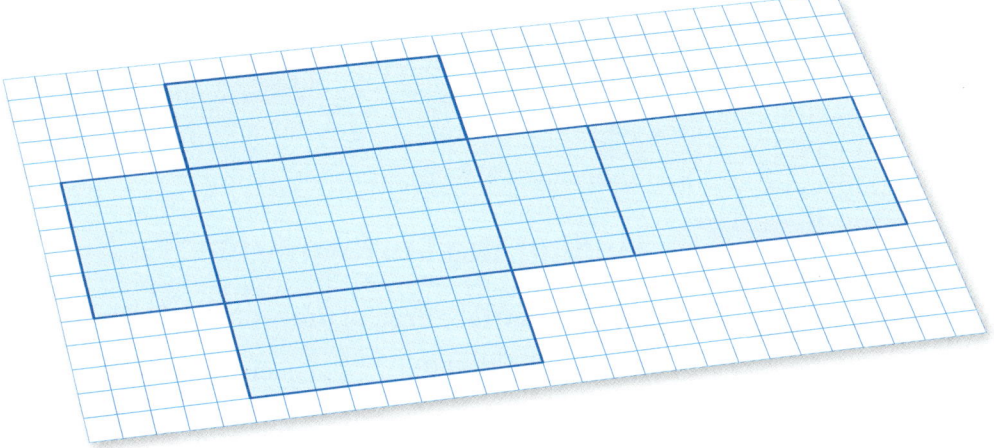

a. Cut out and fold the figure to form a solid. What type of solid does the figure form?

b. What is the area of the entire surface of the solid?

EXPLORATION 2

Finding the Area of the Entire Surface

Work with a partner. Find the area of the entire surface of each solid. Explain your reasoning.

Math Practice

Repeat Calculations

When finding the area of the entire surface, what calculations do you repeat?

a.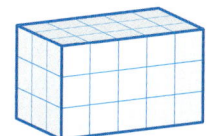

b.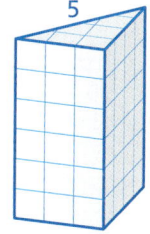

Section 7.5   Surface Areas of Prisms   311

7.5 Lesson

Key Vocabulary
surface area, p. 312
net, p. 312

The **surface area** of a solid is the sum of the areas of all of its faces. You can use a two-dimensional representation of a solid, called a **net**, to find the surface area of the solid. Surface area is measured in *square units*.

Key Idea

Net of a Rectangular Prism

A *rectangular prism* is a prism with rectangular bases.

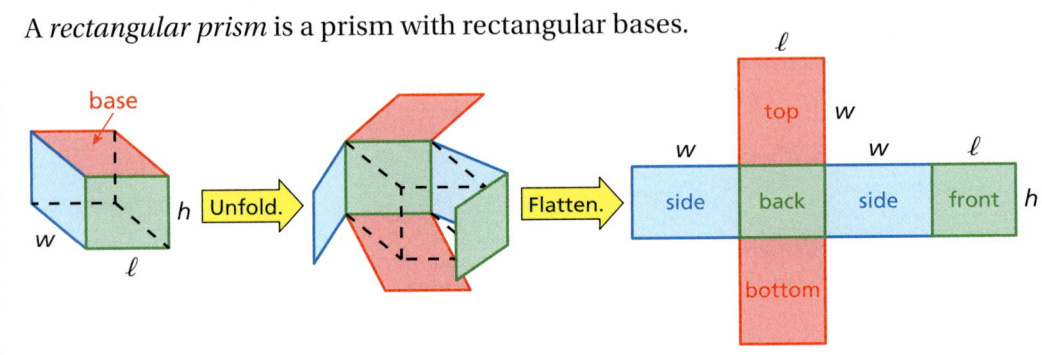

EXAMPLE 1 Finding the Surface Area of a Rectangular Prism

Find the surface area of the rectangular prism.

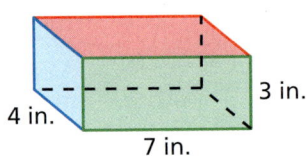

Use a net to find the area of each face.

Top: $7 \cdot 4 = 28$
Bottom: $7 \cdot 4 = 28$
Front: $7 \cdot 3 = 21$
Back: $7 \cdot 3 = 21$
Side: $4 \cdot 3 = 12$
Side: $4 \cdot 3 = 12$

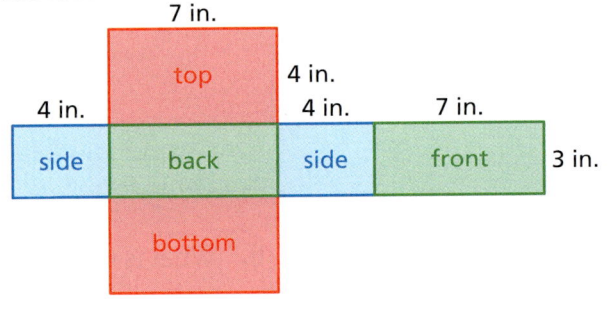

Find the sum of the areas of the faces.

Surface Area = Area of top + Area of bottom + Area of front + Area of back + Area of a side + Area of a side

$S = 28 + 28 + 21 + 21 + 12 + 12 = 122$

▶ So, the surface area is 122 square inches.

Try It Find the surface area of the rectangular prism.

1.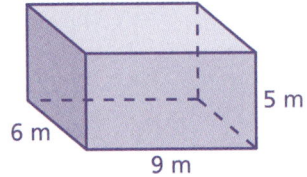

2.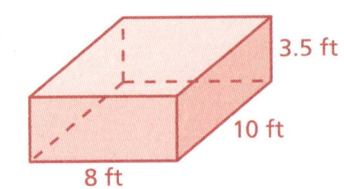

312 Chapter 7 Area, Surface Area, and Volume

Key Idea

Net of a Triangular Prism

A *triangular prism* is a prism with triangular bases.

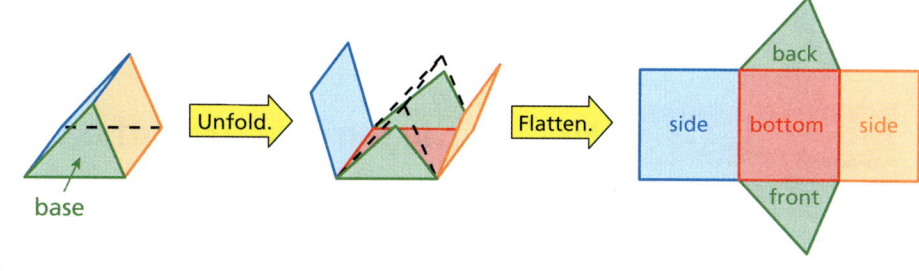

EXAMPLE 2   Finding the Surface Area of a Triangular Prism

Find the surface area of the triangular prism.

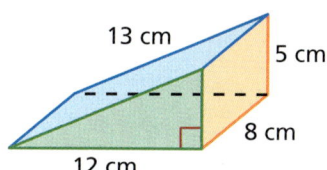

Use a net to find the area of each face.

Bottom: $12 \cdot 8 = 96$

Front: $\frac{1}{2} \cdot 12 \cdot 5 = 30$

Back: $\frac{1}{2} \cdot 12 \cdot 5 = 30$

Side: $13 \cdot 8 = 104$

Side: $8 \cdot 5 = 40$

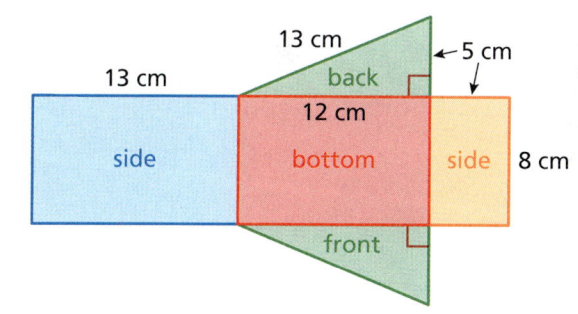

Find the sum of the areas of the faces.

Surface Area = Area of bottom + Area of front + Area of back + Area of a side + Area of a side

$S = 96 + 30 + 30 + 104 + 40$

$= 300$

So, the surface area is 300 square centimeters.

Try It Find the surface area of the triangular prism.

3.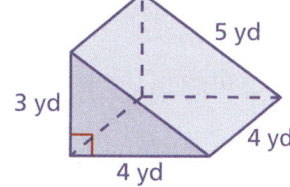

4.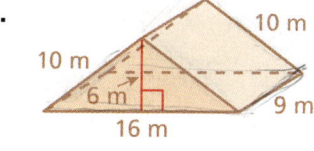

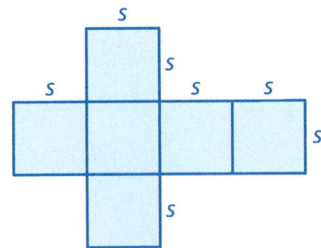
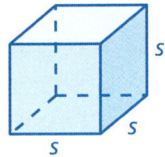

When all the edges of a rectangular prism have the same length s, the rectangular prism is a cube. The net of a cube shows that each of the 6 identical square faces has an area of s^2. So, a formula for the surface area of a cube is

$S = 6s^2$.   Formula for surface area of a cube

EXAMPLE 3 Finding the Surface Area of a Cube

Find the surface area of the cube.

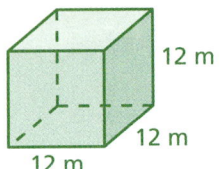

$S = 6s^2$ Write formula for surface area of a cube.

$= 6(12)^2$ Substitute 12 for s.

$= 6(144)$ Evaluate power.

$= 864$ Multiply.

▸ The surface area of the cube is 864 square meters.

Try It Find the surface area of the cube.

5.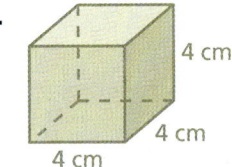

6.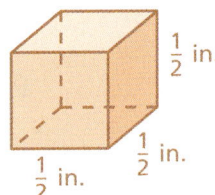

Self-Assessment for Concepts & Skills

Solve each exercise. Then rate your understanding of the success criteria in your journal.

7. **FINDING SURFACE AREA** Find the surface area of a cube with edge lengths of 9 centimeters.

8. **DIFFERENT WORDS, SAME QUESTION** Which is different? Find "both" answers.

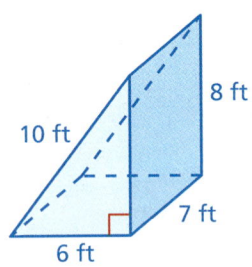

What is the sum of the areas of all of the faces of the prism?

What is the area of the entire surface of the prism?

What is the combined area of the triangular faces of the prism?

What is the surface area of the prism?

314   Chapter 7   Area, Surface Area, and Volume

EXAMPLE 4 Modeling Real Life

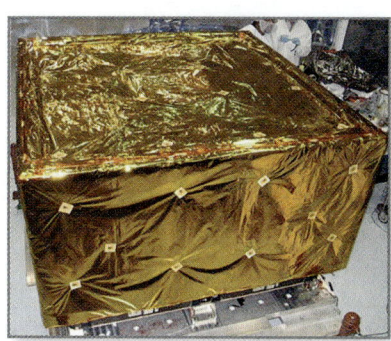

Space instruments are often wrapped in gold-colored multi-layer insulation (MLI) to reflect radiation from the Sun. What is the least amount of MLI needed to wrap an instrument in the shape of a rectangular prism with a length of 5 feet, a width of 5 feet, and a height of 3 feet?

Draw the prism. The least amount of MLI needed is represented by the surface area of the prism. Use a net to find the surface area.

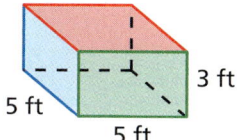

Top: $5 \cdot 5 = 25$
Bottom: $5 \cdot 5 = 25$
Front: $5 \cdot 3 = 15$
Back: $5 \cdot 3 = 15$
Side: $5 \cdot 3 = 15$
Side: $5 \cdot 3 = 15$

$S = 25 + 25 + 15 + 15 + 15 + 15 = 110$

▸ So, the least amount of MLI needed is 110 square feet.

Check Reasonableness The surface area of a 5 ft × 5 ft × 3 ft prism should be less than the surface area of a 5 ft × 5 ft × 5 ft cube. The cube has a surface area of $6(5)^2 = 150$ square feet. Because $110 \text{ ft}^2 < 150 \text{ ft}^2$, the answer is reasonable. ✓

Self-Assessment for Problem Solving

Solve each exercise. Then rate your understanding of the success criteria in your journal.

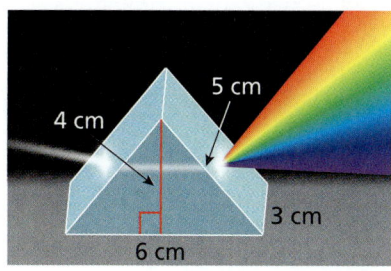

9. Light shines through a glass prism and forms a rainbow. What is the surface area of the prism?

10. One pint of chalkboard paint covers 60 square feet. What is the least number of pints of paint needed to paint the walls of a room in the shape of a rectangular prism with a length of 15 feet, a width of 13 feet, and a height of 10 feet? Explain.

11. **DIG DEEPER!** A flexible *metamaterial* is developed for use in robotics and prosthetics. A block of metamaterial is in the shape of a cube with a surface area of 600 square centimeters. What is the edge length of the block of metamaterial?

7.5 Practice

Review & Refresh

Draw the front, side, and top views of the solid.

1.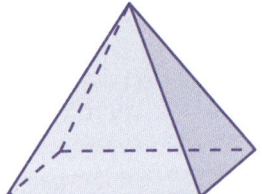
2.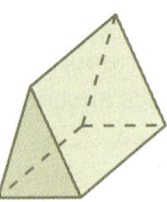
3.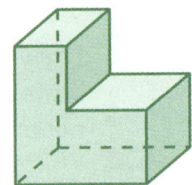

Find the GCF of the numbers.

4. 18, 72
5. 44, 110
6. 78, 93
7. 60, 96, 156

Solve the equation.

8. $s - 5 = 12$
9. $x + 9 = 20$
10. $48 = 6r$
11. $\dfrac{m}{5} = 13$

Divide.

12. $496 \div 16$
13. $765 \div 45$
14. $1173 \div 23$

Concepts, Skills, & Problem Solving

USING TOOLS Use a net to find the area of the entire surface of the solid. Explain your reasoning. *(See Exploration 2, p. 311.)*

15.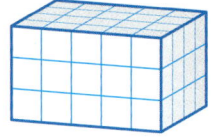
16.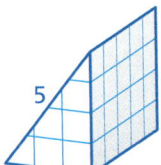
17.

FINDING SURFACE AREA Find the surface area of the rectangular prism.

18.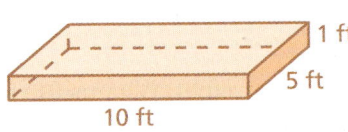
19.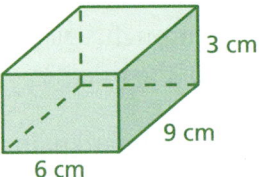
20.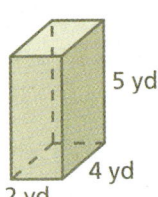
21.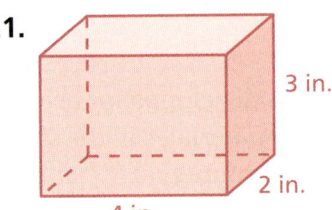
22.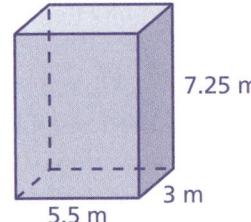
23.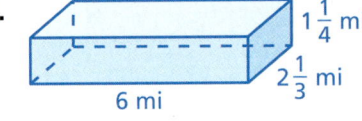

FINDING SURFACE AREA Find the surface area of the triangular prism.

24.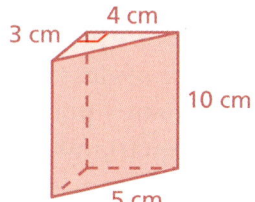

25.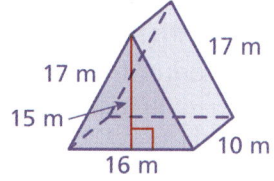

26.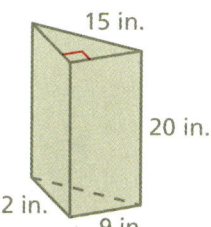

27.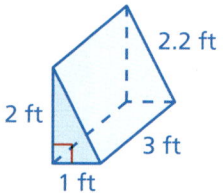

28.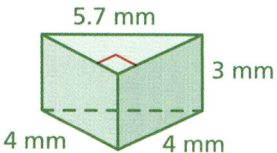

29.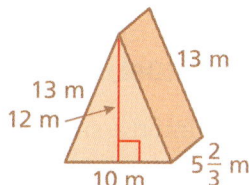

30. **MODELING REAL LIFE** A gift box in the shape of a rectangular prism measures 8 inches by 8 inches by 10 inches. What is the least amount of wrapping paper needed to wrap the gift box? Explain.

31. **MODELING REAL LIFE** What is the least amount of fabric needed to make the tent?

FINDING SURFACE AREA Find the surface area of the cube.

32.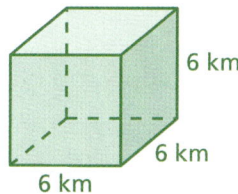

33.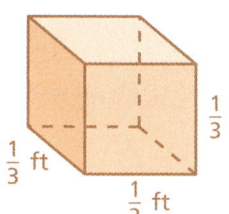

34.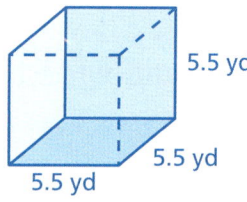

35. **MODELING REAL LIFE** A piece of dry ice is in the shape of a cube with edge lengths of 7 centimeters. Find the surface area of the dry ice.

36. **YOU BE THE TEACHER** Your friend finds the surface area of the prism. Is your friend correct? Explain your reasoning.

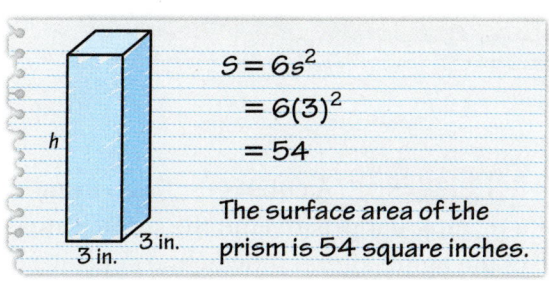

$S = 6s^2$
$= 6(3)^2$
$= 54$

The surface area of the prism is 54 square inches.

Section 7.5   Surface Areas of Prisms   317

37. CRITICAL THINKING A public library has the aquarium shown. The front piece of glass has an area of 24 square feet. How many square feet of glass were used to build the aquarium? (The top of the aquarium is open and the bottom is glass.)

38. MP PROBLEM SOLVING A cereal box has the dimensions shown.

a. Find the surface area of the cereal box.

b. The manufacturer decides to decrease the size of the box by reducing each of the dimensions by 1 inch. Find the decrease in surface area.

39. MP REASONING The material used to make a storage box costs $1.25 per square foot. The boxes have the same volume. Which box might a company prefer to make? Explain your reasoning.

	Length	Width	Height
Box 1	20 in.	6 in.	4 in.
Box 2	15 in.	4 in.	8 in.

40. MP LOGIC Which of the following are nets of a cube? Select all that apply.

A. 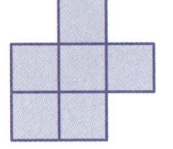   B. 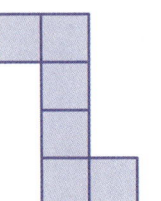   C. 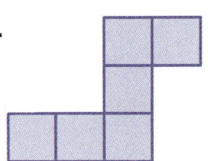   D.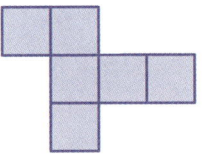

41. MODELING REAL LIFE A quart of stain covers 100 square feet. How many quarts should you buy to stain the wheelchair ramp? (Assume you do not have to stain the bottom of the ramp.)

42. DIG DEEPER! A cube is removed from a rectangular prism. Find the surface area of the figure after removing the cube.

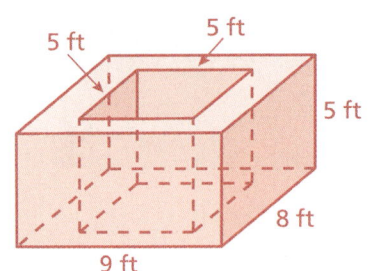

7.6 Surface Areas of Pyramids

Learning Target: Represent pyramids using nets and use nets to find surface areas of pyramids.

Success Criteria:
- I can draw nets to represent pyramids.
- I can use nets to find surface areas of pyramids.
- I can apply surface areas of pyramids to solve real-life problems.

EXPLORATION 1

Using a Net to Construct a Solid

Work with a partner. Copy the net shown below onto grid paper.

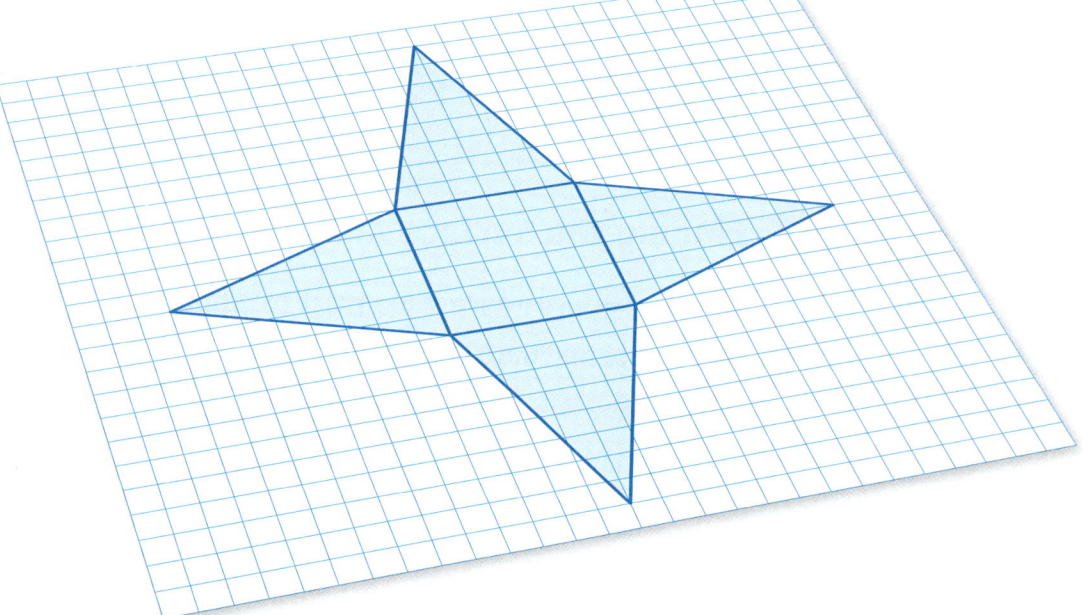

a. Cut out and fold the net to form a solid. What type of solid does the net form?

b. What is the surface area of the solid?

EXPLORATION 2

Finding Surface Areas of Solids

Work with a partner. Find the surface area of each solid. Explain your reasoning.

Math Practice

Analyze Givens
What information can you determine from the diagram? How does this help you find the surface area of the solid?

a.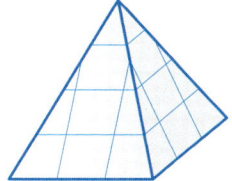

b.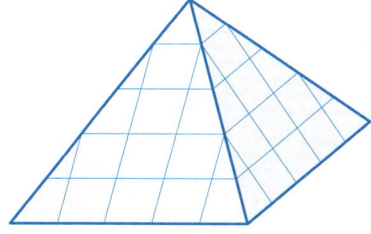

Section 7.6 Surface Areas of Pyramids 319

7.6 Lesson

Key Idea

Net of a Pyramid

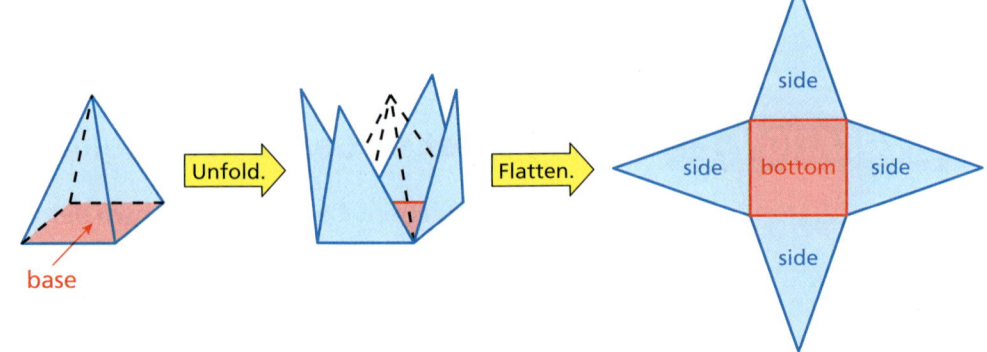

Remember

A *square pyramid* is a pyramid with a square base. A *triangular pyramid* is a pyramid with a triangular base.

In this book, the base of every pyramid is either a square or an equilateral triangle. So, the lateral faces are identical triangles.

EXAMPLE 1   Finding the Surface Area of a Square Pyramid

Find the surface area of the square pyramid.

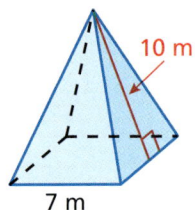

Use a net to find the area of each face.

Bottom: $7 \cdot 7 = 49$

Side: $\frac{1}{2} \cdot 7 \cdot 10 = 35$

Side: $\frac{1}{2} \cdot 7 \cdot 10 = 35$

Side: $\frac{1}{2} \cdot 7 \cdot 10 = 35$

Side: $\frac{1}{2} \cdot 7 \cdot 10 = 35$

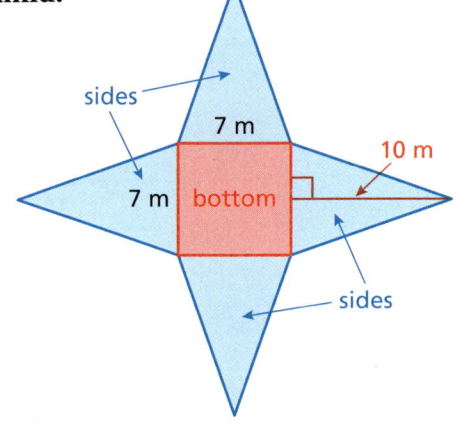

Find the sum of the areas of the faces.

Surface Area = Area of bottom + Area of a side + Area of a side + Area of a side + Area of a side

$S = 49 + 35 + 35 + 35 + 35 = 189$

▸ So, the surface area is 189 square meters.

Try It   Find the surface area of the square pyramid.

1.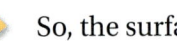

2.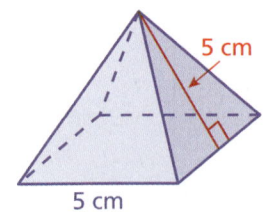

320   Chapter 7   Area, Surface Area, and Volume

EXAMPLE 2 Finding the Surface Area of a Triangular Pyramid

Find the surface area of the triangular pyramid.

Use a net to find the area of each face.

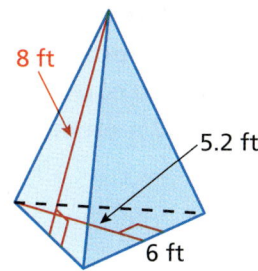
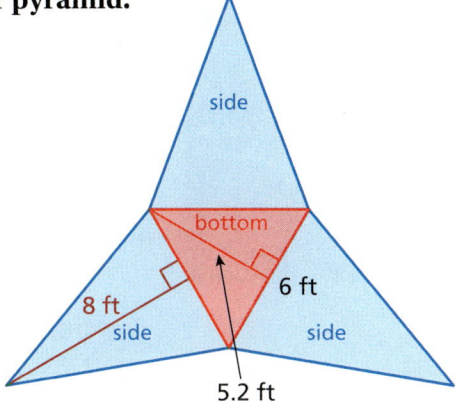

Bottom: $\frac{1}{2} \cdot 6 \cdot 5.2 = 15.6$

Side: $\frac{1}{2} \cdot 6 \cdot 8 = 24$

Side: $\frac{1}{2} \cdot 6 \cdot 8 = 24$

Side: $\frac{1}{2} \cdot 6 \cdot 8 = 24$

Find the sum of the areas of the faces.

$$\text{Surface Area} = \boxed{\text{Area of bottom}} + \boxed{\text{Area of a side}} + \boxed{\text{Area of a side}} + \boxed{\text{Area of a side}}$$

$$S = 15.6 + 24 + 24 + 24$$

$$= 87.6$$

▸ So, the surface area is 87.6 square feet.

Try It Find the surface area of the triangular pyramid.

3.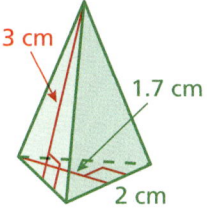

4.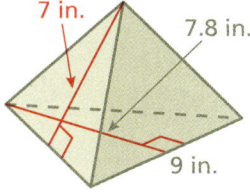

Self-Assessment for Concepts & Skills

Solve each exercise. Then rate your understanding of the success criteria in your journal.

5. **MP PRECISION** Explain how to find the surface area of a pyramid.

FINDING SURFACE AREA Find the surface area of the pyramid.

6.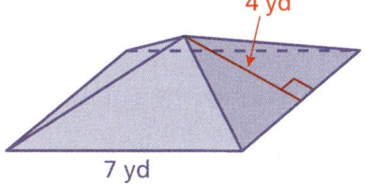

7.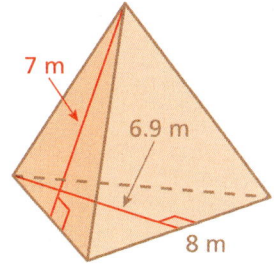

EXAMPLE 3 Modeling Real Life

The uppermost piece of an ancient Egyptian pyramid is called a pyramidion. These square pyramid-shaped pieces were sometimes covered with gold. What is the least amount of gold needed to cover a pyramidion in which each triangular face has a height of 1.2 meters and a base of 1.5 meters?

Draw the pyramid. The least amount of gold needed is represented by the surface area of the pyramid. Use a net to find the surface area.

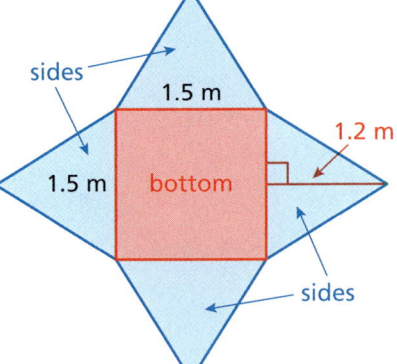

Bottom: $1.5 \cdot 1.5 = 2.25$

Side: $\frac{1}{2} \cdot 1.5 \cdot 1.2 = 0.9$

Side: $\frac{1}{2} \cdot 1.5 \cdot 1.2 = 0.9$

Side: $\frac{1}{2} \cdot 1.5 \cdot 1.2 = 0.9$

Side: $\frac{1}{2} \cdot 1.5 \cdot 1.2 = 0.9$

$S = 2.25 + 0.9 + 0.9 + 0.9 + 0.9 = 5.85$

▸ So, the least amount of gold needed is 5.85 square meters.

Self-Assessment for Problem Solving

Solve each exercise. Then rate your understanding of the success criteria in your journal.

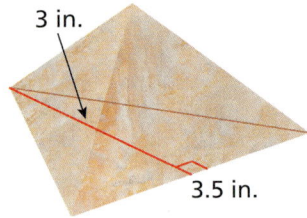

Surface Area = 15.75 in.²

8. A salt lamp is shaped like a triangular pyramid. Find the surface area of each triangular face.

9. **DIG DEEPER!** Originally, each triangular face of the Great Pyramid of Giza had a height of 612 feet and a base of 756 feet. Today, the height of each triangular face of the square pyramid is 592 feet. Find the change in the total surface area of the four triangular faces of the Great Pyramid of Giza.

7.6 Practice

Review & Refresh

Find the surface area of the prism.

1.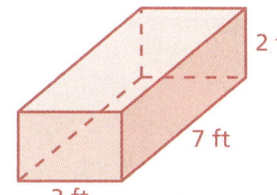
2.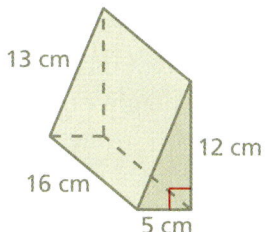
3.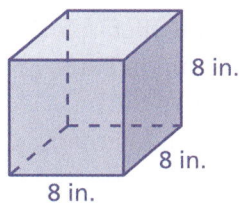

Match the expression with an equivalent expression.

4. $3(4n + 2)$ 5. $6(2n + 3)$ 6. $4(3n + 4)$ 7. $12(n + 1)$

   A. $2(6n + 6)$ **B.** $12n + 18$ **C.** $2(6n + 3)$ **D.** $12n + 16$

Write the fraction or mixed number as a percent.

8. $\frac{17}{25}$ 9. $\frac{19}{20}$ 10. $6\frac{7}{8}$ 11. $\frac{3}{400}$

Concepts, Skills, & Problem Solving

USING TOOLS Use a net to find the surface area of the solid. Explain your reasoning. (See Exploration 2, p. 319.)

12.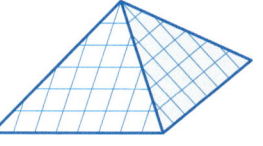
13.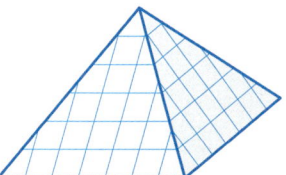
14.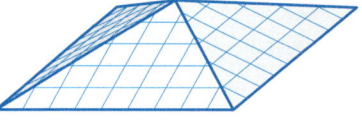

FINDING SURFACE AREA Find the surface area of the pyramid.

15.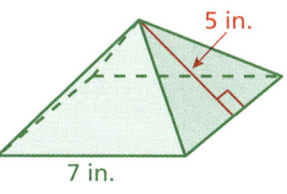
16.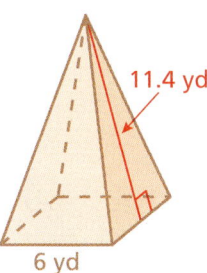
17.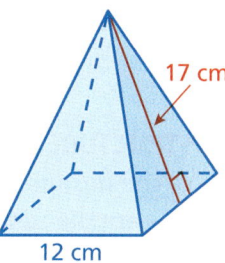

18.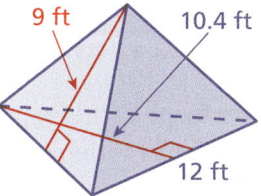
19.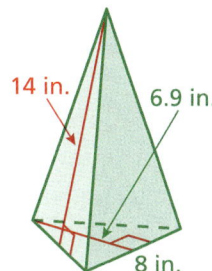
20.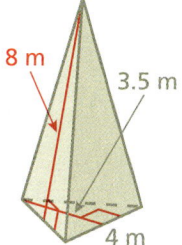

Section 7.6 Surface Areas of Pyramids 323

21. **MODELING REAL LIFE** A paperweight is shaped like a triangular pyramid. Find the surface area of the paperweight.

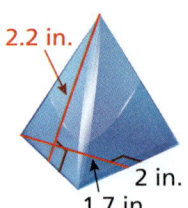

22. **PROBLEM SOLVING** The entrance to the Louvre Museum in Paris, France, is a square pyramid. The side length of the base is 116 feet, and the height of one of the triangular faces is 91.7 feet. Find the surface area of the four triangular faces of the entrance to the Louvre Museum.

23. **MODELING REAL LIFE** A silicon wafer is textured to minimize light reflection. This results in a surface made up of square pyramids. Each triangular face of one of the pyramids has a base of 5 micrometers and a height of 5.6 micrometers. Find the surface area of the pyramid, including the base.

24. **REASONING** A hanging light cover made of glass is shaped like a square pyramid. The cover does not have a bottom. One square foot of the glass weighs 2.45 pounds. The chain can support 35 pounds. Will the chain support the light cover? Explain.

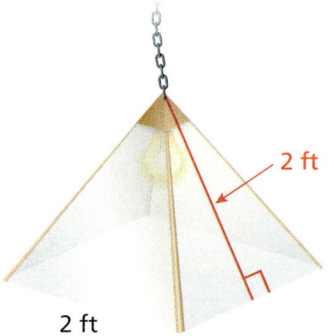

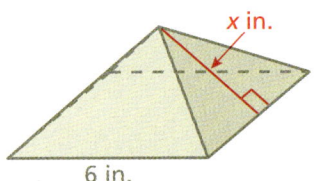

25. **GEOMETRY** The surface area of the square pyramid shown is 84 square inches. What is the value of x?

26. **STRUCTURE** In the diagram of the base of the hexagonal pyramid, all the triangles are the same. Find the surface area of the hexagonal pyramid.

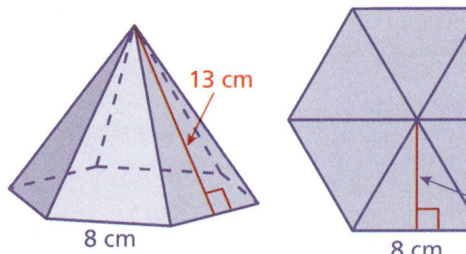

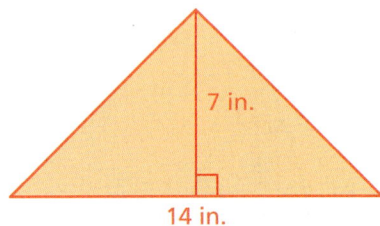

27. **CRITICAL THINKING** Can you form a square pyramid using a square with side lengths of 14 inches and four of the triangles shown? Explain your reasoning.

324   Chapter 7   Area, Surface Area, and Volume

7.7 Volumes of Rectangular Prisms

Learning Target: Find volumes and missing dimensions of rectangular prisms.

Success Criteria:
- I can use a formula to find the volume of a rectangular prism.
- I can use a formula to find the volume of a cube.
- I can use the volume of a rectangular prism and two of its dimensions to find the other dimension.
- I can apply volumes of rectangular prisms to solve real-life problems.

Recall that the **volume** of a three-dimensional figure is a measure of the amount of space that it occupies. Volume is measured in *cubic units*.

EXPLORATION 1

Using a Unit Cube

Work with a partner. A *unit cube* is a cube with an edge length of 1 unit. The parallel edges of the unit cube have been divided into 2, 3, and 4 equal parts to create smaller rectangular prisms that are identical.

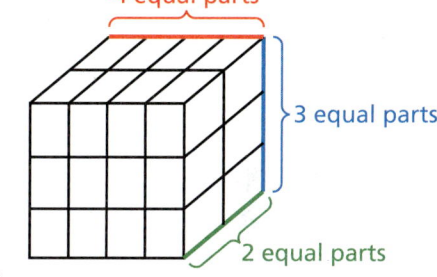

a. The volumes of the identical prisms are equal. What else can you determine about the volumes of the prisms? Explain.

b. Use the identical prisms in part (a) to find the volume of the prism below. Explain your reasoning.

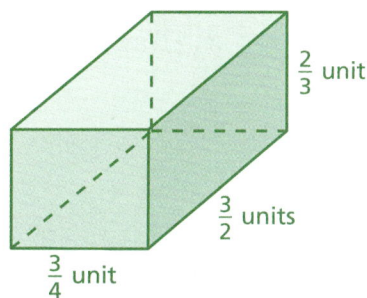

c. How can you use a unit cube to find the volume of the prism below? Explain.

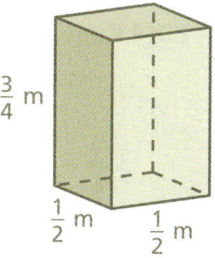

Math Practice

Communicate Precisely

In part (c), explain why you decided to divide the unit cube in the way you did.

d. Do the formulas $V = Bh$ and $V = \ell wh$ work for rectangular prisms with fractional edge lengths? Give examples to support your answer.

7.7 Lesson

Key Vocabulary
volume, p. 325

Key Idea

Volume of a Rectangular Prism

Words The volume V of a rectangular prism is the product of the area of the base and the height of the prism.

Algebra $V = Bh$ or $V = \ell wh$

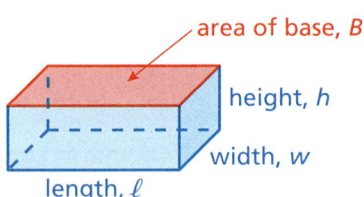

When a rectangular prism is a cube with an edge length of s, you can also use the formula $V = s^3$ to find the volume V of the cube.

EXAMPLE 1 Finding Volumes of Rectangular Prisms

Find the volume of each prism.

a.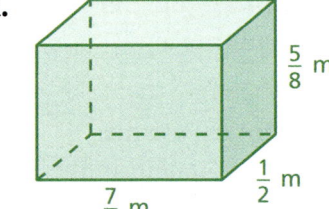

b.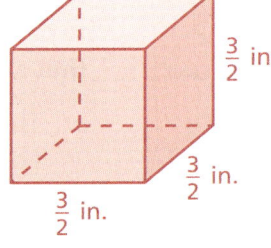

> In Example 1(b), the rectangular prism is a cube. Notice that you obtain the same answer when you use the formula $V = \ell wh$ to find the volume.

a.
$V = \ell wh$

$= \dfrac{7}{8}\left(\dfrac{1}{2}\right)\left(\dfrac{5}{8}\right)$

$= \dfrac{35}{128}$

▸ So, the volume is $\dfrac{35}{128}$ cubic meter.

b.
$V = s^3$

$= \left(\dfrac{3}{2}\right)^3$

$= \dfrac{3}{2}\left(\dfrac{3}{2}\right)\left(\dfrac{3}{2}\right)$

$= \dfrac{27}{8}$, or $3\dfrac{3}{8}$

▸ So, the volume is $3\dfrac{3}{8}$ cubic inches.

Try It Find the volume of the prism.

1.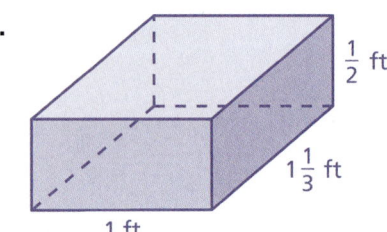

2.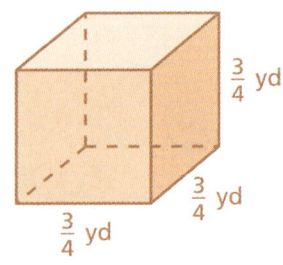

EXAMPLE 2 **Finding a Missing Dimension of a Rectangular Prism**

16 in.   7 in.
Volume = 1792 in.³

Find the height of the computer tower.

$V = \ell wh$		Write formula for volume.
$1792 = 16(7)h$		Substitute values.
$1792 = 112h$		Simplify.
$\dfrac{1792}{112} = \dfrac{112h}{112}$		Division Property of Equality
$16 = h$		Simplify.

▸ So, the height of the computer tower is 16 inches.

Try It Find the missing dimension of the prism.

3. Volume = 72 in.³

2 in.   6 in.
ℓ

4. Volume = 1375 cm³

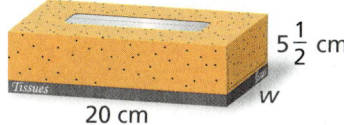

$5\frac{1}{2}$ cm
20 cm w

Self-Assessment for Concepts & Skills

Solve each exercise. Then rate your understanding of the success criteria in your journal.

5. **CRITICAL THINKING** Explain how volume and surface area are different.

6. **FINDING A MISSING DIMENSION** The base of a rectangular prism has an area of 24 square millimeters. The volume of the prism is 144 cubic millimeters. Make a sketch of the prism. Then find the height of the prism.

FINDING VOLUME Find the volume of the prism.

7.
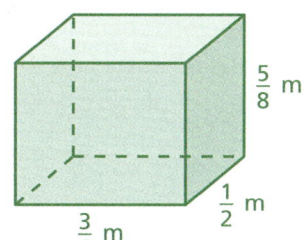
$\frac{5}{8}$ m
$\frac{1}{2}$ m
$\frac{3}{4}$ m

8.
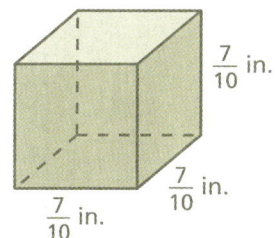
$\frac{7}{10}$ in.
$\frac{7}{10}$ in.
$\frac{7}{10}$ in.

Section 7.7 Volumes of Rectangular Prisms

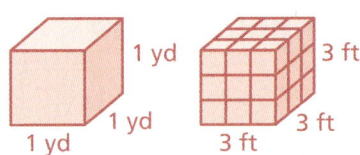

When finding volumes, you may need to convert cubic units. The diagrams at the left show that there are 27 cubic feet per cubic yard.

$$1 \text{ yd}^3 = (1 \text{ yd})(1 \text{ yd})(1 \text{ yd}) = (3 \text{ ft})(3 \text{ ft})(3 \text{ ft}) = 27 \text{ ft}^3$$

You can use a similar procedure to convert other cubic units.

EXAMPLE 3 Modeling Real Life

The dump truck shown delivers dirt for $18 per cubic yard. About how much does a full load of dirt cost?

Find the volume of a full load of dirt in cubic feet.

$V = \ell w h$ Write formula for volume.

$= 17(8)\left(4\dfrac{3}{4}\right)$ Substitute values.

$= 646$ Multiply.

A full load of dirt is 646 cubic feet. Because $1 \text{ yd}^3 = 27 \text{ ft}^3$, convert the volume to cubic yards using a conversion factor.

$$646 \text{ ft}^3 = 646 \text{ ft}^3 \times \dfrac{1 \text{ yd}^3}{27 \text{ ft}^3} \approx 24 \text{ yd}^3$$

 So, a full load of dirt costs about $24 \text{ yd}^3 \times \dfrac{\$18}{1 \text{ yd}^3} = \$432$.

 Self-Assessment for Problem Solving

Solve each exercise. Then rate your understanding of the success criteria in your journal.

9. **DIG DEEPER!** The shark cage is in the shape of a rectangular prism and has a volume of 315 cubic feet. Find a set of reasonable dimensions for the base of the cage. Justify your answer.

10. The hot tub is in the shape of a rectangular prism. How many pounds of water can the hot tub hold? One cubic foot of water weighs about 62.4 pounds.

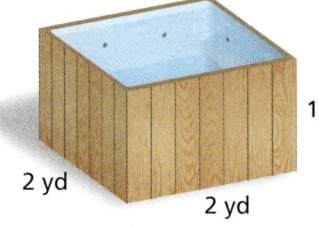

7.7 Practice

Go to BigIdeasMath.com to get HELP with solving the exercises.

▶ Review & Refresh

Find the surface area of the pyramid.

1.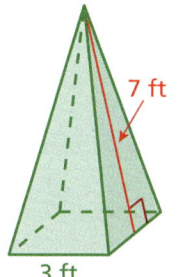

2.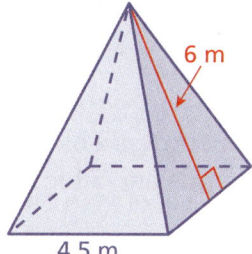

3.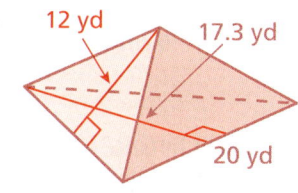

Write the phrase as an expression. Then evaluate the expression when $x = 2$ and $y = 12$.

4. 8 more than a number x

5. the difference of a number y and 9

▶ Concepts, Skills, & Problem Solving

MP STRUCTURE The unit cube is divided into identical rectangular prisms. What is the volume of one of the identical prisms? *(See Exploration 1, p. 325.)*

6.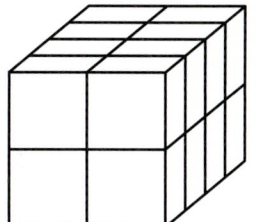

7.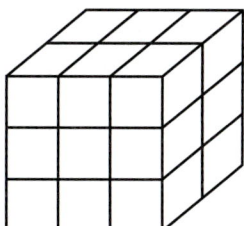

8.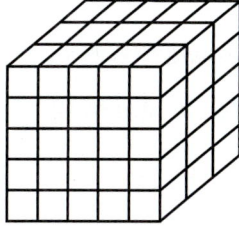

FINDING VOLUME Find the volume of the prism.

9.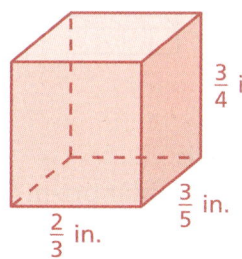

10.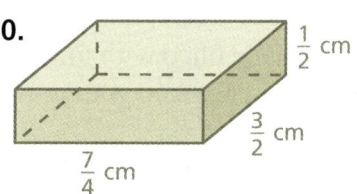

11.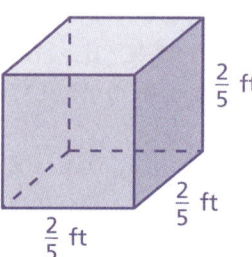

12.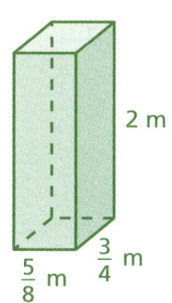

13.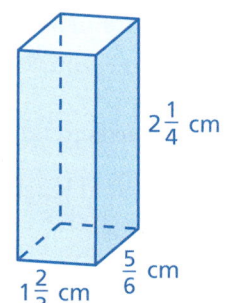

14.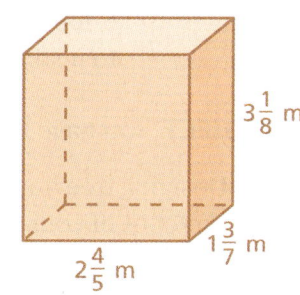

Section 7.7 Volumes of Rectangular Prisms 329

FINDING A MISSING DIMENSION Find the missing dimension of the prism.

15. Volume = 1620 cm³   **16.** Volume = 220.5 cm³   **17.** Volume = 532 in.³

 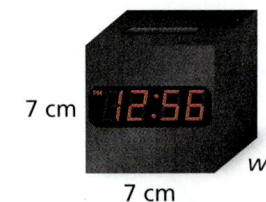 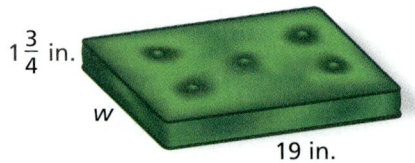

18. MODELING REAL LIFE An FBI agent orders a block of ballistics gel. The gel weighs 54 pounds per cubic foot. What is the weight of the block of gel?

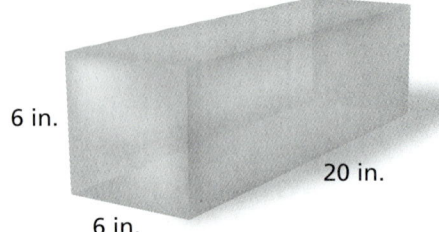

19. MODELING REAL LIFE

a. Estimate the amount of casserole left in the dish.

b. Will the casserole fit in the storage container? Explain your reasoning.

20. GEOMETRY How many $\frac{3}{4}$-centimeter cubes do you need to create a cube with an edge length of 12 centimeters?

21. MP REASONING How many one-millimeter cubes do you need to fill a cube that has an edge length of 1 centimeter? How can this result help you convert a volume from cubic millimeters to cubic centimeters? from cubic centimeters to cubic millimeters?

22. MP LOGIC The container is partially filled with unit cubes. How many unit cubes fit in the container? Explain your reasoning.

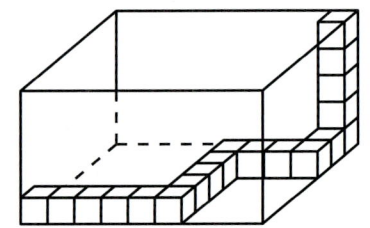

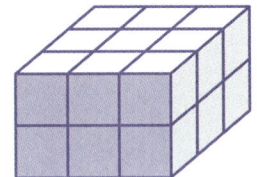

23. MP PROBLEM SOLVING The area of the shaded face is 96 square centimeters. What is the volume of the rectangular prism?

24. DIG DEEPER! Is the combined volume of a 4-foot cube and a 6-foot cube equal to the volume of a 10-foot cube? Use a diagram to justify your answer.

25. PROJECT You have 1400 square feet of boards to use for a new tree house.

a. Design a tree house that has a volume of at least 250 cubic feet. Include sketches of your tree house.

b. Are your dimensions reasonable? Explain your reasoning.

7 Connecting Concepts

Using the Problem-Solving Plan

1. A sports complex has two swimming pools that are shaped like rectangular prisms. The amount of water in the smaller pool is what percent of the amount of water in the larger pool?

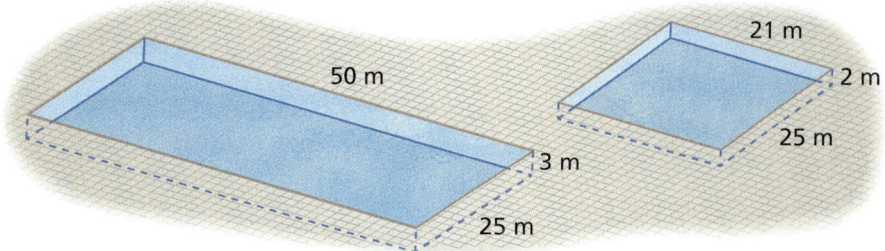

Understand the problem. You know the shape and the dimensions of the two swimming pools. You are asked to find the amount of water in the smaller pool as a percent of the amount of water in the larger pool.

Make a plan. First, find the volume of each pool. Then represent the amount of water in the smaller pool as a fraction of the amount of water in the larger pool. Find an equivalent fraction whose denominator is 100 to find the percent.

Solve and check. Use the plan to solve the problem. Then check your solution.

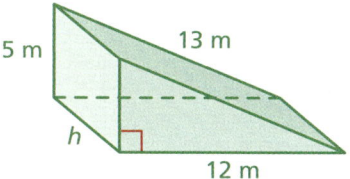

2. Use a graph to represent the relationship between the surface area S (in square meters) and the height h (in meters) of the triangular prism. Then find the height when the surface area is 260 square meters.

3. A toy company sells two different toy chests. The toy chests have different dimensions, but the same volume. What is the width w of Toy Chest 2?

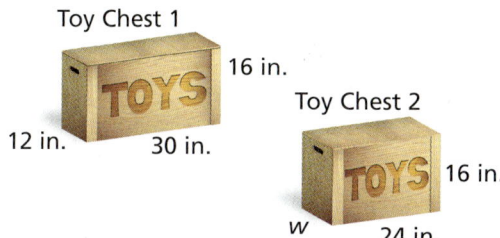

Performance Task

Maximizing the Volumes of Boxes

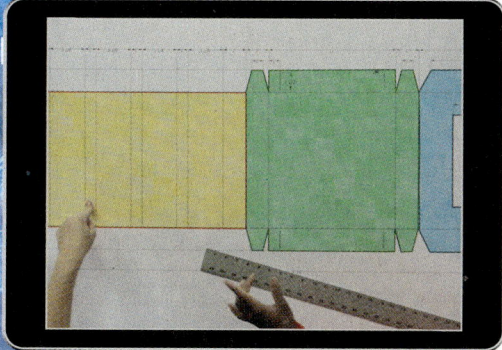

At the beginning of this chapter, you watched a STEAM Video called "Packaging Design." You are now ready to complete the performance task related to this video, available at *BigIdeasMath.com*. Be sure to use the problem-solving plan as you work through the performance task.

7 Chapter Review

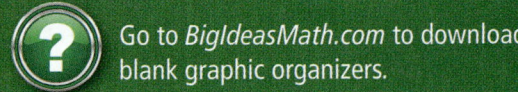

Go to *BigIdeasMath.com* to download blank graphic organizers.

▶ Review Vocabulary

Write the definition and give an example of each vocabulary term.

polygon, p. 285
composite figure, p. 294
kite, p. 298
solid, p. 306
polyhedron, p. 306
face, p. 306
edge, p. 306
vertex, p. 306
prism, p. 306
pyramid, p. 306
surface area, p. 312
net, p. 312
volume, p. 325

▶ Graphic Organizers

You can use a **Four Square** to organize information about a concept. Each of the four squares can be a category, such as definition, vocabulary, example, non-example, words, algebra, table, numbers, visual, graph, or equation. Here is an example of a Four Square for the *area of a parallelogram*.

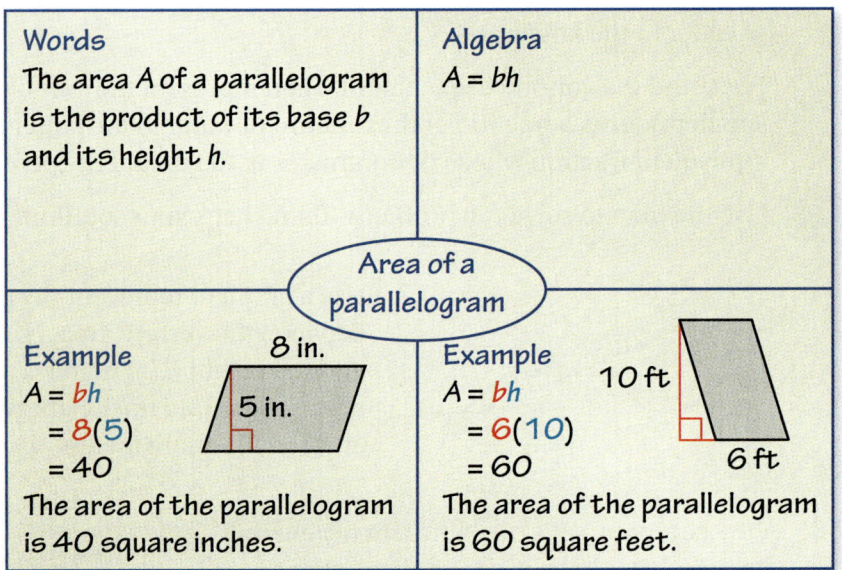

Choose and complete a graphic organizer to help you study the concept.

1. area of a triangle

2. area of a trapezoid

3. area of a composite figure

4. polyhedron

5. surface area of a prism

6. surface area of a pyramid

7. volume of a rectangular prism

"Here is my Four Square to organize information about rattlesnakes. How do you like it?"

332   Chapter 7   Area, Surface Area, and Volume

Chapter Self-Assessment

As you complete the exercises, use the scale below to rate your understanding of the success criteria in your journal.

1	2	3	4
I do not understand.	I can do it with help.	I can do it on my own.	I can teach someone else.

7.1 Areas of Parallelograms (pp. 285–290)

Learning Target: Find areas and missing dimensions of parallelograms.

Find the area of the parallelogram.

1. (20 yd height, 25 yd base)
2. (22 mm height, 11 mm base)
3. (9 cm height, 5 cm base)

4. Find the area (in square inches) of the parallelogram. (2 ft, 3 ft)

5. The billboard shown is in the shape of a parallelogram with a base of 48 feet. What is the height of the billboard?

   ADVERTISE HERE
   672 square feet of space available
   Call: 1-800-555-0100

6. The freeway noise barrier shown is made of identical parallelogram-shaped sections. The area of each section is 7.5 square meters, and the height of the barrier is 5 meters. How many meters wide is each section of the noise barrier?

7. Draw a parallelogram that has an area between 58 and 60 square centimeters.

7.2 Areas of Triangles (pp. 291–296)

Learning Target: Find areas and missing dimensions of triangles, and find areas of composite figures.

Find the area of the triangle.

8.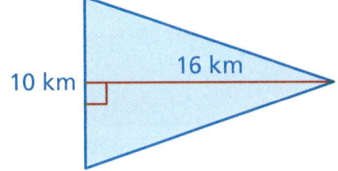

9.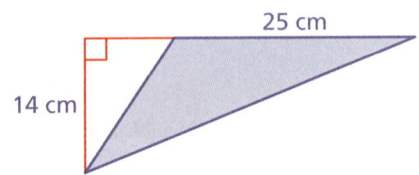

Find the missing dimension of the triangle.

10. Area = 35 mi²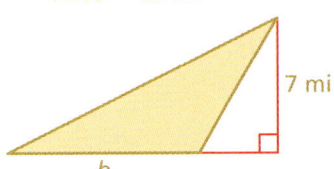

11. Area = 5 cm²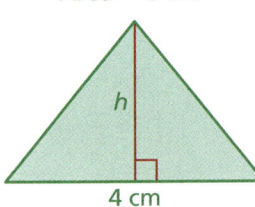

Find the area of the figure.

12.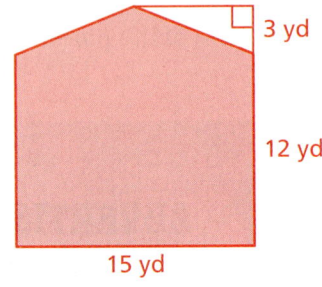

13.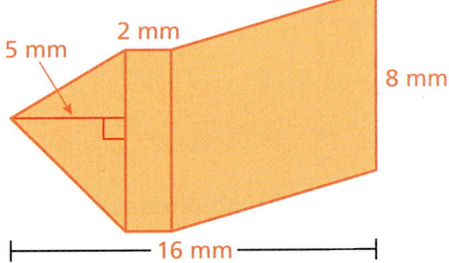

14. Draw a composite figure that has an area less than 35 square inches.

15. The triangle-shaped entrance to the cavern is $2\frac{1}{2}$ feet tall and 4 feet wide. What is the area of the entrance?

7.3 Areas of Trapezoids and Kites (pp. 297–304)

Learning Target: Find areas of trapezoids, kites, and composite figures.

16. Use decomposition to find the area of the kite.

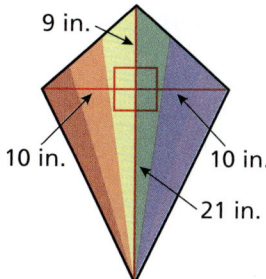

Find the area of the trapezoid.

17.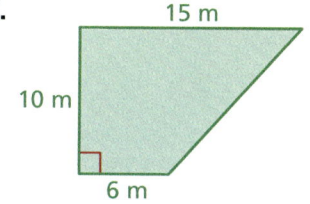

18.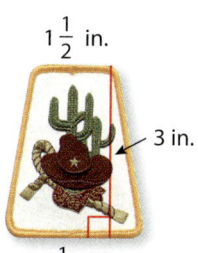

19.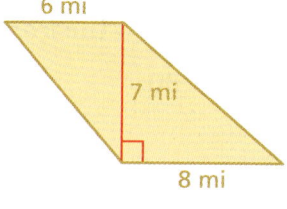

Find the area of the figure.

20.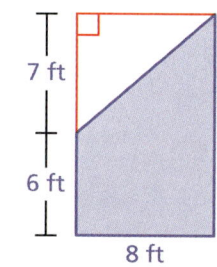

21.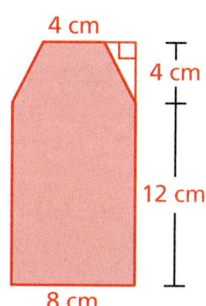

22.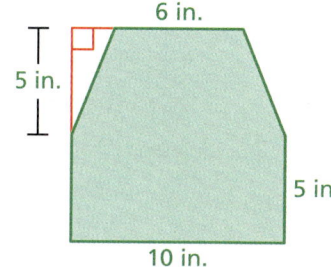

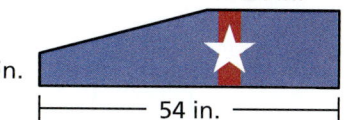

23. You are creating a design for the side of the soapbox car. How much area do you have for the design?

24. Find the area (in square centimeters) of a trapezoid with a height of 2 meters and base lengths of 3 meters and 5 meters.

Chapter Review   **335**

7.4 Three-Dimensional Figures (pp. 305–310)

Learning Target: Describe and draw three-dimensional figures.

Find the numbers of faces, edges, and vertices of the solid.

25.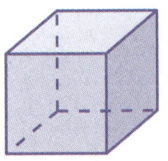
26.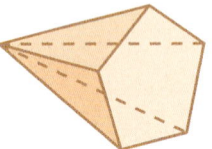
27.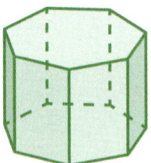

Draw the solid.

28. square pyramid
29. hexagonal prism

Draw the front, side, and top views of the solid.

30.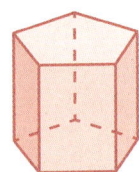
31.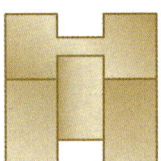
32.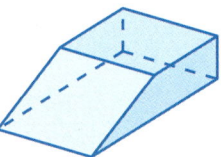

7.5 Surface Areas of Prisms (pp. 311–318)

Learning Target: Represent prisms using nets and use nets to find the surface areas of prisms.

Find the surface area of the prism.

33.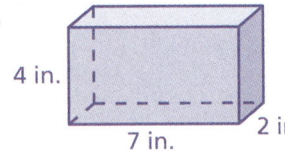
4 in., 7 in., 2 in.

34.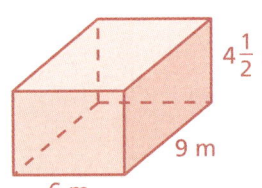
$4\frac{1}{2}$ m, 9 m, 6 m

35.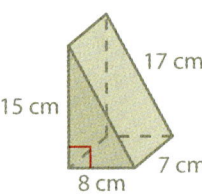
17 cm, 15 cm, 7 cm, 8 cm

36.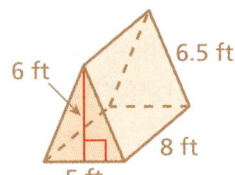
6 ft, 6.5 ft, 8 ft, 5 ft

37.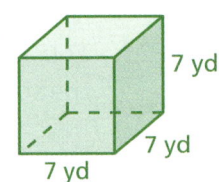
7 yd, 7 yd, 7 yd

38.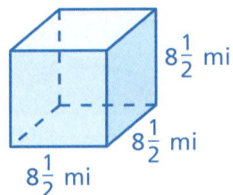
$8\frac{1}{2}$ mi, $8\frac{1}{2}$ mi, $8\frac{1}{2}$ mi

39. One quart of water-resistant paint covers 75 square feet. A swimming pool is in the shape of a rectangular prism with a length of 20 feet, a width of 10 feet, and a height of 5 feet. How many quarts should you buy to paint the swimming pool with two coats of paint?

7.6 Surface Areas of Pyramids (pp. 319–324)

Learning Target: Represent pyramids using nets and use nets to find the surface areas of pyramids.

Find the surface area of the pyramid.

40.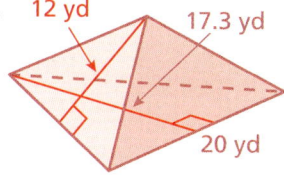

41.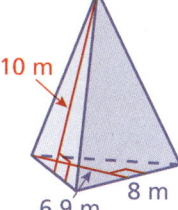

42.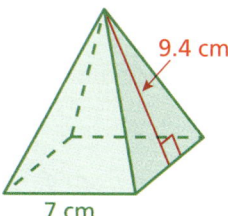

43. You make a square pyramid for a school project. Find the surface area of the pyramid.

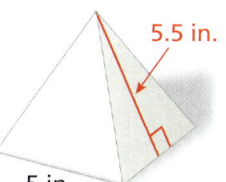

7.7 Volumes of Rectangular Prisms (pp. 325–330)

Learning Target: Find volumes and missing dimensions of rectangular prisms.

Find the volume of the prism.

44.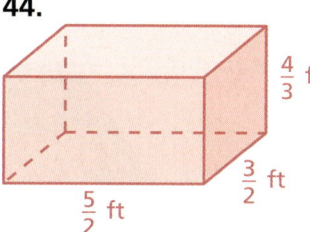

45.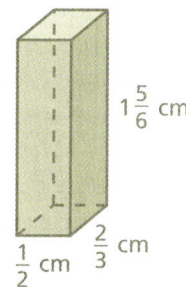

46.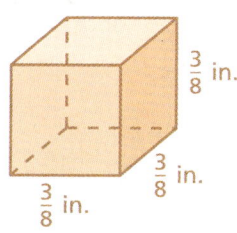

47. The prism has a volume of 150 cubic feet. Find the length of the prism.

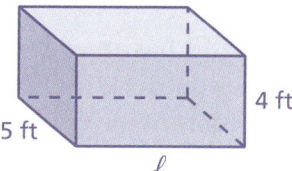

48. How many cubic inches of tissues can the box hold?

49. Draw a rectangular prism that has a volume less than 1 cubic inch.

Chapter Review 337

7 Practice Test

Find the area of the figure.

1.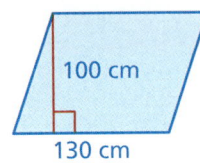
2.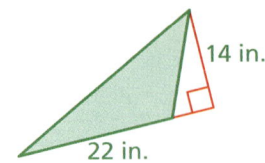
3.

Find the surface area of the solid.

4.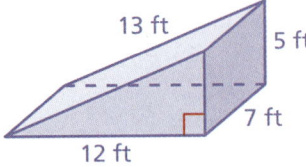
5.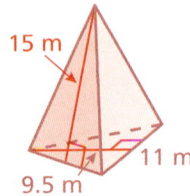

Find the volume of the prism.

6.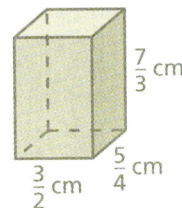
7.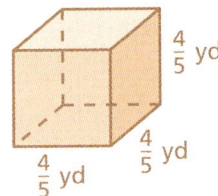

8. Draw an octagonal prism.

9. Find the numbers of faces, edges, and vertices of the solid.

10. The area of a parallelogram is 156 square meters. What is the height of the parallelogram when the base is 13 meters?

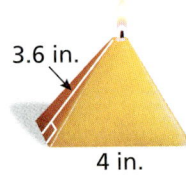

11. A candle is shaped like a square pyramid. Find the surface area of the candle.

12. You are wrapping the boxed DVD collection as a present. What is the least amount of wrapping paper needed to wrap the box?

13. A cube has an edge length of 4 inches. You double the edge lengths. How many times greater is the volume of the new cube?

14. The Pentagon in Arlington, Virginia, is the headquarters of the U.S. Department of Defense. The building's center contains a pentagon-shaped courtyard with an area of about 5 acres. Find the land areas (in square feet) of the courtyard and the building.

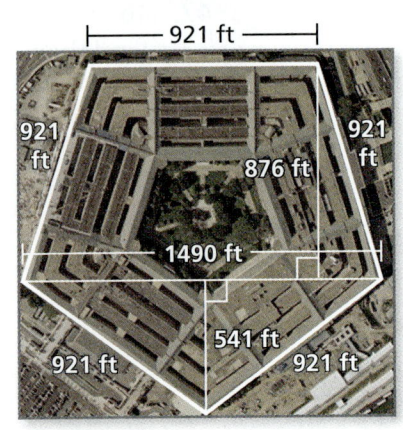

7 Cumulative Practice

Test-Taking Strategy: Answer Easy Questions First

1. A cruise ship is carrying a total of 4971 people. Each lifeboat can hold a maximum of 150 people. What is the minimum number of lifeboats needed to evacuate everyone on the cruise ship?

   A. 33 lifeboats
   B. 34 lifeboats
   C. 54 lifeboats
   D. 332 lifeboats

2. Which number is equivalent to the expression?

   $$3 \cdot 4^2 + 6 \div 2$$

   F. 27
   G. 33
   H. 51
   I. 75

3. What is the volume of the package?

   A. 240 in.^{3}
   B. 376 in.^{3}
   C. 480 in.^{3}
   D. 960 in.^{3}

4. A housing community started with 60 homes. In each of the following years, 8 more homes were built. Let y represent the number of years that have passed since the first year, and let n represent the number of homes. Which equation describes the relationship between n and y?

   F. $n = 8y + 60$
   G. $n = 68y$
   H. $n = 60y + 8$
   I. $n = 60 + 8 + y$

5. What is the value of *m* that makes the equation true?

$$4m = 6$$

6. What is the surface area of the square pyramid?

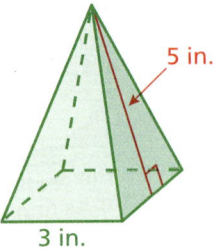

- **A.** 30 in.²
- **B.** 31.5 in.²
- **C.** 39 in.²
- **D.** 69 in.²

7. A wooden box has a length of 12 inches, a width of 6 inches, and a height of 8 inches.

Part A Draw and label a rectangular prism with the dimensions of the wooden box.

Part B What is the surface area, in square inches, of the wooden box? Show your work.

Part C You have a two-fluid ounce sample of wood stain that covers 900 square inches. Is this enough to give the entire box two coats of stain? Show your work and explain your reasoning.

8. On Saturday, you earned $35 mowing lawns. This was *x* dollars more than you earned on Thursday. Which expression represents the amount, in dollars, you earned mowing lawns on Thursday?

- **F.** 35*x*
- **G.** *x* + 35
- **H.** *x* − 35
- **I.** 35 − *x*

9. What is the area, in square yards, of the triangle?

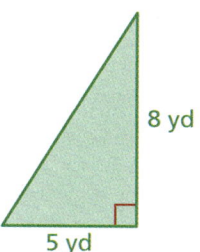

10. Which expression is equivalent to $\frac{12}{35}$?

   A. $\frac{5}{6} \div \frac{2}{7}$

   B. $\frac{2}{7} \div \frac{6}{5}$

   C. $\frac{2}{7} \div \frac{5}{6}$

   D. $\frac{5}{6} \div \frac{7}{2}$

11. The description below represents the area of which polygon?

   "one-half the product of its height and the sum of its bases"

   F. rectangle
   G. parallelogram
   H. trapezoid
   I. triangle

12. What is the missing quantity in the double number line?

   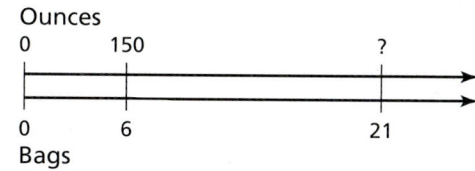

   A. 25 ounces
   B. 165 ounces
   C. 525 ounces
   D. 600 ounces

Cumulative Practice 341

8 Integers, Number Lines, and the Coordinate Plane

- **8.1** Integers
- **8.2** Comparing and Ordering Integers
- **8.3** Rational Numbers
- **8.4** Absolute Value
- **8.5** The Coordinate Plane
- **8.6** Polygons in the Coordinate Plane
- **8.7** Writing and Graphing Inequalities
- **8.8** Solving Inequalities

Chapter Learning Target:
Understand integers.

Chapter Success Criteria:
- ■ I can write integers to represent quantities.
- ■ I can describe quantities.
- ■ I can order and compare quantities.
- ■ I can apply integers to model real-life problems.

STEAM Video: "Designing a CubeSat"

STEAM Video

Designing a CubeSat

A *CubeSat* is a type of miniature satellite that is used for space research. Each CubeSat has the dimensions shown and a mass of no more than 1.33 kilograms.

Watch the STEAM Video "Designing a CubeSat." Then answer the following questions.

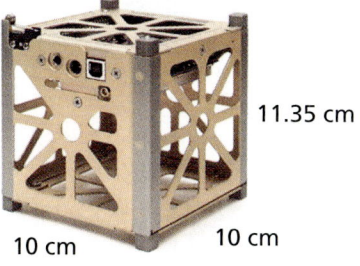

1. For what fields of study do you think CubeSats can be used?

2. Tony says g-forces are a measure of how heavy you feel. The table shows the g-forces on a CubeSat at three points in time. Why can g-forces be as high as 6 during a rocket launch and as low as 0 in space?

Time	Before launch	During launch	After entering space
G-Force	1	6	0

3. What would happen to a CubeSat that cannot withstand a g-force of 6? a g-force of 0?

Performance Task

Launching a CubeSat

After completing this chapter, you will be able to use the concepts you learned to answer the questions in the *STEAM Video Performance Task*. You will be given information about three different types of Cubesats that you can purchase.

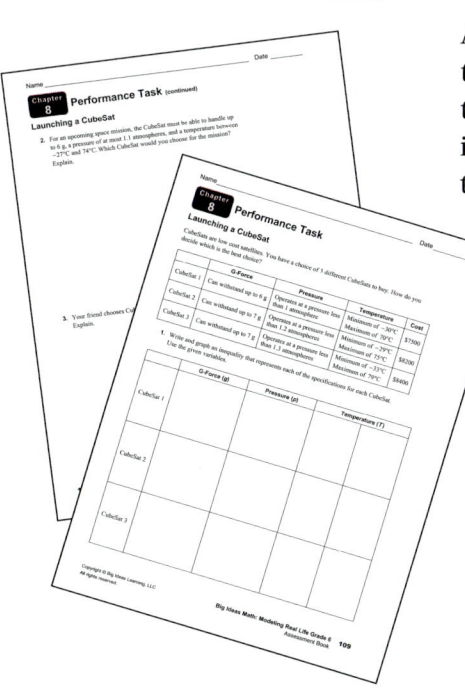

CubeSat 1: $7500

CubeSat 2: $8200

CubeSat 3: $8400

You will determine which of the three CubeSats is the best option for a mission. Why might g-force, pressure, and temperature be important considerations for making your decision?

Getting Ready for Chapter 8

Chapter Exploration

1. Work with a partner. Plot and connect the points to make a picture.

1(6, 9)	**2**(4, 11)	**3**(2, 12)	**4**(0, 11)	**5**(−2, 9)
6(−6, 2)	**7**(−9, 1)	**8**(−11, −3)	**9**(−7, 0)	**10**(−5, −1)
11(−5, −5)	**12**(−4, −8)	**13**(−6, −10)	**14**(−3, −9)	**15**(−3, −10)
16(−4, −11)	**17**(−4, −12)	**18**(−3, −11)	**19**(−2, −12)	**20**(−2, −11)
21(−1, −12)	**22**(−1, −11)	**23**(−2, −10)	**24**(−2, −9)	**25**(1, −9)
26(2, −8)	**27**(2, −10)	**28**(1, −11)	**29**(1, −12)	**30**(2, −11)
31(3, −12)	**32**(3, −11)	**33**(4, −12)	**34**(4, −11)	**35**(3, −10)
36(3, −8)	**37**(4, −6)	**38**(6, 0)	**39**(9, −3)	**40**(9, −1)
41(8, 1)	**42**(5, 3)	**43**(3, 6)	**44**(3, 7)	**45**(4, 8)

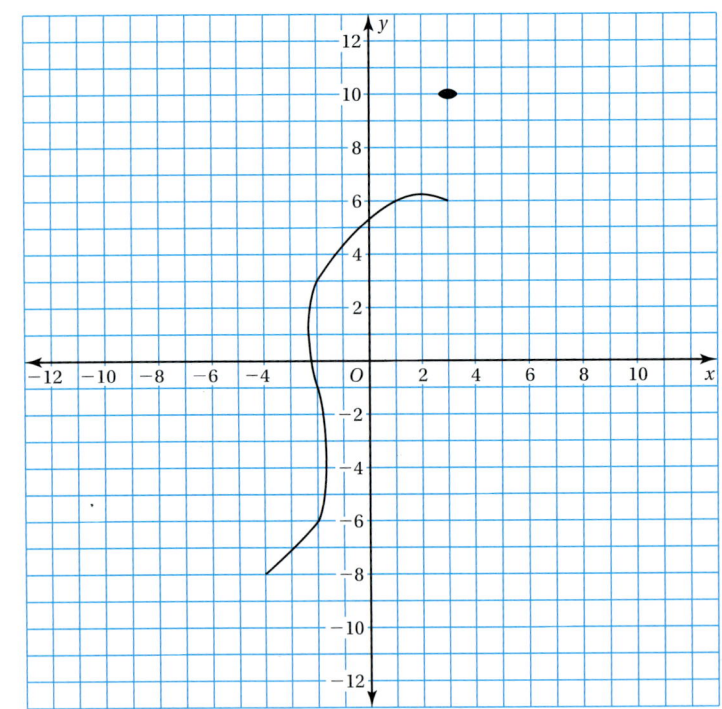

2. Create your own "dot-to-dot" picture. Use at least 20 points.

Vocabulary

The following vocabulary terms are defined in this chapter. Think about what each term might mean and record your thoughts.

negative numbers   opposites   inequality   quadrants

8.1 Integers

Learning Target: Understand the concept of negative numbers and that they are used along with positive numbers to describe quantities.

Success Criteria:
- I can write integers to represent quantities in real life.
- I can graph integers on a number line.
- I can find the opposite of an integer.
- I can apply integers to model real-life problems.

EXPLORATION 1

Reading and Describing Temperatures

Work with a partner. The thermometers show the temperatures in four cities.

    Honolulu, Hawaii   *Anchorage, Alaska*

    Death Valley, California   *Seattle, Washington*

a. Match each temperature with its most appropriate location.

i. 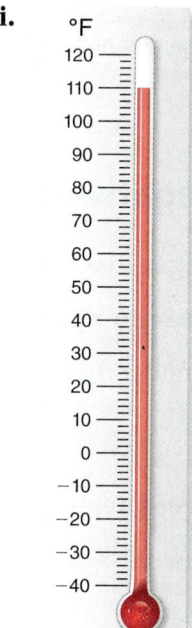 ii. 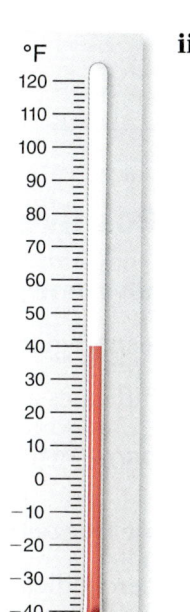 iii. 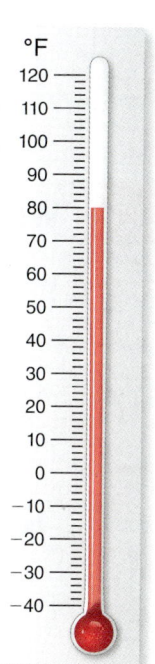 iv.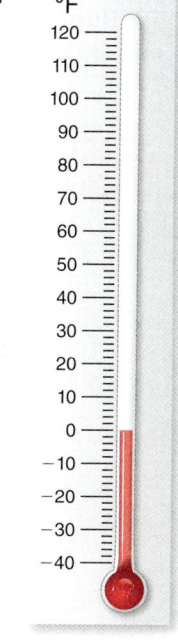

b. What do all of the temperatures have in common?

c. What does it mean for a temperature to be *below* zero? Provide an example. Can you think of any other situations in which numbers may be less than zero?

d. The thermometers show temperatures on a vertical number line. How else can you represent numbers less than zero? Provide an example.

Math Practice

Maintain Oversight

How does this exploration help you represent numbers less than 0?

Section 8.1   Integers   **345**

8.1 Lesson

Key Vocabulary
positive numbers, p. 346
negative numbers, p. 346
opposites, p. 346
integers, p. 346

Positive numbers are greater than 0. They can be written with or without a positive sign (+).

+1   5   +20   10,000

Negative numbers are less than 0. They are written with a negative sign (−).

−1   −5   −20   −10,000

Two numbers that are the same distance from 0 on a number line, but on opposite sides of 0, are called **opposites**. The opposite of 0 is 0.

The Meaning of a Word
Opposite

When you sit across from your friend at the lunch table, you sit **opposite** your friend.

Key Idea

Integers

Words **Integers** are the set of whole numbers and their opposites.

Graph
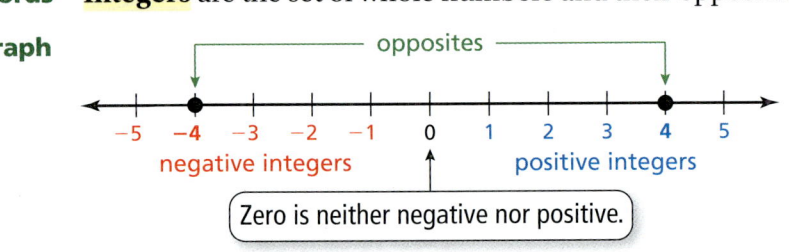

Zero is neither negative nor positive.

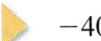 **EXAMPLE 1**   **Writing Positive and Negative Integers**

Write a positive or negative integer that represents each situation.

a. **A contestant gains 250 points on a game show.**

   Gains indicates a number greater than 0. So, use a positive integer.

   ▸ +250, or 250

b. **Gasoline freezes at 40 degrees below zero.**

   Below zero indicates a number less than 0. So, use a negative integer.

   ▸ −40

Try It   Write a positive or negative integer that represents the situation.

1. A hiker climbs 900 feet up a mountain.
2. You have a debt of $24.
3. A student loses 5 points for not showing work on a quiz.
4. A savings account earns $10.

EXAMPLE 2 — Graphing Integers

Graph each integer and its opposite.

a. 3

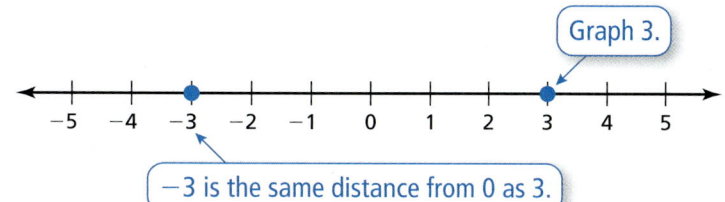

b. −2

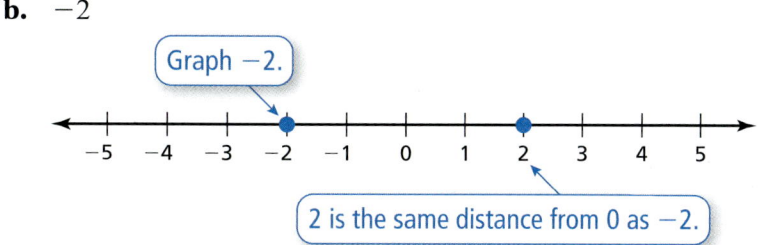

Reading

You can think of the negative sign (−) as referring to the opposite of a number. In Example 2(b), you can read −2 as "the opposite of 2."

Try It Graph the integer and its opposite.

5. 6
6. −4
7. −12
8. 1

Self-Assessment for Concepts & Skills

Solve each exercise. Then rate your understanding of the success criteria in your journal.

WRITING INTEGERS Write a positive or negative integer that represents the situation.

9. A baseball is thrown at a speed of 78 miles per hour.

10. A submarine is 3750 feet below sea level.

GRAPHING INTEGERS Graph the integer and its opposite.

11. 8
12. −7
13. 11

14. **VOCABULARY** Which of the following numbers are integers?

$$8, -4.1, -9, \frac{1}{6}, 1.75, 22$$

15. **VOCABULARY** List three words or phrases used in real life that indicate negative integers.

16. **WRITING** Describe the opposite of a positive integer, the opposite of a negative integer, and the opposite of zero.

EXAMPLE 3 Modeling Real Life

You deliver flowers to an office building. You enter at ground level and go down 2 floors to make the first delivery. Then you go up 7 floors to make the second delivery.

a. Write an integer that represents each position.

Position	Integer
You enter at ground level.	0
You go down 2 floors.	−2
You go up 7 floors.	+7

b. Write an integer that represents how you return to ground level.

Use a number line to model your movement, as shown.

The second delivery is on the fifth floor. You must go down 5 floors to return to ground level.

 The integer representing "down 5 floors" is −5.

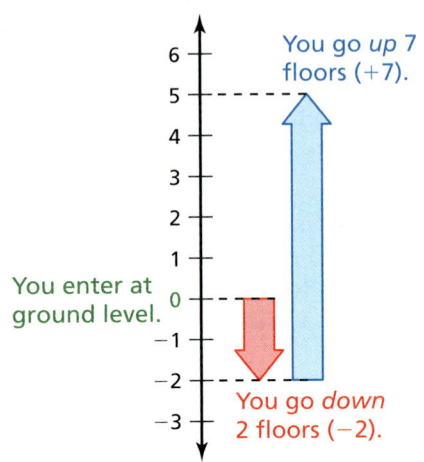

Self-Assessment for Problem Solving

Solve each exercise. Then rate your understanding of the success criteria in your journal.

17. The world record for scuba diving is 332 meters below sea level. Write an integer that represents a new world record. Explain.

18. The indoor and outdoor temperatures are shown. The freezing point of water is 32°F. Write integers that represent how each temperature must change to reach the freezing point of water. Explain.

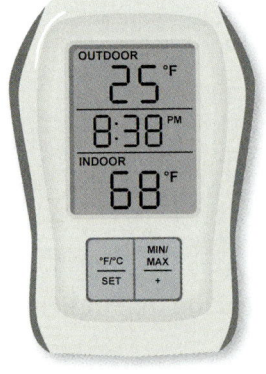

19. An *ion* is an atom that has a positive or negative electric charge. When an ion has more protons than electrons, it has a positive charge. When an ion has fewer protons than electrons, it has a negative charge. Explain what it means for an atom to have an electric charge of zero.

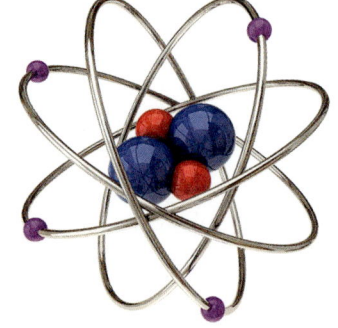

8.1 Practice

Go to **BigIdeasMath.com** to get HELP with solving the exercises.

► Review & Refresh

Find the volume of the prism.

1. A prism with dimensions $\frac{4}{5}$ mm, $\frac{1}{2}$ mm, and $\frac{3}{8}$ mm.

2. 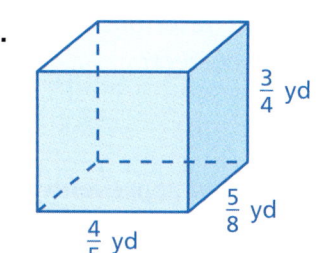 A prism with dimensions $\frac{4}{5}$ yd, $\frac{5}{8}$ yd, and $\frac{3}{4}$ yd.

3. 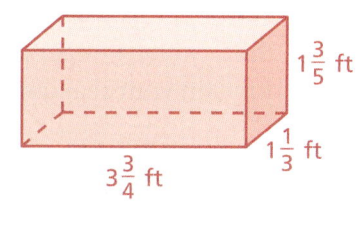 A prism with dimensions $3\frac{3}{4}$ ft, $1\frac{1}{3}$ ft, and $1\frac{3}{5}$ ft.

Factor the expression using the GCF.

4. $4m + 32$
5. $18z - 22$
6. $38x + 80$
7. $42n - 27s$

8. The height of a statue is 276 inches. What is the height of the statue in meters? Round your answer to the nearest hundredth.

   A. 1.09 m   **B.** 7.01 m   **C.** 108.66 m   **D.** 701.04 m

► Concepts, Skills, & Problem Solving

OPEN-ENDED Describe a situation that can be represented by the integer. (See Exploration 1, p. 345.)

9. -6
10. 12
11. -45

WRITING INTEGERS Write a positive or negative integer that represents the situation.

12. A football team loses 3 yards.
13. The temperature is 6 degrees below zero.
14. You earn $15 raking leaves.
15. A person climbs 600 feet up a mountain.
16. You withdraw $42 from an account.
17. An airplane climbs to 37,500 feet.
18. The temperature rises 17 degrees.
19. You lose 56 points in a video game.
20. A ball falls 350 centimeters.
21. You receive 5 bonus points in class.

22. **MODELING REAL LIFE** On December 17, 1903, the Wright brothers accomplished the first powered flight. The plane traveled a distance of 120 feet. Write this distance as an integer.

23. **MODELING REAL LIFE** A stock market gains 83 points. The next day, the stock market loses 47 points. Write each amount as an integer.

Section 8.1 Integers 349

GRAPHING INTEGERS Graph the integer and its opposite.

24. −5   25. −8   26. 14   27. 9
28. −14   29. 20   30. −26   31. 18
32. 30   33. −150   34. −32   35. 400

36. **YOU BE THE TEACHER** Your friend describes the positive integers. Is your friend correct? Explain your reasoning.

> The positive integers are 0, 1, 2, 3, ...

USING A NUMBER LINE Identify the integer represented by the point on the number line.

37. A 38. B 39. C 40. D

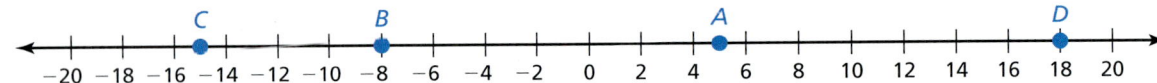

41. **DIG DEEPER!** Low tide, represented by the integer −1, is 1 foot below the average water level. High tide is 5 feet higher than low tide.

    a. What does 0 represent in this situation?

    b. Write an integer that represents the average water level relative to high tide.

42. **REPEATED REASONING** Consider an integer n.

    a. Is the opposite of n always less than 0? Explain your reasoning.

    b. What can you conclude about the opposite of the opposite of n? Justify your answer.

    c. Describe the meaning of $-[-(-n)]$. What is it equal to?

43. **NUMBER SENSE** In a game of tug-of-war, a team wins by pulling the flag over its goal line. The flag begins at 0. During a game, the flag moves 8 feet to the right, 12 feet to the left, and 13 feet back to the right. Did a team win? Explain. If not, what does each team need to do in order to win?

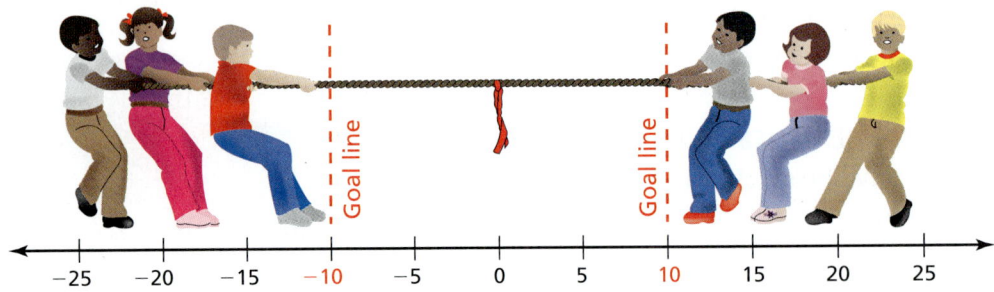

350   Chapter 8   Integers, Number Lines, and the Coordinate Plane

8.2 Comparing and Ordering Integers

Learning Target: Compare and order integers.

Success Criteria:
- I can explain how to determine which of two integers is greater.
- I can order a set of integers from least to greatest.
- I can interpret statements about order in real-life problems.

EXPLORATION 1

Seconds to Liftoff

Work with a partner. You are listening to a command center before the liftoff of a rocket.

You hear the following:

"T minus 10 seconds . . . go for main engine start . . . T minus 9 . . . 8 . . . 7 . . . 6 . . . 5 . . . 4 . . . 3 . . . 2 . . . 1 . . . we have liftoff."

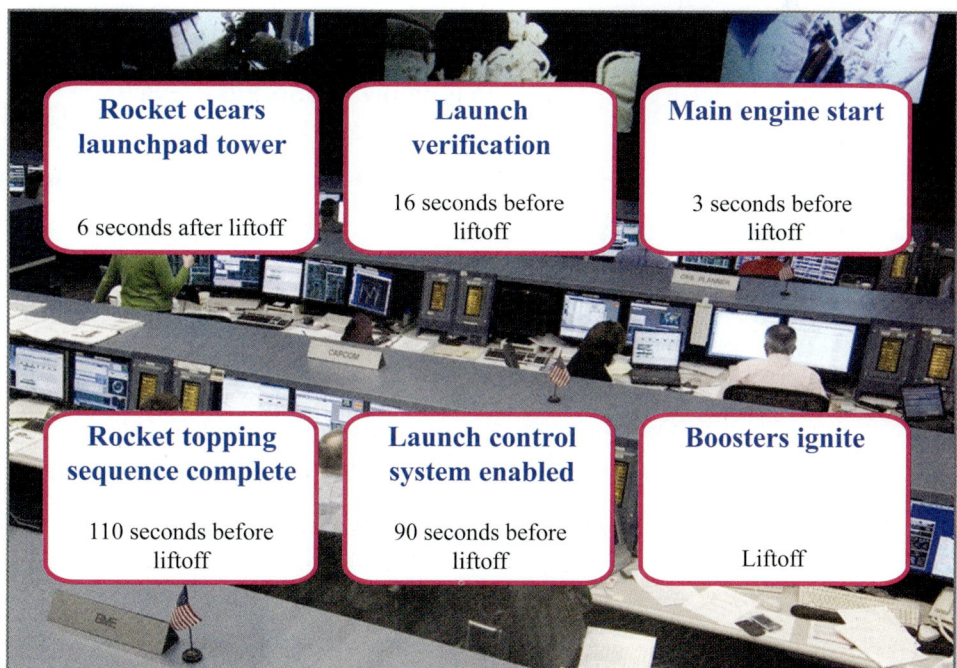

Rocket clears launchpad tower	Launch verification	Main engine start
6 seconds after liftoff	16 seconds before liftoff	3 seconds before liftoff

Rocket topping sequence complete	Launch control system enabled	Boosters ignite
110 seconds before liftoff	90 seconds before liftoff	Liftoff

Math Practice

Use Other Resources
Which sources would give you the most accurate information? How do you know you can trust the information you find?

a. Represent these events on a number line.

b. List the events in the order they occurred. Explain your reasoning.

c. Extend the number line in part (a) to show events in an astronaut's day. Include at least five events before liftoff and at least five events after liftoff. Use the Internet or another reference source to gather information.

8.2 Lesson

Recall that on a horizontal number line, numbers to the left are less than numbers to the right. Numbers to the right are greater than numbers to the left.

On a vertical number line, numbers below are less than numbers above. Numbers above are greater than numbers below.

EXAMPLE 1 **Comparing Integers**

a. Compare 2 and −6.

Graph each number on a horizontal number line.

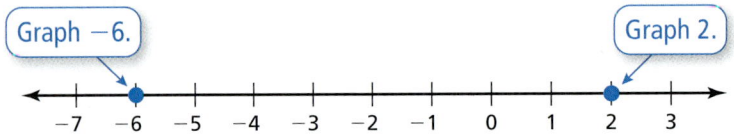

▶ 2 is to the right of −6. So, 2 > −6.

b. Compare −5 and −3.

Graph each number on a vertical number line.

▶ −5 is below −3. So, −5 < −3.

Try It Copy and complete the statement using < or >.

1. 0 ▢ −4
2. −5 ▢ 5
3. −8 ▢ −7

EXAMPLE 2 **Ordering Integers**

Order −4, 3, 0, −1, −2 from least to greatest.

Graph each integer on a number line.

Write the integers as they appear on the number line from left to right.

▶ So, the order from least to greatest is −4, −2, −1, 0, 3.

Try It Order the integers from least to greatest.

4. −2, −3, 3, 1, −1
5. 4, −7, −8, 6, 1

EXAMPLE 3 Reasoning with Integers

A number is greater than −8 and less than 0. What is the greatest possible integer value of this number?

A. −10   **B.** −7   **C.** −1   **D.** 2

The number is greater than −8 and less than 0. So, the number must be to the right of −8 and to the left of 0 on a horizontal number line.

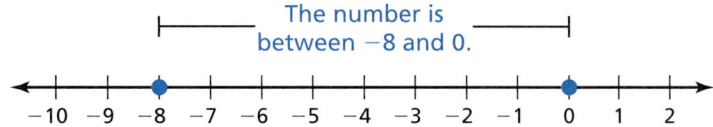

In Example 3, you can eliminate Choices A and D because −10 is to the left of −8 and 2 is to the right of 0.

The greatest possible integer value between −8 and 0 is the integer farthest to the right on the number line between these values, which is −1.

▶ So, the correct answer is **C**.

Try It

6. In Example 3, what is the least possible integer value of the number?

Self-Assessment for Concepts & Skills

Solve each exercise. Then rate your understanding of the success criteria in your journal.

ORDERING INTEGERS Order the integers from least to greatest.

7. 6, −4, −1, 3, 5

8. −7, −9, 0, 8, −2

9. WRITING Explain how to determine which of two integers is greater.

10. MP REASONING The positions of four fish are shown.

  a. Use red, blue, yellow, and green dots to graph the positions of the fish on a horizontal number line and a vertical number line.

  b. Explain how to use the number lines from part (a) to order the positions from least to greatest.

11. MP NUMBER SENSE a and b are negative integers. Compare a and b. Explain your reasoning.

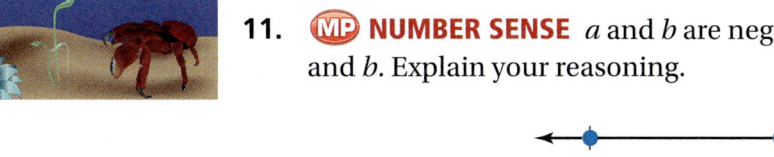

EXAMPLE 4 Modeling Real Life

The diagram shows the coldest recorded temperatures for several cities in North Carolina.

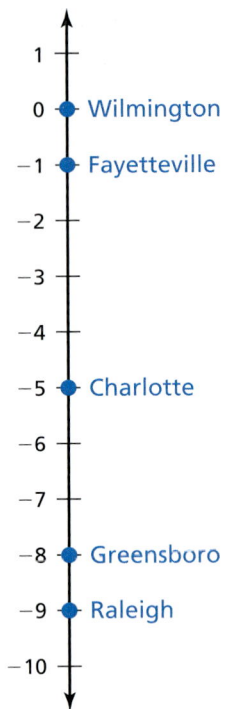
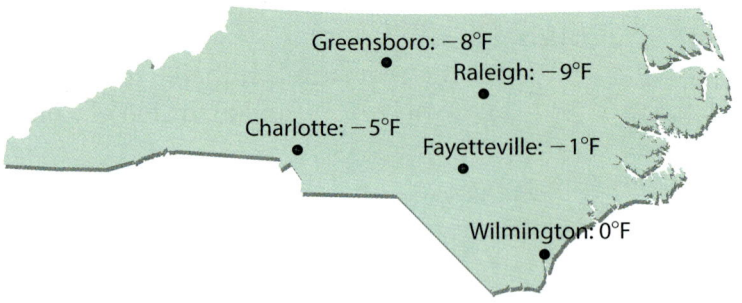

a. **Which city has the coldest recorded temperature?**

Graph each integer on a vertical number line.

▶ −9 is the lowest on the number line. So, Raleigh has the coldest recorded temperature.

b. **Has a negative Fahrenheit temperature ever been recorded in Wilmington? Explain.**

▶ The coldest recorded temperature in Wilmington is 0°F, which is greater than every negative Fahrenheit temperature. So, a negative Fahrenheit temperature has never been recorded in Wilmington.

Self-Assessment for Problem Solving

Solve each exercise. Then rate your understanding of the success criteria in your journal.

12. The freezing temperature of nitrogen is −210°C, and the freezing temperature of oxygen is −219°C. A container of nitrogen and a container of oxygen are both cooled to −215°C. Do the contents of each container freeze? Explain.

13. **DIG DEEPER!** The diagram shows the daily high temperatures during a school week. Was a positive Celsius temperature recorded on Tuesday? on Friday? Explain.

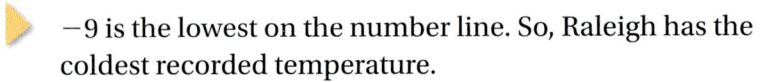

8.2 Practice

> Go to BigIdeasMath.com to get HELP with solving the exercises.

Review & Refresh

Write a positive or negative integer that represents the situation.

1. You walk up 83 stairs.
2. A whale is 17 yards below sea level.
3. An organization receives a $75 donation.
4. A rock falls 250 feet off a cliff.

5. What is the area of the trapezoid?

   A. 6.3 ft^2 **B.** 44.1 ft^2
   C. 50.4 ft^2 **D.** 88.2 ft^2

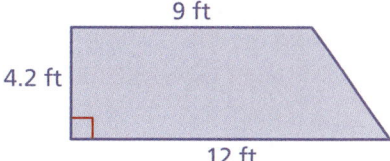

Divide. Write the answer in simplest form.

6. $\dfrac{1}{5} \div \dfrac{1}{9}$
7. $\dfrac{2}{5} \div \dfrac{1}{3}$
8. $\dfrac{1}{4} \div 3$
9. $\dfrac{4}{7} \div 8$

Concepts, Skills, & Problem Solving

OPEN-ENDED Name an event that could occur at the given time (in seconds) in Exploration 1. Describe when the event occurs in the order of events from the exploration. *(See Exploration 1, p. 351.)*

10. -300
11. -150
12. 10

COMPARING INTEGERS Copy and complete the statement using < or >.

13. $3 \;\square\; 0$
14. $-2 \;\square\; 0$
15. $6 \;\square\; -6$
16. $3 \;\square\; -4$
17. $-1 \;\square\; 4$
18. $-7 \;\square\; -8$
19. $-3 \;\square\; -2$
20. $-5 \;\square\; -10$

YOU BE THE TEACHER Your friend compares two integers. Is your friend correct? Explain your reasoning.

21. Compare -3 and -1.
    $3 > 1$. So, $-3 > -1$.

22. Compare -7 and -3.
    Because -7 is to the left of -3 on a number line, $-7 < -3$.

ORDERING INTEGERS Order the integers from least to greatest.

23. $0, -1, 2, 3, -3$
24. $-4, -2, -3, 2, 1$
25. $-2, 3, -3, -4, 4$
26. $5, -11, -9, 3, -4$
27. $-3, 8, 4, 0, -13$
28. $-7, 2, 6, -4, 3$
29. $12, -8, -16, 7, 1$
30. $10, -10, 30, -30, -50$
31. $-5, 15, -10, -20, 25$

Section 8.2 Comparing and Ordering Integers 355

32. **MODELING REAL LIFE** An archaeologist discovers the two artifacts shown.

    a. What integer represents ground level?

    b. A dinosaur bone is 42 centimeters below ground level. Is it deeper than both of the artifacts? Explain.

33. **MP REASONING** A number is between −2 and −10. What is the least possible integer value of this number? What is the greatest possible integer value of this number?

34. **MP NUMBER SENSE** Describe the locations of the integers m and n on a number line for each situation.

    a. $m < n$ b. $m > n$ c. $n > m$ d. $n < m$

CRITICAL THINKING Tell whether the statement is *always*, *sometimes*, or *never* true. Explain.

35. A positive integer is greater than its opposite.

36. An integer is less than its opposite and greater than 0.

37. **MODELING REAL LIFE** The table shows the highest and lowest elevations for five states.

    a. Order the states by their highest elevations, from least to greatest.

    b. Order the states by their lowest elevations, from least to greatest.

    c. What does the lowest elevation for Florida represent?

State	Highest Elevation (feet)	Lowest Elevation (feet)
Arkansas	2753	55
California	14,494	−282
Florida	345	0
Louisiana	535	−8
Tennessee	6643	178

38. **MP NUMBER SENSE** Point A is on a number line halfway between −17 and 5. Point B is halfway between Point A and 0. What integer does Point B represent?

39. **MP REASONING** Eleven Fahrenheit temperatures are shown on a map during a weather report. When the temperatures are ordered from least to greatest, the middle temperature is below 0°F. Do you know exactly how many of the temperatures are represented by negative numbers? Explain.

40. **PUZZLE** Nine students each choose one integer. Here are seven of them:

    $$5, -8, 10, -1, -12, -20, \text{ and } 1.$$

    a. When all nine integers are ordered from least to greatest, the middle integer is 1. Describe the integers chosen by the other two students.

    b. When all nine integers are ordered from least to greatest, the middle integer is −3. Describe the integers chosen by the other two students.

8.3 Rational Numbers

Learning Target: Compare and order rational numbers.

Success Criteria:
- I can explain the meaning of a rational number.
- I can graph rational numbers on a number line.
- I can determine which of two rational numbers is greater.
- I can order a set of rational numbers from least to greatest.

EXPLORATION 1

Locating Fractions on a Number Line

Work with a partner. Represent the events on a number line using a fraction or a mixed number.

a. Radio Transmission: 11:30 A.M.

b. Space Walk: 7:30 P.M.

c. Physical Exam: 4:45 A.M.

d. Take Photograph: 3:15 A.M.

e. Float in the Cabin: 6:20 P.M.

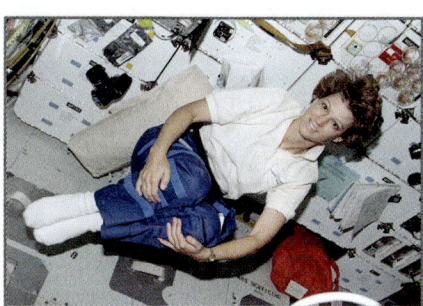

f. Eat Dinner: 8:40 P.M.

Math Practice

Use a Diagram
How can you graph values between negative integers on a number line?

Section 8.3 Rational Numbers

8.3 Lesson

Key Vocabulary
rational number, p. 358

Integers, fractions, and decimals make up the set of *rational numbers*. A **rational number** is a number that can be written as $\frac{a}{b}$, where a and b are integers and $b \neq 0$.

EXAMPLE 1 — Graphing Rational Numbers

Graph each number and its opposite.

You can also graph rational numbers on vertical number lines.

a. $\frac{3}{4}$

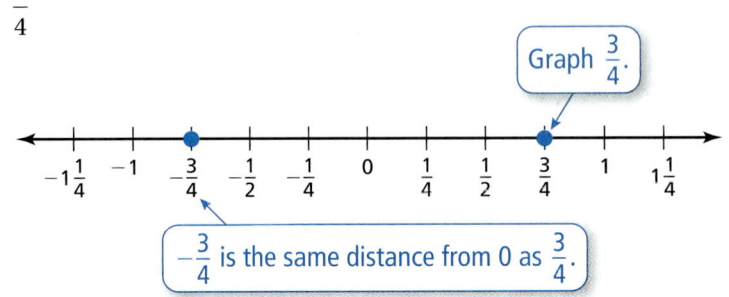

b. -1.6

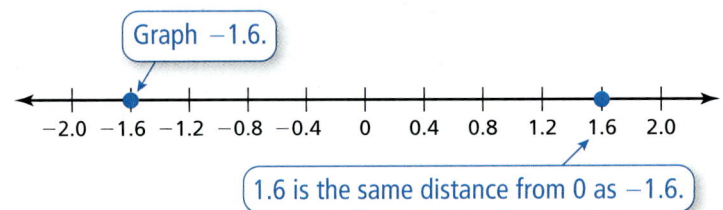

Try It Graph the number and its opposite.

1. $2\frac{1}{2}$
2. $-\frac{4}{5}$
3. -3.5
4. 5.25

EXAMPLE 2 — Comparing Fractions and Mixed Numbers

a. Compare $-\frac{1}{2}$ and $-\frac{3}{4}$.

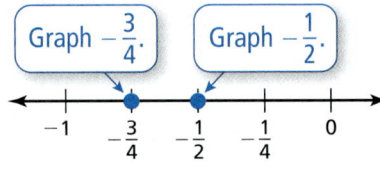

$-\frac{1}{2}$ is to the right of $-\frac{3}{4}$.

▶ So, $-\frac{1}{2} > -\frac{3}{4}$.

b. Compare $-4\frac{5}{6}$ and $-4\frac{1}{6}$.

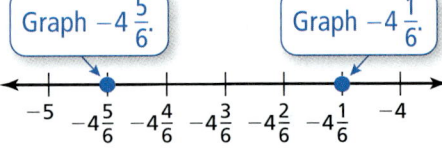

$-4\frac{5}{6}$ is to the left of $-4\frac{1}{6}$.

▶ So, $-4\frac{5}{6} < -4\frac{1}{6}$.

Try It Copy and complete the statement using < or >.

5. $-\frac{4}{7}$ 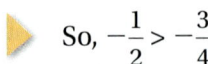 $-\frac{1}{7}$

6. $-1\frac{2}{3}$ ▢ $-1\frac{5}{6}$

358   Chapter 8   Integers, Number Lines, and the Coordinate Plane   Multi-Language Glossary at BigIdeasMath.com

EXAMPLE 3 Comparing Decimals

a. Compare −2.3 and −1.5.

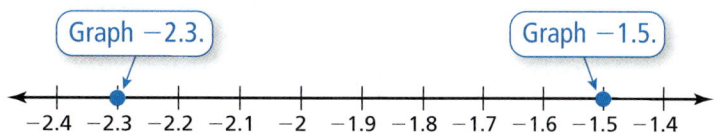

−2.3 is to the left of −1.5.

▸ So, −2.3 < −1.5.

b. Compare −3.08 and −3.8.

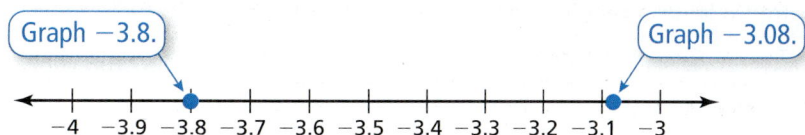

−3.08 is to the right of −3.8.

▸ So, −3.08 > −3.8.

Try It Copy and complete the statement using < or >.

7. −0.5 ▢ 0.3

8. −6.5 ▢ −6.75

Self-Assessment for Concepts & Skills

Solve each exercise. Then rate your understanding of the success criteria in your journal.

COMPARING RATIONAL NUMBERS Copy and complete the statement using < or >.

9. $-\dfrac{2}{3}$ ▢ $-\dfrac{5}{9}$

10. $-2\dfrac{1}{4}$ ▢ $-2\dfrac{3}{8}$

11. -1.7 ▢ -2.4

12. **NUMBER SENSE** Which statement is *not* true?

   A. On a number line, $-2\dfrac{1}{6}$ is to the left of $-2\dfrac{2}{3}$.

   B. $-2\dfrac{2}{3}$ is less than $-2\dfrac{1}{6}$.

   C. On a number line, $-2\dfrac{2}{3}$ is to the left of $-2\dfrac{1}{6}$.

13. **WRITING** Explain how to determine whether a number is a rational number.

EXAMPLE 4 Modeling Real Life

A *Chinook wind* is a warm mountain wind that can cause rapid temperature changes. The table shows three of the greatest temperature drops ever recorded after a Chinook wind occurred. On which date did the temperature drop the fastest? Explain.

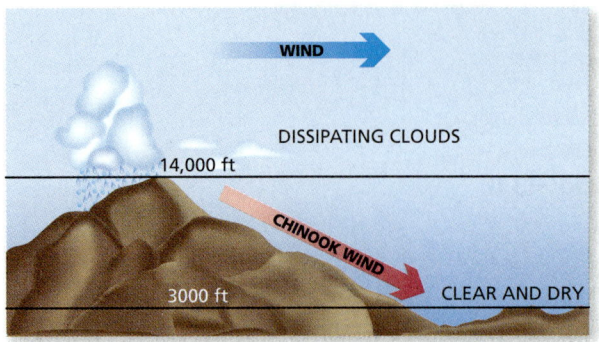

Date	Temperature Change
January 10, 1911	$-3\frac{1}{10}$ °F per minute
November 10, 1911	$-\frac{5}{8}$ °F per minute
January 22, 1943	$-2\frac{1}{5}$ °F per minute

Graph the numbers on a number line.

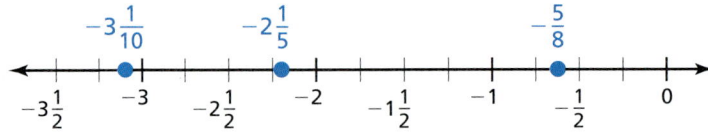

$-3\frac{1}{10}$ is farthest to the left.

 So, the temperature dropped the fastest on January 10, 1911.

 **Self-Assessment** for Problem Solving

Solve each exercise. Then rate your understanding of the success criteria in your journal.

14. You and your friend rappel down a cliff. Your friend descends 0.11 mile and then waits for you to catch up. You descend and your current change in elevation is -0.12 mile. Have you reached your friend? Explain.

15. The table shows the changes in the value of a stock over a period of three days. On which day does the value of the stock change the most? Explain your reasoning.

Day	1	2	3
Change (dollars)	-0.42	0	-0.45

8.3 Practice

▶ Review & Refresh

Copy and complete the statement using < or >.

1. 5 ▢ 8
2. −4 ▢ −7
3. 2 ▢ −5
4. 0 ▢ −3

5. You pay $48 for 8 pounds of chicken. Which is an equivalent rate?
   - **A.** $44 for 4 pounds
   - **B.** $28 for 4 pounds
   - **C.** $15 for 3 pounds
   - **D.** $30 for 5 pounds

Find the whole.

6. 40% of what number is 24?
7. 12% of what number is 9?
8. 48% of what number is 84?
9. 140% of what number is 98?

Multiply.

10. 0.53×3
11. 0.06×0.7
12. 3.7×4.854
13. 2.9×8.8609

▶ Concepts, Skills, & Problem Solving

USING TOOLS Use a fraction or a mixed number to represent the time on a number line. Let 0 represent noon. (See Exploration 1, p. 357.)

14. 8:30 A.M.
15. 12:15 P.M.
16. 3:12 P.M.

GRAPHING RATIONAL NUMBERS Graph the number and its opposite.

17. $\frac{2}{3}$
18. -4.3
19. 2.15
20. $-\frac{3}{7}$
21. -0.4
22. $5\frac{1}{3}$
23. $-2\frac{1}{4}$
24. $-5\frac{3}{10}$

COMPARING RATIONAL NUMBERS Copy and complete the statement using < or >.

25. $-3\frac{1}{3}$ ▢ $-3\frac{2}{3}$
26. $-\frac{1}{2}$ ▢ $-\frac{1}{6}$
27. $-\frac{3}{4}$ ▢ $\frac{5}{8}$
28. $-2\frac{2}{3}$ ▢ $-2\frac{1}{2}$
29. $-1\frac{5}{6}$ ▢ $-1\frac{3}{4}$
30. -4.6 ▢ -4.8
31. -0.12 ▢ -0.05
32. 2.41 ▢ -3.16
33. -3.524 ▢ -3.542

ORDERING RATIONAL NUMBERS Order the numbers from least to greatest.

34. $1.3, -2, -1.8, 0, -1.75$
35. $-4, -4.35, -4.9, -5, -4.3$
36. $1.6, 1.2, 0, 0.8, -0.1$
37. $-\frac{1}{2}, \frac{1}{8}, \frac{3}{4}, -1, -\frac{1}{4}$
38. $-2\frac{3}{10}, -2\frac{2}{5}, -2, -2\frac{1}{2}, -3$
39. $-\frac{1}{20}, -\frac{5}{8}, 0, -1, -\frac{3}{4}$

Section 8.3 Rational Numbers 361

40. **MODELING REAL LIFE** In rough water, a small sand dollar burrows $-\frac{1}{2}$ centimeter into the sand. A larger sand dollar burrows $-1\frac{1}{4}$ centimeters into the sand. Which sand dollar burrowed deeper?

41. **MODELING REAL LIFE** Two golfers calculate their average scores relative to par over several rounds of golf. Golfer A has an average score of $-1\frac{1}{4}$. Golfer B has an average score of $-1\frac{3}{8}$. Who has the lesser average score?

42. **MODELING REAL LIFE** The *apparent magnitude* of a star measures how bright the star appears as seen from Earth. The brighter the star, the lesser the number. Which star is the brightest?

Star	Alpha Centauri	Antares	Canopus	Deneb	Sirius
Apparent Magnitude	−0.27	0.96	−0.72	1.25	−1.46

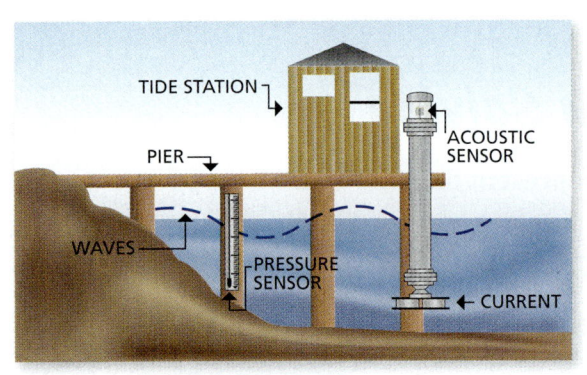

43. **MP REPEATED REASONING** The daily water level relative to the pier is recorded for seven straight days at a tide station on the Big Marco River in Florida. On which days is the water level higher than on the previous day? On which days is it lower? Explain.

Day	Sun	Mon	Tues	Wed	Thurs	Fri	Sat
Water Level (feet)	$-\frac{3}{25}$	$-\frac{7}{20}$	$-\frac{27}{50}$	$-\frac{13}{20}$	$-\frac{16}{25}$	$-\frac{53}{100}$	$-\frac{1}{3}$

44. **DIG DEEPER!** A guitar tuner allows you to tune a guitar string to its correct pitch. The units on a tuner are measured in *cents*. The units tell you how far the string tone is above or below the correct pitch.

Guitar String	6	5	4	3	2	1
Number of Cents Away from the Correct Pitch	−0.3	1.6	−2.3	2.8	2.4	−3.6

a. What number on the tuner represents a correctly tuned guitar string?

b. Which strings have a pitch below the correct pitch?

c. Which string has a pitch closest to its correct pitch?

d. Which string has a pitch farthest from its correct pitch?

e. The tuner is rated to be accurate to within 0.5 cent of the true pitch. Which string could possibly be correct? Explain your reasoning.

45. **MP NUMBER SENSE** What integer values of x make the statement $-\frac{3}{x} < -\frac{x}{3}$ true?

8.4 Absolute Value

Learning Target: Understand the concept of absolute value.

Success Criteria:
- I can find the absolute value of a number.
- I can make comparisons that involve absolute values of numbers.
- I can apply absolute value in real-life problems.

EXPLORATION 1

Comparing Positions of Objects

Work with a partner. The diagram shows the positions of several objects.

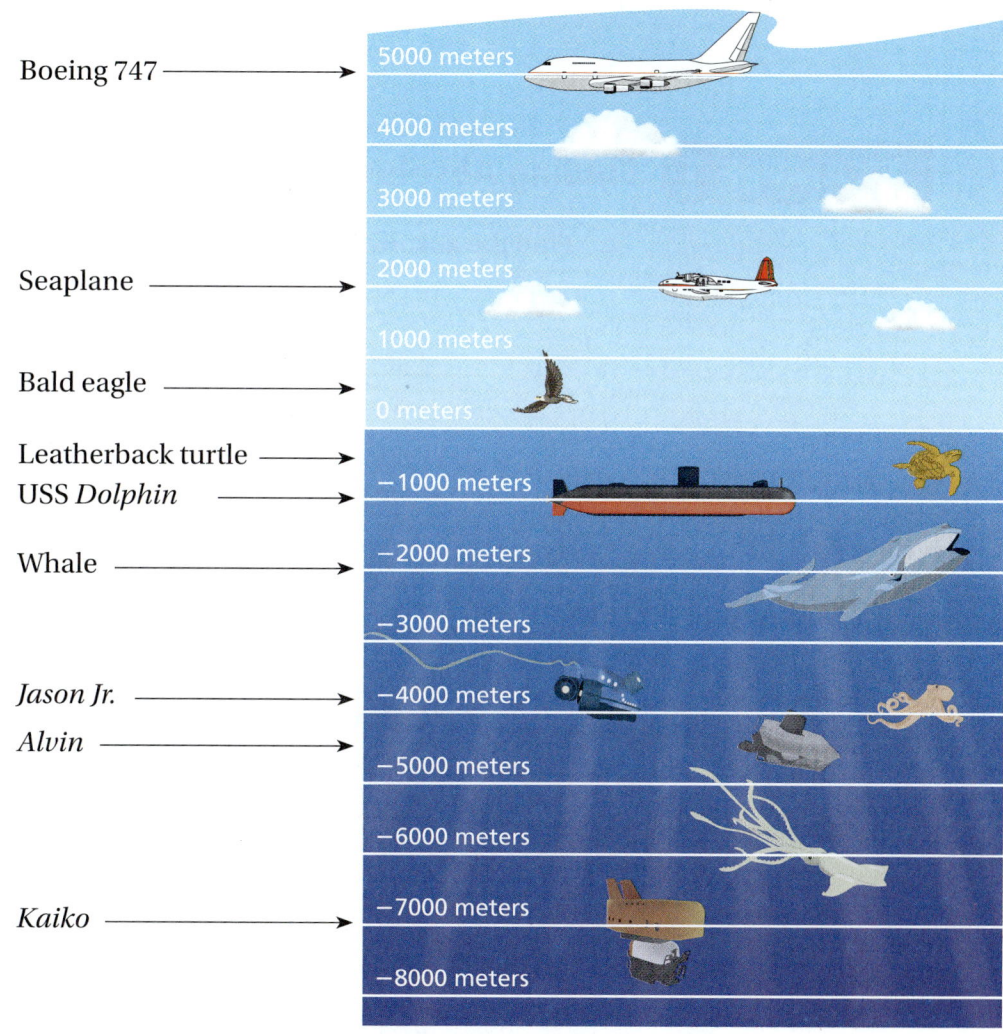

Math Practice

Understand Quantities
What do positive numbers represent in the problem? What do negative numbers represent?

a. What integer represents sea level? How can you compare the positions of objects relative to sea level?

b. Which pairs of objects are the same distance from sea level? How do you know?

c. The vessels *Kaiko*, *Alvin*, and *Jason Jr.* move to be the same distance from sea level as the Boeing 747. About how many meters did each vessel travel?

Section 8.4 Absolute Value 363

8.4 Lesson

Key Vocabulary
absolute value, p. 364

Key Idea

Absolute Value

Words The **absolute value** of a number is the distance between the number and 0 on a number line. The absolute value of a number a is written as $|a|$.

Numbers $\quad |-2| = 2 \quad |2| = 2$

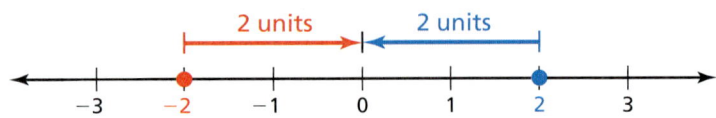

EXAMPLE 1 Finding Absolute Value

a. Find the absolute value of 3.

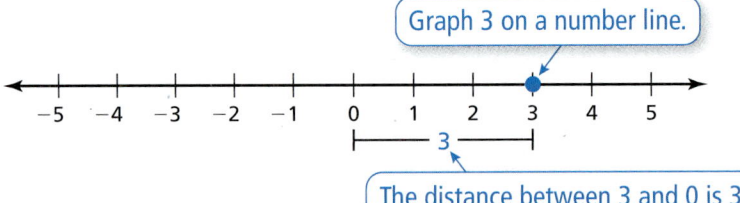

➤ So, $|3| = 3$.

b. Find the absolute value of $-2\frac{1}{2}$.

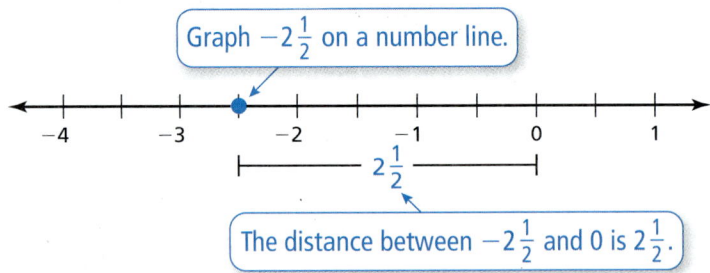

➤ So, $\left|-2\frac{1}{2}\right| = 2\frac{1}{2}$.

Try It Find the absolute value.

1. $|8|$
2. $|-6|$
3. $|0|$
4. $\left|\frac{1}{4}\right|$
5. $\left|-7\frac{1}{3}\right|$
6. $|-12.9|$

EXAMPLE 2 Comparing Values

a. Compare 2 and $|-5|$.

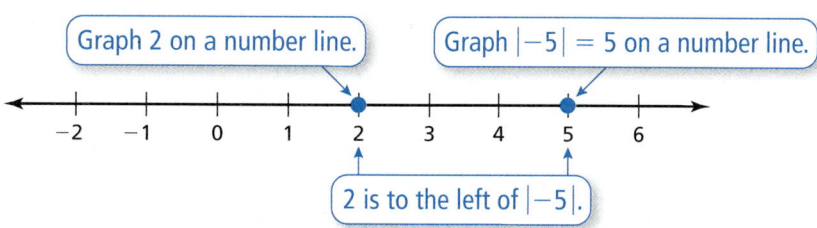

So, $2 < |-5|$.

b. Compare $|3.5|$ and -1.

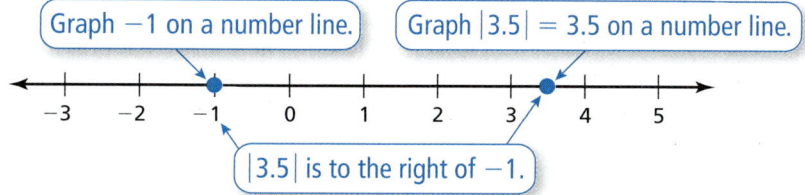

So, $|3.5| > -1$.

Try It Copy and complete the statement using <, >, or =.

7. $|-4|$ ⬚ -2

8. -5 ⬚ $|5|$

9. $|9|$ ⬚ 10

10. 3.9 ⬚ $|-3.9|$

Self-Assessment for Concepts & Skills

Solve each exercise. Then rate your understanding of the success criteria in your journal.

COMPARING VALUES Copy and complete the statement using <, >, or =.

11. $|-6|$ ⬚ 3

12. $|-3.5|$ ⬚ 4

13. $3\frac{1}{2}$ ⬚ $\left|-4\frac{3}{4}\right|$

14. **DIFFERENT WORDS, SAME QUESTION** Which is different? Find "both" answers.

> How far is -3 from 0?

> What integer is 3 units to the left of 0?

> What is the absolute value of -3?

> What is the distance between -3 and 0?

Section 8.4 Absolute Value

EXAMPLE 3 Modeling Real Life

The table shows the elevations of several animals.

Animal	Elevation (feet)
Shark	−4
Sea lion	5
Seagull	56
Shrimp	−65
Turtle	−22

a. **Which animal is the deepest? Explain.**

Graph each elevation.

The lowest elevation represents the animal that is the deepest. The integer that is lowest on the number line is −65.

▸ So, the shrimp is the deepest.

b. **Is the shark or the sea lion closer to sea level?**

Because sea level is at 0 feet, use absolute values.

Shark: $|-4| = 4$ **Sea lion:** $|5| = 5$

▸ Because 4 is less than 5, the shark is closer to sea level than the sea lion.

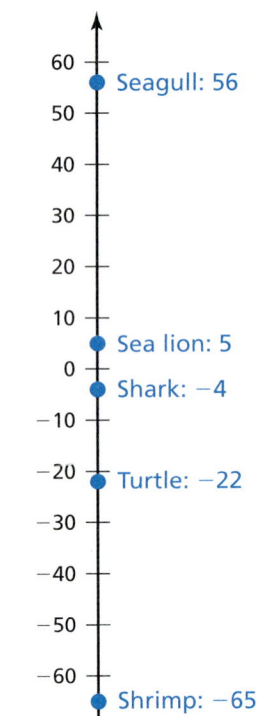

Self-Assessment for Problem Solving

Solve each exercise. Then rate your understanding of the success criteria in your journal.

15. Describe the position of an object in your classroom using a negative rational number. Then describe the position of a second object using a positive rational number. Which number has a greater absolute value? What does this mean?

16. **DIG DEEPER!** The table shows the elevations of several checkpoints along a hiking trail.

Checkpoint	1	2	3	4	5
Elevation (feet)	110	38	−24	12	−142

a. Which checkpoint is farthest from sea level?

b. Which checkpoint is closest to sea level?

c. Between which checkpoints do you reach sea level? Explain your reasoning.

8.4 Practice

Go to *BigIdeasMath.com* to get HELP with solving the exercises.

Review & Refresh

Order the numbers from least to greatest.

1. $2.4, -3.2, -1.8, 0.6, -1.3$
2. $-0.3, 0.7, -1.5, 0, 2.2$
3. $\frac{3}{4}, \frac{1}{2}, \frac{2}{3}, 2, \frac{1}{4}$
4. $\frac{1}{5}, 1\frac{2}{5}, -2\frac{3}{4}, \frac{4}{5}, -2\frac{1}{2}$

Represent the ratio relationship using a graph.

5.
Beats	9	18	27
Seconds	5	10	15

6.
Yogurt (ounces)	7	14	21
Granola (ounces)	3	6	9

Evaluate the expression when $a = 2$, $b = 5$, and $c = 8$.

7. $5 + c$
8. $b - 4$
9. $\frac{c}{a}$
10. $b \cdot c$

Concepts, Skills, & Problem Solving

COMPARING POSITIONS OF OBJECTS Tell which object is farther from sea level. Explain your reasoning. (See Exploration 1, p. 363.)

11. Scuba diver: -15 m
    Dolphin: -22 m
12. Seagull: 12 m
    School of fish: -4 m
13. Shark: -40 m
    Flag on a ship: 32 m

FINDING ABSOLUTE VALUE Find the absolute value.

14. $|-2|$
15. $|23|$
16. $|11|$
17. $|-68|$
18. $|-8.35|$
19. $\left|\frac{1}{6}\right|$
20. $|14.06|$
21. $\left|-\frac{5}{8}\right|$
22. $\left|-3\frac{2}{5}\right|$
23. $|1.026|$
24. $\left|1\frac{1}{3}\right|$
25. $|-6.308|$

26. **REASONING** Write two integers that have an absolute value of 10.

27. **YOU BE THE TEACHER** Your friend finds the absolute value of 14. Is your friend correct? Explain your reasoning.

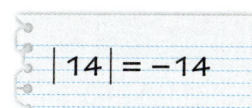
$|14| = -14$

COMPARING VALUES Copy and complete the statement using <, >, or =.

28. 6 ☐ $|-8|$
29. $|-3|$ ☐ 3
30. $|-4.3|$ ☐ 3.4
31. $\frac{1}{5}$ ☐ $\left|-\frac{2}{9}\right|$
32. $|-0.05|$ ☐ 0
33. $|-5.5|$ ☐ $|-3.1|$
34. $\frac{3}{4}$ ☐ $\left|-\frac{2}{5}\right|$
35. $|-6.8|$ ☐ $|8.25|$
36. -12 ☐ $|12|$

37. MODELING REAL LIFE The table shows the change in the balance of a bank account after each of three transactions. Which transaction has the greatest effect on the balance of the account? Which transaction has the least effect on the balance of the account?

Transaction	Change (dollars)
1	14.72
2	−15.36
3	−38.75

ORDERING VALUES Order the values from least to greatest.

38. 5, 0, $|-1|$, $|4|$, -2

39. $|-3|$, $|5|$, -3, -4, $|-4|$

40. 10, $|-6|$, 9, $|3|$, -11, 0

41. -18, $|30|$, -19, $|-22|$, -20, $|-18|$

SIMPLIFYING EXPRESSIONS Simplify the expression.

42. $-|2|$

43. $-|6|$

44. $-|-1|$

45. MP REASONING The coldest possible temperature is called *absolute zero*. It is represented by 0 K on the Kelvin temperature scale.

a. Which temperature is closer to 0 K: 32°F or −50°C?

b. What do absolute values and temperatures on the Kelvin scale have in common?

Absolute Zero
Thermometers compare Fahrenheit, Celsius, and Kelvin scales.

Water boils. 212°F 100°C 373 K
Water freezes. 32°F 0°C 273 K
Absolute Zero −459°F −273°C 0 K
Fahrenheit Celsius Kelvin

CRITICAL THINKING Tell whether the statement is *always*, *sometimes*, or *never* true. Explain.

46. The absolute value of a number is greater than the number.

47. The absolute value of a negative number is positive.

48. The absolute value of a positive number is its opposite.

MATCHING Match the account balance with the debt that it represents. Explain your reasoning.

49. account balance = −$25

50. account balance < −$25

51. account balance > −$25

A. debt > $25

B. debt = $25

C. debt < $25

52. MP PATTERNS A *palindrome* is a word or sentence that reads the same forward as it does backward.

a. Graph and label the following points on a number line: $A = -2$, $C = -1$, $E = 0$, $R = -3$. Then, using the same letters as the original points, graph and label the absolute value of each point on the *same* number line.

b. What word do the letters spell? Is this a palindrome?

c. Assign letters to points on a number line to make up your own palindrome using the process in part (a).

53. CRITICAL THINKING Find values of x and y so that $|x| < |y|$ and $x > y$.

8.5 The Coordinate Plane

Learning Target: Plot and reflect ordered pairs in all four quadrants of a coordinate plane.

Success Criteria:
- I can identify ordered pairs in a coordinate plane.
- I can plot ordered pairs in a coordinate plane and describe their locations.
- I can reflect points in the *x*-axis, the *y*-axis, or both axes.
- I can apply plotting points in all four quadrants to solve real-life problems.

EXPLORATION 1

Extending the Coordinate Plane

Work with a partner. Previously, you plotted points with positive coordinates in a coordinate plane like the one shown at the right.

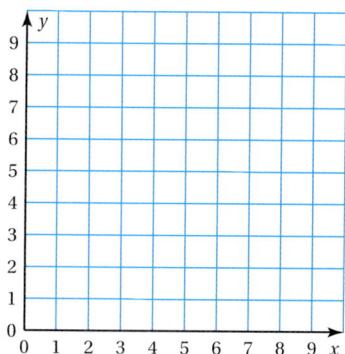

a. You can also plot points in which one or both of the coordinates are negative numbers. Create ordered pairs with different combinations of positive and negative coordinates, as described below. Then plot the ordered pairs and explain how you extended the coordinate plane shown.

   (positive, positive)   (negative, positive)

   (negative, negative)   (positive, negative)

b. How many regions of the coordinate plane are created by the *x*-axis and *y*-axis? What do the points in each of these regions have in common?

c. The photo shows the *reflection*, or mirror image, of a mountain in a lake. When you fold the photo on its axis, the mountain and its reflection align.

Actual mountain

Axis

Reflection of mountain

Math Practice

Check Progress
How can you check your progress to make sure you are reflecting your point correctly?

Plot a point and its *reflection* in one of the axes. Explain your reasoning. What do you notice about the coordinates of the points?

Section 8.5   The Coordinate Plane   369

8.5 Lesson

Key Vocabulary
coordinate plane, p. 370
origin, p. 370
quadrants, p. 370

Previously, you plotted points with positive coordinates. Now, you will plot points with positive and negative coordinates.

Key Idea

The Coordinate Plane

A **coordinate plane** is formed by the intersection of a horizontal number line and a vertical number line. The number lines intersect at the **origin** and separate the coordinate plane into four regions called **quadrants**.

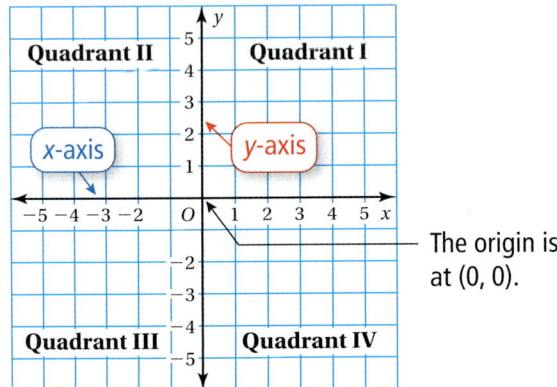

An *ordered pair* is used to locate a point in a coordinate plane.

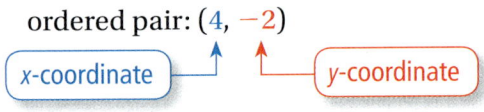

EXAMPLE 1 Identifying an Ordered Pair

Which ordered pair corresponds to Point T?

A. $(-3, -3)$ **B.** $(-3, 3)$

C. $(3, -3)$ **D.** $(3, 3)$

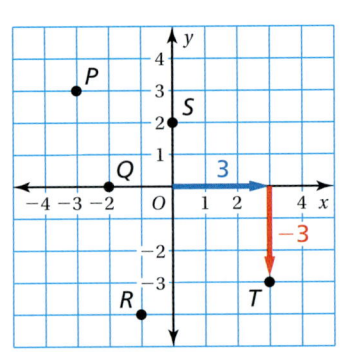

Point T is 3 units to the **right** of the origin and 3 units **down**. So, the x-coordinate is 3 and the y-coordinate is -3.

▶ The ordered pair $(3, -3)$ corresponds to Point T. The correct answer is **C**.

Try It Use the graph in Example 1 to write an ordered pair corresponding to the point.

1. Point P 2. Point Q 3. Point R 4. Point S

370 Chapter 8 Integers, Number Lines, and the Coordinate Plane 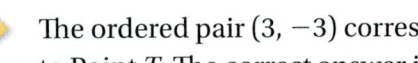 Multi-Language Glossary at *BigIdeasMath.com*

EXAMPLE 2 Plotting Ordered Pairs

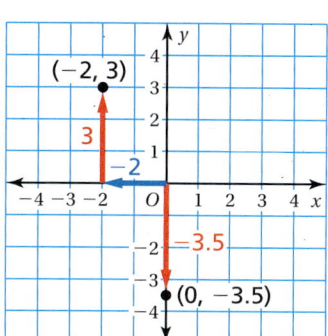

Plot (a) $(-2, 3)$ and (b) $(0, -3.5)$ in a coordinate plane. Describe the location of each point.

a. Start at the origin. Move 2 units left and 3 units up. Then plot the point.

▶ The point is in Quadrant II.

b. Start at the origin. Move 3.5 units down. Then plot the point.

▶ The point is on the y-axis.

Try It Plot the ordered pair in a coordinate plane. Describe the location of the point.

5. $(3, -1)$ 6. $(-5, 0)$ 7. $(-2.5, -1)$ 8. $\left(-1\frac{1}{2}, \frac{1}{2}\right)$

Key Idea

Reflecting a Point in the Coordinate Plane

- To reflect a point in the x-axis, use the same x-coordinate and take the opposite of the y-coordinate.
- To reflect a point in the y-axis, use the same y-coordinate and take the opposite of the x-coordinate.

EXAMPLE 3 Reflecting Points in One Axis

a. Reflect $(-2, 4)$ in the x-axis.

Plot $(-2, 4)$.

To reflect $(-2, 4)$ in the x-axis, use the same x-coordinate, -2, and take the opposite of the y-coordinate. The opposite of 4 is -4.

▶ So, the reflection of $(-2, 4)$ in the x-axis is $(-2, -4)$.

b. Reflect $(-3, -1)$ in the y-axis.

Plot $(-3, -1)$.

To reflect $(-3, -1)$ in the y-axis, use the same y-coordinate, -1, and take the opposite of the x-coordinate. The opposite of -3 is 3.

▶ So, the reflection of $(-3, -1)$ in the y-axis is $(3, -1)$.

Try It Reflect the point in (a) the x-axis and (b) the y-axis.

9. $(3, -2)$ 10. $(4, 0)$ 11. $(-5, 1.5)$

EXAMPLE 4 Reflecting a Point in Both Axes

Reflect (2, 1) in the *x*-axis followed by the *y*-axis.

Step 1: Plot (2, 1).

Step 2: Reflect (2, 1) in the *x*-axis.
Use the same *x*-coordinate, 2, and take the opposite of the *y*-coordinate. The opposite of 1 is −1.

The reflection of (2, 1) in the *x*-axis is (2, −1).

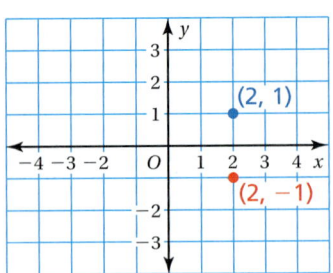

Step 3: Reflect (2, −1) in the *y*-axis.
Use the same *y*-coordinate, −1, and take the opposite of the *x*-coordinate. The opposite of 2 is −2.

The reflection of (2, −1) in the *y*-axis is (−2, −1).

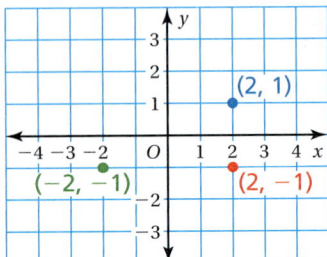

Common Error
When reflecting a second time, be sure to use the reflected point and not the original point.

▶ So, the reflection of (2, 1) in the *x*-axis followed by the *y*-axis is (−2, −1).

Try It Reflect the point in the *x*-axis followed by the *y*-axis.

12. (3, 2) **13.** (−1, 2) **14.** (−4, −3) **15.** (5, −2.5)

Self-Assessment for Concepts & Skills

Solve each exercise. Then rate your understanding of the success criteria in your journal.

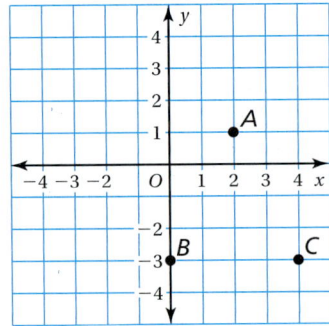

WRITING ORDERED PAIRS Write an ordered pair corresponding to the point shown in the coordinate plane.

16. Point *A* **17.** Point *B* **18.** Point *C*

PLOTTING ORDERED PAIRS Plot the ordered pair in a coordinate plane. Describe the location of the point.

19. $J(2, 5)$ **20.** $K(4, -6)$ **21.** $L\left(-3, -2\frac{1}{2}\right)$

REFLECTING POINTS Reflect the point in the given axis or axes.

22. (9, 8); *x*-axis **23.** (−7, 3); *y*-axis

24. (6, −4); *x*-axis then *y*-axis **25.** (2.5, −4); *y*-axis then *x*-axis

You can use line graphs to display data that are collected over a period of time. Graphing and connecting the ordered pairs can show patterns or trends in the data. This type of line graph is also called a *time series graph*.

EXAMPLE 5 Modeling Real Life

A blizzard hits a town at midnight. The table shows the hourly temperatures from midnight to 8:00 A.M. Display the data in a line graph. Then describe the change in temperature over time.

Hours after Midnight, x	0	1	2	3	4	5	6	7	8
Temperature, y	7°F	5°F	3°F	0°F	−1°F	−4°F	−5°F	−2°F	2°F

Write the ordered pairs.

(0, 7)   (1, 5)   (2, 3)

(3, 0)   (4, −1) (5, −4)

(6, −5) (7, −2) (8, 2)

Plot and label the ordered pairs. Then connect the ordered pairs with line segments.

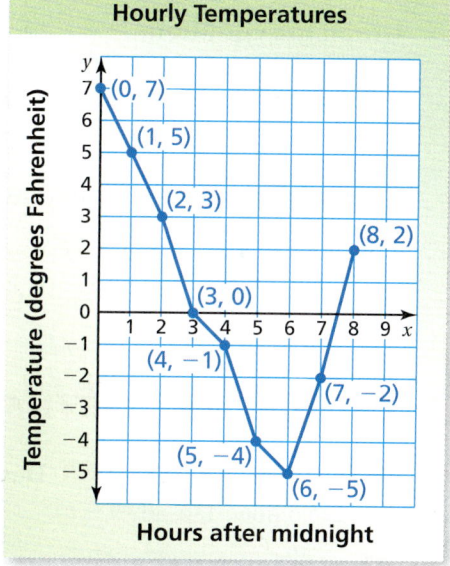

 The hourly temperatures decrease from midnight to 6:00 A.M. and then increase from 6:00 A.M. to 8:00 A.M.

Self-Assessment for Problem Solving

Solve each exercise. Then rate your understanding of the success criteria in your journal.

26. **DIG DEEPER!** At a park, the welcome center is located at (0, 0), the theater is located at (2, 4), and the restrooms are located at (−4.5, 6). The snack bar is exactly halfway between the welcome center and the theater. Graph each location in a coordinate plane.

27. The table shows the elevations of a submarine each hour from noon to 5:00 P.M. Display the data in a line graph. Then describe the change in elevation over time.

Hours after Noon, x	0	1	2	3	4	5
Elevation (kilometers), y	−4.5	−3	−2.5	−2	−3.5	−4

Section 8.5 The Coordinate Plane

8.5 Practice

Go to **BigIdeasMath.com** to get HELP with solving the exercises.

▶ Review & Refresh

Find the absolute value.

1. $|35|$
2. $|-18|$
3. $|4.7|$
4. $\left|-6\frac{7}{12}\right|$

5. What is the ratio of ducks to swans?

   A. $4:9$
   B. $4:5$
   C. $5:4$
   D. $5:9$

Graph the equation.

6. $y = 8x$
7. $y = 3x + 7$
8. $y = \frac{2}{5}x + 2$

Tell which property the statement illustrates.

9. $(2 \cdot p) \cdot 3 = 2 \cdot (p \cdot 3)$
10. $m + 0 = m$
11. $w \cdot 1 = w$
12. $15 + k = k + 15$

▶ Concepts, Skills, & Problem Solving

DESCRIBING REFLECTIONS Describe the reflection shown in the image. (See Exploration 1, p. 369.)

13.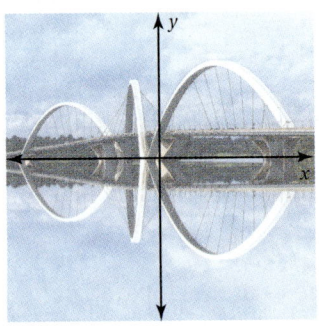

14.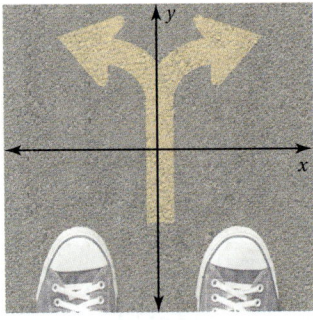

15.

WRITING ORDERED PAIRS Write an ordered pair corresponding to the point.

16. Point A
17. Point B
18. Point C
19. Point D
20. Point E
21. Point F
22. Point G
23. Point H
24. Point I
25. Point J

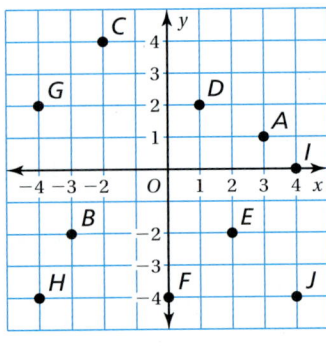

374   Chapter 8   Integers, Number Lines, and the Coordinate Plane

PLOTTING ORDERED PAIRS Plot the ordered pair in a coordinate plane. Describe the location of the point.

26. $K(4, 3)$
27. $L(-1, 2)$
28. $M(0, -6)$
29. $N(3, -7)$
30. $P(-5, -9)$
31. $R(8, 0)$
32. $S(-1.5, 0)$
33. $T(3.5, -1.5)$
34. $U(2, -4)$
35. $V(-4, 1)$
36. $W\left(2\frac{1}{2}, 0\right)$
37. $Z(-4, -5)$

YOU BE THE TEACHER Your friend describes how to plot the point. Is your friend correct? Explain your reasoning.

38.
> To plot (4, 5), start at (0, 0) and move 4 units up and 5 units right.

39.
> To plot (−6, 3), start at (0, 0) and move 6 units left and 3 units up.

MODELING REAL LIFE In Exercises 40–44, use the map of the zoo.

40. Which exhibit is located at (2, 1)?

41. Name an attraction on the positive y-axis.

42. Is parking available in Quadrant II? If not, name a quadrant in which you can park.

43. Write two different ordered pairs that represent the location of the Rain Forest.

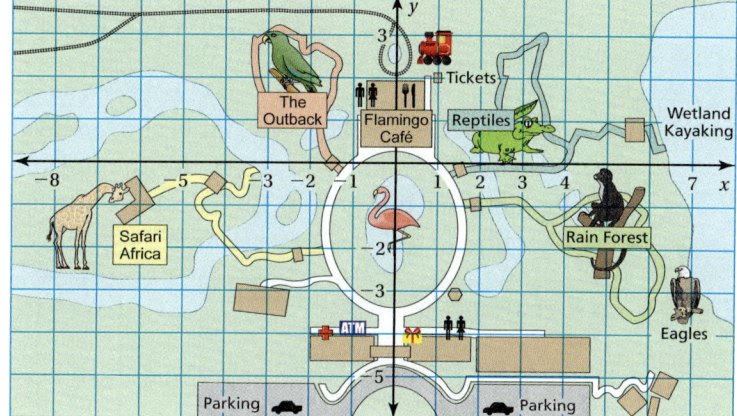

44. Which exhibit is closest to $(-8, -3)$?

REFLECTING POINTS IN ONE AXIS Reflect the point in (a) the x-axis and (b) the y-axis.

45. (3, 2)
46. (−4, 4)
47. (−5, −6)
48. (4, −7)
49. (−9, 3)
50. (6, −2)
51. (0, −1)
52. (−8, 0)
53. (−3.5, 2)
54. (2.5, 4.5)
55. $\left(-5\frac{1}{2}, 3\right)$
56. $\left(\frac{1}{4}, -\frac{7}{8}\right)$

REFLECTING POINTS IN BOTH AXES Reflect the point in the x-axis followed by the y-axis.

57. (4, 5)
58. (−1, 7)
59. (−2, −2)
60. (6, −7)
61. (−8, 8)
62. (5, 9)
63. (0, −2)
64. (−9, 0)
65. (6.5, −10.5)
66. (−0.4, 0.7)
67. $\left(\frac{1}{3}, -\frac{2}{3}\right)$
68. $\left(-1\frac{2}{5}, -1\frac{4}{5}\right)$

69. **MP STRUCTURE** Reflect a point in the x-axis followed by the y-axis. Then reflect the original point in the y-axis followed by the x-axis. Do you get the same results? Explain.

MP REASONING Describe the possible location(s) of the point (x, y).

70. $x > 0, y > 0$ **71.** $x < 0, y < 0$ **72.** $x > 0, y < 0$

73. $x > 0$ **74.** $y < 0$ **75.** $x = 0, y = 0$

CRITICAL THINKING Tell whether the statement is *always*, *sometimes*, or *never* true. Explain your reasoning.

76. The *x*-coordinate of a point on the *x*-axis is zero.

77. The *y*-coordinates of points in Quadrant III are positive.

78. The *x*-coordinate of a point in Quadrant II has the same sign as the *y*-coordinate of a point in Quadrant IV.

79. MODELING REAL LIFE The table shows the number of people who participate in a blood drive each year for 9 years. Display the data in a line graph. Then describe the change in the number of participants over time.

Year, x	1	2	3	4	5	6	7	8	9
Participants, y	140	136	134	132	131	135	136	142	145

80. MODELING REAL LIFE The table shows the amount of carbon dioxide emissions of a country, relative to an environmental standard, each year for 7 years. Display the data in a line graph. Then describe the change in carbon dioxide emissions over time.

Year, x	1	2	3	4	5	6	7
Carbon Dioxide Emissions (millions of metric tons), y	0.6	−0.2	−1.2	1.2	0.8	1	−0.6

81. MP PATTERNS The table shows the total miles run through each of 18 weeks for a marathon training program.

Week	1	2	3	4	5	6	7	8	9
Total Miles	22	46	72	96	124	151	181	211	244

Week	10	11	12	13	14	15	16	17	18
Total Miles	279	317	357	397	437	473	506	530	544

a. Create a table for the distance run during each week of training.

b. Display the data from part (a) in a line graph.

c. Explain the pattern shown in the graph.

82. MP LOGIC Two points are plotted in the coordinate plane. Plot each of the following ordered pairs in the same coordinate plane.

a. $P(a, -b)$ b. $Q(-a, b)$ c. $R(c, -d)$

d. $S(-c, -d)$ e. $T(c, -a)$ f. $U(-d, -b)$

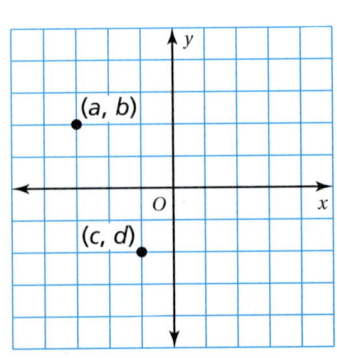

376 Chapter 8   Integers, Number Lines, and the Coordinate Plane

8.6 Polygons in the Coordinate Plane

Learning Target: Draw polygons in the coordinate plane and find distances between points in the coordinate plane.

Success Criteria:
- I can draw polygons in the coordinate plane.
- I can find distances between points in the coordinate plane with the same x-coordinates or the same y-coordinates.
- I can find horizontal and vertical side lengths of polygons in the coordinate plane.
- I can draw polygons in the coordinate plane to solve real-life problems.

EXPLORATION 1

Drawing Polygons in the Coordinate Plane

Work with a partner.

a. Write three ordered pairs that meet the following requirements. Then plot the ordered pairs in a coordinate plane, like the one shown.

- Two of the ordered pairs have the same x-coordinates.
- Two of the ordered pairs have the same y-coordinates.
- Two of the points are in the same quadrant. The other point is in a different quadrant.

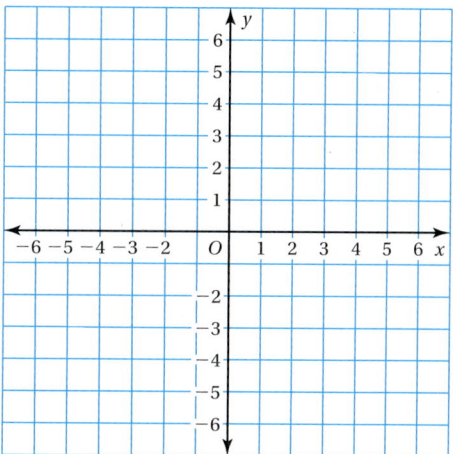

Math Practice

Find General Methods

How can you use absolute values of coordinates to find lengths of horizontal and vertical line segments in the coordinate plane?

b. The points represent the vertices of a polygon. What conclusions can you make about the polygon?

c. Can you plot another point to form a rectangle? a trapezoid? If so, what measures of the quadrilateral can you calculate?

Section 8.6 Polygons in the Coordinate Plane

8.6 Lesson

You can use ordered pairs to represent vertices of polygons. To draw a polygon in a coordinate plane, plot and connect the vertices.

EXAMPLE 1 Drawing a Polygon in a Coordinate Plane

The vertices of a quadrilateral are $A(-1, 1)$, $B(0, 6)$, $C(4, 5)$, and $D(5, -2)$. Draw the quadrilateral in a coordinate plane.

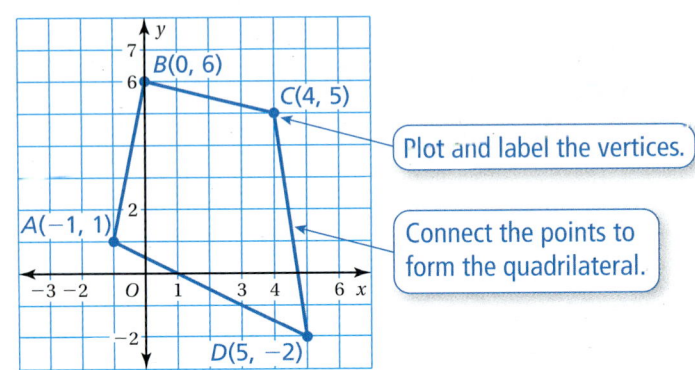

After you plot the vertices, connect them in order to draw the polygon.

Plot and label the vertices.

Connect the points to form the quadrilateral.

Try It Draw the polygon with the given vertices in a coordinate plane.

1. $A(0, 0)$, $B(5, 7)$, $C(4, -3)$
2. $W(4, 4)$, $X(7, 4)$, $Y\left(2\frac{1}{2}, -2\right)$, $Z\left(-\frac{1}{2}, -2\right)$

🔑 Key Idea

Finding Distances between Points in a Coordinate Plane

You can find distances between points in a coordinate plane with the same x-coordinates or the same y-coordinates using the absolute values of the coordinates that are different.

When finding distances between points in the same quadrant, be sure to subtract the lesser absolute value from the greater absolute value.

Points in the same quadrant:

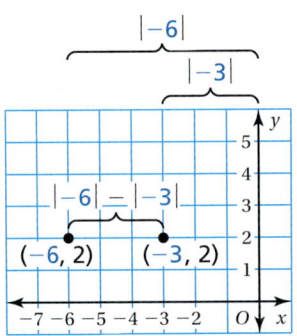

Points in different quadrants:

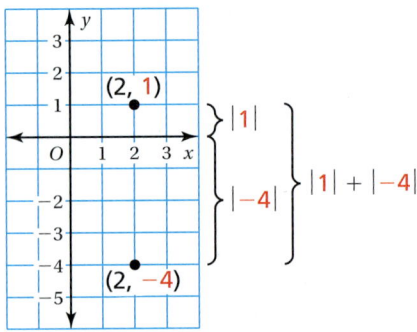

378   Chapter 8   Integers, Number Lines, and the Coordinate Plane

EXAMPLE 2 Finding Distances between Points

a. Find the distance between $(-3, -5)$ and $(2, -5)$.

   Plot the points.

   The points are in different quadrants and have the same y-coordinates. The distance between the points is the sum of the absolute values of the x-coordinates.

   $|-3| + |2| = 3 + 2 = 5$

   ▸ So, the distance between $(-3, -5)$ and $(2, -5)$ is 5.

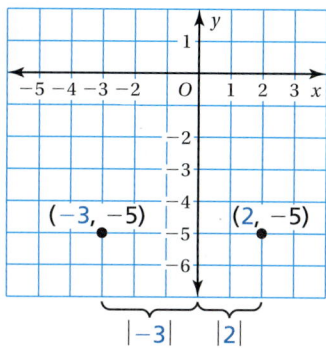

b. Find the distance between $(3, -2.5)$ and $(3, -5)$.

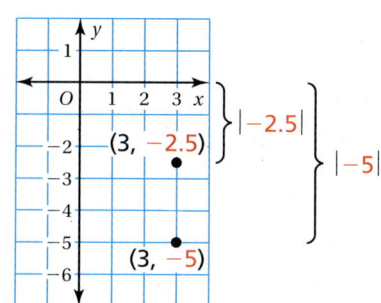

   Plot the points.

   The points are in the same quadrant and have the same x-coordinates. The distance between the points is the difference of the absolute values of the y-coordinates.

   $|-5| - |-2.5| = 5 - 2.5 = 2.5$

   ▸ So, the distance between $(3, -2.5)$ and $(3, -5)$ is 2.5.

Try It Find the distance between the points.

3. $(-6, 6.5), (-2, 6.5)$
4. $(-4, 2), (-4, -5)$

Self-Assessment for Concepts & Skills

Solve each exercise. Then rate your understanding of the success criteria in your journal.

DRAWING A POLYGON Draw the polygon with the given vertices in a coordinate plane.

5. $A(-5, -7), B(-2, 4), C(5, -1)$
6. $D\left(-\dfrac{1}{2}, 6\right), E(3, 1), F\left(0, -4\dfrac{1}{2}\right)$

FINDING DISTANCES Find the distance between the points.

7. $(2, 7), (2, 9)$
8. $(-3, -8), (6, -8)$

9. **WHICH ONE DOESN'T BELONG?** Which pair of points does *not* belong with the other three? Explain your reasoning.

$(-2, 5), (4, 5)$ $(6, -3), (6, 3)$

$(-7, -1), (-7, 4)$ $(2, -1), (-4, -1)$

Section 8.6 Polygons in the Coordinate Plane

EXAMPLE 3   Modeling Real Life

An archaeologist divides an area using a coordinate plane in which the coordinates are measured in meters. The vertices of a secret chamber are $(-8, 10)$, $(4, 10)$, $(4, 2)$, and $(-8, 2)$. Find the perimeter and the area of the secret chamber.

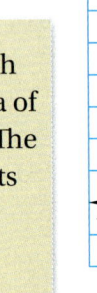

You are given the vertices of a secret chamber. You are asked to find the perimeter and the area of the chamber.

Plot and connect the vertices to draw the polygon that represents the secret chamber. Identify the polygon and find its dimensions. Then find the perimeter and the area of the polygon.

Draw the polygon.

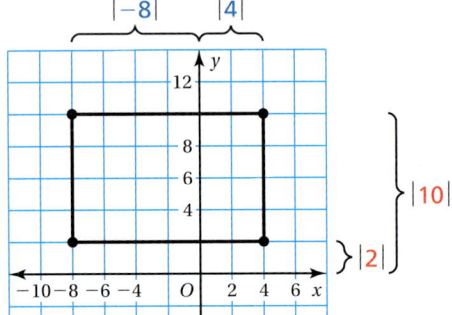

The secret chamber is rectangular. Find the length and the width of the rectangle.

Length: $|-8| + |4| = 8 + 4 = 12$

Width: $|10| - |2| = 10 - 2 = 8$

Another Method Each grid square has an area of $2^2 = 4$ square meters. The secret chamber consists of 24 grid squares. So, the area is $24(4) = 96$ square meters. ✓

The perimeter of the secret chamber is $2(12) + 2(8) = 40$ meters, and the area is $12(8) = 96$ square meters.

Self-Assessment for Problem Solving

Solve each exercise. Then rate your understanding of the success criteria in your journal.

10. A digital map of your hometown is shown in a coordinate plane in which the coordinates are measured in miles. The map shows your house at $(-2, -7)$, your school at $(5, -7)$, and your friend's house at $(-2, 1)$. How far is your house from your school? How far is your house from your friend's house?

11. You design a tree house using a coordinate plane in which the coordinates are measured in feet. The vertices of the floor are $(-2, -3)$, $(-2, 4)$, $(5, 4)$, and $(5, -3)$. Find the perimeter (in yards) and the area (in square yards) of the floor.

8.6 Practice

Review & Refresh

Write an ordered pair corresponding to the point.

1. Point A
2. Point B
3. Point C
4. Point D

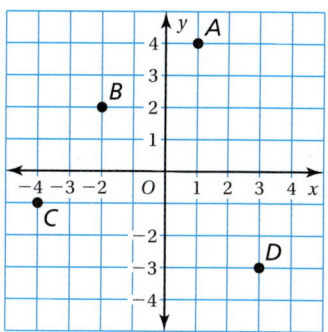

Write the percent as a decimal.

5. 62%
6. 7%
7. 133%
8. 0.45%

9. The tape diagram represents the ratio of the time you spend online to the time your friend spends online. You are online for 6 hours. How many hours does your friend spend online?

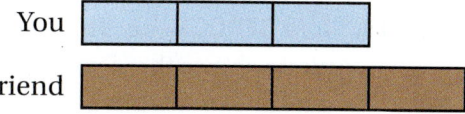

Concepts, Skills, & Problem Solving

STRUCTURE Plot the ordered pairs in a coordinate plane. Then plot another point to form a rectangle. *(See Exploration 1, p. 377.)*

10. $(3, 2), (3, 6), (-5, 2)$
11. $(-4, 7), (-1, 7), (-4, -2)$

DRAWING A POLYGON Draw the polygon with the given vertices in a coordinate plane.

12. $A(4, 7), B(6, 2), C(0, 0)$
13. $D\left(\frac{1}{2}, 2\right), E(-5, 5), F(-4, 1)$
14. $G\left(1\frac{1}{2}, 4\right), H\left(1\frac{1}{2}, -8\right), J(5, -8), K(5, 4)$
15. $L(-3, 2), M(-3, 5), N(2, 2), P(2, -1)$
16. $Q(0, 4), R(-3, 8), S(-7, 4), T(-1, -2), U(7, -2)$
17. $V(-4, -2), W\left(-3, 3\frac{1}{2}\right), X\left(2, 3\frac{1}{2}\right), Y(4, 0), Z(1, -4)$

18. **YOU BE THE TEACHER** Your friend draws a triangle with vertices $A(3, -1), B(4, 3),$ and $C(-1, 2)$. Is your friend correct? Explain your reasoning.

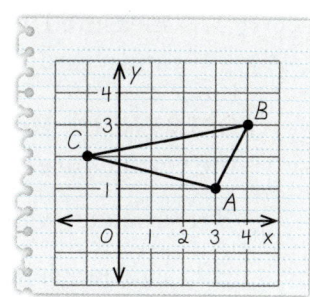

FINDING DISTANCES Find the distance between the points.

19. $(4, 6), (9, 6)$
20. $(5, 10), (5, 4)$
21. $(3, 0), (3, -2)$
22. $(5, -2), (-6, -2)$
23. $(-1, 12), (-1, -3)$
24. $(-7, 8), (7, 8)$
25. $(-6, 5), (-6, -3.5)$
26. $(-2.5, 3), (5, 3)$
27. $(4.5, -1.5), (4.5, 7.25)$

Section 8.6 Polygons in the Coordinate Plane 381

GEOMETRY Find the perimeter and the area of the polygon with the given vertices.

28. $C(1, 1), D(1, 4), E(4, 4), F(4, 1)$

29. $J(-1, -2), K(-6, -2), L(-6, -8), M(-1, -8)$

30. $N(-4, 2), P(5, 2), Q(5, 5), R(-4, 5)$

31. $S(-11, -8), T(-11, 0), U(0, 0), V(0, -8)$

32. **MODELING REAL LIFE** The coordinates of several stars drawn in a coordinate plane are $(8, 0), (7, -3), (3, -2.5), (3.5, 0.5), (-1, 3), (-3, 5)$, and $(-7, 6)$. Plot the locations of the stars. Draw a constellation by connecting the points.

33. **MP STRUCTURE** The coordinate plane shows three vertices of a parallelogram. Find two possible points that could represent the fourth vertex.

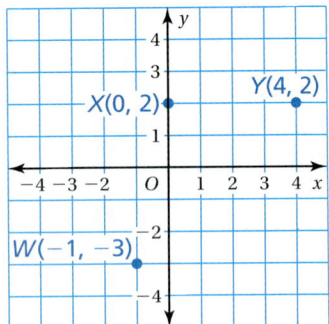

34. **MP PROBLEM SOLVING** Polygon JKLMNP represents a bus route. Each grid square represents 9 square miles. What is the shortest distance, in miles, from Station P to Station L using the bus route? Explain.

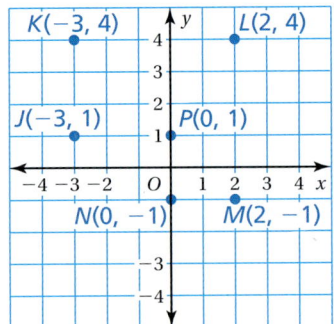

35. **MODELING REAL LIFE** In a topographical map of a city, the vertices of the city limits are $A(-7, 3), B(1, 3), C(1, -4), D(-3, -1.5)$, and $E(-7, -1.5)$. The coordinates are measured in miles. What is the area of the city?

36. **DIG DEEPER!** A map shows that the vertices of a backyard are $W(-100, -70), X(-100, 0), Y(0, 0)$, and $Z(-60, -70)$. The coordinates are measured in feet. The line segment XZ separates the backyard into a lawn and a garden. How many times larger is the lawn than the garden?

OPEN-ENDED Draw a polygon with the given conditions in a coordinate plane where the vertices are not all in the same quadrant.

37. a square with a perimeter of 20 units

38. a rectangle with a perimeter of 18 units

39. a rectangle with an area of 24 units^2

40. a triangle with an area of 15 units^2

41. **MP PRECISION** The vertices of a rectangle are $(1, 0), (1, a), (5, a)$, and $(5, 0)$. The vertices of a parallelogram are $(1, 0), (2, b), (6, b)$, and $(5, 0)$. The values of a and b are both positive and $a > b$. Which polygon has a greater area? Explain.

8.7 Writing and Graphing Inequalities

Learning Target: Write inequalities and represent solutions of inequalities on number lines.

Success Criteria:
- I can write word sentences as inequalities.
- I can determine whether a value is a solution of an inequality.
- I can graph the solutions of inequalities.

EXPLORATION 1

Understanding Inequality Statements

Work with a partner. Create a number line on the floor with both positive and negative integers.

a. For each statement, stand at a number on your number line that makes the statement true. On what other numbers can you stand?

- Class starts more than 3 minutes late.

- You need at least 3 peaches for a recipe.

- The temperature is at most 3 degrees Celsius.

- After playing a video game for 5 minutes, you have fewer than 3 points.

Math Practice

State the Meaning of Symbols

You know the inequality symbols < and >. What do the symbols ≤ and ≥ mean?

b. How can you represent the solutions of each statement in part (a) on a number line?

Section 8.7 Writing and Graphing Inequalities 383

8.7 Lesson

Key Vocabulary
inequality, *p. 384*
solution of an inequality, *p. 385*
solution set, *p. 385*
graph of an inequality, *p. 386*

An **inequality** is a mathematical sentence that compares expressions. It contains the symbols $<$, $>$, $\leq$, or $\geq$. To write a word sentence as an inequality, look for the following phrases to determine where to place the inequality symbol.

Inequality Symbols

Symbol	$<$	$>$	$\leq$	$\geq$
Key Phrases	• is less than • is fewer than	• is greater than • is more than	• is less than or equal to • is at most • is no more than	• is greater than or equal to • is at least • is no less than

EXAMPLE 1 Writing Inequalities

Write each word sentence as an inequality.

a. A number c is less than -4.

A number c | is less than | -4.
c $\quad$ $<$ $\quad$ -4

▸ An inequality is $c < -4$.

b. A number k plus 5 is greater than or equal to 8.

A number k plus 5 | is greater than or equal to | 8.
$k + 5$ $\quad$ $\geq$ $\quad$ 8

▸ An inequality is $k + 5 \geq 8$.

c. Four times a number q is at most 16.

Four times a number q | is at most | 16.
$4q$ $\quad$ $\leq$ $\quad$ 16

▸ An inequality is $4q \leq 16$.

Try It Write the word sentence as an inequality.

1. A number n is greater than 1.
2. Twice a number p is fewer than 7.
3. A number w minus 3 is less than or equal to 10.
4. A number z divided by 2 is at least -6.

384 Chapter 8 Integers, Number Lines, and the Coordinate Plane

A **solution of an inequality** is a value that makes the inequality true. An inequality can have more than one solution. The set of all solutions of an inequality is called the **solution set**.

Value of x	x + 3 ≤ 7	Is the inequality true?
3	$3 + 3 \stackrel{?}{\leq} 7$ $6 \leq 7$ ✓	yes
4	$4 + 3 \stackrel{?}{\leq} 7$ $7 \leq 7$ ✓	yes
5	$5 + 3 \stackrel{?}{\leq} 7$ $8 \not\leq 7$ ✗	no

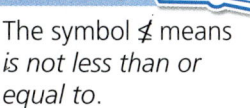

Reading

The symbol $\not\leq$ means *is not less than or equal to*.

EXAMPLE 2 Checking Solutions

Tell whether the given value is a solution of the inequality.

a. $x + 1 > 7$; $x = 8$

$x + 1 > 7$ — Write the inequality.
$8 + 1 \stackrel{?}{>} 7$ — Substitute 8 for x.
$9 > 7$ ✓ — Add. 9 is greater than 7.

▸ So, 8 is a solution of the inequality.

b. $7y < 27$; $y = 4$

$7y < 27$ — Write the inequality.
$7(4) \stackrel{?}{<} 27$ — Substitute 4 for y.
$28 \not< 27$ ✗ — Multiply. 28 is *not* less than 27.

▸ So, 4 is *not* a solution of the inequality.

c. $5 \geq \dfrac{z}{3}$; $z = 15$

$5 \geq \dfrac{z}{3}$ — Write the inequality.
$5 \stackrel{?}{\geq} \dfrac{15}{3}$ — Substitute 15 for z.
$5 \geq 5$ ✓ — Divide. 5 is greater than or equal to 5.

▸ So, 15 is a solution of the inequality.

Try It Tell whether 3 is a solution of the inequality.

5. $b + 4 < 6$ **6.** $9 - n \geq 6$ **7.** $10 \leq 18 \div x$

The **graph of an inequality** shows all the solutions of the inequality on a number line. An open circle, ○, is used when a number is *not* a solution. A closed circle, ●, is used when a number is a solution. An arrow to the left or right shows that the graph continues in that direction.

EXAMPLE 3 Graphing an Inequality

Graph $g > 2$.

Reading

The inequality $g > 2$ is the same as $2 < g$.

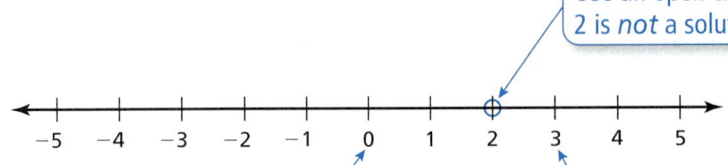

Use an open circle because 2 is *not* a solution.

Test a number to the left of 2. $g = 0$ is *not* a solution.

Test a number to the right of 2. $g = 3$ is a solution.

Shade the number line on the side where you found the solution. Every number on the shaded arrow is a solution of the inequality. So, there are *infinitely many* solutions.

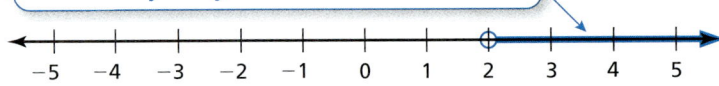

Try It Graph the inequality on a number line.

8. $a < 4$ **9.** $f \leq 7$ **10.** $n > 0$ **11.** $-3 \leq p$

Self-Assessment for Concepts & Skills

Solve each exercise. Then rate your understanding of the success criteria in your journal.

12. DIFFERENT WORDS, SAME QUESTION Which is different? Write "both" inequalities.

| A number n is at most 3. | A number n is no more than 3. |

| A number n is less than or equal to 3. | A number n is at least 3. |

CHECKING SOLUTIONS Tell whether the given value is a solution of the inequality.

13. $p + 5 \leq 12; p = 6$ **14.** $w - 12 < 4; w = 16$

GRAPHING AN INEQUALITY Graph the inequality on a number line.

15. $n > 8$ **16.** $q \leq -4$ **17.** $5 < s$

EXAMPLE 4   Modeling Real Life

The NASA *Solar Probe Plus* can withstand temperatures up to and including 2600°F. Write and graph an inequality that represents the temperatures the probe can withstand.

Words   temperatures   up to and including   2600°F

Variable   Let t be the temperatures (in degrees Fahrenheit) that the probe can withstand.

Inequality t $\leq$ 2600

▶ An inequality is $t \leq 2600$.

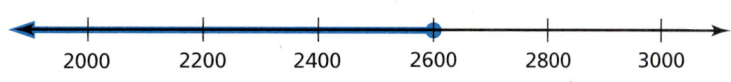

Check The graph shows that 2000°F is a solution. Check this in the inequality.

$$t \leq 2600$$
$$2000 \leq 2600 \checkmark$$

 **Self-Assessment** for Problem Solving

Solve each exercise. Then rate your understanding of the success criteria in your journal.

18. To obtain a babysitting license, you still need to train for at least 6 hours and 45 minutes. Describe the amounts of time that you can train and still not obtain a license. Use a graph to justify your answer.

19. **DIG DEEPER!** The farthest away a drone can fly is 200 meters. A building is 380 meters tall. You control the drone from a floor that is halfway to the top of the building. Can the drone reach the top of the building? Explain your reasoning.

20. Each visit to a water park costs $19.95. An annual pass to the park costs $89.95. Write an inequality that represents the numbers of times you would need to visit the park for the pass to be a better deal.

Section 8.7   Writing and Graphing Inequalities   387

8.7 Practice

Go to *BigIdeasMath.com* to get HELP with solving the exercises.

▶ Review & Refresh

Find the distance between the points.

1. (2, 8), (6, 8)
2. (−5, 9), (7, 9)
3. (−3, 6), (−3, −2)

Solve the equation. Check your solution.

4. $x + 3 = 12$
5. $x - 6 = 8$
6. $\dfrac{t}{12} = 4$
7. $8x = 72$

8. A stack of boards is 24 inches high. The thickness of each board is $\dfrac{3}{8}$ inch. How many boards are in the stack?

   A. $\dfrac{1}{9}$ B. $\dfrac{1}{6}$ C. 9   D. 64

Find the area of the parallelogram.

9.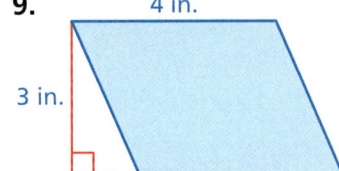
4 in.
3 in.

10.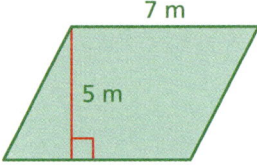
7 m
5 m

11.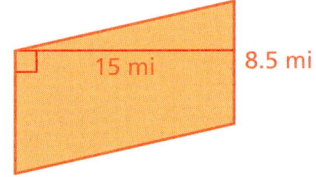
15 mi
8.5 mi

▶ Concepts, Skills, & Problem Solving

UNDERSTANDING INEQUALITY STATEMENTS Choose a number that makes the statement true. What other numbers make the statement true? *(See Exploration 1, p. 383.)*

12. You are less than 3 miles from home.
13. You need at least $5 for lunch.
14. You buy more than 2 movie tickets.
15. A game lasts no more than 10 minutes.

WRITING INEQUALITIES Write the word sentence as an inequality.

16. A number k is less than 10.
17. A number a is more than 6.
18. A number z is fewer than $\dfrac{3}{4}$.
19. A number b is at least −3.
20. One plus a number y is no more than −13.
21. A number x divided by 3 is at most 5.

CHECKING SOLUTIONS Tell whether the given value is a solution of the inequality.

22. $x - 1 \leq 7$; $x = 6$
23. $y + 5 < 13$; $y = 17$
24. $3z > 6$; $z = 3$
25. $6 \leq \dfrac{b}{2}$; $b = 10$
26. $c + 2.5 < 4.3$; $c = 1.8$
27. $a \leq 0$; $a = -5$

388 Chapter 8 Integers, Number Lines, and the Coordinate Plane

MATCHING Match the inequality with its graph.

28. $x \geq -2$ 29. $x < -2$ 30. $x > -2$ 31. $x \leq -2$

A.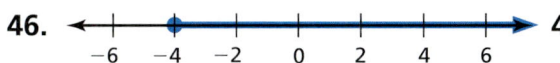

B.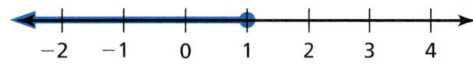

C. (number line with closed circle at -2, shaded left)

D.

GRAPHING AN INEQUALITY Graph the inequality on a number line.

32. $a > 4$ 33. $n \geq 8$ 34. $3 \geq x$ 35. $y < \dfrac{1}{2}$

36. $x < \dfrac{2}{9}$ 37. $-3 \geq c$ 38. $m > -5$ 39. $0 \leq b$

40. $1.5 > f$ 41. $t \geq -\dfrac{1}{2}$ 42. $p > -1.6$ 43. $\dfrac{7}{3} \geq z$

OPEN-ENDED Write an inequality and a word sentence that represent the graph.

44. (number line from -3 to 3, open circle at 1, shaded left)

45. (number line from -2 to 4, closed circle at 1, shaded left)

46.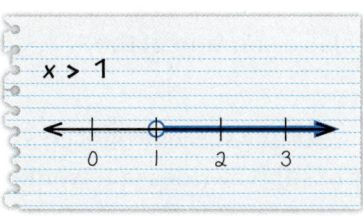

47.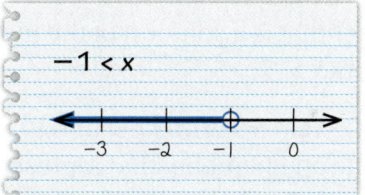

YOU BE THE TEACHER Your friend graphs the inequality. Is your friend correct? Explain your reasoning.

48. $x > 1$
(number line 0 to 3, open circle at 1, shaded right)

49. $-1 < x$
(number line -3 to 0, open circle at -1, shaded left)

50. **MODELING REAL LIFE** The world record for the farthest flight by hoverboard is 2252.4 meters. Write and graph an inequality that represents the distances that would set a new world record.

51. **MODELING REAL LIFE** You are fishing and are allowed to keep at most 3 striped bass. Each striped bass must be no less than 18 inches long.

    a. Write and graph an inequality that represents the numbers of striped bass you are allowed to keep.

    b. Write and graph an inequality that represents the lengths of striped bass you are allowed to keep.

Section 8.7 Writing and Graphing Inequalities 389

52. **MP REASONING** You have $33. You want to buy a necklace and one other item from the list.

    a. Write an inequality that represents the situation.

    b. Can the other item be a T-shirt? Explain.

    c. Can the other item be a book? Explain.

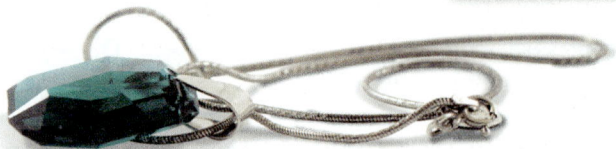

53. **MP LOGIC** For a food to be labeled *low sodium*, there must be no more than 140 milligrams of sodium per serving.

    a. Write and graph an inequality that represents the amount of sodium in a low-sodium serving.

    b. Write and graph an inequality that represents the amount of sodium in a serving that does *not* qualify as low sodium.

    c. Does the food represented by the nutrition facts label qualify as a low-sodium food? Explain.

CRITICAL THINKING Determine whether the statement is *always*, *sometimes*, or *never* true. Explain your reasoning.

54. A number that is a solution of the inequality $x > 5$ is also a solution of the inequality $x \geq 5$.

55. A number that is a solution of the inequality $5 \leq x$ is also a solution of the inequality $x > 5$.

56. **MP PROBLEM SOLVING** A subway ride costs $1.50. A 30-day subway pass costs $36. Write an inequality that represents the numbers of subway rides you would need to take for the pass to be a better deal.

57. **MP PROBLEM SOLVING** Fifty people are seated in a movie theater. The maximum capacity of the theater is 425 people. Write an inequality that represents the numbers of additional people who can be seated.

58. **CRITICAL THINKING** The map shows the elevations above sea level for an area of land.

    a. Graph the possible elevations of *A*. Write the set of elevations as two inequalities.

    b. Graph the possible elevations of *C*. How can you write this set of elevations as a single inequality? Explain.

    c. What is the elevation of *B*? Explain.

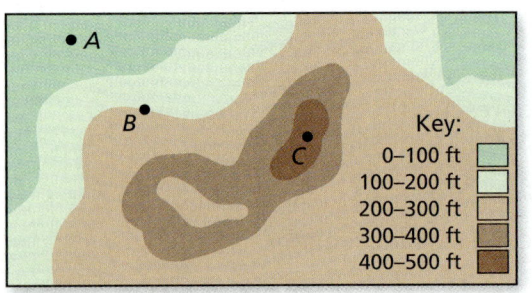

8.8 Solving Inequalities

Learning Target: Write and solve inequalities.

Success Criteria:
- I can apply the properties of inequality to generate equivalent inequalities.
- I can solve inequalities using addition or subtraction.
- I can solve inequalities using multiplication or division.
- I can write and solve inequalities that represent real-life problems.

EXPLORATION 1

Using Tape Diagrams

Work with a partner. In Section 6.2 Exploration 1, the tape diagram below was used to model the equation $x + 4 = 12$.

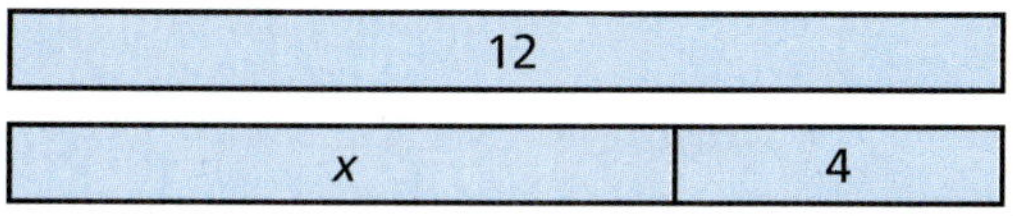

a. Suppose that $x + 4$ is greater than 12. How can you change the equation to represent the new relationship between $x + 4$ and 12?

b. A student finds the possible values of x using the tape diagrams below. What is the solution? How can you find the solution algebraically?

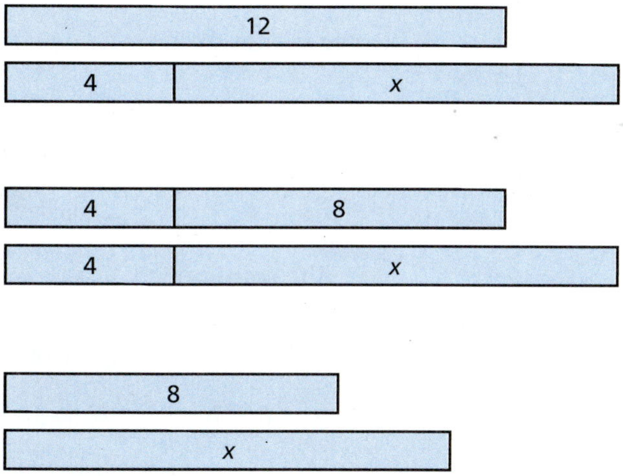

Math Practice

Interpret Results
How is the solution in part (b) different from the solution of an equation?

c. Describe the relationship between $4x$ and 20 as shown by the tape diagram below. What can you conclude about x?

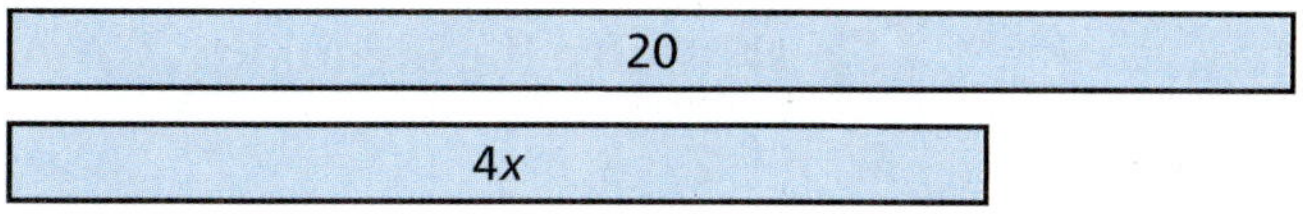

Section 8.8 Solving Inequalities

8.8 Lesson

 **Key Ideas**

Addition Property of Inequality

Words When you add the same number to each side of an inequality, the inequality remains true.

Numbers
$$3 < 5$$
$$+2 \quad +2$$
$$\overline{5 < 7}$$

Algebra
$$x - 4 > 5$$
$$+4 \quad +4$$
$$\overline{x > 9}$$

Subtraction Property of Inequality

Words When you subtract the same number from each side of an inequality, the inequality remains true.

Numbers
$$3 < 5$$
$$-2 \quad -2$$
$$\overline{1 < 3}$$

Algebra
$$x + 4 > 5$$
$$-4 \quad -4$$
$$\overline{x > 1}$$

These properties are also true for ≤ and ≥.

> You can solve inequalities the same way you solve equations. Use inverse operations to get the variable by itself.

EXAMPLE 1   Solving Inequalities Using Addition or Subtraction

a. Solve $x - 3 > 1$. Graph the solution.

$$x - 3 > 1 \quad \text{Write the inequality.}$$
$$+3 \quad +3 \quad \text{Addition Property of Inequality}$$
$$x > 4 \quad \text{Simplify.}$$

(Undo the subtraction.)

▶ The solution is $x > 4$.

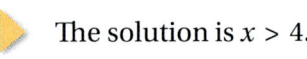

b. Solve $15 \geq 6 + h$. Graph the solution.

$$15 \geq 6 + h \quad \text{Write the inequality.}$$
$$-6 \quad -6 \quad \text{Subtraction Property of Inequality}$$
$$9 \geq h \quad \text{Simplify.}$$

(Undo the addition.)

▶ The solution is $h \leq 9$.

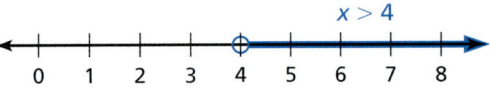

Reading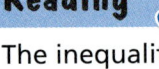
The inequality $h \leq 9$ is the same as $9 \geq h$.

Try It Solve the inequality. Graph the solution.

1. $x - 2 < 3$ **2.** $10 \geq z - 1$ **3.** $y + 2 \geq 17$

Remember
Multiplication and division are inverse operations.

 Key Ideas

Multiplication Property of Inequality

Words When you multiply each side of an inequality by the same *positive* number, the inequality remains true.

Numbers $8 > 6$

$8 \times 2 > 6 \times 2$

$16 > 12$

Algebra $\dfrac{x}{4} < 2$

$\dfrac{x}{4} \cdot 4 < 2 \cdot 4$

$x < 8$

Division Property of Inequality

Words When you divide each side of an inequality by the same *positive* number, the inequality remains true.

Numbers $8 > 6$

$8 \div 2 > 6 \div 2$

$4 > 3$

Algebra $4x < 8$

$\dfrac{4x}{4} < \dfrac{8}{4}$

$x < 2$

These properties are also true for $\leq$ and $\geq$.

EXAMPLE 2 Solving Inequalities Using Multiplication or Division

a. Solve $\dfrac{x}{5} < 2$. Graph the solution.

$\dfrac{x}{5} < 2$ Write the inequality.

Undo the division. → $\dfrac{x}{5} \cdot 5 < 2 \cdot 5$ Multiplication Property of Inequality

$x < 10$ Simplify.

 The solution is $x < 10$.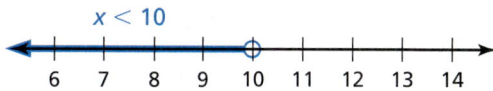

b. Solve $4n \geq 32$. Graph the solution.

$4n \geq 32$ Write the inequality.

Undo the multiplication. → $\dfrac{4n}{4} \geq \dfrac{32}{4}$ Division Property of Inequality

$n \geq 8$ Simplify.

 The solution is $n \geq 8$.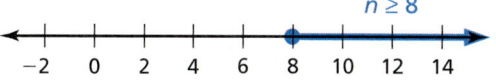

Try It Solve the inequality. Graph the solution.

4. $p \div 3 \geq 2$ **5.** $1 < \dfrac{s}{7}$ **6.** $11k \leq 33$

Section 8.8 Solving Inequalities

EXAMPLE 3 Solving an Inequality Using a Reciprocal

Solve $\dfrac{2x}{3} \leq 4$. Graph the solution.

$\dfrac{2x}{3} \leq 4$ Write the inequality.

$\dfrac{2}{3}x \leq 4$ Rewrite $\dfrac{2x}{3}$ as $\dfrac{2}{3}x$.

Multiply each side by the reciprocal of $\dfrac{2}{3}$. $\rightarrow$ $\dfrac{3}{2} \cdot \dfrac{2}{3}x \leq \dfrac{3}{2} \cdot 4$ Multiplication Property of Inequality

$x \leq 6$ Simplify.

▶ The solution is $x \leq 6$.

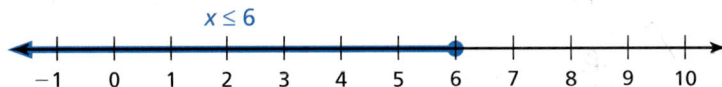

Try It Solve the inequality. Graph the solution.

7. $\dfrac{3}{2}m > 1$ 8. $\dfrac{3}{5}q \leq 6$ 9. $5 > \dfrac{5t}{6}$

Self-Assessment for Concepts & Skills

Solve each exercise. Then rate your understanding of the success criteria in your journal.

SOLVING INEQUALITIES Solve the inequality. Graph the solution.

10. $n + 6 < 10$ 11. $h - 13 \geq 7$

12. $5g > 45$ 13. $\dfrac{3}{4}k \leq 6$

14. **OPEN-ENDED** Write an inequality that the graph represents. Then use the Addition Property of Inequality to write another inequality that the graph represents.

15. **MP REASONING** How is the graph of the solution of $2x \geq 10$ different from the graph of the solution of $2x = 10$?

16. **OPEN-ENDED** Write two inequalities that have the same solution set: one that you can solve using division and one that you can solve using subtraction.

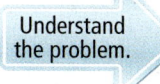 **Modeling Real Life**

A one-way bus ride costs $1.75. A 30-day bus pass costs $42. When is the 30-day pass a better deal?

Understand the problem. You are given the cost of a one-way bus ride and the cost of a 30-day bus pass. You are asked to determine when the pass is a better deal.

Make a plan. For the pass to be a better deal, the total cost of one-way bus rides in a 30-day period must be greater than the cost of the pass. Use a verbal model to write an inequality. Then solve the inequality.

Solve and check.

Verbal Model	Cost of a one-way ride (dollars)	•	Number of one-way rides	>	Cost of a 30-day bus pass (dollars)

Variable Let r be the number of one-way rides.

Inequality $\quad 1.75 \quad • \quad r \quad > \quad 42$

$1.75r > 42$ — Write the inequality.

$\dfrac{1.75r}{1.75} > \dfrac{42}{1.75}$ — Division Property of Inequality

$r > 24$ — Simplify.

▸ So, the 30-day pass is a better deal when you take more than 24 one-way rides in a 30-day period.

Check Verify that 23 one-way rides are less than $42 and 25 one-way rides are greater than $42.

$23(1.75) = \$40.25$ ✓
$25(1.75) = \$43.75$ ✓

 **Self-Assessment** for Problem Solving

Solve each exercise. Then rate your understanding of the success criteria in your journal.

17. A small pizza costs $4.50, and a salad costs $3.75. You plan to buy two small pizzas and four salads. Write and solve an inequality to find the additional amounts you can spend to get free delivery.

18. **DIG DEEPER!** Students at a playground are divided into 5 groups with at least 6 students in each group.

 a. Find the possible numbers of students at the playground.

 b. Suppose the students are divided into 5 *equal* groups. How does this change your answer in part (a)?

Section 8.8 Solving Inequalities

8.8 Practice

> Go to BigIdeasMath.com to get HELP with solving the exercises.

▶ Review & Refresh

Tell whether the given value is a solution of the inequality.

1. $n + 4 > 15$; $n = 9$
2. $s - 12 \leq 8$; $s = 20$
3. $\dfrac{z}{4} \geq 7$; $z = 32$
4. $6g < 48$; $g = 8$

Find the area of the triangle.

5.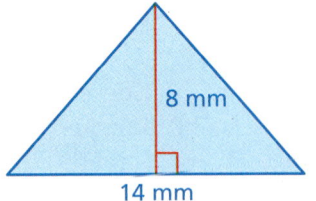
6.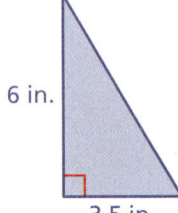
7.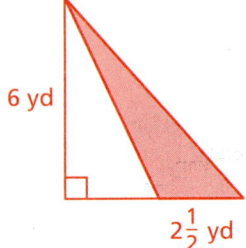

Write the product as a power.

8. 7×7
9. $12 \times 12 \times 12 \times 12$
10. $1.4 \times 1.4 \times 1.4$

▶ Concepts, Skills, & Problem Solving

USING TOOLS Describe the relationship shown by the tape diagram. What can you conclude about x? *(See Exploration 1, p. 391.)*

11.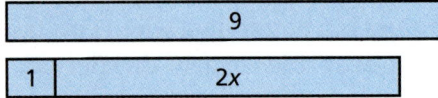
12.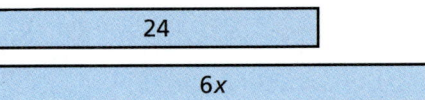

SOLVING INEQUALITIES Solve the inequality. Graph the solution.

13. $x - 4 < 5$
14. $5 + h > 7$
15. $3 \geq y - 2$
16. $y - 21 < 85$

17. $18 > 12 + x$
18. $\dfrac{m}{8} < 4$
19. $n \div 6 > 2$
20. $12x < 96$

21. $\dfrac{1}{11}c \geq 9$
22. $8 \cdot w \leq 72$
23. $7.2 < x + 4.2$
24. $12.7 \geq s - 5.3$

25. $\dfrac{3}{4} \leq \dfrac{1}{2} + n$
26. $7.5p \leq 45$
27. $\dfrac{5}{9}v \leq 45$
28. $\dfrac{5x}{8} \geq 30$

YOU BE THE TEACHER Your friend solves the inequality. Is your friend correct? Explain your reasoning.

29.
```
28 ≥ t - 9
 -9    -9
 19 ≥ t
```

30.
```
 x
 ─ ≤ 30
 6
 x
 ─ · 6 ≤ 30 · 6
 6
     x ≤ 180
```

396 Chapter 8 Integers, Number Lines, and the Coordinate Plane

WRITING INEQUALITIES Write the word sentence as an inequality. Then solve the inequality.

31. Five more than a number *p* is less than 17.

32. Three less than a number *b* is more than 15.

33. Eight times a number *n* is less than 72.

34. A number *t* divided by 32 is at most 4.25.

35. 225 is no less than $\frac{3}{4}$ times a number *w*.

36. MODELING REAL LIFE Your carry-on bag can weigh at most 40 pounds. Write and solve an inequality that represents how much more weight you can add to the bag and still meet the requirement.

37. MODELING REAL LIFE It costs $x for a round-trip bus ticket to the mall. You have $24. Write and solve an inequality that represents how much money you can spend for the bus ticket and still have enough to buy a hat that costs $18.99.

38. GEOMETRY The length of a rectangle is 8 feet, and its area is less than 168 square feet. Write and solve an inequality that represents the possible widths of the rectangle.

39. MODELING REAL LIFE A ticket to a dinosaur exhibit costs $7.50. A one-year pass to the exhibit costs $30. When is the one-year pass a better deal? Explain.

40. MP REASONING A thrill ride at an amusement park holds a maximum of 12 people per ride.

    a. Find the possible numbers of rides needed for 15,000 people.

    b. Is it reasonable for 15,000 people to ride the thrill ride in one day? Explain.

41. OPEN-ENDED Give an example of a real-life situation in which you can list all the solutions of an inequality. Give an example of a real-life situation in which you cannot list all the solutions of an inequality.

Park Hours
10:00 A.M.–10:00 P.M.

42. MP LOGIC Describe the solution of $7x < 7x$. Explain your reasoning.

43. **MP NUMBER SENSE** The possible values of x are given by $x - 3 \geq 2$. What is the least possible value of $5x$?

SOLVING INEQUALITIES Solve the inequality. Graph the solution.

44. $x + 9 - 3 \leq 14$

45. $44 > 7 + s + 26$

46. $6.1 - 0.3 \geq c + 1$

47. $2n < 4.6 \times 12$

48. $32 \geq 2h + 6h$

49. $2\frac{2}{5}b - 1\frac{3}{10}b \leq 6\frac{3}{5}$

50. **MP PROBLEM SOLVING** The high score for a video game is 36,480. Your current score is 34,280. Each dragonfly you catch is worth 1 point. You also get a 1000-point bonus for reaching 35,000 points. Find the possible numbers of dragonflies you can catch to earn a new high score.

51. **MP REASONING** A winning football team more than doubled the offensive yards gained by its opponent. The opponent gained 272 offensive yards. The winning team had 80 offensive plays. Find the possible numbers of yards per play for the winning team. Justify your answer.

52. **DIG DEEPER!** You complete two events of a triathlon. Your goal is to finish with an overall time of less than 100 minutes.

    a. Find the possible numbers of minutes you can take to finish the running event and still meet your goal.

    b. The running event is 3.1 miles long. Estimate how many minutes it would take you to run 3.1 miles. Would this time allow you to reach your goal? Explain your reasoning.

Triathlon	
Event	Your Time (minutes)
Swimming	18.2
Biking	45.4
Running	?

SOLVING INEQUALITIES Graph the numbers that are solutions of both inequalities.

53. $x + 7 > 9$ and $8x \leq 64$

54. $z - 3 \leq 8$ and $6z < 72$

55. $w + 5 \geq 8$ and $4w > 20$

56. $g - 6 \leq 1$ and $3g \geq 21$

57. $2.7 + k \geq 5.3$ and $0.8k \leq 3.36$

58. $p + \frac{3}{4} < 3$ and $\frac{1}{4}p > \frac{3}{8}$

59. **MP PROBLEM SOLVING** You are selling items from a catalog for a school fundraiser. Find the range of sales that will earn you at least $40 and at most $50.

CRITICAL THINKING Let $a > b > 0$ and $x > y > 0$. Tell whether the statement is *always* true. Explain your reasoning.

60. $a + x > b + y$

61. $a - x > b - y$

62. $ax > by$

63. $\frac{a}{x} > \frac{y}{b}$

Connecting Concepts

▶ Using the Problem-Solving Plan

1. You use a coordinate plane to design a kite for a competition. The vertices of the design are $A(0, 0)$, $B(13.5, 9)$, $C(27, 0)$, and $D(13.5, -36)$. The coordinates are measured in inches. Find the least number of square yards of fabric you need to make the kite.

   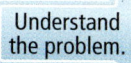
   You know the vertices of your kite design in a coordinate plane, where the coordinates are measured in inches. You are asked to find the least number of square yards of fabric needed to make the kite.

   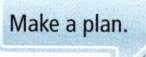
   First, draw a diagram of the design in a coordinate plane. Then decompose the figure into two triangles to find the area of the kite in square inches. Finally, convert the area from square inches to square yards.

   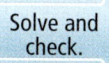
   Use the plan to solve the problem. Then check your solution.

2. You have $240 in a savings account. You deposit $60 per month. The tape diagram represents the ratio of money deposited to money withdrawn each month. Find the monthly change in your account balance. How long will it take for the account to have a balance of $0? Justify your answer.

3. A cord made of synthetic fiber can support 630 pounds, which is at least 450% of the weight that can be supported by a cord made of steel. Graph the possible weights that can be supported by the steel cord.

Performance Task

Launching a CubeSat

At the beginning of this chapter, you watched a STEAM Video called "Designing a CubeSat." You are now ready to complete the performance task related to this video, available at *BigIdeasMath.com*. Be sure to use the problem-solving plan as you work through the performance task.

8 Chapter Review

Review Vocabulary

Write the definition and give an example of each vocabulary term.

positive numbers, *p. 346*
negative numbers, *p. 346*
opposites, *p. 346*
integers, *p. 346*
rational number, *p. 358*
absolute value, *p. 364*
coordinate plane, *p. 370*
origin, *p. 370*
quadrants, *p. 370*
inequality, *p. 384*
solution of an inequality, *p. 385*
solution set, *p. 385*
graph of an inequality, *p. 386*

Graphic Organizers

You can use a **Summary Triangle** to explain a concept. Here is an example of a Summary Triangle for *integers*.

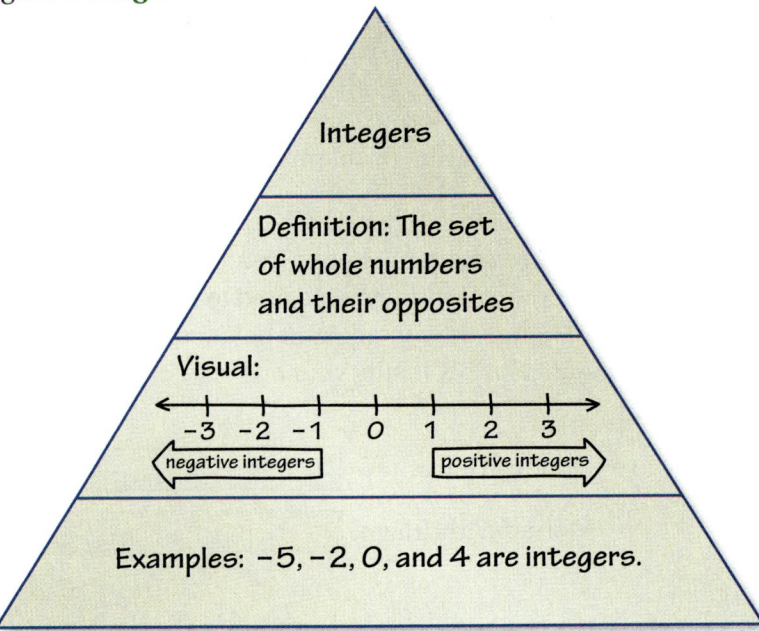

Choose and complete a graphic organizer to help you study the concept.

1. opposites
2. rational number
3. absolute value
4. coordinate plane
5. inequalities
6. solving inequalities using addition or subtraction
7. solving inequalities using multiplication or division

"I'm posting my new **Summary Triangle** on my daily blog. Do you think it will get me more hits?"

400 Chapter 8 Integers, Number Lines, and the Coordinate Plane

Chapter Self-Assessment

As you complete the exercises, use the scale below to rate your understanding of the success criteria in your journal.

1	2	3	4
I do not understand.	I can do it with help.	I can do it on my own.	I can teach someone else.

8.1 Integers (pp. 345–350)

Learning Target: Understand the concept of negative numbers and that they are used along with positive numbers to describe quantities.

Write a positive or negative integer that represents the situation.

1. An elevator goes down 8 floors.
2. You earn $12.

Graph the integer and its opposite.

3. -16
4. 13
5. 4
6. -100

Identify the integer represented by the point on the number line.

7. A
8. B
9. C
10. D

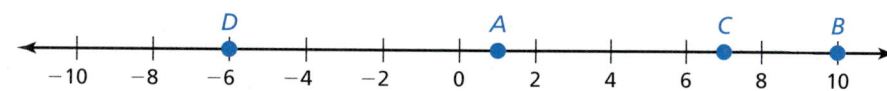

8.2 Comparing and Ordering Integers (pp. 351–356)

Learning Target: Compare and order integers.

Copy and complete the statement using < or >.

11. $4 \;\square\; -7$
12. $-1 \;\square\; 0$
13. $-5 \;\square\; -8$

Order the integers from least to greatest.

14. $-5, 4, 2, -3, -1$
15. $5, -20, -10, 10, 15$
16. $-7, -12, 9, 2, -8$

17. Order the temperatures $-3°C, 8°C, -12°C, -7°C,$ and $0°C$ from coldest to warmest.

18. Your teacher writes five different integers on a note card that are between -10 and 14. When the integers are ordered from least to greatest, the middle number is 1. How many of the integers are positive? negative? Explain.

Chapter Review 401

8.3 Rational Numbers (pp. 357–362)

Learning Target: Compare and order rational numbers.

Graph the number and its opposite.

19. $-\dfrac{2}{5}$
20. $1\dfrac{3}{4}$
21. -1.2
22. 2.75

Copy and complete the statement using < or >.

23. $-2\dfrac{1}{6}\ \square\ -2\dfrac{5}{6}$
24. $-\dfrac{1}{3}\ \square\ -\dfrac{1}{8}$
25. $-3.27\ \square\ -2.68$

Order the numbers from least to greatest.

26. $-2.04,\ -3,\ -2.4,\ -2.19,\ -5.8$
27. $-3\dfrac{7}{8},\ 4,\ -3\dfrac{3}{4},\ \dfrac{1}{2},\ \dfrac{1}{6}$

28. Write a number that is greater than -7.81 and less than -7.

29. A dog buries a bone $-1\dfrac{5}{6}$ inches into the dirt. The dog buries a larger bone $-1\dfrac{3}{4}$ inches into the dirt. Which bone is buried deeper?

8.4 Absolute Value (pp. 363–368)

Learning Target: Understand the concept of absolute value.

Find the absolute value.

30. $|-8|$
31. $|13|$
32. $\left|3\dfrac{6}{7}\right|$
33. $|-1.34|$

Copy and complete the statement using <, >, or =.

34. $|-2|\ \square\ 2$
35. $|4.4|\ \square\ |-2.8|$
36. $\left|\dfrac{1}{6}\right|\ \square\ \left|-\dfrac{2}{9}\right|$

Order the values from least to greatest.

37. $-15,\ |-21|,\ |19|,\ -20,\ 25$
38. $0,\ |-1|,\ -2,\ |2|,\ -3$

39. Simplify $-|-35|$.

40. The latitude of Erie, Pennsylvania, is 42.129. The latitude of Sydney, Australia, is -33.865. Positive values of latitude are north of the equator, negative values of latitude are south of the equator, and the latitude of the equator is 0. Which city is closest to the equator?

8.5 The Coordinate Plane (pp. 369–376)

Learning Target: Plot and reflect ordered pairs in all four quadrants of a coordinate plane.

Write an ordered pair corresponding to the point.

41. Point J

42. Point K

43. Point L

44. Point M

45. Point N

46. Point P

Plot the ordered pair in a coordinate plane. Describe the location of the point.

47. $A(1, 3)$

48. $B(0, -3)$

49. $C(-4, -2)$

50. $D(-3, 1)$

Reflect the point in (a) the x-axis and (b) the y-axis.

51. $(4, 1)$

52. $(-2, 3)$

53. $(2, -5)$

54. $(-3.5, -2.5)$

Reflect the point in the x-axis followed by the y-axis.

55. $(1, 2)$

56. $(-4, 6)$

57. $(3, -4)$

58. $(-3, -3)$

59. Use the map of the town.

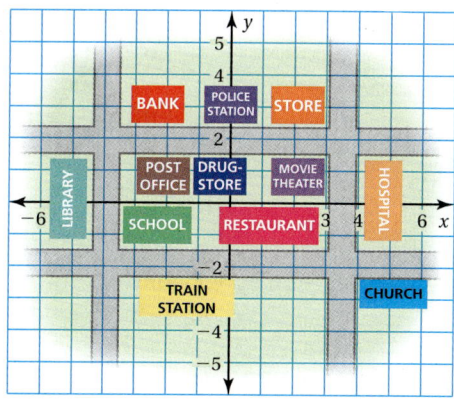

   a. Which building is located at $(-1, 1)$?

   b. Name a building on the positive x-axis.

   c. In which quadrant is the bank located?

   d. Write two different ordered pairs that represent the location of the train station.

   e. You can find the original location of the movie theater by reflecting its location in the y-axis. What building is now in that location?

60. Name the ordered pair that is 5 units right and 2 units down from $(-3, 4)$.

61. A point is reflected in the x-axis. The reflected point is $(3, -9)$. What is the original point?

8.6 Polygons in the Coordinate Plane (pp. 377–382)

Learning Target: Draw polygons in the coordinate plane and find distances between points in the coordinate plane.

Draw the polygon with the given vertices in a coordinate plane.

62. $A(3, 2), B(4, 7), C(6, 0)$

63. $A(1, 2), B(1, -7), C(5, -7), D(8, 2)$

64. $E\left(-1, -3\frac{1}{2}\right), F(1, 0), G(-2, 0), H\left(-4, -3\frac{1}{2}\right)$

Find the distance between the points.

65. $(4, -2), (4, -5)$

66. $(7, 2), (-4, 2)$

67. $(-1, 6), (-1, -3)$

68. $(-5, -8), (-9, -8)$

Find the perimeter and the area of the polygon with the given vertices.

69. $T(2, 7), U(2, 9), V(5, 9), W(5, 7)$

70. $P(4, -3), Q(4, 2), R(9, 2), S(9, -3)$

71. $W(-12, -2), X(-12, 13), Y(5, 13), Z(5, -2)$

72. You design the quilt shown using a coordinate plane in which the coordinates are measured in inches. The vertices of the quilt are $(-3, 5)$, $(-3, -7)$, $(9, 5)$, and $(9, -7)$.

    a. Find the perimeter and the area of the quilt.

    b. The quilt is made of identical-sized square pieces. What is the area of one of the square pieces?

73. Draw a rectangle with a perimeter of 14 units in a coordinate plane where the vertices are in two quadrants.

74. Draw a triangle with an area of 21 square units in a coordinate plane where the vertices are not all in the same quadrant.

8.7 Writing and Graphing Inequalities (pp. 383–390)

Learning Target: Write inequalities and represent solutions of inequalities on a number line.

Write the word sentence as an inequality.

75. A number m is less than 5.

76. Three times a number h is at least -12.

Tell whether $x = 8$ is a solution of the inequality.

77. $\dfrac{x}{2} \geq 3$

78. $13 - x > 5$

79. $19 > 2x$

Graph the inequality on a number line.

80. $x < 0$

81. $a \geq 3$

82. $n \leq -1$

83. The speed limit on a road is 35 miles per hour. Write and graph an inequality that represents the legal speeds on the road.

84. Write an inequality and a word sentence that represent the graph.

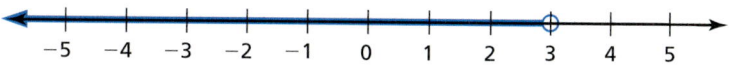

8.8 Solving Inequalities (pp. 391–398)

Learning Target: Write and solve inequalities.

Solve the inequality. Graph the solution.

85. $x + 1 > 3$

86. $y + 8 \geq 9$

87. $k - 7 \leq 0$

88. $9n \geq 63$

89. $24 < 11 + x$

90. $x \div 2 < 4$

91. $4 \leq n - 4$

92. $10p > 40$

93. $s - 1.5 < 2.5$

94. $\dfrac{5}{3}x \leq 10$

95. $\dfrac{1}{4} + m \leq \dfrac{1}{2}$

96. $\dfrac{3}{11}k < 15$

97. Write two inequalities that have the same solution set and can be solved using different operations.

98. You have $15 to spend on a ticket to a movie and snacks. Find the possible amounts you can spend on snacks.

99. You want to use a square section of your yard for a chicken pen. You have at most 52 feet of fencing to form the pen. Find the possible lengths of each side of the chicken pen.

Chapter Review 405

8 Practice Test

Order the values from least to greatest.

1. $0, -2, 3, 1, -4$
2. $-8, |-3|, |5|, 4, -5$
3. $-2.46, -2.5, -2, 1, -2.293$

Graph the number and its opposite.

4. 23
5. $-1\frac{1}{3}$

Find the absolute value.

6. $|7|$
7. $|-11|$

Copy and complete the statement using <, >, or =.

8. $-\frac{2}{3} \ \boxed{} \ -\frac{3}{5}$
9. $2.5 \ \boxed{} \ |2.5|$

Plot the ordered pair in a coordinate plane. Describe the location of the point.

10. $J(4, 0)$
11. $L(1.5, -3.5)$
12. $M(-2, -3)$

13. Reflect $(-5, 1)$ in (a) the x-axis, (b) the y-axis, and (c) the x-axis followed by the y-axis.

Graph the inequality on a number line.

14. $x \geq 5$
15. $m \leq -2$

Solve the inequality. Graph the solution.

16. $x - 3 < 7$
17. $12 \geq n + 6$
18. $\frac{4}{3}b \leq 12$
19. $72 < 12p$

20. A hurricane has wind speeds that are greater than or equal to 74 miles per hour. Write an inequality that represents the possible wind speeds during a hurricane.

21. Two vertices of a triangle are $F(1, -4)$ and $G(6, -4)$. Find two possible points that represent the third vertex so that the triangle has an area of 20 square units.

22. The table shows the melting points (in degrees Celsius) of several elements. Compare the melting point of mercury to the melting point of each of the other elements.

Element	Mercury	Radon	Bromine	Cesium	Francium
Melting Point (°C)	-38.83	-71	-7.2	28.5	27

23. A map shows the vertices of a campsite are $(25, 10)$, $(25, -5)$, $(-5, -5)$, and $(-5, 10)$. The vertices of your tent are $(0, -3)$, $(0, 6)$, $(10, 6)$, and $(10, -3)$. The coordinates are measured in feet. What percent of the campsite is *not* covered by your tent?

8 Cumulative Practice

1. What is the value of the expression when $a = 6$, $b = 5$, and $c = 4$?

   $$8a - 3c + 5b$$

   A. 11
   B. 53
   C. 61
   D. 107

2. Point P is plotted in the coordinate plane.

   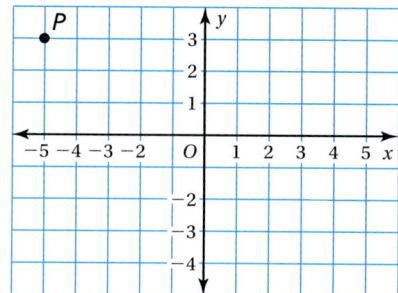

   What are the coordinates of point P?

   F. $(-5, -3)$
   G. $(-5, 3)$
   H. $(-3, -5)$
   I. $(3, -5)$

Test-Taking Strategy
Read All Choices before Answering

"Reading all choices before answering can get you a lot more yummy treats!"

3. What is the value of a that makes the equation true?

   $$a + 6 = 18$$

4. Which list of values is in order from least to greatest?

   A. $2, |-3|, |4|, -6$
   B. $-6, |4|, 2, |-3|$
   C. $-6, |-3|, 2, |4|$
   D. $-6, 2, |-3|, |4|$

5. What is the height of the parallelogram?

   F. 6 meters
   G. 12 meters
   H. 75 meters
   I. 1350 meters

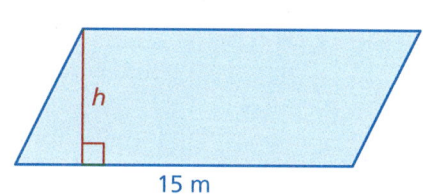

Area = 90 m²
h
15 m

Cumulative Practice 407

6. Which property is illustrated by the statement?

$$4 + (6 + n) = (4 + 6) + n$$

   A. Associative Property of Addition
   B. Commutative Property of Addition
   C. Associative Property of Multiplication
   D. Distributive Property

7. Which number line shows the graph of $x \geq 5$?

   F.

   G.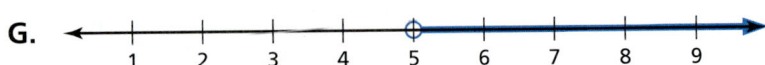

   H.

   I.

8. Which number is the greatest?

   A. $\frac{7}{8}$
   B. 0.86
   C. $\frac{22}{25}$
   D. 85%

9. What is the area of the shaded region?

   F. 23 units^2
   G. 40 units^2
   H. 48 units^2
   I. 60 units^2

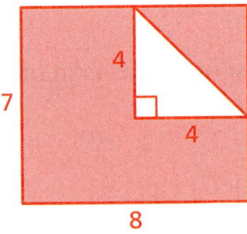

10. Write 23.5% as a decimal.

11. Use grid paper to complete the following.

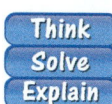

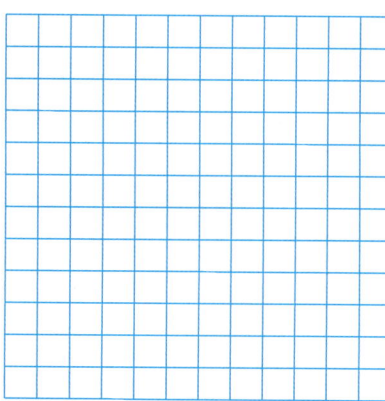

Part A Draw an *x*-axis and a *y*-axis of a coordinate plane. Then plot and label the point $(2, -3)$.

Part B Plot and label *four* points that are 3 units away from $(2, -3)$.

12. What is the perimeter of the rectangle with the vertices shown below?

$$A(-4, -1), B(-4, 7), C(1, 7), D(1, -1)$$

A. 8 units

B. 13 units

C. 26 units

D. 40 units

13. Which net does *not* form a cube?

F.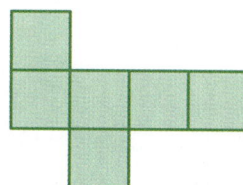

G.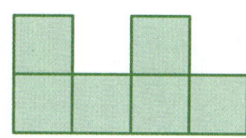

H.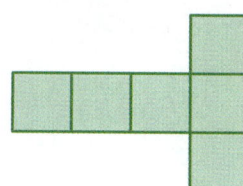

I.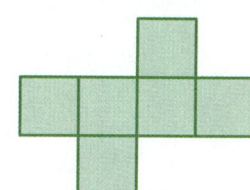

14. Which value of *y* makes the equation true?

$$\frac{3}{4}y = 12$$

A. 9

B. $11\frac{1}{4}$

C. $12\frac{3}{4}$

D. 16

9 Statistical Measures

9.1 Introduction to Statistics
9.2 Mean
9.3 Measures of Center
9.4 Measures of Variation
9.5 Mean Absolute Deviation

Chapter Learning Target:
Understand statistical measures.

Chapter Success Criteria:
- I can construct a data set.
- I can explain how a data set can be interpreted.
- I can find and interpret the measures of center and the measures of variation for a data set.
- I can compare the measures of center and the measures of variation for data sets.

STEAM Video: "Daylight in the Big City"

STEAM Video

Daylight in the Big City

Averages can be used to compare different sets of data. How can you use averages to compare the amounts of daylight in different cities? Can you think of any other real-life situations where averages are useful?

Watch the STEAM Video "Daylight in the Big City." Then answer the following questions.

1. Why do different cities have different amounts of daylight throughout the year?

2. Robert's table includes the difference of the greatest amount of daylight and the least amount of daylight in Lagos, Nigeria, and in Moscow, Russia.

   Lagos: 44 minutes

   Moscow: 633 minutes

   Use these values to make a prediction about the difference of the greatest amount of daylight and the least amount of daylight in a city in Alaska.

Performance Task

Which Measure of Center Is Best: Mean, Median, or Mode?

After completing this chapter, you will be able to use the concepts you learned to answer the questions in the *STEAM Video Performance Task*. You will be given the greatest and least amounts of daylight in the 15 cities in the United States with the greatest populations.

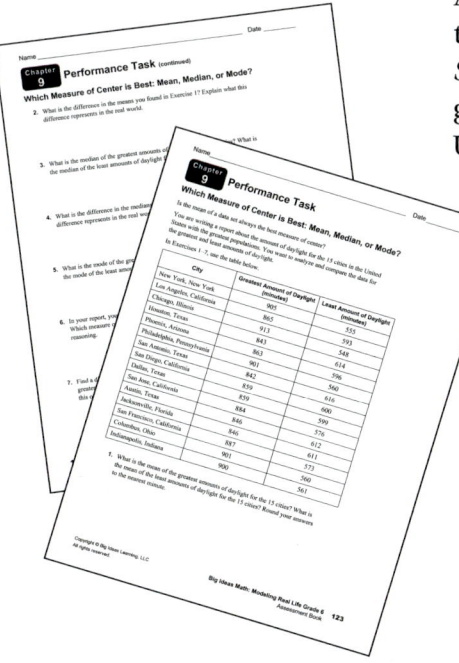

	Greatest	Least
New York:	905 minutes	555 minutes
Los Angeles:	865 minutes	593 minutes
Chicago:	913 minutes	548 minutes

You will determine which measure of center best represents the data. Why might someone be interested in the amounts of daylight throughout the year in a city?

Getting Ready for Chapter

Chapter Exploration

Work with a partner. Write the number of letters in each of your first names on the board.

1. Write all of the numbers on a piece of paper. The collection of numbers is called *data*.

2. Talk with your partner about how you can organize the data. What conclusions can you make about the numbers of letters in the first names of the students in your class?

3. Draw a grid like the one shown below. Then use the grid to draw a graph of the data.

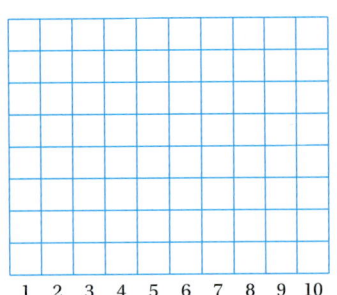

4. **THE CENTER OF THE DATA** Use the graph of the data in Exercise 3 to answer the following.

   a. Is there one number that occurs more than any of the other numbers? If so, write a sentence that interprets this number in the context of your class.

   b. Complete the sentence, "In my class, the average number of letters in a student's first name is _____." Justify your reasoning.

   c. Organize your data using a different type of graph. Describe the advantages or disadvantages of this graph.

Vocabulary

The following vocabulary terms are defined in this chapter. Think about what each term might mean and record your thoughts.

statistical question	measure of center	measure of variation
mean	median	range

9.1 Introduction to Statistics

Learning Target: Identify statistical questions and use data to answer statistical questions.

Success Criteria:
- I can recognize questions that anticipate a variety of answers.
- I can construct and interpret a dot plot.
- I can use data to answer a statistical question.

EXPLORATION 1
Using Data to Answer a Question

Work with a partner.

a. Use your pulse to find your heart rate in beats per minute.

Places to check your pulse:
- on your wrist
- inside your elbow
- on the side of your neck
- on top of your foot

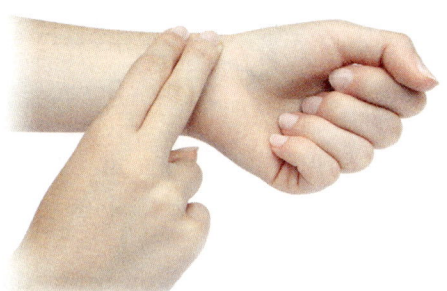

b. Collect the recorded heart rates of the students in your class, including yourself. How spread out are the data? Use a diagram to justify your answer.

c. **MP REASONING** How would you answer the following question by using only one value? Explain your reasoning.

"What is the heart rate of a sixth-grade student?"

EXPLORATION 2
Identifying Types of Questions

Work with a partner.

a. Answer each question on your own. Then compare your answers with your partner's answers. For which questions should your answers be the same? For which questions might your answers be different?

1. How many states are in the United States?
2. How much does a movie ticket cost?
3. What color fur do bears have?
4. How tall is your math teacher?

b. **CONJECTURE** Some of the questions in part (a) are considered *statistical* questions. Which ones are they? Explain.

Math Practice

Build Arguments
How can comparing your answers help you support your conjecture?

Section 9.1   Introduction to Statistics   413

9.1 Lesson

Key Vocabulary
statistics, p. 414
statistical question, p. 414

Statistics is the science of collecting, organizing, analyzing, and interpreting data. A **statistical question** is one for which you do not expect to get a single answer. Instead, you expect a variety of answers, and you are interested in the *distribution* and *tendency* of those answers.

EXAMPLE 1 Identifying Statistical Questions

Determine whether the question is a statistical question. Explain.

a. **How many countries start with the letter Z?**

   ▸ Because there is only one answer, it is not a statistical question.

b. **How much do bags of pretzels cost at the grocery store?**

   ▸ Because you can anticipate that the prices will vary, it is a statistical question. The table at the right may represent prices of several bags of pretzels at a grocery store.

Prices	
$0.99	$2.99
$1.99	$2.99
$1.99	$4.29

c. **How many days does your school have off for spring break this year?**

   ▸ Because there is only one answer, it is not a statistical question.

d. **What are the hair colors of students in your class?**

   ▸ Because you can anticipate that the colors will vary, it is a statistical question. The table below may represent hair colors of several students in a class.

Hair Colors	
Brown	Red
Black	Blonde
Blonde	Pink
Red	Brown

Try It Determine whether the question is a statistical question. Explain.

1. What types of cell phones do students have in your class?
2. How many desks are in your classroom?
3. How much do virtual-reality headsets cost?
4. How many minutes are in your lunch period?

A *dot plot* uses a number line to show the number of times each value in a data set occurs. Dot plots show the *spread* and the *distribution* of a data set.

EXAMPLE 2 Using a Dot Plot

You record the high temperature every day while at summer camp in August. Then you create the vertical dot plot.

a. **Find and interpret the number of data values on the dot plot.**

▸ There are 28 data values on the dot plot. So, you were at camp 28 days, or 4 weeks.

b. **How can you collect these data? What are the units?**

▸ You can collect these data with a thermometer. The units are degrees Fahrenheit (°F).

c. **Write a statistical question that you can answer using the dot plot. Then answer the question.**

One possible statistical question is,

"What is the daily high temperature in August?"

▸ The high temperatures are spread out with about half of the temperatures around 81°F and half of the temperatures around 86°F.

Try It

5. Repeat parts (a)–(c) using the dot plot below that shows the times of students in a 100-meter race.

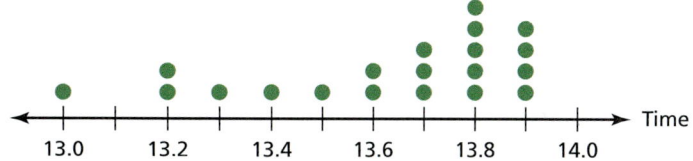

Self-Assessment for Concepts & Skills

Solve each exercise. Then rate your understanding of the success criteria in your journal.

6. **VOCABULARY** What is a statistical question? Give an example and a non-example.

7. **OPEN-ENDED** Write and answer a statistical question using the dot plot. Then find and interpret the number of data values.

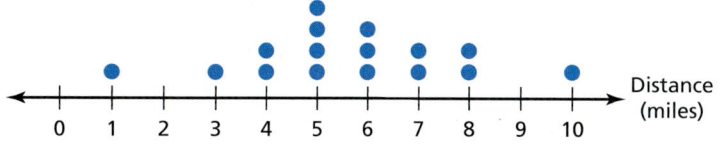

EXAMPLE 3 Modeling Real Life

Your teacher asks you, "What is the mass of a typical mouse?" You conduct a science experiment on house mice. Use the data in the table to answer the question.

Masses (grams)			
20	19	21	20
18	20	27	21
28	23	20	19
20	21	18	27
19	22	21	20

Understand the problem. You know the masses of several mice. You are asked to use the data to answer a statistical question.

Make a plan. Display the data in a dot plot. Identify any clusters, peaks, or gaps in the data. Then use the distribution of the data to answer the question.

Solve and check. Draw a number line that includes the least value, 18, and the greatest value, 28. Then place a dot above the number line for each data value.

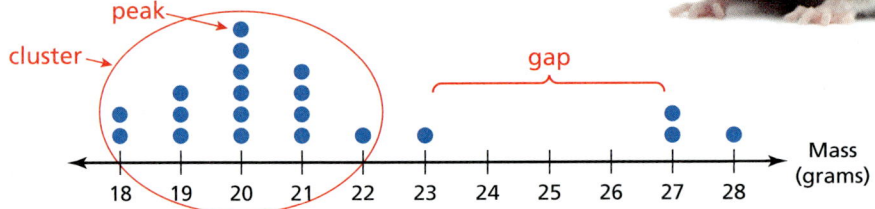

Check Reasonableness
65% of the mice have a mass of 19, 20, or 21 grams. So, it is reasonable to say that a typical mouse has a mass of about 20 grams. ✓

Most of the data are clustered around 20. There is a peak at 20 and a gap between 23 and 27.

▸ A typical mouse has a mass of about 20 grams.

Self-Assessment for Problem Solving

Solve each exercise. Then rate your understanding of the success criteria in your journal.

8. You record the amount of snowfall each day for several days. Then you create the dot plot.

  a. Find and interpret the number of data values on the dot plot.

  b. How can you collect these data? What are the units?

  c. Write a statistical question that you can answer using the dot plot. Then answer the question.

9. You conduct a survey to answer, "How many hours does a typical sixth-grade student spend exercising during a week?" Use the data in the table to answer the question.

Hours of Exercise				
5	1	5	3	5
4	5	2	5	4
3	4	6	5	6

9.1 Practice

Go to *BigIdeasMath.com* to get HELP with solving the exercises.

▶ Review & Refresh

Solve the inequality. Graph the solution.

1. $x - 16 > 8$
2. $p + 6 \leq 8$
3. $54 > 6k$
4. $\dfrac{m}{12} \geq 3$

Tell whether the ordered pair is a solution of the equation.

5. $y = 4x$; (2, 8)
6. $y = 3x + 5$; (3, 15)
7. $y = 6x - 15$; (4, 9)

8. A point is reflected in the *x*-axis. The reflected point is (4, −3). What is the original point?

   A. (−3, 4)   **B.** (−4, 3)   **C.** (−4, −3)   **D.** (4, 3)

Order the numbers from least to greatest.

9. $24\%, \dfrac{1}{4}, 0.2, \dfrac{7}{20}, 0.32$

10. $\dfrac{7}{8}, 85\%, 0.88, \dfrac{3}{4}, 78\%$

▶ Concepts, Skills, & Problem Solving

IDENTIFYING TYPES OF QUESTIONS Answer the question. Tell whether your answer should be the same as your classmates'. *(See Exploration 2, p. 413.)*

11. How many inches are in 1 foot?

12. How many pets do you have?

13. On what day of the month were you born?

14. How many senators are in Congress?

IDENTIFYING STATISTICAL QUESTIONS Determine whether the question is a statistical question. Explain.

15. What are the eye colors of sixth-grade students?

16. At what temperature (in degrees Fahrenheit) does water freeze?

17. How many pages are in the favorite books of students your age?

18. How many hours do sixth-grade students use the Internet each week?

19. **MODELING REAL LIFE** The vertical dot plot shows the heights of the players on a recent NBA championship team.

    a. Find and interpret the number of data values on the dot plot.

    b. How can you collect these data? What are the units?

    c. Write a statistical question that you can answer using the dot plot. Then answer the question.

Section 9.1 Introduction to Statistics 417

20. **MODELING REAL LIFE** The dot plot shows the lengths of earthworms.

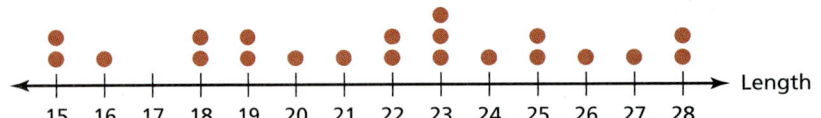

   a. Find and interpret the number of data values on the dot plot.

   b. How can you collect these data? What are the units?

   c. Write a statistical question that you can answer using the dot plot. Then answer the question.

DESCRIBING DATA Display the data in a dot plot. Identify any clusters, peaks, or gaps in the data.

21.

Camper Registrations				
21	25	25	22	21
23	24	26	25	16
24	26	22	25	22

22.

Test Scores				
85	80	83	90	88
82	83	81	80	89
89	84	86	87	83

INTERPRETING DATA The dot plot shows the speeds of cars in a traffic study. Estimate the speed limit. Explain your reasoning.

23.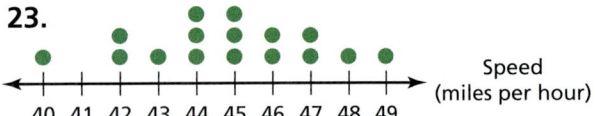

24.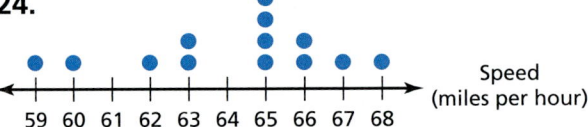

25. **DIG DEEPER!** You conduct a survey to answer, "How many hours does a sixth-grade student spend on homework during a school night?" The table shows the results.

Hours of Homework			
2	4	3	2
1	2	2	1
2	3	5	2

   a. Is this a statistical question? Explain.

   b. Identify any clusters, peaks, or gaps in the data.

   c. Use the distribution of the data to answer the question.

RESEARCH Use the Internet to research and identify the method of measurement and the units used when collecting data about the topic.

26. wind speed

27. amount of rainfall

28. earthquake intensity

29. **MP REASONING** Write a question about letters in the English alphabet that is *not* a statistical question. Then write a question about letters that is a statistical question. Explain your reasoning.

30. **MP REASONING** A bar graph shows the favorite colors of 30 people. Does it make sense to describe clusters in the data? peaks? gaps? Explain.

9.2 Mean

Learning Target: Find and interpret the mean of a data set.

Success Criteria:
- I can explain how the mean summarizes a data set with a single number.
- I can find the mean of a data set.
- I can use the mean of a data set to answer a statistical question.

EXPLORATION 1
Finding a Balance Point

Work with a partner. The diagrams show the numbers of tokens brought to a batting cage. Where on the number line is the data set *balanced*? Is this a good representation of the average? Explain.

a.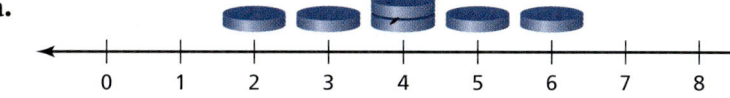

b.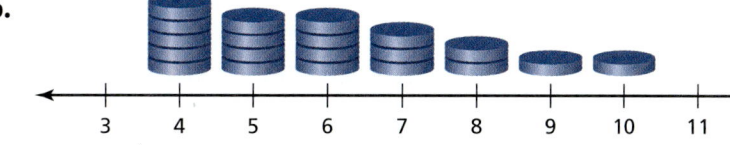

EXPLORATION 2
Finding a Fair Share

Work with a partner. One token lets you hit 12 baseballs in a batting cage. The table shows the numbers of tokens six friends bring to the batting cage.

Tokens					
John	Lisa	Miguel	Matt	Cheryl	Anika
6	3	4	5	2	4

Math Practice

Use Clear Definitions
What does it mean for data to have an average? How does this help you answer the question?

a. Regroup the tokens so that everyone has the same amount. How many times can each friend use the batting cage? Explain how this represents a "fair share."

b. How can you find the answer in part (a) algebraically?

c. Write a statistical question that can be answered using the value in part (a).

9.2 Lesson

Key Vocabulary
mean, *p. 420*
outlier, *p. 422*

 Key Idea

Mean

Words The **mean** of a data set is the sum of the data divided by the number of data values. The mean is a type of average.

Numbers Data: 8, 5, 6, 9   Mean: $\dfrac{8+5+6+9}{4} = \dfrac{28}{4} = 7$

4 data values

EXAMPLE 1   Finding the Mean

Text Messages Sent
Mark: 120
Laura: 95
Stacy: 101
Josh: 125
Kevin: 82
Maria: 108
Manny: 90

The table shows the numbers of text messages sent by a group of friends over 1 week. What is the mean number of messages sent?

A. 100   **B.** 102
C. 103   **D.** 104

$$\text{Mean} = \dfrac{120 + 95 + 101 + 125 + 82 + 108 + 90}{7}$$

sum of the data
number of values

$= \dfrac{721}{7}$ Add values in numerator.

$= 103$ Divide.

▶ The mean number of text messages sent is 103. The correct answer is **C**.

Try It Find the mean of the data.

1.
Dog Weights (pounds)
Sparky: 18
Spot: 9
Rover: 60
Newton: 89
Diego: 44
Ruby: 13
Mookie: 54
Fido: 45

2.
Airbag Backpack Costs				
$600	$450	$350	$650	$800
$300	$550	$500	$600	$750

420   Chapter 9   Statistical Measures   Multi-Language Glossary at *BigIdeasMath.com*

EXAMPLE 2 Comparing Means

The double bar graph shows the monthly rainfall amounts for two cities over a six-month period. Compare the mean monthly rainfalls.

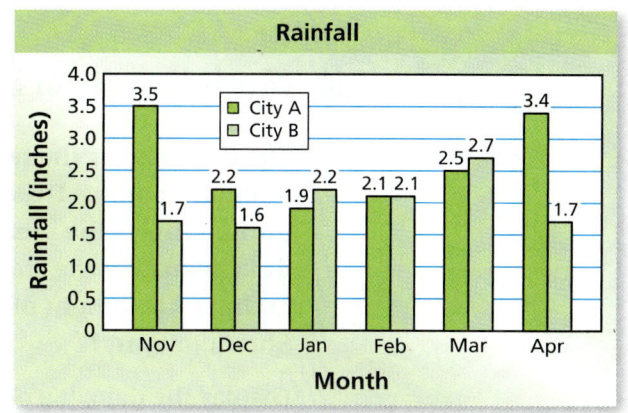

City A mean: $\dfrac{3.5 + 2.2 + 1.9 + 2.1 + 2.5 + 3.4}{6} = \dfrac{15.6}{6} = 2.6$

City B mean: $\dfrac{1.7 + 1.6 + 2.2 + 2.1 + 2.7 + 1.7}{6} = \dfrac{12}{6} = 2$

 Because 2.6 is greater than 2, City A averaged more rainfall.

Try It

3. **WHAT IF?** The monthly rainfall in May was 0.5 inch in City A and 2 inches in City B. Does this affect your answer in Example 2? Explain.

Self-Assessment for Concepts & Skills

Solve each exercise. Then rate your understanding of the success criteria in your journal.

4. **MP NUMBER SENSE** Is the mean always equal to a value in the data set? Explain.

5. **WRITING** Explain why the mean describes a typical value in a data set.

6. **MP NUMBER SENSE** What can you determine when the mean of one data set is greater than the mean of another data set? Explain your reasoning.

7. **COMPARING MEANS** Compare the means of the data sets.

   Data set A: 43, 32, 16, 41, 24, 19, 30, 27

   Data set B: 44, 18, 29, 24, 36, 22, 26, 21

An **outlier** is a data value that is much greater or much less than the other values. When included in a data set, it can affect the mean.

EXAMPLE 3 Modeling Real Life

The table shows the heights of several Shetland ponies. Describe how the outlier affects the mean. Then use the data to answer the statistical question, "What is the height of a typical Shetland Pony?"

Shetland Pony Heights (inches)				
40	37	39	40	42
38	38	37	28	40

Display the data in a dot plot to see the distribution of the data.

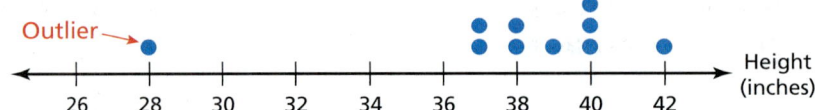

The height of 28 inches is much less than the other heights. So, it is an outlier. Find the mean with and without the outlier.

Mean with outlier:

$$\frac{40 + 37 + 39 + 40 + 42 + 38 + 38 + 37 + 28 + 40}{10} = \frac{379}{10} = 37.9$$

Mean without outlier:

$$\frac{40 + 37 + 39 + 40 + 42 + 38 + 38 + 37 + 40}{9} = \frac{351}{9} = 39$$

▶ With the outlier, the mean is less than all but three of the heights. Without the outlier, the mean better represents the heights. So, the height of a typical Shetland pony is about 39 inches.

Self-Assessment for Problem Solving

Solve each exercise. Then rate your understanding of the success criteria in your journal.

8. **DIG DEEPER!** The monthly numbers of customers at a store in the first half of a year are 282, 270, 320, 351, 319, and 252. The monthly numbers of customers in the second half of the year are 211, 185, 192, 216, 168, and 144. Compare the mean monthly customers in the first half of the year with the mean monthly customers in the second half of the year.

9. The table shows tournament finishes for a golfer. What place does the golfer typically finish in tournaments? Explain how you found your answer.

Tournament Finishes			
5	1	2	5
21	2	8	3
1	5	9	10
3	4	8	9

9.2 Practice

Go to BigIdeasMath.com to get HELP with solving the exercises.

Review & Refresh

Determine whether the question is a statistical question. Explain.

1. How tall are sixth-grade students?
2. How many minutes are in 1 year?
3. How many counties are in Tennessee?
4. What is a student's favorite sport?

Write the percent as a fraction or mixed number in simplest form.

5. 84%
6. 71%
7. 353%
8. 0.2%

Divide. Check your answer.

9. $11.7 \div 9$
10. $5\overline{)72.8}$
11. $6.8\overline{)28.56}$
12. $93 \div 3.75$

Concepts, Skills, & Problem Solving

FINDING A FAIR SHARE Regroup the amounts so that each person has the same amount. What is the amount? (See Exploration 2, p. 419.)

13. Dollars brought by friends to a fair: 11, 12, 12, 12, 12, 12, 13
14. Tickets earned by friends playing an arcade game: 0, 0, 0, 1, 1, 2, 3

FINDING THE MEAN Find the mean of the data.

15.
Pets Owned	
Brandon	I
Jill	III
Mark	II
Nicole	IIII
Steve	0

16.
Brothers and Sisters	
Amanda	♀
Eve	♀♀♀♀♀
Joseph	♀♀♀♀
Michael	♀♀

17.
Sit-Ups		
108	85	94
103	112	115
98	119	126
105	82	89

18. Visits to Your Website

(Bar graph: Day 1: 12, Day 2: 16, Day 3: 0, Day 4: 8, Day 5: 31, Day 6: 28, Day 7: 17)

19. **MODELING REAL LIFE** You and your friends are watching a television show. One of your friends asks, "How long are the commercial breaks during this show?"

Break Times (minutes)				
4.2	3.5	4.55	2.75	2.25

a. Is this a statistical question? Explain.
b. Use the mean of the values in the table to answer the question.

Section 9.2 Mean 423

20. **MODELING REAL LIFE** The table shows the monthly rainfall amounts at a measuring station.

Month	Jan	Feb	Mar	Apr	May	Jun	Jul	Aug	Sep	Oct	Nov	Dec
Rainfall (inches)	2.22	1.51	1.86	2.06	3.48	4.47	3.37	5.40	5.45	4.34	2.64	2.14

a. What is the mean monthly rainfall?

b. Compare the mean monthly rainfall for the first half of the year with the mean monthly rainfall for the second half of the year.

21. **OPEN-ENDED** Create two different data sets that have six values and a mean of 21.

22. **MODELING REAL LIFE** The bar graph shows your cell phone data usage for five months. Describe how the outlier affects the mean. Then use the data to answer the statistical question, "How much cell phone data do you use in a month?"

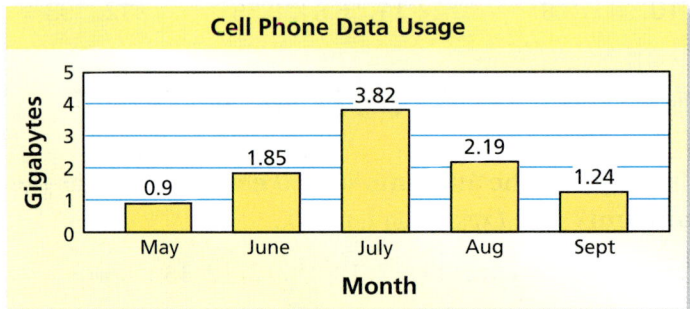

23. **MODELING REAL LIFE** The table shows the heights of the volleyball players on two teams. Compare the mean heights of the two teams. Do outliers affect either mean? Explain.

Player Heights (inches)											
Dolphins	59	65	53	56	58	61	64	68	51	56	54	57
Tigers	63	68	66	58	54	55	61	62	53	70	64	64

24. **REASONING** Use a dot plot to explain why the mean of the data set below is the point where the data set is balanced.

$$11, 13, 17, 15, 12, 18, 12$$

25. **DIG DEEPER!** In your class, 7 students do not receive a weekly allowance, 5 students receive $3, 7 students receive $5, 3 students receive $6, and 2 students receive $8.

a. What is the mean weekly allowance? Explain how you found your answer.

b. A new student who joins your class receives a weekly allowance of $3.50. Without calculating, explain how this affects the mean.

26. **PRECISION** A collection of 8 geodes has a mean weight of 14 ounces. A different collection of 12 geodes has a mean weight of 9 ounces. What is the mean weight of the 20 geodes? Explain how you found your answer.

9.3 Measures of Center

Learning Target: Find and interpret the median and mode of a data set.

Success Criteria:
- I can explain how the median and mode summarize a data set with a single number.
- I can find the median and mode of a data set.
- I can explain how changes to a data set affect the measures of center.
- I can use a measure of center to answer a statistical question.

EXPLORATION 1

Finding the Median

Work with a partner.

a. Write the total numbers of letters in the first and last names of 15 celebrities, historical figures, or people you know. One person is already listed for you.

Person	Number of Letters in First and Last Name
Abraham Lincoln	14

b. Order the values in your data set from least to greatest. Then write the data on a strip of grid paper with 15 boxes.

c. The *middle value* of the data set is called the *median*. The value (or values) that occur most often is called the *mode*. Find the median and the mode of your data set. Explain how you found your answers.

Math Practice

Use a Graph
How can you use a dot plot to find the mode?

d. Why are the median and the mode considered averages of a data set?

Section 9.3   Measures of Center   425

9.3 Lesson

Key Vocabulary 🔊
measure of center, p. 426
median, p. 426
mode, p. 426

A **measure of center** is a measure that describes the typical value of a data set. The mean is one type of measure of center. Here are two others.

Key Ideas

Median

Words Order the data. For a set with an odd number of values, the **median** is the middle value. For a set with an even number of values, the **median** is the mean of the two middle values.

Numbers Data: 5, 8, 9, 12, 14   The median is 9.

Data: 2, 3, 5, 7, 10, 11   The median is $\frac{5+7}{2} = 6$.

Mode

Words The **mode** of a data set is the value or values that occur most often. Data can have one mode, more than one mode, or no mode. When all values occur only once, there is no mode.

Numbers Data: 11, 13, 15, 15, 18, 21, 24, 24   The modes are 15 and 24.

The mode is the only measure of center that you can use to describe a set of data that is not made up of numbers.

EXAMPLE 1 Finding the Median and Mode

Bowling Scores

120	135	160	125	90
205	160	175	105	145

Find the median and mode of the bowling scores.

90, 105, 120, 125, 135, 145, 160, 160, 175, 205   Order the data.

Median: $\frac{135 + 145}{2} = \frac{280}{2} = 140$ Add the two middle values and divide by 2.

Mode: 90, 105, 120, 125, 135, 145, 160, 160, 175, 205

The value 160 occurs most often.

▶ The median is 140. The mode is 160.

Try It Find the median and mode of the data.

1. 20, 4, 17, 8, 12, 9, 5, 20, 13
2. 100, 75, 90, 80, 110, 102

EXAMPLE 2 Finding the Mode

The list shows the favorite types of movies for students in a class. Organize the data in a table. Then find the mode.

Favorite Types of Movies

Comedy	Drama	Horror
Horror	Drama	Horror
Comedy	Comedy	Action
Action	Comedy	Action
Horror	Drama	Comedy
Comedy	Comedy	Horror
Horror	Comedy	Action
Horror	Action	Drama

Type	Number of Students
Action	5
Comedy	8
Drama	4
Horror	7

Comedy received the most votes.

 So, the mode is comedy.

Try It

3. One member of the class was absent and ends up voting for horror. Does this change the mode? Explain.

EXAMPLE 3 Removing an Outlier

Seven competitors eat 26, 33, 34, 2, 32, 34, and 42 bugs in a bug-eating competition. Find the mean, median, and mode of the data with and without the outlier. Which measure does the outlier affect the most?

The competitor with 2 bugs ate much less than any other competitor. So, the outlier is 2.

	Mean	Median	Mode
With Outlier	29	33	34
Without Outlier	33.5	33.5	34

Removing the outlier increases the mean more than the median. The mode is not affected.

 So, the mean is affected the most by the outlier.

Try It

4. The times (in minutes) it takes six students to travel to school are 8, 10, 10, 15, 20, and 45. Find the mean, median, and mode of the data with and without the outlier. Which measure does the outlier affect the most?

EXAMPLE 4 Changing the Values of a Data Set

The prices of six video games at an online store are shown in the table. The price of each game increases by $4.98 when a shipping charge is included. How does this increase affect the mean, median, and mode?

Video Game Prices

$53.42	$35.69
$18.99	$25.13
$27.97	$53.42

Make a new table by adding $4.98 to each price. Then find the mean, median, and mode of both data sets.

Video Game Prices with Shipping Charge

$58.40	$40.67
$23.97	$30.11
$32.95	$58.40

	Mean	Median	Mode
Original Price	35.77	31.83	53.42
Price with Shipping Charge	40.75	36.81	58.4

Compare the measures of center of both data sets.

Mean: $40.75 - 35.77 = 4.98$

Median: $36.81 - 31.83 = 4.98$

Mode: $58.4 - 53.42 = 4.98$

▸ By increasing each video game price by $4.98 for shipping, the mean, median, and mode all increase by $4.98.

Try It

5. **WHAT IF?** The store decreases the price of each video game by $3. How does this decrease affect the mean, median, and mode?

Self-Assessment for Concepts & Skills

Solve each exercise. Then rate your understanding of the success criteria in your journal.

6. **FINDING MEASURES OF CENTER** Consider the data set below.

    15, 18, 13, 11, 12, 21, 9, 11

    a. Find the mean, median, and mode of the data.

    b. Each value in the data set is decreased by 7. How does this change affect the mean, median, and mode?

7. **WRITING** Explain why a typical value in a data set can be described by the median or the mode.

428 Chapter 9 Statistical Measures

EXAMPLE 5 Modeling Real Life

Use the data to answer the statistical question, "How much do climbing shoes cost?"

Find the mean, median, and mode of the data. Then answer the question by creating a dot plot to determine which measure best represents the data.

Mean: $\dfrac{40 + 51 + 142 + 68 + 57 + 40 + 65 + 85}{8} = \dfrac{548}{8} = 68.5$

Median: 40, 40, 51, 57, 65, 68, 85, 142   Order the data.

$$\dfrac{57 + 65}{2} = \dfrac{122}{2} = 61$$

Mode: 40, 40, 51, 57, 65, 68, 85, 142   The value 40 occurs most often.

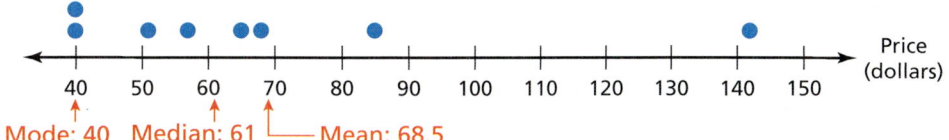

Mode: 40   Median: 61   Mean: 68.5

 The mode is less than most of the data, and the mean is greater than most of the data. The median best represents the data. So, climbing shoes cost about $61.

Self-Assessment for Problem Solving

Solve each exercise. Then rate your understanding of the success criteria in your journal.

8. How does removing the outlier affect your answer in Example 5?

9. It takes 10 contestants on a television show 43, 41, 62, 40, 44, 43, 44, 46, 45, and 41 seconds to cross a canyon on a zipline. Find the mean, median, and mode of the data with and without the outlier. Which measure does the outlier affect the most?

10. The table shows the weights of several great white sharks. Use the data to answer the statistical question, "What is the weight of a great white shark?"

Great White Shark Weights (pounds)			
1700	1500	1700	2100
1600	1700	1900	3900
1700	2200	1800	1600

9.3 Practice

Review & Refresh

Find the mean of the data.

1. 1, 5, 8, 4, 5, 7, 6, 6, 2, 3
2. 9, 12, 11, 11, 10, 7, 4, 8
3. 26, 42, 31, 50, 29, 37, 44, 31
4. 53, 45, 43, 55, 28, 21, 61, 29, 24, 40, 27, 42

5. A shelf in your room can hold at most 30 pounds. There are 12 pounds of books already on the shelf. Which inequality represents the numbers of pounds you can add to the shelf?

   A. $x < 18$ **B.** $x \geq 18$ **C.** $x \leq 42$ **D.** $x \leq 18$

Find the missing values in the ratio table. Then write the equivalent ratios.

6.
Turkey (pounds)	1	2	3
Price (dollars)	5.50		

7.
Guests	5	30	6
Cost (dollars)	100		

Find the surface area of the prism.

8.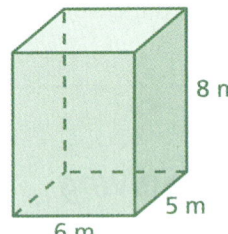

9.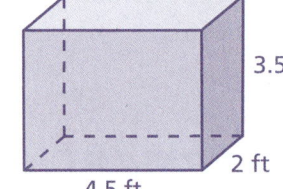

10.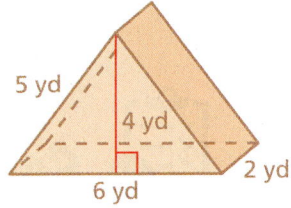

Concepts, Skills, & Problem Solving

FINDING THE MEDIAN Use grid paper to find the median of the data. (See Exploration 1, p. 425.)

11. 9, 7, 2, 4, 3, 5, 9, 6, 8, 0, 3, 8
12. 16, 24, 13, 36, 22, 26, 22, 28, 25

FINDING THE MEDIAN AND MODE Find the median and mode of the data.

13. 3, 5, 7, 9, 11, 3, 8
14. 14, 19, 16, 13, 16, 14
15. 93, 81, 94, 71, 89, 92, 94, 99
16. 44, 13, 36, 52, 19, 27, 33
17. 12, 33, 18, 28, 29, 12, 17, 4, 2
18. 55, 44, 40, 55, 48, 44, 58, 67

19. **YOU BE THE TEACHER** Your friend finds the median of the data. Is your friend correct? Explain your reasoning.

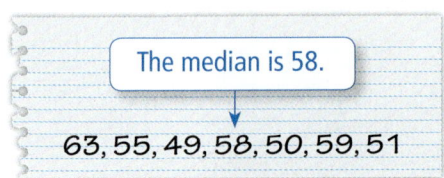

The median is 58.

63, 55, 49, 58, 50, 59, 51

430 Chapter 9 Statistical Measures

FINDING THE MODE Find the mode of the data.

20.
Shirt Colors

Black	Blue	Red
Pink	Black	Black
Gray	Green	Blue
Blue	Blue	Red
Yellow	Blue	Blue
Black	Orange	Black
Black		

21.
Talent Show Acts

Singing	Dancing	Comedy
Singing	Singing	Dancing
Juggling	Dancing	Singing
Singing	Poetry	Dancing
Comedy	Magic	Dancing
Poetry	Singing	Singing

22. **MP REASONING** In Exercises 20 and 21, can you find the mean and median of the data? Explain.

FINDING MEASURES OF CENTER Find the mean, median, and mode of the data.

23. 4.7, 8.51, 6.5, 7.42, 9.64, 7.2, 9.3

24. $8\frac{1}{2}, 6\frac{5}{8}, 3\frac{1}{8}, 5\frac{3}{4}, 6\frac{5}{8}, 5\frac{1}{4}, 10\frac{5}{8}, 4\frac{1}{2}$

25. **MODELING REAL LIFE** The weights (in ounces) of several moon rocks are shown in the table. Find the mean, median, and mode of the weights.

Moon Rock Weights (ounces)		
2.2	2.2	3.2
2.4	2.8	3.4
2.6	3.0	2.5

REMOVING AN OUTLIER Find the mean, median, and mode of the data with and without the outlier. Which measure does the outlier affect the most?

26. 45, 52, 17, 63, 57, 42, 54, 58

27. 85, 77, 211, 88, 91, 84, 85

28. 23, 73, 45, 27, 23, 25, 43, 45

29. 101, 110, 99, 100, 64, 112, 110, 111, 102

30. **MP REASONING** The table shows the monthly salaries for employees at a company.

Monthly Salaries (dollars)				
1940	1660	1860	2100	1720
1540	1760	1940	1820	1600

   a. Find the mean, median, and mode of the data.

   b. Each employee receives a 5% raise. Find the mean, median, and mode of the data with the raise. How does this increase affect the mean, median, and mode of the data?

   c. How are the mean, median, and mode of the monthly salaries related to the mean, median, and mode of the annual salaries?

CHOOSING A MEASURE OF CENTER Find the mean, median, and mode of the data. Choose the measure that best represents the data. Explain your reasoning.

31. 48, 12, 11, 45, 48, 48, 43, 32

32. 12, 13, 40, 95, 88, 7, 95

33. 2, 8, 10, 12, 56, 9, 5, 2, 4

34. 126, 62, 144, 81, 144, 103

35. **MODELING REAL LIFE** The weather forecast for a week is shown. Which measure of center best represents the high temperatures? the low temperatures? Explain your reasoning.

	Sun	Mon	Tue	Wed	Thu	Fri	Sat
High	90°F	91°F	89°F	97°F	101°F	99°F	91°F
Low	74°F	78°F	77°F	77°F	83°F	78°F	72°F

36. **RESEARCH** Find the costs of 10 different boxes of cereal. Choose one cereal whose cost will be an outlier.

    a. Which measure of center does the outlier affect the most? Justify your answer.

    b. Use the data to answer the statistical question, "How much does a box of cereal cost?"

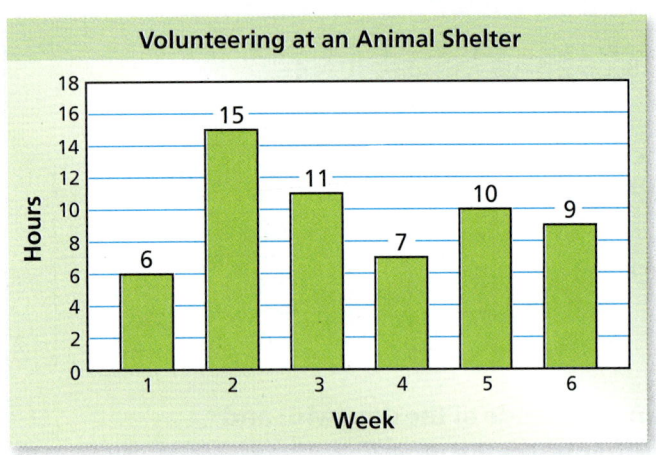

37. **PROBLEM SOLVING** The bar graph shows the numbers of hours you volunteered at an animal shelter. What is the minimum number of hours you need to volunteer in the seventh week to justify that you volunteered an average of 10 hours per week for the 7 weeks? Explain your answer using measures of center.

38. **REASONING** Why is the mode the least frequently used measure of center to describe a data set? Explain.

39. **DIG DEEPER!** The data are the prices of several fitness wristbands at a store.

    $130 $170 $230 $130
    $250 $275 $130 $185

    a. Does the price shown in the advertisement represent the prices well? Explain.

    b. Why might the store use this advertisement?

    c. In this situation, why might a person want to know the mean? the median? the mode? Explain.

40. **CRITICAL THINKING** The expressions $3x$, $9x$, $4x$, $23x$, $6x$, and $3x$ form a data set. Assume $x > 0$.

    a. Find the mean, median, and mode of the data.

    b. Is there an outlier? If so, what is it?

9.4 Measures of Variation

Learning Target: Find and interpret the range and interquartile range of a data set.

Success Criteria:
- I can explain how the range and interquartile range describe the variability of a data set with a single number.
- I can find the range and interquartile range of a data set.
- I can use the interquartile range to identify outliers.

EXPLORATION 1

Interpreting Statements

Work with a partner. There are 24 students in your class. Your teacher makes the following statements.

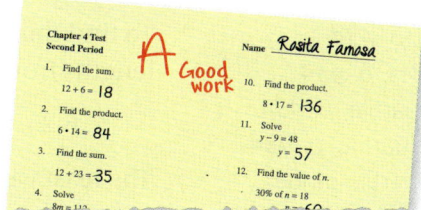

Math Practice

Analyze Givens
How can you use the given information to determine how spread out the data are?

- "The exam scores range from 75% to 96%."
- "Most of the students received high scores."

a. What does each statement mean? Explain.

b. Use your teacher's statements to make a dot plot that can represent the distribution of the exam scores of the class.

c. Compare your dot plot with other groups'. How are they alike? different?

EXPLORATION 2

Grouping Data

Work with a partner. The numbers of U.S. states visited by students in a sixth-grade class are shown.

Numbers of States Visited							
1	7	5	2	4	18	1	6
11	6	3	20	2	7	1	8
10	2	12	5	3	21		

a. Represent the data using a dot plot. Between what values do the data range?

b. Use the dot plot to make observations about the data.

c. How can you describe the *middle half* of the data?

Section 9.4 Measures of Variation 433

9.4 Lesson

A **measure of variation** is a measure that describes the distribution of a data set. A simple measure of variation to find is the *range*. The **range** of a data set is the difference of the greatest value and the least value.

EXAMPLE 1 Finding the Range

Key Vocabulary
measure of variation, *p. 434*
range, *p. 434*
quartiles, *p. 434*
first quartile, *p. 434*
third quartile, *p. 434*
interquartile range, *p. 434*

The table shows the lengths of several Burmese pythons captured for a study. Find and interpret the range of the lengths.

To find the least and the greatest values, order the lengths from least to greatest.

5, 6.25, 8, 10, 11, 12.5, 14, 15.5, 16.25, 18.5

The least value is 5. The greatest value is 18.5.

Lengths (feet)	
18.5	8
11	10
14	15.5
12.5	6.25
16.25	5

▶ So, the range of the lengths is 18.5 − 5 = 13.5 feet. This means that the lengths vary by no more than 13.5 feet.

Try It

1. The ages of people in line for a roller coaster are 15, 17, 21, 32, 41, 30, 25, 52, 16, 39, 11, and 24. Find and interpret the range of the ages.

Key Ideas

Quartiles

The **quartiles** of a data set divide the data into four equal parts. Recall that the median (second quartile) divides the data set into two halves.

Reading
The first quartile can also be called the *lower quartile*. The third quartile can also be called the *upper quartile*.

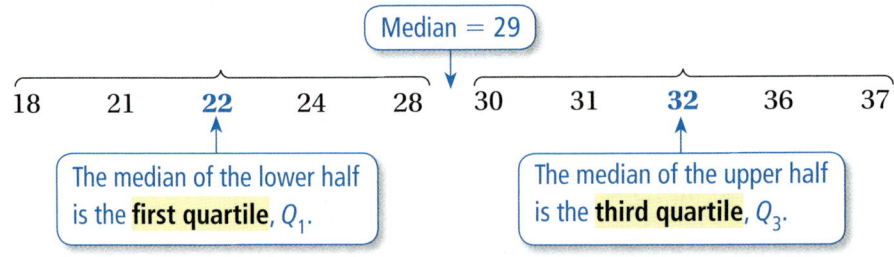

Median = 29

18   21   **22**   24   28   30   31   **32**   36   37

The median of the lower half is the **first quartile**, Q_1.

The median of the upper half is the **third quartile**, Q_3.

Interquartile Range (IQR)

The difference of the third quartile and the first quartile is called the **interquartile range**. The IQR represents the range of the middle half of the data and is another measure of variation.

18   21   **22**   24   28   30   31   **32**   36   37

$$IQR = Q_3 - Q_1$$
$$= 32 - 22$$
$$= 10$$

◀)) Multi-Language Glossary at *BigIdeasMath.com*

EXAMPLE 2 Finding the Interquartile Range

The dot plot shows the top speeds of 12 sports cars. Find and interpret the interquartile range of the data.

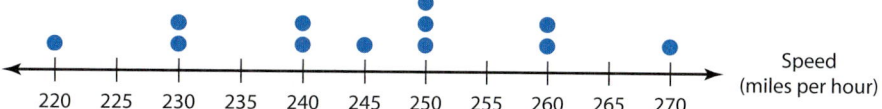

Order the speeds from slowest to fastest. Find the quartiles.

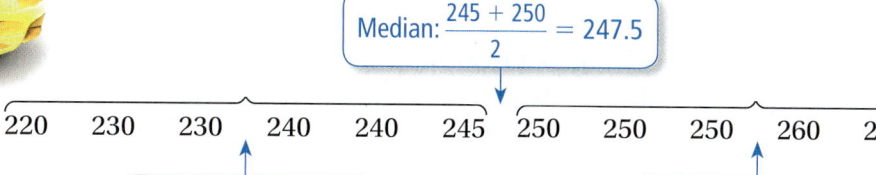

Median: $\dfrac{245 + 250}{2} = 247.5$

$Q_1: \dfrac{230 + 240}{2} = 235$

$Q_3: \dfrac{250 + 260}{2} = 255$

▶ So, the interquartile range is $255 - 235 = 20$. This means that the middle half of the speeds vary by no more than 20 miles per hour.

Try It

2. The data are the number of pages in each of an author's novels. Find and interpret the interquartile range of the data.

   356, 364, 390, 468, 400, 382, 376, 396, 350

Self-Assessment for Concepts & Skills

Solve each exercise. Then rate your understanding of the success criteria in your journal.

3. **WRITING** Explain why the variability of a data set can be described by the range or the interquartile range.

4. **DIFFERENT WORDS, SAME QUESTION** Which is different? Find "both" answers.

   53, 47, 60, 45, 62, 59, 65, 50, 56, 48

   - What is the interquartile range of the data shown?
   - What is the range of the data shown?
   - What is the range of the middle half of the data shown?
   - What is $Q_3 - Q_1$ for the data shown?

Section 9.4 Measures of Variation

You can use the quartiles and the interquartile range to check for outliers.

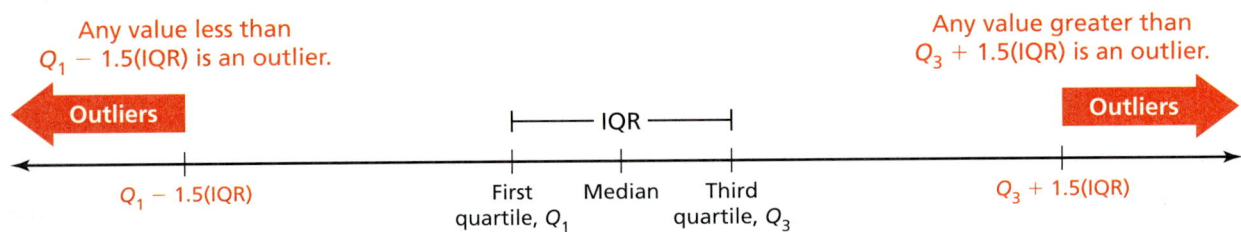

EXAMPLE 3 Modeling Real Life

Bugs Eaten			
26	34	32	42
33	2	34	

Section 9.3 Example 3 identifies 2 as an outlier of the data in the table shown. Use the IQR to determine whether it is the only outlier.

Order the data values from least to greatest. Find the quartiles.

2   26   32   33   34   34   42

$Q_1 = 26$ Median $= 33$ $Q_3 = 34$

The IQR is $34 - 26 = 8$. Use the IQR to find the outlier boundaries.

$Q_1 - 1.5(\text{IQR}) = 26 - 1.5(8)$ $Q_3 + 1.5(\text{IQR}) = 34 + 1.5(8)$
$= 14$ $= 46$

 The only data value less than 14 is 2. There are no data values greater than 46. So, the only outlier is 2.

 Self-Assessment for Problem Solving

Solve each exercise. Then rate your understanding of the success criteria in your journal.

Distances (feet)			
$13\frac{1}{2}$	$21\frac{1}{2}$	21	$16\frac{3}{4}$
$10\frac{1}{4}$	19	32	$26\frac{1}{2}$
29	$16\frac{1}{4}$	$28\frac{1}{2}$	$18\frac{1}{2}$

5. The table shows the distances traveled by a paper airplane. Find and interpret the range and interquartile range of the distances.

6. The table shows the years of teaching experience of math teachers at a school. How do the outlier or outliers affect the variability of the data?

Teaching Experience (years)	5	10	7	8	10
	11	22	8	6	35

9.4 Practice

▶ Review & Refresh

Find the mean, median, and mode of the data.

1. 4, 8, 11, 6, 4, 5, 9, 10, 10, 4
2. 74, 78, 86, 67, 80
3. 15, 18, 17, 17, 15, 16, 14
4. 31, 14, 18, 26, 17, 32

Copy and complete the statement using < or >.

5. 6 ▢ −7
6. −3 ▢ 0
7. 14 ▢ −14
8. 8 ▢ −10

Find the surface area of the pyramid.

9.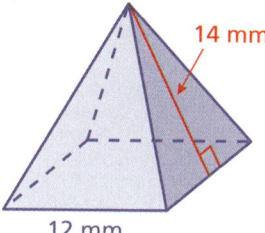

10.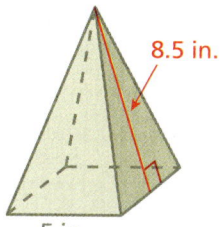

11.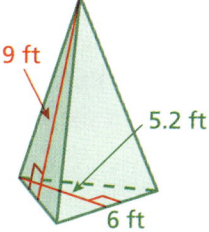

▶ Concepts, Skills, & Problem Solving

INTERPRETING STATEMENTS There are 20 students in your class. Your teacher makes the two statements shown. Use your teacher's statements to make a dot plot that can represent the distribution of the scores of the class. *(See Exploration 1, p. 433.)*

12. "The quiz scores range from 65% to 95%."
    "The scores were evenly spread out."

13. "The project scores range from 78% to 93%."
    "Most of the students received low scores."

FINDING THE RANGE Find the range of the data.

14. 4, 8, 2, 9, 5, 3
15. 28, 42, 36, 23, 14, 47, 40
16. 26, 21, 27, 33, 24, 29
17. 52, 40, 49, 48, 62, 54, 44, 58, 39
18. 133, 117, 152, 127, 168, 146, 174
19. 4.8, 5.5, 4.2, 8.9, 3.4, 7.5, 1.6, 3.8

20. **YOU BE THE TEACHER** Your friend finds the range of the data. Is your friend correct? Explain your reasoning.

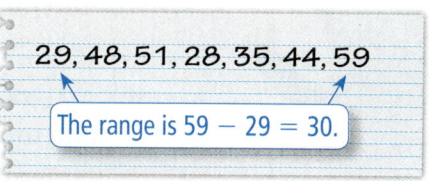

FINDING THE INTERQUARTILE RANGE Find the interquartile range of the data.

21. 4, 6, 4, 2, 9, 1, 12, 7
22. 18, 22, 15, 16, 15, 13, 19, 18
23. 40, 33, 37, 54, 41, 34, 27, 39, 35
24. 84, 75, 90, 87, 99, 91, 85, 88, 76, 92, 94
25. 132, 127, 106, 140, 158, 135, 129, 138
26. 38, 55, 61, 56, 46, 67, 59, 75, 65, 58

27. **MODELING REAL LIFE** The table shows the number of tornadoes in Alabama each year for several years. Find and interpret the range and interquartile range of the data. Then determine whether there are any outliers.

Numbers of Tornadoes			
65	32	54	23
55	145	37	80
94	42	69	77

28. **WRITING** Consider a data set that has no mode. Which measure of variation is greater, the range or the interquartile range? Explain your reasoning.

29. **CRITICAL THINKING** Is it possible for the range of a data set to be equal to the interquartile range? Explain your reasoning.

30. **MP REASONING** How does an outlier affect the range of a data set? Explain.

31. **MODELING REAL LIFE** The table shows the numbers of points scored by players on a sixth-grade basketball team in a season.

Points Scored					
21	53	74	82	84	93
103	108	116	122	193	

   a. Find the range and interquartile range of the data.

   b. Identify the outlier(s) in the data set. Find the range and interquartile range of the data set without the outlier(s). Which measure does the outlier or outliers affect more?

32. **DIG DEEPER!** Two data sets have the same range. Can you assume that the interquartile ranges of the two data sets are about the same? Give an example to justify your answer.

33. **MODELING REAL LIFE** The tables show the ages of the finalists for two reality singing competitions.

   a. Find the mean, median, range, and interquartile range of the ages for each show. Compare the results.

   b. A 21-year-old is voted off Show A, and the 36-year-old is voted off Show B. How do these changes affect the measures in part (a)? Explain.

Ages for Show A		Ages for Show B	
18	17	21	20
15	21	23	13
22	16	15	18
18	28	17	22
24	21	36	25

34. **OPEN-ENDED** Create a set of data with 7 values that has a mean of 30, a median of 26, a range of 50, and an interquartile range of 36.

9.5 Mean Absolute Deviation

Learning Target: Find and interpret the mean absolute deviation of a data set.

Success Criteria:
- I can explain how the mean absolute deviation describes the variability of a data set with a single number.
- I can find the mean absolute deviation of a data set.
- I can compare data sets using the mean absolute deviation to draw conclusions.

The Meaning of a Word ▶ Deviate

When you **deviate** from something, you stray or depart from the normal course of action.

EXPLORATION 1

Finding Distances from the Mean

Work with a partner. The table shows the exam scores of 14 students in your class.

Exam Scores					
Ben	89	Omar	95	Dan	94
Emma	86	Hong	96	Lucy	89
Jeremy	80	Rob	92	Priya	84
Pete	80	Amy	90	Heather	85
Malik	96	Sue	76		

a. Which exam score *deviates* the most from the mean? Which exam score *deviates* the least from the mean? Explain how you found your answers.

b. How far is each data value from the mean?

c. Divide the sum of the values in part (b) by the number of values. In your own words, what does this represent?

d. **MP REASONING** In a data set, what does it mean when the value you found in part (c) is close to 0? Explain.

Math Practice

Use Operations
What operation can you use to find the distance from the mean? Explain.

Section 9.5 Mean Absolute Deviation 439

9.5 Lesson

Key Vocabulary
mean absolute deviation, p. 440

Another measure of variation is the *mean absolute deviation*. The **mean absolute deviation** is an average of how much data values differ from the mean.

🔑 Key Idea

Finding the Mean Absolute Deviation (MAD)

Step 1: Find the mean of the data.

Step 2: Find the distance between each data value and the mean.

Step 3: Find the sum of the distances in Step 2.

Step 4: Divide the sum in Step 3 by the total number of data values.

EXAMPLE 1 Finding the Mean Absolute Deviation

Find and interpret the mean absolute deviation of the data.

$$1, 2, 2, 2, 4, 4, 4, 5$$

Step 1: Mean $= \dfrac{1 + 2 + 2 + 2 + 4 + 4 + 4 + 5}{8} = \dfrac{24}{8} = 3$

Step 2: You can use a dot plot to organize the data. Replace each dot with its distance from the mean.

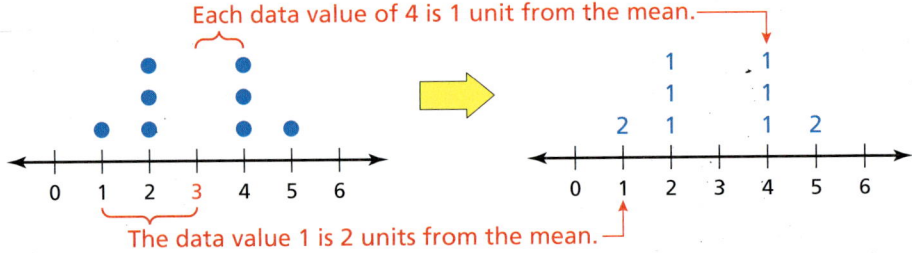

Step 3: The sum of the distances is $2 + 1 + 1 + 1 + 1 + 1 + 1 + 2 = 10$.

Step 4: The mean absolute deviation is $\dfrac{10}{8} = 1.25$.

▸ So, the data values differ from the mean by an average of 1.25.

Try It

1. Find and interpret the mean absolute deviation of the data.

   $$5, 8, 8, 10, 13, 14, 16, 22$$

EXAMPLE 2 **Finding the Mean Absolute Deviation**

The smartphone shows the numbers of runs allowed by a pitcher in his last 10 starts. Find the mean, median, and mean absolute deviation of the data.

Order the runs allowed:

0, 0, 0, 2, 4, 4, 5, 6, 6, 8.

$$\text{Mean} = \frac{35}{10} = 3.5 \qquad \text{Median} = \frac{4+4}{2} = 4$$

Mean absolute deviation:

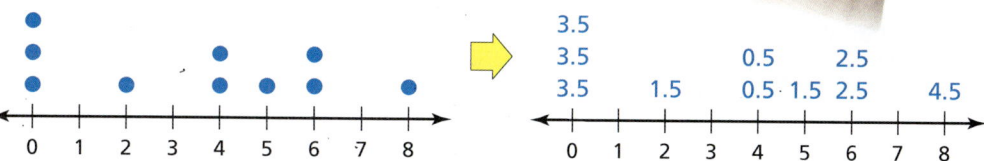

The mean absolute deviation is $\frac{24}{10} = 2.4$.

▶ The mean is 3.5, the median is 4, and the mean absolute deviation is 2.4.

Try It

2. **WHAT IF?** The pitcher allows 4 runs in the next game. How would you expect the mean absolute deviation to change? Explain.

Self-Assessment for Concepts & Skills

Solve each exercise. Then rate your understanding of the success criteria in your journal.

3. **WRITING** Explain why the variability of a data set can be described by the mean absolute deviation.

4. **FINDING THE MEAN ABSOLUTE DEVIATION** Find and interpret the mean absolute deviation of the data.

    8, 12, 4, 3, 14, 1, 9, 13

5. **WHICH ONE DOESN'T BELONG?** Which one does *not* belong with the other three? Explain your reasoning.

range	interquartile range
mean	mean absolute deviation

EXAMPLE 3 Modeling Real Life

Find the mean, median, and mean absolute deviation of the numbers of runs allowed by Pitcher B in his last 10 starts. Which measure can you use to distinguish these data from the data in Example 2? What can you conclude?

Order the runs allowed for Pitcher B: 0, 2, 2, 3, 4, 4, 4, 5, 5, 6.

$$\text{Mean} = \frac{35}{10} = 3.5 \qquad \text{Median} = \frac{4+4}{2} = 4$$

Mean absolute deviation:

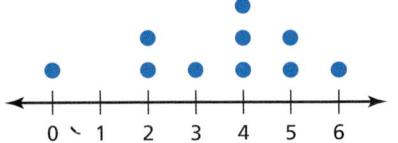 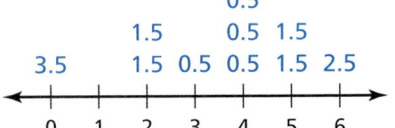

The mean absolute deviation is $\frac{14}{10} = 1.4$.

You cannot use the measures of center to distinguish the data because they are the same for each data set. The measure of variation, MAD, is 2.4 for Pitcher A and 1.4 for Pitcher B. This indicates that the data for Pitcher B has less variation.

The greater the mean absolute deviation, the greater the variation of the data.

▸ Using the MAD to distinguish the data, you can conclude that Pitcher B is more consistent than Pitcher A.

 **Self-Assessment** for Problem Solving

Solve each exercise. Then rate your understanding of the success criteria in your journal.

Tiger Sharks	
Allison	3
Fito	6
Chuck	5
Sumila	4
Lauren	4
Antonio	2

Bear Cats	
Cherie	6
Carlos	1
Dominic	4
Jack	1
Gloria	8
Hannah	4

6. The tables show the numbers of questions answered correctly by members of two teams on a game show. Compare the mean, median, and mean absolute deviation of the numbers of correct answers for each team. What can you conclude?

7. DIG DEEPER! The data set shows the numbers of books that students in your book club read last summer.

8, 6, 11, 12, 14, 12, 11, 6, 15, 9, 7, 10, 9, 13, 5, 8

A new student who read 18 books last summer joins the club. Is 18 an outlier? How does including this value in the data set affect the measures of center and variation? Explain.

9.5 Practice

Review & Refresh

Find the range and interquartile range of the data.

1. 23, 45, 39, 34, 28, 41, 26, 33
2. 63, 53, 48, 61, 69, 63, 57, 72, 46

Graph the integer and its opposite.

3. −15
4. 17
5. 16
6. −22

7. Find the numbers of faces, edges, and vertices of the solid.

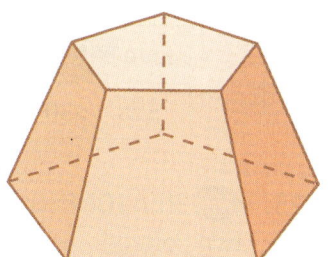

Write the word sentence as an equation.

8. 17 plus a number q is 40.
9. The product of a number s and 14 is 49.
10. The difference of a number b and 9 is 32.
11. The quotient of 36 and a number g is 9.

Concepts, Skills, & Problem Solving

FINDING DISTANCES FROM THE MEAN Find the average distance of each data value in the set from the mean. *(See Exploration 1, p. 439.)*

12. Model years of used cars on a lot: 2014, 2006, 2009, 2011, 2005
13. Prices of kites at a shop: $7, $20, $9, $35, $12, $15, $7, $10, $20, $25

FINDING THE MEAN ABSOLUTE DEVIATION Find and interpret the mean absolute deviation of the data.

14. 69, 51, 71, 77, 71, 80, 75, 63, 73
15. 94, 86, 95, 99, 88, 90
16. 46, 54, 43, 57, 50, 62, 78, 42
17. 25, 28, 20, 22, 32, 28, 35, 34, 30, 36
18. 101, 115, 124, 125, 173, 165, 170
19. 1.1, 7.5, 4.9, 0.4, 2.2, 3.3, 5.1
20. $\frac{1}{4}, \frac{5}{8}, \frac{3}{8}, \frac{3}{4}, \frac{1}{2}$
21. 4.6, 8.5, 7.2, 6.6, 5.1, 6.2, 8.1, 10.3

22. **YOU BE THE TEACHER** Your friend finds and interprets the mean absolute deviation of the data set 35, 40, 38, 32, 42, and 41. Is your friend correct? Explain your reasoning.

> Mean = $\dfrac{35 + 40 + 38 + 32 + 42 + 41}{6}$ = 38
>
> MAD = $\dfrac{3 + 2 + 0 + 6 + 4 + 3}{6}$ = 3
>
> So, the data values differ from the mean by an average of 3.

23. **MODELING REAL LIFE** The data set shows the admission prices at several glass-blowing workshops.

    $20, $20, $16, $12, $15, $25, $11

    Find and interpret the range, interquartile range, and mean absolute deviation of the data.

24. **MODELING REAL LIFE** The table shows the prices of the five most-expensive and least-expensive dishes on a menu. Find the MAD of each data set. Then compare their variations.

Five Most-Expensive Dishes	Five Least-Expensive Dishes
$28 $30 $28 $39 $25	$7 $7 $10 $8 $12

25. **MP REASONING** The data sets show the years of the coins in two collections.

    Your collection: 1950, 1952, 1908, 1902, 1955, 1954, 1901, 1910

    Your friend's collection: 1929, 1935, 1928, 1930, 1925, 1932, 1933, 1920

    Compare the measures of center and the measures of variation for each data set. What can you conclude?

Movies Watched			
7	5	14	5
6	9	10	12
15	4	5	8
11	10	9	2

26. **MODELING REAL LIFE** You survey students in your class about the numbers of movies they watched last month. A new student joins the class who watched 22 movies last month. Is 22 an outlier? How does including this value affect the measures of center and the measures of variation? Explain.

MP REASONING Which data set would have the greater mean absolute deviation? Explain your reasoning.

27. guesses for number of gumballs in a jar
    guesses for number of baseballs in a jar

28. monthly rainfall amounts in a city
    monthly amounts of water used in a home

29. **MP REASONING** Range, interquartile range, and mean absolute deviation are all measures of variation. Which measure of variation is most reliable? Explain your reasoning.

30. **DIG DEEPER!** Add and subtract the MAD from the mean in the original data set in Exercise 26.

    a. What percent of the values are within one MAD of the mean? two MADs of the mean? Which values are more than twice the MAD from the mean?

    b. What do you notice as you get more and more MADs away from the mean? Explain.

9 Connecting Concepts

▶ Using the Problem-Solving Plan

1. Six friends play a carnival game in which a person throws darts at balloons. Each person throws the same number of darts and then records the portion of the balloons that pop. Find and interpret the mean, median, and MAD of the data.

Whitney   16%
Chen $\frac{2}{25}$
Bjorn   0.06
Dustin $\frac{1}{50}$
Philip   0.12
Maria   0.04

Understand the problem. You know that each person throws the same number of darts. You are given the portion of balloons popped by each person as a fraction, a decimal, or a percent.

Make a plan. First, write each fraction and each decimal as a percent. Next, order the percents from least to greatest. Then find and interpret the mean, median, and MAD of the data.

Solve and check. Use the plan to solve the problem. Then check your solution.

2. The cost c (in dollars) to rent skis at a resort for n days is represented by the equation $c = 22n$. The durations of several ski rentals are shown in the table. Find the range and interquartile range of the costs of the ski rentals. Then determine whether any of the costs are outliers.

Duration of Rentals (days)							
1	5	1	3	1	2	5	4
3	12	1	12	5	7	4	1

Performance Task

Which Measure of Center Is Best: Mean, Median, or Mode?

At the beginning of this chapter, you watched a STEAM Video called "Daylight in the Big City." You are now ready to complete the performance task related to this video, available at *BigIdeasMath.com*. Be sure to use the problem-solving plan as you work through the performance task.

9 Chapter Review

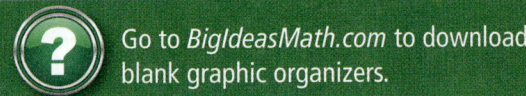

Go to *BigIdeasMath.com* to download blank graphic organizers.

Review Vocabulary

Write the definition and give an example of each vocabulary term.

statistics, *p. 414*
statistical question, *p. 414*
mean, *p. 420*
outlier, *p. 422*
measure of center, *p. 426*
median, *p. 426*
mode, *p. 426*
measure of variation, *p. 434*
range, *p. 434*
quartiles, *p. 434*
first quartile, *p. 434*
third quartile, *p. 434*
interquartile range, *p. 434*
mean absolute deviation, *p. 440*

Graphic Organizers

You can use a **Definition and Example Chart** to organize information about a concept. Here is an example of a Definition and Example Chart for the vocabulary term *statistical question*.

Statistical Question: a question in which you do not expect to get a single answer

Example
What are the heights of sixth-grade students?

Example
What are the ages of people in the auditorium?

Example
What are the numbers of letters in the first names of sixth-grade students?

Choose and complete a graphic organizer to help you study the concept.

1. mean
2. outlier
3. median
4. mode
5. range
6. quartiles
7. interquartile range
8. mean absolute deviation

"Here is my **Definition and Example Chart**. I'll toss one of my frisbees into the surf and you see if you can fetch it."

Chapter Self-Assessment

As you complete the exercises, use the scale below to rate your understanding of the success criteria in your journal.

1	2	3	4
I do not understand.	I can do it with help.	I can do it on my own.	I can teach someone else.

9.1 Introduction to Statistics (pp. 413–418)

Learning Target: Identify statistical questions and use data to answer statistical questions.

Determine whether the question is a statistical question. Explain.

1. How many positive integers are less than 20?

2. In what month were the students in a sixth-grade class born?

3. The dot plot shows the number of televisions owned by each family on a city block.

    a. Find and interpret the number of data values on the dot plot.

    b. Write a statistical question that you can answer using the dot plot. Then answer the question.

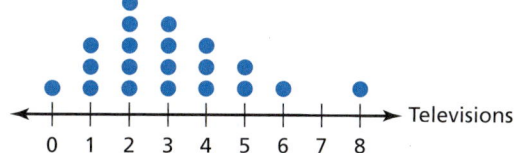

Display the data in a dot plot. Identify any clusters, peaks, or gaps in the data.

4.
Distances (feet)			
56	55	56	57
58	54	51	55
51	56	49	56

5.
Weights (pounds)				
83	88	89	90	89
91	89	84	90	92
90	88	89	83	88

6. You conduct a survey to answer, "What is the heart rate of a typical sixth-grade student?" The table shows the results. Use the distribution of the data to answer the question.

Heart Rates (beats per minute)				
68	69	74	68	64	67
66	70	67	68	74	69
68	69	74	70	68	70
70	68	68	69	67	64
69	65	66	68	69	67

Chapter Review

9.2 Mean (pp. 419–424)

Learning Target: Find and interpret the mean of a data set.

7. Find the mean of the data.

Boiling Points (°F)	Arsenic	Cadmium	Mercury	Phosphorus	Potassium	Sodium
	1117	1409	675	536	1398	1621

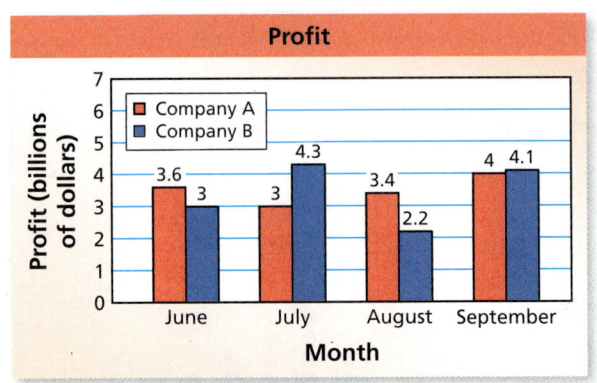

8. The double bar graph shows the monthly profit for two toy companies over a four-month period. Compare the mean monthly profits.

9. The table shows the test scores for a class of sixth-grade students. Describe how the outlier affects the mean. Then use the data to answer the statistical question, "What is the typical test score for a student in the class?"

Test Scores		
91	81	100
72	83	70
85	97	75
80	90	36

9.3 Measures of Center (pp. 425–432)

Learning Target: Find and interpret the median and mode of a data set.

Find the median and mode of the data.

10. 8, 8, 6, 8, 4, 5, 6

11. 24, 74, 61, 29, 38, 27, 68, 54

12. Find the mean, median, and mode of the data set 67, 52, 50, 99, 66, 50, and 57 with and without the outlier. Which measure does the outlier affect the most?

13. The table shows the lengths of several movies. Which measure of center best represents the data? Explain your reasoning.

14. Give an example of a data set that does not have a median. Explain why the data set does not have a median.

Movie Lengths (minutes)		
91	112	126
142	113	112
92	144	148

9.4 Measures of Variation (pp. 433–438)

Learning Target: Find and interpret the range and interquartile range of a data set.

Find the range of the data.

15. 45, 76, 98, 21, 52, 39

16. 95, 63, 52, 8, 93, 16, 42, 37, 62

Find the interquartile range of the data.

17. 28, 46, 25, 76, 18, 25, 47, 83, 44

18. 14, 25, 97, 55, 66, 28, 92, 38, 94

19. The table shows the weights of several adult emperor penguins. Find and interpret the range and interquartile range of the data. Then determine whether there are any outliers.

20. Two data sets have the same interquartile range. Can you assume that the ranges of the two data sets are about the same? Give an example to justify your answer.

Weights (kilograms)	
25	27
36	23.5
33.5	31.25
30.75	32
24	29.25

9.5 Mean Absolute Deviation (pp. 439–444)

Learning Target: Find and interpret the mean absolute deviation of a data set.

Find and interpret the mean absolute deviation of the data.

21.

Shoe Sizes			
6	8.5	6	9
10	7	8	9.5

22.

Prices of Tablets (dollars)				
130	150	190	100	175
120	165	140	180	190

23. The table shows the prices of the five most-expensive and least-expensive manicures given by a salon technician on a particular day. Find the MAD of each data set. Then compare their variations.

Five Most-Expensive Manicures	Five Least-Expensive Manicures
$58 $52 $70 $49 $56	$10 $10 $15 $10 $15

24. You record the lengths of songs you stream. The next song is 276 seconds long. Is 276 an outlier? How does including this value affect the measures of center and the measures of variation? Explain.

Song Lengths (seconds)					
233	219	163	213	224	208
225	220	222	240	228	219
260	249	209	236	206	

9 Practice Test

Find the mean, median, mode, range, and interquartile range of the data.

1. 5, 6, 4, 24, 10, 6, 9, 8
2. 46, 27, 94, 56, 53, 65, 43
3. 32, 58, 19, 36, 44, 57, 11, 26, 74
4. 36, 24, 49, 32, 37, 28, 38, 40, 39

Find and interpret the mean absolute deviation of the data.

5.

Distances Driven (miles)			
312	286	196	201
158	225	206	192

6.

Prices of Sunglasses (dollars)				
15	8	19	20	18
20	22	14	10	15

7. You conduct a survey to answer, "How many minutes does it take a typical sixth-grade student to run a mile?" The table shows the results. Use the distribution of the data to answer the question.

Times (minutes)			
8.25	9	10.25	8.75
8.5	8.25	9.25	8.5
7.75	8.5	8.75	7.5

Weights (pounds)				
81	81	80	82	81
83	76	83	76	80
75	83	94	82	81

8. The table shows the weights of Alaskan malamute dogs at a veterinarian's office. Which measure of center best represents the weight of an Alaskan malamute? Explain your reasoning.

9. The table shows the numbers of guests at a hotel on different days.

Numbers of Guests					
66	58	90	57	63	55
60	62	56	54	72	

   a. Find the range and interquartile range of the data.

   b. Use the interquartile range to identify the outlier(s) in the data set. Find the range and interquartile range of the data set without the outlier(s). Which measure did the outlier or outliers affect more?

10. The data sets show the numbers of hours worked each week by two people for several weeks.

    Person A: 9, 18, 12, 6, 9, 21, 3, 12

    Person B: 12, 18, 15, 16, 14, 12, 15, 18

    Compare the measures of center and the measures of variation for each data set. What can you conclude?

11. The table shows the lengths of several bearded dragons captured for a study. Find the mean, median, and mode of the data in centimeters and in inches. How does converting to inches affect the mean, median, and mode?

Lengths (centimeters)					
36	58	42	43	55	57
52	46	41	52	56	50

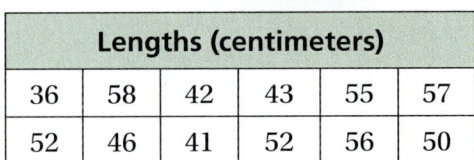

9 Cumulative Practice

1. Which statement can be represented by a negative integer?

    A. The temperature rises 15 degrees.

    B. A hot-air balloon ascends 450 yards.

    C. You earn $50 completing chores.

    D. A submarine submerges 260 feet.

Test-Taking Strategy
Use Intelligent Guessing

What is the mean length of these hyena fangs: 4 in., 3 in., 3 in., 4 in., 5 in., 5 in.?
Ⓐ 6 in.   Ⓑ $\frac{1}{3}$ ft   Ⓒ 2 in.   Ⓓ 5 in.

MEOW!

"The mean can't be 6 or 2 or 5 inches. So, you can use intelligent guessing to find that the answer is $\frac{1}{3}$ ft, or 4 in."

2. What is the height h (in inches) of the prism?

Volume = 5880 in.^{3}

h

$12\frac{1}{4}$ in.

30 in.

3. Which is the solution of the inequality $\frac{2}{3}x < 6$?

    F. $x < 4$

    G. $x < 5\frac{1}{3}$

    H. $x < 6\frac{2}{3}$

    I. $x < 9$

4. The number of hours that each of six students spent reading last week is shown in the bar graph.

Hours Spent Reading

Arnie: 7, Betsy: 10, Chan: 12, Donna: 3, Elena: 10, Frankie: 6

For the data in the bar graph, which measure is the *least*?

    A. mean

    B. median

    C. mode

    D. range

5. Which list of numbers is in order from least to greatest?

   F. $-5.41, -3.6, -3.2, -3.06, -1$
   G. $-1, -3.06, -3.2, -3.6, -5.41$
   H. $-5.41, -3.06, -3.2, -3.6, -1$
   I. $-1, -3.6, -3.2, -3.06, -5.41$

6. What is the mean absolute deviation of the data shown in the dot plot, rounded to the nearest tenth?

   A. 1.4
   B. 3
   C. 3.2
   D. 5

7. A family wants to buy tickets to a theme park. There are separate ticket prices for adults and children.

   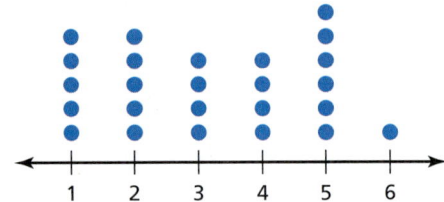

   Roller-Coaster World!
   Tickets: $30 per adult
   $20 per child

   Which expression represents the total cost (in dollars) for a adult tickets and c child tickets?

   F. $600(a + c)$
   G. $50(a \times c)$
   H. $30a + 20c$
   I. $30a \times 20c$

8.  The dot plot shows the leap distances (in feet) of a tree frog. How many leaps were recorded?

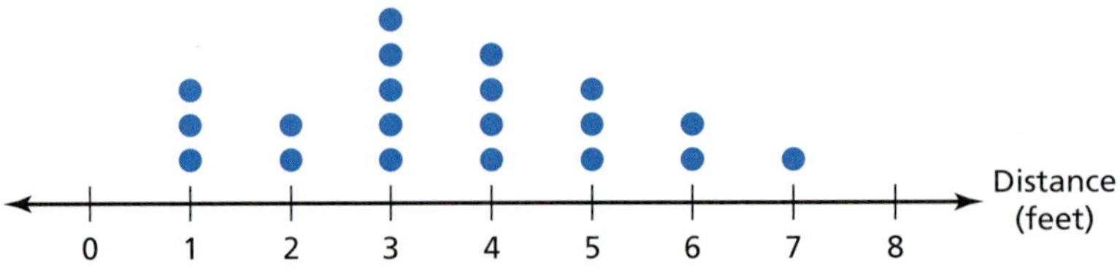

452 Chapter 9 Statistical Measures

9. What is the value of the expression when $a = 6$ and $b = 14$?

$$0.8a + 0.02b$$

- **A.** 0.4828
- **B.** 0.8814
- **C.** 5.08
- **D.** 16.4

10. Which property was *not* used to simplify the expression?

$$0.3 \times y + y \times 0.7 = y \times 0.3 + y \times 0.7$$
$$= y \times (0.3 + 0.7)$$
$$= y \times 1$$
$$= y$$

- **F.** Distributive Property
- **G.** Associative Property of Addition
- **H.** Multiplication Property of One
- **I.** Commutative Property of Multiplication

11. What are the coordinates of Point P?

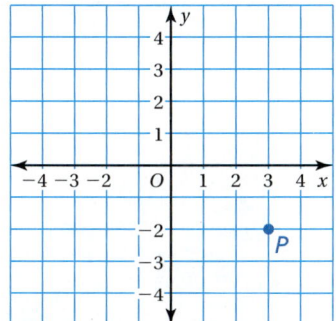

- **A.** $(-3, -2)$
- **B.** $(3, -2)$
- **C.** $(-2, -3)$
- **D.** $(-2, 3)$

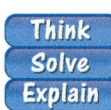

12. Create a data set with 5 numbers that has the following measures.

- a mean of 7
- a median of 9

Explain how you created your data set.

Cumulative Practice 453

10 Data Displays

- **10.1** Stem-and-Leaf Plots
- **10.2** Histograms
- **10.3** Shapes of Distributions
- **10.4** Choosing Appropriate Measures
- **10.5** Box-and-Whisker Plots

Chapter Learning Target:
Understand data displays.

Chapter Success Criteria:
- ■ I can construct a data display.
- ■ I can interpret data in a data display.
- ■ I can choose the appropriate measures of center and variation to describe a data set.
- ■ I can compare data sets.

STEAM Video: "Choosing a Dog"

STEAM Video

Choosing a Dog

Different animals grow at different rates. Given a group of puppies, describe an experiment that you can perform to compare their growth rates. Describe a real-life situation where knowing an animal's growth rate can be useful.

Watch the STEAM Video "Choosing a Dog." Then answer the following questions.

1. Using Alex and Tony's stem-and-leaf plots below, describe the weights of most dogs at 3 months of age and 6 months of age.

3 months

Stem	Leaf
2	9
3	4
4	0 0 1 2 4 6 7 8 8
5	3

Key: 3 | 4 = 34% of adult weight

6 months

Stem	Leaf
5	7 8
6	1 1 3 4 5 5 5 6 7
7	3

Key: 6 | 4 = 64% of adult weight

2. Make predictions about how the stem-and-leaf plot will look after 9 months and after 1 year.

Performance Task

Classifying Dog Breeds by Size

After completing this chapter, you will be able to use the concepts you learned to answer the questions in the *STEAM Video Performance Task*. You will be given names, breeds, and weights of full-grown dogs at a shelter. For example:

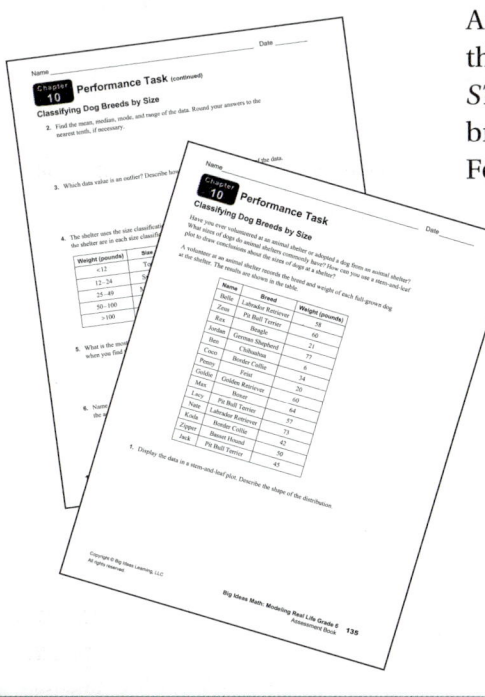

Name	Breed	Weight
Jordan	German shepherd	77 pounds
Ben	Chihuahua	6 pounds
Koda	Border collie	42 pounds

You will use a data display to make conclusions about the sizes of dogs at the shelter. Why might someone be interested in knowing the sizes of dogs at a shelter?

455

Getting Ready for Chapter 10

Chapter Exploration

Work with a partner. A famous data set was collected in Scotland in the mid-1800s. It contains the chest sizes (in inches) of 5738 men in the Scottish Militia.

Chest Size	Number of Men
33	3
34	18
35	81
36	185
37	420
38	749
39	1073
40	1079
41	934
42	658
43	370
44	92
45	50
46	21
47	4
48	1

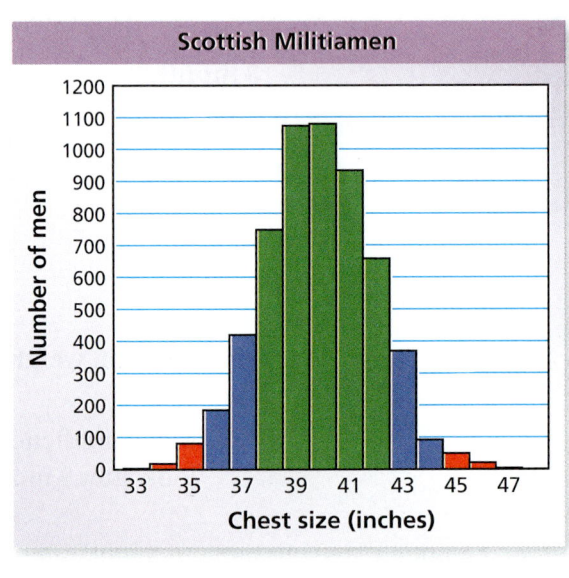

1. Describe the shape of the bar graph shown above.

2. Which of the following data sets have a bar graph that is similar in shape to the bar graph shown above? Assume the sample is selected randomly from the population. Explain your reasoning.

    a. the heights of 500 women
    b. the ages of 500 dogs
    c. the last digit of 500 phone numbers
    d. the weights of 500 newborn babies

3. Describe two other real-life data sets, one that is similar in shape to the bar graph shown above and one that is not.

Vocabulary

The following vocabulary terms are defined in this chapter. Think about what each term might mean and record your thoughts.

stem-and-leaf plot   box-and-whisker plot
frequency table   five-number summary

10.1 Stem-and-Leaf Plots

Learning Target: Display and interpret data in stem-and-leaf plots.

Success Criteria:
- I can explain how to choose stems and leaves of a data set.
- I can make and interpret a stem-and-leaf plot.
- I can use a stem-and-leaf plot to describe the distribution of a data set.

EXPLORATION 1 Making a Data Display

Work with a partner. The list below gives the ages of women when they became first ladies of the United States.

THE WHITE HOUSE
WASHINGTON, D.C.

Frances Cleveland – 21   Jacqueline Kennedy – 31
Caroline Harrison – 56   Claudia Johnson – 50
Ida McKinley – 49   Pat Nixon – 56
Edith Roosevelt – 40   Elizabeth Ford – 56
Helen Taft – 47   Rosalynn Carter – 49
Ellen Wilson – 52   Nancy Reagan – 59
Florence Harding – 60   Barbara Bush – 63
Grace Coolidge – 44   Hillary Clinton – 45
Lou Hoover – 54   Laura Bush – 54
Eleanor Roosevelt – 48   Michelle Obama – 45
Elizabeth Truman – 60   Melania Trump – 46
Mamie Eisenhower – 56

a. The incomplete data display shows the ages of the first ladies in the left column of the list above. What do the numbers on the left represent? What do the numbers on the right represent?

Ages of First Ladies

2	1
3	
4	0 4 7 8 9
5	2 4 6 6
6	0 0

b. This data display is called a *stem-and-leaf plot*. What numbers do you think represent the *stems*? *leaves*? Explain your reasoning.

c. Complete the stem-and-leaf plot using the remaining ages.

d. **MP REASONING** Write a question about the ages of first ladies that is easier to answer using a stem-and-leaf plot than a dot plot.

Math Practice

Listen and Ask Questions

Listen to other students' questions in part (d) and decide if they make sense. If not, ask for clarification.

Section 10.1 Stem-and-Leaf Plots 457

10.1 Lesson

Key Vocabulary
stem-and-leaf plot, p. 458
stem, p. 458
leaf, p. 458

The leaves of a stem-and-leaf plot are usually written in ascending order.

Key Idea

Stem-and-Leaf Plots

A **stem-and-leaf plot** uses the digits of data values to organize a data set. Each data value is broken into a **stem** (digit or digits on the left) and a **leaf** (digit or digits on the right).

A stem-and-leaf plot shows how data are distributed.

Stem	Leaf
2	0 0 1 2 5 7
3	1 4 8
4	2
5	8 9

Key: 2 | 0 = 20

The key explains what the stems and leaves represent.

EXAMPLE 1 — Making a Stem-and-Leaf Plot

	A	B
1	DATE	MINUTES
2	JULY 9	55
3	JULY 9	3
4	JULY 9	6
5	JULY 10	14
6	JULY 10	18
7	JULY 10	5
8	JULY 10	23
9	JULY 11	30
10	JULY 11	23
11	JULY 11	10
12	JULY 11	2
13	JULY 11	36

Make a stem-and-leaf plot of the lengths of the 12 phone calls.

Step 1: Order the data.

2, 3, 5, 6, 10, 14, 18, 23, 23, 30, 36, 55

Step 2: Choose the stems and the leaves. Because the data values range from 2 to 55, use the *tens* digits for the stems and the *ones* digits for the leaves. Be sure to include the key.

Step 3: Write the stems to the *left* of the vertical line.

Step 4: Write the leaves for each stem to the *right* of the vertical line.

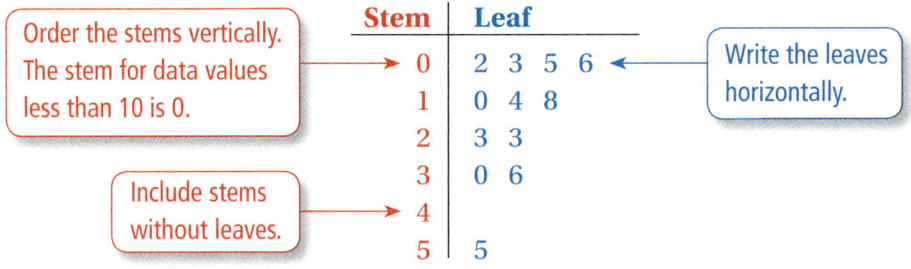

Phone Call Lengths

Stem	Leaf
0	2 3 5 6
1	0 4 8
2	3 3
3	0 6
4	
5	5

Key: 1 | 4 = 14 minutes

Order the stems vertically. The stem for data values less than 10 is 0.

Write the leaves horizontally.

Include stems without leaves.

Try It

1. Make a stem-and-leaf plot of the hair lengths.

Hair Lengths (centimeters)									
5	1	20	12	27	2	30	5	7	38
40	47	1	2	1	32	4	44	33	23

EXAMPLE 2 Interpreting a Stem-and-Leaf Plot

Quiz Scores

Stem	Leaf
6	6
7	0 5 7 8
8	1 1 3 4 4 6 8 8 9
9	0 2 9
10	0

Key: 9 | 2 = 9.2 points

The stem-and-leaf plot shows student quiz scores. (a) How many students scored less than 8 points? (b) How many students scored at least 9 points? (c) How are the data distributed?

a. There are five scores less than 8 points: 6.6, 7.0, 7.5, 7.7, and 7.8.

▶ Five students scored less than 8 points.

b. There are four scores of at least 9 points: 9.0, 9.2, 9.9, and 10.0.

▶ Four students scored at least 9 points.

c. There are few low quiz scores and few high quiz scores. So, most of the scores are in the middle, from 8.1 to 8.9 points.

Try It

2. Use the grading scale at the right.

   a. How many students received a B on the quiz?

   b. How many students received a C on the quiz?

   A: 9.0–10.0
   B: 8.0–8.9
   C: 7.0–7.9
   D: 6.0–6.9
   F: 5.9 and below

Self-Assessment for Concepts & Skills

Solve each exercise. Then rate your understanding of the success criteria in your journal.

3. **MAKING A STEM-AND-LEAF PLOT** Make a stem-and-leaf plot of the data values 14, 22, 9, 13, 30, 8, 25, and 29.

4. **WRITING** How does a stem-and-leaf plot show the distribution of a data set?

5. **MP REASONING** Consider the stem-and-leaf plot shown.

   a. How many data values are at most 10?

   b. How many data values are at least 30?

   c. How are the data distributed?

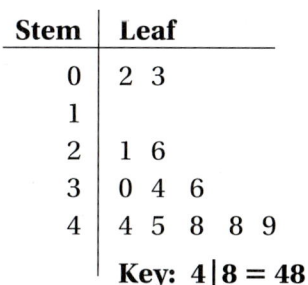

6. **CRITICAL THINKING** How can you display data whose values range from 82 through 129 in a stem-and-leaf plot?

EXAMPLE 3 Modeling Real Life

The stem-and-leaf plot shows the heights of several houseplants. Use the data to answer the question, "What is a typical height of a houseplant?"

Plant Heights

Stem	Leaf
0	1 2 4 5 6 8 9
1	1 1 1 2 3 5
2	2
3	2

Key: $1 \mid 5 = 15$ inches

Find the mean, median, and mode of the data. Use the measure that best represents the data to answer the statistical question.

Mean: $\dfrac{162}{15} = 10.8$

Median: 11

Mode: 11

The mean is slightly less than the median and mode, but all three measures can be used to represent the data.

 So, the typical height of a houseplant is about 11 inches.

Self-Assessment for Problem Solving

Solve each exercise. Then rate your understanding of the success criteria in your journal.

7. **DIG DEEPER!** Work with a partner. Use two number cubes to conduct the following experiment. Then use a stem-and-leaf plot to organize your results and describe the distribution of the data.

- Toss the cubes and find the product of the resulting numbers. Record your results.
- Repeat this process 30 times.

8. The stem-and-leaf plot shows the weights (in pounds) of several puppies at a pet store. Use the data to answer the question, "How much does a puppy at the pet store weigh?"

Puppy Weights

Stem	Leaf
0	8
1	2 5 7 8
2	4 4
3	1

Key: $2 \mid 4 = 24$ pounds

460 Chapter 10 Data Displays

10.1 Practice

Go to *BigIdeasMath.com* to get HELP with solving the exercises.

Review & Refresh

Find and interpret the mean absolute deviation of the data.

1. 8, 6, 8, 5, 3, 10, 11, 5, 7
2. 55, 46, 39, 62, 55, 51, 48, 60, 39, 45
3. 37, 54, 41, 18, 28, 32
4. 12, 25, 8, 22, 6, 1, 10, 4

Use the Distributive Property to simplify the expression.

5. $5(n + 8)$
6. $7(y - 6)$
7. $14(2b + 3)$
8. $11(9 + s)$

Solve the equation.

9. $\dfrac{p}{3} = 8$
10. $28 = 6g$
11. $3d \div 4 = 9$
12. $10 = \dfrac{2z}{3}$

Concepts, Skills, & Problem Solving

REASONING Write a question that is easier to answer using the stem-and-leaf plot than a dot plot. *(See Exploration 1, p. 457.)*

13. **Numbers of Customers**

Stem	Leaf
1	2 3 6 7
2	0 1 1 3 3 8 8
3	2 3 4 4 5 5 6 9 9
4	0 1 1 2 4 6 7 8 9 9

Key: 1 | 3 = 13 customers

14. **Text Messages Received**

Stem	Leaf
4	0 0 2 6 6 9
5	1 1 3 3 7 7 7 9 9 9
6	1 2 2 5 5 6 7 8 8
7	0 2 2 3 4

Key: 5 | 1 = 51 text messages

MAKING A STEM-AND-LEAF PLOT Make a stem-and-leaf plot of the data.

15. **Books Read**

26	15	20	9
31	25	29	32
17	26	19	40

16. **Hours Online**

8	12	21	14
18	6	15	24
12	17	2	0

17. **Test Scores (%)**

87	82	95	91	69
88	68	87	65	81
97	85	80	90	62

18. **Points Scored**

58	50	42	71	75
45	51	43	38	71
42	70	56	58	43

19. **Bikes Sold**

78	112	105	99
86	96	115	100
79	81	99	108

20. **Minutes in Line**

4.0	2.6	1.9	3.1
3.6	2.2	2.7	3.8
1.6	2.0	3.1	2.9

21. **YOU BE THE TEACHER** Your friend makes a stem-and-leaf plot of the data. Is your friend correct? Explain your reasoning.

    51, 25, 47, 42, 55, 26, 50, 44, 55

Stem	Leaf
2	5 6
4	2 4 7
5	0 1 5 5

    Key: 4|2 = 42

MODELING REAL LIFE The stem-and-leaf plot shows the numbers of confirmed cases of a virus in 15 countries.

22. How many of the countries have more than 60 confirmed cases?

23. Find the mean, median, mode, range, and interquartile range of the data.

24. How are the data distributed?

25. Which data value is an outlier? Describe how the outlier affects the mean.

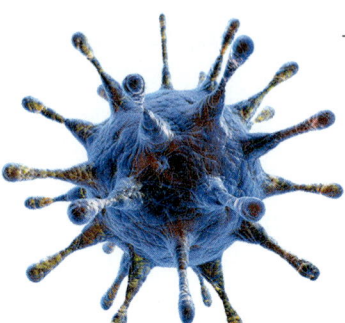

Stem	Leaf
4	1 1 3 3 5
5	0 2 3 4
6	2 3 3 7
7	5
8	
9	7

Key: 5|0 = 50 cases

26. **MP REASONING** Each stem-and-leaf plot below has a mean of 39. Without calculating, determine which stem-and-leaf plot has the lesser mean absolute deviation. Explain your reasoning.

Stem	Leaf
2	3 7
3	0 2 6 9
4	1 2 5 8
5	1 4

    Key: 4|1 = 41

Stem	Leaf
2	2 4 5 8 9
3	3 8
4	5
5	3 6 7 8

    Key: 5|3 = 53

27. **DIG DEEPER!** The stem-and-leaf plot shows the daily high temperatures (in degrees Fahrenheit) for the first 15 days of June.

Stem	Leaf
6	7 8
7	0 0 3 4 6 8 9
8	2 3 6 7 8 9

    Key: 6|7 = 67°F

    a. When you include the daily high temperatures for the rest of the month, the mean absolute deviation increases. Draw a stem-and-leaf plot that could represent all of the daily high temperatures for the month.

    b. Use your stem-and-leaf plot from part (a) to answer the question, "What is a typical daily high temperature in June?"

28. **CRITICAL THINKING** The back-to-back stem-and-leaf plot shows the 9-hole golf scores for two golfers. Only one of the golfers can compete in a tournament as your teammate. Use measures of center and measures of variation to support choosing either golfer.

Rich		Will
7 5	3	
8 5 4 3 2 1	4	2 3 4 4 6 7 7 8 9
5 0	5	0

    Key: 1|4|2 = 41 and 42 strokes

10.2 Histograms

Learning Target: Display and interpret data in histograms.

Success Criteria:
- I can explain how to draw a histogram.
- I can make and interpret a histogram.
- I can determine whether a question can be answered using a histogram.

EXPLORATION 1

Performing an Experiment

Work with a partner.

a. Make the airplane shown from a single sheet of $8\frac{1}{2}$-by-11-inch paper. Then design and make your own paper airplane.

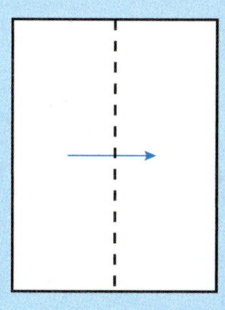
1. Fold in half. Then unfold.

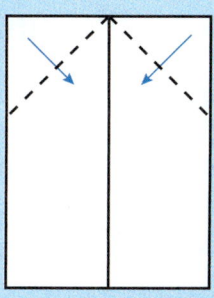
2. Fold corners.

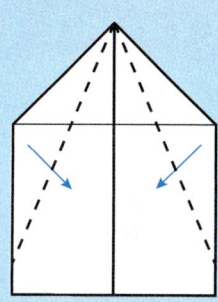
3. Fold corners again.

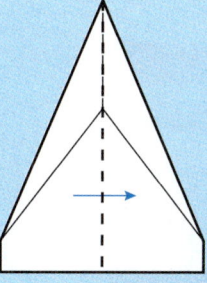
4. Fold in half.

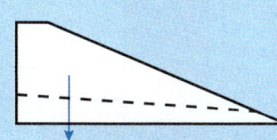
5. Fold wings out on both sides.

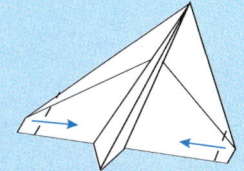
6. Fold wing edges up.

Math Practice

Specify Units
What units will you use to measure the distance flown? Will the units you use affect the results in your frequency table? Explain.

b. **MP PRECISION** Fly each airplane 20 times. Keep track of the distance flown each time.

c. A **frequency table** groups data values into intervals. The **frequency** is the number of values in an interval. Use a frequency table to organize the results for each airplane.

d. **MODELING** Represent the data in the frequency tables graphically. Which airplane flies farther? Explain your reasoning.

10.2 Lesson

Key Vocabulary
frequency table, *p. 463*
frequency, *p. 463*
histogram, *p. 464*

Key Idea

Histograms

A **histogram** is a bar graph that shows the frequencies of data values in intervals of the same size.

The height of a bar represents the frequency of the values in the interval.

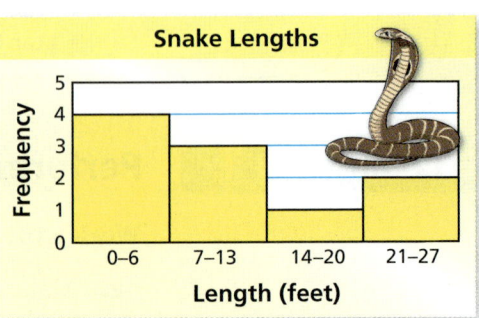

EXAMPLE 1 Making a Histogram

The frequency table shows the numbers of laps that people in a swimming class completed today. Display the data in a histogram.

Number of Laps	Frequency
1–3	11
4–6	4
7–9	0
10–12	3
13–15	6

Step 1: Draw and label the axes.

Step 2: Draw a bar to represent the frequency of each interval.

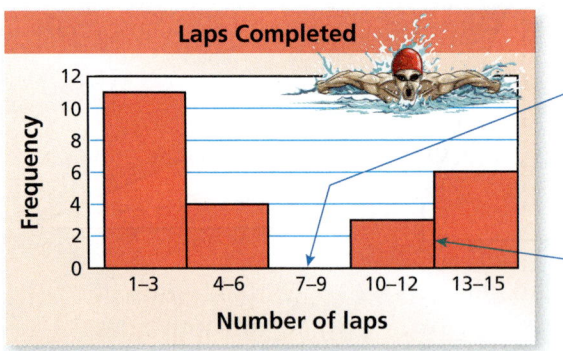

Include any interval with a frequency of 0. The bar height is 0.

There is no space between the bars of a histogram.

Try It

1. The frequency table shows the ages of people riding a roller coaster. Display the data in a histogram.

Age	10–19	20–29	30–39	40–49	50–59
Frequency	16	11	5	2	4

464 Chapter 10 Data Displays

EXAMPLE 2 Using a Histogram

The histogram shows winning speeds at the Daytona 500.
(a) Which interval contains the most data values?
(b) How many of the winning speeds are less than 140 miles per hour?
(c) How many of the winning speeds are at least 160 miles per hour?

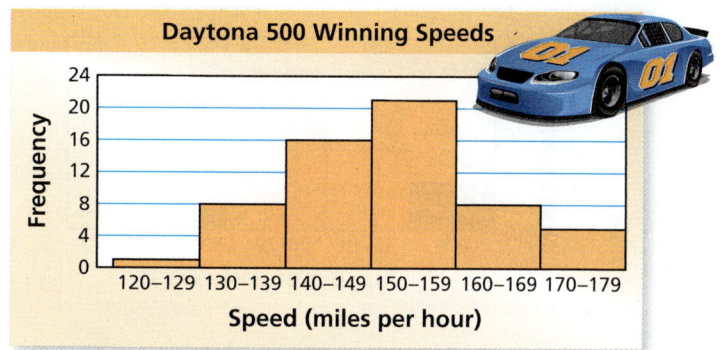

a. The interval with the tallest bar contains the most data values.

▸ So, the 150−159 miles per hour interval contains the most data values.

b. One winning speed is in the 120−129 miles per hour interval, and eight winning speeds are in the 130−139 miles per hour interval.

▸ So, 1 + 8 = 9 winning speeds are less than 140 miles per hour.

c. Eight winning speeds are in the 160−169 miles per hour interval, and five winning speeds are in the 170−179 miles per hour interval.

▸ So, 8 + 5 = 13 winning speeds are at least 160 miles per hour.

Try It

2. The histogram shows the numbers of hours that students in a class slept last night.

   a. How many students slept at least 8 hours?

   b. How many students slept less than 12 hours?

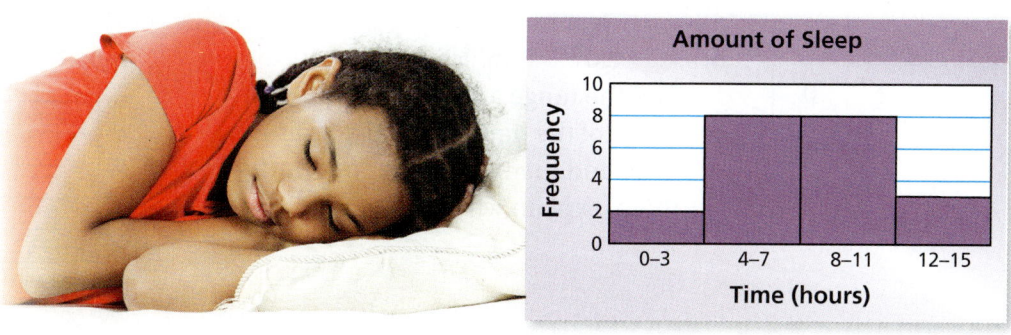

Section 10.2   Histograms   465

EXAMPLE 3 Comparing Data Displays

The data displays show how many push-ups students in a class completed for a physical fitness test. Which data display can you use to find how many students are in the class? Explain.

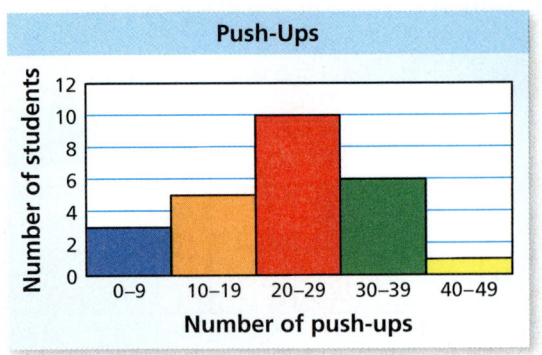
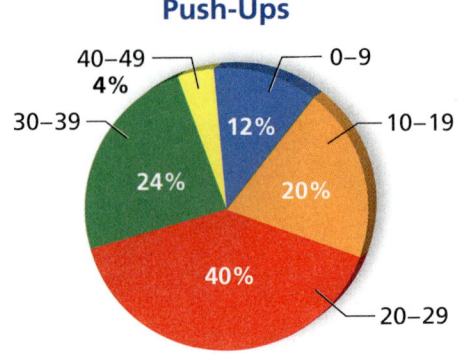

▸ You can use the histogram because it shows the number of students in each interval. The sum of these values represents the number of students in the class. You cannot use the circle graph because it does not show the number of students in each interval.

Try It

3. Which data display should you use to describe the portion of the entire class that completed 30−39 push-ups? Explain.

Self-Assessment for Concepts & Skills

Solve each exercise. Then rate your understanding of the success criteria in your journal.

Numbers of Siblings			
0	2	3	6
3	2	1	2
1	4	0	3
7	3	1	5
4	3	2	0

4. **MAKING A HISTOGRAM** The table shows the numbers of siblings of students in a class.

   a. Display the data in a histogram.

   b. Explain how you chose reasonable intervals for your histogram in part (a).

5. **NUMBER SENSE** Can you find the range and the interquartile range of the data in the histogram? If so, find them. If you cannot find them, explain why not.

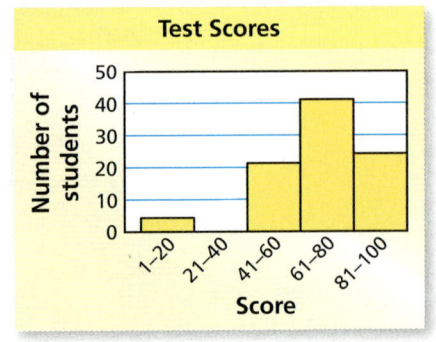

EXAMPLE 4 Modeling Real Life

Which statements *cannot* be made using the data displays in Example 3?

A. Twelve percent of the class completed 9 push-ups.

B. Five students completed at least 10 and at most 19 push-ups.

C. At least one student completed more than 39 push-ups.

D. Less than $\frac{1}{4}$ of the class completed 30 or more push-ups.

The circle graph shows that 12% completed 0–9 push-ups, but you cannot determine how many completed exactly 9. So, Statement A cannot be made.

In the histogram, the bar height for the 10–19 interval is 5, and the bar height for the 40–49 interval is 1. So, Statements B and C can be made.

The circle graph shows that 24% completed 30–39 push-ups, and 4% completed 40–49 push-ups. So, 24% + 4% = 28% completed 30 or more push-ups. Because $\frac{1}{4}$ = 25% and 28% > 25%, Statement D cannot be made.

▸ The correct answers are **A** and **D**.

Self-Assessment for Problem Solving

Solve each exercise. Then rate your understanding of the success criteria in your journal.

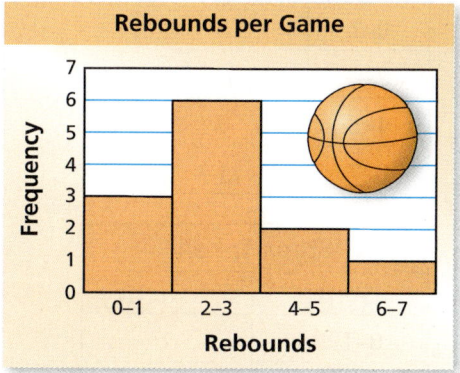

6. The histogram shows the numbers of rebounds per game for a middle school basketball player in a season.

   a. Which interval contains the most data values?

   b. How many games did the player play during the season?

   c. In what percent of the games did the player have 4 or more rebounds?

7. Determine whether you can make each statement by using the histogram in the previous exercise. Explain.

   a. The basketball player had 2 rebounds in 6 different games.

   b. The basketball player had more than 1 rebound in 9 different games.

10.2 Practice

Review & Refresh

Make a stem-and-leaf plot of the data.

1.
Blog Posts			
7	5	12	4
20	11	9	15
4	8	6	12

2.
Social Media Comments			
4	18	1	32
10	36	16	7
44	3	7	15

Find the percent of the number.

3. 25% of 180
4. 30% of 90
5. 16% of 140
6. 64% of 80

7. What is the least common multiple of 7 and 12?

   A. 28   **B.** 42   **C.** 84   **D.** 168

Concepts, Skills, & Problem Solving

MAKING A FREQUENCY TABLE Organize the data using a frequency table. (See Exploration 1, p. 463.)

8.
Members of Book Clubs			
6	17	13	19
13	9	18	24
11	15	21	14

9.
Points Scored				
42	45	57	39	55
38	48	36	48	46
51	29	45	54	42

MAKING A HISTOGRAM Display the data in a histogram.

10.
States Visited	
States	Frequency
1–5	12
6–10	14
11–15	6
16–20	3

11.
Chess Team	
Wins	Frequency
10–13	3
14–17	4
18–21	4
22–25	2

12.
Movies Watched	
Movies	Frequency
0–1	5
2–3	11
4–5	8
6–7	1

13.
Ages of Celebrities	
Ages	Frequency
10–19	2
20–29	6
30–39	8
40–49	4

14.
Shoes Owned	
Pairs of Shoes	Frequency
1–3	3
4–6	8
7–9	10
10–12	0
13–15	2

15.
Steps Taken	
Steps	Frequency
0–1999	1
2000–3999	4
4000–5999	9
6000–7999	12
8000–9999	11

16. **YOU BE THE TEACHER** Your friend displays the data in a histogram. Is your friend correct? Explain your reasoning.

Snow Days per School	
Snow Days	Frequency
0–1	9
2–3	6
4–5	2
6–7	1

17. **MODELING REAL LIFE** The histogram shows the numbers of magazines read last month by the students in a class.

    a. Which interval contains the fewest data values?

    b. How many students are in the class?

    c. What percent of the students read fewer than six magazines?

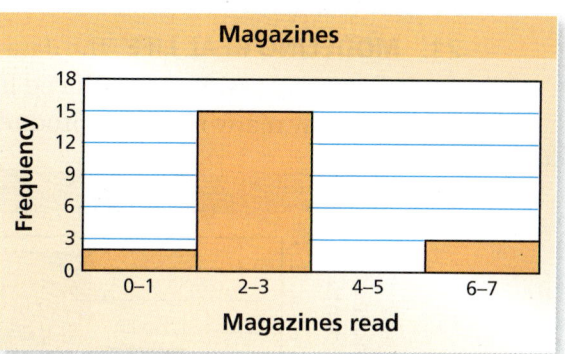

18. **YOU BE THE TEACHER** Your friend interprets the histogram. Is your friend correct? Explain your reasoning.

    12% of the songs took 5–8 seconds to download.

19. **REASONING** The histogram shows the percent of the voting-age population in each state who voted in a presidential election. Explain whether the graph supports each statement.

    a. Only 40% of one state voted.

    b. In most states, between 50% and 64.9% voted.

    c. The mode of the data is between 55% and 59.9%.

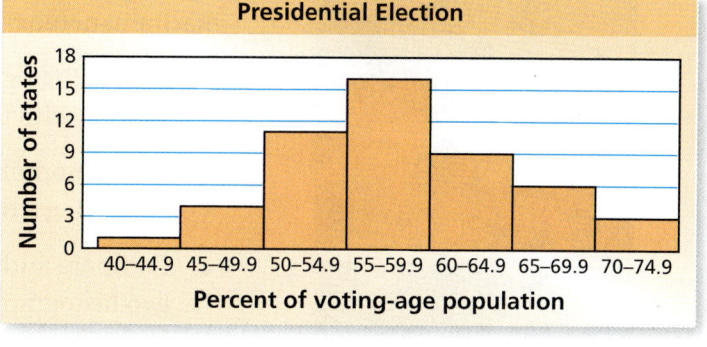

20. **MP PROBLEM SOLVING** The histograms show the areas of counties in Pennsylvania and Indiana. Which state do you think has the greater area? Explain.

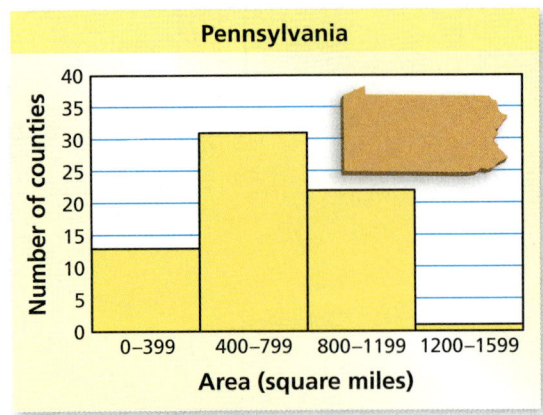
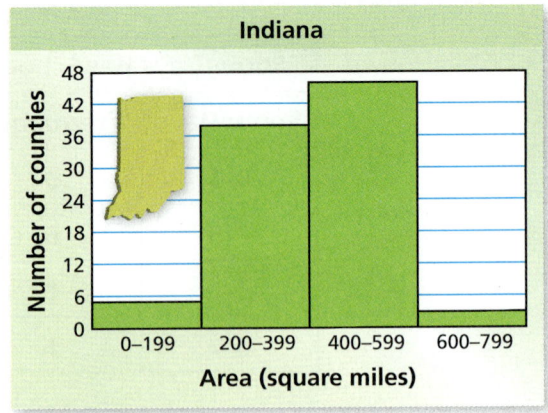

21. **MODELING REAL LIFE** The data displays show how many pounds of garbage apartment residents produced in 1 week. Which data display can you use to find how many residents produced more than 25 pounds of garbage? Explain.

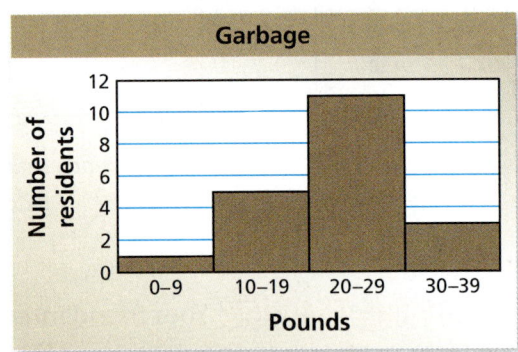

Garbage

Stem	Leaf
0	9
1	0 5 8 8 9
2	1 2 5 5 6 7 7 7 9 9 9
3	2 3 3

Key: 1 | 5 = 15 pounds

22. **MP REASONING** Determine whether you can make each statement by using the data displays in Exercise 21. Explain your reasoning.

   a. One resident produced 10 pounds of garbage.

   b. Twelve residents produced between 20 and 29 pounds of garbage.

23. **DIG DEEPER!** The table shows the lengths of some whales in a marine sanctuary.

   a. Make a histogram of the data starting with the interval 51–55.

   b. Make another histogram of the data using a different-sized interval.

   c. Compare and contrast the two histograms.

Lengths (feet)				
81	88	57	82	70
71	51	82	77	79
83	80	54	80	81
59	84	75	76	68
83	78	55	67	85
85	77	73	78	79

24. **MP LOGIC** Can you find the mean or the median of the data in Exercise 17? Explain.

10.3 Shapes of Distributions

Learning Target: Describe and compare shapes of distributions.

Success Criteria:
- I can explain what it means for a distribution to be skewed left, skewed right, or symmetric.
- I can use data displays to describe shapes of distributions.
- I can use shapes of distributions to compare data sets.

Math Practice

Apply Mathematics

How can the word *skewed* be applied in mathematics?

The Meaning of a Word ▶ Skewed

When something is **skewed**,  it has a slanted direction or position.

EXPLORATION 1

Describing Shapes of Distributions

Work with a partner. The lists show the first three digits and last four digits of several phone numbers in the contact list of a cell phone.

a. Compare and contrast the distribution of the last digit of each phone number to the distribution of the first digit of each phone number. Describe the shapes of the distributions.

b. Describe the shape of the distribution of the data in the table below. Compare it to the distributions in part (a).

Ages of Cell Phones (years)					
0	1	0	6	4	0
2	3	5	1	1	2
0	1	2	3	1	0
0	0	1	1	1	1
7	1	4	2	2	2

Section 10.3 Shapes of Distributions 471

10.3 Lesson

You can use dot plots and histograms to identify shapes of distributions.

Key Ideas

Symmetric and Skewed Distributions

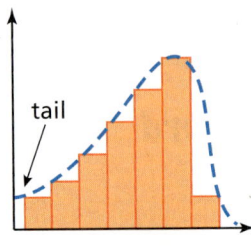
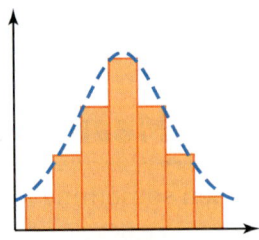
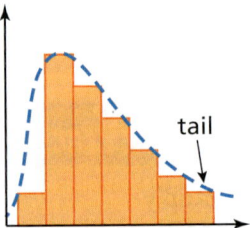

Skewed left   *Symmetric*   *Skewed right*

- The "tail" of the graph extends to the left.
- Most data are on the right.

- The left side of the graph is a mirror image of the right side of the graph.

- The "tail" of the graph extends to the right.
- Most data are on the left.

> If all the dots of a dot plot or bars of a histogram are about the same height, then the distribution is a *flat*, or *uniform*, distribution. A uniform distribution is also symmetric.

EXAMPLE 1 Describing Shapes of Distributions

Describe the shape of each distribution.

a. Daily Snowfall Amounts

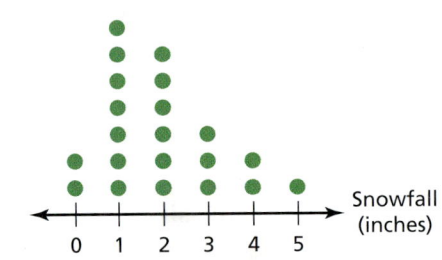

b.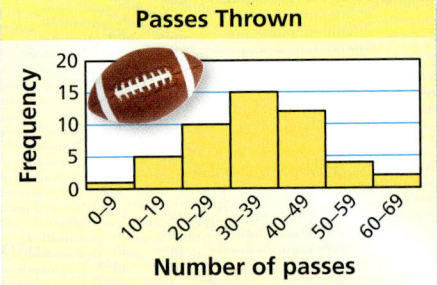

Most of the data are on the left, and the tail extends to the right.

▸ So, the distribution is skewed right.

The left side of the graph is approximately a mirror image of the right side of the graph.

▸ So, the distribution is symmetric.

Try It

1. Describe the shape of the distribution.

Daily Spam Emails Received

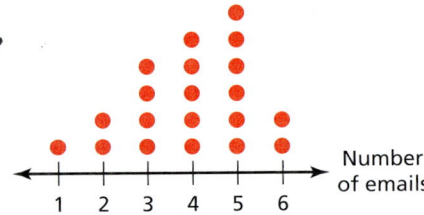

472 Chapter 10 Data Displays

EXAMPLE 2 Describing the Shape of a Distribution

Ages	Frequency
10–13	1
14–17	3
18–21	7
22–25	12
26–29	20
30–33	18
34–37	3

The frequency table shows the ages of people watching a comedy in a theater. Display the data in a histogram. Then describe the shape of the distribution.

Draw and label the axes. Then draw a bar to represent the frequency of each interval.

Most of the data are on the right, and the tail extends to the left.

▸ So, the distribution is skewed left.

Try It

2. The frequency table shows the ages of people watching a historical movie in a theater. Display the data in a histogram. Describe the shape of the distribution.

Ages	10–19	20–29	30–39	40–49	50–59	60–69
Frequency	3	18	36	40	14	5

Self-Assessment for Concepts & Skills

Solve each exercise. Then rate your understanding of the success criteria in your journal.

3. **WRITING** Explain in your own words what it means for a distribution to be (a) skewed left, (b) symmetric, and (c) skewed right.

4. **DESCRIBING A DISTRIBUTION** Display the data shown in a histogram. Describe the shape of the distribution.

Calories	Frequency
1–100	2
101–200	8
201–300	10
301–400	5
401–500	3

5. **WHICH ONE DOESN'T BELONG?** Which histogram does *not* belong with the other three? Explain your reasoning.

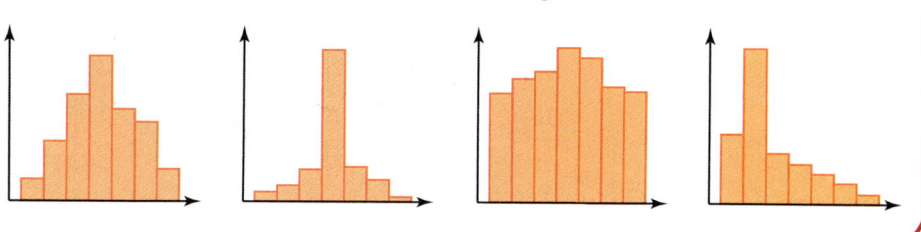

Section 10.3 Shapes of Distributions

EXAMPLE 3 Modeling Real Life

The histogram shows the ages of people watching an animated movie in the same theater as in Example 2. Which movie has an older audience?

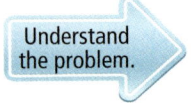 **Understand the problem.** You are given histograms that display the ages of people watching two movies. You are asked to determine which movie has an older audience.

 **Make a plan.** Use the intervals and distributions of the data to determine which movie has an older audience.

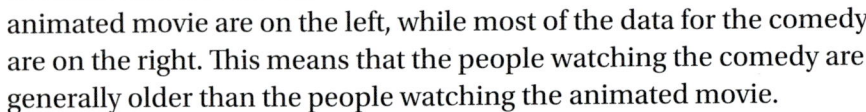

Animated Movie Attendance

 **Solve and check.** The intervals in the histograms are the same. Most of the data for the animated movie are on the left, while most of the data for the comedy are on the right. This means that the people watching the comedy are generally older than the people watching the animated movie.

▶ So, the comedy has an older audience.

Check Reasonableness The movies have similar attendance. However, only 4 people watching the comedy are 17 or under. A total of 35 people watching the animated movie are 17 or under. So, it is reasonable to conclude that the comedy has an older audience. ✓

Self-Assessment for Problem Solving

Solve each exercise. Then rate your understanding of the success criteria in your journal.

Visitors	Aurora	Grover
1–20	3	6
21–40	5	11
41–60	6	7
61–80	10	4
81–100	7	3

6. The frequency table shows the numbers of visitors each day to parks in Aurora and Grover in one month. Which park generally has more daily visitors? Justify your answer.

7. **DIG DEEPER!** The frequency tables below show the ages of guests on two cruises. Can you make accurate comparisons of the ages of the guests? Explain your reasoning.

Ages	Frequency
18–24	26
25–31	26
32–38	22
39–45	14
46–52	8

Ages	Frequency
18–22	16
23–27	22
28–32	26
33–37	20
38–42	12

10.3 Practice

Review & Refresh

Display the data in a histogram.

1.
Goals per Game	
Goals	Frequency
0–1	5
2–3	4
4–5	0
6–7	1

2.
Minutes Practiced	
Minutes	Frequency
0–19	8
20–39	10
40–59	11
60–79	2

3.
Poems Written for Class	
Poems	Frequency
0–4	6
5–9	16
10–14	4
15–19	2

Write a unit rate for the situation.

4. $200 per 8 days

5. 60 kilometers for every 1.5 hours

Concepts, Skills, & Problem Solving

DESCRIBING SHAPES OF DISTRIBUTIONS Describe the shape of the distribution of the data in the table. (See Exploration 1, p. 471.)

6.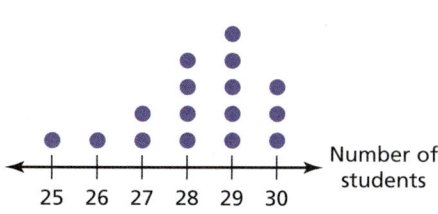

7.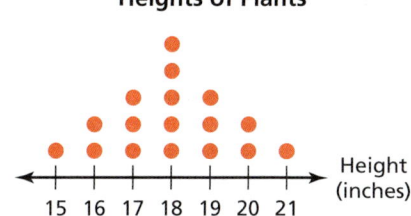

DESCRIBING SHAPES OF DISTRIBUTIONS Describe the shape of the distribution.

8.

9.

10.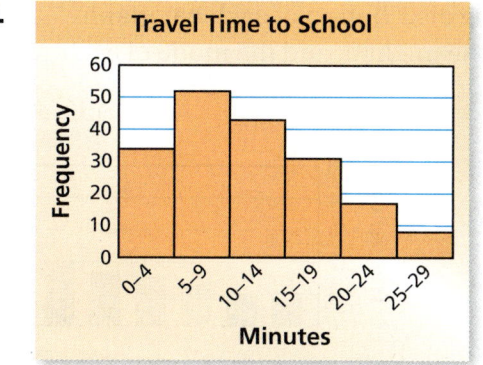

11.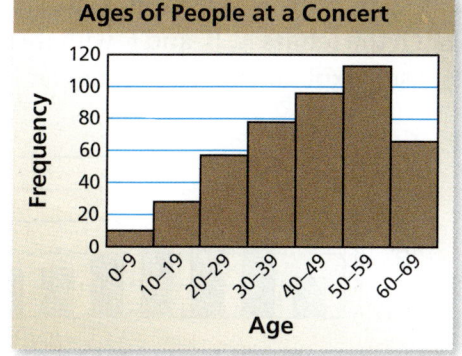

12. **MODELING REAL LIFE** The frequency table shows the years of experience for the medical staffs in Jones County and Pine County. Display the data for each county in a histogram. Which county's medical staff has less experience? Explain.

Years of Experience	0–3	4–7	8–11	12–15	16–19	20–23	24–27
Frequency for Jones County	7	15	17	12	8	5	3
Frequency for Pine County	3	5	9	14	10	6	2

13. **MP REASONING** What is the shape of the distribution of the restaurant waiting times? Explain your reasoning.

14. **MP LOGIC** Are all distributions either approximately symmetric or skewed? Explain. If not, give an example.

15. **MP REASONING** Can you use a stem-and-leaf plot to describe the shape of a distribution? Explain your reasoning.

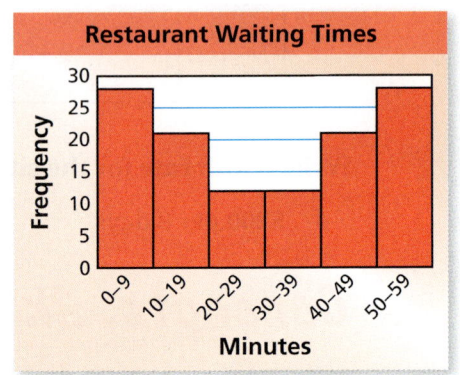

16. **DIG DEEPER!** The table shows the donation amounts received by a charity in one day.

Donations (dollars)												
20	15	40	70	20	5	25	50	47	20	62	55	40
10	50	18	20	100	40	80	60	20	80	3	30	50
25	30	10	33	20	50	7	35	40	25	70		

a. Make a histogram of the data starting with the interval 0–14. Describe the shape of the distribution.

b. A company adds $5 to each donation. Make another histogram starting with the same interval as in part (a). Compare the shape of this distribution with the distribution in part (a). Explain any differences in the distributions.

17. **CRITICAL THINKING** Describe the shape of the distribution of each bar graph. Match the letters A, B, and C with the mean, the median, and the mode of each data set. Explain your reasoning.

a.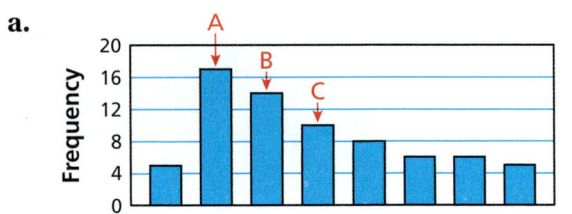

b.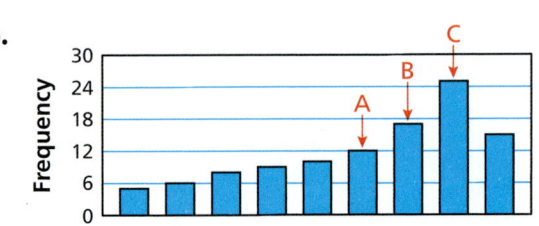

10.4 Choosing Appropriate Measures

Learning Target: Determine which measures of center and variation best describe a data set.

Success Criteria:
- I can describe the shape of a distribution.
- I can use the shape of a distribution to determine which measure of center best describes the data.
- I can use the shape of a distribution to determine which measure of variation best describes the data.

EXPLORATION 1

Using Shapes of Distributions

Work with a partner.

a. In Section 10.3 Exploration 1(a), you described the distribution of the first digits of the numbers at the right. In Exploration 1(b), you described the distribution of the data set below.

Ages of Cell Phones (years)					
0	1	0	6	4	0
2	3	5	1	1	2
0	1	2	3	1	0
0	0	1	1	1	1
7	1	4	2	2	2

538-
438-
664-
761-
868-
735-
694-
599-
725-
556-
555-
456-
736-
664-
576-

664-
664-
538-
855-
664-
538-
654-
654-
725-
538-
799-
764-
664-
664-
725-

What do you notice about the measures of center, measures of variation, and the shapes of the distributions? Explain.

b. Which measure of center best describes each data set? Explain your reasoning.

c. Which measure of variation best describes each data set? Explain your reasoning.

Math Practice

Construct Arguments
Explain why the shapes of the distributions in Exploration 1 affect which measures best describe the data.

Section 10.4 Choosing Appropriate Measures 477

10.4 Lesson

You can use a measure of center and a measure of variation to describe the distribution of a data set. The shape of the distribution can help you choose which measures are the most appropriate to use.

Key Idea

Choosing Appropriate Measures

The mean absolute deviation (MAD) uses the mean in its calculation. So, when a data distribution is *symmetric*,

- use the mean to describe the center and
- use the MAD to describe the variation.

The interquartile range (IQR) uses quartiles in its calculation. So, when a data distribution is *skewed*,

- use the median to describe the center and
- use the IQR to describe the variation.

EXAMPLE 1 — Choosing Appropriate Measures

Bordering States	Frequency
0–1	3
2–3	13
4–5	21
6–7	11
8–9	2

The frequency table shows the number of states that border each state in the United States. What are the most appropriate measures to describe the center and the variation?

To see the distribution of the data, display the data in a histogram.

The left side of the graph is approximately a mirror image of the right side of the graph. The distribution is symmetric.

▸ So, the mean and the mean absolute deviation are the most appropriate measures to describe the center and the variation.

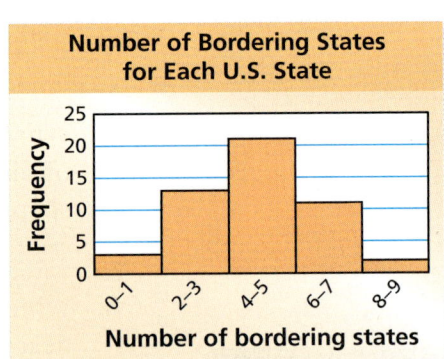

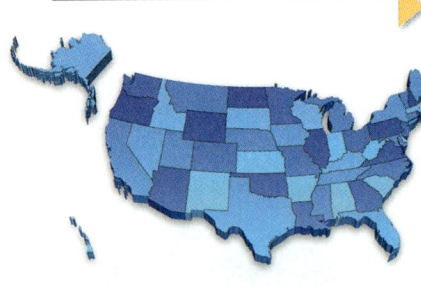

Try It

1. The frequency table shows the gas mileages of several motorcycles made by a company. What are the most appropriate measures to describe the center and the variation?

Mileage (miles per gallon)	40–44	45–49	50–54	55–59	60–64	65–69
Frequency	2	1	6	8	10	3

478   Chapter 10   Data Displays

EXAMPLE 2 Describing a Data Set

The dot plot shows the average numbers of hours students in a class sleep each night. Describe the center and the variation of the data set.

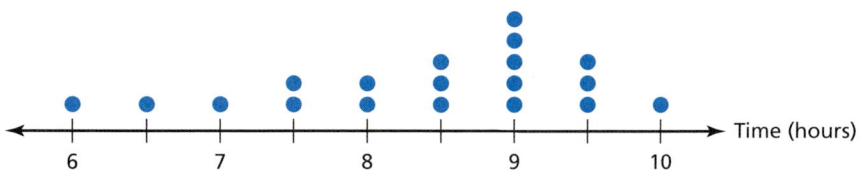

Amount of Sleep

Most of the data values are on the right, clustered around 9, and the tail extends to the left. The distribution is skewed left, so the median and the interquartile range are the most appropriate measures to describe the center and the variation.

The median is 8.5 hours. The first quartile is 7.5, and the third quartile is 9. So, the interquartile range is $9 - 7.5 = 1.5$ hours.

▸ The data are centered around 8.5 hours. The middle half of the data varies by no more than 1.5 hours.

Weekly Gym Time

Try It

2. The dot plot shows the numbers of hours people spent at the gym last week. Describe the center and the variation of the data set.

Self-Assessment for Concepts & Skills

Solve each exercise. Then rate your understanding of the success criteria in your journal.

3. **OPEN-ENDED** Construct a dot plot for which the mean is the most appropriate measure to describe the center of the distribution.

CHOOSING APPROPRIATE MEASURES Choose the most appropriate measures to describe the center and the variation. Explain your reasoning. Then find the measures you chose.

4.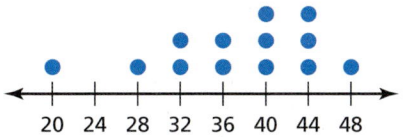

5.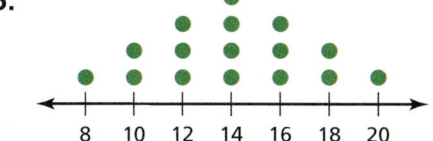

6. **WRITING** Explain why the most appropriate measures to describe the center and the variation of a data set are determined by the shape of the distribution.

EXAMPLE 3 Modeling Real Life

Basket A		Basket B	
$70	$40	$45	$30
$60	$90	$45	$55
$10	$0	$55	$40
$70	$40	$50	$60
$30	$100	$45	$50
$50	$60	$55	$55
$40	$50	$60	$45
$60	$30	$40	$70

Two baskets each have 16 envelopes with money inside, as shown in the tables. How much does a typical envelope in each basket contain? Why might a person want to pick from Basket B instead of Basket A?

To answer each question, display the data in dot plots to see the distributions of the data.

In each graph, the left side is a mirror image of the right side. Because both distributions are symmetric, the mean and the mean absolute deviation are the most appropriate measures to describe the center and the variation.

The mean of each data set is $\frac{800}{16} = \$50$. The MAD of Basket A is $\frac{320}{16} = \$20$, and the MAD of Basket B is $\frac{120}{16} = \$7.50$. So, Basket A has more variability.

▸ A typical envelope in each basket contains about $50. A person may choose from Basket B instead of Basket A because there is less variability. This means it is more likely to get an amount near $50 by choosing an envelope from Basket B than by choosing an envelope from Basket A.

Self-Assessment for Problem Solving

Solve each exercise. Then rate your understanding of the success criteria in your journal.

Room A		Room B	
5	25	25	15
30	45	30	20
20	15	5	10
50	20	20	50
25	30	45	5
10	25	25	50

7. Why might a person want to pick from Basket A instead of Basket B in Example 3? Explain your reasoning.

8. In a video game, two rooms each have 12 treasure chests containing gold coins. The tables show the numbers of coins in each chest. You pick one chest and are rewarded with the coins inside. From which room would you choose? Explain your reasoning.

9. Create a dot plot of the numbers of pets that students in your class own. Describe the center and the variation of the data set.

10.4 Practice

Review & Refresh

Describe the shape of the distribution.

1. Hours Worked

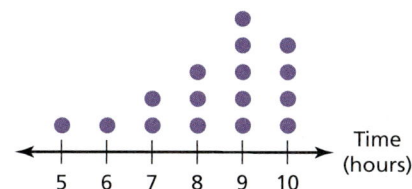

2.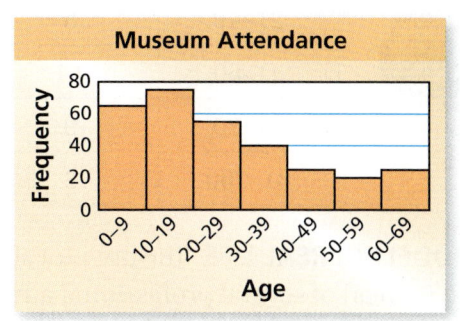

Find the median, first quartile, third quartile, and interquartile range of the data.

3. 68, 74, 67, 72, 63, 70, 78, 64, 76

4. 39, 48, 33, 24, 30, 44, 36, 41, 28, 53

Divide. Write the answer in simplest form.

5. $4\frac{2}{5} \div 2$

6. $5\frac{1}{8} \div \frac{7}{8}$

7. $2\frac{3}{7} \div 1\frac{1}{7}$

8. $\frac{4}{5} \div 7\frac{1}{2}$

Concepts, Skills, & Problem Solving

USING SHAPES OF DISTRIBUTIONS Find the mean and the median of the data set. Which measure of center best describes the data set? Explain your reasoning. *(See Exploration 1, p. 477.)*

9. 9, 3, 7, 7, 9, 2, 8, 9, 6, 7, 8, 9

10. 24, 25, 27, 27, 23, 29, 26, 26, 26, 25, 28

CHOOSING APPROPRIATE MEASURES Choose the most appropriate measures to describe the center and the variation.

11. Prices of Jeans

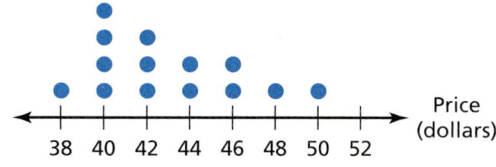

12. Time Volunteering

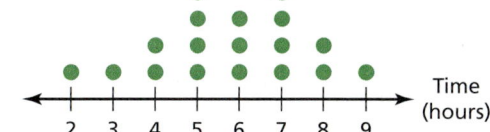

13. Ages of Cats at a Shelter

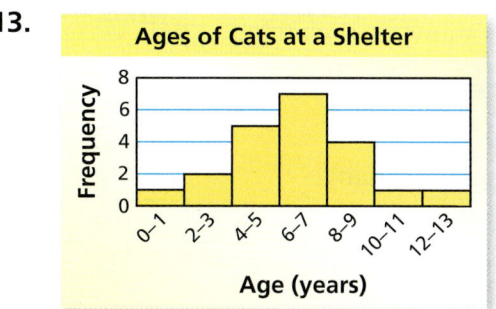

14. Practice Exercises Completed

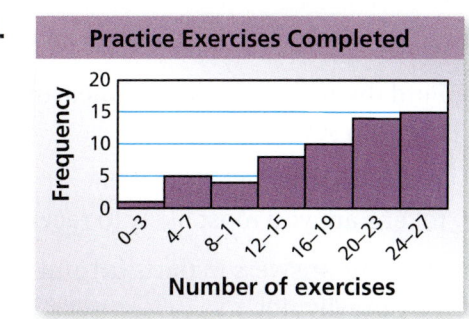

15. **DESCRIBING DATA SETS** Describe the centers and the variations of the data sets in Exercises 11 and 12.

Eggs (thousands)	Frequency
1–20	5
21–40	8
41–60	14
61–80	9
81–100	3

16. **MODELING REAL LIFE** The frequency table shows the numbers of eggs laid by several octopi. What are the most appropriate measures to describe the center and the variation? Explain your reasoning.

17. **MODELING REAL LIFE** The dot plot shows the vertical jump heights (in inches) of several professional athletes. Describe the center and the variation of the data set.

18. **OPEN-ENDED** Describe a real-life situation where the median and the interquartile range are likely the best measures of center and variation to describe the data. Explain your reasoning.

Pile A		Pile B	
5	7	7	5
4	3	5	2
6	4	3	8
5	6	5	6
4	6	5	4
3	5	0	8
5	7	2	10

19. **MP PROBLEM SOLVING** You play a board game in which you draw from one of two piles of cards. Each card has a number that says how many spaces you will move your piece forward on the game board. The tables show the numbers on the cards in each pile. From which pile would you choose? Explain your reasoning.

20. **DIG DEEPER!** The frequency table shows the numbers of words that several students can form in 1 minute using the letters P, S, E, D, A. What are the most appropriate measures to describe the center and variation? Can you find the exact values of the measures of center and variation for the data? Explain.

Number of Words	1–3	4–6	7–9	10–12	13–15
Frequency	7	9	5	3	1

21. **MP REASONING** A bag contains 20 vouchers that can be redeemed for different numbers of tokens at an arcade, as shown in the table.

   a. Find the most appropriate measure to describe the center of the data set.

   b. You randomly select a voucher from the bag. How many tokens are you most likely to receive? Explain.

   c. Are your answers in parts (a) and (b) the same? Explain why or why not.

Numbers of Tokens				
1	10	2	1	4
9	3	10	2	3
4	2	1	4	1
2	1	10	3	7

482   Chapter 10   Data Displays

10.5 Box-and-Whisker Plots

Learning Target: Display and interpret data in box-and-whisker plots.

Success Criteria:
- I can find the five-number summary of a data set.
- I can make a box-and-whisker plot.
- I can explain what the box and the whiskers of a box-and-whisker plot represent.
- I can compare data sets represented by box-and-whisker plots.

EXPLORATION 1

Drawing a Box-and-Whisker Plot

Work with a partner. Each student in a sixth-grade class is asked to choose a number from 1 to 20. The results are shown below.

Numbers Chosen

4	5	14	16
5	16	17	8
18	13	17	18
17	14	19	11
15	8	2	18
13	19	8	7

Math Practice

View as Components
What do the different components of a box-and-whisker plot represent?

a. The *box-and-whisker plot* below represents the data set. Which part represents the *box*? the *whiskers*? Explain.

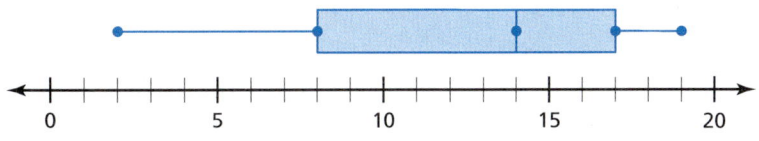

b. What does each of the five plotted points represent?

c. In your own words, describe what a box-and-whisker plot is and what it tells you about a data set.

d. Conduct a survey in your class. Have each student write a number from 1 to 20 on a piece of paper. Collect all of the data and draw a box-and-whisker plot that represents the data. Compare the data with the box-and-whisker plot in part (a).

10.5 Lesson

Key Vocabulary 🔊
box-and-whisker plot, p. 484
five-number summary, p. 484

Key Idea

Box-and-Whisker Plot

A **box-and-whisker plot** represents a data set along a number line by using the least value, the greatest value, and the quartiles of the data. A box-and-whisker plot shows the *variability* of a data set.

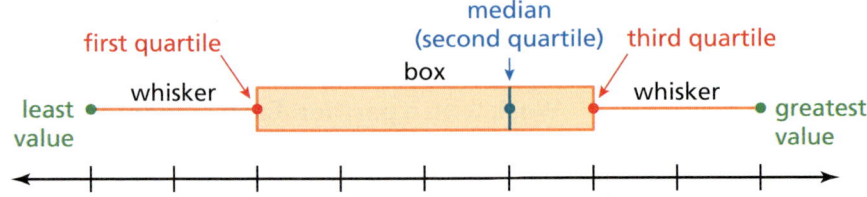

The five numbers that make up the box-and-whisker plot are called the **five-number summary** of the data set.

EXAMPLE 1   Making a Box-and-Whisker Plot

Make a box-and-whisker plot for the ages (in years) of the spider monkeys at a zoo.

15, 20, 14, 38, 30, 36, 30, 30, 27, 26, 33, 35

Step 1: Order the data. Find the quartiles.

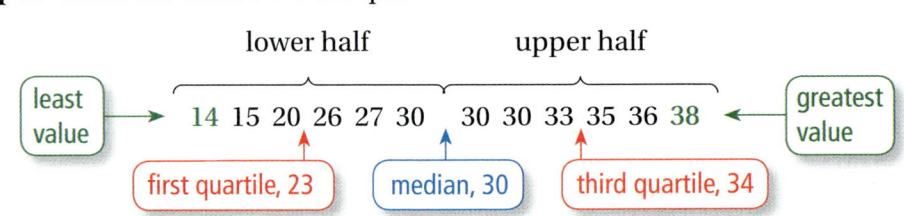

Step 2: Draw a number line that includes the least and greatest values. Graph points above the number line that represent the five-number summary.

Step 3: Draw a box using the quartiles. Draw a line through the median. Draw whiskers from the box to the least and the greatest values.

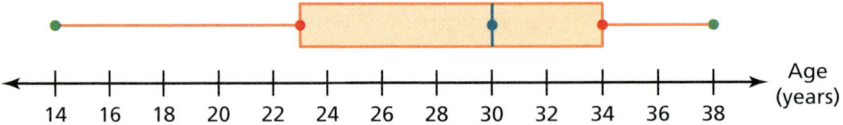

Try It

1. A group of friends spent 1, 0, 2, 3, 4, 3, 6, 1, 0, 1, 2, and 2 hours online last night. Make a box-and-whisker plot for the data.

484   Chapter 10   Data Displays

The figure shows how data are distributed in a box-and-whisker plot.

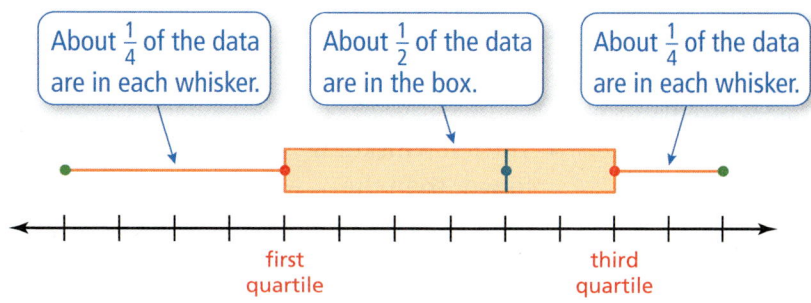

A long whisker or box indicates that the data are more spread out.

About $\frac{1}{4}$ of the data are in each whisker. About $\frac{1}{2}$ of the data are in the box. About $\frac{1}{4}$ of the data are in each whisker.

first quartile   third quartile

EXAMPLE 2 Analyzing a Box-and-Whisker Plot

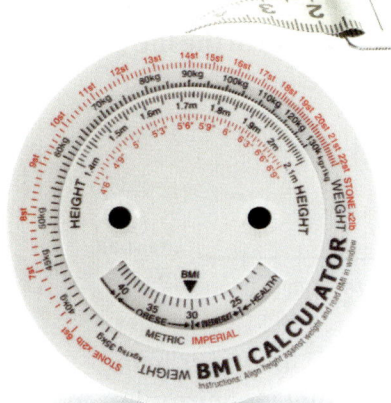

The box-and-whisker plot shows the body mass index (BMI) of a sixth-grade class.

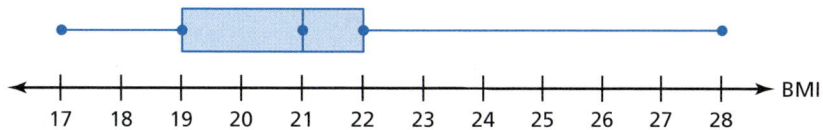

a. **What fraction of the students have a BMI of at least 22?**

   The right whisker represents students who have a BMI of at least 22.

   ▸ So, about $\frac{1}{4}$ of the students have a BMI of at least 22.

b. **Are the data more spread out below the first quartile or above the third quartile? Explain.**

   The right whisker is longer than the left whisker.

   ▸ So, the data are more spread out above the third quartile than below the first quartile.

c. **Find and interpret the interquartile range of the data.**

   interquartile range = third quartile − first quartile

   $= 22 - 19 = 3$

   ▸ So, the middle half of the students' BMIs varies by no more than 3.

Try It

2. The box-and-whisker plot shows the heights of the roller coasters at an amusement park. (a) What fraction of the roller coasters are between 120 feet tall and 220 feet tall? (b) Are the data more spread out below or above the median? Explain. (c) Find and interpret the interquartile range of the data.

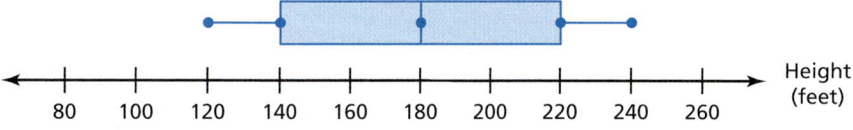

A box-and-whisker plot also shows the shape of a distribution.

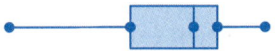

Skewed left   *Symmetric*   *Skewed right*

> If you can draw a line through the median of a box-and-whisker plot, and each side is a mirror image of the other, then the distribution is symmetric.

- The left whisker is longer than the right whisker.
- Most data are on the right.

- The whiskers are about the same length.
- The median is in the middle of the box.

- The right whisker is longer than the left whisker.
- Most data are on the left.

EXAMPLE 3 — Identifying Shapes of Distributions

The double box-and-whisker plot represents the life spans of crocodiles and alligators at a zoo. Identify the shape of the distribution of the life spans of alligators.

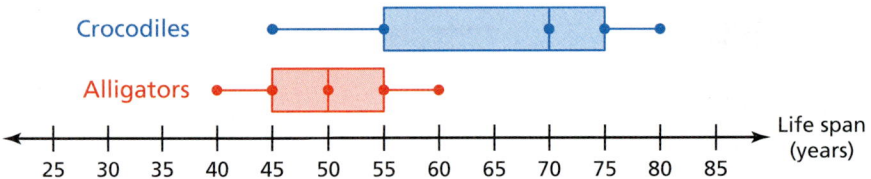

For alligator life spans, the whisker lengths are equal. The median is in the middle of the box. The left side of the box-and-whisker plot is a mirror image of the right side of the box-and-whisker plot.

▶ So, the distribution is symmetric.

Try It

3. Identify the shape of the distribution of the life spans of crocodiles.

Self-Assessment for Concepts & Skills

Solve each exercise. Then rate your understanding of the success criteria in your journal.

4. **VOCABULARY** Explain how to find the five-number summary of a data set.

MAKING A BOX-AND-WHISKER PLOT Make a box-and-whisker plot for the data. Identify the shape of the distribution.

5. Ticket prices (dollars): 39, 42, 40, 47, 38, 39, 44, 55, 44, 58, 45

6. Number of sit-ups: 20, 20, 23, 25, 25, 26, 27, 29, 30, 30, 32, 34, 37, 38

7. **MP NUMBER SENSE** In a box-and-whisker plot, what fraction of the data is greater than the first quartile?

EXAMPLE 4   Modeling Real Life

The double box-and-whisker plot represents the prices of snowboards at two stores.

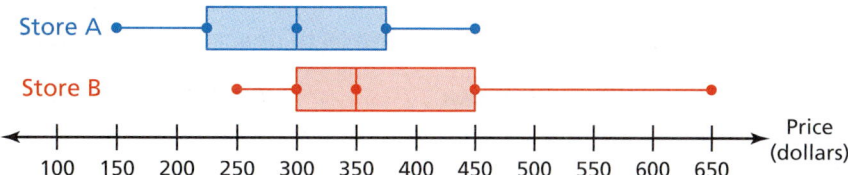

a. **Which store's prices are more spread out? Explain.**

Both boxes appear to be the same length. So, the interquartile range of each data set is equal. The range of the prices in Store B, however, is greater than the range of the prices in Store A.

 So, the prices in Store B are more spread out.

b. **Which store's prices are generally higher? Explain.**

For Store A, the distribution is symmetric with about one-half of the prices above $300.

For Store B, the distribution is skewed right with about three-fourths of the prices above $300.

 So, the prices in Store B are generally higher.

Self-Assessment for Problem Solving

Solve each exercise. Then rate your understanding of the success criteria in your journal.

Test Scores	
75	65
64	79
100	75
94	52
73	80

Test Scores	
56	70
47	100
83	44
45	58
54	30

8. The tables at the left show the test scores of two sixth-grade achievement tests. The same group of students took both tests. The students took one test in the fall and the other in the spring.

   a. Analyze each distribution. Then compare and contrast the test results.

   b. Which table likely represents the results of which test? Explain your reasoning.

9. Make a box-and-whisker plot that represents the heights of the boys in your class. Then make a box-and-whisker plot that represents the heights of the girls in your class. Compare and contrast the distributions.

10.5 Practice

Go to **BigIdeasMath.com** to get HELP with solving the exercises.

▶ Review & Refresh

Choose the most appropriate measures to describe the center and the variation.

1.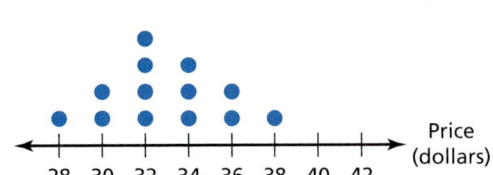
   Prices of Games

2.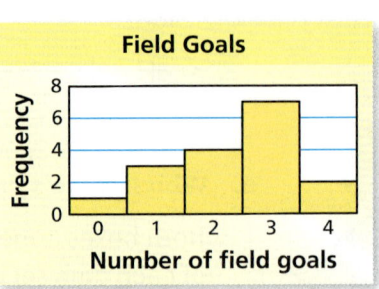

Copy and complete the statement using < or >.

3. $-\dfrac{2}{3}$ ☐ $-\dfrac{3}{4}$

4. $-2\dfrac{1}{5}$ ☐ $-2\dfrac{1}{6}$

5. -5.3 ☐ -5.5

Factor the expression using the GCF.

6. $42 + 14$

7. $12x - 18$

8. $28n + 20$

9. $60g - 25h$

▶ Concepts, Skills, & Problem Solving

COMPARING DATA Compare the data in the box-and-whisker plots. (See Exploration 1, p. 483.)

10.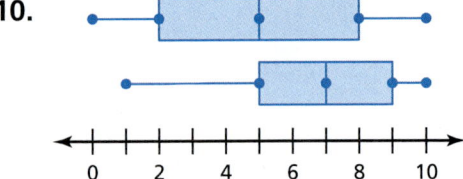

11.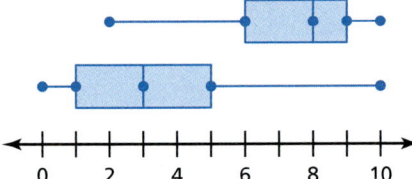

MAKING A BOX-AND-WHISKER PLOT Make a box-and-whisker plot for the data.

12. Ages of teachers (in years): 30, 62, 26, 35, 45, 22, 49, 32, 28, 50, 42, 35

13. Quiz scores: 8, 12, 9, 10, 12, 8, 5, 9, 7, 10, 8, 9, 11

14. Donations (in dollars): 10, 30, 5, 15, 50, 25, 5, 20, 15, 35, 10, 30, 20

15. Science test scores: 85, 76, 99, 84, 92, 95, 68, 100, 93, 88, 87, 85

16. Shoe sizes: 12, 8.5, 9, 10, 9, 11, 11.5, 9, 9, 10, 10, 10.5, 8

17. Ski lengths (in centimeters): 180, 175, 205, 160, 210, 175, 190, 205, 190, 160, 165, 195

18. **YOU BE THE TEACHER** Your friend makes a box-and-whisker plot for the data shown. Is your friend correct? Explain your reasoning.

    2, 6, 4, 3, 7, 4, 6, 9, 6, 8, 5, 7

    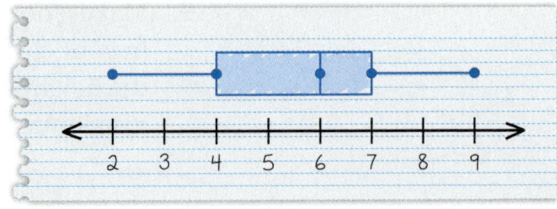

19. **MODELING REAL LIFE** The numbers of days 12 friends went camping during the summer are 6, 2, 0, 10, 3, 6, 6, 4, 12, 0, 6, and 2. Make a box-and-whisker plot for the data. What is the range of the data?

20. **ANALYZING A BOX-AND-WHISKER PLOT** The box-and-whisker plot represents the numbers of gallons of water needed to fill different types of dunk tanks offered by a company.

    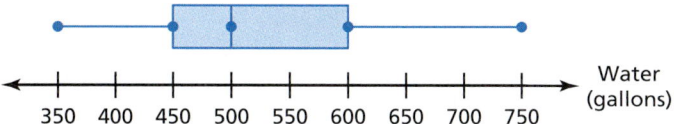

    a. What fraction of the dunk tanks requires at least 500 gallons of water?

    b. Are the data more spread out below the first quartile or above the third quartile? Explain.

    c. Find and interpret the interquartile range of the data.

21. **MODELING REAL LIFE** The box-and-whisker plot represents the heights (in meters) of the tallest buildings in Chicago.

    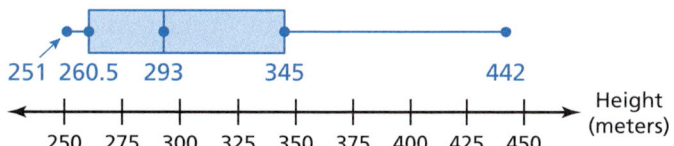

    a. What percent of the buildings are no taller than 345 meters?

    b. Is there more variability in the heights above 345 meters or below 260.5 meters? Explain.

    c. Find and interpret the interquartile range of the data.

22. **CRITICAL THINKING** The numbers of spots on several frogs in a jungle are shown in the dot plot.

    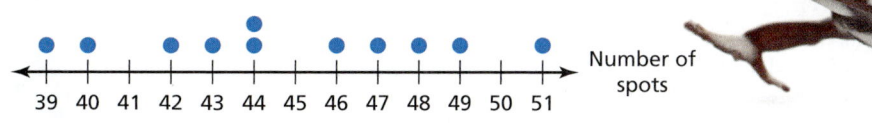

    a. Make a box-and-whisker plot for the data.

    b. Compare the dot plot and the box-and-whisker plot. Describe the advantages and disadvantages of each data display.

SHAPES OF BOX-AND-WHISKER PLOTS Identify the shape of the distribution. Explain.

23.

24.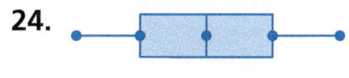

25.

26.

27. **MODELING REAL LIFE** The double box-and-whisker plot represents the start times of recess for classes at two schools.

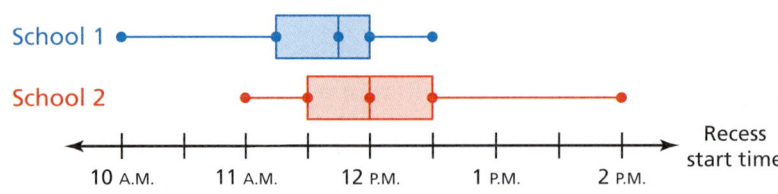

a. Identify the shape of each distribution.

b. Which school's start times for recess are more spread out? Explain.

c. You randomly pick one class from each school. Which class is more likely to have recess before lunch? Explain.

MAKING A BOX-AND-WHISKER PLOT Make a box-and-whisker plot for the data.

28. Temperatures (in °C): 15, 11, 14, 10, 19, 10, 2, 15, 12, 14, 9, 20, 17, 5

29. Checking account balances (in dollars): 30, 0, 50, 20, 90, −15, 40, 100, 45, −20, 70, 0

30. **REASONING** The data set in Exercise 28 has an outlier. Describe how removing the outlier affects the box-and-whisker plot.

31. **OPEN-ENDED** Write a data set with 12 values that has a symmetric box-and-whisker plot.

32. **CRITICAL THINKING** When does a box-and-whisker plot *not* have one or both whiskers?

33. **STRUCTURE** Draw a histogram that could represent the distribution shown in Exercise 25.

34. **DIG DEEPER!** The double box-and-whisker plot represents the goals scored per game by two lacrosse teams during a 16-game season.

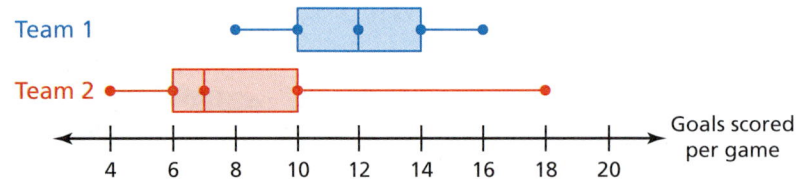

a. Which team is more consistent? Explain.

b. Team 1 played Team 2 once during the season. Which team do you think won? Explain.

c. Can you determine the number of games in which Team 2 scored 10 goals or less? Explain your reasoning.

35. **CHOOSE TOOLS** A market research company wants to summarize the variability of the SAT scores of graduating seniors in the United States. Should the company use a stem-and-leaf plot, a histogram, or a box-and-whisker plot? Explain.

10 Connecting Concepts

▶ Using the Problem-Solving Plan

1. The locations of pitches in an at-bat are shown in the coordinate plane, where the coordinates are measured in inches. Describe the location of a typical pitch in the at-bat.

   Understand the problem. You know the locations of the pitches. You are asked to find the location of a typical pitch in the at-bat.

   Make a plan. First, use the coordinates of the pitches to create two data sets, one for the x-coordinates of the pitches and one for the y-coordinates of the pitches. Next, make a box-and-whisker plot for each data set. Then use the most appropriate measure of center for each data set to find the location of a typical pitch.

   Solve and check. Use the plan to solve the problem. Then check your solution.

2. A set of 20 data values is described below. Sketch a histogram that could represent the data set. Explain.

   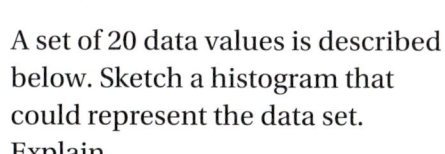

   - least value: 10
   - third quartile: 34
   - first quartile: 25
   - greatest value: 48
   - mean: 29
   - MAD: 7

3. The chart shows the dimensions (in inches) of several flat-rate shipping boxes. Each box is in the shape of a rectangular prism. Describe the distribution of the volumes of the boxes. Then find the most appropriate measures to describe the center and the variation of the volumes.

SHIPPING RATES

Dimensions	Price
5 × 5 × 4	$6.80
8 × 5 × 3	$8.30
9 × 8 × 5	$9.75
10 × 6 × 6	$10.75
10 × 10 × 4	$10.75
8 × 7 × 5	$10.75
12 × 10 × 3	$11.25
15 × 10 × 3	$12.25
12 × 12 × 5	$17.40

Performance Task

Classifying Dog Breeds by Size

At the beginning of this chapter, you watched a STEAM Video called "Choosing a Dog." You are now ready to complete the performance task related to this video, available at *BigIdeasMath.com*. Be sure to use the problem-solving plan as you work through the performance task.

10 Chapter Review

Go to *BigIdeasMath.com* to download blank graphic organizers.

Review Vocabulary

Write the definition and give an example of each vocabulary term.

stem-and-leaf plot, *p. 458*
stem, *p. 458*
leaf, *p. 458*
frequency table, *p. 463*
frequency, *p. 463*
histogram, *p. 464*
box-and-whisker plot, *p. 484*
five-number summary, *p. 484*

Graphic Organizers

You can use an **Information Frame** to help you organize and remember concepts. Here is an example of an Information Frame for the vocabulary term *histogram*.

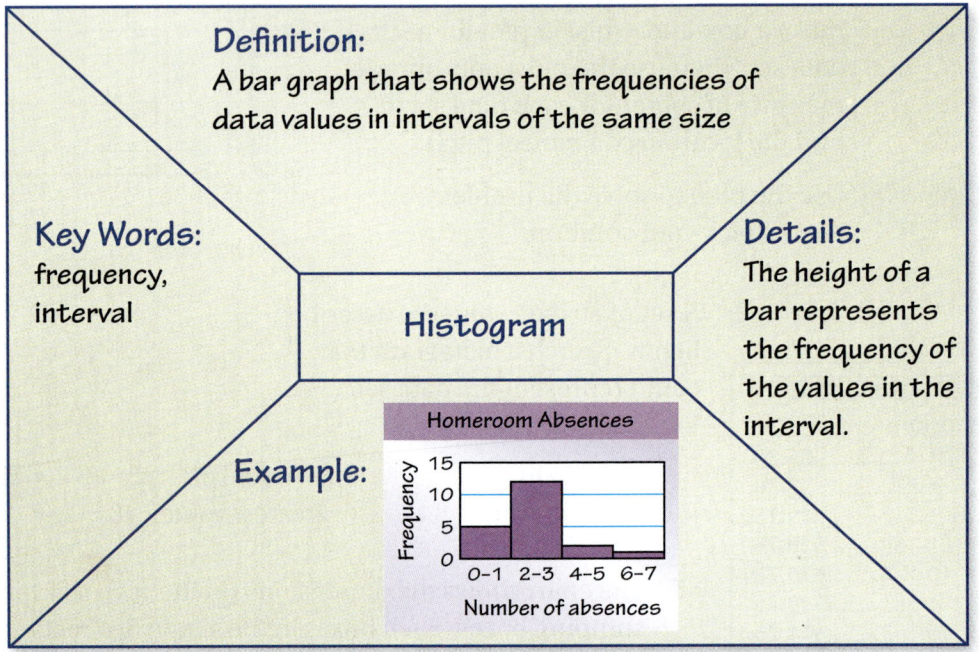

Choose and complete a graphic organizer to help you study the concept.

1. stem-and-leaf plot
2. frequency table
3. shapes of distributions
4. box-and-whisker plot

"I finished my **Information Frame** about dog sweaters. Why don't you make your information frame about barn cats?"

492 Chapter 10 Data Displays

Chapter Self-Assessment

As you complete the exercises, use the scale below to rate your understanding of the success criteria in your journal.

1	2	3	4
I do not understand.	I can do it with help.	I can do it on my own.	I can teach someone else.

10.1 Stem-and-Leaf Plots (pp. 457–462)

Learning Target: Display and interpret data in stem-and-leaf plots.

Make a stem-and-leaf plot of the data.

1. **Hats Sold Each Day**

5	18	12	15
21	30	8	12
13	9	14	25

2. **Ages of Park Volunteers**

13	17	40	15
48	21	19	52
13	55	60	20

3. The stem-and-leaf plot shows the weights (in pounds) of yellowfin tuna caught during a fishing contest.

Stem	Leaf
7	6 6 8
8	0 2 5 5 5 7 9
9	3 5 6 6
10	2

    Key: 8 | 5 = 85 pounds

    a. How many tuna weigh less than 90 pounds?
    b. Find the mean, median, mode, range, and interquartile range of the data.
    c. How are the data distributed?

Stem	Leaf
1	6 8 9
2	0 2 3 4 4 5 7 9
3	0 5

    Key: 2 | 7 = 27

4. The stem-and-leaf plot shows the body mass index (BMI) for adults at a recreation center. Use the data to answer the question, "What is the typical BMI for an adult at the recreation center?" Explain.

5. Write a statistical question that can be answered using the stem-and-leaf plot.

    Songs Downloaded per Day

Stem	Leaf
0	0 2 4 5 6 6 6 7 8
1	3 4 8 9
2	
3	2

    Key: 1 | 9 = 19 songs

Chapter Review 493

10.2 Histograms (pp. 463–470)

Learning Target: Display and interpret data in histograms.

Display the data in a histogram.

6.

Heights of Gymnasts	
Heights (inches)	Frequency
50–54	1
55–59	8
60–64	5
65–69	2

7.

Minutes Studied	
Minutes	Frequency
0–19	5
20–39	9
40–59	12
60–79	3

8. The histogram shows the number of crafts each member of a craft club made for a fundraiser.

    a. Which interval contains the most data values?

    b. How many members made at least 6 crafts?

    c. Can you use the histogram to determine the total number of crafts made? Explain.

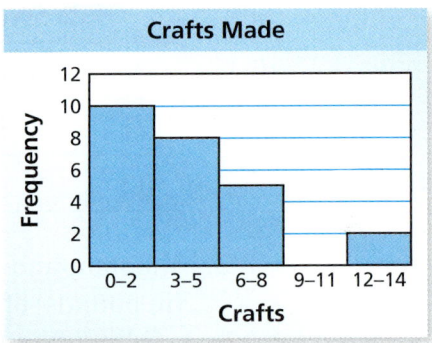

10.3 Shapes of Distributions (pp. 471–476)

Learning Target: Describe and compare shapes of distributions.

9. Describe the shape of the distribution.

10. The frequency table shows the math test scores for the same class of students as Exercise 9. Display the data in a histogram. Which test has higher scores?

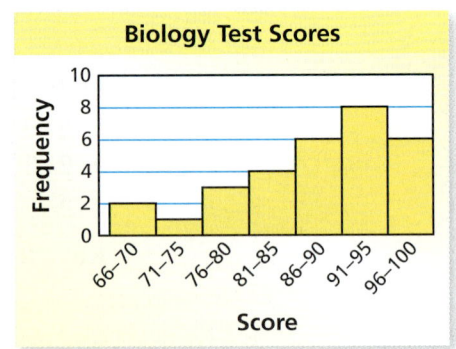

Score	66–70	71–75	76–80	81–85
Frequency	1	5	10	8

Score	86–90	91–95	96–100
Frequency	2	2	2

11. The table shows the numbers of neutrons for several elements in the nonmetal group of the periodic table. Make a histogram of the data starting with the interval 0–9. Describe the shape of the distribution.

Number of Neutrons			
0	45	16	8
7	6	16	

494   Chapter 10   Data Displays

10.4 Choosing Appropriate Measures (pp. 477–482)

Learning Target: Use the shape of the distribution of a data set to determine which measures of center and variation best describe the data.

Choose the most appropriate measures to describe the center and the variation.

12.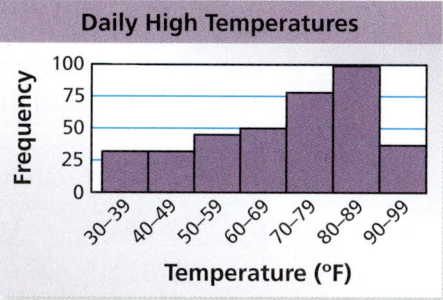

13.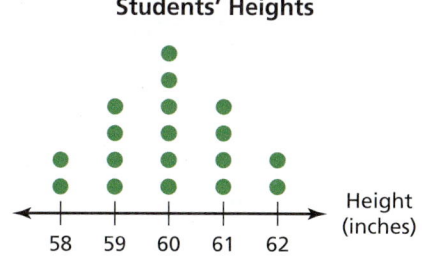

14. Describe the center and the variation of the data set in Exercise 13.

10.5 Box-and-Whisker Plots (pp. 483–490)

Learning Target: Display and interpret data in box-and-whisker plots.

Make a box-and-whisker plot for the data.

15. Ages of volunteers at a hospital:
    14, 17, 20, 16, 17, 14, 21, 18

16. Masses (in kilograms) of lions:
    120, 200, 180, 150, 200, 200, 230, 160

17. The box-and-whisker plot represents the lengths (in minutes) of movies being shown at a theater.

    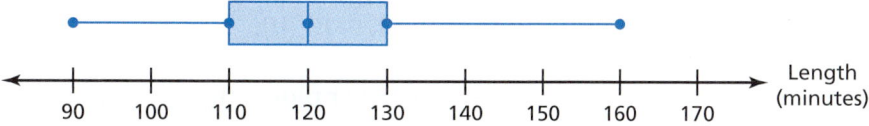

    a. What percent of the movies are no longer than 120 minutes?

    b. Is there more variability in the movie lengths longer than 130 minutes or shorter than 110 minutes? Explain.

    c. Find and interpret the interquartile range of the data.

18. The double box-and-whisker plot represents the heights of students in two math classes.

    a. Identify the shape of each distribution.

    b. Which class has heights that are more spread out? Explain.

    c. You randomly pick one student from each class. Which student is more likely to be taller than 170 centimeters? Explain.

10 Practice Test

Make a stem-and-leaf plot of the data.

1.

Quiz Scores (%)			
96	88	80	72
80	94	92	100
76	80	68	90

2.

Songs Downloaded Each Day				
45	31	29	38	38
67	40	62	45	60
40	39	60	43	48

3. Find the mean, median, mode, range, and interquartile range of the data.

Cooking Time

Stem	Leaf
3	5 8
4	0 1 8
5	0 4 4 4 5 9
6	0

Key: 4 | 1 = 41 minutes

4. Display the data in a histogram. How many people watched less than 20 hours of television per week?

Television Watched per Week	
Hours	Frequency
0–9	14
10–19	16
20–29	10
30–39	8

5. The dot plot shows the numbers of glasses of water that the students in a class drink in one day.

    a. Describe the shape of the distribution.

    b. Choose the most appropriate measures to describe the center and the variation. Find the measures you chose.

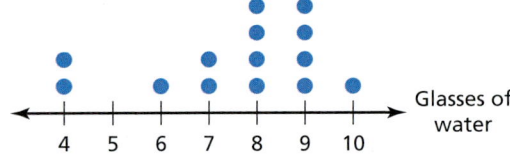

6. Make a box-and-whisker plot for the lengths (in inches) of fish in a pond: 12, 13, 7, 8, 14, 6, 13, 10.

7. The double box-and-whisker plot compares the battery lives (in hours) of two brands of cell phones.

    a. What is the range of the upper 75% of battery life for each brand of cell phone?

    b. Which brand of cell phone typically has a longer battery life? Explain.

    c. In the box-and-whisker plot, there are 190 cell phones of Brand A that have at most 10.5 hours of battery life. About how many cell phones are represented in the box-and-whisker plot for Brand A?

10 Cumulative Practice

1. Research scientists are measuring the numbers of days lettuce seeds take to germinate. In a study, 500 seeds were planted. Of these, 473 seeds germinated. The box-and-whisker plot summarizes the numbers of days it took the seeds to germinate. What can you conclude from the box-and-whisker plot?

   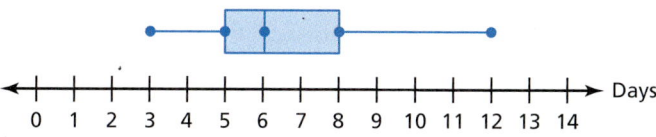

   A. The median number of days for the seeds to germinate is 12.

   B. 50% of the seeds took more than 8 days to germinate.

   C. 50% of the seeds took less than 5 days to germinate.

   D. The median number of days for the seeds to germinate was 6.

2. Find the interquartile range of the data.

   15   7   5   8   9   20   12   7   11   7   15

   F. 8   **G.** 11

   H. 12   **I.** 20

3. There are seven different integers in a set. When they are listed from least to greatest, the middle integer is -1. Which statement below must be true?

   A. There are three negative integers in the set.

   B. There are three positive integers in the set.

   C. There are four negative integers in the set.

   D. The integer in the set after -1 is positive.

4. What is the mean number of seats?

- F. 2.4 seats
- G. 5 seats
- H. 6.5 seats
- I. 7 seats

5. On Wednesday, a town received 17 millimeters of rain. This was x millimeters more rain than the town received on Tuesday. Which expression represents the amount of rain, in millimeters, the town received on Tuesday?

- A. $17x$
- B. $17 - x$
- C. $x + 17$
- D. $x - 17$

6. One of the leaves is missing in the stem-and-leaf plot.

   The median of the data set represented by the stem-and-leaf plot is 38. What is the value of the missing leaf?

Stem	Leaf
1	3 4
2	
3	4 5 7 7 ? 9
4	0 1 1 4
5	0 2 3

Key: $1\,|\,4 = 14$

7. Which property is demonstrated by the equation?

$$723 + (y + 277) = 723 + (277 + y)$$

- F. Associative Property of Addition
- G. Commutative Property of Addition
- H. Distributive Property
- I. Addition Property of Zero

498 Chapter 10 Data Displays

8. A student took five tests and had a mean score of 92. Her scores on the first 4 tests were 90, 96, 86, and 92. What was her score on the fifth test?

   A. 92
   B. 93
   C. 96
   D. 98

9. At the end of the school year, your teacher counted the number of absences for each student. The results are shown in the histogram. How many students had fewer than 10 absences?

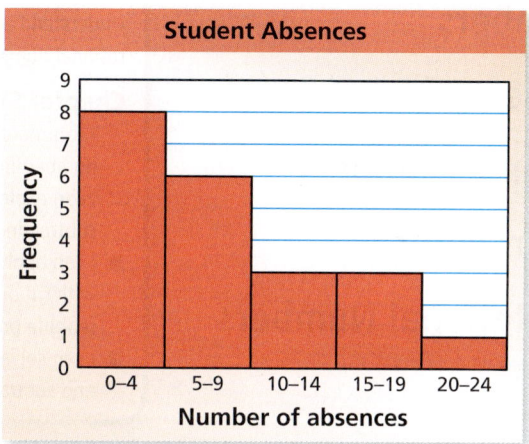

10. The ages of the 16 members of a camera club are listed below.

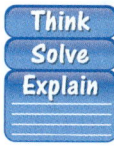

40, 22, 24, 58, 30, 31, 37, 25, 62, 40, 39, 37, 28, 28, 51, 44

Part A Order the ages from youngest to oldest.

Part B Find the median of the ages.

Part C Make a box-and-whisker plot for the ages.

A Adding and Subtracting Rational Numbers

- **A.1** Rational Numbers
- **A.2** Adding Integers
- **A.3** Adding Rational Numbers
- **A.4** Subtracting Integers
- **A.5** Subtracting Rational Numbers

Chapter Learning Target:
Understand adding and subtracting rational numbers.

Chapter Success Criteria:
- I can represent rational numbers on a number line.
- I can explain the rules for adding and subtracting integers using absolute value.
- I can apply addition and subtraction with rational numbers to model real-life problems.
- I can solve problems involving addition and subtraction of rational numbers.

STEAM Video: "Freezing Solid"

STEAM Video

Freezing Solid

The Celsius temperature scale is defined using the freezing point, 0°C, and the boiling point, 100°C, of water. Why do you think the scale is defined using these two points?

Watch the STEAM Video "Freezing Solid." Then answer the following questions.

1. In the video, Tony says that the freezing point of wax is 53°C and the boiling point of wax is 343°C.

    a. Describe the temperature of wax that has just changed from liquid form to solid form. Explain your reasoning.

    b. After Tony blows out the candle, he demonstrates that there is still gas in the smoke. What do you know about the temperature of the gas that is in the smoke?

    c. In what form is wax when the temperature is at 100°C, the boiling point of water?

2. Consider wax in solid, liquid, and gaseous forms. Which is hottest? coldest?

Performance Task

Melting Matters

After completing this chapter, you will be able to use the concepts you learned to answer the questions in the *STEAM Video Performance Task*. You will answer questions using the melting points of the substances below.

Ice	Tin
Beeswax	Ethanol
Mercury	Acetone
Plastic	Chocolate

You will graph the melting points of the substances on a number line to make comparisons. How is the freezing point of a substance related to its melting point? What is meant when someone says it is below freezing outside? Explain.

501

Getting Ready for Chapter A

Chapter Exploration

1. Work with a partner. Plot and connect the points to make a picture.

   1(1, 11) **2**(4, 10) **3**(7, 10) **4**(11, 9) **5**(13, 8)
   6(15, 5) **7**(15, 3) **8**(16, 1) **9**(16, −1) **10**(15, −1)
   11(11, 1) **12**(9, 2) **13**(7, 1) **14**(5, −1) **15**(1, −1)
   16(0, 0) **17**(3, 1) **18**(1, 1) **19**(−2, 0) **20**(−6, −2)
   21(−9, −6) **22**(−9, −7) **23**(−7, −9) **24**(−7, −11) **25**(−8, −12)
   26(−9, −11) **27**(−11, −10) **28**(−13, −11) **29**(−15, −11) **30**(−17, −12)
   31(−17, −10) **32**(−15, −7) **33**(−12, −6) **34**(−11, −6) **35**(−10, −3)
   36(−8, 2) **37**(−5, 6) **38**(−3, 9) **39**(−4, 10) **40**(−5, 10)
   41(−2, 12)

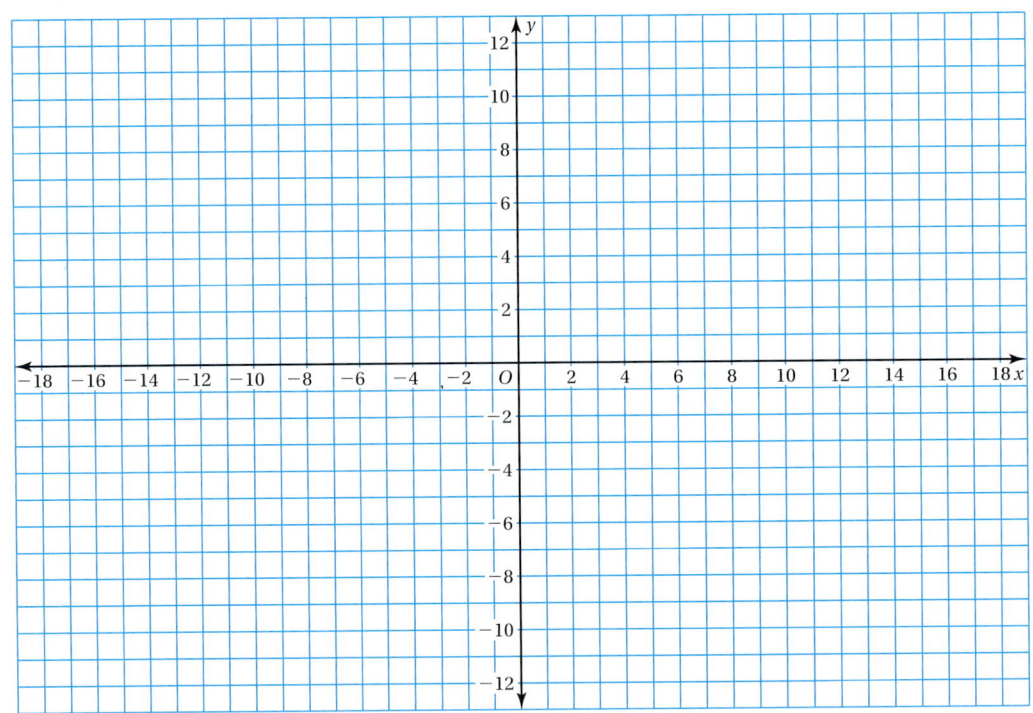

2. Create your own "dot-to-dot" picture. Use at least 20 points.

Vocabulary

The following vocabulary terms are defined in this chapter. Think about what each term might mean and record your thoughts.

integers   absolute value   rational number   additive inverse

A.1 Rational Numbers

Learning Target: Understand absolute values and ordering of rational numbers.

Success Criteria:
- I can graph rational numbers on a number line.
- I can find the absolute value of a rational number.
- I can use a number line to compare rational numbers.

Recall that **integers** are the set of whole numbers and their opposites. A **rational number** is a number that can be written as $\frac{a}{b}$, where a and b are integers and $b \neq 0$.

EXPLORATION 1

Using a Number Line

Work with a partner. Make a number line on the floor. Include both negative numbers and positive numbers.

a. Stand on an integer. Then have your partner stand on the opposite of the integer. How far are each of you from 0? What do you call the distance between a number and 0 on a number line?

b. Stand on a rational number that is not an integer. Then have your partner stand on any other number. Which number is greater? How do you know?

c. Stand on any number other than 0 on the number line. Can your partner stand on a number that is:
- greater than your number and farther from 0?
- greater than your number and closer to 0?
- less than your number and the same distance from 0?
- less than your number and farther from 0?

For each case in which it was not possible to stand on a number as directed, explain why it is not possible. In each of the other cases, how can you decide where your partner can stand?

Math Practice

Find Entry Points

What are some ways to determine which of two numbers is greater?

A.1 Lesson

Key Vocabulary
integers, *p. 503*
rational number, *p. 503*
absolute value, *p. 504*

🔑 Key Idea

Absolute Value

Words The **absolute value** of a number is the distance between the number and 0 on a number line. The absolute value of a number a is written as $|a|$.

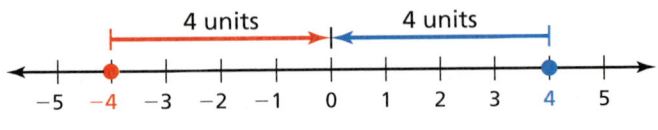

Numbers $|-4| = 4 \qquad |4| = 4$

EXAMPLE 1 — Finding Absolute Values of Rational Numbers

a. Find the absolute value of -3.

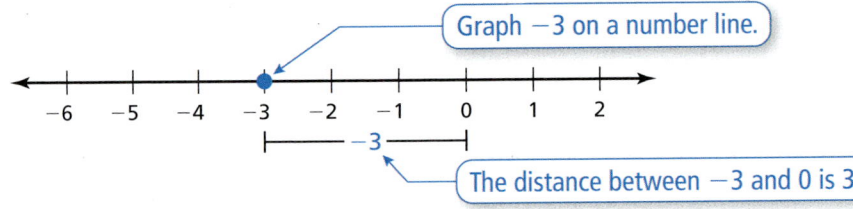

▸ So, $|-3| = 3$.

b. Find the absolute value of $1\frac{1}{4}$.

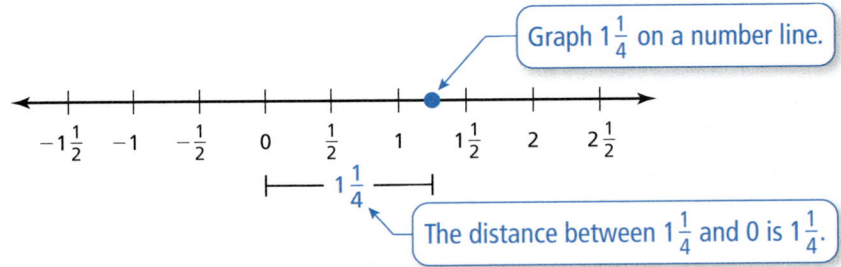

▸ So, $\left|1\frac{1}{4}\right| = 1\frac{1}{4}$.

Try It Find the absolute value.

1. $|7|$
2. $\left|-\frac{5}{3}\right|$
3. $|-2.6|$

EXAMPLE 2 Comparing Rational Numbers

Compare $\left|-2.5\right|$ and $\frac{3}{2}$.

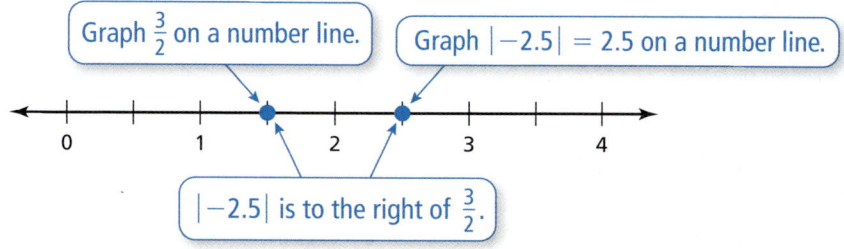

So, $\left|-2.5\right| > \frac{3}{2}$.

Try It Copy and complete the statement using <, >, or =.

4. $\left|9\right|$ ▢ $\left|-9\right|$
5. $-\left|\frac{1}{2}\right|$ ▢ $-\frac{1}{4}$
6. 7 ▢ $-\left|-4.5\right|$

> **Remember**
> Two numbers that are the same distance from 0 on a number line, but on opposite sides of 0, are called *opposites*. The opposite of a number a is $-a$.

Self-Assessment for Concepts & Skills

Solve each exercise. Then rate your understanding of the success criteria in your journal.

7. **VOCABULARY** Which of the following numbers are integers?

    $9, 3.2, -1, \frac{1}{2}, -0.25, 15$

8. **VOCABULARY** What is the absolute value of a number?

COMPARING RATIONAL NUMBERS Copy and complete the statement using <, >, or =. Use a number line to justify your answer.

9. 3.5 ▢ $\left|-\frac{7}{2}\right|$
10. $\left|\frac{11}{4}\right|$ ▢ $\left|-2.8\right|$

11. **WRITING** You compare two numbers, a and b. Explain how $a > b$ and $\left|a\right| < \left|b\right|$ can both be true statements.

12. **WHICH ONE DOESN'T BELONG?** Which expression does *not* belong with the other three? Explain your reasoning.

    $\left|6\right|$ 6 -6 $\left|-6\right|$

Section A.1 Rational Numbers 505

EXAMPLE 3 Modeling Real Life

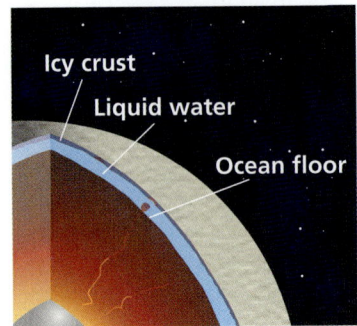

A moon has an ocean underneath its icy surface. Scientists run tests above and below the surface. The table shows the elevations of each test. Which test is deepest? Which test is closest to the surface?

Test	Temperature	Salinity	Atmosphere	Organics	Ice
Elevation (miles)	−3.8	−5.15	0.3	−4.5	−0.25

To determine which test is deepest, find the least elevation. Graph the elevations on a vertical number line.

The number line shows that the salinity test is deepest. The number line also shows that the atmosphere test and the ice test are closest to the surface. To determine which is closer to the surface, identify which elevation has a lesser absolute value.

Atmosphere: $|0.3| = 0.3$ **Ice:** $|-0.25| = 0.25$

 So, the salinity test is deepest and the ice test is closest to the surface.

Self-Assessment for Problem Solving

Solve each exercise. Then rate your understanding of the success criteria in your journal.

13. An airplane is at an elevation of 5.5 miles. A submarine is at an elevation of −10.9 kilometers. Which is closer to sea level? Explain.

14. The image shows the corrective powers (in *diopters*) of contact lenses for eight people. The farther the number of diopters is from 0, the greater the power of the lens. Positive diopters correct *farsightedness* and negative diopters correct *nearsightedness*. Who is the most nearsighted? the most farsighted? Who has the best eyesight?

Patient	1	2	3	4	5	6	7	8
Power (diopters)	−1.25	0.75	2.5	−3.75	−2.5	−4.75	−7.5	1.5

A.1 Practice

Go to *BigIdeasMath.com* to get HELP with solving the exercises.

▶ Review & Refresh

Write the ratio.

1. deer to bears
2. bears to deer
3. bears to animals
4. animals to deer

Find the GCF of the numbers.

5. 8, 20
6. 12, 30
7. 7, 28
8. 48, 72

▶ Concepts, Skills, & Problem Solving

MP NUMBER SENSE Determine which number is greater and which number is farther from 0. Explain your reasoning. *(See Exploration 1, p. 503.)*

9. $4, -6$
10. $-3.25, \dfrac{7}{2}$
11. $-\dfrac{4}{5}, -1.3$

FINDING ABSOLUTE VALUES Find the absolute value.

12. $|8|$
13. $|-2|$
14. $|-10|$
15. $|10|$

16. $|0|$
17. $\left|\dfrac{1}{3}\right|$
18. $\left|\dfrac{7}{8}\right|$
19. $\left|-\dfrac{5}{9}\right|$

20. $\left|\dfrac{11}{8}\right|$
21. $|3.8|$
22. $|-5.3|$
23. $\left|-\dfrac{15}{4}\right|$

24. $|7.64|$
25. $|-18.26|$
26. $\left|4\dfrac{2}{5}\right|$
27. $\left|-5\dfrac{1}{6}\right|$

COMPARING RATIONAL NUMBERS Copy and complete the statement using <, >, or =.

28. $2 \;\square\; |-5|$
29. $|-1| \;\square\; |-8|$
30. $|5| \;\square\; |-5|$

31. $|-2| \;\square\; 0$
32. $0.4 \;\square\; \left|-\dfrac{7}{8}\right|$
33. $|4.9| \;\square\; |-5.3|$

34. $-|4.7| \;\square\; \dfrac{1}{2}$
35. $\left|-\dfrac{3}{4}\right| \;\square\; -\left|\dfrac{3}{4}\right|$
36. $-\left|1\dfrac{1}{4}\right| \;\square\; -\left|-1\dfrac{3}{8}\right|$

YOU BE THE TEACHER Your friend compares two rational numbers. Is your friend correct? Explain your reasoning.

37. $|-10| = -10$

38. $-\dfrac{4}{5} > -\dfrac{1}{2}$

Section A.1 Rational Numbers 507

39. OPEN-ENDED Write a negative number whose absolute value is greater than 3.

40. MODELING REAL LIFE The *summit elevation* of a volcano is the elevation of the top of the volcano relative to sea level. The summit elevation of Kilauea, a volcano in Hawaii, is 1277 meters. The summit elevation of Loihi, an underwater volcano in Hawaii, is −969 meters. Which summit is higher? Which summit is closer to sea level?

41. MODELING REAL LIFE The *freezing point* of a liquid is the temperature at which the liquid becomes a solid.

a. Which liquid in the table has the lowest freezing point?

b. Is the freezing point of mercury or butter closer to the freezing point of water, 0°C?

Liquid	Freezing Point (°C)
Butter	35
Airplane fuel	−53
Honey	−3
Mercury	−39
Candle wax	53

ORDERING RATIONAL NUMBERS Order the values from least to greatest.

42. $8, |3|, -5, |-2|, -2$

43. $|-6.3|, -7.2, 8, |5|, -6.3$

44. $|3.5|, |-1.8|, 4.6, 3\frac{2}{5}, |2.7|$

45. $\left|-\frac{3}{4}\right|, \frac{5}{8}, \left|\frac{1}{4}\right|, -\frac{1}{2}, \left|-\frac{7}{8}\right|$

46. MP PROBLEM SOLVING The table shows golf scores, relative to *par*.

a. The player with the lowest score wins. Which player wins?

b. Which player is closest to par?

c. Which player is farthest from par?

Player	Score
1	+5
2	0
3	−4
4	−1
5	+2

47. DIG DEEPER! You use the table below to record the temperature at the same location each hour for several hours. At what time is the temperature coldest? At what time is the temperature closest to the freezing point of water, 0°C?

Time	10:00 A.M.	11:00 A.M.	12:00 P.M.	1:00 P.M.	2:00 P.M.	3:00 P.M.
Temperature (°C)	−2.6	−2.7	−0.15	1.6	−1.25	−3.4

MP REASONING Determine whether $n \geq 0$ or $n \leq 0$.

48. $n + |-n| = 2n$

49. $n + |-n| = 0$

TRUE OR FALSE? Determine whether the statement is *true* or *false*. Explain your reasoning.

50. If $x < 0$, then $|x| = -x$.

51. The absolute value of every rational number is positive.

A.2 Adding Integers

Learning Target: Find sums of integers.

Success Criteria:
- I can explain how to model addition of integers on a number line.
- I can find sums of integers by reasoning about absolute values.
- I can explain why the sum of a number and its opposite is 0.

EXPLORATION 1   Using Integer Counters to Find Sums

$\boxed{+} = +1$
$\boxed{-} = -1$

Work with a partner. You can use the integer counters shown at the left to find sums of integers.

a. How can you use integer counters to model a sum? a sum that equals 0?

b. What expression is being modeled below? What is the value of the sum?

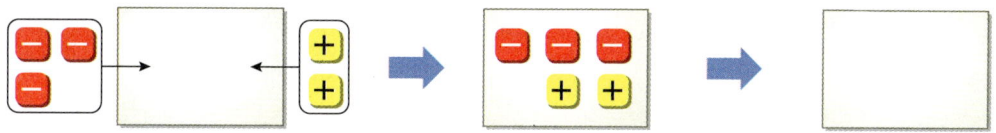

c. **INDUCTIVE REASONING** Use integer counters to complete the table.

Expression	Type of Sum	Sum	Sum: Positive, Negative, or Zero
$-3 + 2$	Integers with different signs		
$-4 + (-3)$			
$5 + (-3)$			
$7 + (-7)$			
$2 + 4$			
$-6 + (-2)$			
$-5 + 9$			
$15 + (-9)$			
$-10 + 10$			
$-6 + (-6)$			
$13 + (-13)$			

Math Practice

Make Conjectures
How can absolute values be used to write a rule about the sum of two integers?

d. How can you tell whether the sum of two integers is *positive*, *negative*, or *zero*?

e. Write rules for adding (i) two integers with the same sign, (ii) two integers with different signs, and (iii) two opposite integers.

Section A.2 Adding Integers 509

A.2 Lesson

Key Vocabulary
additive inverse, p. 511

You have used number lines to find sums of positive numbers, which involve movement to the right. Now you will find sums with negative numbers, which involve movement to the left.

EXAMPLE 1 Using Number Lines to Find Sums

a. Find $4 + (-4)$.

Draw an arrow from 0 to 4 to represent 4. Then draw an arrow 4 units to the left to represent adding -4.

The length of each arrow is the absolute value of the number it represents.

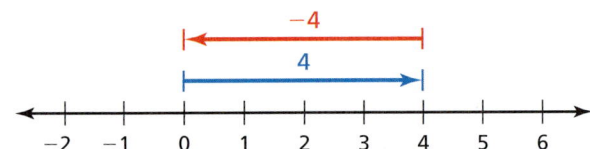

▸ So, $4 + (-4) = 0$.

b. Find $-1 + (-3)$.

Draw an arrow from 0 to -1 to represent -1. Then draw an arrow 3 units to the left to represent adding -3.

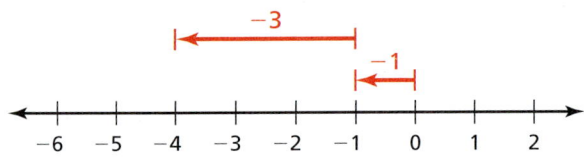

▸ So, $-1 + (-3) = -4$.

c. Find $-2 + 6$.

Draw an arrow from 0 to -2 to represent -2. Then draw an arrow 6 units to the right to represent adding 6.

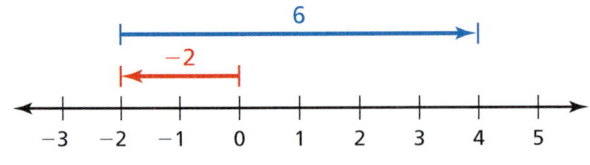

▸ So, $-2 + 6 = 4$.

Try It Use a number line to find the sum.

1. $-2 + 2$
2. $4 + (-5)$
3. $-3 + (-3)$

Using integer counters and number lines leads to the following rules for adding integers.

Key Ideas

Adding Integers with the Same Sign

Words Add the absolute values of the integers. Then use the common sign.

Numbers $2 + 5 = 7$ $-2 + (-5) = -7$

Adding Integers with Different Signs

Words Subtract the lesser absolute value from the greater absolute value. Then use the sign of the integer with the greater absolute value.

Numbers $8 + (-10) = -2$ $-13 + 17 = 4$

Additive Inverse Property

Words The sum of a number and its **additive inverse,** or opposite, is 0.

Numbers $6 + (-6) = 0$ $-25 + 25 = 0$

Algebra $a + (-a) = 0$

> Notice that Example 1(a) shows the Additive Inverse Property.

EXAMPLE 2 Adding Integers with the Same Sign

Find $-4 + (-2)$.

$-4 + (-2) = -6$ Add $|-4|$ and $|-2|$.

Use the common sign.

▸ The sum is -6.

Check Use integer counters.

$-4 \quad + \quad -2 \quad = \quad -6$ ✓

Try It Find the sum.

4. $7 + 13$ **5.** $-8 + (-5)$ **6.** $-2 + (-15)$

Section A.2   Adding Integers   511

EXAMPLE 3 Adding Integers with Different Signs

a. Find $5 + (-10)$.

$$5 + (-10) = -5$$

$|-10| > |5|$. So, subtract $|5|$ from $|-10|$.

Use the sign of -10.

▸ The sum is -5.

b. Find $-3 + 7$.

$$-3 + 7 = 4$$

$|7| > |-3|$. So, subtract $|-3|$ from $|7|$.

Use the sign of 7.

▸ The sum is 4.

c. Find $-12 + 12$.

$$-12 + 12 = 0$$

The sum is 0 by the Additive Inverse Property.

-12 and 12 are opposites.

▸ The sum is 0.

Try It Find the sum.

7. $-2 + 11$
8. $9 + (-10)$
9. $-31 + 31$

Self-Assessment for Concepts & Skills

Solve each exercise. Then rate your understanding of the success criteria in your journal.

10. **WRITING** Explain how to use a number line to find the sum of two integers.

ADDING INTEGERS Find the sum. Use a number line to justify your answer.

11. $-8 + 20$
12. $30 + (-30)$
13. $-10 + (-18)$

14. **MP NUMBER SENSE** Is $3 + (-4)$ the same as $-4 + 3$? Explain.

MP LOGIC Tell whether the statement is *true* or *false*. Explain your reasoning.

15. The sum of two negative integers is always negative.

16. The sum of an integer and its absolute value is always 0.

512   Chapter A   Adding and Subtracting Rational Numbers

You can use the Commutative and Associative Properties of Addition to find sums of integers.

EXAMPLE 4 Modeling Real Life

The list shows four account transactions. Find the change in the account balance.

JULY TRANSACTIONS	
Withdrawal	-$40
Deposit	$50
Deposit	$75
Withdrawal	-$50

Understand the problem. You are given amounts of two withdrawals and two deposits. You are asked to find how much the balance in the account changed.

Make a plan. Find the sum of the transactions. Notice that 50 and −50 are opposites and combine to make 0. So, use properties of addition to first group those terms.

Solve and check.

$$-40 + 50 + 75 + (-50) = -40 + 75 + 50 + (-50) \quad \text{Comm. Prop. of Add.}$$
$$= -40 + 75 + [50 + (-50)] \quad \text{Assoc. Prop. of Add.}$$
$$= -40 + 75 + 0 \quad \text{Add. Inv. Prop.}$$
$$= 35 + 0 \quad \text{Add } -40 \text{ and } 75.$$
$$= 35 \quad \text{Add. Prop. of Zero}$$

 So, the account balance increased $35.

> **Another Method** Find the sum by grouping the first two terms and the last two terms.
> $$-40 + 50 + 75 + (-50) = (-40 + 50) + [75 + (-50)]$$
> $$= 10 + 25 = 35 \checkmark$$

Self-Assessment for Problem Solving

Solve each exercise. Then rate your understanding of the success criteria in your journal.

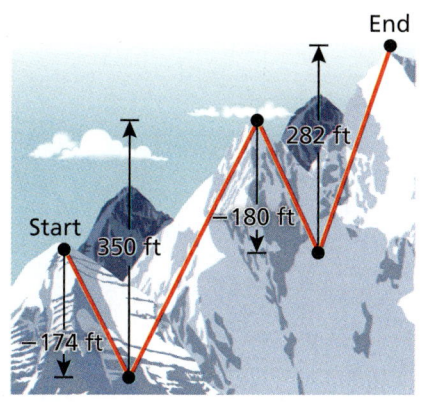

17. At 12:00 P.M., the water pressure on a submarine is 435 pounds per square inch. From 12:00 P.M. to 12:30 P.M., the water pressure increases 58 pounds per square inch. From 12:30 P.M. to 1:00 P.M., the water pressure decreases 116 pounds per square inch. What is the water pressure at 1:00 P.M.?

18. **DIG DEEPER!** The diagram shows the elevation changes between checkpoints on a trail. The trail begins at an elevation of 8136 feet. What is the elevation at the end of the trail?

A.2 Practice

Review & Refresh

Copy and complete the statement using <, >, or =.

1. $5 \;\square\; |-7|$
2. $|-2.6| \;\square\; |-2.06|$
3. $\left|-\dfrac{3}{5}\right| \;\square\; -\left|\dfrac{5}{8}\right|$

Add.

4. $8.43 + 5.21$
5. $2.316 + 4.09$
6. $\dfrac{5}{9} + \dfrac{3}{9}$
7. $\dfrac{1}{2} + \dfrac{1}{8}$

8. The regular price of a photograph printed on a canvas is $18. You have a coupon for 15% off. How much is the discount?

   A. $2.70   **B.** $3   **C.** $15   **D.** $15.30

9. Represent the ratio relationship using a graph.

Time (hours)	1	2	3
Distance (miles)	55	110	165

Concepts, Skills, & Problem Solving

USING INTEGER COUNTERS Use integer counters to complete the table. (See Exploration 1, p. 509.)

	Expression	Type of Sum	Sum	Sum: Positive, Negative, or Zero
10.	$-5 + 8$			
11.	$-3 + (-7)$			

USING NUMBER LINES Write an addition expression represented by the number line. Then find the sum.

12.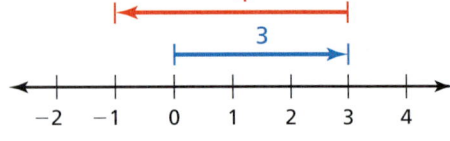

13.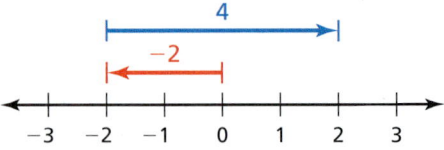

14.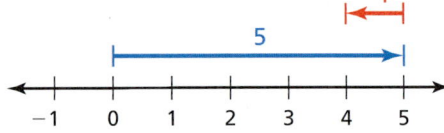

15.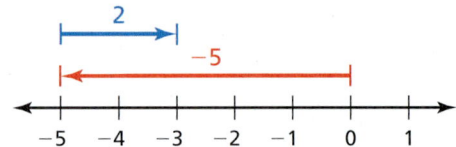

514   Chapter A   Adding and Subtracting Rational Numbers

ADDING INTEGERS Find the sum. Use integer counters or a number line to verify your answer.

16. $6 + 4$ **17.** $-4 + (-6)$ **18.** $-2 + (-3)$ **19.** $-5 + 12$

20. $5 + (-7)$ **21.** $8 + (-8)$ **22.** $9 + (-11)$ **23.** $-3 + 13$

24. $-4 + (-16)$ **25.** $-3 + (-1)$ **26.** $14 + (-5)$ **27.** $0 + (-11)$

28. $-10 + (-15)$ **29.** $-13 + 9$ **30.** $18 + (-18)$ **31.** $-25 + (-9)$

YOU BE THE TEACHER Your friend finds the sum. Is your friend correct? Explain your reasoning.

32. $9 + (-6) = 3$

33. $-10 + (-10) = 0$

34. MODELING REAL LIFE The temperature is −3°F at 7:00 A.M. During the next 4 hours, the temperature increases 21°F. What is the temperature at 11:00 A.M.?

35. MODELING REAL LIFE Your bank account has a balance of −$12. You deposit $60. What is your new balance?

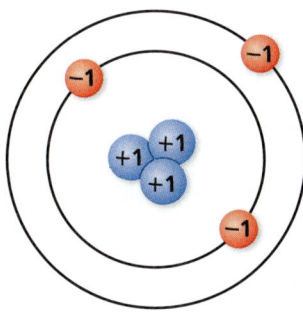

Lithium Atom

36. PROBLEM SOLVING A lithium atom has positively charged protons and negatively charged electrons. The sum of the charges represents the charge of the lithium atom. Find the charge of the atom.

37. OPEN-ENDED Write two integers with different signs that have a sum of −25. Write two integers with the same sign that have a sum of −25.

USING PROPERTIES Tell how the Commutative and Associative Properties of Addition can help you find the sum using mental math. Then find the sum.

38. $9 + 6 + (-6)$ **39.** $-8 + 13 + (-13)$ **40.** $9 + (-17) + (-9)$

41. $7 + (-12) + (-7)$ **42.** $-12 + 25 + (-15)$ **43.** $6 + (-9) + 14$

ADDING INTEGERS Find the sum.

44. $13 + (-21) + 16$ **45.** $22 + (-14) + (-12)$ **46.** $-13 + 27 + (-18)$

47. $-19 + 26 + 14$ **48.** $-32 + (-17) + 42$ **49.** $-41 + (-15) + (-29)$

DESCRIBING A SUM Describe the location of the sum, relative to p, on a number line.

50. $p + 3$ **51.** $p + (-7)$ **52.** $p + 0$ **53.** $p + q$

ALGEBRA Evaluate the expression when $a = 4$, $b = -5$, and $c = -8$.

54. $a + b$ **55.** $-b + c$ **56.** $|a + b + c|$

Section A.2   Adding Integers   515

57. MODELING REAL LIFE The table shows the income and expenses for a school carnival. The school's goal was to raise $1100. Did the school reach its goal? Explain.

Games	Concessions	Donations	Flyers	Decorations
$650	$530	$52	−$28	−$75

OPEN-ENDED Write a real-life story using the given topic that involves the sum of an integer and its additive inverse.

58. income and expenses

59. the amount of water in a bottle

60. the elevation of a blimp

MENTAL MATH Use mental math to solve the equation.

61. $d + 12 = 2$ **62.** $b + (-2) = 0$ **63.** $-8 + m = -15$

64. DIG DEEPER! Starting at point A, the path of a dolphin jumping out of the water is shown.

   a. Is the dolphin deeper at point C or point E? Explain your reasoning.

   b. Is the dolphin higher at point B or point D? Explain your reasoning.

   c. What is the change in elevation of the dolphin from point A to point E?

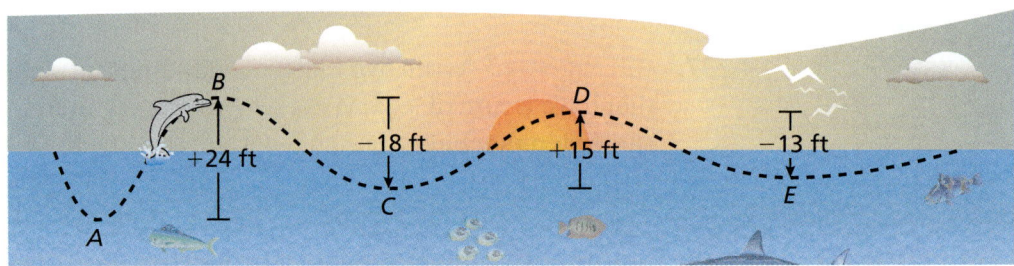

65. NUMBER SENSE Consider the integers p and q. Describe all of the possible values of p and q for each circumstance. Justify your answers.

   a. $p + q = 0$ **b.** $p + q < 0$ **c.** $p + q > 0$

66. PUZZLE According to a legend, the Chinese Emperor Yu-Huang saw a magic square on the back of a turtle. In a *magic square*, the numbers in each row and in each column have the same sum. This sum is called the *magic sum*.

Copy and complete the magic square so that each row and each column has a magic sum of 0. Use each integer from −4 to 4 exactly once.

A.3 Adding Rational Numbers

Learning Target: Find sums of rational numbers.

Success Criteria:
- I can explain how to model addition of rational numbers on a number line.
- I can find sums of rational numbers by reasoning about absolute values.
- I can use properties of addition to efficiently add rational numbers.

EXPLORATION 1

Adding Rational Numbers

Work with a partner.

a. Choose a unit fraction to represent the space between the tick marks on each number line. What addition expressions are being modeled? What are the sums?

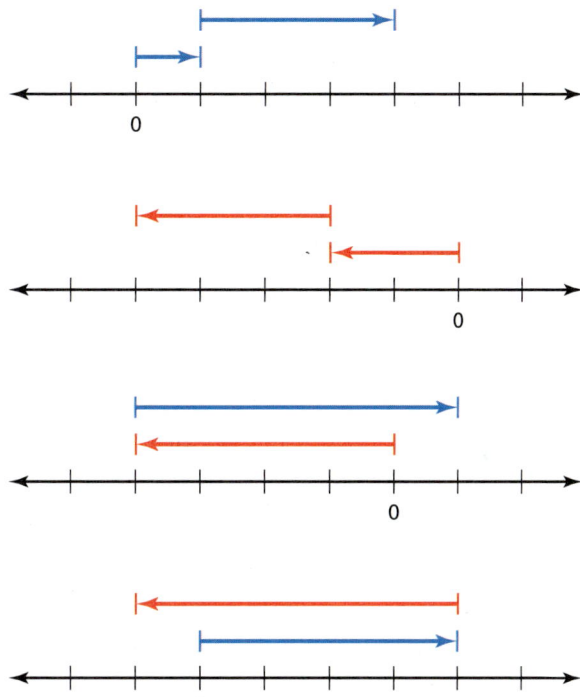

Math Practice

Look for Structure
How do the lengths and directions of the arrows determine the sign of the sum?

b. Do the rules for adding integers apply to all rational numbers? Explain your reasoning.

c. You have used the following properties to add integers. Do these properties apply to all rational numbers? Explain your reasoning.

- Commutative Property of Addition
- Associative Property of Addition
- Additive Inverse Property

A.3 Lesson

 Key Idea

Adding Rational Numbers

Words To add rational numbers, use the same rules as you used for adding integers.

Numbers $\dfrac{3}{5} + \left(-\dfrac{1}{5}\right) = \left|\dfrac{3}{5}\right| - \left|-\dfrac{1}{5}\right|$

$= \dfrac{3}{5} - \dfrac{1}{5}$

$= \dfrac{2}{5}$

Model

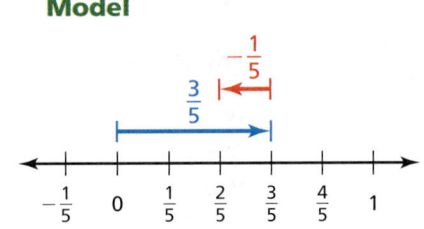

EXAMPLE 1 Adding Rational Numbers

Find $-\dfrac{8}{3} + \dfrac{5}{6}$. **Estimate** $-3 + 1 = -2$

Because the signs are different and $\left|-\dfrac{8}{3}\right| > \left|\dfrac{5}{6}\right|$, subtract $\left|\dfrac{5}{6}\right|$ from $\left|-\dfrac{8}{3}\right|$.

$\left|-\dfrac{8}{3}\right| - \left|\dfrac{5}{6}\right| = \dfrac{8}{3} - \dfrac{5}{6}$ Find the absolute values.

$= \dfrac{16}{6} - \dfrac{5}{6}$ Rewrite $\dfrac{8}{3}$ as $\dfrac{16}{6}$.

$= \dfrac{16 - 5}{6}$ Write the difference of the numerators over the common denominator.

$= \dfrac{11}{6}$, or $1\dfrac{5}{6}$ Simplify.

Because $\left|-\dfrac{8}{3}\right| > \left|\dfrac{5}{6}\right|$, use the sign of $-\dfrac{8}{3}$.

▶ So, $-\dfrac{8}{3} + \dfrac{5}{6} = -1\dfrac{5}{6}$. **Reasonable?** $-1\dfrac{5}{6} \approx -2$ ✓

Try It Find the sum. Write your answer in simplest form.

1. $-\dfrac{1}{2} + \left(-\dfrac{3}{2}\right)$
2. $-1\dfrac{3}{8} + \dfrac{3}{4}$
3. $4 + \left(-\dfrac{7}{2}\right)$

EXAMPLE 2 Adding Rational Numbers

Find $-0.75 + (-1.5)$.   **Estimate** $-1 + (-1.5) = -2.5$

Because the signs are the same, add $|-0.75|$ and $|-1.5|$.

$$|-0.75| + |-1.5| = 0.75 + 1.5 \qquad \text{Find the absolute values.}$$
$$= 2.25 \qquad \text{Add.}$$

Because -0.75 and -1.5 are both negative, use a negative sign in the sum.

 So, $-0.75 + (-1.5) = -2.25$.   **Reasonable?** $-2.25 \approx -2.5$

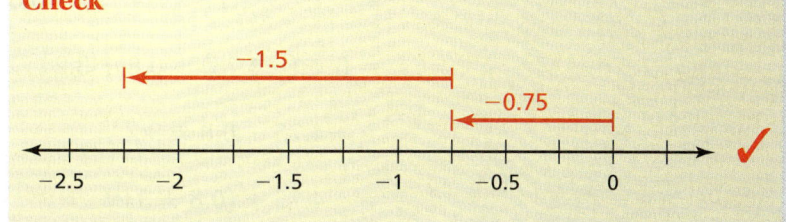

Try It   Find the sum.

4. $-3.3 + (-2.7)$ 5. $-5.35 + 4$ 6. $1.65 + (-0.9)$

 ## Self-Assessment for Concepts & Skills

Solve each exercise. Then rate your understanding of the success criteria in your journal.

7. **WRITING** Explain how to use a number line to find the sum of two rational numbers.

ADDING RATIONAL NUMBERS Find the sum.

8. $-\dfrac{7}{10} + \dfrac{1}{5}$ 9. $-\dfrac{3}{4} + \left(-\dfrac{1}{3}\right)$ 10. $-2.6 + 4.3$

11. **DIFFERENT WORDS, SAME QUESTION** Which is different? Find "both" answers.

> Add -4.5 and 3.5.   What is the distance between -4.5 and 3.5?
>
> What is -4.5 increased by 3.5?   Find the sum of -4.5 and 3.5.

Section A.3   Adding Rational Numbers   519

EXAMPLE 3 Modeling Real Life

The table shows the annual profits (in millions of dollars) of an online gaming company from 2013 to 2017. Positive numbers represent *gains*, and negative numbers represent *losses*. Which statement describes the profit over the five-year period?

Year	Profit (millions of dollars)
2013	−1.7
2014	−4.75
2015	1.7
2016	0.8
2017	3.2

A. gain of $0.75 million **B.** gain of $75,000

C. loss of $75,000 **D.** loss of $750,000

To determine the amount of the gain or loss, find the sum of the profits.

$$\begin{aligned}
\text{five-year profit} &= -1.7 + (-4.75) + 1.7 + 0.8 + 3.2 &&\text{Write the sum.}\\
&= -1.7 + 1.7 + (-4.75) + 0.8 + 3.2 &&\text{Comm. Prop. of Add.}\\
&= 0 + (-4.75) + 0.8 + 3.2 &&\text{Additive Inv. Prop.}\\
&= -4.75 + 0.8 + 3.2 &&\text{Add. Prop. of Zero}\\
&= -4.75 + (0.8 + 3.2) &&\text{Assoc. Prop. of Add.}\\
&= -4.75 + 4 &&\text{Add 0.8 and 3.2.}\\
&= -0.75 &&\text{Add −4.75 and 4.}
\end{aligned}$$

The five-year profit is −$0.75 million. So, the company has a five-year loss of $0.75 million, or $750,000.

 The correct answer is **D**.

> The Commutative and Associative Properties of Addition are true for all rational numbers.

Self-Assessment for Problem Solving

Solve each exercise. Then rate your understanding of the success criteria in your journal.

12. A bottle contains 10.5 cups of orange juice. You drink 1.2 cups of the juice each morning and 0.9 cup of the juice each afternoon. How much total juice do you drink each day? When will you run out of juice?

13.  **DIG DEEPER!** The table shows the changes in elevation of a hiker each day for three days. How many miles of elevation must the hiker gain on the fourth day to gain $\frac{1}{4}$ mile of elevation over the four days?

Day	Change in elevation (miles)
1	$-\frac{1}{4}$
2	$\frac{1}{2}$
3	$-\frac{1}{5}$
4	?

A.3 Practice

> Go to **BigIdeasMath.com** to get HELP with solving the exercises.

▶ Review & Refresh

Find the sum. Use a number line to verify your answer.

1. $3 + 12$
2. $5 + (-7)$
3. $-4 + (-1)$
4. $-6 + 6$

Subtract.

5. $69 - 38$
6. $82 - 74$
7. $177 - 63$
8. $451 - 268$

9. What is the range of the numbers below?

   12, 8, 17, 12, 15, 18, 30

   A. 12 **B.** 15 **C.** 18 **D.** 22

▶ Concepts, Skills, & Problem Solving

USING TOOLS Choose a unit fraction to represent the space between the tick marks on the number line. Write the addition expression being modeled. Then find the sum. *(See Exploration 1, p. 517.)*

10.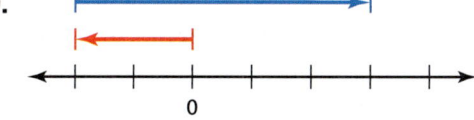

11.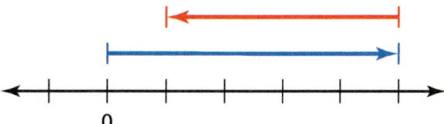

ADDING RATIONAL NUMBERS Find the sum. Write fractions in simplest form.

12. $\dfrac{11}{12} + \left(-\dfrac{7}{12}\right)$

13. $-1\dfrac{1}{5} + \left(-\dfrac{3}{5}\right)$

14. $-4.2 + 3.3$

15. $-\dfrac{9}{14} + \dfrac{2}{7}$

16. $12.48 + (-10.636)$

17. $-2\dfrac{1}{6} + \left(-\dfrac{2}{3}\right)$

18. $-20.25 + 15.711$

19. $-32.306 + (-24.884)$

20. $\dfrac{15}{4} + \left(-4\dfrac{1}{3}\right)$

21. **YOU BE THE TEACHER** Your friend finds the sum. Is your friend correct? Explain your reasoning.

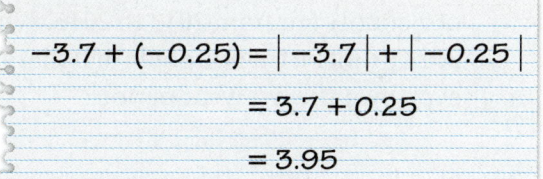

$$-3.7 + (-0.25) = |-3.7| + |-0.25|$$
$$= 3.7 + 0.25$$
$$= 3.95$$

OPEN-ENDED Describe a real-life situation that can be represented by the addition expression modeled on the number line.

22.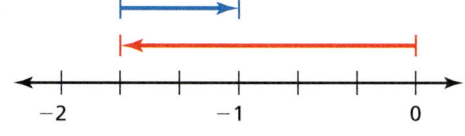

23.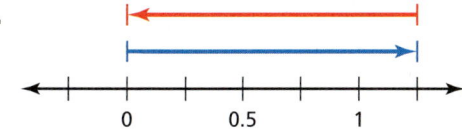

Section A.3 Adding Rational Numbers 521

24. **MODELING REAL LIFE** You eat $\frac{3}{10}$ of a coconut. Your friend eats $\frac{1}{5}$ of the coconut. What fraction of the coconut do you and your friend eat?

25. **MODELING REAL LIFE** Your bank account balance is −$20.85. You deposit $15.50. What is your new balance?

26. **MP NUMBER SENSE** When is the sum of two negative mixed numbers an integer?

27. **WRITING** You are adding two rational numbers with different signs. How can you tell if the sum will be *positive*, *negative*, or *zero*?

28. **DIG DEEPER!** The table at the left shows the water level (in inches) of a reservoir for three months compared to the yearly average. Is the water level for the three-month period greater than or less than the yearly average? Explain.

June	July	August
$-2\frac{1}{8}$	$1\frac{1}{4}$	$-\frac{7}{8}$

USING PROPERTIES Tell how the Commutative and Associative Properties of Addition can help you find the sum using mental math. Then find the sum.

29. $4.5 + (-6.21) + (-4.5)$

30. $\frac{1}{3} + \left(\frac{2}{3} + \frac{5}{8}\right)$

31. $8\frac{1}{2} + \left[4\frac{1}{10} + \left(-8\frac{1}{2}\right)\right]$

ADDING RATIONAL NUMBERS Find the sum. Explain each step.

32. $6 + 4\frac{3}{4} + (-2.5)$

33. $-4.3 + \frac{4}{5} + 12$

34. $5\frac{1}{3} + 7.5 + \left(-3\frac{1}{6}\right)$

35. **MP PROBLEM SOLVING** The table at the right shows the annual profits (in thousands of dollars) of a county fair from 2013 to 2016. What must the 2017 profit be (in hundreds of dollars) to break even over the five-year period?

Year	Profit (thousands of dollars)
2013	2.5
2014	1.4
2015	−3.3
2016	−1.4
2017	?

36. **MP REASONING** Is $|a+b| = |a| + |b|$ true for all rational numbers a and b? Explain.

37. **MP REPEATED REASONING** Evaluate the expression.

$$\frac{19}{20} + \left(-\frac{18}{20}\right) + \frac{17}{20} + \left(-\frac{16}{20}\right) + \cdots + \left(-\frac{4}{20}\right) + \frac{3}{20} + \left(-\frac{2}{20}\right) + \frac{1}{20}$$

A.4 Subtracting Integers

Learning Target: Find differences of integers.

Success Criteria:
- I can explain how subtracting integers is related to adding integers.
- I can explain how to model subtraction of integers on a number line.
- I can find differences of integers by reasoning about absolute values.

EXPLORATION 1

Using Integer Counters to Find Differences

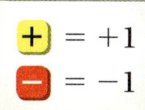

Work with a partner.

a. Use integer counters to find the following sum and difference. What do you notice?

$$4 + (-2) \qquad 4 - 2$$

b. In part (a), you *removed* zero pairs to find the sums. How can you use integer counters and zero pairs to find $-3 - 1$?

c. **INDUCTIVE REASONING** Use integer counters to complete the table.

Expression	Operation: Add or Subtract	Answer
$4 - 2$	Subtract 2.	
$4 + (-2)$		
$-3 - 1$		
$-3 + (-1)$		
$3 - 8$		
$3 + (-8)$		
$9 - 13$		
$9 + (-13)$		
$-6 - (-3)$		
$-6 + 3$		
$-5 - (-12)$		
$-5 + 12$		

Math Practice

Interpret Results
What do the results tell you about the relationship between subtracting integers and adding integers?

d. Write a general rule for subtracting integers.

A.4 Lesson

 **Key Idea**

Subtracting Integers

Words To subtract an integer, add its opposite.

Numbers $3 - 4 = 3 + (-4) = -1$

Models

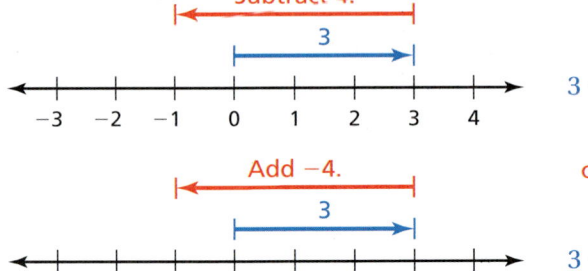

EXAMPLE 1   Using Number Lines to Find Differences

a. **Find $-2 - 4$.**

Draw an arrow from 0 to -2 to represent -2. Then draw an arrow 4 units to the left to represent subtracting 4, or adding -4.

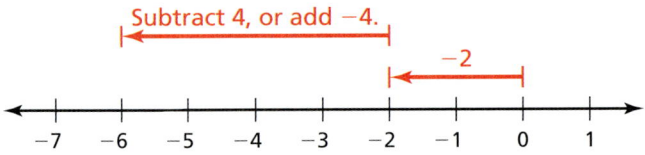

▶ So, $-2 - 4 = -6$.

b. **Find $-3 - (-7)$.**

Draw an arrow from 0 to -3 to represent -3. Then draw an arrow 7 units to the right to represent subtracting -7, or adding 7.

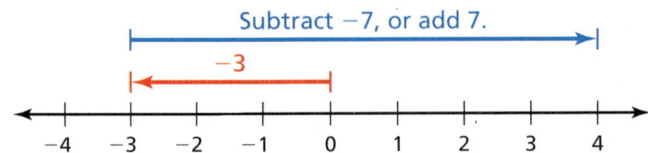

▶ So, $-3 - (-7) = 4$.

Try It Use a number line to find the difference.

1. $1 - 4$
2. $-5 - 2$
3. $6 - (-5)$

524   Chapter A   Adding and Subtracting Rational Numbers

EXAMPLE 2 Subtracting Integers

a. Find $3 - 12$.

$$3 - 12 = 3 + (-12) \quad \text{Add the opposite of 12.}$$
$$= -9 \quad \text{Add.}$$

▸ The difference is -9.

b. Find $-8 - (-13)$.

$$-8 - (-13) = -8 + 13 \quad \text{Add the opposite of } -13.$$
$$= 5 \quad \text{Add.}$$

▸ The difference is 5.

c. Find $5 - (-4)$.

$$5 - (-4) = 5 + 4 \quad \text{Add the opposite of } -4.$$
$$= 9 \quad \text{Add.}$$

▸ The difference is 9.

Try It Find the difference.

4. $8 - 3$
5. $9 - 17$
6. $-3 - 3$
7. $-14 - 9$
8. $10 - (-8)$
9. $-12 - (-12)$

Self-Assessment for Concepts & Skills

Solve each exercise. Then rate your understanding of the success criteria in your journal.

10. **WRITING** Explain how to use a number line to find the difference of two integers.

MATCHING Match the subtraction expression with the corresponding addition expression. Explain your reasoning.

11. $9 - (-5)$
12. $-9 - 5$
13. $-9 - (-5)$
14. $9 - 5$

A. $-9 + 5$
B. $9 + (-5)$
C. $-9 + (-5)$
D. $9 + 5$

SUBTRACTING INTEGERS Find the difference. Use a number line to justify your answer.

15. $10 - 12$
16. $6 - (-8)$
17. $-7 - (-4)$

EXAMPLE 3 Modeling Real Life

Which continent has the greater range of elevations?

	North America	Africa
Highest Elevation	6198 m	5895 m
Lowest Elevation	−86 m	−155 m

Understand the problem. You are given the highest and lowest elevations in North America and Africa. You are asked to find the continent with the greater difference between its highest and lowest elevations.

Make a plan. Find the range of elevations for each continent by subtracting the lowest elevation from the highest elevation. Then compare the ranges.

Solve and check.

North America
$$\text{range} = 6198 - (-86)$$
$$= 6198 + 86$$
$$= 6284 \text{ m}$$

Africa
$$\text{range} = 5895 - (-155)$$
$$= 5895 + 155$$
$$= 6050 \text{ m}$$

▶ Because 6284 meters is greater than 6050 meters, North America has the greater range of elevations.

Another Method North America's highest elevation is $6198 - 5895 = 303$ meters higher than Africa's highest elevation. Africa's lowest elevation is $|-155| - |-86| = 69$ meters lower than North America's lowest elevation. Because $303 > 69$, North America has the greater range of elevations. ✓

Self-Assessment for Problem Solving

Solve each exercise. Then rate your understanding of the success criteria in your journal.

18. A polar vortex causes the temperature to decrease from 3°C at 3:00 P.M. to −2°C at 4:00 P.M. The temperature continues to change by the same amount each hour until 8:00 P.M. Find the total change in temperature from 3:00 P.M. to 8:00 P.M.

19. **DIG DEEPER!** While on vacation, you map several locations using a coordinate plane in which each unit represents 1 mile. A cove is at $(3, -7)$, an island is at $(-5, 4)$, and you are currently at $(3, 4)$. Are you closer to the cove or the island? Justify your answer.

A.4 Practice

Review & Refresh

Find the sum. Write fractions in simplest form.

1. $\frac{5}{9} + \left(-\frac{2}{9}\right)$
2. $-8.75 + 2.43$
3. $-3\frac{1}{8} + \left(-2\frac{3}{8}\right)$

Add.

4. $2.48 + 6.711$
5. $12.807 + 7.116$
6. $18.7126 + 14.033$

Write an addition expression represented by the number line. Then find the sum.

7.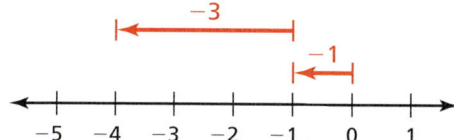
8.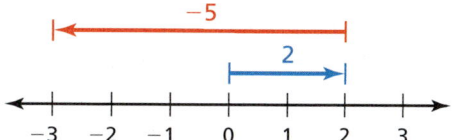

Concepts, Skills, & Problem Solving

USING INTEGER COUNTERS Use integer counters to find the difference. (See Exploration 1, p. 523.)

9. $5 - 3$
10. $1 - 4$
11. $-2 - (-6)$

USING NUMBER LINES Write an addition expression and write a subtraction expression represented by the number line. Then evaluate the expressions.

12.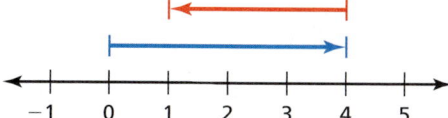
13.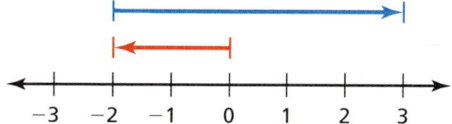

SUBTRACTING INTEGERS Find the difference. Use a number line to verify your answer.

14. $4 - 7$
15. $8 - (-5)$
16. $-6 - (-7)$
17. $-2 - 3$
18. $5 - 8$
19. $-4 - 6$
20. $-8 - (-3)$
21. $10 - 7$
22. $-8 - 13$
23. $15 - (-2)$
24. $-9 - (-13)$
25. $-7 - (-8)$
26. $-6 - (-6)$
27. $-10 - 12$
28. $32 - (-6)$
29. $0 - (-20)$

30. **YOU BE THE TEACHER** Your friend finds the difference. Is your friend correct? Explain your reasoning.

    $7 - (-12) = 7 + 12 = 19$

31. **MP STRUCTURE** A scientist records the water temperature and the air temperature in Antarctica. The water temperature is −2°C. The air is 9°C colder than the water. Which expression can be used to find the air temperature? Explain your reasoning.

$$-2 + 9 \qquad -2 - 9 \qquad 9 - 2$$

32. **MODELING REAL LIFE** A shark is 80 feet below the surface of the water. It swims up and jumps out of the water to a height of 15 feet above the surface. Find the vertical distance the shark travels. Justify your answer.

33. **MODELING REAL LIFE** The figure shows a diver diving from a platform. The diver reaches a depth of 4 meters. What is the change in elevation of the diver?

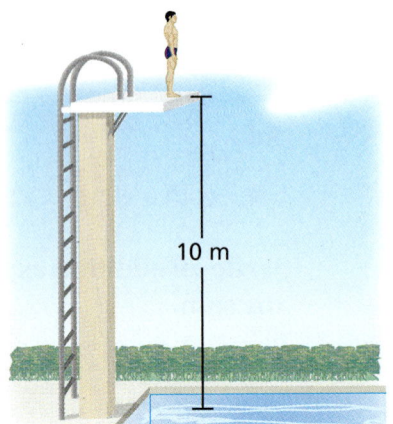

34. **OPEN-ENDED** Write two different pairs of negative integers, x and y, that make the statement $x - y = -1$ true.

USING PROPERTIES Tell how the Commutative and Associative Properties of Addition can help you evaluate the expression using mental math. Then evaluate the expression.

35. $2 - 7 + (-2)$
36. $-6 - 8 + 6$
37. $8 + (-8 - 5)$
38. $-39 + 46 - (-39)$
39. $[13 + (-28)] - 13$
40. $-2 + (-47 - 8)$

ALGEBRA Evaluate the expression when $k = -3$, $m = -6$, and $n = 9$.

41. $4 - n$
42. $m - (-8)$
43. $-5 + k - n$
44. $|m - k|$

45. **MODELING REAL LIFE** The table shows the record monthly high and low temperatures for a city in Alaska.

	Jan	Feb	Mar	Apr	May	Jun	Jul	Aug	Sep	Oct	Nov	Dec
High (°F)	56	57	56	72	82	92	84	85	73	64	62	53
Low (°F)	−35	−38	−24	−15	1	29	34	31	19	−6	−21	−36

a. Which month has the greatest range of temperatures?

b. What is the range of temperatures for the year?

MP REASONING Tell whether the difference of the two integers is *always*, *sometimes*, or *never* positive. Explain your reasoning.

46. two positive integers
47. a positive integer and a negative integer
48. two negative integers
49. a negative integer and a positive integer

MP NUMBER SENSE For what values of a and b is the statement true?

50. $|a - b| = |b - a|$
51. $|a - b| = |a| - |b|$

A.5 Subtracting Rational Numbers

Learning Target: Find differences of rational numbers and find distances between numbers on a number line.

Success Criteria:
- I can explain how to model subtraction of rational numbers on a number line.
- I can find differences of rational numbers by reasoning about absolute values.
- I can find distances between numbers on a number line.

EXPLORATION 1

Subtracting Rational Numbers

Work with a partner.

a. Choose a unit fraction to represent the space between the tick marks on each number line. What expressions involving subtraction are being modeled? What are the differences?

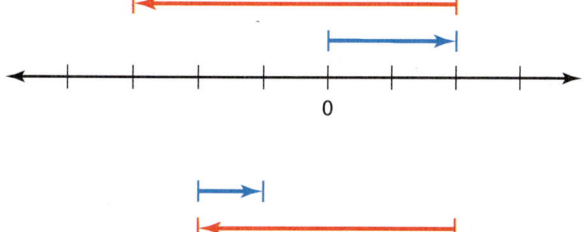

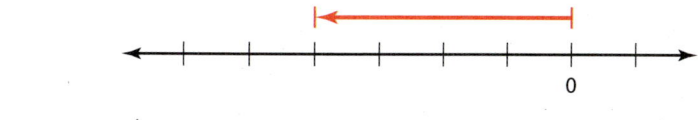

b. Do the rules for subtracting integers apply to all rational numbers? Explain your reasoning.

c. You have used the commutative and associative properties to add integers. Do these properties apply in expressions involving subtraction? Explain your reasoning.

EXPLORATION 2

Finding Distances on a Number Line

Work with a partner.

Math Practice

Find General Methods
How can you find the distance between any two rational numbers on a number line?

a. Find the distance between 3 and −2 on a number line.

b. The distance between 3 and 0 is the absolute value of 3, because $|3 - 0| = |3| = 3$. How can you use absolute values to find the distance between 3 and −2? Justify your answer.

c. Choose any two rational numbers. Use your method in part (b) to find the distance between the numbers. Use a number line to check your answer.

A.5 Lesson

🗝 Key Idea

Subtracting Rational Numbers

Words To subtract rational numbers, use the same rules as you used for subtracting integers.

Numbers $\frac{1}{5} - \left(-\frac{4}{5}\right) = \frac{1}{5} + \frac{4}{5}$

$= \frac{5}{5}$

$= 1$

Model

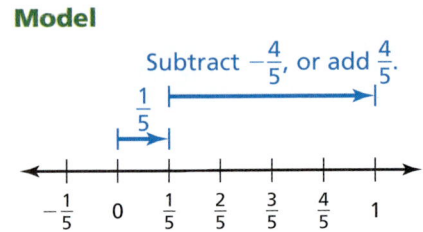

EXAMPLE 1   Subtracting Rational Numbers

Find $-4\frac{1}{7} - \frac{5}{7}$.   **Estimate** $-4 - 1 = -5$

Rewrite the difference as a sum by adding the opposite.

$$-4\frac{1}{7} - \frac{5}{7} = -4\frac{1}{7} + \left(-\frac{5}{7}\right)$$

Because the signs are the same, add $\left|-4\frac{1}{7}\right|$ and $\left|-\frac{5}{7}\right|$.

$\left|-4\frac{1}{7}\right| + \left|-\frac{5}{7}\right| = 4\frac{1}{7} + \frac{5}{7}$ Find the absolute values.

$= 4 + \frac{1}{7} + \frac{5}{7}$ Write $4\frac{1}{7}$ as $4 + \frac{1}{7}$.

$= 4 + \frac{6}{7}$, or $4\frac{6}{7}$ Add fractions and simplify.

Because $-4\frac{1}{7}$ and $-\frac{5}{7}$ are both negative, use a negative sign in the difference.

▶ So, $-4\frac{1}{7} - \frac{5}{7} = -4\frac{6}{7}$.   **Reasonable?** $-4\frac{6}{7} \approx -5$ ✓

Try It Find the difference. Write your answer in simplest form.

1. $\frac{1}{3} - \left(-\frac{1}{3}\right)$
2. $-3\frac{1}{3} - \frac{2}{3}$
3. $4 - 5\frac{1}{2}$

EXAMPLE 2 Subtracting Rational Numbers

Find 2.4 − 5.6.

Rewrite the difference as a sum by adding the opposite.

$$2.4 - 5.6 = 2.4 + (-5.6)$$

Because the signs are different and $|-5.6| > |2.4|$, subtract $|2.4|$ from $|-5.6|$.

$$|-5.6| - |2.4| = 5.6 - 2.4 \qquad \text{Find the absolute values.}$$
$$= 3.2 \qquad \text{Subtract.}$$

Because $|-5.6| > |2.4|$, use the sign of -5.6.

▶ So, $2.4 - 5.6 = -3.2$.

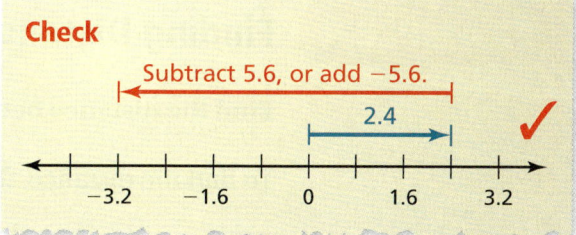
Check

Try It Find the difference.

4. $-2.1 - 3.9$
5. $-8.8 - (-8.8)$
6. $0.45 - (-0.05)$

EXAMPLE 3 Using Properties of Addition

Evaluate $-1\frac{3}{8} - 8\frac{1}{2} - \left(-6\frac{7}{8}\right)$.

Use properties of addition to group the mixed numbers that include fractions with the same denominator.

$$-1\frac{3}{8} - 8\frac{1}{2} - \left(-6\frac{7}{8}\right) = -1\frac{3}{8} + \left(-8\frac{1}{2}\right) + 6\frac{7}{8} \qquad \text{Rewrite as a sum of terms.}$$

$$= -1\frac{3}{8} + 6\frac{7}{8} + \left(-8\frac{1}{2}\right) \qquad \text{Comm. Prop. of Add.}$$

$$= 5\frac{1}{2} + \left(-8\frac{1}{2}\right) \qquad \text{Add } -1\frac{3}{8} \text{ and } 6\frac{7}{8}.$$

$$= -3 \qquad \text{Add } 5\frac{1}{2} \text{ and } -8\frac{1}{2}.$$

▶ So, $-1\frac{3}{8} - 8\frac{1}{2} - \left(-6\frac{7}{8}\right) = -3$.

Try It Evaluate the expression. Write fractions in simplest form.

7. $-2 - \frac{2}{5} + 1\frac{3}{5}$
8. $7.8 - 3.3 - (-1.2) + 4.3$

Key Idea

Distance between Numbers on a Number Line

Words The distance between any two numbers on a number line is the absolute value of the difference of the numbers.

Model

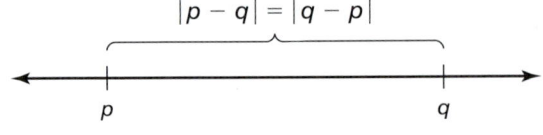

EXAMPLE 4 Finding Distance on a Number Line

Find the distance between $-\frac{1}{3}$ and -2 on a number line.

To find the distance, find the absolute value of the difference of the numbers.

$$\left| -2 - \left(-\frac{1}{3} \right) \right| = \left| -2 + \frac{1}{3} \right| \quad \text{Add the opposite of } -\frac{1}{3}.$$

$$= \left| -1\frac{2}{3} \right| \quad \text{Add } -2 \text{ and } \frac{1}{3}.$$

$$= 1\frac{2}{3} \quad \text{Find the absolute value.}$$

▶ So, the distance between $-\frac{1}{3}$ and -2 is $1\frac{2}{3}$.

Try It Find the distance between the two numbers on a number line.

9. -3 and 9
10. -7.5 and -15.3
11. $1\frac{1}{2}$ and $-\frac{2}{3}$

Self-Assessment for Concepts & Skills

Solve each exercise. Then rate your understanding of the success criteria in your journal.

12. **WRITING** Explain how to use a number line to find the difference of two rational numbers.

SUBTRACTING RATIONAL NUMBERS Find the difference. Use a number line to justify your answer.

13. $4.9 - 1.6$
14. $\frac{7}{8} - \left(-\frac{3}{4} \right)$
15. $\frac{1}{3} - 2\frac{1}{6}$

EXAMPLE 5 Modeling Real Life

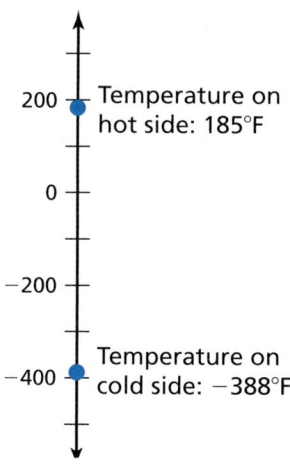

The number line shows the temperatures on each side of the James Webb telescope when in Earth's orbit. Find and interpret the distance between the points.

The number line shows that the temperature on the hot side is 185°F and the temperature on the cold side is −388°F.

To find the distance between the points, find the absolute value of the difference of the numbers.

$$|185 - (-388)| = |185 + 388| \quad \text{Add the opposite of } -388.$$
$$= |573| \quad \text{Add 185 and 388.}$$
$$= 573 \quad \text{Find the absolute value.}$$

 The temperatures are 573°F apart on the number line. So, the hot side is 573°F hotter than the cold side.

 ## Self-Assessment for Problem Solving

Solve each exercise. Then rate your understanding of the success criteria in your journal.

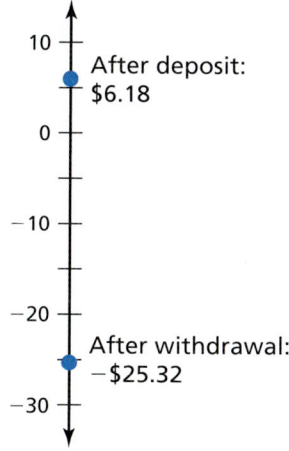

16. A parasail is $\frac{3}{100}$ mile above the water. After 5 minutes, the parasail is $\frac{1}{50}$ mile above the water. Find and interpret the change in height of the parasail.

17. **DIG DEEPER!** You withdraw $55 from a bank account to purchase a game. Then you make a deposit. The number line shows the balances of the account after each transaction.

  a. Find and interpret the distance between the points.

  b. How much money was in your account before buying the game?

A.5 Practice

Review & Refresh

Find the difference. Use a number line to verify your answer.

1. $9 - 5$
2. $-8 - (-8)$
3. $-12 - 7$
4. $12 - (-3)$

Find the volume of the prism.

5.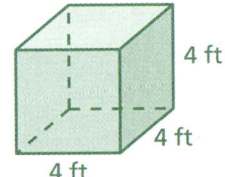
6.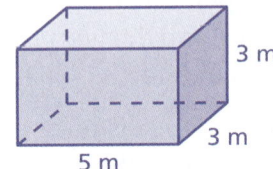

Order the values from least to greatest.

7. $6, |3|, |-4|, 1, -2$
8. $|4.5|, -3.6, 2, |-1.8|, 1.2$

Concepts, Skills, & Problem Solving

USING TOOLS Choose a unit fraction to represent the space between the tick marks on the number line. Write an expression involving subtraction that is being modeled. Then find the difference. *(See Exploration 1, p. 529.)*

9.
10.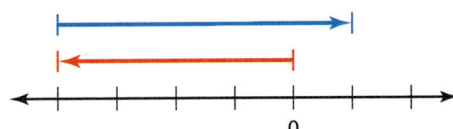

SUBTRACTING RATIONAL NUMBERS Find the difference. Write fractions in simplest form.

11. $\dfrac{5}{8} - \left(-\dfrac{7}{8}\right)$

12. $-1\dfrac{1}{3} - 1\dfrac{2}{3}$

13. $-1 - 2.5$

14. $\dfrac{4}{5} - \left(-\dfrac{3}{10}\right)$

15. $5.5 - 8.1$

16. $-5 - \dfrac{5}{3}$

17. $-8\dfrac{3}{8} - 10\dfrac{1}{6}$

18. $-4.62 - 3.51$

19. $-\dfrac{1}{2} - \left(-\dfrac{5}{9}\right)$

20. $-7.34 - (-5.51)$

21. $6.673 - (-8.29)$

22. $12\dfrac{2}{5} - 17\dfrac{1}{3}$

23. **YOU BE THE TEACHER** Your friend finds the difference. Is your friend correct? Explain your reasoning.

$$\dfrac{3}{2} - \dfrac{9}{2} = \left|\dfrac{3}{2}\right| + \left|\dfrac{9}{2}\right| = \dfrac{12}{2} = 6$$

OPEN-ENDED Describe a real-life situation that can be represented by the subtraction expression modeled on the number line.

24. 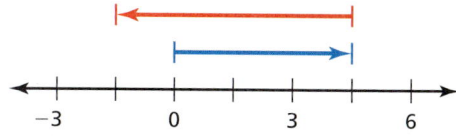   25.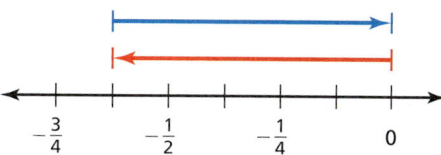

26. **MODELING REAL LIFE** Your water bottle is $\frac{5}{6}$ full. After tennis practice, the bottle is $\frac{3}{8}$ full. How much of the water did you drink?

27. **MODELING REAL LIFE** You have $2\frac{2}{3}$ ounces of sodium chloride. You want to replicate an experiment that uses $2\frac{3}{4}$ ounces of sodium chloride. Do you have enough sodium chloride? If not, how much more do you need?

28. **MP REASONING** When is the difference of two decimals an integer? Explain.

USING PROPERTIES Tell how the Commutative and Associative Properties of Addition can help you evaluate the expression. Then evaluate the expression.

29. $\frac{3}{4} + \frac{2}{3} - \frac{3}{4}$

30. $\frac{2}{5} - \frac{7}{10} - \left(-\frac{3}{5}\right)$

31. $8.5 + 3.4 - 6.5 - (-1.6)$

32. $-1\frac{3}{4} - \left(-8\frac{1}{3}\right) - \left(-4\frac{1}{4}\right)$

33. $2.1 + (5.8 - 4.1)$

34. $2\frac{3}{8} - 4\frac{1}{2} + 3\frac{1}{8} - \left(-\frac{1}{2}\right)$

FINDING DISTANCE ON A NUMBER LINE Find the distance between the two numbers on a number line.

35. 2.7 and 5.9

36. $-\frac{7}{9}$ and $-\frac{2}{9}$

37. -2.2 and 8.4

38. $\frac{3}{4}$ and $\frac{1}{8}$

39. -1.85 and 7.36

40. -7 and $-3\frac{2}{3}$

41. 2.491 and -3.065

42. $-2\frac{1}{2}$ and $-5\frac{3}{4}$

43. $-1\frac{1}{3}$ and $12\frac{7}{12}$

44. **MODELING REAL LIFE** The number line shows the temperatures at 2:00 A.M. and 2:00 P.M. in the Gobi Desert. Find and interpret the distance between the points.

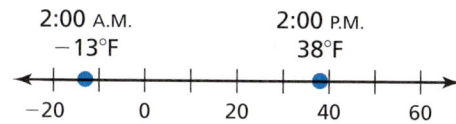

Section A.5 Subtracting Rational Numbers 535

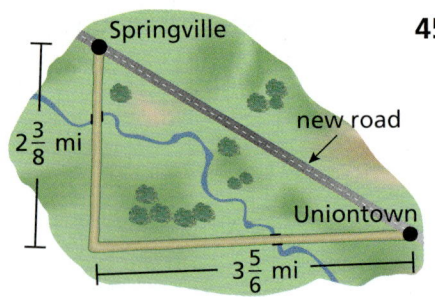

45. PROBLEM SOLVING A new road that connects Uniontown to Springville is $4\frac{1}{3}$ miles long. What is the change in distance when using the new road instead of the dirt roads?

FINDING DISTANCE IN A COORDINATE PLANE Find the distance between the points in a coordinate plane.

46. $(-4, 7.8), (-4, -3.5)$ **47.** $(-2.63, 7), (1.85, 7)$ **48.** $\left(-\frac{1}{2}, -1\right), \left(\frac{5}{8}, -1\right)$

49. $\left(6, 2\frac{1}{3}\right), \left(6, -5\frac{2}{9}\right)$ **50.** $(-6.2, 1.4), (8.9, 1.4)$ **51.** $\left(7\frac{1}{7}, 1\frac{4}{5}\right), \left(7\frac{1}{7}, -\frac{9}{10}\right)$

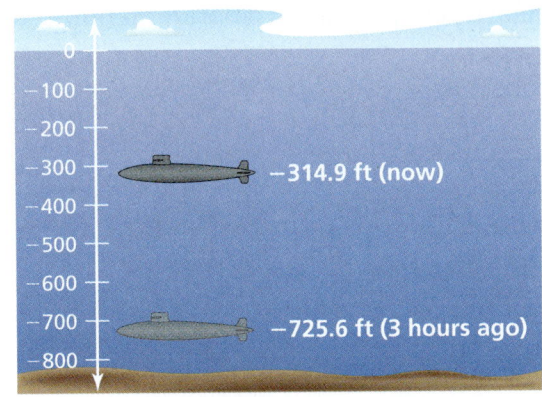

52. DIG DEEPER! The figure shows the elevations of a submarine.

  a. Find the vertical distance traveled by the submarine.

  b. Find the mean hourly vertical distance traveled by the submarine.

53. MP LOGIC The bar graph shows how each month's rainfall compares to the historical average.

  a. What is the difference in rainfall of the wettest month and the driest month?

  b. What do you know about the total amount of rainfall for the year?

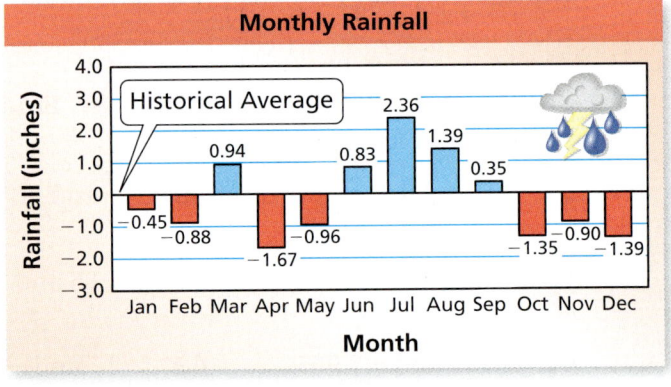

54. OPEN-ENDED Write two different pairs of negative decimals, x and y, that make the statement $x - y = 0.6$ true.

MP REASONING Tell whether the difference of the two numbers is *always*, *sometimes*, or *never* positive. Explain your reasoning.

55. two negative fractions

56. a positive decimal and a negative decimal

57. MP STRUCTURE Fill in the blanks to complete the decimals.

$$5.\boxed{}4 - \boxed{}.\boxed{} = -3.61$$

A Connecting Concepts

Problem-Solving Strategies

Using an appropriate strategy will help you make sense of problems as you study the mathematics in this course. You can use the following strategies to solve problems that you encounter.

- Use a verbal model.
- Draw a diagram.
- Write an equation.
- Solve a simpler problem.
- Sketch a graph or number line.
- Make a table.
- Make a list.
- Break the problem into parts.

Using the Problem-Solving Plan

1. A land surveyor uses a coordinate plane to draw a map of a park, where each unit represents 1 mile. The park is in the shape of a parallelogram with vertices $(-2.5, 1.5)$, $(-1.5, -2.25)$, $(2.75, -2.25)$, and $(1.75, 1.5)$. Find the area of the park.

   Understand the problem. You know the vertices of the parallelogram-shaped park and that each unit represents 1 mile. You are asked to find the area of the park.

   Make a plan. Use a coordinate plane to draw a map of the park. Then find the height and base length of the park. Find the area by using the formula for the area of a parallelogram.

   Solve and check. Use the plan to solve the problem. Then check your solution.

2. The diagram shows the height requirement for driving a go-cart. You are $5\frac{1}{4}$ feet tall. Write and solve an inequality to represent how much taller you must be to drive a go-cart.

Performance Task

Melting Matters

At the beginning of this chapter, you watched a STEAM Video called "Freezing Solid." You are now ready to complete the performance task related to this video, available at *BigIdeasMath.com*. Be sure to use the problem-solving plan as you work through the performance task.

Chapter Review

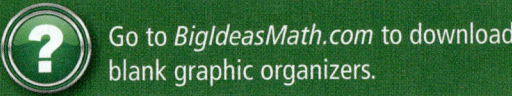

Go to *BigIdeasMath.com* to download blank graphic organizers.

Review Vocabulary

Write the definition and give an example of each vocabulary term.

integers, *p. 503*
rational number, *p. 503*
absolute value, *p. 504*
additive inverse, *p. 511*

Graphic Organizers

You can use a **Definition and Example Chart** to organize information about a concept. Here is an example of a Definition and Example Chart for the vocabulary term *absolute value*.

Absolute value: the distance between a number and 0 on a number line

Example: $|3| = 3$

Example: $|-5| = 5$

Example: $|0| = 0$

Choose and complete a graphic organizer to help you study the concept.

1. integers
2. rational numbers
3. adding integers
4. Additive Inverse Property
5. adding rational numbers
6. subtracting integers
7. subtracting rational numbers

"I made a **Definition and Example Chart** to give my owner ideas for my birthday next week."

538   Chapter A   Adding and Subtracting Rational Numbers

Chapter Self-Assessment

As you complete the exercises, use the scale below to rate your understanding of the success criteria in your journal.

1	2	3	4
I do not understand.	I can do it with help.	I can do it on my own.	I can teach someone else.

A.1 Rational Numbers (pp. 503–508)

Learning Target: Understand absolute values and ordering of rational numbers.

Find the absolute value.

1. $|3|$
2. $|-9|$
3. $\left|\dfrac{3}{4}\right|$
4. $|-5.2|$
5. $\left|-\dfrac{6}{7}\right|$
6. $|4.15|$

Copy and complete the statement using <, >, or =.

7. $|-2|\ \square\ -2$
8. $\left|-\dfrac{1}{3}\right|\ \square\ \left|-\dfrac{5}{6}\right|$
9. $-|1.7|\ \square\ -1.7$

10. Order $|2.25|$, $|-1.5|$, $1\dfrac{1}{4}$, $\left|2\dfrac{1}{2}\right|$, and -2 from least to greatest.

11. Your friend is in Death Valley, California, at an elevation of −282 feet. You are near the Mississippi River in Illinois at an elevation of 279 feet. Who is closer to sea level?

12. Give values for a and b so that $a < b$ and $|a| > |b|$.

13. The map shows the longitudes (in degrees) for Salvador, Brazil, and Nairobi, Kenya. Which city is closer to the Prime Meridian?

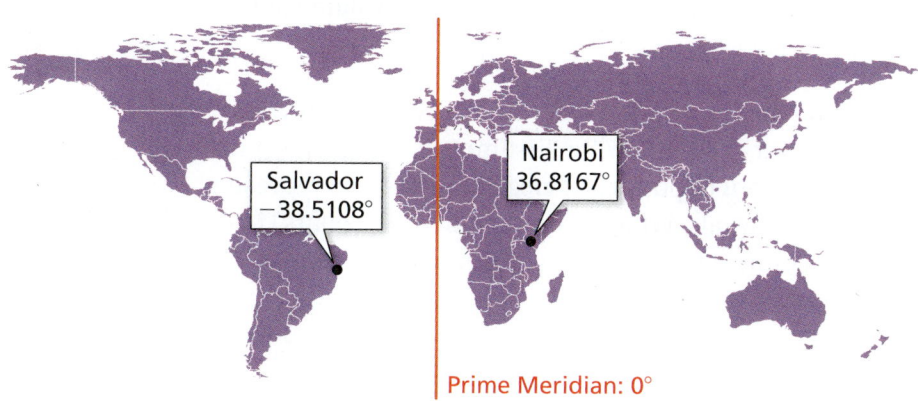

Salvador −38.5108°
Nairobi 36.8167°
Prime Meridian: 0°

A.2 Adding Integers (pp. 509–516)

Learning Target: Find sums of integers.

14. Write an addition expression represented by the number line. Then find the sum.

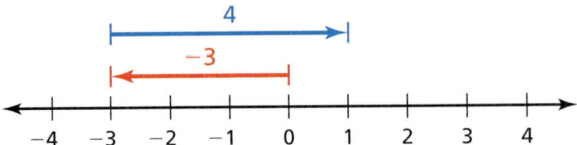

Find the sum. Use a number line to verify your answer.

15. $-16 + (-11)$

16. $-15 + 5$

17. $100 + (-75)$

18. $-32 + (-2)$

19. $-2 + (-7) + 15$

20. $9 + (-14) + 3$

21. During the first play of a football game, you lose 3 yards. You gain 7 yards during the second play. What is your total gain of yards for these two plays?

22. Write an addition expression using integers that equals -2. Use a number line to justify your answer.

23. Describe a real-life situation that uses the sum of the integers -8 and 12.

A.3 Adding Rational Numbers (pp. 517–522)

Learning Target: Find sums of rational numbers.

Find the sum. Write fractions in simplest form.

24. $\dfrac{9}{10} + \left(-\dfrac{4}{5}\right)$

25. $-4\dfrac{5}{9} + \dfrac{8}{9}$

26. $-1.6 + (-2.4)$

27. Find the sum of $-4 + 6\dfrac{2}{5} + (-2.7)$. Explain each step.

28. You open a new bank account. The table shows the activity of your account for the first month. Positive numbers represent deposits and negative numbers represent withdrawals. What is your balance (in dollars) in the account at the end of the first month?

Date	Amount (dollars)
3/5	100
3/12	-12.25
3/16	25.82
3/21	14.95
3/29	-18.56

A.4 Subtracting Integers (pp. 523–528)

Learning Target: Find differences of integers.

Find the difference. Use a number line to verify your answer.

29. $8 - 18$ **30.** $-16 - (-5)$ **31.** $-18 - 7$ **32.** $-12 - (-27)$

33. Your score on a game show is -300. You answer the final question incorrectly, so you lose 400 points. What is your final score?

34. Oxygen has a boiling point of $-183°C$ and a melting point of $-219°C$. What is the temperature difference of the melting point and the boiling point?

35. In one month, you earn $16 for mowing the lawn, $15 for babysitting, and $20 for allowance. You spend $12 at the movie theater. How much more money do you need to buy a $45 video game?

36. Write a subtraction expression using integers that equals -6.

37. Write two negative integers whose difference is positive.

A.5 Subtracting Rational Numbers (pp. 529–536)

Learning Target: Find differences of rational numbers and find distances between numbers on a number line.

Find the difference. Write fractions in simplest form.

38. $-\dfrac{5}{12} - \dfrac{3}{10}$ **39.** $3\dfrac{3}{4} - \dfrac{7}{8}$ **40.** $3.8 - (-7.45)$

41. Find the distance between -3.71 and -2.59 on a number line.

42. A turtle is $20\dfrac{5}{6}$ inches below the surface of a pond. It dives to a depth of $32\dfrac{1}{4}$ inches. What is the change in the turtle's position?

43. The lowest temperature ever recorded on Earth was $-89.2°C$ at Soviet Vostok Station in Antarctica. The highest temperature ever recorded was $56.7°C$ at Greenland Ranch in California. What is the difference between the highest and lowest recorded temperatures?

A Practice Test

Find the absolute value.

1. $\left|-\dfrac{4}{5}\right|$
2. $|6.43|$
3. $|-22|$

Copy and complete the statement using <, >, or =.

4. $4\ \square\ |-8|$
5. $|-7|\ \square\ -12$
6. $-7\ \square\ |3|$

Add or subtract. Write fractions in simplest form.

7. $-6 + (-11)$
8. $2 - (-9)$
9. $-\dfrac{4}{9} + \left(-\dfrac{23}{18}\right)$
10. $\dfrac{17}{12} - \left(-\dfrac{1}{8}\right)$
11. $9.2 + (-2.8)$
12. $2.86 - 12.1$

13. Write an addition expression and write a subtraction expression represented by the number line. Then evaluate the expressions.

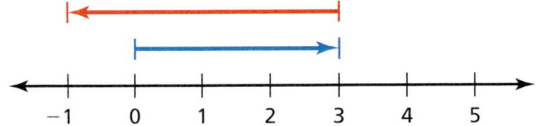

14. The table shows your scores, relative to *par*, for nine holes of golf. What is your total score for the nine holes?

Hole	1	2	3	4	5	6	7	8	9
Score	+1	−2	−1	0	−1	+3	−1	−3	+1

15. The elevation of a fish is −27 feet. The fish descends 32 feet, and then rises 14 feet. What is its new elevation?

16. The table shows the rainfall (in inches) for three months compared to the yearly average. Is the total rainfall for the three-month period greater than or less than the yearly average? Explain.

October	November	December
−0.86	2.56	−1.24

17. Bank Account A has $750.92, and Bank Account B has $675.44. Account A changes by −$216.38, and Account B changes by −$168.49. Which account has the greater balance? Explain.

18. On January 1, you recorded the lowest temperature as 23°F and the highest temperature as 6°C. A formula for converting from degrees Fahrenheit F to degrees Celsius C is $C = \dfrac{5}{9}F - \dfrac{160}{9}$. What is the temperature range (in degrees Celsius) for January 1?

A Cumulative Practice

1. A football team gains 2 yards on the first play, loses 5 yards on the second play, loses 3 yards on the third play, and gains 4 yards on the fourth play. What is the team's total gain or loss?

   A. a gain of 14 yards B. a gain of 2 yards

   C. a loss of 2 yards D. a loss of 14 yards

2. Which expression is *not* equal to 0?

   F. $5 - 5$ G. $-7 + 7$

   H. $6 - (-6)$ I. $-8 - (-8)$

3. What is the value of the expression?

   $$|-2 - (-2.5)|$$

   A. -4.5 B. -0.5

   C. 0.5 D. 4.5

4. What is the value of the expression?

   $$17 - (-8)$$

5. What is the distance between the two numbers on the number line?

   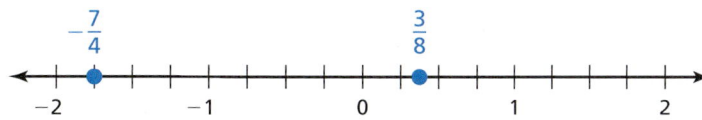

   F. $-2\frac{1}{8}$ G. $-1\frac{3}{8}$

   H. $1\frac{3}{8}$ I. $2\frac{1}{8}$

6. What is the value of the expression when $a = 8$, $b = 3$, and $c = 6$?

$$\left| a^2 - 2ac + 5b \right|$$

A. -65 **B.** -17

C. 17 **D.** 65

7. What is the value of the expression?

$$-9.74 + (-2.23)$$

8. Four friends are playing a game using the spinner shown. Each friend starts with a score of 0 and then spins four times. When you spin blue, you add the number to your score. When you spin red, you subtract the number from your score. The highest score after four spins wins. Each friend's spins are shown. Which spins belong to the winner?

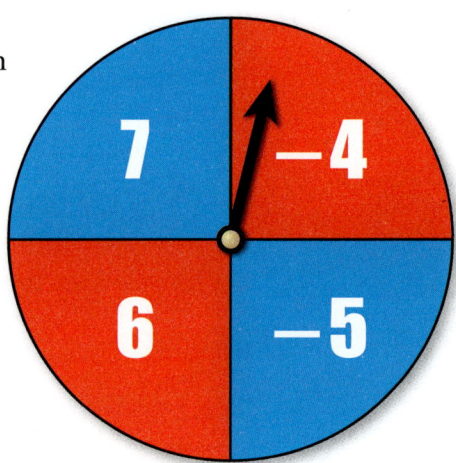

F. $6, 7, 7, 6$

G. $-4, -4, 7, -5$

H. $6, -5, -4, 7$

I. $-5, 6, -5, 6$

9. What number belongs in the box to make the equation true?

$$3\tfrac{1}{2} \div 5\tfrac{2}{3} = \tfrac{7}{2} \times \boxed{}$$

A. $\dfrac{3}{17}$ **B.** $\dfrac{3}{2}$

C. $\dfrac{17}{3}$ **D.** $\dfrac{13}{2}$

10. What is the value of the expression?

$$\dfrac{5.2 - 2.25}{0.05}$$

F. -346 **G.** 0.59

H. 5.9 **I.** 59

544 Chapter A Adding and Subtracting Rational Numbers

11. You leave school and walk 1.237 miles west. Your friend leaves school and walks 0.56 mile east. How far apart are you and your friend?

  A. 0.677 mile **B.** 0.69272 mile

  C. 1.293 miles **D.** 1.797 miles

12. Which property does the equation represent?

$$-80 + 30 + (-30) = -80 + [30 + (-30)]$$

  F. Commutative Property of Addition

  G. Associative Property of Addition

  H. Additive Inverse Property

  I. Addition Property of Zero

13. The values of which two points have the greatest sum?

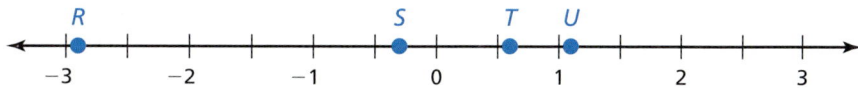

  A. R and S **B.** R and U

  C. S and T **D.** T and U

14. Consider the number line shown.

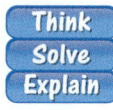
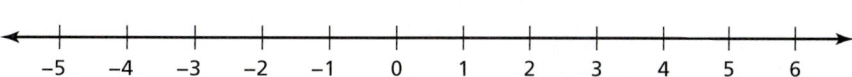

  Part A Use the number line to explain how to add -2 and -3.

  Part B Use the number line to explain how to subtract 5 from 2.

15. Which expression represents a *negative* value?

  F. $2 - |-7 + 3|$ **G.** $|-12 + 9|$

  H. $|5| + |11|$ **I.** $|8 - 14|$

B Multiplying and Dividing Rational Numbers

- **B.1** Multiplying Integers
- **B.2** Dividing Integers
- **B.3** Converting Between Fractions and Decimals
- **B.4** Multiplying Rational Numbers
- **B.5** Dividing Rational Numbers

Chapter Learning Target:
Understand multiplying and dividing rational numbers.

Chapter Success Criteria:
- ■ I can explain the rules for multiplying integers.
- ■ I can explain the rules for dividing integers.
- ■ I can evaluate expressions involving rational numbers.
- ■ I can solve real-life problems involving multiplication and division of rational numbers.

STEAM Video: "Carpenter or Joiner"

STEAM Video

Carpenter or Joiner

Carpenters and joiners must be precise with their measurements when building structures. In what other real-life situations must measurements be precise?

Watch the STEAM Video "Carpenter or Joiner." Then answer the following questions.

1. Robert says that changes in water content cause wood to shrink or expand *across* the grain more than *along* the grain. What does this mean?

2. Describe how you can cut a log so that the pieces shrink in different ways as they dry out.

Performance Task

Precisely Perfect

After completing this chapter, you will be able to use the concepts you learned to answer the questions in the *STEAM Video Performance Task*. You will be given the accuracies of seven telescopes. For example:

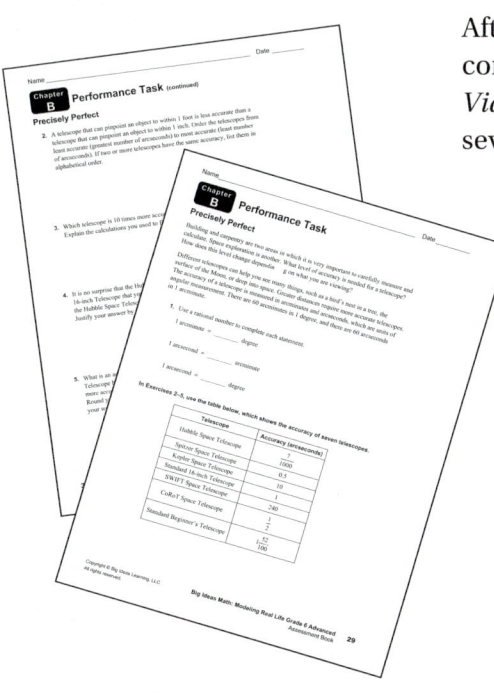

Accuracy (arcseconds)

Hubble Space Telescope: $\frac{7}{1000}$

Kepler Space Telescope: 10

Standard Beginner's Telescope: $1\frac{52}{100}$

You will be asked to compare the accuracies of the telescopes. Why do different telescopes have different accuracies?

547

Getting Ready for Chapter

Chapter Exploration

1. Work with a partner. Use integer counters to find each product.   + = +1   − = −1

   a. $(+3) \times (-2)$

   $(+3) \times (-2)$
   Add 3 groups of −2. $(+3) \times (-2) = $ ▢

   b. $(-2) \times (-2)$

   $(-2) \times (-2)$
   Remove 2 groups of −2. $(-2) \times (-2) = $ ▢

   Start with enough zero pairs so you can remove 2 groups of −2.

   c. $(-2) \times (+3)$

   $(-2) \times (+3)$
   Remove 2 groups of 3. $(-2) \times (+3) = $ ▢

   Start with enough zero pairs so you can remove 2 groups of 3.

   Work with a partner. Use integer counters to find the product.

   2. $(+3) \times (+2)$ 3. $(+3) \times (-1)$ 4. $(+2) \times (-4)$
   5. $(-3) \times (+2)$ 6. $(-2) \times (-3)$ 7. $(-1) \times (-4)$
   8. $(-1) \times (-2)$ 9. $(+3) \times (+1)$ 10. $(-3) \times (-2)$
   11. $(-2) \times (+2)$ 12. $(-2) \times (+4)$ 13. $(-4) \times (-2)$

14. **MAKE A CONJECTURE** Use your results in Exercises 1–13 to determine the sign of each product.

    a. negative integer and a positive integer
    b. two negative integers
    c. two positive integers

Vocabulary

The following vocabulary terms are defined in this chapter. Think about what each term might mean and record your thoughts.

terminating decimal   repeating decimal   complex fraction

B.1 Multiplying Integers

Learning Target: Find products of integers.

Success Criteria:
- I can explain the rules for multiplying integers.
- I can find products of integers with the same sign.
- I can find products of integers with different signs.

EXPLORATION 1

Understanding Products Involving Negative Integers

Work with a partner.

a. The number line and integer counters model the product $3 \cdot 2$. How can you find $3 \cdot (-2)$? Explain.

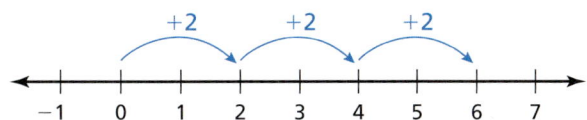

b. Use the tables to find $-3 \cdot 2$ and $-3 \cdot (-2)$. Explain your reasoning.

2	•	2	=	4
1	•	2	=	2
0	•	2	=	0
−1	•	2	=	
−2	•	2	=	
−3	•	2	=	

−3	•	3	=	−9
−3	•	2	=	−6
−3	•	1	=	−3
−3	•	0	=	
−3	•	−1	=	
−3	•	−2	=	

c. **INDUCTIVE REASONING** Complete the table. Then write general rules for multiplying (i) two integers with the same sign and (ii) two integers with different signs.

Expression	Type of Product	Product	Product: Positive or Negative
$3 \cdot 2$	Integers with the same sign		
$3 \cdot (-2)$			
$-3 \cdot 2$			
$-3 \cdot (-2)$			
$6 \cdot 3$			
$2 \cdot (-5)$			
$-6 \cdot 5$			
$-5 \cdot (-3)$			

Math Practice

Construct Arguments

Construct an argument that you can use to convince a friend of the rules you wrote in Exploration 1(c).

B.1 Lesson

Consider the following methods for evaluating $3(-2 + 4)$.

Evaluate in parentheses:
$3(-2 + 4) = 3(2)$
$= 6$

Use the Distributive Property:
$3(-2 + 4) = 3(-2) + 3(4)$
$= ? + 12$

For the Distributive Property to be true, $3(-2)$ must equal -6. This leads to the following rules for multiplying integers.

 Key Ideas

Multiplying Integers with the Same Sign

Words The product of two integers with the same sign is positive.

Numbers $2 \cdot 3 = 6$ $\qquad -2 \cdot (-3) = 6$

Multiplying Integers with Different Signs

Words The product of two integers with different signs is negative.

Numbers $2 \cdot (-3) = -6$ $\qquad -2 \cdot 3 = -6$

EXAMPLE 1 Multiplying Integers

Find each product.

a. $-5 \cdot (-6)$

The integers have the same sign.

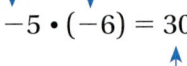

$-5 \cdot (-6) = 30$

The product is positive.

▸ The product is 30.

b. $3(-4)$

The integers have different signs.

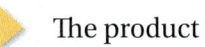

$3(-4) = -12$

The product is negative.

▸ The product is -12.

Try It Find the product.

1. $5 \cdot 5$
2. $-1(-9)$
3. $-7 \cdot (-8)$
4. $12 \cdot (-2)$
5. $4(-6)$
6. $-25(0)$

EXAMPLE 2 Evaluating Expressions

a. Find $(-2)^2$.

$(-2)^2 = (-2) \cdot (-2)$ Write $(-2)^2$ as repeated multiplication.

$= 4$ Multiply.

> The expression $(-2)^2$ indicates to multiply the number in parentheses, -2, by itself.
> The expression -2^2, however, indicates to find the opposite of 2^2.

b. Find -2^2.

$-2^2 = -(2 \cdot 2)$ Write 2^2 as repeated multiplication.

$= -4$ Multiply 2 and 2.

c. Find $-2 \cdot 17 \cdot (-5)$.

$-2 \cdot 17 \cdot (-5) = -2 \cdot (-5) \cdot 17$ Commutative Property of Multiplication

$= 10 \cdot 17$ Multiply -2 and -5.

$= 170$ Multiply 10 and 17.

Remember
Use order of operations when evaluating an expression.

d. Find $-6(-3 + 4) + 6$.

$-6(-3 + 4) + 6 = -6(1) + 6$ Perform operation in parentheses.

$= -6 + 6$ Multiplication Property of 1

$= 0$ Additive Inverse Property

Try It Evaluate the expression.

7. $8 \cdot (-15) \cdot 0$
8. $24 - 3^3$
9. $10 - 7(3 - 5)$

Self-Assessment for Concepts & Skills

Solve each exercise. Then rate your understanding of the success criteria in your journal.

10. **WRITING** What can you conclude about two integers whose product is (a) positive and (b) negative?

EVALUATING AN EXPRESSION Evaluate the expression.

11. $4(-8)$
12. $-5(-7)$
13. $12 - 3^2 \cdot (-2)$

MP REASONING Tell whether the statement is *true* or *false*. Explain your reasoning.

14. The product of three positive integers is positive.

15. The product of three negative integers is positive.

Section B.1 Multiplying Integers 551

EXAMPLE 3 Modeling Real Life

You solve a number puzzle on your phone. You start with 250 points. You finish the puzzle in 8 minutes 45 seconds and make 3 mistakes. What is your score?

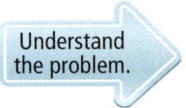 **Understand the problem.** You are given ways to gain points and lose points when completing a puzzle. You are asked to find your score after finishing the puzzle.

 **Make a plan.** Use a verbal model to solve the problem. Find the sum of the starting points, mistake penalties, and time bonus.

 **Solve and check.**

Score = Starting points + Number of mistakes · Penalty per mistake + Time bonus

$= 250 + 3(-50) + 75$ ← 10 min − 8 min 45 sec = 1 min 15 sec = 75 sec

$= 250 + (-150) + 75$

$= 100 + 75$

$= 175$

So, your score is 175 points.

Another Method
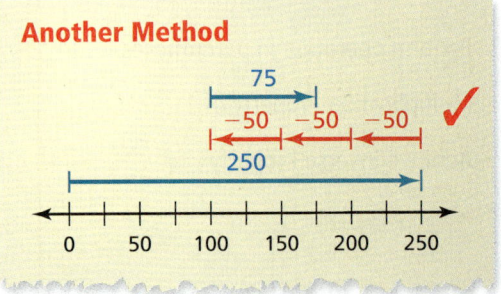

Self-Assessment for Problem Solving

Solve each exercise. Then rate your understanding of the success criteria in your journal.

16. On a mountain, the temperature decreases by 18°F for each 5000-foot increase in elevation. At 7000 feet, the temperature is 41°F. What is the temperature at 22,000 feet? Justify your answer.

17. Players in a racing game earn 3 points for each coin they collect. Each player loses 5 points for each second that he or she finishes after the first-place finisher. The table shows the results of a race. List the players in order from greatest to least number of points.

Player	Coins	Time
1	31	0:02:03
2	18	0:01:55
3	24	0:01:58
4	27	0:02:01

B.1 Practice

 Go to *BigIdeasMath.com* to get HELP with solving the exercises.

▶ Review & Refresh

Find the distance between the two numbers on a number line.

1. -4.3 and 0.8
2. -7.7 and -6.4
3. $-2\frac{3}{5}$ and -1

Divide.

4. $27 \div 9$
5. $48 \div 6$
6. $56 \div 4$
7. $153 \div 8$

8. What is the prime factorization of 84?

   A. $2^2 \times 3^2$
   B. $2^3 \times 7$
   C. $3^3 \times 7$
   D. $2^2 \times 3 \times 7$

▶ Concepts, Skills, & Problem Solving

MP CHOOSE TOOLS Use a number line or integer counters to find the product. (See Exploration 1, p. 549.)

9. $2(-4)$
10. $-6(3)$
11. $4(-5)$

MULTIPLYING INTEGERS Find the product.

12. $6 \cdot 4$
13. $7(-3)$
14. $-2(8)$
15. $-3(-4)$
16. $-6 \cdot 7$
17. $3 \cdot 9$
18. $8 \cdot (-5)$
19. $-1 \cdot (-12)$
20. $-5(10)$
21. $-13(0)$
22. $-9 \cdot 9$
23. $15(-2)$
24. $-10 \cdot 11$
25. $-6 \cdot (-13)$
26. $7(-14)$
27. $-11 \cdot (-11)$

28. **MODELING REAL LIFE** You burn 10 calories each minute you jog. What integer represents the change in your calories after you jog for 20 minutes?

29. **MODELING REAL LIFE** In a four-year period, about 80,000 acres of coastal wetlands in the United States are lost each year. What integer represents the total change in coastal wetlands?

EVALUATING EXPRESSIONS Evaluate the expression.

30. $(-4)^2$
31. -6^2
32. $-5 \cdot 3 \cdot (-2)$
33. $3 \cdot (-12) \cdot 0$
34. $-5(-7)(-20)$
35. $5 - 8^2$
36. $-5^2 \cdot 4$
37. $-2 \cdot (-3)^3$
38. $2 + 1 \cdot (-7 + 5)$
39. $4 - (-2)^3$
40. $4 \cdot (25 \cdot 3^2)$
41. $-4(3^2 - 8) + 1$

YOU BE THE TEACHER Your friend evaluates the expression. Is your friend correct? Explain your reasoning.

42.
$-2(-7) = -14$

43.
$-10^2 = -100$

MP PATTERNS Find the next two numbers in the pattern.

44. $-12, 60, -300, 1500, \ldots$

45. $7, -28, 112, -448, \ldots$

46. **MP PROBLEM SOLVING** In a scavenger hunt, each team earns 25 points for each item that they find. Each team loses 15 points for every minute after 4:00 P.M. that they report to the city park. The table shows the number of items found by each team and the time that each team reported to the park. Which team wins the scavenger hunt? Justify your answer.

Team	Items	Time
A	13	4:03 P.M.
B	15	4:07 P.M.
C	11	3:56 P.M.
D	12	4:01 P.M.

47. **MP REASONING** The height of an airplane during a landing is given by $22{,}000 + (-480t)$, where t is the time in minutes. Estimate how many minutes it takes the plane to land. Explain your reasoning.

48. **MP PROBLEM SOLVING** The table shows the price of a bluetooth speaker each month for 4 months.

Month	Price (dollars)
June	165
July	$165 + (-12)$
August	$165 + 2(-12)$
September	$165 + 3(-12)$

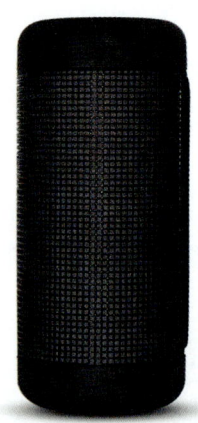

a. Describe the change in the price of the speaker.

b. The table at the right shows the amount of money you save each month. When do you have enough money saved to buy the speaker? Explain your reasoning.

Amount Saved	
June	$35
July	$55
August	$45
September	$18

49. **DIG DEEPER!** Two integers, a and b, have a product of 24. What is the least possible sum of a and b?

50. **MP NUMBER SENSE** Consider two integers p and q. Explain why $p \times (-q) = (-p) \times q = -pq$.

B.2 Dividing Integers

Learning Target: Find quotients of integers.

Success Criteria:
- I can explain the rules for dividing integers.
- I can find quotients of integers with the same sign.
- I can find quotients of integers with different signs.

EXPLORATION 1

Understanding Quotients Involving Negative Integers

Work with a partner.

a. Discuss the relationship between multiplication and division with your partner.

b. **INDUCTIVE REASONING** Complete the table. Then write general rules for dividing (i) two integers with the same sign and (ii) two integers with different signs.

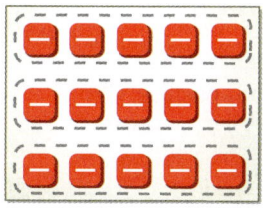

Expression	Type of Quotient	Quotient	Quotient: Positive, Negative, or Zero
$-15 \div 3$	Integers with different signs		
$12 \div (-6)$			
$10 \div (-2)$			
$-6 \div 2$			
$-12 \div (-12)$			
$-21 \div (-7)$			
$0 \div (-15)$			
$0 \div 4$			
$-5 \div 4$			
$5 \div (-4)$			

Math Practice

Recognize Usefulness of Tools

Can you use number lines or integer counters to reach the same conclusions as in part (b)? Explain why or why not.

c. Find the values of $-\dfrac{8}{4}$, $\dfrac{-8}{4}$, and $\dfrac{8}{-4}$. What do you notice? Is this true for $-\dfrac{a}{b}$, $\dfrac{-a}{b}$, and $\dfrac{a}{-b}$ when a and b are integers? Explain.

d. Is every quotient of integers a rational number? Explain your reasoning.

B.2 Lesson

 **Key Ideas**

Remember
Division by 0 is undefined.

Dividing Integers with the Same Sign

Words The quotient of two integers with the same sign is positive.

Numbers $8 \div 2 = 4$ $-8 \div (-2) = 4$

Dividing Integers with Different Signs

Words The quotient of two integers with different signs is negative.

Numbers $8 \div (-2) = -4$ $-8 \div 2 = -4$

EXAMPLE 1 — Dividing Integers with the Same Sign

Find $-18 \div (-6)$.

The integers have the same sign.

$-18 \div (-6) = 3$

The quotient is positive.

 The quotient is 3.

Try It Find the quotient.

1. $14 \div 2$
2. $-32 \div (-4)$
3. $-40 \div (-8)$

EXAMPLE 2 — Dividing Integers with Different Signs

Find each quotient.

a. $75 \div (-25)$

b. $\dfrac{-54}{6}$

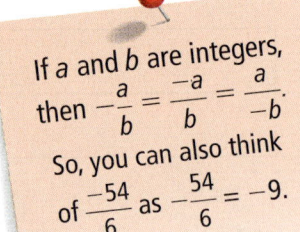

If a and b are integers, then $-\dfrac{a}{b} = \dfrac{-a}{b} = \dfrac{a}{-b}$.
So, you can also think of $\dfrac{-54}{6}$ as $-\dfrac{54}{6} = -9$.

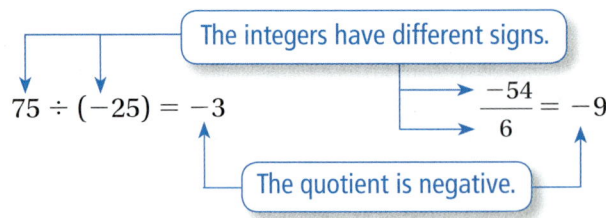

The integers have different signs.

$75 \div (-25) = -3$ $\dfrac{-54}{6} = -9$

The quotient is negative.

▶ The quotient is -3.   ▶ The quotient is -9.

Try It Find the quotient.

4. $0 \div (-6)$
5. $\dfrac{-49}{7}$
6. $\dfrac{21}{-3}$

EXAMPLE 3 Evaluating Expressions

Find the value of each expression when $x = 8$ and $y = -4$.

a. $\dfrac{x}{2y}$

$\dfrac{x}{2y} = \dfrac{8}{2(-4)}$ Substitute 8 for x and -4 for y.

$= \dfrac{8}{-8}$ Multiply 2 and -4.

$= -1$ Divide 8 by -8.

▶ The value of the expression is -1.

b. $-x^2 + 12 \div y$

$-x^2 + 12 \div y = -8^2 + 12 \div (-4)$ Substitute 8 for x and -4 for y.

$= -(8 \cdot 8) + 12 \div (-4)$ Write 8^2 as repeated multiplication.

$= -64 + 12 \div (-4)$ Multiply 8 and 8.

$= -64 + (-3)$ Divide 12 by -4.

$= -67$ Add.

▶ The value of the expression is -67.

Try It Evaluate the expression when $a = -18$ and $b = -6$.

7. $a \div b$ **8.** $\dfrac{a+6}{3}$ **9.** $\dfrac{b^2}{a} + 4$

Self-Assessment for Concepts & Skills

Solve each exercise. Then rate your understanding of the success criteria in your journal.

10. WRITING What can you conclude about two integers whose quotient is (a) positive, (b) negative, or (c) zero?

DIVIDING INTEGERS Find the quotient.

11. $-12 \div 4$ **12.** $\dfrac{-6}{-2}$ **13.** $15 \div (-3)$

14. WHICH ONE DOESN'T BELONG? Which expression does *not* belong with the other three? Explain your reasoning.

$\dfrac{10}{-5}$ $\dfrac{-10}{5}$ $\dfrac{-10}{-5}$ $-\dfrac{10}{5}$

Section B.2 Dividing Integers 557

EXAMPLE 4 Modeling Real Life

You measure the height of the tide using the support beams of a pier. What is the mean hourly change in the height?

59 inches at 2 P.M.
8 inches at 8 P.M.

To find the mean hourly change in the height of the tide, divide the change in the height by the elapsed time.

$$\text{mean hourly change} = \frac{\text{final height} - \text{initial height}}{\text{elapsed time}}$$

The elapsed time from 2 P.M. to 8 P.M. is 6 hours.

$$= \frac{8 - 59}{6} \quad \text{Substitute.}$$

$$= \frac{-51}{6} \quad \text{Subtract.}$$

$$= -8\frac{1}{2} \quad \text{Divide.}$$

 The mean change in the height of the tide is $-8\frac{1}{2}$ inches per hour.

Self-Assessment for Problem Solving

Solve each exercise. Then rate your understanding of the success criteria in your journal.

15. A female grizzly bear weighs 500 pounds. After hibernating for 6 months, she weighs only 350 pounds. What is the mean monthly change in weight?

16. The table shows the change in the number of crimes committed in a city each year for 4 years. What is the mean yearly change in the number of crimes?

Year	2014	2015	2016	2017
Change in Crimes	215	−321	−185	95

17. **DIG DEEPER!** At a restaurant, when a customer buys 4 pretzels, the fifth pretzel is free. Soft pretzels cost $3.90 each. You order 12 soft pretzels. What is your mean cost per pretzel?

B.2 Practice

Review & Refresh

Find the product.

1. $8 \cdot 10$
2. $-6(9)$
3. $4(7)$
4. $-9(-8)$

Order the numbers from least to greatest.

5. $28\%, \frac{1}{4}, 0.24$
6. $42\%, 0.45, \frac{2}{5}$
7. $\frac{7}{10}, 0.69, 71\%, \frac{9}{10}, 0.84$

Write an addition expression and write a subtraction expression represented by the number line. Then evaluate the expressions.

8.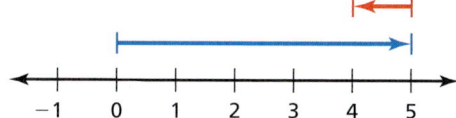

9.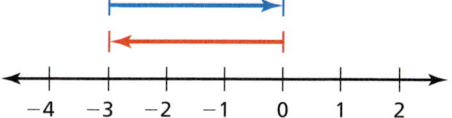

Concepts, Skills, & Problem Solving

MP CHOOSE TOOLS Complete the table. (See Exploration 1, p. 555.)

	Expression	Type of Quotient	Quotient	Quotient: Positive, Negative, or Zero
10.	$14 \div (-2)$			
11.	$-24 \div 12$			
12.	$-55 \div (-5)$			

DIVIDING INTEGERS Find the quotient, if possible.

13. $4 \div (-2)$
14. $21 \div (-7)$
15. $-20 \div 4$
16. $-18 \div (-3)$
17. $\frac{-14}{2}$
18. $\frac{0}{6}$
19. $\frac{-15}{-5}$
20. $\frac{54}{-9}$
21. $-\frac{33}{11}$
22. $-49 \div (-7)$
23. $0 \div (-2)$
24. $\frac{60}{-6}$
25. $\frac{-56}{14}$
26. $\frac{18}{0}$
27. $-\frac{65}{5}$
28. $\frac{-84}{-7}$

YOU BE THE TEACHER Your friend finds the quotient. Is your friend correct? Explain your reasoning.

29. $\dfrac{-63}{-9} = -7$

30. $0 \div (-5) = -5$

31. MODELING REAL LIFE You read 105 pages of a novel over 7 days. What is the mean number of pages you read each day?

USING ORDER OF OPERATIONS Evaluate the expression.

32. $-8 - 14 \div 2 + 5$

33. $24 \div (-4) + (-2) \cdot (-5)$

EVALUATING EXPRESSIONS Evaluate the expression when $x = 10$, $y = -2$, and $z = -5$.

34. $x \div y$

35. $12 \div 3y$

36. $\dfrac{2z}{y}$

37. $\dfrac{-x + y}{6}$

38. $100 \div (-z^2)$

39. $\dfrac{10y^2}{z}$

40. $\left| \dfrac{xz}{-y} \right|$

41. $\dfrac{-x^2 + 6z}{y}$

42. MP PATTERNS Find the next two numbers in the pattern $-128, 64, -32, 16, \ldots$. Explain your reasoning.

43. MODELING REAL LIFE The Detroit-Windsor Tunnel is an underwater highway that connects the cities of Detroit, Michigan, and Windsor, Ontario. How many times deeper is the roadway than the bottom of the ship?

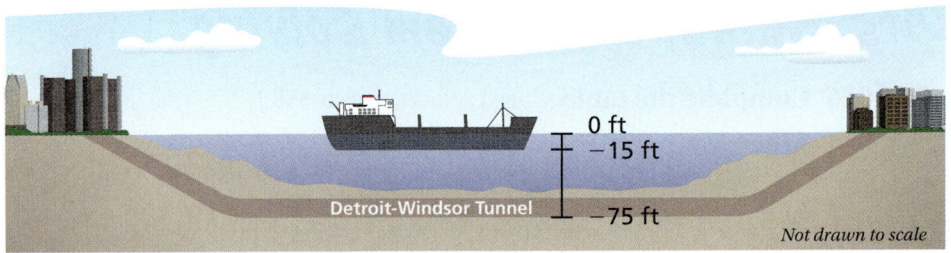

44. MODELING REAL LIFE A snowboarder descends from an elevation of 2253 feet to an elevation of 1011 feet in 3 minutes. What is the mean change in elevation per minute?

45. MP REASONING The table shows a golfer's scores relative to *par* for three out of four rounds of a tournament.

a. What was the golfer's mean score per round for the first 3 rounds?

b. The golfer's goal for the tournament is to have a mean score no greater than -3. Describe how the golfer can achieve this goal.

Scorecard	
Round 1	$+1$
Round 2	-4
Round 3	-3
Round 4	?

46. MP PROBLEM SOLVING The regular admission price for an amusement park is $72. For a group of 15 or more, the admission price is reduced by $25 per person. How many people need to be in a group to save $500?

47. DIG DEEPER! Write a set of five different integers that has a mean of -10. Explain how you found your answer.

B.3 Converting Between Fractions and Decimals

Learning Target: Convert between different forms of rational numbers.

Success Criteria:
- I can explain the difference between terminating and repeating decimals.
- I can write fractions and mixed numbers as decimals.
- I can write decimals as fractions and mixed numbers.

EXPLORATION 1

Analyzing Denominators of Decimal Fractions

Work with a partner.

a. Write each decimal as a fraction or mixed number.

   0.7   1.29   12.831   0.0041

b. What do the factors of the denominators of the fractions you wrote have in common? Is this always true for decimal fractions?

EXPLORATION 2

Exploring Decimal Representations

Work with a partner.

a. A fraction $\dfrac{a}{b}$ can be interpreted as $a \div b$. Use a calculator to convert each unit fraction to a decimal. Do some of the decimals look different than the others? Explain.

$$\frac{1}{2} \qquad \frac{1}{3} \qquad \frac{1}{4} \qquad \frac{1}{5}$$

$$\frac{1}{6} \qquad \frac{1}{7} \qquad \frac{1}{8}$$

$$\frac{1}{9} \qquad \frac{1}{10} \qquad \frac{1}{11} \qquad \frac{1}{12}$$

> **Math Practice**
>
> **Use Technology to Explore**
> How do calculators help you learn about different types of decimals? How can you find decimal forms of fractions without using a calculator?

b. Compare and contrast the fractions in part (a) with the fractions you wrote in Exploration 1. What conclusions can you make?

c. Does every fraction have a decimal form that either *terminates* or *repeats*? Explain your reasoning.

B.3 Lesson

Key Vocabulary
terminating decimal, p. 562
repeating decimal, p. 562

Because you can divide any integer by any nonzero integer, you can use long division to write fractions and mixed numbers as decimals. These decimals are rational numbers and will either *terminate* or *repeat*.

A **terminating decimal** is a decimal that ends.

$$1.5, \ -0.25, \ 10.824$$

A **repeating decimal** is a decimal that has a pattern that repeats.

$$-1.333\ldots = -1.\overline{3}$$
$$0.151515\ldots = 0.\overline{15}$$

Use *bar notation* to show which of the digits repeat.

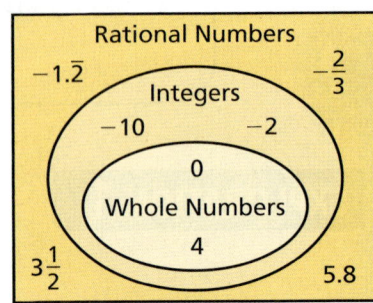

EXAMPLE 1 Writing Fractions and Mixed Numbers as Decimals

a. Write $-2\frac{1}{4}$ as a decimal.

Notice that $-2\frac{1}{4} = -\frac{9}{4}$.

Use long division to divide 9 by 4.

Divide 9 by 4.

$$\begin{array}{r} 2.25 \\ 4{\overline{\smash{\big)}\,9.00}} \\ -8 \\ \hline 1\,0 \\ -\,8 \\ \hline 20 \\ -20 \\ \hline 0 \end{array}$$

The remainder is 0. So, it is a terminating decimal.

Another Method
Use equivalent fractions.
$$\frac{1}{4} = \frac{1 \times 25}{4 \times 25} = \frac{25}{100}$$
So, $-2\frac{1}{4} = -2\frac{25}{100}$
$= -2.25.$ ✓

So, $-2\frac{1}{4} = -2.25$.

b. Write $\frac{5}{11}$ as a decimal.

Use long division to divide 5 by 11.

Divide 5 by 11.

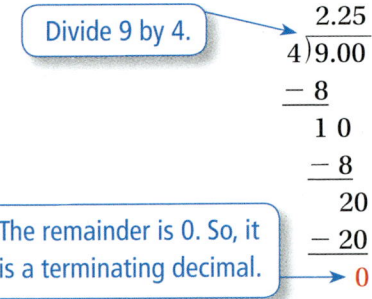

The remainder repeats. So, it is a repeating decimal.

So, $\frac{5}{11} = 0.\overline{45}$.

Try It Write the fraction or mixed number as a decimal.

1. $-\frac{6}{5}$ 2. $-7\frac{3}{8}$ 3. $-\frac{3}{11}$ 4. $1\frac{5}{27}$

This shows two representations for simplifying fractions. Use whichever you prefer.

Any terminating decimal can be written as a fraction whose denominator is a power of 10. You can often simplify the resulting fraction by *dividing out* any common factors, which is the same as removing the common factor from the numerator and denominator.

$$0.48 = \frac{48}{100} = \frac{48 \div 4}{100 \div 4} = \frac{12}{25} \quad \text{or} \quad 0.48 = \frac{48}{100} = \frac{12 \cdot \cancel{4}}{25 \cdot \cancel{4}} = \frac{12}{25}$$

EXAMPLE 2 Writing a Terminating Decimal as a Fraction

Write −0.26 as a fraction in simplest form.

$-0.26 = -\dfrac{26}{100}$ ← Write the digits after the decimal point in the numerator.
← The last digit is in the hundredths place. So, use 100 in the denominator.

$ = -\dfrac{13 \cdot \cancel{2}}{50 \cdot \cancel{2}}$ Divide out the common factor, 2.

$ = -\dfrac{13}{50}$ Simplify.

▶ So, $-0.26 = -\dfrac{13}{50}$.

Reading
−0.26 is read as "negative twenty-six hundredths."

Try It Write the decimal as a fraction or mixed number in simplest form.

5. −0.3   **6.** 0.125   **7.** −3.1   **8.** −10.25

Self-Assessment for Concepts & Skills

Solve each exercise. Then rate your understanding of the success criteria in your journal.

9. WRITING Compare and contrast terminating decimals and repeating decimals.

WRITING A FRACTION OR MIXED NUMBER AS A DECIMAL Write the fraction or mixed number as a decimal.

10. $\dfrac{3}{16}$ **11.** $-\dfrac{7}{15}$ **12.** $6\dfrac{17}{20}$

WRITING A DECIMAL AS A FRACTION OR MIXED NUMBER Write the decimal as a fraction or mixed number in simplest form.

13. 0.6   **14.** −12.48   **15.** 0.408

EXAMPLE 3 Modeling Real Life

The table shows the elevations of four sea creatures relative to sea level. Which of the sea creatures are deeper than the whale? Explain.

Creature	Elevation (kilometers)
Anglerfish	$-\dfrac{13}{10}$
Shark	$-\dfrac{2}{11}$
Squid	$-2\dfrac{1}{5}$
Whale	-0.8

One way to compare the depths of the creatures is to use a number line. First, write each fraction or mixed number as a decimal.

$$-\dfrac{13}{10} = -1.3$$

$$-\dfrac{2}{11} = -0.\overline{18}$$ ⟶ Divide 2 by 11.

$$-2\dfrac{1}{5} = -2\dfrac{2}{10} = -2.2$$

```
       0.1818
   11)2.0000
      -1 1
         90
        -88
         20
        -11
          90
         -88
           2
```

The remainder repeats. So, it is a repeating decimal.

Then graph each decimal on a number line.

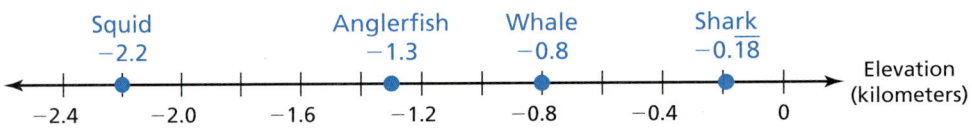

▸ Both -2.2 and -1.3 are less than -0.8. So, the squid and the anglerfish are deeper than the whale.

 **Self-Assessment** for Problem Solving

Solve each exercise. Then rate your understanding of the success criteria in your journal.

16. A box turtle hibernates in sand at an elevation of -1.625 feet. A spotted turtle hibernates at an elevation of $-1\dfrac{7}{12}$ feet. Which turtle hibernates deeper in the sand? How much deeper?

17. A *red sprite* is an electrical flash that occurs in Earth's upper atmosphere. The table shows the elevations of four red sprites. What is the range of the elevations?

Elevation (miles)	
50.6	$50\dfrac{13}{25}$
$50\dfrac{8}{15}$	$\dfrac{155}{3}$

564   Chapter B   Multiplying and Dividing Rational Numbers

B.3 Practice

Go to **BigIdeasMath.com** to get HELP with solving the exercises.

Review & Refresh

Find the quotient.

1. $12 \div (-6)$
2. $-48 \div 8$
3. $-42 \div (-7)$
4. $-33 \div (-3)$

Find the product.

5. 5.88×6
6. $2.0035 \cdot 4$
7. 5.49×13.509
8. 1.0006×0.003

9. Find the missing values in the ratio table. Then write the equivalent ratios.

Hours	2		$\frac{4}{3}$
Dollars Earned	18	72	

Concepts, Skills, & Problem Solving

MP STRUCTURE Without dividing, determine whether the decimal form of the fraction *terminates* or *repeats*. Explain. (See Explorations 1 & 2, p. 561.)

10. $\frac{3}{8}$
11. $\frac{5}{7}$
12. $\frac{11}{40}$
13. $\frac{5}{24}$

WRITING A FRACTION OR MIXED NUMBER AS A DECIMAL Write the fraction or mixed number as a decimal.

14. $\frac{7}{8}$
15. $\frac{1}{11}$
16. $-3\frac{1}{2}$
17. $-\frac{7}{9}$
18. $-\frac{17}{40}$
19. $1\frac{5}{6}$
20. $4\frac{2}{15}$
21. $\frac{25}{24}$
22. $-\frac{13}{11}$
23. $-2\frac{17}{18}$
24. $-5\frac{7}{12}$
25. $8\frac{15}{22}$

26. **YOU BE THE TEACHER** Your friend writes $-\frac{7}{11}$ as a decimal. Is your friend correct? Explain your reasoning.

$$-\frac{7}{11} = -0.6\overline{3}$$

WRITING A DECIMAL AS A FRACTION OR MIXED NUMBER Write the decimal as a fraction or mixed number in simplest form.

27. -0.9
28. 0.45
29. -0.258
30. -0.312
31. -2.32
32. -1.64
33. 6.012
34. -12.405

35. **MODELING REAL LIFE** You find one quarter, two dimes, and two nickels.

a. Write the dollar amount as a decimal.

b. Write the dollar amount as a fraction or mixed number in simplest form.

Section B.3 Converting Between Fractions and Decimals

COMPARING RATIONAL NUMBERS Copy and complete the statement using < or >.

36. $-4\frac{6}{10}$ ▨ -4.65 **37.** $-5\frac{3}{11}$ ▨ $-5.\overline{2}$ **38.** $-2\frac{13}{16}$ ▨ $-2\frac{11}{14}$

39. MODELING REAL LIFE Is the half pipe deeper than the skating bowl? Explain.

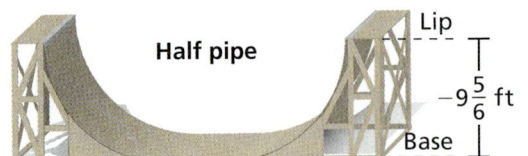
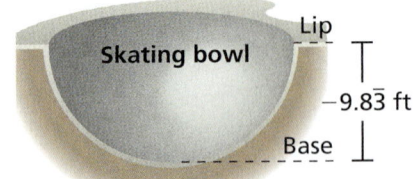

Player	Hits	At Bats
1	42	90
2	38	80

40. MODELING REAL LIFE In softball, a batting average is the number of hits divided by the number of times at bat. Does Player 1 or Player 2 have the greater batting average?

ORDERING RATIONAL NUMBERS Order the numbers from least to greatest.

41. $-\frac{3}{4}, 0.5, \frac{2}{3}, -\frac{7}{3}, 1.2$ **42.** $\frac{9}{5}, -2.5, -1.1, -\frac{4}{5}, 0.8$ **43.** $-1.4, -\frac{8}{5}, 0.6, -0.9, \frac{1}{4}$

44. $2.1, -\frac{6}{10}, -\frac{9}{4}, -0.75, \frac{5}{3}$ **45.** $-\frac{7}{2}, -2.8, -\frac{5}{4}, \frac{4}{3}, 1.3$ **46.** $-\frac{11}{5}, -2.4, 1.6, \frac{15}{10}, -2.25$

47. MODELING REAL LIFE The table shows the changes in the water level of a pond over several weeks. Order the numbers from least to greatest.

Week	1	2	3	4
Change (inches)	$-\frac{7}{5}$	$-1\frac{5}{11}$	-1.45	$-1\frac{91}{200}$

48. OPEN-ENDED Find one terminating decimal and one repeating decimal between $-\frac{1}{2}$ and $-\frac{1}{3}$.

49. PROBLEM SOLVING You miss 3 out of 10 questions on a science quiz and 4 out of 15 questions on a math quiz. On which quiz did you have a greater percentage of correct answers?

50. CRITICAL THINKING A hackberry tree has roots that reach a depth of $6\frac{5}{12}$ meters. The top of the tree is $18.2\overline{8}$ meters above the ground. Find the total height from the bottom of the roots to the top of the tree.

51. DIG DEEPER! Let a and b be integers.

   a. When can $-\frac{1}{a}$ be written as a positive, repeating decimal?

   b. When can $\frac{1}{ab}$ be written as a positive, terminating decimal?

B.4 Multiplying Rational Numbers

Learning Target: Find products of rational numbers.

Success Criteria:
- I can explain the rules for multiplying rational numbers.
- I can find products of rational numbers with the same sign.
- I can find products of rational numbers with different signs.

EXPLORATION 1

Finding Products of Rational Numbers

Work with a partner.

a. Write a multiplication expression represented by each area model. Then find the product.

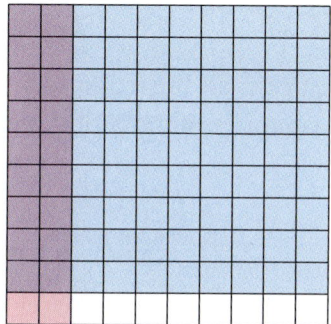 

b. Complete the table.

	Expression	Product	Expression	Product
i.	0.2×0.9		-0.2×0.9	
ii.	$0.3(0.5)$		$0.3(-0.5)$	
iii.	$\dfrac{1}{4} \cdot \dfrac{1}{2}$		$\dfrac{1}{4} \cdot \left(-\dfrac{1}{2}\right)$	
iv.	$1.2(0.4)$		$-1.2(-0.4)$	
v.	$\dfrac{3}{10}\left(\dfrac{2}{5}\right)$		$-\dfrac{3}{10}\left(-\dfrac{2}{5}\right)$	
vi.	0.6×1.8		-0.6×1.8	
vii.	$1\dfrac{1}{4} \cdot 2\dfrac{1}{2}$		$-1\dfrac{1}{4} \cdot \left(-2\dfrac{1}{2}\right)$	

Math Practice

Consider Similar Problems

How is multiplying integers similar to multiplying other rational numbers? How is it different?

c. Do the rules for multiplying integers apply to all rational numbers? Explain your reasoning.

B.4 Lesson

> When the signs of two numbers are different, their product is negative. When the signs of two numbers are the same, their product is positive.

Key Idea

Multiplying Rational Numbers

Words To multiply rational numbers, use the same rules for signs as you used for multiplying integers.

Numbers $\quad -\dfrac{2}{7} \cdot \dfrac{1}{3} = -\dfrac{2}{21} \qquad\qquad -\dfrac{2}{7} \cdot \left(-\dfrac{1}{3}\right) = \dfrac{2}{21}$

EXAMPLE 1 Multiplying Rational Numbers

a. Find -2.5×3.6.   **Estimate** $-2.5 \cdot 4 = -10$

Because the decimals have different signs, the product is negative. So, find the opposite of the product of 2.5 and 3.6.

$$
\begin{array}{r}
2.5 \\
\times\ 3.6 \\
\hline
1\ 5\ 0 \\
7\ 5\ 0\ \\
\hline
9.0\ 0
\end{array}
$$

← 1 decimal place
← + 1 decimal place

← 2 decimal places

▸ So, $-2.5 \times 3.6 = -9$.   **Reasonable?** $-9 \approx -10$ ✓

b. Find $-\dfrac{1}{3}\left(-2\dfrac{3}{4}\right)$.   **Estimate** $-\dfrac{1}{3} \cdot (-3) = 1$

Because the numbers have the same sign, the product is positive. So, find the product of $\dfrac{1}{3}$ and $2\dfrac{3}{4}$.

$\dfrac{1}{3}\left(2\dfrac{3}{4}\right) = \dfrac{1}{3}\left(\dfrac{11}{4}\right)$ Write the mixed number as an improper fraction.

$\phantom{\dfrac{1}{3}\left(2\dfrac{3}{4}\right)} = \dfrac{11}{12}$ Multiply the numerators and the denominators.

▸ So, $-\dfrac{1}{3}\left(-2\dfrac{3}{4}\right) = \dfrac{11}{12}$.   **Reasonable?** $\dfrac{11}{12} \approx 1$ ✓

Try It Find the product. Write fractions in simplest form.

1. -5.1×1.8 2. $-6.3(-0.6)$ 3. $-\dfrac{4}{5}\left(-\dfrac{2}{3}\right)$ 4. $4\dfrac{1}{2} \cdot \left(-2\dfrac{1}{3}\right)$

The properties of multiplication you have used apply to all rational numbers. You can also write $-\frac{a}{b}$ as $\frac{-a}{b}$ or $\frac{a}{-b}$ when performing operations with rational numbers.

EXAMPLE 2 Using Properties to Multiply Rational Numbers

Find $\left(-\frac{1}{7} \cdot \frac{4}{5}\right) \cdot (-7) \cdot \left(-\frac{1}{2}\right)$.

You can use properties of multiplication to find the product.

$\left(-\frac{1}{7} \cdot \frac{4}{5}\right) \cdot (-7) \cdot \left(-\frac{1}{2}\right) = -7 \cdot \left(-\frac{1}{7} \cdot \frac{4}{5}\right) \cdot \left(-\frac{1}{2}\right)$ Commutative Property of Multiplication

$= \left[-7 \cdot \left(-\frac{1}{7}\right)\right] \cdot \frac{4}{5} \cdot \left(-\frac{1}{2}\right)$ Associative Property of Multiplication

$= 1 \cdot \frac{4}{5} \cdot \left(-\frac{1}{2}\right)$ Multiplicative Inverse Property

$= \frac{4}{5} \cdot \left(\frac{-1}{2}\right)$ Multiplication Property of One

$= \frac{\overset{2}{\cancel{4}} \cdot (-1)}{5 \cdot \cancel{2}_1}$ Multiply. Divide out the common factor, 2.

$= \frac{-2}{5}$, or $-\frac{2}{5}$ Simplify.

Notice that Example 2 uses different notation to demonstrate the following.

$\dfrac{4 \cdot (-1)}{5 \cdot 2} = \dfrac{2 \cdot \overset{1}{\cancel{2}} \cdot (-1)}{5 \cdot \cancel{2}_1}$

Try It Find the product. Write fractions in simplest form.

5. $-\dfrac{2}{3} \cdot 7\dfrac{7}{8} \cdot \dfrac{3}{2}$

6. $-7.02(0.1)(100)(-10)$

Self-Assessment for Concepts & Skills

Solve each exercise. Then rate your understanding of the success criteria in your journal.

7. **WRITING** Explain how to determine whether a product of two rational numbers is *positive* or *negative*.

MULTIPLYING RATIONAL NUMBERS Find the product. Write fractions in simplest form.

8. $-\dfrac{3}{10} \times \left(-\dfrac{8}{15}\right)$

9. $-\dfrac{2}{3} \cdot 1\dfrac{1}{3}$

10. $-2.8(-1.7)$

11. $1\dfrac{3}{5} \cdot \left(-3\dfrac{3}{4}\right)$

Section B.4 Multiplying Rational Numbers

EXAMPLE 3 Modeling Real Life

A school record for the 40-meter dash is 15.24 seconds. Predict the school record after 15 years when the school record decreases by about 0.06 second per year.

Use a verbal model to solve the problem. Because the school record *decreases* by about 0.06 second per year, the change in the school record each year is −0.06 second.

$$\boxed{\text{School record after 15 years}} = \boxed{\text{Current school record}} + \boxed{\text{Number of years}} \cdot \boxed{\text{Average yearly change}}$$

$= 15.24 + 15(-0.06)$ Substitute.

$= 15.24 + (-0.9)$ Multiply 15 and −0.06.

$= 14.34$ Add 15.24 and −0.9.

▶ You can predict that the school record will be about 14.34 seconds after 15 years.

> **Check Reasonableness**
> Because 0.06 < 0.1, the school record decreases by less than $0.1 \cdot 15 = 1.5$ seconds. So, the school record is greater than $15.24 - 1.5 = 13.74$ seconds.
>
> Because 14.34 > 13.74, the answer is reasonable. ✓

Self-Assessment for Problem Solving

Solve each exercise. Then rate your understanding of the success criteria in your journal.

12. A swimmer's best time in an event is 53.87 seconds. On average, his best time decreases by 0.28 second each of the next five times he swims the event. Does he accomplish his goal of swimming the event in less than 52.5 seconds?

13. **DIG DEEPER!** *Terminal velocity* is the fastest speed that an object can fall through the air. A skydiver reaches a terminal velocity of 120 miles per hour. What is the change in elevation of the skydiver after falling at terminal velocity for 15 seconds? Justify your answer.

B.4 Practice

Go to *BigIdeasMath.com* to get HELP with solving the exercises.

▶ Review & Refresh

Write the fraction or mixed number as a decimal.

1. $\dfrac{5}{16}$
2. $-\dfrac{9}{22}$
3. $6\dfrac{8}{11}$
4. $-\dfrac{26}{24}$

Find the area of the figure.

5.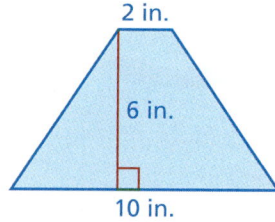

6.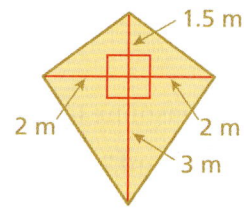

7.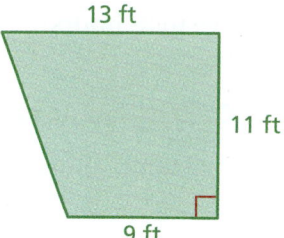

▶ Concepts, Skills, & Problem Solving

FINDING PRODUCTS OF RATIONAL NUMBERS Write a multiplication expression represented by the area model. Then find the product. *(See Exploration 1, p. 567.)*

8.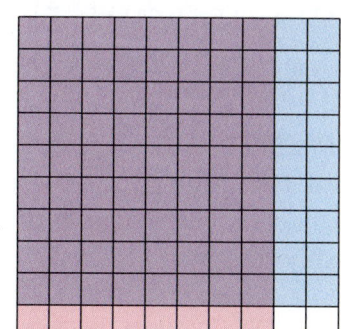

9.

MP REASONING Without multiplying, tell whether the value of the expression is positive or negative. Explain your reasoning.

10. $-1\left(\dfrac{4}{5}\right)$
11. $\dfrac{4}{7} \cdot \left(-3\dfrac{1}{2}\right)$
12. $-0.25(-3.659)$

MULTIPLYING RATIONAL NUMBERS Find the product. Write fractions in simplest form.

13. $-\dfrac{1}{4} \times \left(-\dfrac{4}{3}\right)$
14. $\dfrac{5}{6}\left(-\dfrac{8}{15}\right)$
15. $-2\left(-1\dfrac{1}{4}\right)$
16. $-3\dfrac{1}{3} \cdot \left(-2\dfrac{7}{10}\right)$
17. $0.4 \times (-0.03)$
18. $-0.05 \times (-0.5)$
19. $-8(0.09)(-0.5)$
20. $\dfrac{5}{6} \cdot \left(-4\dfrac{1}{2}\right) \cdot \left(-2\dfrac{1}{5}\right)$
21. $\left(-1\dfrac{2}{3}\right)^3$

Section B.4 Multiplying Rational Numbers

YOU BE THE TEACHER Your friend evaluates the expression. Is your friend correct? Explain your reasoning.

22.
$$-\frac{1}{4} \times \frac{3}{2} = \frac{-1}{4} \times \frac{3}{2}$$
$$= \frac{-3}{8}$$

23.
$$-2.2 \times (-3.7) = -8.14$$

24. **MODELING REAL LIFE** The hour hand of a clock moves $-30°$ every hour. How many degrees does it move in $2\frac{1}{5}$ hours?

25. **MODELING REAL LIFE** A 14.5-gallon gasoline tank is $\frac{3}{4}$ full. How many gallons will it take to fill the tank?

26. **OPEN-ENDED** Write two fractions whose product is $-\frac{3}{5}$.

USING PROPERTIES Find the product. Write fractions in simplest form.

27. $\frac{1}{5} \cdot \frac{3}{8} \cdot (-5)$

28. $0.01(4.6)(-200)$

29. $(-17.2 \times 2.5) \times 4$

30. $\left(-\frac{5}{9} \times \frac{2}{7}\right) \times \left(-\frac{7}{2}\right)$

31. $\left[-\frac{2}{3} \cdot \left(-\frac{5}{7}\right)\right] \cdot \left(-\frac{9}{4}\right)$

32. $(-4.5 \cdot 8.61) \cdot \left(-\frac{2}{9}\right)$

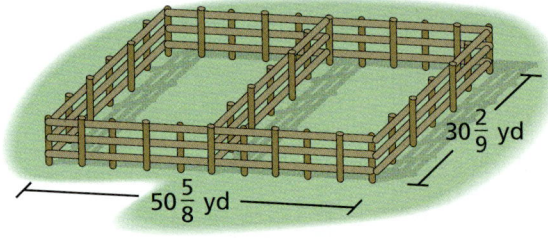

33. **MP PROBLEM SOLVING** Fencing costs $25.80 per yard. How much does it cost to enclose two adjacent rectangular pastures as shown? Justify your answer.

$30\frac{2}{9}$ yd

$50\frac{5}{8}$ yd

ALGEBRA Evaluate the expression when $x = -2$, $y = 3$, and $z = -\frac{1}{5}$.

34. $x \cdot z$

35. xyz

36. $\frac{1}{3} + x \cdot z$

37. $\frac{1}{2}z - \frac{2}{3}y$

EVALUATING AN EXPRESSION Evaluate the expression. Write fractions in simplest form.

38. $-4.2 + 8.1 \times (-1.9)$

39. $-3\frac{3}{4} \times \frac{5}{6} - 2\frac{1}{3}$

40. $\left(-\frac{2}{3}\right)^2 - \frac{3}{4}\left(2\frac{1}{3}\right)$

41. **DIG DEEPER!** Use positive or negative integers to fill in the blanks so that the product is $\frac{1}{4}$. Justify your answer.

$$\frac{}{2} \times \left(-\frac{5}{}\right) \times \frac{}{}$$

B.5 Dividing Rational Numbers

Learning Target: Find quotients of rational numbers.

Success Criteria:
- I can explain the rules for dividing rational numbers.
- I can find quotients of rational numbers with the same sign.
- I can find quotients of rational numbers with different signs.

EXPLORATION 1

Finding Quotients of Rational Numbers

Work with a partner.

a. Write two division expressions represented by the area model. Then find the quotients.

b. Complete the table.

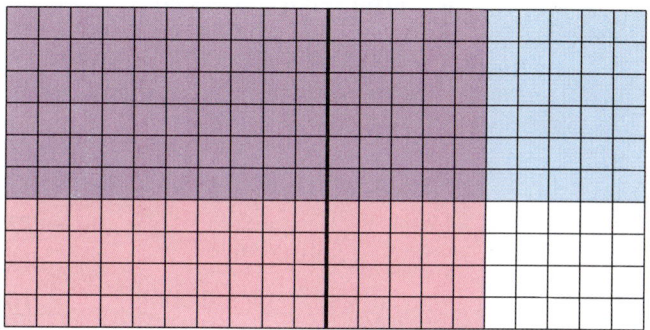

	Expression	Quotient	Expression	Quotient
i.	$0.9 \div 1.5$		$-0.9 \div 1.5$	
ii.	$1 \div \frac{1}{2}$		$-1 \div \frac{1}{2}$	
iii.	$2 \div 0.25$		$2 \div (-0.25)$	
iv.	$0 \div \frac{4}{5}$		$0 \div \left(-\frac{4}{5}\right)$	
v.	$1\frac{1}{2} \div 3$		$-1\frac{1}{2} \div (-3)$	
vi.	$0.8 \div 0.1$		$-0.8 \div (-0.1)$	

Math Practice

Applying Mathematics

How does interpreting a division expression in a real-life story help you make sense of the quotient?

c. Do the rules for dividing integers apply to all rational numbers? Explain your reasoning.

d. Write a real-life story involving the quotient $-0.75 \div 3$. Interpret the quotient in the context of the story.

B.5 Lesson

Key Vocabulary
complex fraction, p. 575

Key Idea

Dividing Rational Numbers

Words To divide rational numbers, use the same rules for signs as you used for dividing integers.

Numbers $-\dfrac{1}{2} \div \dfrac{4}{9} = -\dfrac{1}{2} \cdot \dfrac{9}{4} = -\dfrac{9}{8}$ $-\dfrac{1}{2} \div \left(-\dfrac{4}{9}\right) = -\dfrac{1}{2} \cdot \left(-\dfrac{9}{4}\right) = \dfrac{9}{8}$

EXAMPLE 1 Dividing Rational Numbers

a. Find $-8.4 \div (-3.6)$.

Because the decimals have the same sign, the quotient is positive. Use long division to divide 8.4 by 3.6.

$$3.6\overline{)8.4} \longrightarrow 36\overline{)84.00} \quad \begin{array}{r} 2.33 \\ \end{array}$$

$$\begin{array}{r} -72 \\ \hline 12\,0 \\ -10\,8 \\ \hline 1\,20 \\ -1\,08 \\ \hline 12 \end{array}$$

The remainder repeats. So, it is a repeating decimal.

Another Method Write the division expression as a fraction.

$$\dfrac{-8.4}{-3.6} = \dfrac{84}{36}$$
$$= \dfrac{7}{3}$$
$$= 2\dfrac{1}{3}, \text{ or } 2.\overline{3} \checkmark$$

So, $-8.4 \div (-3.6) = 2.\overline{3}$.

b. Find $\dfrac{6}{5} \div \left(-\dfrac{4}{3}\right)$.

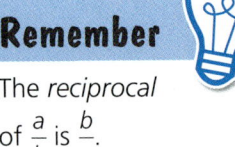

Remember
The *reciprocal* of $\dfrac{a}{b}$ is $\dfrac{b}{a}$.

$\dfrac{6}{5} \div \left(-\dfrac{4}{3}\right) = \dfrac{6}{5} \cdot \left(-\dfrac{3}{4}\right)$ Multiply by the reciprocal of $-\dfrac{4}{3}$.

$= \dfrac{\overset{3}{\cancel{6}} \cdot (-3)}{5 \cdot \underset{2}{\cancel{4}}}$ Multiply the numerators and the denominators. Divide out the common factor, 2.

$= \dfrac{-9}{10}, \text{ or } -\dfrac{9}{10}$ Simplify.

So, $\dfrac{6}{5} \div \left(-\dfrac{4}{3}\right) = -\dfrac{9}{10}$.

Try It Find the quotient. Write fractions in simplest form.

1. $-2.4 \div 3.2$ **2.** $-6 \div (-1.1)$ **3.** $-\dfrac{6}{5} \div \left(-\dfrac{1}{2}\right)$ **4.** $-\dfrac{1}{3} \div 2\dfrac{2}{3}$

You can represent division involving fractions using *complex fractions*. A **complex fraction** has at least one fraction in the numerator, denominator, or both.

EXAMPLE 2 Evaluating a Complex Fraction

Evaluate $\dfrac{-\dfrac{10}{9}}{-\dfrac{1}{6}+1}$.

Rewrite the complex fraction as a division expression.

Notice how $-\dfrac{1}{6} = \dfrac{-1}{6}$ is used to find the sum in parentheses.

$$-\dfrac{10}{9} \div \left(-\dfrac{1}{6} + 1\right) = -\dfrac{10}{9} \div \left(\dfrac{-1}{6} + \dfrac{6}{6}\right) \qquad \text{Rewrite } -\dfrac{1}{6} \text{ as } \dfrac{-1}{6} \text{ and 1 as } \dfrac{6}{6}.$$

$$= -\dfrac{10}{9} \div \dfrac{5}{6} \qquad \text{Add fractions.}$$

$$= -\dfrac{10}{9} \cdot \dfrac{6}{5} \qquad \text{Multiply by the reciprocal of } \dfrac{5}{6}.$$

$$= -\dfrac{\overset{2}{\cancel{10}} \cdot \overset{2}{\cancel{6}}}{\underset{3}{\cancel{9}} \cdot \underset{1}{\cancel{5}}} \qquad \text{Multiply. Divide out common factors.}$$

$$= -\dfrac{4}{3} \qquad \text{Simplify.}$$

Try It Evaluate the expression. Write fractions in simplest form.

5. $\dfrac{-\dfrac{1}{2}}{6}$

6. $\dfrac{-2\dfrac{1}{2}}{-\dfrac{3}{4}}$

7. $\dfrac{-1\dfrac{2}{3} \cdot \left(-\dfrac{3}{5}\right)}{\left(\dfrac{1}{3}\right)^2}$

Self-Assessment for Concepts & Skills

Solve each exercise. Then rate your understanding of the success criteria in your journal.

8. **WRITING** Explain how to determine whether a quotient of two rational numbers is *positive* or *negative*.

EVALUATING AN EXPRESSION Evaluate the expression. Write fractions in simplest form.

9. $\dfrac{3}{8} \div \left(-\dfrac{9}{5}\right)$

10. $-6.8 \div (-3.6)$

11. $\dfrac{-\dfrac{2}{9}}{2\dfrac{2}{5}}$

EXAMPLE 3 Modeling Real Life

A restaurant launches a mobile app that allows customers to rate their food on a scale from −5 to 5. So far, customers have given the lasagna scores of 2.25, −3.5, 0, −4.5, 1.75, −1, 3.5, and −2.5. Should the restaurant consider changing the recipe? Explain.

Understand the problem.
You are given eight scores for lasagna. You are asked to determine whether the restaurant should make changes to the lasagna recipe.

Make a plan.
Use the mean score to determine whether people generally like the lasagna. Then decide whether the recipe should change.

Solve and check.
Divide the sum of the scores by the number of scores. Group together scores that are convenient to add.

$$\text{mean} = \frac{0 + (-3.5 + 3.5) + (2.25 + 1.75) + [(-4.5) + (-2.5) + (-1)]}{8}$$

$$= \frac{0 + 0 + 4 + (-8)}{8}$$

$$= \frac{-4}{8}, \text{ or } -0.5$$

The mean score is below the "mediocre" score of 0.

▶ So, the restaurant should consider changing the recipe.

Look Back
Only 3 of the 8 scores were better than "mediocre." So, it makes sense to conclude that the restaurant should change the recipe. ✓

Self-Assessment for Problem Solving

Solve each exercise. Then rate your understanding of the success criteria in your journal.

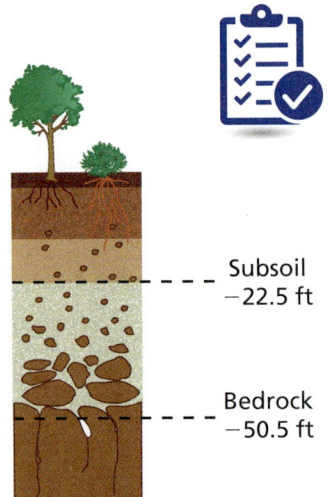

Subsoil −22.5 ft

Bedrock −50.5 ft

12. **DIG DEEPER!** Soil is composed of several layers. A geologist measures the depths of the *subsoil* and the *bedrock*, as shown. Find and interpret two quotients involving the depths of the subsoil and the bedrock.

13. The restaurant in Example 3 receives additional scores of −0.75, −1.5, −1.25, 4.75, −0.25, −0.5, 5, and −0.5 for the lasagna. Given the additional data, should the restaurant consider changing the recipe? Explain.

B.5 Practice

Go to **BigIdeasMath.com** to get HELP with solving the exercises.

▶ Review & Refresh

Find the product. Write fractions in simplest form.

1. $-0.5(1.31)$
2. $\dfrac{9}{10}\left(-1\dfrac{1}{4}\right)$
3. $-\dfrac{7}{12}\left(-\dfrac{3}{14}\right)$

Identify the terms, coefficients, and constants in the expression.

4. $3b + 12$
5. $14 + z + 6f$
6. $8g + 14 + 5c + 7$
7. $42m + 18 + 12c^2$

▶ Concepts, Skills, & Problem Solving

USING TOOLS Write two division expressions represented by the area model. Then find the quotients. *(See Exploration 1, p. 573.)*

8.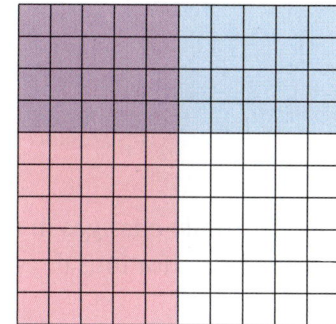
9.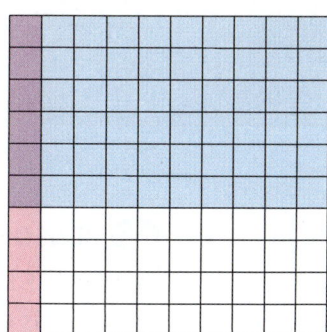

DIVIDING RATIONAL NUMBERS Find the quotient. Write fractions in simplest form.

10. $-\dfrac{7}{10} \div \dfrac{2}{5}$
11. $-0.18 \div 0.03$
12. $-3.45 \div (-15)$
13. $-8 \div (-2.2)$
14. $\dfrac{1}{4} \div \left(-\dfrac{3}{8}\right)$
15. $8.722 \div (-3.56)$
16. $12.42 \div (-4.8)$
17. $-2\dfrac{4}{5} \div (-7)$
18. $-10\dfrac{2}{7} \div \left(-4\dfrac{4}{11}\right)$

YOU BE THE TEACHER Your friend evaluates the expression. Is your friend correct? Explain your reasoning.

19.
$$-\dfrac{2}{3} \div \dfrac{4}{5} = \dfrac{-3}{2} \times \dfrac{4}{5}$$
$$= \dfrac{-12}{10}$$
$$= -\dfrac{6}{5}$$

20.
$$-4.25 \div 1.7 = 2.5$$

Section B.5 Dividing Rational Numbers 577

21. **MODELING REAL LIFE** How many 0.75-pound packages can you make with 4.5 pounds of sunflower seeds?

EVALUATING AN EXPRESSION Evaluate the expression. Write fractions in simplest form.

22. $\dfrac{\frac{14}{9}}{-\frac{1}{3}-\frac{1}{6}}$

23. $\dfrac{-\frac{12}{5}+\frac{3}{10}}{\frac{11}{14}-\left(-\frac{9}{14}\right)}$

24. $-0.42 \div 0.8 + 0.2$

25. $2.85 - 6.2 \div 2^2$

26. $\dfrac{3}{4} + \dfrac{7}{10} - \dfrac{1}{8} \div \left(-\dfrac{1}{2}\right)$

27. $\dfrac{\frac{7}{6}}{\left(-\frac{11}{5}\right)\left(10\frac{1}{2}\right)\left(-\frac{5}{11}\right)}$

28. **MP PROBLEM SOLVING** The section of the boardwalk shown is made using boards that are each $9\frac{1}{4}$ inches wide. The spacing between each board is equal. What is the width of the spacing between each board?

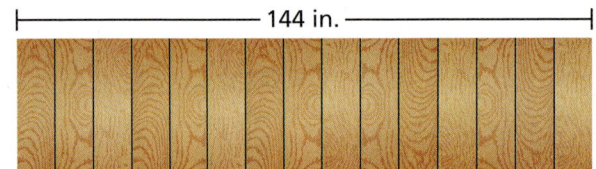

Day	Change in pressure
Monday	−0.05
Tuesday	0.09
Wednesday	−0.04
Thursday	−0.08

29. **MP REASONING** The table shows the daily changes in the barometric pressure (in inches of mercury) for four days.
  a. What is the mean change?
  b. The mean change for Monday through Friday is −0.01 inch. What is the change in the barometric pressure on Friday? Explain.

30. **MP LOGIC** In an online survey, gym members react to the statement shown by adjusting the position of the needle. The responses have values of −4.2, 1.6, 0.4, 0, 2.1, −5.0, −4.7, 0.6, 1.1, 0.8, 0.4, and 2.1. Explain how two people can use the results of the survey to reach different conclusions about whether the gym should adjust its membership prices.

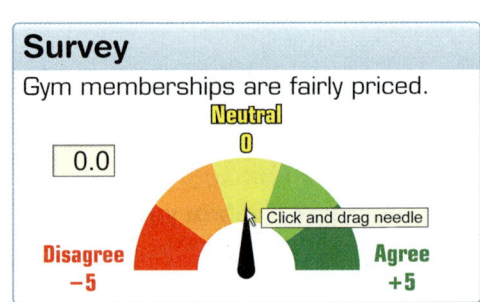

31. **CRITICAL THINKING** Determine whether the statement is *sometimes*, *always*, or *never* true. Explain your reasoning.
  a. The product of two terminating decimals is a terminating decimal.
  b. The quotient of two terminating decimals is a terminating decimal.

Connecting Concepts

Using the Problem-Solving Plan

1. You feed several adult hamsters equal amounts of a new food recipe over a period of 1 month. You record the changes in the weights of the hamsters in the table. Use the data to answer the question "What is the typical weight change of a hamster that is fed the new recipe?"

Weight Change (ounces)				
−0.07	−0.03	−0.11	−0.04	−0.08
0.02	−0.08	−0.08	−0.06	−0.05
−0.11	−0.1	0	−0.07	−0.08

Understand the problem. You know the weight changes of 15 hamsters. You want to use this information to find the typical weight change.

Make a plan. Display the data in a dot plot to see the distribution of the data. Then use the distribution to determine the most appropriate measure of center.

Solve and check. Use the plan to solve the problem. Then check your solution.

2. Evaluate the expression shown at the right. Write your answer in simplest form.

$$-\frac{1}{2} + \frac{2}{3}$$
$$\frac{3}{5}\left(\frac{3}{4} - \frac{11}{8}\right)$$

3. You drop a racquetball from a height of 60 inches. On each bounce, the racquetball bounces to a height that is 70% of its previous height. What is the change in the height of the racquetball after 3 bounces?

Performance Task

Precisely Perfect

At the beginning of this chapter, you watched a STEAM Video called "Carpenter or Joiner." You are now ready to complete the performance task related to this video, available at **BigIdeasMath.com**. Be sure to use the problem-solving plan as you work through the performance task.

Chapter Review

Go to BigIdeasMath.com to download blank graphic organizers.

Review Vocabulary

Write the definition and give an example of each vocabulary term.

terminating decimal, *p. 562*   repeating decimal, *p. 562*   complex fraction, *p. 575*

Graphic Organizers

You can use an **Information Frame** to help organize and remember a concept. Here is an example of an Information Frame for *multiplying integers*.

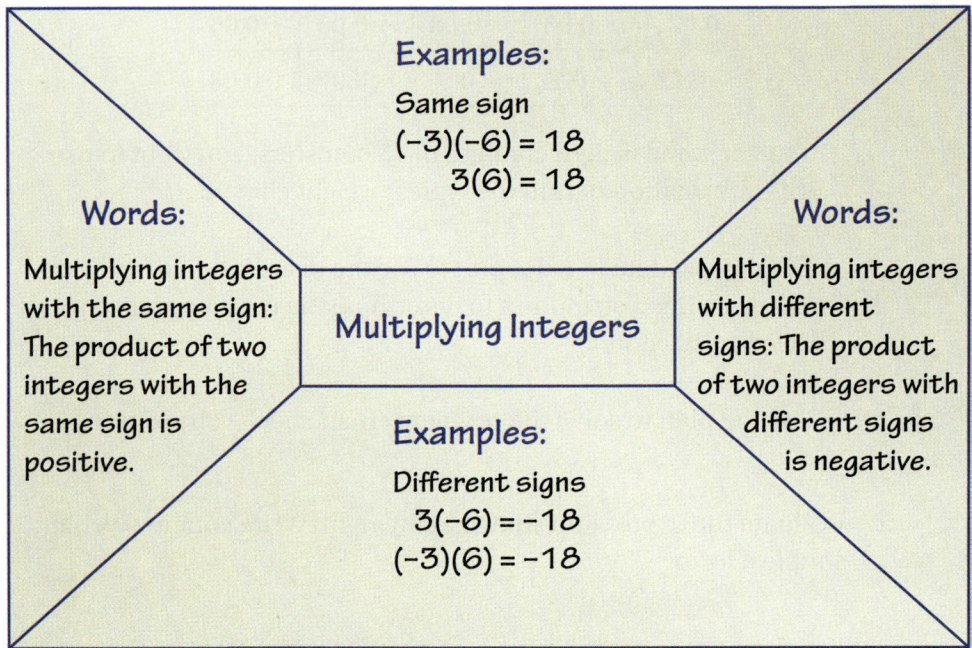

Choose and complete a graphic organizer to help you study the concept.

1. dividing integers

2. writing fractions or mixed numbers as decimals

3. writing decimals as fractions or mixed numbers

4. multiplying rational numbers

5. dividing rational numbers

"I finished my **Information Frame** about rainforests. It makes me want to visit Costa Rica. How about you?"

Chapter B   Multiplying and Dividing Rational Numbers

Chapter Self-Assessment

As you complete the exercises, use the scale below to rate your understanding of the success criteria in your journal.

1	2	3	4
I do not understand.	I can do it with help.	I can do it on my own.	I can teach someone else.

B.1 Multiplying Integers (pp. 549–554)

Learning Target: Find products of integers.

Find the product.

1. $-8 \cdot 6$
2. $10(-7)$
3. $-3 \cdot (-6)$

4. You and a group of friends participate in a game where you must use clues to escape from a room. You have a limited amount of time to escape and are allowed 3 free clues. Additional clues may be requested, but each removes 5 minutes from your remaining time. What integer represents the total change in the time when you use 5 clues?

Evaluate the expression.

5. $(-3)^3$
6. $(-3)(-4)(10)$
7. $24 - 3(2 - 4^2)$

8. Write three integers whose product is negative.

9. You are playing laser tag. The table shows how many points you gain or lose when you tag or are tagged by another player in different locations. You are tagged three times on the back, twice on the shoulder, and twice on the laser. You tag two players on the front, four players on the back, and one player on the laser. What is your score?

Tag Locations	Points Gained	Points Lost
Front	200	50
Back	100	25
Shoulder	50	12
Laser	50	12

10. The product of three integers is positive. How many of the integers can be negative? Explain.

11. Two integers, c and d, have a product of -6. What is the *greatest* possible sum of c and d?

Chapter Review 581

B.2 Dividing Integers (pp. 555–560)

Learning Target: Find quotients of integers.

Find the quotient.

12. $-18 \div 9$
13. $\dfrac{-42}{-6}$
14. $\dfrac{-30}{6}$
15. $84 \div (-7)$

Evaluate the expression when $x = 3$, $y = -4$, and $z = -6$.

16. $z \div x$
17. $\dfrac{xy}{z}$
18. $\dfrac{z - 2x}{y}$

Find the mean of the integers.

19. $-3, -8, 12, -15, 9$
20. $-54, -32, -70, -25, -65, -42$

21. The table shows the weekly profits of a fruit vendor. What is the mean profit for these weeks?

Week	1	2	3	4
Profit	−$125	−$86	$54	−$35

B.3 Converting Between Fractions and Decimals (pp. 561–566)

Learning Target: Convert between different forms of rational numbers.

Write the fraction or mixed number as a decimal.

22. $-\dfrac{8}{15}$
23. $\dfrac{5}{8}$
24. $-\dfrac{13}{6}$
25. $1\dfrac{7}{16}$

Write the decimal as a fraction or mixed number in simplest form.

26. -0.6
27. -0.35
28. -5.8
29. 24.23

30. The table shows the changes in the average yearly precipitation (in inches) in a city for several months. Order the numbers from least to greatest.

February	March	April	May
-1.75	$\dfrac{3}{11}$	0.3	$-1\dfrac{7}{9}$

B.4 Multiplying Rational Numbers (pp. 567–572)

Learning Target: Find products of rational numbers.

Find the product. Write fractions in simplest form.

31. $-\dfrac{4}{9}\left(-\dfrac{7}{9}\right)$

32. $\dfrac{8}{15}\left(-\dfrac{2}{3}\right)$

33. $-5.9(-9.7)$

34. $4.5(-5.26)$

35. $-\dfrac{2}{3}\left(2\dfrac{1}{2}\right)(-3)$

36. $-1.6(0.5)(-20)$

37. The elevation of a sunken ship is -120 feet. You are in a submarine at an elevation that is $\dfrac{5}{8}$ of the ship's elevation. What is your elevation?

38. Write two fractions whose product is between $\dfrac{1}{5}$ and $\dfrac{1}{2}$, and whose sum is negative.

B.5 Dividing Rational Numbers (pp. 573–578)

Learning Target: Find quotients of rational numbers.

Find the quotient. Write fractions in simplest form.

39. $\dfrac{9}{10} \div \left(-\dfrac{6}{5}\right)$

40. $-\dfrac{4}{11} \div \dfrac{2}{7}$

41. $-\dfrac{7}{8} \div \left(-\dfrac{5}{12}\right)$

42. $6.4 \div (-3.2)$

43. $-15.4 \div (-2.5)$

44. $-23.8 \div 5.6$

45. You use a debit card to purchase several shirts. Your account balance after buying the shirts changes by $-\$30.60$. For each shirt you purchased, the change in your account balance was $-\$6.12$. How many shirts did you buy?

46. Evaluate $\dfrac{z}{y - \dfrac{3}{4} + x}$ when $x = 4$, $y = -3$, and $z = -\dfrac{1}{8}$.

B Practice Test

Evaluate the expression. Write fractions in simplest form.

1. $-9 \cdot 2$
2. $-72 \div (-3)$
3. $3\frac{9}{10} \times \left(-\frac{8}{3}\right)$
4. $-1\frac{5}{6} \div 4\frac{1}{6}$
5. $-4.4 \times (-6.02)$
6. $-5 \div 1.5$

Write the fraction or mixed number as a decimal.

7. $\frac{7}{40}$
8. $-\frac{1}{9}$
9. $-1\frac{5}{16}$

Write the decimal as a fraction or mixed number in simplest form.

10. -0.122
11. 0.33
12. -7.09

Evaluate the expression when $x = 5$, $y = -3$, and $z = -2$.

13. $\dfrac{y + z}{x}$
14. $\dfrac{x - 5z}{y}$
15. $\dfrac{\frac{1}{3}x}{\frac{y}{z}}$

16. Find the mean of 11, -7, -14, 10, and -5.

17. A driver receives -25 points for each rule violation. What integer represents the change in points after 4 rule violations?

18. How many 2.25-pound containers can you fill with 24.75 pounds of almonds?

19. In a recent 10-year period, the change in the number of visitors to U.S. national parks was about $-11{,}150{,}000$ visitors.

    a. What was the mean yearly change in the number of visitors?

    b. During the seventh year, the change in the number of visitors was about 10,800,000. Explain how the change for the 10-year period can be negative.

20. You have a $50 gift card to go shopping for school supplies. You buy 2 packs of pencils, 5 notebooks, 6 folders, 1 pack of pens, 3 packs of paper, 1 pack of highlighters, and 2 binders.

    a. What number represents the change in the value of the gift card after buying your school supplies?

    b. What percentage of the value remains on your gift card?

Cumulative Practice

1. When José and Sean were each 5 years old, José was $1\frac{1}{2}$ inches taller than Sean. Then José grew at an average rate of $2\frac{3}{4}$ inches per year until he was 13 years old. José was 63 inches tall when he was 13 years old. How tall was Sean when he was 5 years old?

   A. $39\frac{1}{2}$ in.
   B. $42\frac{1}{2}$ in.
   C. $44\frac{3}{4}$ in.
   D. $47\frac{3}{4}$ in.

   Test-Taking Strategy
   Estimate the Answer

   "One-fourth of the 36 cats in our town are tabbies. How many are not tabbies?
   Ⓐ 9 Ⓑ 18 Ⓒ 27 Ⓓ 36"

   "Using **estimation** you can see that there are about 10 tabbies. So about 30 are not tabbies."

2. What is the value of $-5 + (-7)$?

   F. -12
   G. -2
   H. 2
   I. 12

3. What is the value of the expression?

   $$-\frac{9}{16} + \frac{9}{8}$$

4. What is the value of $|a^2 - 2ac + 5b|$ when $a = -2$, $b = 3$, and $c = -5$?

   A. -9
   B. -1
   C. 1
   D. 9

5. Your friend evaluated the expression.

   $$2 - 3 - (-5) = -5 - (-5)$$
   $$= -5 + 5$$
   $$= 0$$

   What should your friend do to correct the error that he made?

   F. Subtract 5 from -5 instead of adding.

   G. Rewrite $2 - 3$ as -1.

   H. Subtract -5 from 3 before subtracting 3 from 2.

   I. Rewrite $-5 + 5$ as -10.

Cumulative Practice 585

6. What is the value of $-1\frac{1}{2} - \left(-1\frac{3}{4}\right)$?

   A. $-3\frac{1}{4}$

   B. $\frac{1}{4}$

   C. $\frac{6}{7}$

   D. $2\frac{5}{8}$

7. What is the value of the expression when $q = -2$, $r = -12$, and $s = 8$?

   $$\frac{-q^2 - r}{s}$$

   F. -2

   G. -1

   H. 1

   I. 2

8. You are stacking wooden blocks with the dimensions shown. How many blocks do you need to stack vertically to build a block tower that is $7\frac{1}{2}$ inches tall?

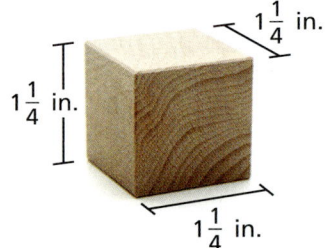

9. Your friend evaluated an expression.

   $$-4\frac{3}{4} + 2\frac{1}{5} = -\frac{19}{4} + \frac{11}{5}$$
   $$= -\frac{95}{20} + \frac{44}{20}$$
   $$= \frac{-95 + 44}{20}$$
   $$= \frac{-139}{20}$$
   $$= -6\frac{19}{20}$$

   What should your friend do to correct the error that she made?

   A. Rewrite $-\frac{19}{4} + \frac{11}{5}$ as $\frac{-19 + 11}{4 + 5}$.

   B. Rewrite $-95 + 44$ as -51.

   C. Rewrite $\frac{-95 + 44}{20}$ as $\frac{51}{20}$.

   D. Rewrite $-4\frac{3}{4}$ as $-\frac{13}{4}$.

10. Which expression has the greatest value when $x = -2$ and $y = -3$?

  F. $-xy$
  G. xy
  H. $x - y$
  I. $-x - y$

11. Four points are graphed on the number line.

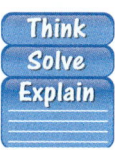

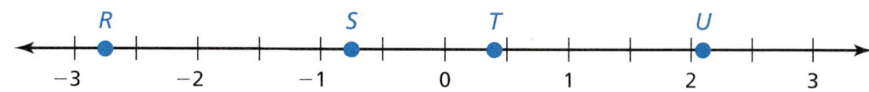

Part A Choose the two points whose values have the greatest sum. Approximate this sum. Explain your reasoning.

Part B Choose the two points whose values have the greatest difference. Approximate this difference. Explain your reasoning.

Part C Choose the two points whose values have the greatest product. Approximate this product. Explain your reasoning.

Part D Choose the two points whose values have the greatest quotient. Approximate this quotient. Explain your reasoning.

12. What number belongs in the box to make the equation true?

$$\frac{-0.4}{\boxed{}} + 0.8 = -1.2$$

  A. -1
  B. -0.2
  C. 0.2
  D. 1

13. Which expression has a negative value when $x = -4$ and $y = 2$?

  F. $-x + y$
  G. $y - x$
  H. $x - y$
  I. $-x - y$

14. What is the area of a triangle with a base of $2\frac{1}{2}$ inches and a height of 2 inches?

  A. $2\frac{1}{4}$ in.^{2}
  B. $2\frac{1}{2}$ in.^{2}
  C. $4\frac{1}{2}$ in.^{2}
  D. 5 in.^{2}

15. Which decimal is equivalent to $\frac{2}{9}$?

  F. 0.2
  G. $0.\overline{2}$
  H. 0.29
  I. 4.5

C Expressions

- **C.1** Algebraic Expressions
- **C.2** Adding and Subtracting Linear Expressions
- **C.3** The Distributive Property
- **C.4** Factoring Expressions

Chapter Learning Target:
Understand algebraic expressions.

Chapter Success Criteria:
- I can identify parts of an algebraic expression.
- I can write algebraic expressions.
- I can solve problems using algebraic expressions.
- I can interpret algebraic expressions in real-life problems.

STEAM Video: "Trophic Status"

STEAM Video

Trophic Status

In an ecosystem, energy and nutrients flow between *biotic* and *abiotic* components. Biotic components are the living parts of an ecosystem. Abiotic components are the non-living parts of an ecosystem. What is an example of an ecosystem?

Watch the STEAM video "Trophic Status." Then answer the following questions.

1. Give examples of both biotic and abiotic components in an ecosystem. Explain.

2. When an organism is eaten, its energy flows into the organism that consumes it. Explain how to use an expression to represent the total energy that a person gains from eating each of the items shown.

Performance Task

Chlorophyll in Plants

After completing this chapter, you will be able to use the concepts you learned to answer the questions in the *STEAM Video Performance Task*. You will be given the numbers of atoms found in molecules involved in photosynthesis.

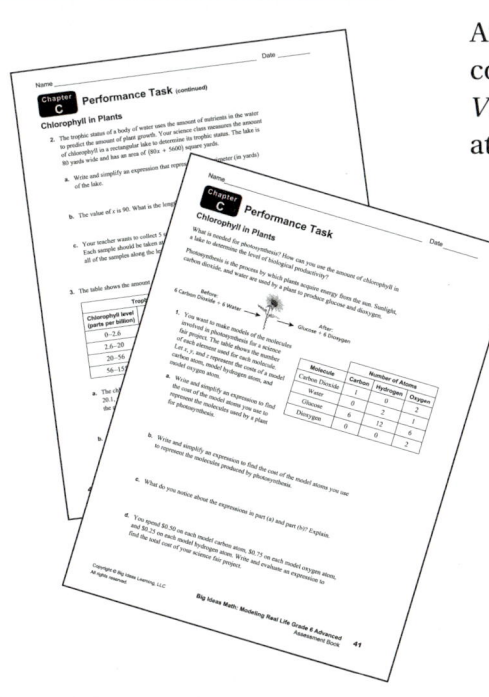

Glucose Molecule

6 carbon atoms

12 hydrogen atoms

6 oxygen atoms

You will be asked to determine the total cost for a model of a molecule given the costs of different types of atom models. How can you find the total cost of purchasing several identical objects?

589

Getting Ready for Chapter C

Chapter Exploration

Work with a partner. Rewrite the algebraic expression so that it has fewer symbols but still has the same value when evaluated for any value of *x*.

	Original Expression	Simplified Expression		Original Expression	Simplified Expression
1.	$2x + 4 + x$		2.	$3(x + 1) - 4$	
3.	$x - (3 - x)$		4.	$5 + 2x - 3$	
5.	$x + 3 + 2x - 4$		6.	$2x + 2 - x + 3$	

7. **WRITING GUIDELINES** Work with a partner. Use your answers in Exercises 1–6 to write guidelines for simplifying an expression.

Simplifying an Algebraic Expression

Key Idea Use the following steps to simplify an algebraic expression.
1.
2.
3.

APPLYING A DEFINITION Work with a partner. Two expressions are equivalent if they have the same value when evaluated for any value of *x*. Decide which two expressions are equivalent. Explain your reasoning.

	Expression A	Expression B	Expression C
8.	$x - (2x + 1)$	$-x + 1$	$-x - 1$
9.	$2x + 3 - x + 4$	$x + 7$	$x - 1$
10.	$3 + x - 2(x + 1)$	$-x + 1$	$-x + 5$
11.	$2 - 2x - (x + 2)$	$-3x$	$-3x + 4$

Vocabulary

The following vocabulary terms are defined in this chapter. Think about what each term might mean and record your thoughts.

like terms   linear expression   factoring an expression

C.1 Algebraic Expressions

Learning Target: Simplify algebraic expressions.

Success Criteria:
- I can identify terms and like terms of algebraic expressions.
- I can combine like terms to simplify algebraic expressions.
- I can write and simplify algebraic expressions to solve real-life problems.

EXPLORATION 1

Simplifying Algebraic Expressions

Work with a partner.

a. Choose a value of x other than 0 or 1 for the last column in the table. Complete the table by evaluating each algebraic expression for each value of x. What do you notice?

	Expression	Value When $x = 0$	$x = 1$	$x = ?$
A.	$-\frac{1}{3} + x + \frac{7}{3}$			
B.	$0.5x + 3 - 1.5x - 1$			
C.	$2x + 6$			
D.	$x + 4$			
E.	$-2x + 2$			
F.	$\frac{1}{2}x - x + \frac{3}{2}x + 4$			
G.	$-4.8x + 2 - x + 3.8x$			
H.	$x + 2$			
I.	$-x + 2$			
J.	$3x + 2 - x + 4$			

> **Math Practice**
>
> **Analyze Conjectures**
>
> A student says that x and x^3 are equivalent because they have the same value when $x = -1$, $x = 0$, and $x = 1$. Explain why the student is or is not correct.

b. How can you use properties of operations to justify your answers in part (a)? Explain your reasoning.

c. To subtract a number, you can add its opposite. Does a similar rule apply to the terms of an algebraic expression? Explain your reasoning.

C.1 Lesson

Key Vocabulary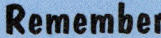
like terms, *p. 592*
simplest form, *p. 592*

In an algebraic expression, **like terms** are terms that have the same variables raised to the same exponents. Constant terms are also like terms. To identify terms and like terms in an expression, first write the expression as a sum of its terms.

EXAMPLE 1 Identifying Terms and Like Terms

Identify the terms and like terms in each expression.

a. $9x - 2 + 7 - x$

Rewrite as a sum of terms.

$$9x + (-2) + 7 + (-x)$$

Terms: $9x, \ -2, \ 7, \ -x$

Like terms: $9x$ and $-x, \ -2$ and 7

b. $z^2 + 5z - 3z^2 + z$

Rewrite as a sum of terms.

$$z^2 + 5z + (-3z^2) + z$$

Terms: $z^2, \ 5z, \ -3z^2, \ z$

Like terms: z^2 and $-3z^2$, $5z$ and z

Try It Identify the terms and like terms in the expression.

1. $y + 10 - \dfrac{3}{2}y$
2. $2r^2 + 7r - r^2 - 9$
3. $7 + 4p - 5 + p + 2q$

An algebraic expression is in **simplest form** when it has no like terms and no parentheses. To *combine* like terms that have variables, use the Distributive Property to add or subtract the coefficients.

EXAMPLE 2 Simplifying Algebraic Expressions

a. **Simplify $6n - 10n$.**

$6n - 10n = (6 - 10)n$ Distributive Property

$\qquad\qquad\quad = -4n$ Subtract.

Remember
The Distributive Property states
$a(b + c) = ab + ac$
and
$a(b - c) = ab - ac$.

b. **Simplify $-8.5w + 5.2w + w$.**

$-8.5w + 5.2w + w = -8.5w + 5.2w + 1w$ Multiplication Property of 1

$\qquad\qquad\qquad\qquad = (-8.5 + 5.2 + 1)w$ Distributive Property

$\qquad\qquad\qquad\qquad = -2.3w$ Add.

Try It Simplify the expression.

4. $-10y + 15y$
5. $\dfrac{3}{8}b - \dfrac{3}{4}b$
6. $2.4g - 2.4g - 9.8g$

EXAMPLE 3 Simplifying Algebraic Expressions

a. Simplify $\frac{3}{4}y + 12 - \frac{1}{2}y - 6$.

$$\frac{3}{4}y + 12 - \frac{1}{2}y - 6 = \frac{3}{4}y + 12 + \left(-\frac{1}{2}y\right) + (-6) \quad \text{Rewrite as a sum.}$$

$$= \frac{3}{4}y + \left(-\frac{1}{2}y\right) + 12 + (-6) \quad \text{Commutative Property of Addition}$$

$$= \left[\frac{3}{4} + \left(-\frac{1}{2}\right)\right]y + 12 + (-6) \quad \text{Distributive Property}$$

$$= \frac{1}{4}y + 6 \quad \text{Combine like terms.}$$

b. Simplify $-3y - 5y + 4z + 9z$.

$$-3y - 5y + 4z + 9z = (-3 - 5)y + (4 + 9)z \quad \text{Distributive Property}$$

$$= -8y + 13z \quad \text{Simplify.}$$

Try It Simplify the expression.

7. $14 - 3z + 8 + z$
8. $2.5x + 4.3x - 5$
9. $2s - 9s + 8t - t$

Self-Assessment for Concepts & Skills

Solve each exercise. Then rate your understanding of the success criteria in your journal.

10. **WRITING** Explain how to identify the terms and like terms of $3y - 4 - 5y$.

SIMPLIFYING ALGEBRAIC EXPRESSIONS Simplify the expression.

11. $7p + 6p$
12. $\frac{4}{5}n - 3 + \frac{7}{10}n$
13. $2w - g - 7w + 3g$

14. **VOCABULARY** Is the expression $3x + 2x - 4$ in simplest form? Explain.

15. **WHICH ONE DOESN'T BELONG?** Which expression does *not* belong with the other three? Explain your reasoning.

$-4 + 6 + 3x$	$3x + 9 - 7$
$5x - 10 - 2x$	$5x - 4 + 6 - 2x$

Section C.1 Algebraic Expressions

EXAMPLE 4 Modeling Real Life

Each person in a group buys an evening ticket, a medium drink, and a large popcorn. How much does the group pay when there are 5 people in the group?

Write an expression that represents the sum of the costs of the items purchased. Use a verbal model.

| Verbal Model | Number of tickets | · | Cost per ticket | + | Number of medium drinks | · | Cost per medium drink | + | Number of large popcorns | · | Cost per large popcorn |

Variable The same number of each item is purchased. So, x can represent the number of tickets, the number of medium drinks, and the number of large popcorns.

Expression $7.50x$ + $2.75x$ + $4x$

$7.50x + 2.75x + 4x = (7.50 + 2.75 + 4)x$ Distributive Property

$= 14.25x$ Add coefficients.

The expression $14.25x$ indicates that the cost per person is $14.25. To find the cost for a group of 5 people, evaluate the expression when $x = 5$.

$$14.25(5) = 71.25$$

▸ The total cost for a group of 5 people is $71.25.

Remember
Variables can be lowercase or uppercase. Make sure you consistently use the same case for a variable when solving a problem.

Self-Assessment for Problem Solving

Solve each exercise. Then rate your understanding of the success criteria in your journal.

16. MODELING REAL LIFE An exercise mat is 3.3 times as long as it is wide. Write expressions in simplest form that represent the perimeter and the area of the exercise mat.

17. DIG DEEPER! A group of friends visits the movie theater in Example 4. Each person buys a daytime ticket and a small drink. The group shares 2 large popcorns. What is the average cost per person when there are 4 people in the group?

C.1 Practice

> Go to **BigIdeasMath.com** to get HELP with solving the exercises.

▶ Review & Refresh

Find the product or quotient. Write fractions in simplest form.

1. $-\dfrac{2}{7} \times \dfrac{7}{4}$
2. $-\dfrac{2}{3}\left(-\dfrac{9}{10}\right)$
3. $1\dfrac{4}{9} \div \left(-\dfrac{2}{9}\right)$

Order the numbers from least to greatest.

4. $\dfrac{7}{8}$, 0.85, 87%, $\dfrac{3}{4}$, 78%
5. 15%, 14.8, $15\dfrac{4}{5}$, 1450%

6. A bird's nest is 12 feet above the ground. A mole's den is 12 inches below the ground. What is the difference in height of these two positions?

   A. 24 in.   **B.** 11 ft   **C.** 13 ft   **D.** 24 ft

▶ Concepts, Skills, & Problem Solving

MP REASONING Determine whether the expressions are equivalent. Explain your reasoning. (See Exploration 1, p. 591.)

7.
Expression 1	$3 - 5x$
Expression 2	$4.25 - 5x - 4.25$

8.
Expression 1	$1.25x + 4 + 0.75x - 3$
Expression 2	$2x + 1$

IDENTIFYING TERMS AND LIKE TERMS Identify the terms and like terms in the expression.

9. $t + 8 + 3t$
10. $3z + 4 + 2 + 4z$
11. $2n - n - 4 + 7n$
12. $-x - 9x^2 + 12x^2 + 7$
13. $1.4y + 5 - 4.2 - 5y^2 + z$
14. $\dfrac{1}{2}s - 4 + \dfrac{3}{4}s + \dfrac{1}{8} - s^3$

15. **YOU BE THE TEACHER** Your friend identifies the terms and like terms in the expression $3x - 5 - 2x + 9x$. Is your friend correct? Explain your reasoning.

    > $3x - 5 - 2x + 9x$
    > Terms: $3x$, 5, $2x$, and $9x$
    > Like Terms: $3x$, $2x$, and $9x$

SIMPLIFYING ALGEBRAIC EXPRESSIONS Simplify the expression.

16. $12g + 9g$
17. $11x + 9 - 7$
18. $8s - 11s + 6s$
19. $4b - 24 + 19$
20. $4p - 5p - 30p$
21. $4.2v - 5 - 6.5v$
22. $8 + 4a + 6.2 - 9a$
23. $\dfrac{2}{5}y - 4 + 7 - \dfrac{9}{10}y$
24. $-\dfrac{2}{3}c - \dfrac{9}{5} + 14c + \dfrac{3}{10}$

25. **MODELING REAL LIFE** On a hike, each hiker carries the items shown. Write and interpret an expression in simplest form that represents the weight carried by x hikers. How much total weight is carried when there are 4 hikers?

3.4 lb   4.6 lb   2.2 lb

26. **MP STRUCTURE** Evaluate the expression $-8x + 5 - 2x - 4 + 5x$ when $x = 2$ before and after simplifying. Which method do you prefer? Explain.

27. **OPEN-ENDED** Write an expression with five different terms that is equivalent to $8x^2 + 3x^2 + 3y$. Justify your answer.

28. **MP STRUCTURE** Which of the following shows a correct way of simplifying $6 + (3 - 5x)$? Explain the errors made in the other choices.

    A. $6 + (3 - 5x) = (6 + 3 - 5)x = 4x$

    B. $6 + (3 - 5x) = 6 + (3 - 5)x = 6 + (-2)x = 6 - 2x$

    C. $6 + (3 - 5x) = (6 + 3) - 5x = 9 - 5x$

    D. $6 + (3 - 5x) = (6 + 3 + 5) - x = 14 - x$

29. **MP PRECISION** Two comets orbit the Sun. One comet travels 30,000 miles per hour and the other comet travels 28,500 miles per hour. What is the most efficient way to calculate the difference of the distances traveled by the comets for any given number of minutes? Justify your answer.

	Car	Truck
Wash	$8	$10
Wax	$12	$15

30. **MODELING REAL LIFE** Find the earnings for washing and waxing 12 cars and 8 trucks. Justify your answer.

31. **CRITICAL THINKING** You apply gold foil to a piece of red poster board to make the design shown.

    a. Find the area of the gold foil when $x = 3$. Justify your answer.

    b. The pattern at the right is called "St. George's Cross." Find a country that uses this pattern as its flag.

32. **GEOMETRY** Two rectangles have different dimensions. Each rectangle has a perimeter of $(7x + 5)$ inches. Draw and label diagrams that represent possible dimensions of the rectangles.

C.2 Adding and Subtracting Linear Expressions

Learning Target: Find sums and differences of linear expressions.

Success Criteria:
- I can explain the difference between linear and nonlinear expressions.
- I can find opposites of terms that include variables.
- I can apply properties of operations to add and subtract linear expressions.

EXPLORATION 1

Using Algebra Tiles

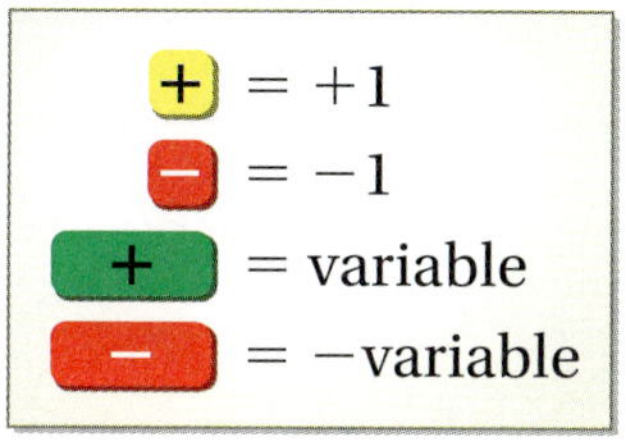

$\boxed{+} = +1$
$\boxed{-} = -1$
$\boxed{+} = $ variable
$\boxed{-} = -$variable

Work with a partner. You can use the algebra tiles shown at the left to find sums and differences of algebraic expressions.

a. How can you use algebra tiles to model a sum of terms that equals 0? Explain your reasoning.

b. Write each sum or difference modeled below. Then use the algebra tiles to simplify the expression.

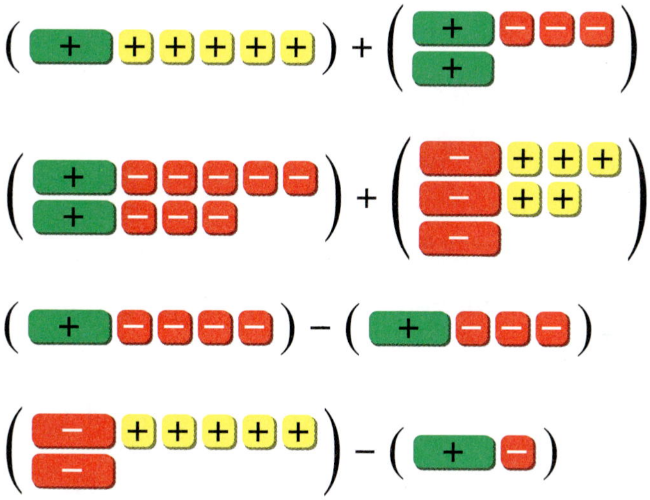

Math Practice

Consider Similar Problems

How is using integer counters to find sums and differences of integers similar to using algebra tiles to find sums and differences of algebraic expressions?

c. Write two algebraic expressions of the form $ax + b$, where a and b are rational numbers. Find the sum and difference of the expressions.

EXPLORATION 2

Using Properties of Operations

Work with a partner.

a. Do algebraic expressions, such as $2x$, $-3y$, and $3z + 1$ have additive inverses? How do you know?

b. How can you find the sums and differences modeled in Exploration 1 without using algebra tiles? Explain your reasoning.

Section C.2 Adding and Subtracting Linear Expressions

C.2 Lesson

Key Vocabulary
linear expression, p. 598

A **linear expression** is an algebraic expression in which the exponent of each variable is 1.

Linear Expressions	$-4x$	$3x + 5y$	$5 - \dfrac{1}{6}x$
Nonlinear Expressions	$\dfrac{1}{2}x^2$	$-7x^3 + x$	$x^5 + 1$

You can use either a vertical or a horizontal method to add linear expressions.

EXAMPLE 1 Adding Linear Expressions

Find each sum.

a. $(x - 2) + (3x + 8)$

Vertical method: Align like terms vertically and add.

$$\begin{array}{r} x - 2 \\ +\ 3x + 8 \\ \hline 4x + 6 \end{array}$$

 The sum is $4x + 6$.

> Linear expressions are usually written with the variable term first.

b. $(-4y + 3) + (11y - 5)$

Horizontal method: Use properties of operations to group like terms and simplify.

$(-4y + 3) + (11y - 5) = -4y + 3 + 11y - 5$	Rewrite the sum.
$= -4y + 11y + 3 - 5$	Commutative Property of Addition
$= (-4y + 11y) + (3 - 5)$	Group like terms.
$= 7y - 2$	Combine like terms.

The sum is $7y - 2$.

Try It Find the sum.

1. $(x + 3) + (2x - 1)$
2. $(-8z + 4) + (8z - 7)$
3. $(4.5 - n) + (-10n + 6.5)$
4. $\left(\dfrac{1}{2}w - 3\right) + \left(\dfrac{1}{4}w + 3\right)$

To subtract one linear expression from another, add the opposite of each term in the expression. You can use a vertical or a horizontal method.

EXAMPLE 2 Subtracting Linear Expressions

Find each difference.

a. $(5x + 6) - (-x + 6)$

Vertical method: Align like terms vertically and subtract.

$$\begin{array}{r}(5x+6)\\-(-x+6)\end{array} \quad \text{Add the opposite.} \quad \begin{array}{r}5x+6\\+x-6\\\hline 6x\end{array}$$

▸ The difference is $6x$.

Common Error

When subtracting an expression, make sure you add the opposite of each term in the expression, not just the first term.

b. $(7y + 5) - (8y - 6)$

Horizontal method: Use properties of operations to group like terms and simplify.

$(7y + 5) - (8y - 6) = (7y + 5) + (-8y + 6)$ Add the opposite.

$ = 7y + (-8y) + 5 + 6$ Commutative Property of Addition

$ = [7y + (-8y)] + (5 + 6)$ Group like terms.

$ = -y + 11$ Combine like terms.

▸ The difference is $-y + 11$.

Try It Find the difference.

5. $(m - 3) - (-m + 12)$

6. $(-2c + 5) - (6.3c + 20)$

Self-Assessment for Concepts & Skills

Solve each exercise. Then rate your understanding of the success criteria in your journal.

7. WRITING Describe how to distinguish a linear expression from a nonlinear expression. Give an example of each.

8. DIFFERENT WORDS, SAME QUESTION Which is different? Find "both" answers.

| What is x more than $3x - 1$? | Find $3x - 1$ decreased by x. |

| What is the difference of $3x - 1$ and x? | Subtract $(x + 1)$ from $3x$. |

Section C.2 Adding and Subtracting Linear Expressions

EXAMPLE 3 Modeling Real Life

Skateboard kits cost d dollars and you have a coupon for $2 off each one you buy. After assembly, you sell each skateboard for $(2d - 4)$ dollars. Find and interpret your profit on each skateboard sold.

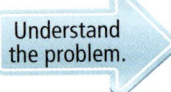

You are given information about purchasing skateboard kits and selling the assembled skateboards. You are asked to find and interpret the profit made on each skateboard sold.

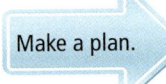

Find the difference of the expressions representing the selling price and the purchase price. Then simplify and interpret the expression.

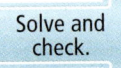

You receive $2 off of d dollars, so you pay $(d - 2)$ dollars for each kit.

$$\text{Profit (dollars)} = \text{Selling price (dollars)} - \text{Purchase price (dollars)}$$

$= (2d - 4) - (d - 2)$	Write the difference.
$= (2d - 4) + (-d + 2)$	Add the opposite.
$= 2d - d - 4 + 2$	Group like terms.
$= d - 2$	Combine like terms.

▸ Your profit on each skateboard sold is $(d - 2)$ dollars. You pay $(d - 2)$ dollars for each kit, so you are doubling your money.

Look Back Assume each kit is $40. Verify that you double your money.

When $d = 40$: You pay $d - 2 = 40 - 2 = \$38$.
You sell it for $2d - 4 = 2(40) - 4 = 80 - 4 = \$76$.
Because $\$38 \cdot 2 = \$76$, you double your money. ✓

Self-Assessment for Problem Solving

Solve each exercise. Then rate your understanding of the success criteria in your journal.

9. **DIG DEEPER!** In a basketball game, the home team scores $(2m + 39)$ points and the away team scores $(3m + 40)$ points, where m is the number of minutes since halftime. Who wins the game? What is the difference in the scores m minutes after halftime? Explain.

10. Electric guitar kits originally cost d dollars online. You buy the kits on sale for 50% of the original price, plus a shipping fee of $4.50 per kit. After painting and assembly, you sell each guitar online for $(1.5d + 4.5)$ dollars. Find and interpret your profit on each guitar sold.

C.2 Practice

Review & Refresh

Simplify the expression.

1. $4f + 11f$
2. $b + 4b - 9b$
3. $-4z - 6 - 7z + 3$

Evaluate the expression when $x = -\dfrac{4}{5}$ and $y = \dfrac{1}{3}$.

4. $x + y$
5. $2x + 6y$
6. $-x + 4y$

7. What is the surface area of a cube that has a side length of 5 feet?

   A. 25 ft^2 **B.** 75 ft^2 **C.** 125 ft^2 **D.** 150 ft^2

Concepts, Skills, & Problem Solving

USING ALGEBRA TILES Write the sum or difference modeled by the algebra tiles. Then use the algebra tiles to simplify the expression. (See Exploration 1, p. 597.)

8. (algebra tiles) + (algebra tiles)

9. (algebra tiles) − (algebra tiles)

ADDING LINEAR EXPRESSIONS Find the sum.

10. $(n + 8) + (n - 12)$
11. $(7 - b) + (3b + 2)$
12. $(2w - 9) + (-4w - 5)$
13. $(2x - 6) + (4x - 12)$
14. $(-3.4k - 7) + (3k + 21)$
15. $\left(-\dfrac{7}{2}z + 4\right) + \left(\dfrac{1}{5}z - 15\right)$
16. $(6 - 2.7h) + (-1.3j - 4)$
17. $\left(\dfrac{7}{4}x - 5\right) + (2y - 3.5) + \left(-\dfrac{1}{4}x + 5\right)$

18. **MODELING REAL LIFE** While catching fireflies, you and a friend decide to have a competition. After m minutes, you have $(3m + 13)$ fireflies and your friend has $(4m + 6)$ fireflies.

    a. How many total fireflies do you and your friend catch? Explain your reasoning.

    b. The competition lasts 3 minutes. Who has more fireflies? Justify your answer.

Section C.2 Adding and Subtracting Linear Expressions

SUBTRACTING LINEAR EXPRESSIONS Find the difference.

19. $(-2g + 7) - (g + 11)$

20. $(6d + 5) - (2 - 3d)$

21. $(4 - 5y) - (2y - 16)$

22. $(2n - 9) - (-2.4n + 4)$

23. $\left(-\dfrac{1}{8}c + 16\right) - \left(\dfrac{3}{8} + 3c\right)$

24. $\left(\dfrac{9}{4}x + 6\right) - \left(-\dfrac{5}{4}x - 24\right)$

25. $\left(\dfrac{1}{3} - 6m\right) - \left(\dfrac{1}{4}n - 8\right)$

26. $(1 - 5q) - (2.5s + 8) - (0.5q + 6)$

27. YOU BE THE TEACHER Your friend finds the difference $(4m + 9) - (2m - 5)$. Is your friend correct? Explain your reasoning.

$(4m + 9) - (2m - 5) = 4m + 9 - 2m - 5$
$= 4m - 2m + 9 - 5$
$= 2m + 4$

28. GEOMETRY The expression $17n + 11$ represents the perimeter of the triangle. What is the length of the third side? Explain your reasoning.

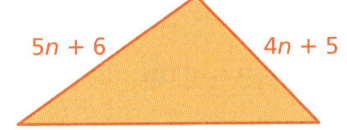

29. MP LOGIC Your friend says the sum of two linear expressions is always a linear expression. Is your friend correct? Explain.

30. MODELING REAL LIFE You burn 265 calories running and then 7 calories per minute swimming. Your friend burns 273 calories running and then 11 calories per minute swimming. You each swim for the same number of minutes. Find and interpret the difference in the amounts of calories burned by you and your friend.

31. DIG DEEPER! You start a new job. After w weeks, you have $(10w + 120)$ dollars in your savings account and $(45w + 25)$ dollars in your checking account.

   a. What is the total amount of money in the accounts? Explain.

   b. How much money did you have before you started your new job? How much money do you save each week?

   c. You want to buy a new phone for $150, and still have $500 left in your accounts afterwards. Explain how to determine when you can buy the phone.

32. MP REASONING Write an expression in simplest form that represents the vertical distance between the two lines shown. What is the distance when $x = 3$? when $x = -3$?

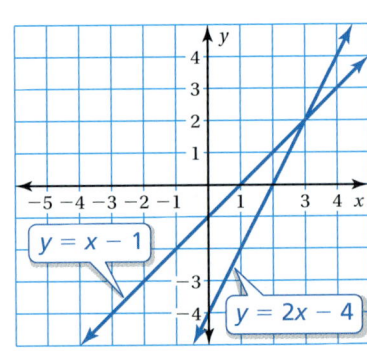

C.3 The Distributive Property

Learning Target: Apply the Distributive Property to generate equivalent expressions.

Success Criteria:
- I can explain how to apply the Distributive Property.
- I can use the Distributive Property to simplify algebraic expressions.

EXPLORATION 1
Using Models to Write Expressions

Work with a partner.

a. Write an expression that represents the area of the shaded region in each figure.

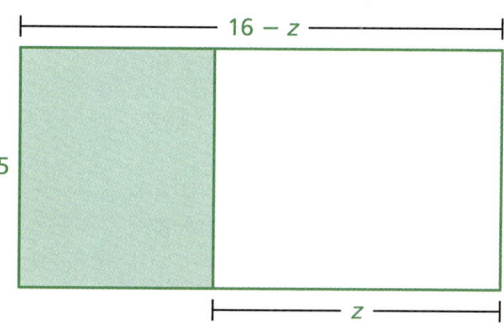

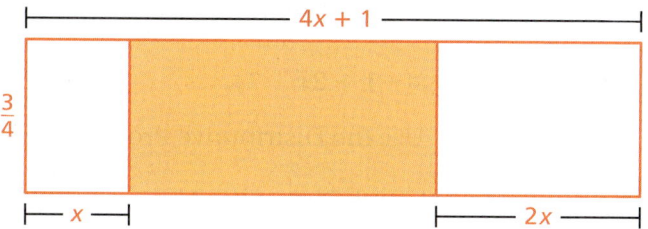

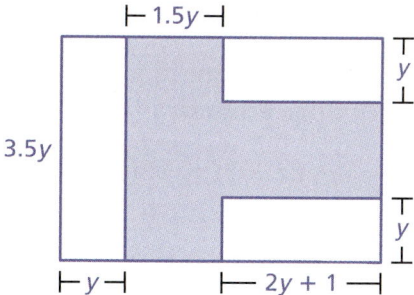

Math Practice

Use Expressions
How can you determine whether expressions that appear to be different are equivalent?

b. Compare your expressions in part (a) with other groups in your class. Did other groups write expressions that look different than yours? If so, determine whether the expressions are equivalent.

Section C.3   The Distributive Property   603

C.3 Lesson

You can use the Distributive Property to simplify expressions involving variable terms and rational numbers.

EXAMPLE 1 Using the Distributive Property

Simplify each expression.

a. $-\frac{1}{3}(3n - 6)$

Remember
The Distributive Property states
$a(b + c) = ab + ac$
and
$a(b - c) = ab - ac$.

$$-\frac{1}{3}(3n - 6) = -\frac{1}{3}(3n) - \left(-\frac{1}{3}\right)(6) \quad \text{Distributive Property}$$
$$= -n - (-2) \quad \text{Multiply.}$$
$$= -n + 2 \quad \text{Add the opposite.}$$

b. $5(-x + 3y)$

$$5(-x + 3y) = 5(-x) + 5(3y) \quad \text{Distributive Property}$$
$$= -5x + 15y \quad \text{Multiply.}$$

Try It Simplify the expression.

1. $-1(x + 9)$
2. $\frac{2}{3}(-3z - 6)$
3. $-1.5(8m - n)$

EXAMPLE 2 Simplifying Expressions

Simplify $-3(-1 + 2x + 7)$.

Method 1: Use the Distributive Property before combining like terms.

$$-3(-1 + 2x + 7) = -3(-1) + (-3)(2x) + (-3)(7) \quad \text{Distributive Property}$$
$$= 3 + (-6x) + (-21) \quad \text{Multiply.}$$
$$= -6x - 18 \quad \text{Combine like terms.}$$

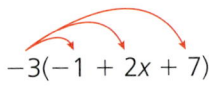

Common Error
Multiply each term in the sum by -3, not 3.

$-3(-1 + 2x + 7)$

Method 2: Combine like terms in parentheses before using the Distributive Property.

$$-3(-1 + 2x + 7) = -3(2x + 6) \quad \text{Combine like terms.}$$
$$= (-3)(2x) + (-3)(6) \quad \text{Distributive Property}$$
$$= -6x - 18 \quad \text{Multiply.}$$

Try It Simplify the expression.

4. $2(-3s + 1 - 5)$
5. $-\frac{3}{2}(a - 4 - 2a)$

EXAMPLE 3 Simplifying Expressions

Simplify each expression.

a. $-\frac{1}{2}(6n + 4) + 2n$

$-\frac{1}{2}(6n + 4) + 2n = -\frac{1}{2}(6n) + \left(-\frac{1}{2}\right)(4) + 2n$ — Distributive Property

$= -3n + (-2) + 2n$ — Multiply.

$= -n - 2$ — Combine like terms.

b. $(6d - 5) - 8\left(\frac{3}{4}d - 1\right)$

$(6d - 5) - 8\left(\frac{3}{4}d - 1\right) = (6d - 5) - \left[8\left(\frac{3}{4}d\right) - 8(1)\right]$ — Distributive Property

$= (6d - 5) - (6d - 8)$ — Multiply.

$= (6d - 5) + (-6d + 8)$ — Add the opposite.

$= [6d + (-6d)] + (-5 + 8)$ — Group like terms.

$= 3$ — Combine like terms.

> You can multiply an expression by −1 to find the opposite of the expression.

Try It Simplify the expression.

6. $3.5m - 1.5(m - 10)$

7. $\frac{4}{5}(10w - 5) - 2(w + 9)$

Self-Assessment for Concepts & Skills

Solve each exercise. Then rate your understanding of the success criteria in your journal.

8. WRITING Explain how to use the Distributive Property when simplifying an expression.

USING THE DISTRIBUTIVE PROPERTY Simplify the expression.

9. $\frac{5}{6}(-2y + 3)$

10. $6(3s - 2.5 - 5s)$

11. $\frac{3}{10}(4m - 8) + 9m$

12. $2.25 - 2(7.5 - 4h)$

13. MP STRUCTURE Use the terms at the left to complete the expression below so that it is equivalent to $9x - 12$. Justify your answer.

☐ (☐ − ☐) + ☐

EXAMPLE 4 Modeling Real Life

A square pool has a side length of *s* feet. How many 1-foot square tiles does it take to tile the border of the pool?

Understand the problem. You are given information about a square pool and square tiles. You are asked to find the number of tiles it takes to tile the border of the pool.

Make a plan. Draw a diagram that represents the situation. Use the diagram to write an expression for the number of tiles needed.

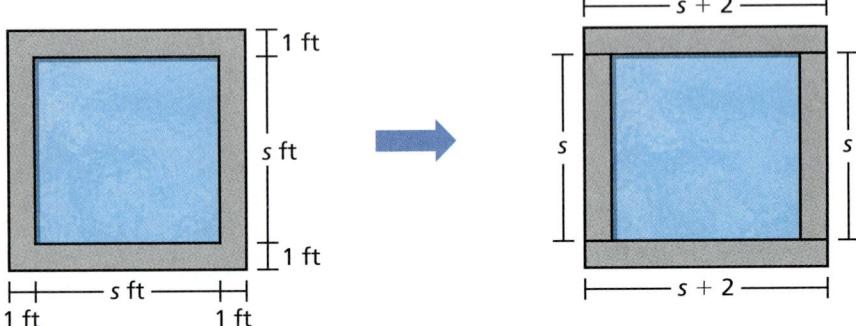

The diagram shows that the tiled border can be divided into two sections that each require $s + 2$ tiles and two sections that each require s tiles. So, the number of tiles can be represented by $2(s + 2) + 2s$. Simplify the expression.

$$2(s + 2) + 2s = 2(s) + 2(2) + 2s \qquad \text{Distributive Property}$$
$$= 4s + 4 \qquad \text{Simplify.}$$

▶ The expression $4s + 4$ represents the number of tiles that are needed.

Another Method
Draw a different diagram.
$$4(s + 1) = 4(s) + 4(1)$$
$$= 4s + 4 \checkmark$$

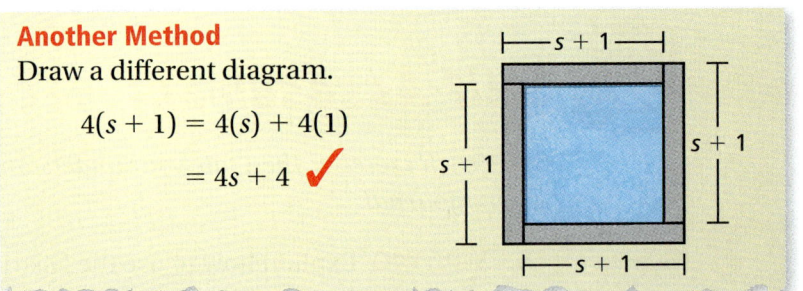

Self-Assessment for Problem Solving

Solve each exercise. Then rate your understanding of the success criteria in your journal.

14. A rectangular room is 10 feet longer than it is wide. How many 1-foot square tiles does it take to tile along the inside walls of the room?

15. How many 2-foot square tiles does it take to tile the border of the pool in Example 4? Explain.

C.3 Practice

Review & Refresh

Find the sum or difference.

1. $(5b - 9) + (b + 8)$
2. $(3m + 5) - (6 - 5m)$
3. $(1 - 9z) + 3(z - 2)$
4. $(7g - 6) - (-3n - 4)$

Evaluate the expression.

5. -6^2
6. $-9^2 \cdot 3$
7. $(-7) \cdot (-2) \cdot (-4)$

Copy and complete the statement using <, >, or =.

8. $11 \; \square \; |-11|$
9. $|3.5| \; \square \; |-5.8|$
10. $|-3.5| \; \square \; \left|\frac{17}{5}\right|$

Concepts, Skills, & Problem Solving

USING MODELS Write two different expressions that represent the area of the shaded region. Show that the expressions are equivalent. *(See Exploration 1, p. 603.)*

11.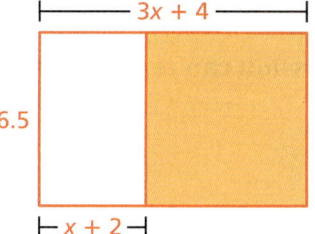

12.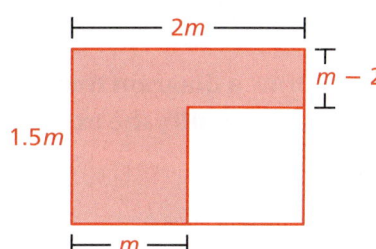

USING THE DISTRIBUTIVE PROPERTY Simplify the expression.

13. $3(a - 7)$
14. $-6(2 + x)$
15. $-5(3m - 4)$
16. $-9(-5 - 4c)$
17. $4.5(3s + 6)$
18. $-1.4(-5 + 7g)$
19. $\frac{2}{5}(6 - 5p)$
20. $-\frac{4}{3}(3q - 10)$
21. $2(3 + 4y + 5)$
22. $-9(8 + 6n - 4)$
23. $-6(-4d - 8.3 + 3d)$
24. $2.3h(6 - k)$
25. $-\frac{3}{8}(-4y + z)$
26. $2(-2w - 1.2 + 7x)$
27. $\frac{5}{3}\left(\frac{4}{3}a + 9b + \frac{2}{3}a\right)$

YOU BE THE TEACHER Your friend simplifies the expression. Is your friend correct? Explain your reasoning.

28.
$$-2(h + 8k) = -2(h) + 2(8k)$$
$$= -2h + 16k$$

29.
$$-3(4 - 5b + 7) = -3(11 - 5b)$$
$$= -3(11) + (-3)(5b)$$
$$= -33 - 15b$$

SIMPLIFYING EXPRESSIONS Simplify the expression.

30. $-3(5g + 1) + 8g$

31. $-6a + 7(-2a - 4)$

32. $9 - 3(5 - 4x)$

33. $-\frac{3}{4}(5p - 12) + 2\left(8 - \frac{1}{4}p\right)$

34. $c(4 + 3c) - 0.75(c + 3)$

35. $-1 - \frac{2}{3}\left(\frac{6}{7} - \frac{3}{7}n\right)$

36. MODELING REAL LIFE The cost (in dollars) of a custom-made sweatshirt is represented by $3.5n + 29.99$, where n is the number of different colors in the design. Write and interpret a simplified expression that represents the cost of 15 sweatshirts.

37. MODELING REAL LIFE A ski resort makes snow using a snow fan that costs $1200. The fan has an average daily operation cost of $9.50. Write and interpret a simplified expression that represents the cost to purchase and operate 6 snow fans.

38. MP NUMBER SENSE Predict whether the instructions below will produce equivalent expressions. Then show whether your prediction is correct.
- Subtract 3 from n, add 3 to the result, and then triple that expression.
- Subtract 3 from n, triple the result, and then add 3 to that expression.

USING A MODEL Draw a diagram that shows how the expression can represent the area of a figure. Then simplify the expression.

39. $5(2 + x + 3)$

40. $(4 + 1)(x + 2x)$

41. DIG DEEPER! A square fire pit with a side length of s feet is bordered by 1-foot square stones as shown.

a. How many stones does it take to border the fire pit with two rows of stones? Use a diagram to justify your answer.

b. You border the fire pit with n rows of stones. How many stones are in the nth row? Explain your reasoning.

42. PUZZLE Your friend asks you to perform the following steps.
1) Pick any number except 0.
2) Add 2 to your number.
3) Multiply the result by 3.
4) Subtract 6 from the result.
5) Divide the result by your original number.

Your friend says, "The final result is 3!" Is your friend correct? If so, explain how your friend knew the final result. If not, explain why not.

C.4 Factoring Expressions

Learning Target: Factor algebraic expressions.

Success Criteria:
- I can identify the greatest common factor of terms, including variable terms.
- I can use the Distributive Property to factor algebraic expressions.
- I can write a term as a product involving a given factor.

EXPLORATION 1

Finding Dimensions

Work with a partner.

a. The models show the areas (in square units) of parts of rectangles. Use the models to find the missing values that complete the expressions. Explain your reasoning.

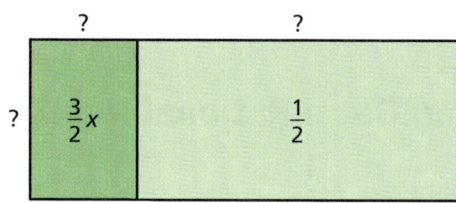

$$\frac{4}{5} + \frac{8}{5} = ?(? + ?)$$

Math Practice

View as Components

How does viewing each rectangle as two distinct parts help you complete the expressions?

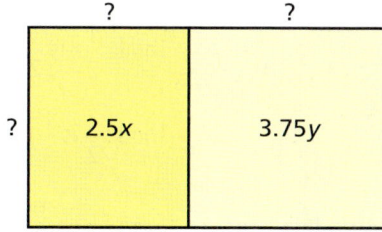

$$\frac{3}{2}x + \frac{1}{2} = ?(? + ?)$$

$$2.5x + 3.75y = ?(? + ?)$$

b. Are the expressions you wrote in part (a) equivalent to the original expressions? Explain your reasoning.

c. Explain how you can use the Distributive Property to find rational number factors of an expression.

C.4 Lesson

Key Vocabulary
factoring an expression, p. 610

When **factoring an expression**, you write the expression as a product of factors. You can use the Distributive Property to factor any rational number from an expression.

EXAMPLE 1 Factoring Out the GCF

Factor $24x - 18$ using the GCF.

Find the GCF of $24x$ and 18.

$$24x = \textcircled{2} \cdot 2 \cdot 2 \cdot \textcircled{3} \cdot x$$
$$18 = \textcircled{2} \cdot \textcircled{3} \cdot 3$$

Circle the common prime factors.

So, the GCF of $24x$ and 18 is $2 \cdot 3 = 6$. Use the GCF to factor the expression.

$$24x - 18 = 6(4x) - 6(3) \qquad \text{Rewrite using GCF.}$$
$$= 6(4x - 3) \qquad \text{Distributive Property}$$

Try It Factor the expression using the GCF.

1. $15x + 25$
2. $4y - 20$
3. $36c + 24d$

EXAMPLE 2 Factoring Out a Rational Number

Factor $\frac{1}{2}$ out of $\frac{1}{2}x + \frac{3}{2}$.

Write each term as a product of $\frac{1}{2}$ and another factor.

$$\frac{1}{2}x = \frac{1}{2} \cdot x \qquad \text{Think: } \frac{1}{2}x \text{ is } \frac{1}{2} \text{ times what?}$$

$$\frac{3}{2} = \frac{1}{2} \cdot 3 \qquad \text{Think: } \frac{3}{2} \text{ is } \frac{1}{2} \text{ times what?}$$

Use the Distributive Property to factor out $\frac{1}{2}$.

$$\frac{1}{2}x + \frac{3}{2} = \frac{1}{2} \cdot x + \frac{1}{2} \cdot 3 \qquad \text{Rewrite the expression.}$$
$$= \frac{1}{2}(x + 3) \qquad \text{Distributive Property}$$

Try It Factor out the coefficient of the variable term.

4. $\frac{1}{2}n - \frac{1}{2}$
5. $\frac{3}{4}p - \frac{3}{2}$
6. $5 + 2.5q$

EXAMPLE 3 Factoring Out a Negative Number

Factor -2 out of $-4p + 10$.

Write each term as a product of -2 and another factor.

$-4p = -2 \cdot 2p$ Think: $-4p$ is -2 times what?

$10 = -2 \cdot (-5)$ Think: 10 is -2 times what?

Use the Distributive Property to factor out -2.

$-4p + 10 = -2 \cdot 2p + (-2) \cdot (-5)$ Rewrite the expression.

$= -2[2p + (-5)]$ Distributive Property

$= -2(2p - 5)$ Simplify.

▷ So, $-4p + 10 = -2(2p - 5)$.

Try It

7. Factor -5 out of $-5d + 30$.

8. Factor -4 out of $-8k - 12$.

Self-Assessment for Concepts & Skills

Solve each exercise. Then rate your understanding of the success criteria in your journal.

FACTORING OUT THE GCF Factor the expression using the GCF.

9. $16n - 24$

10. $42a + 14b$

FACTORING OUT A RATIONAL NUMBER Factor out the coefficient of the variable term.

11. $\frac{1}{10}k - \frac{7}{10}$

12. $42 + 3.5h$

FACTORING OUT A NEGATIVE NUMBER Factor out the indicated number.

13. Factor -8 out of $-32d + 56$.

14. Factor -12 out of $-24k + 120$.

15. **WRITING** Describe the relationship between using the Distributive Property to simplify an expression and to factor an expression. Give an example to justify your answer.

EXAMPLE 4   Modeling Real Life

A rectangular landing platform for a rocket is 60 yards wide and has an area of (60x + 3600) square yards. Write an expression that represents the perimeter (in yards) of the platform.

Factor the width of 60 yards out of the given area expression to find an expression that represents the length (in yards) of the platform.

60x + 3600 = 60 · x + 60 · 60	Rewrite the expression.
= 60(x + 60)	Distributive Property

So, the length (in yards) of the platform can be represented by x + 60. Use the perimeter formula to write an expression that represents the perimeter of the platform.

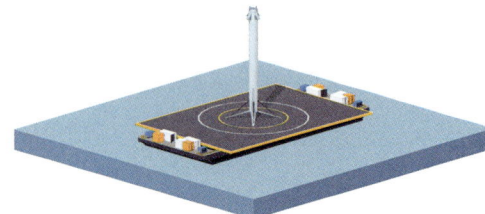

$P = 2\ell + 2w$	Perimeter of a rectangle
$= 2(x + 60) + 2(60)$	Substitute for ℓ and w.
$= 2x + 120 + 120$	Multiply.
$= 2x + 240$	Add.

 So, an expression that represents the perimeter (in yards) of the platform is $2x + 240$.

Self-Assessment for Problem Solving

Solve each exercise. Then rate your understanding of the success criteria in your journal.

16. An organization drills 3 wells to provide access to clean drinking water. The cost (in dollars) to drill and maintain the wells for n years is represented by 34,500 + 540n. Write and interpret an expression that represents the cost to drill and maintain one well for n years.

17. A photograph is 16 inches long and has an area of (16x + 96) square inches. A custom-made frame is 2 inches wide and costs $0.50 per square inch. Write an expression that represents the cost of the frame.

612   Chapter C   Expressions

C.4 Practice

Go to *BigIdeasMath.com* to get HELP with solving the exercises.

Review & Refresh

Simplify the expression.

1. $8(k-5)$
2. $-4.5(-6+2d)$
3. $-\dfrac{1}{4}(3g-6-5g)$

Find the difference. Write fractions in simplest form.

4. $\dfrac{2}{3} - \left(-\dfrac{5}{3}\right)$
5. $-4.7 - 5.6$
6. $-4\dfrac{3}{8} - \left(-2\dfrac{1}{4}\right)$

Evaluate the expression when $x = 4$, $y = -6$, and $z = -3$.

7. $y \div z$
8. $\dfrac{4y}{2x}$
9. $\dfrac{3x - 2y}{z}$

Concepts, Skills, & Problem Solving

FINDING DIMENSIONS The model shows the area (in square units) of each part of a rectangle. Use the model to find the missing values that complete the expression. Explain your reasoning. *(See Exploration 1, p. 609.)*

10. $2.25x + 3 = \boxed{}\left(\boxed{} + \boxed{}\right)$

11. $\dfrac{5}{6}m + \dfrac{2}{3}n = \boxed{}\left(\boxed{} + \boxed{}\right)$

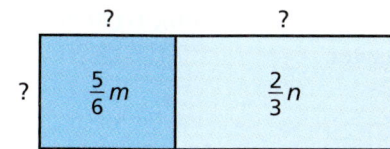

FACTORING OUT THE GCF Factor the expression using the GCF.

12. $9b + 21$
13. $32z - 48$
14. $8x + 2$
15. $3y - 24$
16. $14p - 28$
17. $6 + 16k$
18. $21 - 14d$
19. $20z - 8$
20. $15w + 65$
21. $36a + 16b$
22. $21m - 49n$
23. $12 + 9g - 30h$

FACTORING OUT A RATIONAL NUMBER Factor out the coefficient of the variable term.

24. $\dfrac{1}{7}a + \dfrac{1}{7}$
25. $\dfrac{1}{3}b - \dfrac{1}{3}$
26. $\dfrac{3}{8}d + \dfrac{3}{4}$
27. $2.2x + 4.4$
28. $1.5y - 6$
29. $0.8w + 3.6$
30. $\dfrac{15}{4} + \dfrac{3}{8}x$
31. $4h - 3$
32. $0.15c - 0.072$
33. $\dfrac{3}{8}z + 1$
34. $6s - \dfrac{3}{4}$
35. $\dfrac{5}{2}k - 2$

Section C.4 Factoring Expressions 613

YOU BE THE TEACHER Your friend factors the expression. Is your friend correct? Explain your reasoning.

36.
$$16p - 28 = 4(4p - 28)$$

37.
$$\frac{2}{3}y - \frac{14}{3} = \frac{2}{3} \cdot y - \frac{2}{3} \cdot 7$$
$$= \frac{2}{3}(y - 7)$$

FACTORING OUT A NEGATIVE NUMBER Factor out the indicated number.

38. Factor -4 out of $-8d + 20$.

39. Factor -6 out of $18z - 15$.

40. Factor -0.25 out of $7g + 3.5$.

41. Factor $-\frac{1}{2}$ out of $-\frac{1}{2}x + 6$.

42. Factor -1.75 out of $-14m - 5.25n$.

43. Factor $-\frac{1}{4}$ out of $-\frac{1}{2}x - \frac{5}{4}y$.

44. **MP STRUCTURE** A rectangle has an area of $(4x + 12)$ square units. Write three multiplication expressions that can represent the product of the length and the width of the rectangle.

45. **MODELING REAL LIFE** A square wrestling mat has a perimeter of $(12x - 32)$ feet. Explain how to use the expression to find the length (in feet) of the mat. Justify your answer.

46. **MODELING REAL LIFE** A table is 6 feet long and 3 feet wide. You extend the length of the table by inserting two identical table *leaves*. The extended table is rectangular with an area of $(18 + 6x)$ square feet. Write and interpret an expression that represents the length (in feet) of the extended table.

47. **DIG DEEPER!** A three-dimensional printing pen uses heated plastic to create three-dimensional objects. A kit comes with one 3D-printing pen and p packages of plastic. An art club purchases 6 identical kits for $(180 + 58.5p)$ dollars. Write and interpret an expression that represents the cost of one kit.

48. **MP STRUCTURE** The area of the trapezoid is $\left(\frac{3}{4}x - \frac{1}{4}\right)$ square centimeters. Write two different pairs of expressions that represent the possible base lengths (in centimeters). Justify your answers.

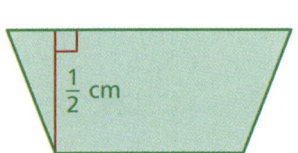

Connecting Concepts

▶ Using the Problem-Solving Plan

1. The runway shown has an area of $(0.05x + 0.125)$ square miles. Write an expression that represents the perimeter (in feet) of the runway.

0.05 mi

Understand the problem. You know the area of the rectangular runway in square miles and the width of the runway in miles. You want to know the perimeter of the runway in feet.

Make a plan. Factor the width of 0.05 mile out of the expression that represents the area to find an expression that represents the length of the runway. Then write an expression that represents the perimeter (in miles) of the runway. Finally, use a measurement conversion to write the expression in terms of feet.

Solve and check. Use the plan to solve the problem. Then check your solution.

2. The populations of two towns after t years can be modeled by $-300t + 7000$ and $-200t + 5500$. What is the combined population of the two towns after t years? The combined population of the towns in Year 10 is what percent of the combined population in Year 0?

Performance Task

Chlorophyll in Plants

At the beginning of this chapter, you watched a STEAM Video called "Tropic Status." You are now ready to complete the performance task related to this video, available at *BigIdeasMath.com*. Be sure to use the problem-solving plan as you work through the performance task.

Chapter Review

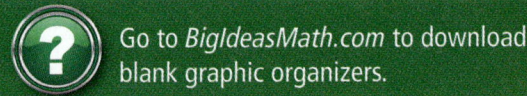
Go to *BigIdeasMath.com* to download blank graphic organizers.

Review Vocabulary

Write the definition and give an example of each vocabulary term.

like terms, *p. 592*
simplest form, *p. 592*
linear expression, *p. 598*
factoring an expression, *p. 610*

Graphic Organizers

You can use an **Example and Non-Example Chart** to list examples and non-examples of a concept. Here is an Example and Non-Example Chart for *like terms*.

Like Terms

Examples	Non-Examples
2 and −3	y and 4
$3x$ and $-7x$	$3x$ and $3y$
x^2 and $6x^2$	$4x$ and $-2x^2$
y and $5y$	$2y$ and 5

Choose and complete a graphic organizer to help you study the concept.

1. simplest form
2. equivalent expressions
3. linear expression
4. Distributive Property
5. factoring an expression

"Here is my **Example and Non-Example Chart** about things that scare cats."

616 Chapter C Expressions

Chapter Self-Assessment

As you complete the exercises, use the scale below to rate your understanding of the success criteria in your journal.

1	2	3	4
I do not understand.	I can do it with help.	I can do it on my own.	I can teach someone else.

C.1 Algebraic Expressions (pp. 591–596)

Learning Target: Simplify algebraic expressions.

Identify the terms and like terms in the expression.

1. $z + 8 - 4z$
2. $3n + 7 - n - 3$
3. $10x^2 - y + 12 - 3x^2$

Simplify the expression.

4. $4h - 8h$
5. $6.4r - 7 - 2.9r$
6. $2m - m - 7m$
7. $6y + 9 + 3y - 7$
8. $\frac{3}{5}x + 19 - \frac{3}{20}x - 7$
9. $\frac{2}{3}y + 14 - \frac{1}{6}y - 8$

10. Write an expression with 4 different terms that is equivalent to $5x^2 - 8$. Justify your answer.

11. Find the earnings for selling the same number of each type of sandwich. Justify your answer.

	Turkey	Ham
Pretzel Roll	2.25	1.55
Bagel	2.00	1.30

12. You buy the same number of brushes, rollers, and paint cans.

    a. Write and interpret an expression in simplest form that represents the total amount of money you spend on painting supplies.

    b. How much do you spend when you buy one set of supplies for each of 3 painters?

Paint $21.79

Brush $3.99

Paint roller $6.89

Chapter Review 617

C.2 Adding and Subtracting Linear Expressions (pp. 597–602)

Learning Target: Find sums and differences of linear expressions.

Find the sum.

13. $(c - 4) + (3c + 9)$

14. $(5z + 4) + (3z - 6)$

15. $(-2.1m - 5) + (3m - 7)$

16. $\left(\frac{5}{4}q + 1\right) + (q - 4) + \left(-\frac{1}{4}q + 2\right)$

Find the difference.

17. $(x - 1) - (3x + 2)$

18. $(4y + 3) - (2y - 9)$

19. $\left(\frac{1}{2}h + 7\right) - \left(\frac{3}{2}h + 9\right)$

20. $(4 - 3.7b) - (-5.4b - 4) - (1.2b + 1)$

21. A basket holds n apples. You pick $(2n - 3)$ apples, and your friend picks $(n + 4)$ apples. How many apples do you and your friend pick together? How many baskets do you need to carry all the apples? Justify your answer.

22. Greenland has a population of x people. Barbados has a population of about 4500 more than 5 times the population of Greenland. Find and interpret the difference in the populations of these two countries.

C.3 The Distributive Property (pp. 603–608)

Learning Target: Apply the Distributive Property to generate equivalent expressions.

Simplify the expression.

23. $2(a - 3)$

24. $-3(4x - 10)$

25. $-2.5(8 - b)$

26. $-7(1 - 3d - 5)$

27. $9(-3w - 6.2 + 2w)$

28. $\frac{3}{4}\left(8g - \frac{1}{4} - \frac{2}{3}g\right)$

29. Mars has m moons. The number of moons of Pluto is one more than twice the number of moons of Mars. The number of moons of Neptune is one less than 3 times the number of moons of Pluto. Write and interpret a simplified expression that represents the number of moons of Neptune.

Simplify the expression.

30. $3(2 + q) + 15$

31. $\frac{1}{8}(16m - 8) - 17$

32. $-1.5(4 - n) + 2.8$

33. $\frac{2}{5}(d - 10) - \frac{2}{3}(d + 6)$

34. The expression for degrees Fahrenheit is $\frac{9}{5}C + 32$, where C represents degrees Celsius. The temperature today is 5 degrees Celsius more than yesterday. Write and simplify an expression for the difference in degrees Fahrenheit for these two days.

C.4 Factoring Expressions (pp. 609–614)

Learning Target: Factor algebraic expressions.

Factor the expression using GCF.

35. $18a - 12$ **36.** $2b + 8$ **37.** $9 - 15x$

Factor out the coefficient of the variable term.

38. $\frac{1}{4}y + \frac{3}{8}$ **39.** $1.7j - 3.4$ **40.** $-5p + 20$

41. Factor $-\frac{3}{4}$ out of $\frac{3}{2}x - \frac{9}{4}y$.

42. You and 4 friends are buying tickets for a concert. The cost to buy one ticket is c dollars. If you buy all the tickets together, there is a discount and the cost is $(5c - 12.5)$ dollars. How much do you save per ticket when you buy the tickets together?

43. The rectangular pupil of an octopus is estimated to be 20 millimeters long with an area of $(20x - 200)$ square millimeters. Write an expression that represents the perimeter (in millimeters) of the octopus pupil.

44. A building block has a square base that has a perimeter of $(12x - 9)$ inches. Explain how to use the expression to find the length (in inches) of the wall shown.

Practice Test

1. Identify the terms and like terms in $4x + 9x^2 - 2x + 2$.

Simplify the expression.

2. $8x - 5 + 2x$

3. $2.5w - 3y + 4w$

4. $\dfrac{5}{7}x + 15 - \dfrac{9}{14}x - 9$

5. $(3j + 11) + (8j - 7)$

6. $(2r - 13) - (-6r + 4)$

7. $-2(4 - 3n)$

8. $3(5 - 2n) + 9n$

9. $\dfrac{1}{3}(6x + 9) - 2$

10. $\dfrac{3}{4}(8p + 12) + \dfrac{3}{8}(16p - 8)$

11. $-2.5(2s - 5) - 3(4.5s - 5.2)$

Factor out the coefficient of the variable term.

12. $6n - 24$

13. $\dfrac{1}{2}q + \dfrac{5}{2}$

14. $-4x + 36$

15. Find the earnings for giving a haircut and a shampoo to m men and w women. Justify your answer.

	Women	Men
Haircut	$45	$15
Shampoo	$12	$7

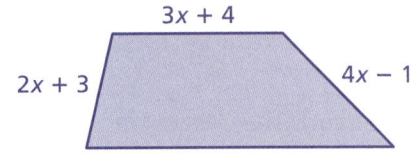

16. The expression $15x + 11$ represents the perimeter of the trapezoid. What is the length of the fourth side? Explain your reasoning.

17. The maximum number of charms that will fit on a bracelet is $3\left(d - \dfrac{2}{3}\right)$, where d is the diameter (in centimeters) of the bracelet.

   a. Write and interpret a simplified expression that represents the maximum number of charms on a bracelet.

   b. What is the maximum number of charms that fit on a bracelet that has a diameter of 6 centimeters?

18. You expand a rectangular garden so the perimeter is now twice the perimeter of the old garden. The expression $12w + 16$ represents the perimeter of the new garden, where w represents the width of the old garden.

   a. Write an expression that represents the perimeter of the old garden. Justify your answer.

   b. Write an expression that represents the area of the old garden.

Cumulative Practice

1. What is the simplified form of the expression?

    $3.7x - 5 - 2.3x$

    A. $-3.6x$

    B. $6x - 5$

    C. $1.4x - 5$

    D. $3.7x - 7.3$

2. What is the value of the expression when $c = 0$ and $d = -6$?

    $$\frac{cd - d^2}{4}$$

3. What is the value of the expression?

    $-38 - (-14)$

    F. -52 **G.** -24

    H. 24 **I.** 52

4. The daily low temperatures for a week are shown.

    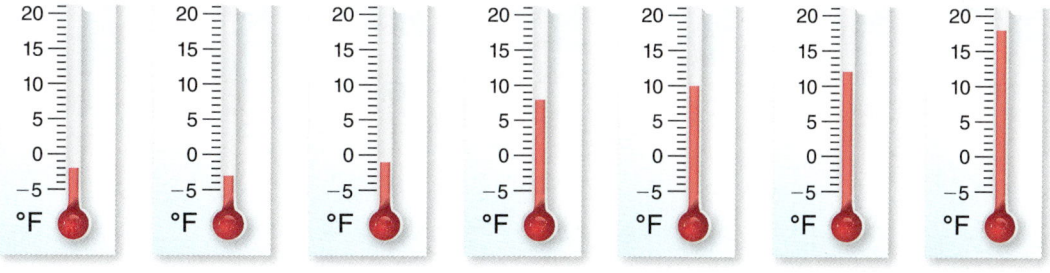

    What is the mean low temperature of the week?

    A. $-2°F$ **B.** $6°F$

    C. $8°F$ **D.** $10°F$

5. You and a friend collect seashells on a beach. After h minutes, you have collected $(11 + 2h)$ seashells and your friend has collected $(5h - 2)$ seashells. How many total seashells have you and your friend collected?

   F. $7h + 9$

   G. $3h - 13$

   H. $16h$

   I. $7h + 13$

6. What is the value of the expression?

   $$-0.28 \div (-0.07)$$

7. Which list is ordered from least to greatest?

   A. $-\left|\frac{3}{4}\right|, -\frac{1}{2}, \left|\frac{3}{8}\right|, -\frac{1}{4}, \left|-\frac{7}{8}\right|$

   B. $-\frac{1}{2}, -\frac{1}{4}, \left|\frac{3}{8}\right|, -\left|\frac{3}{4}\right|, \left|-\frac{7}{8}\right|$

   C. $\left|-\frac{7}{8}\right|, \left|\frac{3}{8}\right|, -\frac{1}{4}, -\frac{1}{2}, -\left|\frac{3}{4}\right|$

   D. $-\left|\frac{3}{4}\right|, -\frac{1}{2}, -\frac{1}{4}, \left|\frac{3}{8}\right|, \left|-\frac{7}{8}\right|$

8. Which number is equivalent to the expression shown?

   $$-2\frac{1}{4} - \left(-8\frac{3}{8}\right)$$

   F. $-10\frac{5}{8}$

   G. $-10\frac{1}{3}$

   H. $6\frac{1}{8}$

   I. $6\frac{1}{2}$

9. What is the simplified form of the expression?

   $$7x - 2(3x + 6)$$

   A. $15x + 30$

   B. $x - 12$

   C. $13x + 12$

   D. $-11x$

10. Which expression is *not* equivalent to the expression?

$$72m - 60$$

F. $6(12m - 10)$

G. $4(18m - 15)$

H. $12m$

I. $12(6m - 5)$

11. You want to buy a bicycle with your friend. You have \$43.50 saved and plan to save an additional \$7.25 every week. Your friend has \$24.50 saved and plans to save an additional \$8.75 every week.

Think Solve Explain

Part A Simplify and interpret an expression that represents the amount of money you and your friend save after w weeks.

Part B After 10 weeks, you and your friend use all of the money and buy the bike. How much does the bike cost? Who pays more towards the cost of the bike? Explain your reasoning.

12. Your friend evaluated $3 + x^2 \div y$ when $x = -2$ and $y = 4$.

$$\begin{aligned} 3 + x^2 \div y &= 3 + (-2^2) \div 4 \\ &= 3 - 4 \div 4 \\ &= 3 - 1 \\ &= 2 \end{aligned}$$

What should your friend do to correct his error?

A. Divide 3 by 4 before subtracting.

B. Square -2, then divide.

C. Divide -2 by 4, then square.

D. Subtract 4 from 3 before dividing.

D Ratios and Proportions

- **D.1** Ratios and Ratio Tables
- **D.2** Rates and Unit Rates
- **D.3** Identifying Proportional Relationships
- **D.4** Writing and Solving Proportions
- **D.5** Graphs of Proportional Relationships
- **D.6** Scale Drawings

Chapter Learning Target:
Understand ratios and proportions.

Chapter Success Criteria:
- I can write and interpret ratios.
- I can describe ratio relationships and proportional relationships.
- I can represent equivalent ratios.
- I can model ratio relationships and proportional relationships to solve real-life problems.

STEAM Video: "Painting a Large Room"

STEAM Video

Painting a Large Room

Shades of paint can be made by mixing other paints. What colors of paints can you mix to make green paint?

Watch the STEAM Video "Painting a Large Room." Then answer the following questions.

1. Enid estimates that they need 2 gallons of paint to apply two coats to the wall shown. How many square feet does she expect $\frac{1}{2}$ gallon of paint will cover?

10 ft

24 ft

2. Describe a room that requires $5\frac{1}{2}$ gallons of paint to apply one coat of paint to each of the four walls.

Performance Task

Mixing Paint

After completing this chapter, you will be able to use the concepts you learned to answer the questions in the *STEAM Video Performance Task*. You will be given the amounts of each tint used to make different colors of paint. For example:

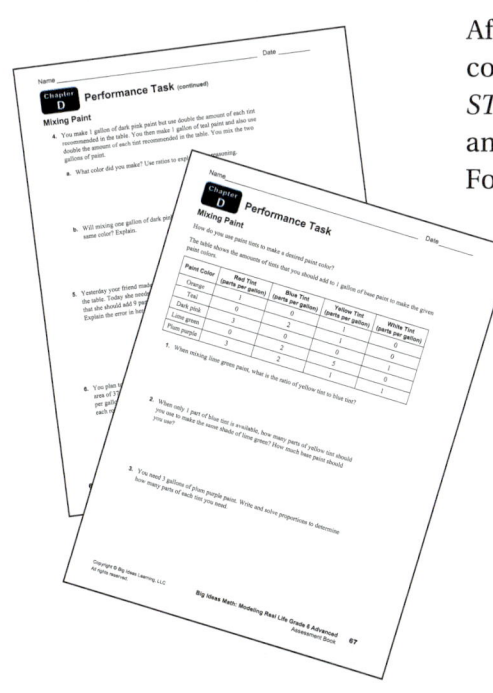

Plum Purple Paint

3 parts red tint per gallon

2 parts blue tint per gallon

1 part yellow tint per gallon

1 part white tint per gallon

You will be asked to solve various ratio problems about mixing paint. Given any color of paint, how can you make the paint slightly lighter in color?

625

Getting Ready for Chapter D

Chapter Exploration

The Meaning of a Word ▶ Rate

When you rent snorkel gear at the beach, you should pay attention to the rental **rate**. The rental rate is in dollars per hour.

1. **Work with a partner. Complete each step.**
   - Match each description with a rate.
   - Match each rate with a fraction.
   - Give a reasonable value for each fraction. Then give an unreasonable value.

Description	Rate	Fraction
Your speed in the 100-meter dash	Dollars per hour	$\dfrac{\boxed{} \text{ inches}}{\text{year}}$
The hourly wage of a worker at a fast-food restaurant	Inches per year	$\dfrac{\boxed{} \text{ pounds}}{\text{square foot}}$
The average annual rainfall in a rain forest	Pounds per square foot	$\dfrac{\$\boxed{}}{\text{hour}}$
The amount of fertilizer spread on a lawn	Meters per second	$\dfrac{\boxed{} \text{ meters}}{\text{second}}$

2. **Work with a partner.** Describe a situation to which the given fraction can apply. Show how to rewrite each expression as a division problem. Then simplify and interpret your result.

   a. $\dfrac{\frac{1}{2} \text{ cup}}{4 \text{ fluid ounces}}$

   b. $\dfrac{2 \text{ inches}}{\frac{3}{4} \text{ second}}$

   c. $\dfrac{\frac{3}{8} \text{ cup sugar}}{\frac{3}{4} \text{ cup flour}}$

   d. $\dfrac{\frac{5}{6} \text{ gallon}}{\frac{2}{3} \text{ second}}$

Vocabulary

The following vocabulary terms are defined in this chapter. Think about what each term might mean and record your thoughts.

proportional   constant of proportionality   scale drawing

D.1 Ratios and Ratio Tables

Learning Target: Understand ratios of rational numbers and use ratio tables to represent equivalent ratios.

Success Criteria:
- I can write and interpret ratios involving rational numbers.
- I can use various operations to create tables of equivalent ratios.
- I can use ratio tables to solve ratio problems.

EXPLORATION 1
Describing Ratio Relationships

Work with a partner. Use the recipe shown.

Chicken Soup
- stewed tomatoes   9 ounces
- chicken broth   15 ounces
- chopped chicken   1 cup
- chopped spinach   9 ounces
- grated parmesan   5 tablespoons

a. Identify several ratios in the recipe.

b. You halve the recipe. Describe your ratio relationships in part (a) using the new quantities. Is the relationship between the ingredients the same as in part (a)? Explain.

EXPLORATION 2
Completing Ratio Tables

Work with a partner. Use the ratio tables shown.

x	5			
y	1			

x	$\frac{1}{4}$			
y	$\frac{1}{2}$			

Math Practice

Communicate Precisely

How can you determine whether the ratios in each table are equivalent?

a. Complete the first ratio table using multiple operations. Use the same operations to complete the second ratio table.

b. Are the ratios in the first table equivalent? the second table? Explain.

c. Do the strategies for completing ratio tables of whole numbers work for completing ratio tables of fractions? Explain your reasoning.

D.1 Lesson

Key Vocabulary
ratio, *p. 628*
value of a ratio, *p. 628*
equivalent ratios, *p. 629*
ratio table, *p. 629*

Key Idea

Ratios

Words A **ratio** is a comparison of two quantities. The **value of the ratio** a to b is the number $\frac{a}{b}$, which describes the multiplicative relationship between the quantities in the ratio.

Examples 2 snails *to* 6 fish

$\frac{1}{2}$ cup of milk *for every* $\frac{1}{4}$ cup of cream

Algebra The ratio of a to b can be written as $a : b$.

Reading
Recall that phrases indicating ratios include *for each*, *for every*, and *per*.

EXAMPLE 1 Writing and Interpreting Ratios

You make *flubber* using the ingredients shown.

Flubber Ingredients
cold water   3/2 cups
hot water   4/3 cups
glue   2 cups
borax   3 teaspoons

a. Write the ratio of cold water to glue.

The recipe uses $\frac{3}{2}$ cups of water per 2 cups of glue.

▶ So, the ratio of cold water to glue is $\frac{3}{2}$ to 2, or $\frac{3}{2} : 2$.

b. Find and interpret the value of the ratio in part (a).

The value of the ratio $\frac{3}{2} : 2$ is

$$\frac{\frac{3}{2}}{2} = \frac{3}{2} \div 2$$
$$= \frac{3}{2} \cdot \frac{1}{2}$$
$$= \frac{3}{4}.$$

So, the multiplicative relationship is $\frac{3}{4}$.

▶ The amount of cold water in the recipe is $\frac{3}{4}$ the amount of glue.

Try It

1. You mix $\frac{2}{3}$ teaspoon of baking soda with 3 teaspoons of salt. Find and interpret the value of the ratio of baking soda to salt.

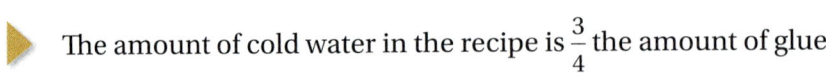

Two ratios that describe the same relationship are **equivalent ratios**. The values of equivalent ratios are equivalent. You can find and organize equivalent ratios in a **ratio table** by:

- adding or subtracting quantities in equivalent ratios.
- multiplying or dividing each quantity in a ratio by the same number.

EXAMPLE 2 Completing A Ratio Table

Find the missing values in the ratio table. Then write the equivalent ratios.

Cups	3	12	15	
Quarts	$\frac{3}{4}$			$\frac{5}{4}$

You can use a combination of operations to find the missing values.

×4   +3   ÷3

Cups	3	12	15	5
Quarts	$\frac{3}{4}$	3	$\frac{15}{4}$	$\frac{5}{4}$

×4   +$\frac{3}{4}$ ÷3

Notice that you obtain the third column by adding the values in the first column to the values in the second column.

$3 + 12 = 15$

$\frac{3}{4} + 3 = \frac{15}{4}$

▶ The equivalent ratios are $3 : \frac{3}{4}$, $12 : 3$, $15 : \frac{15}{4}$, and $5 : \frac{5}{4}$.

Try It Find the missing values in the ratio table. Then write the equivalent ratios.

2.

Kilometers	$\frac{5}{2}$		5
Hours		4	16

3.

Gallons	0.4	1.2	1.6
Days	0.75		

Self-Assessment for Concepts & Skills

Solve each exercise. Then rate your understanding of the success criteria in your journal.

4. **WRITING AND INTERPRETING RATIOS** You include $\frac{1}{2}$ tablespoon of essential oils in a solution for every 12 tablespoons of jojoba oil. Find and interpret the value of the ratio of jojoba oil to essential oils.

5. **MP NUMBER SENSE** Find the missing values in the ratio table. Then write the equivalent ratios.

Pounds	$\frac{3}{2}$		$\frac{21}{2}$
Years	$\frac{1}{12}$	$\frac{2}{3}$	

Section D.1 Ratios and Ratio Tables

EXAMPLE 3 Modeling Real Life

You mix $\frac{1}{2}$ cup of yellow paint for every $\frac{3}{4}$ cup of blue paint to make 15 cups of green paint. How much yellow paint do you use?

Method 1: The ratio of yellow paint to blue paint is $\frac{1}{2}$ to $\frac{3}{4}$. Use a ratio table to find an equivalent ratio in which the total amount of yellow paint and blue paint is 15 cups.

Yellow (cups)	Blue (cups)	Total (cups)
$\frac{1}{2}$	$\frac{3}{4}$	$\frac{1}{2} + \frac{3}{4} = \frac{5}{4}$
2	3	5
6	9	15

×4, ×3

 So, you use 6 cups of yellow paint.

Method 2: You can use the ratio of yellow paint to blue paint to find the fraction of the green paint that is made from yellow paint. You use $\frac{1}{2}$ cup of yellow paint for every $\frac{3}{4}$ cup of blue paint, so the fraction of the green paint that is made from yellow paint is

yellow → $\dfrac{\frac{1}{2}}{\frac{1}{2} + \frac{3}{4}} = \dfrac{\frac{1}{2}}{\frac{5}{4}} = \frac{1}{2} \cdot \frac{4}{5} = \frac{2}{5}$. ← green

 So, you use $\frac{2}{5} \cdot 15 = 6$ cups of yellow paint.

Self-Assessment for Problem Solving

Solve each exercise. Then rate your understanding of the success criteria in your journal.

6. **DIG DEEPER!** A satellite orbiting Earth travels $14\frac{1}{2}$ miles every 3 seconds. How far does the satellite travel in $\frac{3}{4}$ minute?

7. An engine runs on a mixture of 0.1 quart of oil for every 3.5 quarts of gasoline. You make 3 quarts of the mixture. How much oil and how much gasoline do you use?

D.1 Practice

Go to *BigIdeasMath.com* to get HELP with solving the exercises.

▶ Review & Refresh

Solve the inequality. Graph the solution.

1. $4p + 7 \geq 19$
2. $14 < -6n - 10$
3. $-3(2 + d) \leq 15$

Find the quotient. Write fractions in simplest form.

4. $\dfrac{2}{9} \div \dfrac{4}{3}$
5. $10.08 \div 12$
6. $-\dfrac{5}{6} \div \dfrac{3}{10}$

7. Which ratio can be represented by the tape diagram?

   A. $3:4$
   B. $4:5$
   C. $4:9$
   D. $8:12$

▶ Concepts, Skills, & Problem Solving

OPEN-ENDED Complete the ratio table using multiple operations. Are the ratios in the table equivalent? Explain. (See Exploration 2, p. 627.)

8.
x	4		
y	10		

9.
x	$\dfrac{4}{5}$		
y	$\dfrac{1}{2}$		

Fruit Punch Ingredients

chopped watermelon	3 cups
sugar	3/4 cup
mint leaves	1/2 cup
white grape juice	2 cups
lime juice	3/4 cup
club soda	4 cups

WRITING AND INTERPRETING RATIOS Find the ratio. Then find and interpret the value of the ratio.

10. club soda : white grape juice
11. mint leaves : chopped watermelon
12. white grape juice to sugar
13. lime juice to mint leaves

14. **YOU BE THE TEACHER** You have blue ribbon and red ribbon in the ratio $\dfrac{1}{2} : \dfrac{1}{5}$. Your friend finds the value of the ratio. Is your friend correct? Explain your reasoning.

   The value of the ratio is
   $\dfrac{\frac{1}{2}}{\frac{1}{5}} = \dfrac{1}{2} \div \dfrac{1}{5} = \dfrac{1}{10}$.

Section D.1 Ratios and Ratio Tables

COMPLETING A RATIO TABLE Find the missing values in the ratio table. Then write the equivalent ratios.

15.
Calories	20		10	90
Miles	$\frac{1}{6}$	$\frac{2}{3}$		

16.
Meters	8	4		
Minutes	$\frac{1}{3}$		$\frac{1}{4}$	$\frac{5}{12}$

17.
Feet	$\frac{1}{24}$		$\frac{1}{8}$	
Inches	$\frac{1}{2}$	1		$\frac{1}{4}$

18.
Tea (cups)	3.75			
Milk (cups)	1.5	1	3.5	2.5

19. **CRITICAL THINKING** Are the two statements equivalent? Explain your reasoning.
   - The ratio of boys to girls is 2 to 3.
   - The ratio of girls to boys is 3 to 2.

20. **MODELING REAL LIFE** A city dumps plastic *shade balls* into a reservoir to prevent water from evaporating during a drought. It costs $5760 for 16,000 shade balls. How much does it cost for 12,000 shade balls?

21. **MODELING REAL LIFE** An oil spill spreads 25 square meters every $\frac{1}{6}$ hour. What is the area of the oil spill after 2 hours?

22. **MODELING REAL LIFE** You mix 0.25 cup of juice concentrate for every 2 cups of water to make 18 cups of juice. How much juice concentrate do you use? How much water do you use?

23. **MODELING REAL LIFE** A store sells $2\frac{1}{4}$ pounds of mulch for every $1\frac{1}{2}$ pounds of gravel sold. The store sells 180 pounds of mulch and gravel combined. How many pounds of each item does the store sell?

24. **DIG DEEPER!** You mix $\frac{1}{4}$ cup of red paint for every $\frac{1}{2}$ cup of blue paint to make 3 gallons of purple paint.

   a. How much red paint do you use? How much blue paint do you use?

   b. You decide that you want to make a lighter purple paint. You make the new mixture by adding $\frac{1}{4}$ cup of white paint for every $\frac{1}{4}$ cup of red paint and $\frac{1}{2}$ cup of blue paint. How much red paint, blue paint, and white paint do you use to make $1\frac{1}{2}$ gallons of the lighter purple paint?

D.2 Rates and Unit Rates

Learning Target: Understand rates involving fractions and use unit rates to solve problems.

Success Criteria:
- I can find unit rates for rates involving fractions.
- I can use unit rates to solve rate problems.

EXPLORATION 1

Writing Rates

Work with a partner.

a. How many degrees does the minute hand on a clock move every 15 minutes? Write a rate that compares the number of degrees moved by the minute hand to the number of hours elapsed.

Math Practice

Recognize Usefulness of Tools

Can you use a protractor to find the number of degrees the minute hand moves in 15 minutes? in 1 hour?

b. Can you use the rate in part (a) to determine how many degrees the minute hand moves in $\frac{1}{2}$ hour? Explain your reasoning.

c. Write a rate that represents the number of degrees moved by the minute hand every hour. How can you use this rate to find the number of degrees moved by the minute hand in $2\frac{1}{2}$ hours?

d. Draw a clock with hour and minute hands. Draw another clock that shows the time after the minute hand moves 900°. How many degrees does the hour hand move in this time? in one hour? Explain your reasoning.

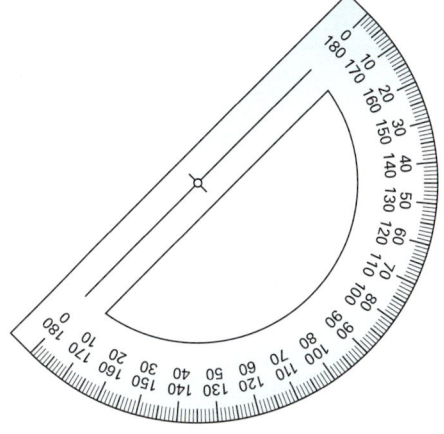

Section D.2 Rates and Unit Rates 633

D.2 Lesson

Key Vocabulary 🔊
rate, p. 634
unit rate, p. 634
equivalent rates, p. 634

Key Idea

Rates and Unit Rates

Words A **rate** is a ratio of two quantities using different units. A **unit rate** compares a quantity to one unit of another quantity. **Equivalent rates** have the same unit rate.

Numbers You pay $350 for every $\frac{1}{4}$ ounce of gold.

| $350 | $350 | $350 | $350 |

Rate: $350 : $\frac{1}{4}$ oz

| $\frac{1}{4}$ oz | $\frac{1}{4}$ oz | $\frac{1}{4}$ oz | $\frac{1}{4}$ oz |

Unit Rate: $1400 : 1 oz

Algebra   Rate: a units : b units   Unit rate: $\frac{a}{b}$ units : 1 unit

---

EXAMPLE 1 Finding Unit Rates

A nutrition label shows that every $\frac{1}{4}$ cup of tuna has $\frac{1}{2}$ gram of fat.

a. How many grams of fat are there for every cup of tuna?

There is $\frac{1}{2}$ gram of fat for every $\frac{1}{4}$ cup of tuna. Find the unit rate.

▶ There are $\dfrac{\frac{1}{2}}{\frac{1}{4}} = 2$ grams of fat for every cup of tuna.

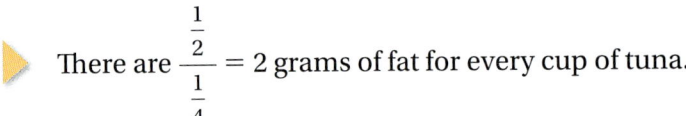

b. How many cups of tuna are there for every gram of fat?

There is $\frac{1}{4}$ cup of tuna for every $\frac{1}{2}$ gram of fat. Find the unit rate.

▶ There is $\dfrac{\frac{1}{4}}{\frac{1}{2}} = \frac{1}{2}$ cup of tuna per gram of fat.

Try It

1. There is $\frac{1}{4}$ gram of fat for every $\frac{1}{3}$ tablespoon of powdered peanut butter. How many grams of fat are there for every tablespoon of the powder?

EXAMPLE 2 Using a Unit Rate to Solve a Rate Problem

A scientist estimates that a jet of liquid iron in the Earth's core travels 9 feet every $\frac{1}{2}$ hour. How far does the liquid iron travel in 1 day?

The ratio of feet to hours is $9 : \frac{1}{2}$. Using a ratio table, divide the quantity by $\frac{1}{2}$ to find the unit rate in feet per hour. Then multiply each quantity by 24 to find the distance traveled in 24 hours, or 1 day.

	×2	×24
Distance (feet)	9	18	432
Time (hours)	$\frac{1}{2}$	1	24
	×2	×24

 So, the liquid iron travels about 432 feet in 1 day.

Try It

2. **WHAT IF?** The scientist later states that the iron travels 3 feet every 10 minutes. Does this change your answer in Example 2? Explain.

Self-Assessment for Concepts & Skills

Solve each exercise. Then rate your understanding of the success criteria in your journal.

3. **VOCABULARY** How can you tell when a rate is a unit rate?

4. **WRITING** Explain why rates are usually written as unit rates.

Find the unit rate.

5. $1.32 for 12 ounces

6. $\frac{1}{4}$ gallon for every $\frac{3}{10}$ mile

7. **USING TOOLS** Find the missing values in the ratio table. Then write the unit rate of grams per cup and the unit rate of cups per gram.

Grams	$\frac{5}{2}$		1	$\frac{15}{4}$	
Cups	$\frac{2}{3}$	$\frac{1}{6}$			4

Section D.2 Rates and Unit Rates 635

EXAMPLE 3 Modeling Real Life

You hike up a mountain trail at a rate of $\frac{1}{4}$ mile every 10 minutes. You hike 5 miles every 2 hours on the way down the trail. How much farther do you hike in 3 hours on the way down than in 3 hours on the way up?

Because 10 minutes is $\frac{1}{6}$ of an hour, the ratio of miles to hours on the way up is $\frac{1}{4} : \frac{1}{6}$. On the way down, the ratio is 5 : 2. Use ratio tables to find how far you hike in 3 hours at each rate.

Hiking Up	
Distance (miles)	Time (hours)
$\frac{1}{4}$	$\frac{1}{6}$
$\frac{3}{2}$	1
$\frac{9}{2}$	3

Hiking Down	
Distance (miles)	Time (hours)
5	2
$\frac{5}{2}$	1
$\frac{15}{2}$	3

Find the unit rate for each part of the hike.

Find the distance you hike in 3 hours on each part of the hike.

▶ So, you hike $\frac{15}{2} - \frac{9}{2} = \frac{6}{2} = 3$ miles farther in 3 hours on the way down than you hike in 3 hours on the way up.

Check Your rate on the way down is $\frac{5}{2} - \frac{3}{2} = \frac{2}{2} = 1$ mile per hour faster than your rate on the way up. So, you hike 3 miles farther in 3 hours on the way down than you hike in 3 hours on the way up. ✓

 Self-Assessment for Problem Solving

Solve each exercise. Then rate your understanding of the success criteria in your journal.

8. Two people compete in a five-mile go-kart race. Person A travels $\frac{1}{10}$ mile every 15 seconds. Person B travels $\frac{3}{8}$ mile every 48 seconds. Who wins the race? What is the difference of the finish times of the competitors?

9. **DIG DEEPER!** A bus travels 0.8 mile east every 45 seconds. A second bus travels 0.55 mile west every 30 seconds. The buses start at the same location. Use two methods to determine how far apart the buses are after 15 minutes. Explain your reasoning.

D.2 Practice

Go to *BigIdeasMath.com* to get HELP with solving the exercises.

▶ Review & Refresh

Find the missing values in the ratio table. Then write the equivalent ratios.

1.
Flour (cups)	$\frac{3}{4}$		3	1
Oats (cups)		$\frac{1}{3}$	$\frac{2}{3}$	

2.
Pages	$\frac{1}{4}$	$\frac{3}{4}$		5
Minutes	$\frac{1}{2}$		3	

Copy and complete the statement using <, >, or =.

3. $\frac{9}{2}$ ▨ $\frac{8}{3}$

4. $-\frac{8}{15}$ ▨ $\frac{10}{18}$

5. $\frac{-6}{24}$ ▨ $\frac{-2}{8}$

▶ Concepts, Skills, & Problem Solving

WRITING RATES Find the number of degrees moved by the minute hand of a clock in the given amount of time. Explain your reasoning. *(See Exploration 1, p. 633.)*

6. $\frac{2}{3}$ hour

7. $\frac{7}{12}$ hour

8. $1\frac{1}{4}$ hours

FINDING UNIT RATES Find the unit rate.

9. 180 miles in 3 hours

10. 256 miles per 8 gallons

11. $\frac{1}{2}$ pound : 5 days

12. 4 grams for every $\frac{3}{4}$ serving

13. $9.60 for 4 pounds

14. $4.80 for 6 cans

15. 297 words in 5.5 minutes

16. $\frac{1}{3}$ kilogram : $\frac{2}{3}$ foot

17. $\frac{5}{8}$ ounce per $\frac{1}{4}$ pint

18. $21\frac{3}{4}$ meters in $2\frac{1}{2}$ hours

USING TOOLS Find the missing values in the ratio table. Then write the equivalent ratios.

19.
Calories	25	50		
Servings		$\frac{1}{3}$	1	$\frac{4}{3}$

20.
Oxygen (liters)	4	$\frac{4}{3}$		16
Time (minute)		$\frac{3}{4}$	1	

21. **MP PROBLEM SOLVING** In January 2012, the U.S. population was about 313 million people. In January 2017, it was about 324 million. What was the average rate of population change per year?

22. **MODELING REAL LIFE** You can sand $\frac{4}{9}$ square yard of wood in $\frac{1}{2}$ hour. How many square yards can you sand in 3.2 hours? Justify your answer.

MP REASONING Tell whether the rates are equivalent. Justify your answer.

23. 75 pounds per 1.5 years
    38.4 ounces per 0.75 year

24. $7\frac{1}{2}$ miles for every $\frac{3}{4}$ hour
    $\frac{1}{2}$ mile for every 3 minutes

25. **MP PROBLEM SOLVING** The table shows nutritional information for three beverages.

Beverage	Serving Size	Calories	Sodium
Whole milk	1 c	146	98 mg
Orange juice	1 pt	210	10 mg
Apple juice	24 fl oz	351	21 mg

    a. Which has the most calories per fluid ounce?

    b. Which has the least sodium per fluid ounce?

26. **MODELING REAL LIFE** A shuttle leaving Earth's atmosphere travels 15 miles every 2 seconds. When entering the Earth's atmosphere, the shuttle travels $2\frac{3}{8}$ miles per $\frac{1}{2}$ second. Find the difference in the distances traveled after 15 seconds when leaving and entering the atmosphere.

27. **RESEARCH** Fire hydrants are one of four different colors to indicate the rate at which water comes from the hydrant.

    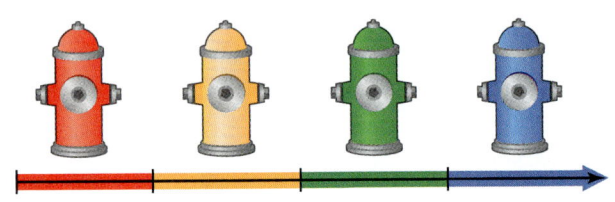

    a. Use the Internet to find the ranges of rates indicated by each color.

    b. Research why a firefighter needs to know the rate at which water comes out of a hydrant.

28. **DIG DEEPER!** You and a friend start riding bikes toward each other from opposite ends of a 24-mile biking route. You ride $2\frac{1}{6}$ miles every $\frac{1}{4}$ hour. Your friend rides $7\frac{1}{3}$ miles per hour.

    a. After how many hours do you meet?

    b. When you meet, who has traveled farther? How much farther?

D.3 Identifying Proportional Relationships

Learning Target: Determine whether two quantities are in a proportional relationship.

Success Criteria:
- I can determine whether ratios form a proportion.
- I can explain how to determine whether quantities are proportional.
- I can distinguish between proportional and nonproportional situations.

EXPLORATION 1

Determining Proportional Relationships

Work with a partner.

a. You can paint 50 square feet of a surface every 40 minutes. How long does it take you to paint the mural shown? Explain how you found your answer.

b. The number of square feet you paint is *proportional* to the number of minutes it takes you. What do you think it means for a quantity to be *proportional* to another quantity?

c. Assume your friends paint at the same rate as you. The table shows how long it takes you and different numbers of friends to paint a fence. Is x proportional to y in the table? Explain.

Painters, x	1	2	3	4
Hours, y	4	2	$\frac{4}{3}$	1

Math Practice

Look for Patterns
How can the table in part (c) help you answer the question in part (d)?

d. How long will it take you and four friends to paint the fence? Explain how you found your answer.

Section D.3 Identifying Proportional Relationships 639

D.3 Lesson

Key Vocabulary 🔊
proportion, *p. 640*
cross products, *p. 641*
proportional, *p. 642*

🔑 Key Idea

Proportions

Words A **proportion** is an equation stating that the values of two ratios are equivalent.

Numbers Equivalent ratios: 2 : 3 and 4 : 6

Proportion: $\dfrac{2}{3} = \dfrac{4}{6}$

EXAMPLE 1 **Determining Whether Ratios Form a Proportion**

Tell whether the ratios form a proportion.

a. 6 : 4 and 8 : 12

Compare the values of the ratios.

$\dfrac{6}{4} = \dfrac{6 \div 2}{4 \div 2} = \dfrac{3}{2}$

$\dfrac{8}{12} = \dfrac{8 \div 4}{12 \div 4} = \dfrac{2}{3}$

← The values of the ratios are *not* equivalent.

▶ Because $\dfrac{3}{2} \neq \dfrac{2}{3}$, the ratios 6 : 4 and 8 : 12 do *not* form a proportion.

b. 10 : 40 and 2.5 : 10

Compare the values of the ratios.

$\dfrac{10}{40} = \dfrac{10 \div 10}{40 \div 10} = \dfrac{1}{4}$

← The values of the ratios are equivalent.

$\dfrac{2.5}{10} = \dfrac{2.5 \times 10}{10 \times 10} = \dfrac{25}{100} = \dfrac{25 \div 25}{100 \div 25} = \dfrac{1}{4}$

▶ Because $\dfrac{1}{4} = \dfrac{1}{4}$, the ratios 10 : 40 and 2.5 : 10 form a proportion.

When you are determining whether ratios form a proportion, you are checking whether the ratios are equivalent.

Try It Tell whether the ratios form a proportion.

1. 1 : 2 and 5 : 10
2. 4 : 6 and 18 : 24
3. 4.5 to 3 and 6 to 9
4. $\dfrac{1}{2}$ to $\dfrac{1}{4}$ and 8 to 4

 **Key Ideas**

Cross Products

In the proportion $\frac{a}{b} = \frac{c}{d}$, the products $a \cdot d$ and $b \cdot c$ are called **cross products**.

Cross Products Property

Words The cross products of a proportion are equal.

Numbers

$\frac{2}{3} = \frac{4}{6}$

$2 \cdot 6 = 3 \cdot 4$

Algebra

$\frac{a}{b} = \frac{c}{d}$

$ad = bc$, where $b \neq 0$ and $d \neq 0$

> You can use the Multiplication Property of Equality to show that the cross products are equal.
>
> $\frac{a}{b} = \frac{c}{d}$
>
> $bd \cdot \frac{a}{b} = bd \cdot \frac{c}{d}$
>
> $ad = bc$

EXAMPLE 2 Using Cross Products

Tell whether the ratios form a proportion.

a. $6:9$ and $12:18$

Use the Cross Products Property to determine whether the ratios form a proportion.

$\frac{6}{9} \stackrel{?}{=} \frac{12}{18}$ Determine whether the values of the ratios are equivalent.

$6 \cdot 18 \stackrel{?}{=} 9 \cdot 12$ Find the cross products.

$108 = 108$ The cross products are equal.

 So, the ratios $6:9$ and $12:18$ form a proportion.

b. $2:3$ and $4:5$

Use the Cross Products Property to determine whether the ratios form a proportion.

$\frac{2}{3} \stackrel{?}{=} \frac{4}{5}$ Determine whether the values of the ratios are equivalent.

$2 \cdot 5 \stackrel{?}{=} 3 \cdot 4$ Find the cross products.

$10 \neq 12$ The cross products are *not* equal.

 So, the ratios $2:3$ and $4:5$ do *not* form a proportion.

Try It Tell whether the ratios form a proportion.

5. $6:2$ and $12:1$

6. $8:12$ and $\frac{2}{3}:1$

Two quantities are **proportional** when all of the ratios relating the quantities are equivalent. These quantities are said to be in a *proportional relationship*.

EXAMPLE 3 Determining Whether Two Quantities are Proportional

Tell whether x and y are proportional.

Compare the values of the ratios x to y.

$$\frac{\frac{1}{2}}{3} = \frac{1}{6} \qquad \frac{1}{6} \qquad \frac{\frac{3}{2}}{9} = \frac{1}{6} \qquad \frac{2}{12} = \frac{1}{6}$$

The values of the ratios are equivalent.

x	y
$\frac{1}{2}$	3
1	6
$\frac{3}{2}$	9
2	12

> You can also compare the values of the ratios y to x. In Example 3, the value of the ratios y to x is equal to 6.

So, x and y are proportional.

Try It Tell whether x and y are proportional.

7.
x	1	2	3	4
y	2	4	6	8

8.
x	2	4	6	8	10
y	4	2	1	$\frac{1}{2}$	$\frac{1}{4}$

Self-Assessment for Concepts & Skills

Solve each exercise. Then rate your understanding of the success criteria in your journal.

PROPORTIONS Tell whether the ratios form a proportion.

9. $4:14$ and $12:40$

10. $9:3$ and $45:15$

11. **VOCABULARY** Explain how to determine whether two quantities are proportional.

12. **WHICH ONE DOESN'T BELONG?** Which ratio does *not* belong with the other three? Explain your reasoning.

$4:10 \qquad 2:5$

$3:5 \qquad 6:15$

EXAMPLE 4 Modeling Real Life

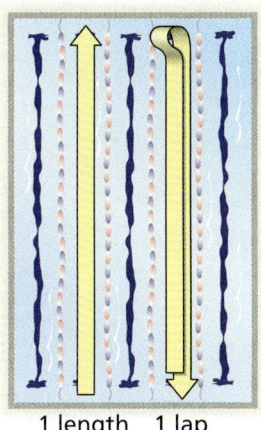

1 length   1 lap

You swim for 16 minutes and complete 20 laps. You swam your first 4 laps in 2.4 minutes. How long does it take you to swim 10 laps?

Compare unit rates to determine whether the number of laps is proportional to your time. If it is, then you can use ratio reasoning to find the time it takes you to swim 10 laps.

2.4 minutes for every 4 laps: $\frac{2.4}{4} = 0.6$ minute per lap

16 minutes for every 20 laps: $\frac{16}{20} = 0.8$ minute per lap

The number of laps is *not* proportional to the time. So, you *cannot* use ratio reasoning to determine the time it takes you to swim 10 laps.

Because you slowed down after your first 4 laps, you can estimate that you swim 10 laps in more than

$$\frac{0.6 \text{ minute}}{1 \text{ lap}} \cdot 10 \text{ laps} = 6 \text{ minutes},$$

but less than

$$\frac{0.8 \text{ minute}}{1 \text{ lap}} \cdot 10 \text{ laps} = 8 \text{ minutes}.$$

▸ So, you can estimate that it takes you about 7 minutes to swim 10 laps.

Self-Assessment for Problem Solving

Solve each exercise. Then rate your understanding of the success criteria in your journal.

13. After making 20 servings of pasta, a chef has used 30 cloves of garlic. The chef used 6 cloves to make the first 4 servings. How many cloves of garlic are used to make 10 servings? Justify your answer.

14. **DIG DEEPER!** A runner completes a 25-mile race in 5 hours. The runner completes the first 7.5 miles in 1.5 hours.

  a. Do these rates form a proportion? Justify your answer.

  b. Can you determine, with certainty, the time it took the runner to complete 10 miles? Explain your reasoning.

D.3 Practice

Go to BigIdeasMath.com to get HELP with solving the exercises.

▶ Review & Refresh

Find the unit rate.

1. 30 inches per 5 years
2. 486 games every 3 seasons
3. 8750 steps every 1.25 hours
4. 3.75 pints out of every 5 gallons

Add or subtract.

5. $-28 + 15$
6. $-6 + (-11)$
7. $-10 - 8$
8. $-17 - (-14)$

Solve the equation.

9. $\frac{x}{6} = 25$
10. $8x = 72$
11. $150 = 2x$
12. $35 = \frac{x}{4}$

▶ Concepts, Skills, & Problem Solving

MP REASONING You can paint 75 square feet of a surface every 45 minutes. Determine how long it takes you to paint a wall with the given dimensions. (See Exploration 1, p. 639.)

13. 8 ft × 5 ft
14. 7 ft × 6 ft
15. 9 ft × 9 ft

PROPORTIONS Tell whether the ratios form a proportion.

16. 1 to 3 and 7 to 21
17. 1 : 5 and 6 : 30
18. 3 to 4 and 24 to 18
19. 3.5 : 2 and 14 : 8
20. 24 : 30 and $3 : \frac{7}{2}$
21. $\frac{21}{2} : 3$ and 16 : 6
22. 0.6 : 0.5 and 12 : 10
23. 2 to 4 and 11 to $\frac{11}{2}$
24. $\frac{5}{8} : \frac{2}{3}$ and $\frac{1}{4} : \frac{1}{3}$

IDENTIFYING PROPORTIONAL RELATIONSHIPS Tell whether x and y are proportional.

25.
x	1	2	3
y	7	8	9

26.
x	2	4	6
y	5	10	15

27.
x	0.25	0.5	0.75
y	4	8	12

28.
x	$\frac{2}{3}$	1	$\frac{4}{3}$
y	$\frac{7}{10}$	$\frac{3}{5}$	$\frac{1}{2}$

YOU BE THE TEACHER Your friend determines whether *x* and *y* are proportional. Is your friend correct? Explain your reasoning.

29.
x	8	9
y	3	4

$$\frac{8+1}{3+1} = \frac{9}{4}$$

The values of the ratios *x* to *y* are equal. So, *x* and *y* are proportional.

30.
x	2	4	8
y	6	12	18

$$\frac{2}{6} = \frac{1}{3} \qquad \frac{4}{12} = \frac{1}{3}$$

The values of the ratios *x* to *y* are equal. So, *x* and *y* are proportional.

PROPORTIONS Tell whether the rates form a proportion.

31. 7 inches in 9 hours; 42 inches in 54 hours

32. 12 players from 21 teams; 15 players from 24 teams

33. 385 calories in 3.5 servings; 300 calories in 3 servings

34. 4.8 laps every 8 minutes; 3.6 laps every 6 minutes

35. $\frac{3}{4}$ pound for every 5 gallons; $\frac{4}{5}$ pound for every $5\frac{1}{3}$ gallons

36. **MODELING REAL LIFE** You do 90 sit-ups in 2 minutes. Your friend does 126 sit-ups in 2.8 minutes. Do these rates form a proportion? Explain.

37. **MODELING REAL LIFE** Find the heart rates of you and your friend. Do these rates form a proportion? Explain.

	Heartbeats	Seconds
You	22	20
Friend	18	15

38. **PROBLEM SOLVING** You earn $56 walking your neighbor's dog for 8 hours. Your friend earns $36 painting your neighbor's fence for 4 hours. Are the pay rates equivalent? Explain.

39. **GEOMETRY** Are the heights and bases of the two triangles proportional? Explain.

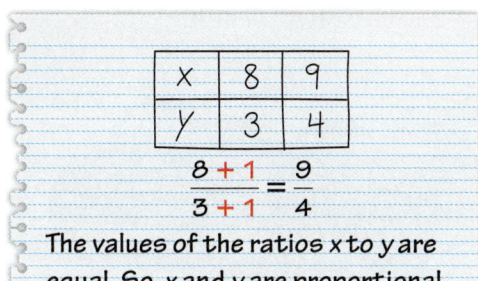

$h = 8$ cm
$h = 12$ cm
$b = 10$ cm
$b = 15$ cm

Session Number, x	Pitches, y	Curveballs, z
1	10	4
2	20	8
3	30	12
4	40	16

40. **REASONING** A pitcher coming back from an injury limits the number of pitches thrown in bullpen sessions as shown.

  a. Which quantities are proportional?

  b. How many pitches that are *not* curveballs will the pitcher likely throw in Session 5?

41. **MP STRUCTURE** You add the same numbers of pennies and dimes to the coins shown. Is the new ratio of pennies to dimes proportional to the original ratio of pennies to dimes? If so, illustrate your answer with an example. If not, show why with a counterexample.

   a.    b.

42. **MP REASONING** You are 13 years old, and your cousin is 19 years old. As you grow older, is your age proportional to your cousin's age? Explain your reasoning.

43. **MODELING REAL LIFE** The shadow of the moon during a solar eclipse travels 2300 miles in 1 hour. In the first 20 minutes, the shadow traveled $766\frac{2}{3}$ miles. How long does it take for the shadow to travel 1150 miles? Justify your answer.

44. **MODELING REAL LIFE** In 60 seconds, a car in a parade travels 0.2 mile. The car traveled the last 0.05 mile in 12 seconds. How long did it take for the car to travel 0.1 mile? Justify your answer.

45. **OPEN-ENDED** Describe (a) a real-life situation where you expect two quantities to be proportional and (b) a real-life situation where you do *not* expect two quantities to be proportional. Explain your reasoning.

46. **MP PROBLEM SOLVING** A specific shade of red nail polish requires 7 parts red to 2 parts yellow. A mixture contains 35 quarts of red and 8 quarts of yellow. Is the mixture the correct shade? If so, justify your answer. If not, explain how you can fix the mixture to make the correct shade of red.

47. **MP LOGIC** The quantities *x* and *y* are proportional. Use each of the integers 1–5 to complete the table. Justify your answer.

x	10		6	
y				0.5

48. **CRITICAL THINKING** Ratio *A* and Ratio *B* form a proportion. Ratio *B* and Ratio *C* also form a proportion. Do Ratio *A* and Ratio *C* form a proportion? Justify your answer.

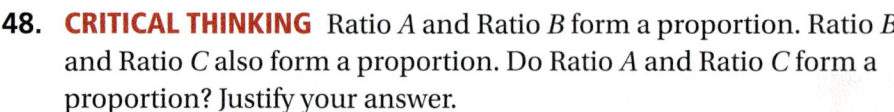

D.4 Writing and Solving Proportions

Learning Target: Use proportions to solve ratio problems.

Success Criteria:
- I can solve proportions using various methods.
- I can find a missing value that makes two ratios equivalent.
- I can use proportions to represent and solve real-life problems.

EXPLORATION 1

Solving a Ratio Problem

Work with a partner. A train travels 50 miles every 40 minutes. To determine the number of miles the train travels in 90 minutes, your friend creates the following table.

Miles	50	x
Minutes	40	90

a. Explain how you can find the value of x.

b. Can you use the information in the table to write a proportion? If so, explain how you can use the proportion to find the value of x. If not, explain why not.

c. How far does the train below travel in 2 hours?

Math Practice

Use Equations
What equation can you use to find the answer in part (c)?

30 miles every $\frac{1}{2}$ hour

d. Share your results in part (c) with other groups. Compare and contrast methods used to solve the problem.

Section D.4 Writing and Solving Proportions 647

D.4 Lesson

You can solve proportions using various methods.

EXAMPLE 1 Solving a Proportion Using Mental Math

Solve $\dfrac{3}{2} = \dfrac{x}{8}$.

Step 1: Think: The product of 2 and what number is 8?

$$\dfrac{3}{2} = \dfrac{x}{8}$$

$2 \times ? = 8$

Step 2: Because the product of 2 and 4 is 8, multiply the numerator by 4 to find x.

$3 \times 4 = 12$

$$\dfrac{3}{2} = \dfrac{x}{8}$$

$2 \times 4 = 8$

 The solution is $x = 12$.

Try It Solve the proportion.

1. $\dfrac{5}{8} = \dfrac{20}{d}$
2. $\dfrac{7}{z} = \dfrac{14}{10}$
3. $\dfrac{21}{24} = \dfrac{x}{8}$

EXAMPLE 2 Solving a Proportion Using Multiplication

Solve $\dfrac{5}{7} = \dfrac{x}{21}$.

$\dfrac{5}{7} = \dfrac{x}{21}$ Write the proportion.

$21 \cdot \dfrac{5}{7} = 21 \cdot \dfrac{x}{21}$ Multiplication Property of Equality

$15 = x$ Simplify.

 The solution is $x = 15$.

Try It Solve the proportion.

4. $\dfrac{w}{6} = \dfrac{6}{9}$
5. $\dfrac{12}{10} = \dfrac{a}{15}$
6. $\dfrac{y}{10} = \dfrac{3}{5}$

EXAMPLE 3 Solving a Proportion Using Cross Products

Solve each proportion.

a. $\dfrac{x}{8} = \dfrac{7}{10}$

$x \cdot 10 = 8 \cdot 7$ Cross Products Property

$10x = 56$ Multiply.

$x = 5.6$ Divide each side by 10.

▸ The solution is $x = 5.6$.

b. $\dfrac{9}{y} = \dfrac{3}{17}$

$9 \cdot 17 = y \cdot 3$ Cross Products Property

$153 = 3y$ Multiply.

$51 = y$ Divide each side by 3.

▸ The solution is $y = 51$.

Try It Solve the proportion.

7. $\dfrac{2}{7} = \dfrac{x}{28}$ **8.** $\dfrac{12}{5} = \dfrac{6}{y}$ **9.** $\dfrac{40}{z+1} = \dfrac{15}{6}$

EXAMPLE 4 Writing and Solving a Proportion

Find the value of x so that the ratios $3:8$ and $x:20$ are equivalent.

For the ratios to be equivalent, the values of the ratios must be equal. So, find the value of x for which $\dfrac{3}{8}$ and $\dfrac{x}{20}$ are equal by solving a proportion.

$\dfrac{3}{8} = \dfrac{x}{20}$ Write a proportion.

$20 \cdot \dfrac{3}{8} = 20 \cdot \dfrac{x}{20}$ Multiplication Property of Equality

$7.5 = x$ Simplify.

▸ So, $3:8$ and $x:20$ are equivalent when $x = 7.5$.

Try It Find the value of x so that the ratios are equivalent.

10. $2:4$ and $x:6$ **11.** $x:5$ and $8:2$ **12.** 4 to 3 and 10 to x

EXAMPLE 5 Writing a Proportion

A chef increases the amounts of ingredients in a recipe to make a proportional recipe. The new recipe has 6 cups of black beans. Which proportion can be used to find the number x of cups of water in the new recipe?

Black Bean Soup
1.5 cups black beans
0.5 cup salsa
2 cups water
1 tomato
2 teaspoons seasoning

A. $\dfrac{2}{1.5} = \dfrac{6}{x}$ **B.** $\dfrac{1.5}{6} = \dfrac{x}{2}$

C. $\dfrac{1.5}{2} = \dfrac{x}{6}$ **D.** $\dfrac{1.5}{2} = \dfrac{6}{x}$

In the original recipe, the ratio of cups of black beans to cups of water is 1.5 : 2. In the new recipe, the ratio is 6 : x.

For the new recipe to be proportional to the original recipe, these ratios must be equivalent. So, the values of the ratios must be equal, $\dfrac{1.5}{2} = \dfrac{6}{x}$.

 The correct answer is **D**.

Try It

13. Write a proportion that can be used to find the number of tomatoes in the new recipe.

Self-Assessment for Concepts & Skills

Solve each exercise. Then rate your understanding of the success criteria in your journal.

SOLVING A PROPORTION Solve the proportion.

14. $\dfrac{5}{12} = \dfrac{b}{36}$ **15.** $\dfrac{6}{p} = \dfrac{42}{35}$

16. WRITING AND SOLVING A PROPORTION Find the value of x so that the ratios $x : 9$ and $5 : 6$ are equivalent.

17. DIFFERENT WORDS, SAME QUESTION Which is different? Find "both" answers.

Solve $\dfrac{3}{x} = \dfrac{12}{8}$.

Find x so that $3 : x$ and $12 : 8$ are equivalent.

Find x so that $3 : 12$ and $x : 8$ are equivalent.

Solve $\dfrac{12}{x} = \dfrac{3}{8}$.

EXAMPLE 6 Modeling Real Life

A titanosaur's heart pumped 50 gallons of blood for every 2 heartbeats. How many heartbeats did it take to pump 1000 gallons of blood?

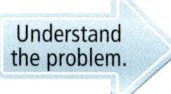

You are given the rate at which a titanosaur's heart pumped blood. Because all of the rates you can write using this relationship are equivalent, the amount of blood pumped is proportional to the number of heartbeats. You are asked to find how many heartbeats it took to pump 1000 gallons of blood.

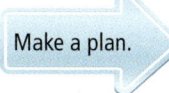

The ratio of heartbeats to gallons of blood is 2 : 50. The number x of heartbeats for every 1000 gallons of blood can be represented by the ratio x : 1000. Use a proportion to find the value of x for which $\frac{2}{50}$ and $\frac{x}{1000}$ are equal.

$$\frac{2}{50} = \frac{x}{1000}$$ Write a proportion.

$40 = x$ Multiply each side by 1000.

 So, it took 40 heartbeats to pump 1000 gallons of blood.

Another Method
You can use a ratio table to solve the problem.

× 20

Heartbeats	2	40	✓
Blood (gallons)	50	1000	

× 20

Self-Assessment for Problem Solving

Solve each exercise. Then rate your understanding of the success criteria in your journal.

18. You burn 35 calories every 3 minutes running on a treadmill. You want to run for at least 15 minutes, but no more than 30 minutes. What are the possible numbers of calories that you will burn? Justify your answer.

19. DIG DEEPER! Two boats travel at the same speed to different destinations. Boat A reaches its destination in 12 minutes. Boat B reaches its destination in 18 minutes. Boat B travels 3 miles farther than Boat A. How fast do the boats travel? Justify your answer.

Section D.4 Writing and Solving Proportions

D.4 Practice

Go to *BigIdeasMath.com* to get HELP with solving the exercises.

Review & Refresh

Tell whether x and y are proportional.

1.
x	4	6	8
y	6	8	10

2.
x	$\frac{2}{5}$	$\frac{4}{5}$	4
y	3	6	30

Plot the ordered pair in a coordinate plane.

3. $A(-5, -2)$
4. $B(-3, 0)$
5. $C(-1, 2)$
6. $D(1, 4)$

7. Which expression is equivalent to $(3w - 8) - 4(2w + 3)$?

  A. $11w + 4$
  B. $-5w - 5$
  C. $-5w + 4$
  D. $-5w - 20$

Concepts, Skills, & Problem Solving

SOLVING A RATIO PROBLEM Determine how far the vehicle travels in 3 hours. (See Exploration 1, p. 647.)

8. A helicopter travels 240 miles every 2 hours.

9. A motorcycle travels 25 miles every 0.5 hour.

10. A train travels 10 miles every $\frac{1}{4}$ hour.

11. A ferry travels 45 miles every $1\frac{1}{2}$ hours.

SOLVING A PROPORTION Solve the proportion. Explain your choice of method.

12. $\frac{1}{4} = \frac{z}{20}$
13. $\frac{3}{4} = \frac{12}{y}$
14. $\frac{35}{k} = \frac{7}{3}$
15. $\frac{b}{36} = \frac{5}{9}$

16. $\frac{x}{8} = \frac{3}{12}$
17. $\frac{3}{4} = \frac{v}{14}$
18. $\frac{15}{8} = \frac{45}{c}$
19. $\frac{35}{28} = \frac{n}{12}$

20. $\frac{a}{6} = \frac{15}{2}$
21. $\frac{y}{9} = \frac{44}{54}$
22. $\frac{4}{24} = \frac{c}{36}$
23. $\frac{20}{16} = \frac{d}{12}$

24. $\frac{10}{7} = \frac{8}{k}$
25. $\frac{5}{n} = \frac{16}{32}$
26. $\frac{9}{10} = \frac{d}{6.4}$
27. $\frac{2.4}{1.8} = \frac{7.2}{k}$

28. **YOU BE THE TEACHER** Your friend solves the proportion $\frac{m}{8} = \frac{15}{24}$. Is your friend correct? Explain your reasoning.

   $\frac{m}{8} = \frac{15}{24}$
   $m \cdot 24 = 8 \cdot 15$
   $m = 5$

652 Chapter D Ratios and Proportions

29. **MP NUMBER SENSE** Without solving, determine whether $\frac{x}{4} = \frac{15}{3}$ and $\frac{x}{15} = \frac{4}{3}$ have the same solution. Explain your reasoning.

WRITING A PROPORTION Use the table to write a proportion.

30.

	Game 1	Game 2
Points	12	18
Shots	14	w

31.

	May	June
Winners	n	34
Entries	85	170

32.

	Today	Yesterday
Miles	15	m
Hours	2.5	4

33.

	Race 1	Race 2
Meters	100	200
Seconds	x	22.4

WRITING AND SOLVING A PROPORTION Find the value of x so that the ratios are equivalent.

34. $1:8$ and $4:x$

35. 4 to 5 and x to 20

36. $3:x$ and $12:40$

37. x to 0.25 and 6 to 1.5

38. $x:\frac{5}{2}$ and $8:10$

39. $\frac{7}{4}$ to 14 and x to 32

40. **WRITING A PROPORTION** Your science teacher has a photograph of the space shuttle *Atlantis*. Every 1 centimeter in the photograph represents 200 centimeters on the actual shuttle. Which of the proportions can you use to find the actual length x of *Atlantis*? Explain.

$\frac{1}{200} = \frac{19.5}{x}$ $\frac{1}{200} = \frac{x}{19.5}$ $\frac{200}{19.5} = \frac{x}{1}$ $\frac{x}{200} = \frac{1}{19.5}$

19.5 cm

41. **MODELING REAL LIFE** In an orchestra, the ratio of trombones to violas is 1 to 3. There are 9 violas. How many trombones are in the orchestra?

42. **MODELING REAL LIFE** A dance team has 80 dancers. The ratio of seventh-grade dancers to all dancers is $5:16$. Find the number of seventh-grade dancers on the team.

43. **MODELING REAL LIFE** There are 144 people in an audience. The ratio of adults to children is 5 to 3. How many are adults?

44. **MP PROBLEM SOLVING** You have $50 to buy T-shirts. You can buy 3 T-shirts for $24. Do you have enough money to buy 7 T-shirts? Justify your answer.

45. **MP PROBLEM SOLVING** You buy 10 vegetarian pizzas and pay with $100. How much change do you receive?

46. **MODELING REAL LIFE** A person who weighs 120 pounds on Earth weighs 20 pounds on the Moon. How much does a 93-pound person weigh on the Moon?

47. **MP PROBLEM SOLVING** Three pounds of lawn seed covers 1800 square feet. How many bags are needed to cover 8400 square feet?

48. **MODELING REAL LIFE** There are 180 white lockers in a school. There are 3 white lockers for every 5 blue lockers. How many lockers are in the school?

CONVERTING MEASURES Use a proportion to complete the statement. Round to the nearest hundredth if necessary.

49. 6 km ≈ ☐ mi

50. 2.5 L ≈ ☐ gal

51. 90 lb ≈ ☐ kg

SOLVING A PROPORTION Solve the proportion.

52. $\dfrac{2x}{5} = \dfrac{9}{15}$

53. $\dfrac{5}{2} = \dfrac{d-2}{4}$

54. $\dfrac{4}{k+3} = \dfrac{8}{14}$

55. **MP LOGIC** It takes 6 hours for 2 people to build a swing set. Can you use the proportion $\dfrac{2}{6} = \dfrac{5}{h}$ to determine the number of hours h it will take 5 people to build the swing set? Explain.

56. **MP STRUCTURE** The ratios $a:b$ and $c:d$ are equivalent. Which of the following equations are proportions? Explain your reasoning.

$\dfrac{b}{a} = \dfrac{d}{c}$ $\dfrac{a}{c} = \dfrac{b}{d}$ $\dfrac{a}{d} = \dfrac{c}{b}$ $\dfrac{c}{a} = \dfrac{d}{b}$

57. **CRITICAL THINKING** Consider the proportions $\dfrac{m}{n} = \dfrac{1}{2}$ and $\dfrac{n}{k} = \dfrac{2}{5}$. What is $\dfrac{m}{k}$? Explain your reasoning.

D.5 Graphs of Proportional Relationships

Learning Target: Represent proportional relationships using graphs and equations.

Success Criteria:
- I can determine whether quantities are proportional using a graph.
- I can find the unit rate of a proportional relationship using a graph.
- I can create equations to represent proportional relationships.

EXPLORATION 1

Representing Relationships Graphically

Work with a partner. The tables represent two different ways that red and blue food coloring are mixed.

Mixture 1

Drops of Blue, x	Drops of Red, y
1	2
2	4
3	6
4	8

Mixture 2

Drops of Blue, x	Drops of Red, y
0	2
2	4
4	6
6	8

a. Represent each table in the same coordinate plane. Which graph represents a proportional relationship? How do you know?

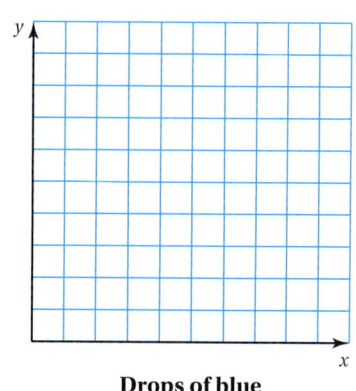

Math Practice

Use a Graph

How is the graph of the proportional relationship different from the other graph?

b. Find the unit rate of the proportional relationship. How is the unit rate shown on the graph?

c. What is the multiplicative relationship between x and y for the proportional relationship? How can you use this value to write an equation that relates y and x?

D.5 Lesson

Key Vocabulary
constant of proportionality, p. 656

The equation $y = kx$ can also be written as $\frac{y}{x} = k$. So, k is equal to the value of the ratio $y : x$.

Key Idea

Graphs of Proportional Relationships

Words Two quantities x and y are proportional when $y = kx$, where k is a number and $k \neq 0$. The number k represents the multiplicative relationship between the quantities and is called the **constant of proportionality**.

Graph The graph of $y = kx$ is a line that passes through the origin.

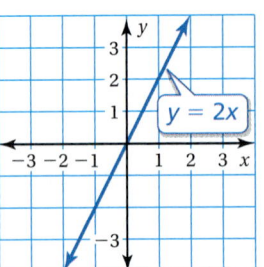

EXAMPLE 1 — Determining Whether Two Quantities are Proportional

Tell whether x and y are proportional. Explain your reasoning.

a.
x	1	2	3	4
y	−2	0	2	4

Plot the points. Draw a line through the points.

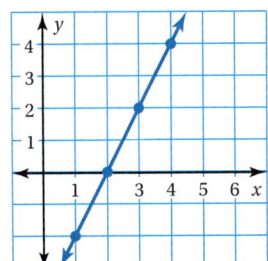

▸ The line does *not* pass through the origin. So, x and y are not proportional.

b.
x	0	2	4	6
y	0	2	4	6

Plot the points. Draw a line through the points.

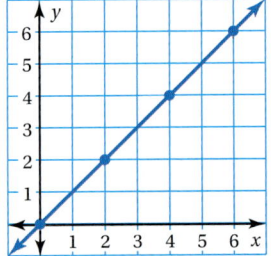

▸ The line passes through the origin. So, x and y are proportional.

Try It — Tell whether x and y are proportional. Explain your reasoning.

1.
x	y
0	−2
1	1
2	4
3	7

2.
x	y
1	4
2	8
3	12
4	16

3.
x	y
−2	4
−1	2
0	0
1	2

EXAMPLE 2 Finding a Unit Rate from a Graph

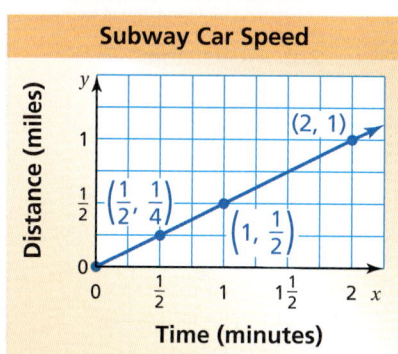

The graph shows the speed of a subway car. Find the speed in miles per minute.

The graph is a line through the origin, so time and distance are proportional. To find the speed in miles per minute, use a point on the graph to find the unit rate.

One Way: Use the point (2, 1) to find the speed.

The point (2, 1) indicates that the subway car travels 1 mile every 2 minutes. So, the unit rate is

$\frac{1}{2}$ mile per minute.

▶ The speed of the subway car is $\frac{1}{2}$ mile per minute.

On the graph of a proportional relationship, the point (1, k) indicates the unit rate, k : 1, and the constant of proportionality, k. This value is a measure of the steepness, or slope, of the line.

Another Way: Use the point $\left(1, \frac{1}{2}\right)$ to find the speed.

The point $\left(1, \frac{1}{2}\right)$ indicates that the subway car travels $\frac{1}{2}$ mile every 1 minute. This is the unit rate.

▶ The speed of the subway car is $\frac{1}{2}$ mile per minute.

Try It

4. **WHAT IF?** Does your answer change when you use the point $\left(\frac{1}{2}, \frac{1}{4}\right)$ to find the speed of the subway car? Explain your reasoning.

Self-Assessment for Concepts & Skills

Solve each exercise. Then rate your understanding of the success criteria in your journal.

5. **IDENTIFYING A PROPORTIONAL RELATIONSHIP** Use the graph shown to tell whether x and y are proportional. Explain your reasoning.

6. **FINDING A UNIT RATE** Interpret each plotted point in the graph. Then identify the unit rate, if possible.

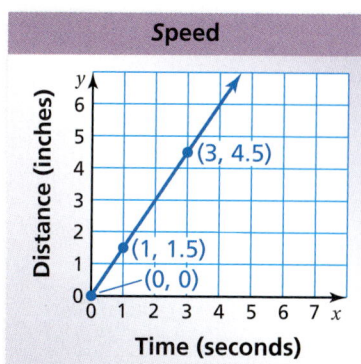

Section D.5 Graphs of Proportional Relationships 657

EXAMPLE 3 Modeling Real Life

The graph shows the area y (in square feet) that a robotic vacuum cleans in x minutes. Find the area cleaned in 10 minutes.

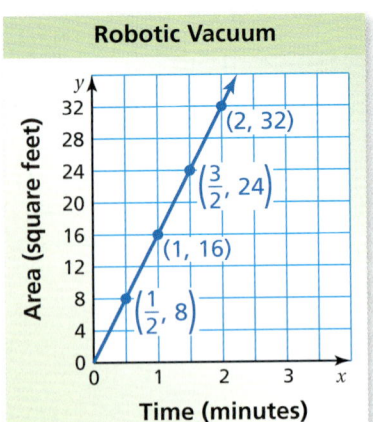

The graph is a line through the origin, so x and y are proportional. You can write an equation to represent the relationship between area and time.

Because the graph passes through the point $(1, 16)$, the unit rate is 16 square feet per minute and the constant of proportionality is $k = 16$. So, an equation of the line is $y = 16x$. Substitute to find the area cleaned in 10 minutes.

$y = 16x$	Write the equation.
$ = 16(10)$	Substitute 10 for x.
$ = 160$	Multiply.

▶ So, the vacuum cleans 160 square feet in 10 minutes.

Self-Assessment for Problem Solving

Solve each exercise. Then rate your understanding of the success criteria in your journal.

7. The table shows the temperature y (in degrees Fahrenheit), x hours after midnight.

Hours, x	0	0.5	1	1.5
Temperature, y (°F)	42	44	46	48

   a. Describe a proportional relationship between time and temperature shown by the table. Explain your reasoning.

   b. Find the temperature 3.5 hours after midnight.

8. **DIG DEEPER!** Show how you can use a proportional relationship to plan the heights of the vertical supports of a waterskiing ramp. Then explain how increasing the steepness of the ramp affects the proportional relationship.

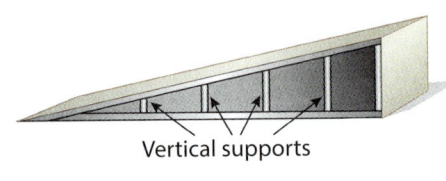
Vertical supports

D.5 Practice

Review & Refresh

Find the value of *x* so that the ratios are equivalent.

1. 2 : 7 and 8 : x
2. 3 to 2 and x to 18
3. 9 : x and 54 : 8

Find the quotient, if possible.

4. $36 \div 4$
5. $42 \div (-6)$
6. $-39 \div 3$
7. $-44 \div (-4)$

Solve the inequality. Graph the solution.

8. $-\dfrac{x}{3} < 2$
9. $\dfrac{1}{3}p \geq 4$
10. $-8 < \dfrac{2}{3}n$
11. $-2w \leq 10$

Concepts, Skills, & Problem Solving

REPRESENTING RELATIONSHIPS GRAPHICALLY Represent the table graphically. Does the graph represent a proportional relationship? How do you know? (See Exploration 1, p. 655.)

12.
Hours, x	Miles, y
0	50
1	100
2	150

13.
Cucumbers, x	Tomatoes, y
2	4
3	6
4	8

IDENTIFYING A PROPORTIONAL RELATIONSHIP Tell whether *x* and *y* are proportional. Explain your reasoning.

14.
x	1	2	3	4
y	2	4	6	8

15.
x	-2	-1	0	1
y	0	2	4	6

16.
x	-1	0	1	2
y	-2	-1	0	1

17.
x	3	6	9	12
y	2	4	6	8

18.
x	1	2	3	4
y	3	4	5	6

19.
x	1	3	5	7
y	0.5	1.5	2.5	3.5

20. **YOU BE THE TEACHER** Your friend uses the graph to determine whether *x* and *y* are proportional. Is your friend correct? Explain your reasoning.

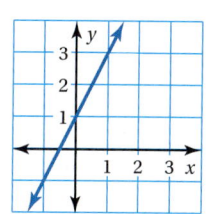

> The graph is a line, so x and y are proportional.

FINDING A UNIT RATE Interpret each plotted point in the graph. Then identify the unit rate.

21.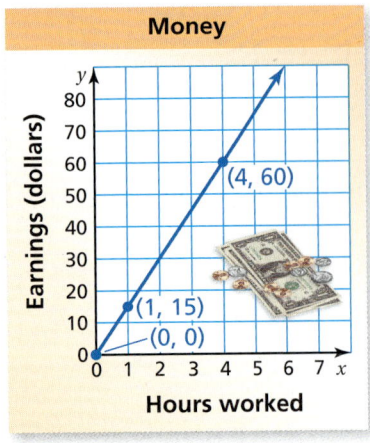

22.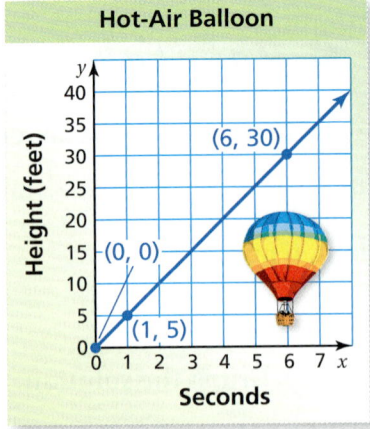

IDENTIFYING A PROPORTIONAL RELATIONSHIP Tell whether x and y are proportional. If so, identify the constant of proportionality. Explain your reasoning.

23. $x - y = 0$

24. $\dfrac{x}{y} = 2$

25. $8 = xy$

26. $x^2 = y$

WRITING AN EQUATION The variables x and y are proportional. Use the values to find the constant of proportionality. Then write an equation that relates x and y.

27. When $y = 72$, $x = 3$.

28. When $y = 20$, $x = 12$.

29. When $y = 45$, $x = 40$.

30. **MODELING REAL LIFE** The table shows the profit y for recycling x pounds of aluminum. Find the profit for recycling 75 pounds of aluminum.

Aluminum (lb), x	10	20	30	40
Profit, y	$4.50	$9.00	$13.50	$18.00

31. **MODELING REAL LIFE** The graph shows the cost of buying concert tickets. Tell whether x and y are proportional. If so, find and interpret the constant of proportionality. Then find the cost of 14 tickets.

32. **REASONING** The graph of a proportional relationship passes through (12, 16) and (1, y). Find y.

33. **PROBLEM SOLVING** The amount of chlorine in a swimming pool is proportional to the volume of water. The pool has 2.5 milligrams of chlorine per liter of water. How much chlorine is in the pool?

8000 gallons

34. **DIG DEEPER!** A vehicle travels 250 feet every 3 seconds. Find the value of the ratio, the unit rate, and the constant of proportionality. How are they related?

D.6 Scale Drawings

Learning Target: Solve problems involving scale drawings.

Success Criteria:
- I can find an actual distance in a scale drawing.
- I can explain the meaning of scale and scale factor.
- I can use a scale drawing to find the actual lengths and areas of real-life objects.

EXPLORATION 1

Creating a Scale Drawing

Work with a partner. Several sections in a zoo are drawn on 1-centimeter grid paper as shown. Each centimeter in the drawing represents 4 meters.

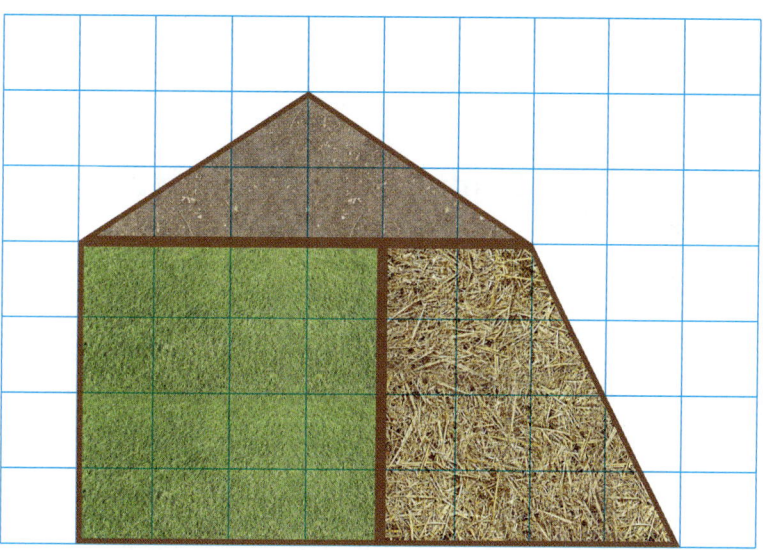

a. Describe the relationship between the lengths of the fences in the drawing and the actual side lengths of the fences.

b. Describe the relationship between the areas of the sections in the drawing and the actual areas of the sections.

c. Are the relationships in parts (a) and (b) the same? Explain your reasoning.

d. Choose a different distance to represent each centimeter on a piece of 1-centimeter grid paper. Then create a new drawing of the sections in the zoo using the distance you chose. Describe any similarities or differences in the drawings.

> **Math Practice**
>
> **Analyze Givens**
>
> How does the information given about the drawing shown help you create an accurate drawing in part (d)?

Section D.6 Scale Drawings **661**

D.6 Lesson

Key Vocabulary
scale drawing, *p. 662*
scale model, *p. 662*
scale, *p. 662*
scale factor, *p. 663*

Key Idea

Scale Drawings and Models
A **scale drawing** is a proportional, two-dimensional drawing of an object.
A **scale model** is a proportional, three-dimensional model of an object.

Scale
The measurements in scale drawings and models are proportional to the measurements of the actual object. The **scale** gives the ratio that compares the measurements of the drawing or model with the actual measurements.

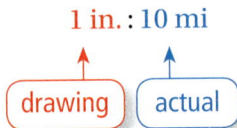

> Recall that a ratio $a : b$ is equivalent to $1 : \frac{b}{a}$.
>
> A scale is usually written as a ratio where the first quantity is 1 unit.

EXAMPLE 1 Finding an Actual Distance

What is the actual distance d between Cadillac and Detroit?

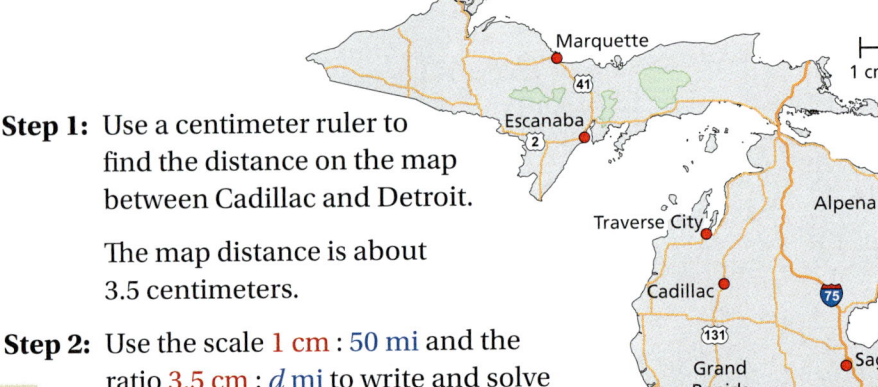

Step 1: Use a centimeter ruler to find the distance on the map between Cadillac and Detroit.

The map distance is about 3.5 centimeters.

Step 2: Use the scale 1 cm : 50 mi and the ratio 3.5 cm : d mi to write and solve a proportion.

$$\frac{1}{50} = \frac{3.5}{d}$$ ← map distance (cm) / actual distance (m)

$d = 50 \cdot 3.5$ Cross Products Property

$d = 175$ Multiply.

▶ So, the distance between Cadillac and Detroit is about 175 miles.

Another Method
You can use a ratio table.

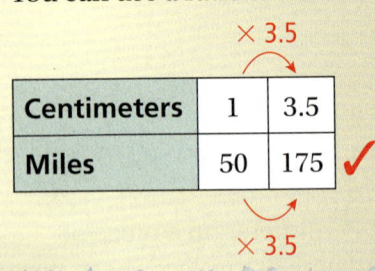

Centimeters	1	3.5
Miles	50	175 ✓

Try It

1. What is the actual distance between Traverse City and Marquette?

A scale can be written without units when the units are the same. The value of this ratio is called the **scale factor**. The scale factor describes the multiplicative relationship between the dimensions of a scale drawing or scale model and the dimensions of the actual object.

EXAMPLE 2 Finding a Scale Factor

A scale model of the Sergeant Floyd Monument is 10 inches tall. The actual monument is 100 feet tall.

a. **What does 1 inch represent in the model? What is the scale?**

The ratio of the model height to the actual height is 10 in. : 100 ft. Divide each quantity by 10 to determine the number of feet represented by 1 inch in the model.

$$\div 10 \begin{pmatrix} 10 \text{ in.} : 100 \text{ ft} \\ 1 \text{ in.} : 10 \text{ ft} \end{pmatrix} \div 10$$

▶ In the model, 1 inch represents 10 feet. So, the scale is 1 in. : 10 ft.

b. **What is the scale factor of the model?**

Write the scale with the same units. Use the fact that 1 ft = 12 in.

$$10 \text{ ft} = 10 \text{ ft} \times \frac{12 \text{ in.}}{1 \text{ ft}} = 120 \text{ in.}$$

▶ The scale is 1 in. : 120 in., or 1 : 120. So, the scale factor is $\frac{1}{120}$.

Try It

2. A drawing has a scale of 1 mm : 20 cm. What is the scale factor of the drawing?

Self-Assessment for Concepts & Skills

Solve each exercise. Then rate your understanding of the success criteria in your journal.

1 cm : 32 ft

3. **VOCABULARY** In your own words, explain the meaning of the scale and scale factor of a drawing or model.

4. **FINDING AN ACTUAL DISTANCE** Consider the scale drawing of Balanced Rock in Arches National Park. What is the actual height of the structure?

5. **FINDING A SCALE FACTOR** A drawing has a scale of 3 in. : 2 ft. What is the scale factor of the drawing?

6. **MP REASONING** Describe the scale factor of a model that is (a) larger than the actual object and (b) smaller than the actual object.

EXAMPLE 3 Modeling Real Life

1 cm : 2 mm

The scale drawing of a square computer chip helps you see the individual components on the chip.

a. **Find the perimeter and the area of the computer chip in the scale drawing.**

   When measured using a centimeter ruler, the scale drawing of the computer chip has a side length of 4 centimeters.

   ▸ So, the perimeter of the computer chip in the scale drawing is $4(4) = 16$ centimeters, and the area is $4^2 = 16$ square centimeters.

b. **Find the actual perimeter and area of the computer chip.**

   Multiplying each quantity in the scale by 4 shows that the actual side length of the computer chip is 8 millimeters.

   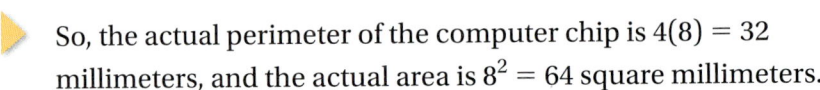

   ▸ So, the actual perimeter of the computer chip is $4(8) = 32$ millimeters, and the actual area is $8^2 = 64$ square millimeters.

c. **Compare the side lengths of the scale drawing with the actual side lengths of the computer chip.**

   Find the scale factor. Use the fact that 1 cm = 10 mm.

   Because the scale can be written as 10 mm : 2 mm, or 10 : 2, the scale factor is $\dfrac{10}{2} = 5$.

   ▸ So, the side lengths of the scale drawing are 5 times the actual side lengths of the computer chip.

 Self-Assessment for Problem Solving

Solve each exercise. Then rate your understanding of the success criteria in your journal.

Scale: 1 ft : 11.2 ft

7. A scale drawing of the Parthenon is shown. Find the actual perimeter and area of the rectangular face of the Parthenon. Then recreate the scale drawing with a scale factor of 0.2. Find the perimeter and area of the rectangular face in your drawing.

8. **DIG DEEPER!** You are in charge of creating a billboard advertisement that is 16 feet long and 8 feet tall. Choose a product. Create a scale drawing of the billboard using words and a picture. What is the scale factor of your design?

D.6 Practice

Go to **BigIdeasMath.com** to get HELP with solving the exercises.

▶ Review & Refresh

Tell whether x and y are proportional. Explain your reasoning.

1. | x | 10 | 9 | 8 | 7 |
    |---|----|---|---|---|
    | y | 5 | 4 | 3 | 2 |

2. | x | 6 | 12 | 18 | 24 |
    |---|---|----|----|----|
    | y | 7 | 14 | 21 | 28 |

Simplify the expression.

3. $7p + 6p$

4. $8 + 3d - 17$

5. $-2 + \frac{2}{5}b - \frac{1}{4}b + 6$

Write the word sentence as an inequality.

6. A number c is less than -3.

7. 7 plus a number z is more than 5.

8. The product of a number m and 6 is no less than 30.

▶ Concepts, Skills, & Problem Solving

CREATING A SCALE DRAWING Each centimeter on the 1-centimeter grid paper represents 8 inches. Create a proportional drawing of the figure that is larger or smaller than the figure shown. (See Exploration 1, p. 661.)

9.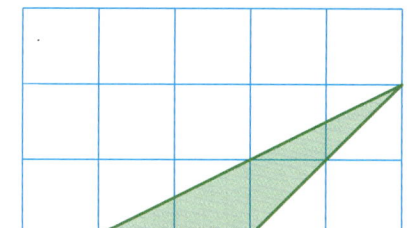

10.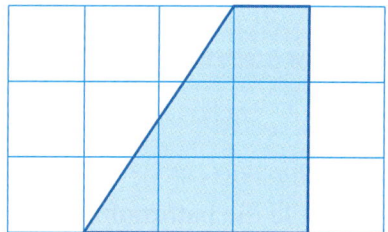

FINDING AN ACTUAL DISTANCE Use the map in Example 1 to find the actual distance between the cities.

11. Kalamazoo and Ann Arbor

12. Lansing and Flint

13. Grand Rapids and Escanaba

14. Saginaw and Alpena

USING A SCALE Find the missing dimension. Use the scale 1 : 12.

	Item	Model	Actual
15.	Mattress	Length: 6.25 in.	Length: in.
16.	Corvette	Length: in.	Length: 15 ft
17.	Water tower	Depth: 32 cm	Depth: m
18.	Wingspan	Width: 5.4 ft	Width: yd
19.	Football helmet	Diameter: mm	Diameter: 21 cm

Section D.5 Scale Drawings 665

FINDING A SCALE FACTOR Use a centimeter ruler to find the scale and the scale factor of the drawing.

20.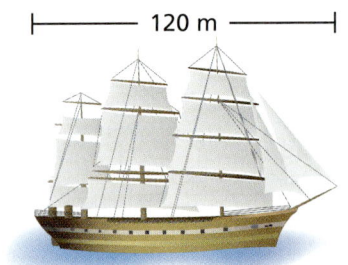

21.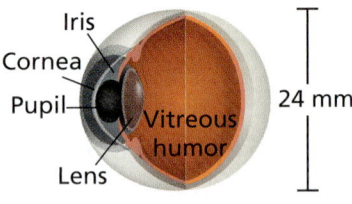

22. **CRITICAL THINKING** You know the length and the width of a scale model. What additional information do you need to know to find the scale of the model? Explain.

23. **MODELING REAL LIFE** Central Park is a rectangular park in New York City.

    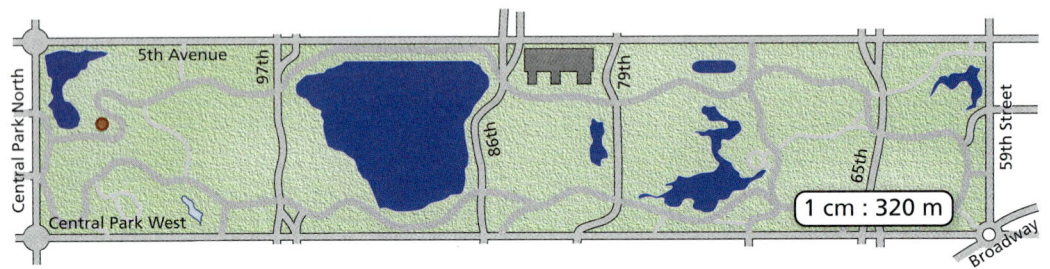

    a. Find the perimeter and the area of the scale drawing of Central Park.

    b. Find the actual perimeter and area of Central Park.

24. **MP PROBLEM SOLVING** In a blueprint, each square has a side length of $\frac{1}{4}$ inch.

    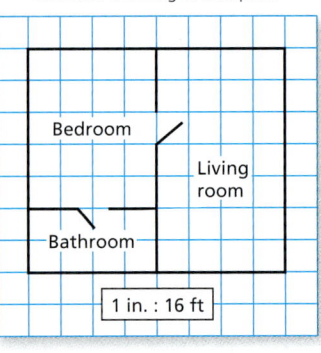

    Reduced Drawing of Blueprint

    a. Ceramic tile costs $5 per square foot. How much does it cost to tile the bathroom?

    b. Carpet costs $18 per square yard. How much does it cost to carpet the bedroom and living room?

REPRODUCING A SCALE DRAWING Recreate the scale drawing so that it has a scale of 1 cm : 4 m.

25.
1 cm : 8 m

26.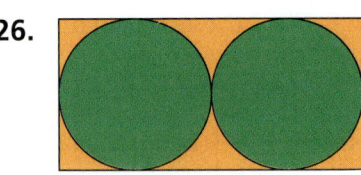
1 cm : 2 m

27. **DIG DEEPER!** Make a conjecture about the relationship between the scale factor of a drawing and the quotients $\frac{\text{drawing perimeter}}{\text{actual perimeter}}$ and $\frac{\text{drawing area}}{\text{actual area}}$. Explain your reasoning.

666 Chapter D Ratios and Proportions

D Connecting Concepts

Using the Problem-Solving Plan

1. The table shows the toll y (in dollars) for traveling x miles on a turnpike. You have $8.25 to pay your toll. How far can you travel on the turnpike?

Distance, x (miles)	25	30	35	40
Toll, y (dollars)	3.75	4.50	5.25	6.00

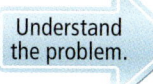

The table shows the tolls for traveling several different distances on a turnpike. You have $8.25 to pay the toll. You are asked to find how far you can travel on the turnpike with $8.25 for tolls.

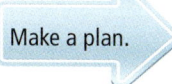

First, determine the relationship between x and y and write an equation to represent the relationship. Then use the equation to determine the distance you can travel.

Use the plan to solve the problem. Then check your solution.

2. A company uses a silo in the shape of a rectangular prism to store bird seed. The base of the silo is a square with side lengths of 20 feet. Are the height and the volume of the silo proportional? Justify your answer.

3. A rectangle is drawn in a coordinate plane as shown. In the same coordinate plane, create a scale drawing of the rectangle that has a vertex at (0, 0) and a scale factor of 3.

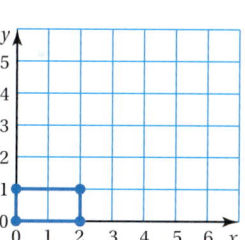

Performance Task

Mixing Paint

At the beginning of this chapter, you watched a STEAM Video called "Painting a Large Room." You are now ready to complete the performance task related to this video, available at *BigIdeasMath.com*. Be sure to use the problem-solving plan as you work through the performance task.

D Chapter Review

Go to *BigIdeasMath.com* to download blank graphic organizers.

▶ Review Vocabulary

Write the definition and give an example of each vocabulary term.

ratio, *p. 628*
value of a ratio, *p. 628*
equivalent ratios, *p. 629*
ratio table, *p. 629*
rate, *p. 634*
unit rate, *p. 634*
equivalent rates, *p. 634*
proportion, *p. 640*
cross products, *p. 641*
proportional, *p. 642*
constant of proportionality, *p. 656*
scale drawing, *p. 662*
scale model, *p. 662*
scale, *p. 662*
scale factor, *p. 663*

▶ Graphic Organizers

You can use an **Example and Non-Example Chart** to list examples and non-examples of a concept. Here is an Example and Non-Example Chart for *scale factor*.

Scale factor

Examples	Non-Examples
$\frac{5}{1}$	1 cm : 2 mm
$\frac{1}{200}$	1 mm : 20 cm
$\frac{1}{1}$	12 in. : 1 ft
$\frac{3}{2}$	3 mi : 2 in.

Choose and complete a graphic organizer to help you study the concept.

1. ratio
2. equivalent ratios
3. rate
4. unit rate
5. equivalent rates
6. proportion
7. cross products
8. proportional
9. scale

"What do you think of my **Example & Non-Example Chart** for popular cat toys?"

Chapter Self-Assessment

As you complete the exercises, use the scale below to rate your understanding of the success criteria in your journal.

1	2	3	4
I do not understand.	I can do it with help.	I can do it on my own.	I can teach someone else.

D.1 Ratios and Ratio Tables (pp. 627–632)

Learning Target: Understand ratios of rational numbers and use ratio tables to represent equivalent ratios.

Write the ratio. Then find and interpret the value of the ratio.

1. salt : flour
2. water to flour
3. salt to water

Modeling Clay
Ingredients:
2 cups flour $\frac{1}{2}$ cup salt $\frac{3}{4}$ cup water

Find the missing values in the ratio table. Then write the equivalent ratios.

4.
Flour (cups)	$\frac{3}{2}$	3		
Milk (cups)	$\frac{1}{2}$		$\frac{3}{2}$	2

5.
Miles	45	135		90
Hours	0.75		3	

6. The cost for 16 ounces of cheese is $3.20. What is the cost for 20 ounces of cheese?

D.2 Rates and Unit Rates (pp. 633–638)

Learning Target: Understand rates involving fractions and use unit rates to solve problems.

Find the unit rate.

7. 289 miles on 10 gallons
8. $6\frac{2}{5}$ revolutions in $2\frac{2}{3}$ seconds
9. You can mow 23,760 square feet in $\frac{1}{2}$ hour. How many square feet can you mow in 2 hours? Justify your answer.

Tell whether the rates are equivalent. Justify your answer.

10. 60 centimeters every 2.5 years
    30 centimeters every 15 months

11. $2.56 per $\frac{1}{2}$ pound
    $0.48 per 6 ounces

Chapter Review   669

D.3 Identifying Proportional Relationships (pp. 639–646)

Learning Target: Determine whether two quantities are in a proportional relationship.

Tell whether the ratios form a proportion.

12. 4 to 9 and 2 to 3

13. 12 : 22 and 18 : 33

14. $\frac{1}{2} : 2$ and $\frac{1}{4} : \frac{1}{10}$

15. 3.2 to 8 and 1.2 to 3

16. Tell whether x and y are proportional.

x	1	3	6	8
y	4	12	24	32

17. You can type 250 characters in 60 seconds. Your friend can type 375 characters in 90 seconds. Do these rates form a proportion? Explain.

D.4 Writing and Solving Proportions (pp. 647–654)

Learning Target: Use proportions to solve ratio problems.

Solve the proportion. Explain your choice of method.

18. $\frac{3}{8} = \frac{9}{x}$

19. $\frac{x}{4} = \frac{2}{5}$

20. $\frac{5}{12} = \frac{y}{15}$

21. $\frac{s+1}{4} = \frac{4}{8}$

Use the table to write a proportion.

22.

	Game 1	Game 2
Penalties	6	8
Minutes	12	m

23.

	Concert 1	Concert 2
Songs	15	18
Hours	2.5	h

24. Find the value of x so that the ratios 8 : 20 and 6 : x are equivalent.

25. Swamp gas consists primarily of methane, a chemical compound consisting of a 1 : 4 ratio of carbon to hydrogen atoms. If a sample of methane contains 1564 hydrogen atoms, how many carbon atoms are present in the sample?

D.5 Graphs of Proportional Relationships (pp. 655–660)

Learning Target: Represent proportional relationships using graphs and equations.

26. Tell whether x and y are proportional. Explain your reasoning.

x	−3	−1	1	3
y	6	2	−2	−6

27. The graph shows the number of visits your website received over the past 6 months. Interpret each plotted point in the graph. Then identify the unit rate.

Tell whether x and y are proportional. If so, identify the constant of proportionality. Explain your reasoning.

28. $x + y = 6$ 29. $y - x = 0$ 30. $\dfrac{x}{y} = 20$ 31. $x = y + 2$

32. The variables x and y are proportional. When $y = 4$, $x = \dfrac{1}{2}$. Find the constant of proportionality. Then write an equation that relates x and y.

D.6 Scale Drawings (pp. 661–666)

Learning Target: Solve problems involving scale drawings.

Find the missing dimension. Use the scale factor 1 : 20.

	Item	Model	Actual
33.	Basketball player	Height: in.	Height: 90 in.
34.	Dinosaur	Length: 3.75 ft	Length: ft

Use a centimeter ruler to find the scale and the scale factor of the drawing.

35. ⊢———— 30 in. ————⊣

36. ⊢— 7.5 in. —⊣

37. A scale model of a lighthouse has a scale of 1 in. : 8 ft. The scale model is 20 inches tall. How tall is the lighthouse?

D Practice Test

Find the unit rate.

1. 84 miles in 12 days
2. $2\frac{2}{5}$ kilometers in $3\frac{3}{4}$ minutes

Tell whether the ratios form a proportion.

3. 1 to 0.4 and 9 to 3.6
4. $2 : \frac{8}{3}$ and $\frac{2}{3} : 6$

Tell whether x and y are proportional. Explain your reasoning.

5.
x	2	4	6	8
y	10	20	30	40

6.
x	1	3	5	7
y	3	7	11	15

7. Use the table to write a proportion.

	Monday	Tuesday
Gallons	6	8
Miles	180	m

Solve the proportion.

8. $\frac{x}{8} = \frac{9}{4}$
9. $\frac{17}{4} = \frac{y}{6}$

Tell whether x and y are proportional. If so, identify the constant of proportionality. Explain your reasoning.

10. $xy - 11 = 5$
11. $\frac{y}{x} = 8$

12. A recipe calls for $\frac{2}{3}$ cup flour for every $\frac{1}{2}$ cup sugar. Write the ratio of sugar to flour. Then find and interpret the value of the ratio.

13. The graph shows the number of cycles of a crosswalk signal during the day and during the night.

    a. Write equations that relate x and y for both the day and night periods.

    b. Find how many more cycles occur during the day than during the night for a six-hour period.

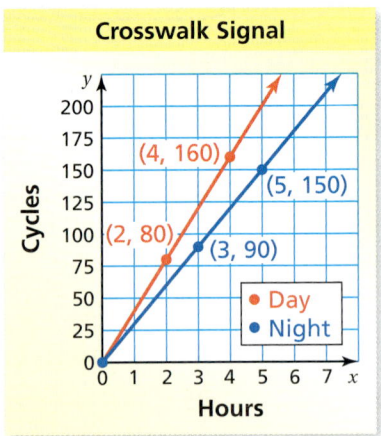

14. An engineer is using computer-aided design (CAD) software to design a component for a space shuttle. The scale of the drawing is 1 cm : 60 in. The actual length of the component is 12.75 feet. What is the length of the component in the drawing?

15. A specific shade of green glaze is made of 5 parts blue glaze to 3 parts yellow glaze. A glaze mixture contains 25 quarts of blue glaze and 9 quarts of yellow glaze. How can you fix the mixture to make the specific shade of green glaze?

D Cumulative Practice

1. The school store sells 4 pencils for $0.80. What is the unit cost of a pencil?

   A. $0.20 B. $0.80

   C. $3.20 D. $5.00

2. What is the simplified form of the expression?

   $$3x - (2x - 5)$$

   F. $x - 5$ G. $x + 5$

   H. $5x - 5$ I. $-x - 5$

3. Which fraction is equivalent to -1.25?

   A. $-12\frac{1}{2}$ B. $-1\frac{1}{4}$

   C. $-\frac{125}{1000}$ D. $1\frac{1}{4}$

4. What is the value of x for the proportion $\frac{8}{12} = \frac{x}{18}$?

5. What inequality is represented by the graph?

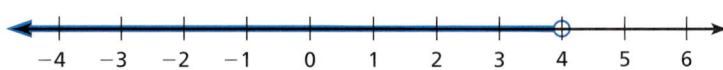

   F. $x - 3 < 7$ G. $x + 6 \leq 10$

   H. $-5 + x < -1$ I. $x - 8 > -4$

6. What is the missing value in the ratio table?

x	$\frac{2}{3}$	$\frac{4}{3}$	$\frac{8}{3}$	$\frac{10}{3}$
y	6	12	24	

   A. $24\frac{2}{3}$ B. 30

   C. 36 D. 48

7. Which expression shows factoring $12x + 54$ using the GCF?

   F. $2(6x + 27)$

   G. $3(4x + 18)$

   H. $6(2x + 9)$

   I. $12\left(x + \dfrac{9}{2}\right)$

8. The distance traveled by a high-speed train is proportional to the number of hours traveled. Which of the following is *not* a valid interpretation of the graph?

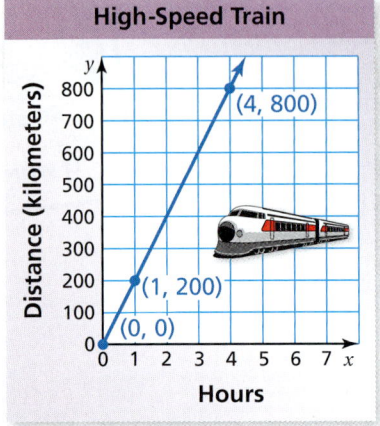

   A. The train travels 0 kilometers in 0 hours.

   B. The unit rate is 200 kilometers per hour.

   C. After 4 hours, the train is traveling 800 kilometers per hour.

   D. The train travels 800 kilometers in 4 hours.

9. Which graph represents a number that is at most -2?

   F.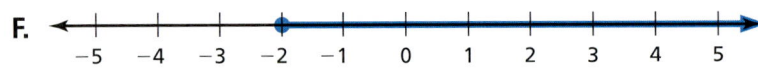

   G.

   H.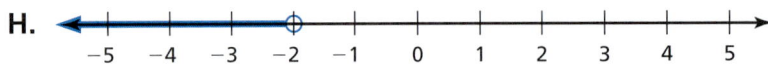

   I.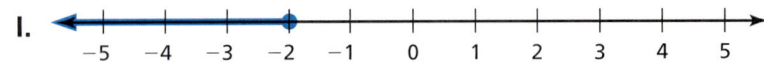

10. A map of the state where your friend lives has the scale $\frac{1}{2}$ in. : 10 mi.

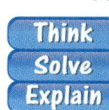

Part A Your friend measured the distance between her town and the state capital on the map. Her measurement was $4\frac{1}{2}$ inches. Based on your friend's measurement, what is the actual distance (in miles) between her town and the state capital? Show your work and explain your reasoning.

Part B Your friend wants to mark her favorite campsite on the map. She knows that the campsite is 65 miles north of her town. What distance on the map (in inches) represents an actual distance of 65 miles? Show your work and explain your reasoning.

11. What is the value of the expression $-56 \div (-8)$?

12. The quantities x and y are proportional. What is the missing value in the table?

x	y
$\frac{5}{7}$	10
$\frac{9}{7}$	18
$\frac{15}{7}$	30
4	

A. 38   **B.** 42

C. 46   **D.** 56

13. To begin a board game, you place a playing piece at START. On your first three turns, you move ahead 8 spaces, move back 3 spaces, and then move ahead 2 spaces. How many spaces are you from START?

F. 2   **G.** 3

H. 7   **I.** 13

E Percents

- **E.1** Fractions, Decimals, and Percents
- **E.2** The Percent Proportion
- **E.3** The Percent Equation
- **E.4** Percents of Increase and Decrease
- **E.5** Discounts and Markups
- **E.6** Simple Interest

Chapter Learning Target:
Understand fractions, decimals, and percents.

Chapter Success Criteria:
- I can rewrite fractions, decimals, and percents.
- I can compare and order fractions, decimals, and percents.
- I can use the percent proportion or percent equation to find a percent, a part, or a whole.
- I can apply percents to solve real-life problems.

STEAM Video: "Tornado!"

STEAM Video

Tornado!

More tornadoes occur each year in the United States than in any other country. How can you use a percent to describe the portion of tornadoes in the United States that occur in your state?

Watch the STEAM Video "Tornado!" Then answer the following questions.

1. The map below shows the average annual number of tornadoes in each state. Which regions have the most tornadoes? the fewest tornadoes?

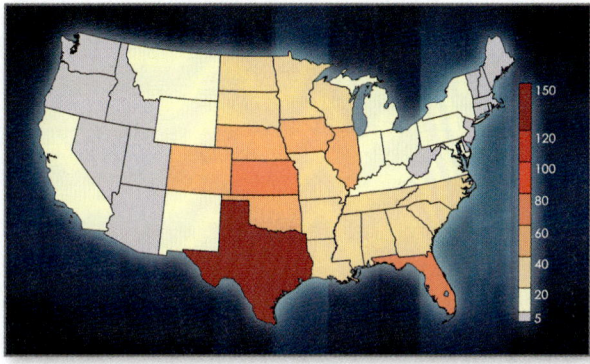

2. Robert says that only Alaska, Hawaii, and Rhode Island average less than 1 tornado per year. What percent of states average *more* than 1 tornado per year?

Performance Task

Tornado Alley

After completing this chapter, you will be able to use the concepts you learned to answer the questions in the *STEAM Video Performance Task*. You will be given information about the average annual numbers of tornadoes in several states over a 25-year period. For example:

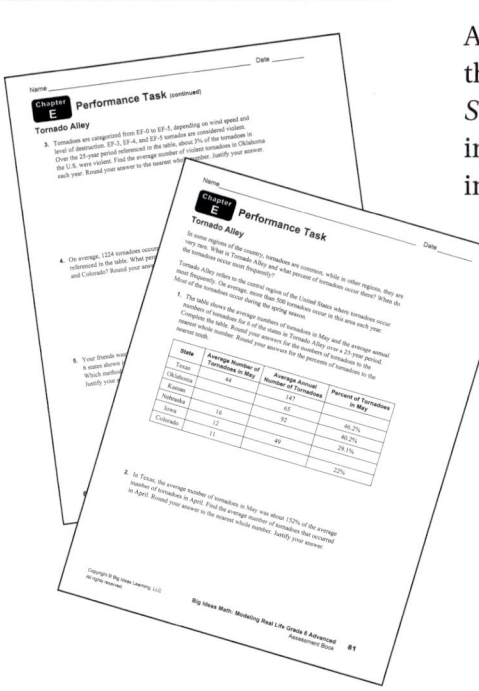

Texas: 147

Kansas: 92

Oklahoma: 65

Iowa: 49

You will be asked to solve various percent problems about tornadoes. Why is it helpful to know the percent of tornadoes that occur in each state?

677

Getting Ready for Chapter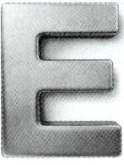

Chapter Exploration

Work with a partner. Write the percent of the model that is shaded. Then write the percent as a decimal.

1.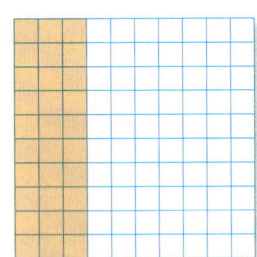

   ☐ % = ☐/☐ ← per / cent

   = ☐/☐   Simplify.

   = ☐   Write the fraction as a decimal.

2. 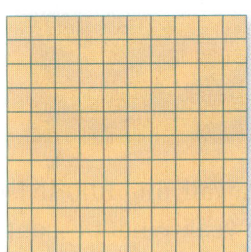   3. 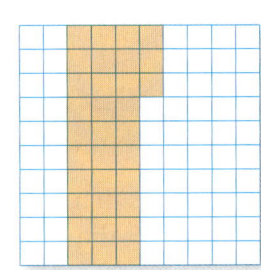   4.

5. 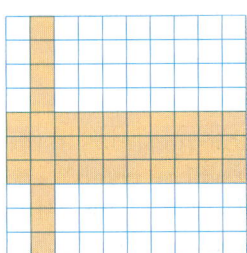   6. 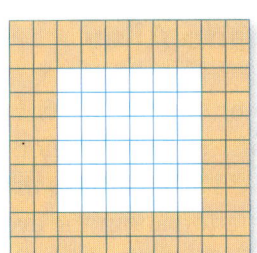   7.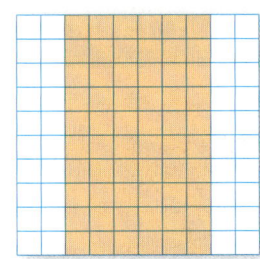

8. **WRITE A PROCEDURE** Work with a partner. Write a procedure for rewriting a percent as a decimal. Use examples to justify your procedure.

Vocabulary

The following vocabulary terms are defined in this chapter. Think about what each term might mean and record your thoughts.

| percent of change | percent of decrease | discount |
| percent of increase | percent error | markup |

E.1 Fractions, Decimals, and Percents

Learning Target: Rewrite fractions, decimals, and percents using different representations.

Success Criteria:
- I can write percents as decimals and decimals as percents.
- I can write fractions as decimals and percents.
- I can compare and order fractions, decimals, and percents.

EXPLORATION 1

Comparing Numbers in Different Forms

Work with a partner. Determine which number is greater. Explain your method.

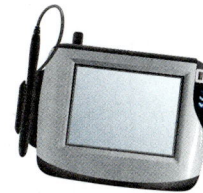

a. 7% sales tax   or $\dfrac{1}{20}$ sales tax

b. 0.37 cup of flour   or $\dfrac{1}{3}$ cup of flour

c. $\dfrac{5}{8}$-inch wrench   or   0.375-inch wrench

d. $12\dfrac{3}{5}$ dollars   or   12.56 dollars

e. $5\dfrac{5}{6}$ fluid ounces   or   5.6 fluid ounces

EXPLORATION 2

Ordering Fractions, Decimals, and Percents

Work with a partner and follow the steps below.

- Write five different numbers on individual slips of paper. Include at least one decimal, one fraction, and one percent.

- On a separate sheet of paper, create an answer key that shows your numbers written from least to greatest.

- Exchange slips of paper with another group and race to order the numbers from least to greatest. Then exchange answer keys to check your orders.

Math Practice

Make a Plan
Make a plan to order the numbers. How might having a plan help you to order numbers quickly?

E.1 Lesson

Key Ideas

Writing Percents as Decimals

Words Remove the percent symbol. Then divide by 100, which moves the decimal point two places to the left.

Numbers $82\% = 82.\cancel{\%} = 0.82$ $2.\overline{45}\% = 02.\overline{45}\cancel{\%} = 0.02\overline{45}$

Writing Decimals as Percents

Words Multiply by 100, which moves the decimal point two places to the right. Then add a percent symbol.

Numbers $0.47 = 0.47 = 47\%$ $0.\overline{2} = 0.222\ldots = 22.\overline{2}\%$

Remember
Bar notation indicates one or more repeating digits.

EXAMPLE 1   Converting Between Percents and Decimals

Write each percent as a decimal or each decimal as a percent. Use a model to represent each number.

a. $61\% = 61.\cancel{\%} = 0.61$

b. $8\% = 08.\cancel{\%} = 0.08$

c. $0.27 = 0.27 = 27\%$

d. $0.\overline{3} = 0.333\ldots = 33.\overline{3}\%$

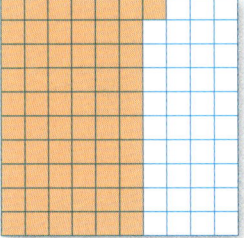
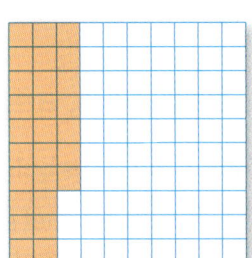
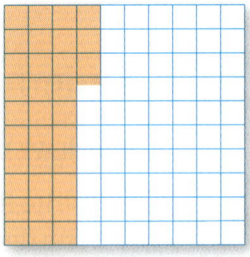

Try It Write the percent as a decimal or the decimal as a percent. Use a model to represent the number.

1. 39%   2. $12.\overline{6}\%$ 3. 0.05   4. 1.25

680   Chapter E   Percents

EXAMPLE 2 Writing Fractions as Decimals and Percents

Write each fraction as a decimal and a percent.

Remember
For a fraction with a denominator of 100, $\frac{n}{100} = n\%$.

a. $\frac{4}{5}$

$$\frac{4}{5} = \frac{4 \times 20}{5 \times 20} = \frac{80}{100} = 80\% = 0.8$$

▶ So, $\frac{4}{5}$ can be written as 0.8 or 80%.

b. $\frac{15}{11}$

Use long division to divide 15 by 11.

$\frac{15}{11} = 1.\overline{36}$

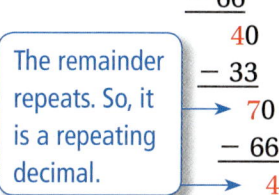

Write $1.\overline{36}$ as a percent.

$1.\overline{36} = 1.3636\ldots = 136.\overline{36}\%$

The remainder repeats. So, it is a repeating decimal.

▶ So, $\frac{15}{11}$ can be written as $1.\overline{36}$ or $136.\overline{36}\%$.

Try It Write the fraction as a decimal and a percent.

5. $\frac{5}{8}$ **6.** $\frac{1}{6}$ **7.** $\frac{11}{3}$ **8.** $\frac{3}{1000}$

Self-Assessment for Concepts & Skills

Solve each exercise. Then rate your understanding of the success criteria in your journal.

CONVERTING BETWEEN PERCENTS AND DECIMALS Write the percent as a decimal or the decimal as a percent. Use a model to represent the number.

9. 46% **10.** $66.\overline{6}\%$ **11.** 0.18 **12.** $2.\overline{3}$

WRITING FRACTIONS AS DECIMALS AND PERCENTS Write the fraction as a decimal and a percent.

13. $\frac{7}{10}$ **14.** $\frac{5}{9}$

15. $\frac{7}{2000}$ **16.** $\frac{17}{15}$

EXAMPLE 3 Modeling Real Life

An ice rink is open December through February. The table shows the attendance each month as a portion of the total attendance. How many times more guests visit the ice rink in the busiest month than in the least busy month?

Month	December	January	February
Portion of Guests	0.72	$\frac{3}{25}$	16%

Write $\frac{3}{25}$ and 16% as decimals.

January: $\frac{3}{25} = \frac{12}{100} = 0.12$ **February:** $16\% = 16.\% = 0.16$

The busiest month was December, the second busiest month was February, and the least busy month was January. So, divide 0.72 by 0.12.

$$0.12\overline{)0.72} \longrightarrow 12\overline{)72.}$$
$$\phantom{0.12\overline{)0.72}\longrightarrow}\underline{-72}$$
$$\phantom{0.12\overline{)0.72}\longrightarrow\,}0$$

Multiply each number by 100.

▸ So, 6 times more guests visit the ice rink in the busiest month than in the least busy month.

Self-Assessment for Problem Solving

Solve each exercise. Then rate your understanding of the success criteria in your journal.

17. An astronaut spends 53% of the day working, 0.1 of the day eating, $\frac{3}{10}$ of the day sleeping, and the rest of the day exercising. Order the events by duration from least to greatest. Justify your answer.

18. **DIG DEEPER!** A band plays one concert in Arizona, one concert in California, and one concert in Georgia. In California, the band earned $\frac{3}{2}$ the profit that they earned in Arizona. Of the total profit earned by the band, 32% is earned in Arizona. How many times more money did the band earn at the most profitable concert than at the least profitable concert? Justify your answer.

E.1 Practice

Go to *BigIdeasMath.com* to get HELP with solving the exercises.

Review & Refresh

Find the missing dimension. Use the scale 1 : 15.

	Item	Model	Actual
1.	Figure skater	Height: ___ in.	Height: 67.5 in.
2.	Pipe	Length: 5 ft	Length: ___ ft

Simplify the expression.

3. $2(3p - 6) + 4p$

4. $5n - 3(4n + 1)$

5. What is the solution of $2n - 4 > -12$?

  A. $n < -10$ **B.** $n < -4$ **C.** $n > -2$ **D.** $n > -4$

Concepts, Skills, & Problem Solving

COMPARING NUMBERS IN DIFFERENT FORMS Determine which number is greater. **Explain your method.** (See Exploration 1, p. 679.)

6. $4\frac{2}{5}$ tons or 4.3 tons

7. 82% success rate or $\frac{5}{6}$ success rate

CONVERTING BETWEEN PERCENTS AND DECIMALS Write the percent as a decimal or the decimal as a percent. Use a model to represent the number.

8. 26%   9. 0.63   10. 9%   11. 0.6

12. 44.7%   13. 55%   14. $39.\overline{2}$%   15. 3.554

16. 123%   17. 0.041   18. 0.122   19. $49.\overline{92}$%

20. **YOU BE THE TEACHER** Your friend writes $4.\overline{8}$% as a decimal. Is your friend correct? Explain your reasoning.

$$4.\overline{8}\% = 4.888\ldots\% = 488.\overline{8}$$

WRITING FRACTIONS AS DECIMALS AND PERCENTS Write the fraction as a decimal and a percent.

21. $\frac{29}{100}$ 22. $\frac{3}{4}$ 23. $\frac{7}{8}$ 24. $\frac{2}{3}$

25. $\frac{7}{9}$ 26. $\frac{12}{5}$ 27. $\frac{9}{2}$ 28. $\frac{1}{1000}$

29. $\frac{17}{6}$ 30. $\frac{3}{11}$ 31. $\frac{1}{750}$ 32. $\frac{22}{9}$

MP PRECISION Order the numbers from least to greatest.

33. 66.1%, 0.66, $\frac{2}{3}$, 0.667

34. $\frac{2}{9}$, 21%, $0.2\overline{1}$, $\frac{11}{50}$

MATCHING Tell which letter shows the graph of the number.

35. $\frac{7}{9}$
36. 0.812
37. $\frac{5}{6}$
38. 79.5%

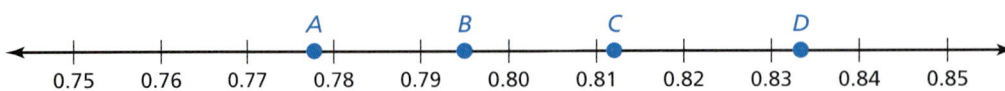

39. **MP PROBLEM SOLVING** The table shows the portion of students in each grade that participate in School Spirit Week. Order the grades by portion of participation from least to greatest.

Grade	Participation
6	0.64
7	$\frac{3}{5}$
8	65%

40. **MODELING REAL LIFE** The table shows the portion of gold medals that were won by the United States in five summer Olympic games. In what year did the United States win the least portion of gold medals? the greatest portion? Justify your answers.

Year	2000	2004	2008	2012	2016
Portion of Gold Medals Won	$12.\overline{3}\%$	$\frac{36}{301}$	$0.\overline{12}$	$\frac{23}{150}$	$\frac{46}{307}$

41. **MP PROBLEM SOLVING** You, your friend, and your cousin have a basketball competition where each person attempts the same number of shots. You make 70% of your shots, your friend makes $\frac{7}{9}$ of her shots, and your cousin makes $0.7\overline{2}$ of his shots. How many times more shots are made by the first place finisher than the third place finisher?

42. **DIG DEEPER!** Three different mixtures contain small amounts of acetic acid. Mixture A is 0.036 acetic acid, Mixture B is 4.2% acetic acid, and Mixture C is $\frac{1}{22}$ acetic acid. Explain how to use this information to determine which mixture contains the greatest amount of acetic acid.

43. **MODELING REAL LIFE** Over 44% of the 30 students in a class read a book last month. What are the possible numbers of students in the class who read a book last month? Justify your answer.

44. **MP NUMBER SENSE** Fill in the blanks using each of the numbers 0–7 exactly once, so that the percent, decimal, and fraction below are ordered from least to greatest. Justify your answer.

 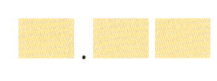 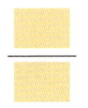

E.2 The Percent Proportion

Learning Target: Use the percent proportion to find missing quantities.

Success Criteria:
- I can write proportions to represent percent problems.
- I can solve a proportion to find a percent, a part, or a whole.

EXPLORATION 1

Using Percent Models

Work with a partner.

a. Complete each model. Explain what each model represents.

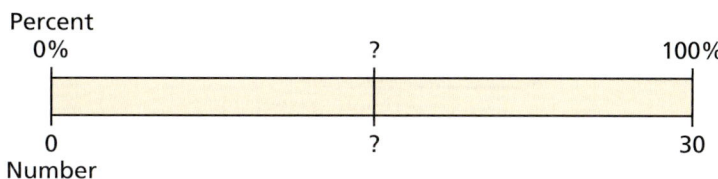

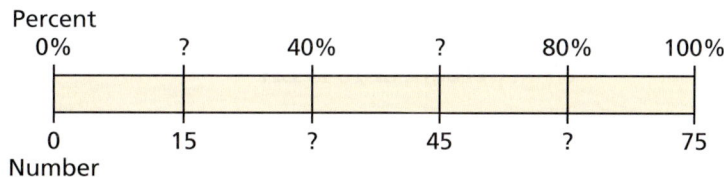

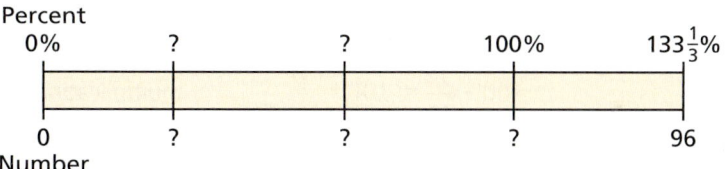

Math Practice

Use a Model
What quantities are given in each model? How can you use these quantities to answer the questions in part (b)?

b. Use the models in part (a) to answer each question.
- What number is 50% of 30?
- 15 is what percent of 75?
- 96 is $133\frac{1}{3}$% of what number?

c. How can you use ratio tables to check your answers in part (b)? How can you use proportions? Provide examples to support your reasoning.

d. Write a question different from those in part (b) that can be answered using one of the models in part (a). Trade questions with another group and find the solution.

Section E.2 The Percent Proportion 685

E.2 Lesson

Key Idea

The Percent Proportion

Words You can represent "a is p percent of w" with the proportion

$$\frac{a}{w} = \frac{p}{100}$$

where a is part of the whole w, and $p\%$, or $\frac{p}{100}$, is the percent.

Numbers   3 out of 4 is 75%.

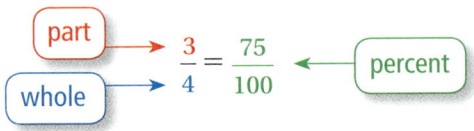

> In percent problems, the word *of* is usually followed by the whole.

EXAMPLE 1   Finding a Percent

What percent of 15 is 12?

$\dfrac{a}{w} = \dfrac{p}{100}$ Write the percent proportion.

$\dfrac{12}{15} = \dfrac{p}{100}$ Substitute 12 for a and 15 for w.

$100 \cdot \dfrac{12}{15} = 100 \cdot \dfrac{p}{100}$ Multiplication Property of Equality

$80 = p$ Simplify.

▶ So, 80% of 15 is 12.

You can also use a ratio table to find the percent.

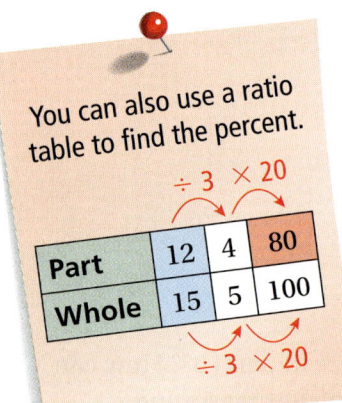

Check Use a model to check your answer.

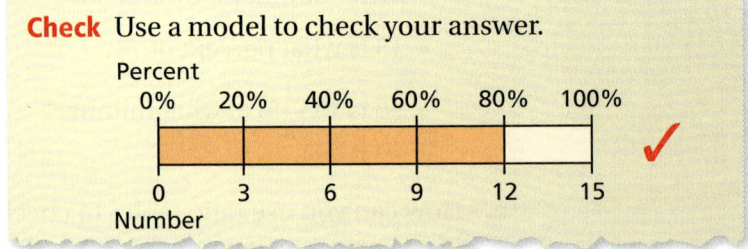

Try It Write and solve a proportion to answer the question.

1. What percent of 5 is 3?
2. 24 is what percent of 20?

686   Chapter E   Percents

EXAMPLE 2 Finding a Part

What number is 0.5% of 200?

$$\frac{a}{w} = \frac{p}{100}$$ Write the percent proportion.

$$\frac{a}{200} = \frac{0.5}{100}$$ Substitute 200 for *w* and 0.5 for *p*.

$$a = 1$$ Multiply each side by 200.

 So, 1 is 0.5% of 200.

Try It Write and solve a proportion to answer the question.

3. What number is 80% of 60?
4. 10% of 40.5 is what number?

EXAMPLE 3 Finding a Whole

150% of what number is 30?

$$\frac{a}{w} = \frac{p}{100}$$ Write the percent proportion.

$$\frac{30}{w} = \frac{150}{100}$$ Substitute 30 for *a* and 150 for *p*.

$$3000 = 150w$$ Cross Products Property

$$20 = w$$ Divide each side by 150.

 So, 150% of 20 is 30.

Try It Write and solve a proportion to answer the question.

5. 0.1% of what number is 4?
6. $\frac{1}{2}$ is 25% of what number?

 **Self-Assessment** for Concepts & Skills

Solve each exercise. Then rate your understanding of the success criteria in your journal.

7. **USING THE PERCENT PROPORTION** Write and solve a proportion to determine what percent of 120 is 54.

8. **MP CHOOSE TOOLS** Use a model to find 60% of 30.

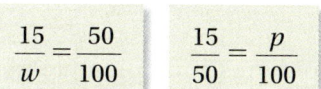

9. **WHICH ONE DOESN'T BELONG?** Which proportion at the left does *not* belong with the other three? Explain your reasoning.

EXAMPLE 4 Modeling Real Life

The bar graph shows the strengths of tornadoes that occurred in a state in a recent year. What percent of the tornadoes were EF1s?

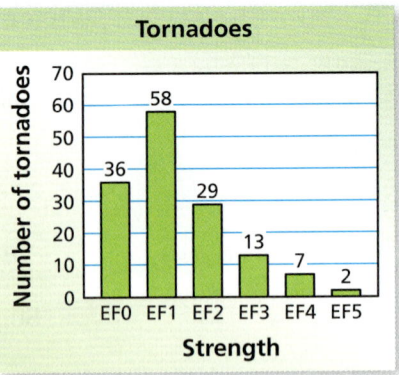

Understand the problem. You are given a bar graph that shows the number of tornadoes in each strength category. You are asked to find the percent of the tornadoes that were EF1s.

Make a plan. The total number of tornadoes, 145, is the whole, and the number of EF1 tornadoes, 58, is the part. Use the percent proportion to find the percent of the tornadoes that were EF1s.

Solve and check.

$$\frac{a}{w} = \frac{p}{100}$$ Write the percent proportion.

$$\frac{58}{145} = \frac{p}{100}$$ Substitute 58 for a and 145 for w.

$$100 \cdot \frac{58}{145} = 100 \cdot \frac{p}{100}$$ Multiplication Property of Equality

$$40 = p$$ Simplify.

▶ So, 40% of the tornadoes were EF1s.

Check Reasonableness
The number of EF1 tornadoes, 58, is less than half the total number of tornadoes, 145. So, the percent of the tornadoes that were EF1s should be less than 50%. Because 40% < 50%, the answer is reasonable. ✓

 **Self-Assessment** for Problem Solving

Solve each exercise. Then rate your understanding of the success criteria in your journal.

10. An arctic woolly-bear caterpillar lives for 7 years and spends 90% of its life frozen. How many days of its life is the arctic woolly-bear frozen?

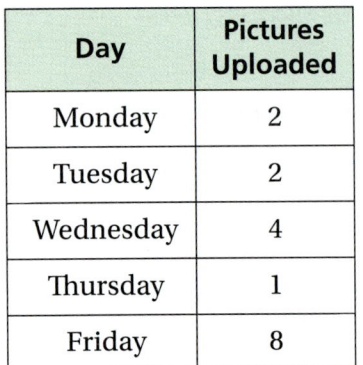

Day	Pictures Uploaded
Monday	2
Tuesday	2
Wednesday	4
Thursday	1
Friday	8

11. **DIG DEEPER!** The table shows the numbers of pictures you upload to a social media website for 5 days in a row. How many total pictures do you upload during the week when 32% of the total pictures are uploaded on Saturday and Sunday?

E.2 Practice

Go to *BigIdeasMath.com* to get HELP with solving the exercises.

▶ Review & Refresh

Write the fraction as a decimal and a percent.

1. $\dfrac{42}{100}$
2. $\dfrac{7}{1000}$
3. $\dfrac{13}{9}$
4. $\dfrac{41}{66}$

Evaluate the expression when $a = -15$ and $b = -5$.

5. $a \div b$
6. $\dfrac{b + 14}{a}$
7. $\dfrac{b^2}{a + 5}$

8. What is the solution of $9x = -1.8$?

   A. $x = -5$ **B.** $x = -0.2$ **C.** $x = 0.2$ **D.** $x = 5$

▶ Concepts, Skills, & Problem Solving

CHOOSE TOOLS Use a model to answer the question. Use a proportion to check your answer. *(See Exploration 1, p. 685.)*

9. What number is 20% of 80?
10. 10 is what percent of 40?
11. 15 is 30% of what number?
12. What number is 120% of 70?
13. 20 is what percent of 50?
14. 48 is 75% of what number?

USING THE PERCENT PROPORTION Write and solve a proportion to answer the question.

15. What percent of 25 is 12?
16. 14 is what percent of 56?
17. 25% of what number is 9?
18. 36 is 0.9% of what number?
19. 75% of 124 is what number?
20. 110% of 90 is what number?
21. What number is 0.4% of 40?
22. 72 is what percent of 45?

23. **YOU BE THE TEACHER** Your friend uses the percent proportion to answer the question below. Is your friend correct? Explain your reasoning.

    $\dfrac{a}{w} = \dfrac{p}{100}$

    $\dfrac{34}{w} = \dfrac{40}{100}$

    $w = 85$

    "40% of what number is 34?"

24. **MODELING REAL LIFE** Of 140 seventh-grade students, 15% earn the Presidential Youth Fitness Award. How many students earn the award?

25. **MODELING REAL LIFE** A salesperson receives a 3% commission on sales. The salesperson receives $180 in commission. What is the amount of sales?

Section E.2 The Percent Proportion 689

USING THE PERCENT PROPORTION Write and solve a proportion to answer the question.

26. 0.5 is what percent of 20?

27. 14.2 is 35.5% of what number?

28. $\frac{3}{4}$ is 60% of what number?

29. What number is 25% of $\frac{7}{8}$?

30. **MODELING REAL LIFE** You are assigned 32 math exercises for homework. You complete 75% of the exercises before dinner. How many exercises do you have left to do after dinner?

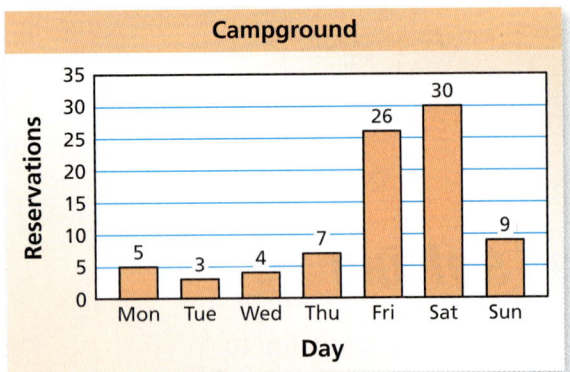

31. **MODELING REAL LIFE** Your friend earns $10.50 per hour, which is 125% of her hourly wage last year. How much did your friend earn per hour last year?

32. **MODELING REAL LIFE** The bar graph shows the numbers of reserved campsites at a campground for one week. What percent of the reservations were for Friday or Saturday?

33. **PROBLEM SOLVING** Your friend displays the results of a survey that asks several people to vote on a new school mascot.

   a. What is missing from the bar graph?

   b. What percent of the votes does the least popular mascot receive? Explain your reasoning.

   c. There are 124 votes total. How many votes does tiger receive?

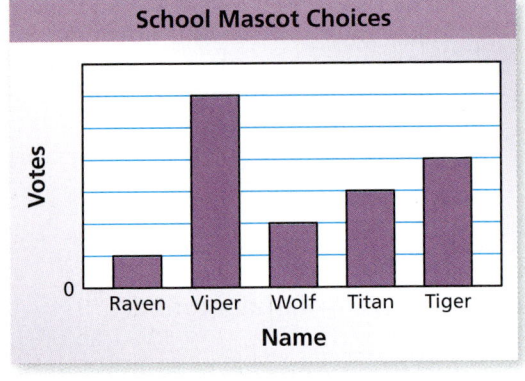

34. **DIG DEEPER!** A quarterback completes 18 of 33 passes during the first three quarters of a football game. He completes every pass in the fourth quarter and 62.5% of his passes for the entire game. How many passes does the quarterback throw in the fourth quarter? Justify your answer.

35. **REASONING** 20% of a number is x. What is 100% of the number? Assume $x > 0$.

36. **STRUCTURE** Answer each question. Assume $x > 0$.

   a. What percent of $8x$ is $5x$?

   b. What is 65% of $80x$?

E.3 The Percent Equation

Learning Target: Use the percent equation to find missing quantities.

Success Criteria:
- I can write equations to represent percent problems.
- I can use the percent equation to find a percent, a part, or a whole.

EXPLORATION 1

Using Percent Equations

Work with a partner.

a. The circle graph shows the number of votes received by each candidate during a school election. So far, only half of the students have voted. Find the percent of students who voted for each candidate. Explain your method.

Votes Received by Each Candidate

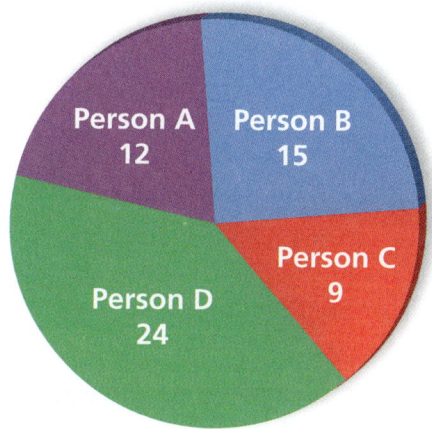

Math Practice

Use Equations

How does the equation you wrote in part (b) compare to the percent proportion? Explain.

b. You have learned that $\dfrac{\text{part}}{\text{whole}} = \text{percent}$. Solve the equation for the "part." Explain your reasoning.

c. The circle graph shows the final results of the election after every student voted. Use the equation you wrote in part (b) to find the number of students who voted for each candidate.

Final Results

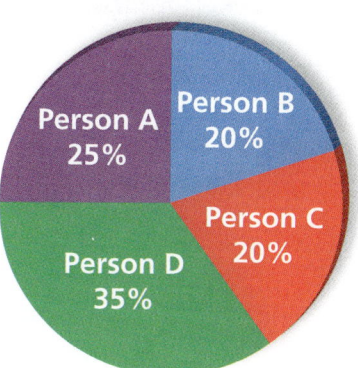

d. Use a different method to check your answers in part (c). Which method do you prefer? Explain.

E.3 Lesson

Key Idea

The Percent Equation

Words To represent "a is p percent of w," use an equation.

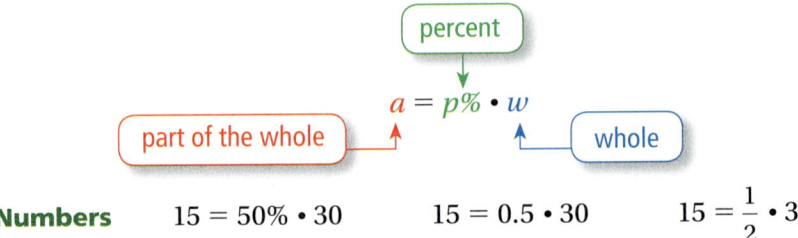

Numbers $15 = 50\% \cdot 30$ $15 = 0.5 \cdot 30$ $15 = \dfrac{1}{2} \cdot 30$

EXAMPLE 1   Finding a Part of a Number

What number is 24% of 50?   **Estimate**

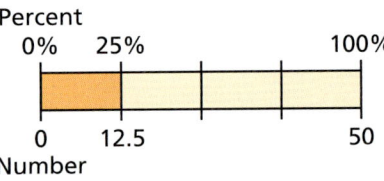

$a = p\% \cdot w$ — Write the percent equation.

$= \dfrac{24}{100} \cdot 50$ — Substitute $\dfrac{24}{100}$ for $p\%$ and 50 for w.

$= 12$ — Simplify.

▸ So, 12 is 24% of 50.   **Reasonable?** $12 \approx 12.5$

Common Error

Remember to convert a percent to a fraction or a decimal when using the percent equation. For Example 1, write 24% as $\dfrac{24}{100}$.

Try It Write and solve an equation to answer the question.

1. What number is 10% of 20?
2. What number is 150% of 40?

EXAMPLE 2   Finding a Percent

9.5 is what percent of 25?

$a = p\% \cdot w$ — Write the percent equation.

$9.5 = p\% \cdot 25$ — Substitute 9.5 for a and 25 for w.

$\dfrac{9.5}{25} = \dfrac{p\% \cdot 25}{25}$ — Division Property of Equality

$0.38 = p\%$ — Simplify.

▸ Because 0.38 equals 38%, 9.5 is 38% of 25.

692   Chapter E   Percents

Try It Write and solve an equation to answer the question.

3. 3 is what percent of 600?

4. 18 is what percent of 20?

EXAMPLE 3 Finding a Whole

39 is 52% of what number?

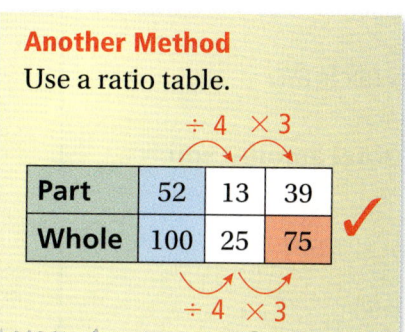

Another Method
Use a ratio table.

	÷4	×3	
Part	52	13	39
Whole	100	25	75

$a = p\% \cdot w$ Write the percent equation.

$39 = 0.52 \cdot w$ Substitute 39 for a and 0.52 for $p\%$.

$\dfrac{39}{0.52} = \dfrac{0.52 \cdot w}{0.52}$ Division Property of Equality

$75 = w$ Simplify.

▶ So, 39 is 52% of 75.

Try It Write and solve an equation to answer the question.

5. 8 is 80% of what number?

6. 90 is 180% of what number?

Self-Assessment for Concepts & Skills

Solve each exercise. Then rate your understanding of the success criteria in your journal.

7. **VOCABULARY** Write the percent equation in words.

USING THE PERCENT EQUATION Write and solve an equation to answer the question.

8. 14 is what percent of 70?

9. What number is 36% of 85?

10. 9 is 12% of what number?

11. 108 is what percent of 72?

12. **DIFFERENT WORDS, SAME QUESTION** Which is different? Find "both" answers.

What number is 20% of 55?	55 is 20% of what number?
20% of 55 is what number?	0.2 • 55 is what number?

Section E.3 The Percent Equation 693

EXAMPLE 4 Modeling Real Life

8th Street Cafe

DATE: MAY04   12:45PM
TABLE: 29
SERVER: JANE

Food Total   27.50
Tax   1.65
Subtotal   29.15
TIP: _____
TOTAL: _____

Thank You

You are paying for lunch and receive the bill shown.

a. Find the percent of sales tax on the food total.

Answer the question: $1.65 is what percent of $27.50?

$a = p\% \cdot w$	Write the percent equation.
$1.65 = p\% \cdot 27.50$	Substitute 1.65 for a and 27.50 for w.
$0.06 = p\%$	Divide each side by 27.50.

  Because 0.06 equals 6%, the percent of sales tax is 6%.

b. You leave a 16% tip on the food total. Find the total amount you pay for lunch.

Answer the question: What tip amount is 16% of $27.50?

$a = p\% \cdot w$	Write the percent equation.
$= 0.16 \cdot 27.50$	Substitute 0.16 for $p\%$ and 27.50 for w.
$= 4.40$	Multiply.

The amount of the tip is $4.40.

  So, you pay a total of $29.15 + $4.40 = $33.55.

Self-Assessment for Problem Solving

Solve each exercise. Then rate your understanding of the success criteria in your journal.

13. DIG DEEPER! A school offers band and chorus classes. The table shows the percents of the 1200 students in the school who are enrolled in band, chorus, or neither class. How many students are enrolled in both classes? Explain.

Class	Enrollment
Band	34%
Chorus	28%
Neither	42%

14. Water Tank A has a capacity of 550 gallons and is 66% full. Water Tank B is 53% full. The ratio of the capacity of Water Tank A to Water Tank B is 11 : 15.

  a. How much water is in each tank?

  b. What percent of the total volume of both tanks is filled with water?

E.3 Practice

Review & Refresh

Write and solve a proportion to answer the question.

1. 30% of what number is 9?
2. 42 is what percent of 80?
3. What percent of 36 is 20?
4. What number is 120% of 80?

Find the distance between the two numbers on a number line.

5. -4 and 10
6. $-\frac{2}{3}$ and $\frac{4}{3}$
7. $-5\frac{2}{5}$ and $-1\frac{3}{10}$
8. -4.3 and 7.5

9. There are 160 people in a grade. The ratio of boys to girls is 3 to 5. Which proportion can you use to find the number x of boys?

   A. $\frac{3}{8} = \frac{x}{160}$
   B. $\frac{3}{5} = \frac{x}{160}$
   C. $\frac{5}{8} = \frac{x}{160}$
   D. $\frac{3}{5} = \frac{160}{x}$

Concepts, Skills, & Problem Solving

USING PERCENT EQUATIONS The circle graph shows the number of votes received by each candidate during a school election. Find the percent of students who voted for the indicated candidate. *(See Exploration 1, p. 691.)*

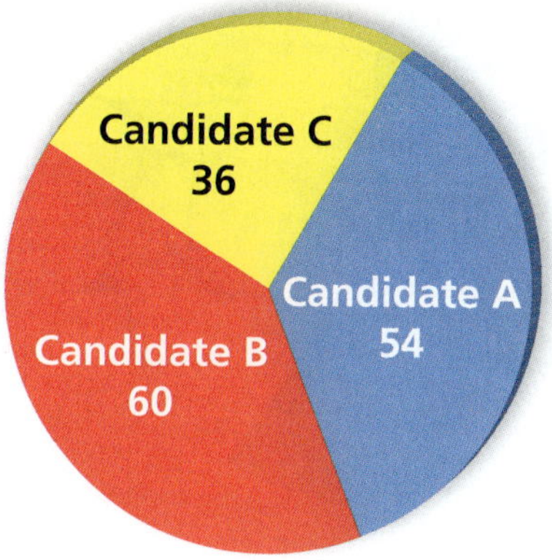

10. Candidate A
11. Candidate B
12. Candidate C

Votes Received by Each Candidate

Candidate C: 36
Candidate A: 54
Candidate B: 60

USING THE PERCENT EQUATION Write and solve an equation to answer the question.

13. 20% of 150 is what number?
14. 45 is what percent of 60?
15. 35% of what number is 35?
16. 0.8% of 150 is what number?
17. 29 is what percent of 20?
18. 0.5% of what number is 12?
19. What percent of 300 is 51?
20. 120% of what number is 102?

YOU BE THE TEACHER Your friend uses the percent equation to answer the question. Is your friend correct? Explain your reasoning.

21. What number is 35% of 20?

    $a = p\% \cdot w$
    $= 0.35 \cdot 20$
    $= 7$

22. 30 is 60% of what number?

    $a = p\% \cdot w$
    $= 0.6 \cdot 30$
    $= 18$

Section E.3 The Percent Equation 695

23. **MODELING REAL LIFE** A salesperson receives a 2.5% commission on sales. What commission does the salesperson receive for $8000 in sales?

24. **MODELING REAL LIFE** Your school raised 125% of its fundraising goal. The school raised $6750. What was the goal?

25. **MODELING REAL LIFE** The sales tax on the model rocket shown is $1.92. What is the percent of sales tax?

PUZZLE There were n signers of the Declaration of Independence. The youngest was Edward Rutledge, who was x years old. The oldest was Benjamin Franklin, who was y years old.

26. x is 25% of 104. What was Rutledge's age?

27. 7 is 10% of y. What was Franklin's age?

28. n is 80% of y. How many signers were there?

Favorite Sport

29. **LOGIC** How can you tell whether a percent of a number will be *greater than*, *less than*, or *equal to* the number? Give examples to support your answer.

30. **PROBLEM SOLVING** In a survey, a group of students is asked their favorite sport. Eighteen students choose "other" sports.
    a. How many students participate in the survey?
    b. How many choose football?

31. **TRUE OR FALSE?** Tell whether the statement is *true* or *false*. Explain your reasoning.

    If W is 25% of Z, then $Z : W$ is $75 : 25$.

32. **DIG DEEPER!** At a restaurant, the amount of your bill before taxes and tip is $19.83. A 6% sales tax is applied to your bill, and you leave a tip equal to 19% of the original amount. Use mental math to estimate the total amount of money you pay. Explain your reasoning. *(Hint: Use 10% of the original amount.)*

33. **REASONING** The table shows your test results in a math class. What score do you need on the last test to earn 90% of the total points on the tests?

Test Score	Point Value
83%	100
91.6%	250
88%	150
?	300

E.4 Percents of Increase and Decrease

Learning Target: Find percents of change in quantities.

Success Criteria:
- I can explain the meaning of percent of change.
- I can find the percent of increase or decrease in a quantity.
- I can find the percent error of a quantity.

EXPLORATION 1

Exploring Percent of Change

Work with a partner.

Each year in the Columbia River Basin, adult salmon swim upriver to streams to lay eggs.

To go up the river, the adult salmon use fish ladders. But to go down the river, the young salmon must pass through several dams.

At one time, there were electric turbines at each of the eight dams on the main stem of the Columbia and Snake Rivers. About 88% of the young salmon pass through a single dam unharmed.

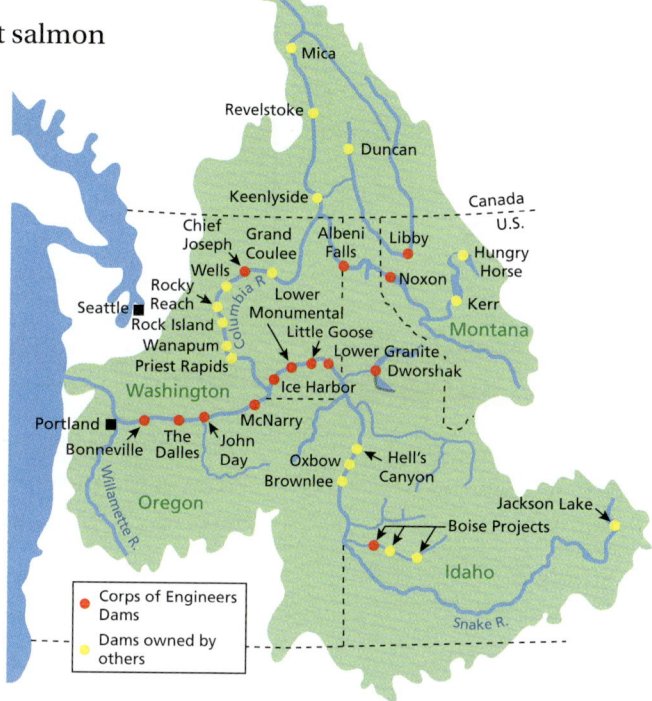

Math Practice

Check Progress

As the number of dams increases, what should be true about the number of young salmon that pass through unharmed?

a. One thousand young salmon pass through a dam. How many pass through unharmed?

b. One thousand young salmon pass through the river basin. How many pass through all 8 dams unharmed?

c. By what percent does the number of young salmon *decrease* when passing through a single dam?

d. Describe a similar real-life situation in which a quantity *increases* by a constant percent each time an event occurs.

Section E.4 Percents of Increase and Decrease 697

E.4 Lesson

Key Vocabulary
percent of change, p. 698
percent of increase, p. 698
percent of decrease, p. 698
percent error, p. 700

A **percent of change** is the percent that a quantity changes from the original amount.

$$\text{percent of change} = \frac{\text{amount of change}}{\text{original amount}}$$

 **Key Idea**

Percents of Increase and Decrease

When the original amount increases, the percent of change is called a **percent of increase**.

$$\text{percent of increase} = \frac{\text{new amount} - \text{original amount}}{\text{original amount}}$$

When the original amount decreases, the percent of change is called a **percent of decrease**.

$$\text{percent of decrease} = \frac{\text{original amount} - \text{new amount}}{\text{original amount}}$$

EXAMPLE 1 — Finding a Percent of Increase

Day	Hours Online
Saturday	2
Sunday	4.5

The table shows the numbers of hours you spent online last weekend. What is the percent of change in your time spent online from Saturday to Sunday?

The time spent online Sunday is greater than the time spent online Saturday. So, the percent of change is a percent of increase.

$$\text{percent of increase} = \frac{\text{new amount} - \text{original amount}}{\text{original amount}}$$

$$= \frac{4.5 - 2}{2} \qquad \text{Substitute.}$$

$$= \frac{2.5}{2} \qquad \text{Subtract.}$$

$$= 1.25, \text{ or } 125\% \qquad \text{Write as a percent.}$$

▶ So, your time spent online increased 125% from Saturday to Sunday.

Try It Find the percent of change. Round to the nearest tenth of a percent if necessary.

1. 10 inches to 25 inches
2. 57 people to 65 people

698   Chapter E   Percents   Multi-Language Glossary at *BigIdeasMath.com*

EXAMPLE 2 Finding a Percent of Decrease

The bar graph shows a softball player's home run totals. What was the percent of change from 2016 to 2017?

The number of home runs decreased from 2016 to 2017. So, the percent of change is a percent of decrease.

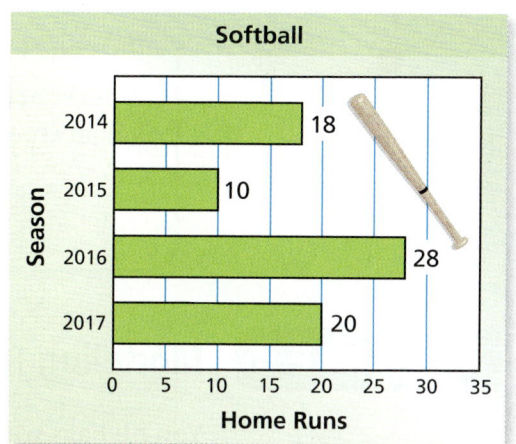

$$\text{percent of decrease} = \frac{\text{original amount} - \text{new amount}}{\text{original amount}}$$

$$= \frac{28 - 20}{28} \qquad \text{Substitute.}$$

$$= \frac{8}{28} \qquad \text{Subtract.}$$

$$\approx 0.286, \text{ or } 28.6\% \qquad \text{Write as a percent.}$$

▸ So, the number of home runs decreased about 28.6% from 2016 to 2017.

Try It

3. In Example 2, what was the percent of change from 2014 to 2015?

Self-Assessment for Concepts & Skills

Solve each exercise. Then rate your understanding of the success criteria in your journal.

4. **VOCABULARY** What does it mean for a quantity to change by $n\%$?

5. **MP NUMBER SENSE** Without calculating, determine which situation has a greater percent of change. Explain.
   - 5 bonus points added to 50 points
   - 5 bonus points added to 100 points

FINDING A PERCENT OF CHANGE Identify the percent of change as an *increase* or a *decrease*. Then find the percent of change.

6. 8 feet to 24 feet

7. 300 miles to 210 miles

Key Idea

Percent Error

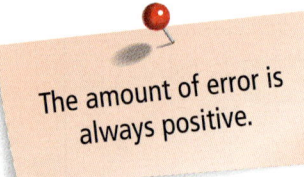

The amount of error is always positive.

A **percent error** is the percent that an estimated amount differs from the actual amount.

$$\text{percent error} = \frac{\text{amount of error}}{\text{actual amount}}$$

EXAMPLE 3 Modeling Real Life

You fill bags with about 16 ounces of homemade dog treats. The acceptable percent error when filling a bag is 5%. Tell whether each bag is acceptable.

Find the percent error for each bag.

Bag A: 15 ounces

Bag B: 16.5 ounces

Bag A: The amount of error is
$16 - 15 = 1$ ounce.

$$\text{percent error} = \frac{\text{amount of error}}{\text{actual amount}}$$

$$= \frac{1}{16}$$

$$\approx 0.0625, \text{ or } 6.25\%$$

Bag B: The amount of error is
$16.5 - 16 = 0.5$ ounce.

$$\text{percent error} = \frac{\text{amount of error}}{\text{actual amount}}$$

$$= \frac{0.5}{16}$$

$$\approx 0.03125, \text{ or } 3.125\%$$

▸ Because 6.25% > 5%, Bag A is not acceptable. Because 3.125% < 5%, Bag B is acceptable.

Self-Assessment for Problem Solving

Solve each exercise. Then rate your understanding of the success criteria in your journal.

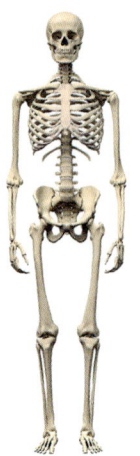

8. In one round of a game, you are asked how many bones are in a human body. If the percent error of your answer is at most 5%, you earn two points. If the percent error is at most 10%, but greater than 5%, you earn one point. You guess 195 bones. The correct answer is 206 bones. How many points do you earn?

9. **DIG DEEPER!** The manager of a restaurant offers a 20% decrease in price to tennis teams. A cashier applies a 10% decrease and then another 10% decrease. Is this the same as applying a 20% decrease? Justify your answer.

E.4 Practice

Go to *BigIdeasMath.com* to get HELP with solving the exercises.

▶ Review & Refresh

Write and solve an equation to answer the question.

1. What number is 25% of 64?
2. 39.2 is what percent of 112?
3. 5 is 5% of what number?
4. 18 is 32% of what number?

Find the sum. Write fractions in simplest form.

5. $\frac{4}{7} + \left(-\frac{6}{7}\right)$
6. $-4.621 + 3.925$
7. $-\frac{5}{12} + \frac{3}{4}$

▶ Concepts, Skills, & Problem Solving

EXPLORING PERCENT CHANGE You are given the percent of salmon that pass through a single dam unharmed. By what percent does the number of salmon decrease when passing through a single dam? (See Exploration 1, p. 697.)

8. 75%
9. 80%
10. 62%
11. 94%

FINDING A PERCENT OF CHANGE Identify the percent of change as an *increase* or a *decrease*. Then find the percent of change. Round to the nearest tenth of a percent if necessary.

12. 12 inches to 36 inches
13. 75 people to 25 people
14. 50 pounds to 35 pounds
15. 24 songs to 78 songs
16. 10 gallons to 24 gallons
17. 72 paper clips to 63 paper clips
18. 16 centimeters to 44.2 centimeters
19. 68 miles to 42.5 miles

20. **YOU BE THE TEACHER** Your friend finds the percent increase from 18 to 26. Is your friend correct? Explain your reasoning.

$\frac{26 - 18}{26} \approx 0.31 = 31\%$

21. **MODELING REAL LIFE** Last week, you finished Level 2 of a video game in 32 minutes. Today, you finish Level 2 in 28 minutes. What is the percent of change?

22. **MODELING REAL LIFE** You estimate that a baby pig weighs 20 pounds. The actual weight of the baby pig is 16 pounds. Find the percent error.

23. **MP PRECISION** A researcher estimates that a fossil is 3200 years old. Using *carbon-14 dating*, a procedure used to determine the age of an object, the researcher discovers that the fossil is 3600 years old.

    a. Find the percent error.
    b. What other estimate gives the same percent error? Explain your reasoning.

FINDING A PERCENT OF CHANGE Identify the percent of change as an *increase* or a *decrease*. Then find the percent of change. Round to the nearest tenth of a percent if necessary.

24. $\frac{1}{4}$ to $\frac{1}{2}$ 25. $\frac{4}{5}$ to $\frac{3}{5}$ 26. $\frac{3}{8}$ to $\frac{7}{8}$ 27. $\frac{5}{4}$ to $\frac{3}{8}$

28. **CRITICAL THINKING** Explain why a change from 20 to 40 is a 100% increase, but a change from 40 to 20 is a 50% decrease.

29. **MODELING REAL LIFE** The table shows population data for a community.

Year	Population
2011	118,000
2017	138,000

    a. What is the percent of change from 2011 to 2017?
    b. Predict the population in 2023. Explain your reasoning.

30. **GEOMETRY** Suppose the length and the width of the sandbox are doubled.

    a. Find the percent of change in the perimeter.
    b. Find the percent of change in the area.

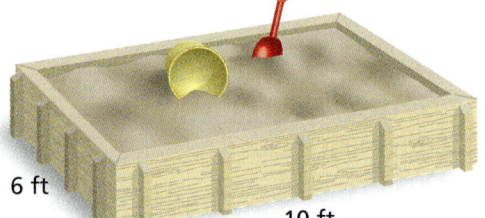

31. **MODELING REAL LIFE** A company fills boxes with about 21 ounces of cereal. The acceptable percent error in filling a box is 2.5%. Box A contains 20.4 ounces of cereal and Box B contains 21.5 ounces of cereal. Tell whether each box is an acceptable weight.

32. **MP PRECISION** Find the percent of change from June to September in the mile-run times shown.

33. **CRITICAL THINKING** A number increases by 10% and then decreases by 10%. Will the result be *greater than*, *less than*, or *equal to* the original number? Explain.

34. **MP PROBLEM SOLVING** You want to reduce your daily calorie consumption by about 9%. You currently consume about 2100 calories per day. Use mental math to estimate the number of calories you should consume in one week to meet your goal. Explain.

35. **DIG DEEPER!** Donations to an annual fundraiser are 15% greater this year than last year. Last year, donations were 10% greater than the year before. The amount raised this year is $10,120. How much was raised two years ago?

36. **MP REASONING** Forty students are in the science club. Of those, 45% are girls. This percent increases to 56% after more girls join the club. How many more girls join?

E.5 Discounts and Markups

Learning Target: Solve percent problems involving discounts and markups.

Success Criteria:
- I can use percent models to solve problems involving discounts and markups.
- I can write and solve equations to solve problems involving discounts and markups.

EXPLORATION 1

Comparing Discounts

Work with a partner.

a. The same pair of earrings is on sale at three stores. Which store has the best price? Use the percent models to justify your answer.

Store A:
Regular price: $45

Store B:
Regular price: $49

Store C:
Regular price: $39

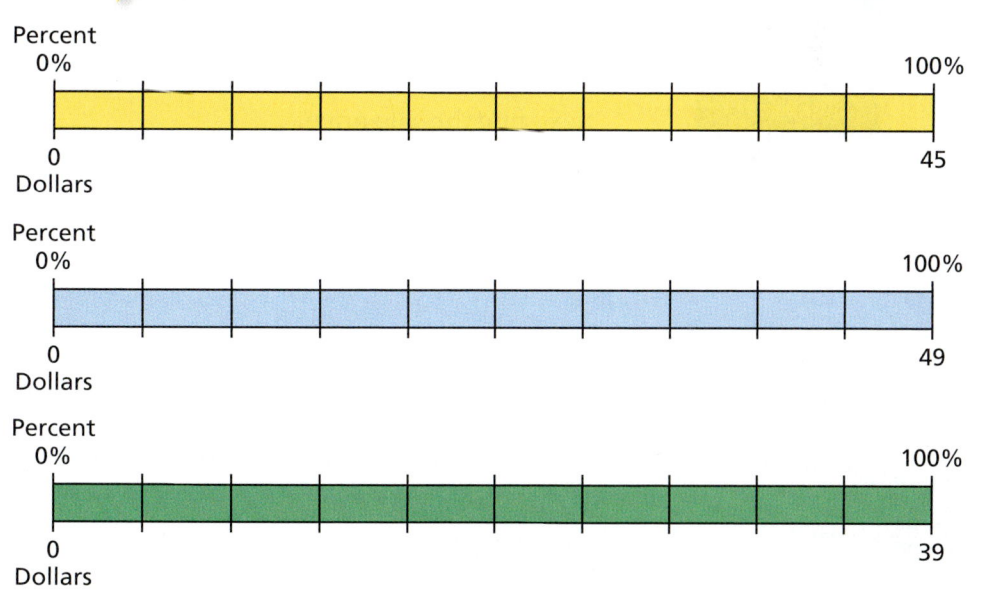

Math Practice

Communicate Precisely

Explain to your partner why 30% of the original price in part (b) is not the same as 30% of what you paid.

b. You buy the earrings on sale for 30% off at a different store. You pay $22.40. What was the original price of the earrings? Use the percent model to justify your answer.

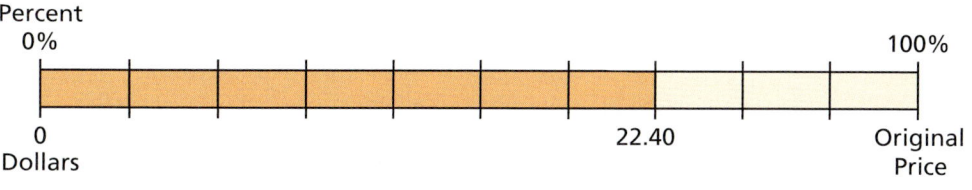

c. You sell the earrings in part (b) to a friend for 60% more than what you paid. What is the selling price? Use a percent model to justify your answer.

Section E.5 Discounts and Markups 703

E.5 Lesson

Key Vocabulary 🔊
discount, p. 704
markup, p. 704

 **Key Ideas**

Discounts
A **discount** is a decrease in the original price of an item.

Markups
To make a profit, stores charge more than what they pay. The increase from what the store pays to the selling price is called a **markup**.

EXAMPLE 1 Finding a Sale Price

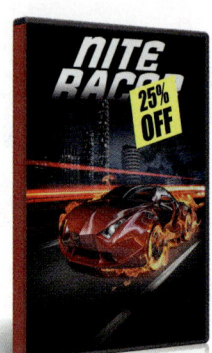

The original price of a video game is $35. What is the sale price?

Method 1: First, find the discount. The discount is 25% of $35.

$$a = p\% \cdot w \qquad \text{Write the percent equation.}$$
$$= 0.25 \cdot 35 \qquad \text{Substitute 0.25 for } p\% \text{ and 35 for } w.$$
$$= 8.75 \qquad \text{Multiply.}$$

Next, find the sale price.

Sale price = Original price − Discount
= 35 − 8.75
= 26.25

▶ So, the sale price is $26.25.

Method 2: Use the fact that the sale price is 100% − 25% = 75% of the original price.

Find the sale price.

Sale price = 75% of $35
= 0.75 • 35
= 26.25

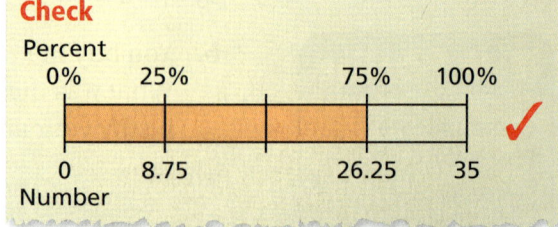

▶ So, the sale price is $26.25.

> For an item with an original price of p, the price after a 25% discount is $p - 0.25p$, or $0.75p$. So, a 25% discount is the same as paying 75% of the original price.

Try It

1. The original price of a skateboard is $50. The skateboard is on sale for 20% off. What is the sale price?

EXAMPLE 2 **Finding an Original Price**

What is the original price of the cleats?

The sale price is 100% − 40% = 60% of the original price.

Answer the question:

33 is 60% of what number?

$a = p\% \cdot w$	Write the percent equation.
$33 = 0.6 \cdot w$	Substitute 33 for a and 0.6 for $p\%$.
$55 = w$	Divide each side by 0.6.

 So, the original price of the cleats is $55.

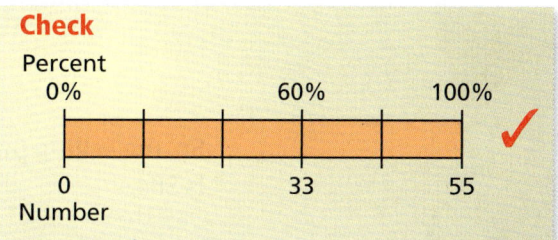

Try It

2. The discount on a DVD is 50%. It is on sale for $10. What is the original price of the DVD?

 **Self-Assessment** for Concepts & Skills

Solve each exercise. Then rate your understanding of the success criteria in your journal.

3. **WRITING** Describe how to find the sale price of an item that has a 15% discount.

FINDING A SALE PRICE Find the sale price. Use a percent model to check your answer.

4. A portable table tennis set costs $30 before a 30% discount.

5. The original price of an easel is $70. The easel is on sale for 20% off.

FINDING AN ORIGINAL PRICE Find the original price. Use a percent model to check your answer.

6. A bracelet costs $36 after a 25% discount.

7. The discount on a toy robot is 40%. The toy robot is on sale for $54.

Section E.5   Discounts and Markups   705

EXAMPLE 3 Modeling Real Life

A store pays $70 for a bicycle. What is the selling price when the markup is 20%?

Method 1: First, find the markup. The markup is 20% of $70.

$$a = p\% \cdot w$$
$$= 0.20 \cdot 70$$
$$= 14$$

Next, find the selling price.

Selling price = Cost to store + Markup

$$= 70 + 14$$
$$= 84$$

▶ So, the selling price is $84.

Method 2: Use a ratio table. The selling price is 120% of the cost to the store.

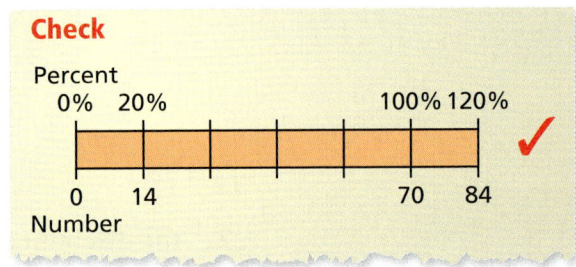

▶ So, the selling price is $84.

Check

Percent

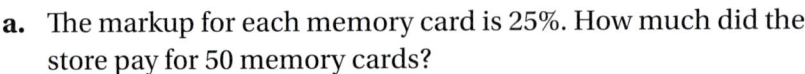

Self-Assessment for Problem Solving

Solve each exercise. Then rate your understanding of the success criteria in your journal.

8. **DIG DEEPER!** You have two coupons for a store. The first coupon applies a $15 discount to a single purchase, and the second coupon applies a 10% discount to a single purchase. You can only use one coupon on a purchase. When should you use each coupon? Explain.

9. A store sells memory cards for $25 each.

    a. The markup for each memory card is 25%. How much did the store pay for 50 memory cards?

    b. The store offers a discount when a customer buys two or more memory cards. A customer pays $47.50 for two memory cards. What is the percent of discount?

    c. How much does a customer pay for three memory cards if the store increases the percent of discount in part (b) by 2%?

E.5 Practice

▶ Review & Refresh

Identify the percent of change as an *increase* or a *decrease*. Then find the percent of change. Round to the nearest tenth of a percent if necessary.

1. 16 meters to 20 meters
2. 9 points to 4 points
3. 15 ounces to 5 ounces
4. 38 staples to 55 staples

Find the product. Write fractions in simplest form.

5. $\dfrac{4}{7}\left(-\dfrac{1}{6}\right)$
6. $-1.58(6.02)$
7. $-3\left(-2\dfrac{1}{8}\right)$

▶ Concepts, Skills, & Problem Solving

COMPARING DISCOUNTS The same item is on sale at two stores. Which one is the better price? Use percent models to justify your answer. *(See Exploration 1, p. 703.)*

8. 60% off $60 or 55% off $50
9. 85% off $90 or 70% off $65

USING TOOLS Copy and complete the table.

	Original Price	Percent of Discount	Sale Price
10.	$80	20%	
11.	$42	15%	
12.	$120	80%	
13.	$112	32%	
14.	$69.80	60%	
15.		25%	$40
16.		5%	$57
17.		80%	$90
18.		64%	$72
19.		15%	$146.54
20.	$60		$45
21.	$82		$65.60
22.	$95		$61.75

FINDING A SELLING PRICE Find the selling price.

23. Cost to store: $50
    Markup: 10%
24. Cost to store: $80
    Markup: 60%
25. Cost to store: $140
    Markup: 25%

Section E.5 Discounts and Markups 707

26. **YOU BE THE TEACHER** A store pays $60 for an item. Your friend finds the selling price when the markup is 20%. Is your friend correct? Explain your reasoning.

> 0.2($60) = $12
> So, the selling price is $12.

27. **MP STRUCTURE** The scooter is being sold at a 10% discount. The original price is shown. Which methods can you use to find the new sale price? Which method do you prefer? Explain.

   - Multiply $42.00 by 0.9.
   - Multiply $42.00 by 0.1, then subtract from $42.00.
   - Multiply $42.00 by 0.9, then add to $42.00.
   - Multiply $42.00 by 0.9, then subtract from $42.00.

28. **MP NUMBER SENSE** The original price of an item is p dollars. Is the price of the item with an 18% markup the same as multiplying the original price by 1.18? Use two expressions to justify your answer.

29. **MP PROBLEM SOLVING** You are shopping for a video game system.

   a. At which store should you buy the system?
   b. Store A has a weekend sale. What discount must Store A offer for you to buy the system there?

Store	Cost to Store	Markup
A	$162	40%
B	$155	30%
C	$160	25%

30. **DIG DEEPER!** A pool manager balances the pH level of a pool. The price of a bucket of chlorine tablets is $90, and the price of a pH test kit is $11. The manager uses a coupon that applies a 40% discount to the total cost of the two items. How much money does the pool manager pay for each item?

31. **MP PRECISION** You buy a pair of jeans at a department store.

   a. What is the percent of discount to the nearest percent?
   b. What is the percent of sales tax to the nearest tenth of a percent?
   c. The price of the jeans includes a 60% markup. After the discount, what is the percent of markup to the nearest percent?

   Department Store
   Jeans   39.99
   Discount -10.00
   Subtotal   29.99
   Sales Tax   1.95
   Total   31.94
   Thank You

32. **CRITICAL THINKING** You buy a bicycle helmet for $22.26, which includes 6% sales tax. The helmet is discounted 30% off the selling price. What is the original price?

33. **MP REASONING** A drone that costs $129.50 is discounted 40%. The next month, the sale price is discounted an additional 60%. Is the drone now "free"? If so, explain. If not, find the sale price.

708   Chapter E   Percents

E.6 Simple Interest

Learning Target: Understand and apply the simple interest formula.

Success Criteria:
- I can explain the meaning of simple interest.
- I can use the simple interest formula to solve problems.

EXPLORATION 1

Understanding Simple Interest

Work with a partner. You deposit $150 in an account that earns 6% *simple interest per year*. You do not make any other deposits or withdrawals. The table shows the balance of the account at the end of each year.

Years	Balance
0	$150
1	$159
2	$168
3	$177
4	$186
5	$195
6	$204

a. Describe any patterns you see in the account balance.

b. How is the amount of interest determined each year?

c. How can you find the amount of simple interest earned when you are given an initial amount, an interest rate, and a period of time?

d. You deposit $150 in a different account that earns simple interest. The table shows the balance of the account each year. What is the interest rate of the account? What is the balance after 10 years?

Years	0	1	2	3
Balance	$150	$165	$180	$195

Math Practice

Look for Patterns
How does the pattern in the balances help you find the simple interest rate?

Section E.6 Simple Interest 709

E.6 Lesson

Key Vocabulary
interest, p. 710
principal, p. 710
simple interest, p. 710

Interest is money paid or earned for using or lending money. The **principal** is the amount of money borrowed or deposited.

 **Key Idea**

Simple Interest

Words **Simple interest** is money paid or earned only on the principal.

Algebra

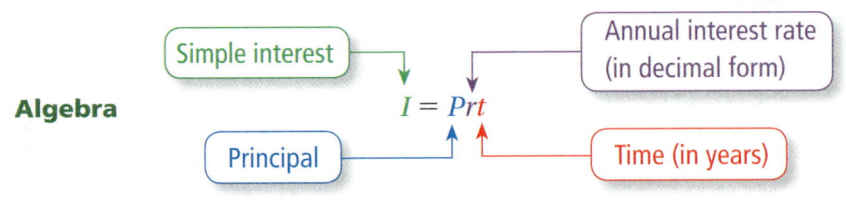

$I = Prt$

Reading
An interest rate per year is also called an annual interest rate.

EXAMPLE 1 Finding a Balance

You deposit $500 in a savings account. The account earns 3% simple interest per year. What is the balance after 3 years?

To find the balance, calculate the interest and add it to the principal.

$I = Prt$ Write the simple interest formula.

$= 500(0.03)(3)$ Substitute 500 for P, 0.03 for r, and 3 for t.

$= 45$ Multiply.

The interest earned is $45 after 3 years.

▶ So, the balance is $500 + $45 = $545 after 3 years.

Try It

1. What is the balance of the account after 9 months?

EXAMPLE 2 Finding an Annual Interest Rate

You deposit $1000 in an account. The account earns $100 simple interest in 4 years. What is the annual interest rate?

$I = Prt$ Write the simple interest formula.

$100 = 1000(r)(4)$ Substitute 100 for I, 1000 for P, and 4 for t.

$100 = 4000r$ Simplify.

$0.025 = r$ Divide each side by 4000.

▶ So, the annual interest rate of the account is 0.025, or 2.5%.

Try It

2. You deposit $350 in an account. The account earns $17.50 simple interest in 2.5 years. What is the annual interest rate?

EXAMPLE 3 Finding an Amount of Time

A bank offers three savings accounts. The simple annual interest rate is determined by the principal. How long does it take an account with a principal of $800 to earn $100 in interest?

The diagram shows that the interest rate for a principal of $800 is 2%.

$I = Prt$	Write the simple interest formula.
$100 = 800(0.02)(t)$	Substitute 100 for I, 800 for P, and 0.02 for r.
$100 = 16t$	Simplify.
$6.25 = t$	Divide each side by 16.

  So, the account earns $100 in interest in 6.25 years.

Try It

3. In Example 3, how long does it take an account with a principal of $10,000 to earn $750 in interest?

Self-Assessment for Concepts & Skills

Solve each exercise. Then rate your understanding of the success criteria in your journal.

4. **VOCABULARY** Explain the meaning of simple interest.

USING THE SIMPLE INTEREST FORMULA Use the simple interest formula.

5. You deposit $20 in a savings account. The account earns 4% simple interest per year. What is the balance after 4 years?

6. You deposit $800 in an account. The account earns $360 simple interest in 3 years. What is the annual interest rate?

7. You deposit $650 in a savings account. How long does it take an account with an annual interest rate of 5% to earn $178.25 in interest?

EXAMPLE 4 Modeling Real Life

You borrow $600 to buy a violin. The simple annual interest rate is 15%. You pay off the loan after 2 years of equal monthly payments. How much is each payment?

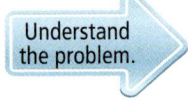 You are given the amount and simple annual interest rate of a loan that you pay back in 2 years. You are asked to find the monthly payment.

 Use the simple interest formula to find the interest you pay on the loan. Then divide the total amount you pay by the number of months in 2 years.

Solve and check.

$I = Prt$	Write the simple interest formula.
$= 600(0.15)(2)$	Substitute 600 for P, 0.15 for r, and 2 for t.
$= 180$	Multiply.

You pay $600 + $180 = $780 for the loan.

 So, each monthly payment is $\dfrac{780}{24} = \$32.50$.

Look Back When you substitute 600 for P and 0.15 for r, you obtain $I = 90t$. This indicates that you pay $90 in interest each year. So, in 2 years you pay $2(90) = \$180$ in interest. ✓

Self-Assessment for Problem Solving

Solve each exercise. Then rate your understanding of the success criteria in your journal.

8. You want to deposit $1000 in a savings account for 3 years. One bank adds a $100 bonus to your principal and offers a 2% simple annual interest rate. Another bank does not add a bonus, but offers 6% simple interest per year. Which bank should you choose? Explain.

9. Your cousin borrows $1125 to repair her car. The simple annual interest rate is 10%. She makes equal monthly payments of $25. How many years will it take to pay off the loan?

10. **DIG DEEPER!** You borrow $900 to buy a laptop. You plan to pay off the loan after 5 years of equal monthly payments. After 10 payments, you have $1200 left to pay. What is the simple annual interest rate of your loan?

E.6 Practice

Go to *BigIdeasMath.com* to get HELP with solving the exercises.

▶ Review & Refresh

Find the selling price.

1. A store pays $8 for a pool noodle. The markup is 20%.
2. A store pays $3 for a magazine. The markup is 5%.

Solve the inequality. Graph the solution.

3. $x + 5 < 2$
4. $b - 2 \geq -1$
5. $w + 6 \leq -3$

▶ Concepts, Skills, & Problem Solving

UNDERSTANDING SIMPLE INTEREST The table shows the balance of an account each year. What is the interest rate of the account? What is the balance after 10 years?
(See Exploration 1, p. 709.)

6.

Years	Balance
0	$40
1	$42
2	$44
3	$46

7.

Years	Balance
0	$175
1	$189
2	$203
3	$217

FINDING INTEREST EARNED An account earns simple annual interest.
(a) Find the interest earned. (b) Find the balance of the account.

8. $600 at 5% for 2 years
9. $1500 at 4% for 5 years
10. $350 at 3% for 10 years
11. $1800 at 6.5% for 30 months
12. $925 at 2.3% for 2.4 years
13. $5200 at 7.36% for 54 months

14. **YOU BE THE TEACHER** Your friend finds the simple interest earned on $500 at 6% for 18 months. Is your friend correct? Explain your reasoning.

$$I = (500)(0.06)(18)$$
$$= \$540$$

FINDING AN ANNUAL INTEREST RATE Find the annual interest rate.

15. $I = \$24$, $P = \$400$, $t = 2$ years
16. $I = \$562.50$, $P = \$1500$, $t = 5$ years
17. $I = \$54$, $P = \$900$, $t = 18$ months
18. $I = \$160$, $P = \$2000$, $t = 8$ months

FINDING AN AMOUNT OF TIME Find the amount of time.

19. $I = \$30$, $P = \$500$, $r = 3\%$
20. $I = \$720$, $P = \$1000$, $r = 9\%$
21. $I = \$54$, $P = \$800$, $r = 4.5\%$
22. $I = \$450$, $P = \$2400$, $r = 7.5\%$

23. **FINDING AN ACCOUNT BALANCE** A savings account earns 5% simple interest per year. The principal is $1200. What is the balance after 4 years?

24. **FINDING AN ANNUAL INTEREST RATE** You deposit $400 in an account. The account earns $18 simple interest in 9 months. What is the annual interest rate?

25. **FINDING AN AMOUNT OF TIME** You deposit $3000 in a CD (certificate of deposit) that earns 5.6% simple annual interest. How long will it take to earn $336 in interest?

FINDING AN AMOUNT PAID Find the amount paid for the loan.

26. $1500 at 9% for 2 years

27. $2000 at 12% for 3 years

28. $2400 at 10.5% for 5 years

29. $4800 at 9.9% for 4 years

USING THE SIMPLE INTEREST FORMULA Copy and complete the table.

	Principal	Annual Interest Rate	Time	Simple Interest
30.	$12,000	4.25%	5 years	
31.		6.5%	18 months	$828.75
32.	$15,500	8.75%		$5425.00
33.	$18,000		54 months	$4252.50

34. **MODELING REAL LIFE** A family borrows money for a rainforest tour. The simple annual interest rate is 12%. The loan is paid after 3 months. What is the total amount paid for the tour?

Rainforest Tour
Tickets $940
Food $170
Supplies $120

35. **MODELING REAL LIFE** You deposit $5000 in an account earning 7.5% simple interest per year. How long will it take for the balance of the account to be $6500?

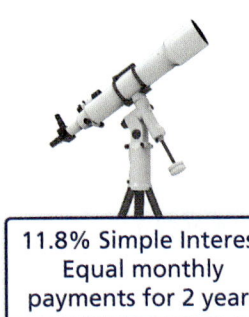

11.8% Simple Interest
Equal monthly payments for 2 years

36. **MODELING REAL LIFE** You borrow $1300 to buy a telescope. What is the monthly payment?

37. **MP REASONING** How many years will it take for $2000 to double at a simple annual interest rate of 8%? Explain how you found your answer.

38. **DIG DEEPER!** You take out two loans. After 2 years, the total interest for the loans is $138. On the first loan, you pay 7.5% simple annual interest on a principal of $800. On the second loan, you pay 3% simple annual interest. What is the principal for the second loan?

39. **MP REPEATED REASONING** You deposit $500 in an account that earns 4% simple annual interest. The interest earned each year is added to the principal to create a new principal. Find the total amount in your account after each year for 3 years.

40. **MP NUMBER SENSE** An account earns r% simple interest per year. Does doubling the initial principal have the same effect on the total interest earned as doubling the amount of time? Justify your answer.

714 Chapter E Percents

E Connecting Concepts

Using the Problem-Solving Plan

1. The table shows the percent of successful shots for each team in a hockey game. A total of 55 shots are taken in the game. The ratio of shots taken by the Blazers to shots taken by the Hawks is 6 : 5. How many goals does each team score?

Team	Percent of Successful Shots
Blazers	10%
Hawks	16%

   Understand the problem. You know that 55 shots are taken in a hockey game and that the Blazers take 6 shots for every 5 shots taken by the Hawks. You also know the percent of successful shots for each team.

   Make a plan. Use a ratio table to determine the number of shots taken by each team. Then use the percent equation to determine the number of successful shots for each team.

   Solve and check. Use the plan to solve the problem. Then check your solution.

2. Fill in the blanks with positive numbers so that the sum of the fractions is 37.5% of the first fraction. Justify your answer.

$$\frac{\boxed{}}{5} + \left(-\frac{\boxed{}}{4}\right)$$

3. The graph shows the distance traveled by a motorcycle on a dirt road. After turning onto a paved road, the motorcycle travels $\frac{1}{5}$ mile every $\frac{1}{4}$ minute. Find the percent of change in the speed of the motorcycle. Round to the nearest tenth of a percent if necessary.

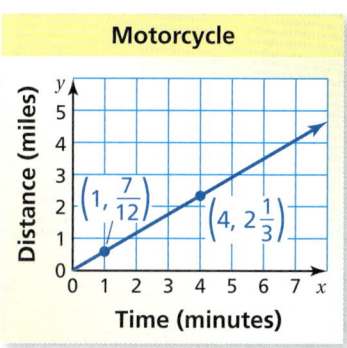

Motorcycle graph showing points $(1, \frac{7}{12})$ and $(4, 2\frac{1}{3})$; Distance (miles) vs Time (minutes).

Performance Task

Tornado Alley

At the beginning of this chapter, you watched a STEAM Video called "Tornado!" You are now ready to complete the performance task related to this video, available at *BigIdeasMath.com*. Be sure to use the problem-solving plan as you work through the performance task.

E Chapter Review

Go to *BigIdeasMath.com* to download blank graphic organizers.

▶ Review Vocabulary

Write the definition and give an example of each vocabulary term.

percent of change, *p. 698*
percent of increase, *p. 698*
percent of decrease, *p. 698*
percent error, *p. 700*
discount, *p. 704*
markup, *p. 704*
interest, *p. 710*
principal, *p. 710*
simple interest, *p. 710*

▶ Graphic Organizers

You can use a **Summary Triangle** to explain a concept. Here is an example of a Summary Triangle for *writing a percent as a decimal*.

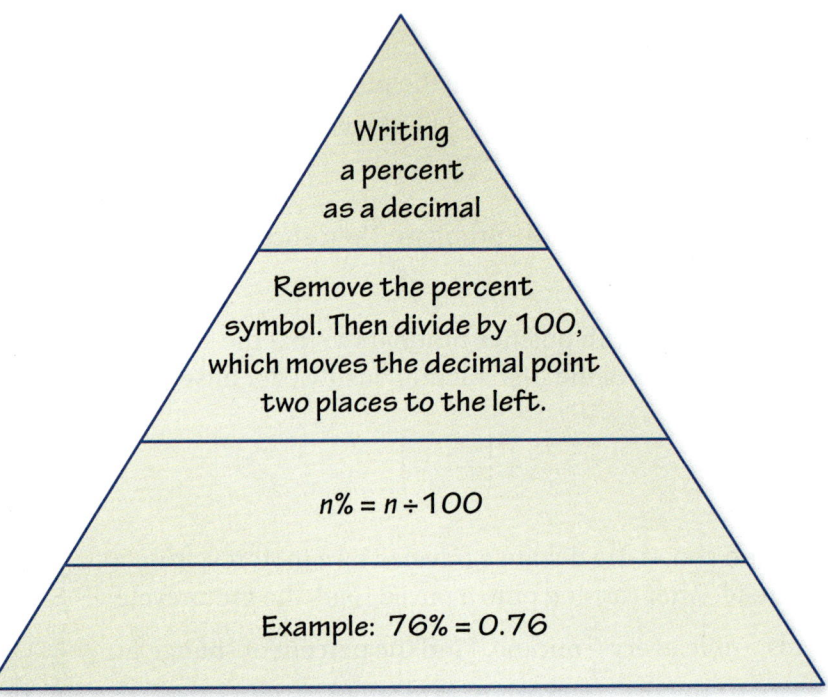

Choose and complete a graphic organizer to help you study the concept.

1. writing a decimal as a percent
2. comparing and ordering fractions, decimals, and percents
3. the percent proportion
4. the percent equation
5. percent of change
6. discount
7. markup

"I found this great **Summary Triangle** in my *Beautiful Beagle Magazine*."

716 Chapter E Percents

Chapter Self-Assessment

As you complete the exercises, use the scale below to rate your understanding of the success criteria in your journal.

1	2	3	4
I do not understand.	I can do it with help.	I can do it on my own.	I can teach someone else.

E.1 Fractions, Decimals, and Percents (pp. 679–684)

Learning Target: Rewrite fractions, decimals, and percents using different representations.

Write the percent as a decimal or the decimal as a percent. Use a model to represent the number.

1. 74%
2. 2%
3. 221%
4. 0.17
5. $4.\overline{3}$
6. 0.079

Write the fraction as a decimal and a percent.

7. $\frac{17}{20}$
8. $\frac{3}{8}$
9. $\frac{14}{9}$

10. For school spirit day, 11.875% of your class wears orange shirts, $\frac{5}{8}$ of your class wears blue shirts, 0.15625 of your class wears white shirts, and the rest of your class wears gold shirts. Order the portions of shirts of each color from least to greatest. Justify your answer.

E.2 The Percent Proportion (pp. 685–690)

Learning Target: Use the percent proportion to find missing quantities.

Write and solve a proportion to answer the question.

11. What percent of 60 is 18?
12. 40 is what percent of 32?
13. What number is 70% of 70?
14. $\frac{3}{4}$ is 75% of what number?
15. About 29% of the Earth's surface is covered by land. The total surface area of the Earth is about 510 million square kilometers. What is the area of the Earth's surface covered by land?

E.3 The Percent Equation (pp. 691–696)

Learning Target: Use the percent equation to find missing quantities.

Write and solve an equation to answer the question.

16. What number is 24% of 25?

17. 9 is what percent of 20?

18. 60.8 is what percent of 32?

19. 91 is 130% of what number?

20. 85% of what number is 10.2?

21. 83% of 20 is what number?

22. 15% of the parking spaces at a school are handicap spaces. The school has 18 handicap spaces. How many parking spaces are there in total?

23. Of the 25 students on a field trip, 16 bring cameras. What percent of the students bring cameras?

E.4 Percents of Increase and Decrease (pp. 697–702)

Learning Target: Find percents of change in quantities.

Identify the percent of change as an *increase* or a *decrease*. Then find the percent of change. Round to the nearest tenth of a percent if necessary.

24. 6 yards to 36 yards

25. 120 meals to 52 meals

26. You estimate that a jar contains 68 marbles. The actual number of marbles is 60. Find the percent error.

27. The table shows the numbers of skim boarders at a beach on Saturday and Sunday. What was the percent of change in boarders from Saturday to Sunday?

Day	Number of Skim Boarders
Saturday	12
Sunday	9

E.5 Discounts and Markups (pp. 703–708)

Learning Target: Solve percent problems involving discounts and markups.

Find the sale price or original price.

28. Original price: $50
    Discount: 15%
    Sale price: ?

29. Original price: ?
    Discount: 20%
    Sale price: $75

30. What is the original price of the tennis racquet?

31. A store pays $50 for a pair of shoes. The markup is 25%.

    a. What is the selling price for the shoes?

    b. What is the total cost for a person to buy the shoes including a 6% sales tax?

E.6 Simple Interest (pp. 709–714)

Learning Target: Understand and apply the simple interest formula.

An account earns simple interest. (a) Find the interest earned. (b) Find the balance of the account.

32. $300 at 4% for 3 years

33. $2000 at 3.5% for 4 years

Find the annual interest rate.

34. $I = \$17$, $P = \$500$, $t = 2$ years

35. $I = \$426$, $P = \$1200$, $t = 5$ years

Find the amount of time.

36. $I = \$60$, $P = \$400$, $r = 5\%$

37. $I = \$237.90$, $P = \$1525$, $r = 2.6\%$

38. You deposit $100 in an account. The account earns $2 simple interest in 6 months. What is the annual interest rate?

39. Bank A is offering a loan with a simple interest rate of 8% for 2 years. Bank B is offering a loan with a simple interest rate of 6.5% for 3 years.

    a. Assuming the monthly payments are equal, what is the monthly payment for the four wheeler from Bank A? from Bank B?

    b. Give reasons for why a person might choose Bank A and why a person might choose Bank B for a loan to buy the four wheeler. Explain your reasoning.

Practice Test

Write the percent as a decimal, or the decimal as a percent. Use a model to represent the number.

1. 0.96%
2. 3%
3. 25.$\overline{5}$%
4. 0.$\overline{6}$
5. 7.88
6. 0.58

Order the numbers from least to greatest.

7. 86%, $\frac{15}{18}$, 0.84, $\frac{8}{9}$, 0.8$\overline{6}$

8. 91.6%, 0.91, $\frac{11}{12}$, 0.917, 9.2%

Write and solve a proportion or equation to answer the question.

9. What percent of 28 is 21?
10. 64 is what percent of 40?
11. What number is 80% of 45?
12. 0.8% of what number is 6?

Identify the percent of change as an *increase* or a *decrease*. Then find the percent of change. Round to the nearest tenth of a percent if necessary.

13. 4 strikeouts to 10 strikeouts
14. $24 to $18

Find the sale price or selling price.

15. Original price: $15
    Discount: 5%
    Sale price: ?

16. Cost to store: $5.50
    Markup: 75%
    Selling price: ?

An account earns simple interest. Find the interest earned or the principal.

17. Interest earned: ?
    Principal: $450
    Interest rate: 6%
    Time: 8 years

18. Interest earned: $27
    Principal: ?
    Interest rate: 1.5%
    Time: 2 years

19. You spend 8 hours each weekday at school. (a) Write the portion of a weekday spent at school as a fraction, a decimal, and a percent. (b) What percent of a week is spent at school if you go to school 4 days that week? Round to the nearest tenth.

20. Research indicates that 90% of the volume of an iceberg is below water. The volume of the iceberg above the water is 160,000 cubic feet. What is the volume of the iceberg below water?

21. You estimate that there are 66 cars in a parking lot. The actual number of cars is 75.

    a. Find the percent error.
    b. What other estimate gives the same percent error? Explain your reasoning.

Cumulative Practice

1. A movie theater offers 30% off the price of a movie ticket to students from your school. The regular price of a movie ticket is $8.50. What is the discounted price that you pay for a ticket?

   A. $2.55
   B. $5.50
   C. $5.95
   D. $8.20

2.  What is the least value of x for which the inequality is true?

   $$16 \geq -2x$$

3. You are building a scale model of a park that is planned for a city. The model uses the scale 1 centimeter = 2 meters. The park will have a rectangular reflecting pool with a length of 20 meters and a width of 12 meters. In your scale model, what will be the area of the reflecting pool?

   F. 60 cm²
   G. 120 cm²
   H. 480 cm²
   I. 960 cm²

4. Which proportion represents the problem?

   "17% of a number is 43. What is the number?"

   A. $\dfrac{17}{43} = \dfrac{n}{100}$
   B. $\dfrac{n}{17} = \dfrac{43}{100}$
   C. $\dfrac{n}{43} = \dfrac{17}{100}$
   D. $\dfrac{43}{n} = \dfrac{17}{100}$

5. Which list of numbers is in order from least to greatest?

   F. $0.8, \dfrac{5}{8}, 70\%, 0.09$
   G. $0.09, \dfrac{5}{8}, 0.8, 70\%$
   H. $\dfrac{5}{8}, 70\%, 0.8, 0.09$
   I. $0.09, \dfrac{5}{8}, 70\%, 0.8$

6. What is the value of $\dfrac{9}{8} \div \left(-\dfrac{11}{4}\right)$?

7. The number of calories you burn by playing basketball is proportional to the number of minutes you play. Which of the following is a valid interpretation of the graph?

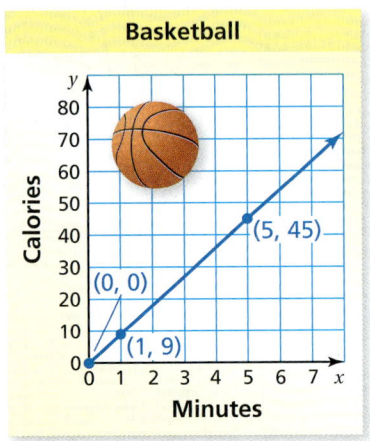

A. The unit rate is $\dfrac{1}{9}$ calorie per minute.

B. You burn 5 calories by playing basketball for 45 minutes.

C. You do not burn any calories if you do not play basketball for at least 1 minute.

D. You burn an additional 9 calories for each minute of basketball you play.

8. A softball team is ordering uniforms. Each player receives one of each of the items shown in the table.

Item	Jersey	Pants	Hat	Socks
Price (dollars)	x	15.99	4.88	3.99

Which expression represents the total cost (in dollars) when there are 15 players on the team?

F. $x + 24.86$

G. $15x + 372.90$

H. $x + 372.90$

I. $x + 387.90$

9. Your friend solves the equation. What should your friend do to correct the error that he made?

$$-3(2 + w) = -45$$
$$2 + w = -15$$
$$w = -17$$

A. Multiply -45 by -3.

B. Add 3 to -45.

C. Add 2 to -15.

D. Divide -45 by -3.

10. You are comparing the costs of a certain model of ladder at a hardware store and at an online store.

Part A What is the total cost of buying the ladder at each of the stores? Show your work and explain your reasoning.

Part B Suppose that the hardware store is offering 10% off the price of the ladder and that the online store is offering free shipping and handling. Which store offers the lower total cost for the ladder? by how much? Show your work and explain your reasoning.

11. Which graph represents the inequality $-5 - 3x \geq -11$.

F.

G.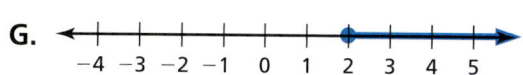

H.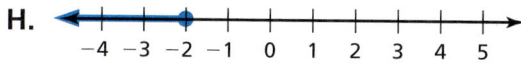

I.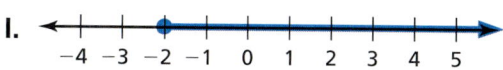

Selected Answers

Chapter 1

Section 1.1
Review & Refresh
1. 300
3. 369
5. $(5 + 8) \times 4$
7. 4.0
9. 3

Concepts, Skills, & Problem Solving
11. 8^2; 64
13. 9^4; 6561
15. 9^2
17. 15^3
19. 14^3
21. 11^5
23. 16^4
25. 167^3
27. 25
29. 36
31. 0
33. 16
35. 343
37. 32
39. 117,649
41. 20,736
43. no; $8^3 = 8 \cdot 8 \cdot 8 = 512$
45. perfect square
47. not a perfect square
49. not a perfect square
51. perfect square
53. 16 in.
55. *Sample answer:* $11^2, 5^3, 2^7$
57.

Power	4^6	4^5	4^4	4^3	4^2	4^1
Value	4096	1024	256	64	16	4

The value of the power is divided by 4. $4^0 = 1$

Section 1.2
Review & Refresh
1. 11^4
3. $h = 8$ in.
5. composite
7. prime

Concepts, Skills, & Problem Solving
9. 10; 22; no
11. 5
13. 24
15. 88
17. 13
19. 41
21. 24
23. 204
25. no; $9 + 3 \times 3^2 = 9 + 27 = 36$
27. 8
29. 20
31. 17
33. 4
35. 12
37. 72
39. 3
41. 621 in.^{2}
43. *Sample answer:* $(5^2 - 3 \times 5) \div 2 + 95$
45. 44 min; Two miles is 3520 yards. Each group can clean $200 \div 5 = 40$ yards each minute, so together the two groups can clean 80 yards each minute and $3520 \div 80 = 44$ minutes.

Section 1.3
Review & Refresh
1. 34
3. 13
5.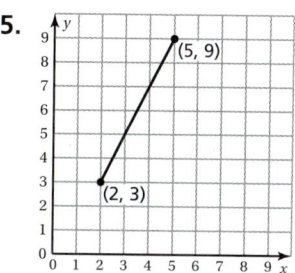
7. 102; $408 \div 4 = (400 + 8) \div 4$
9. 323; $969 \div 3 = (900 + 60 + 9) \div 3$
11. isosceles, obtuse

Concepts, Skills, & Problem Solving
13. $2 \cdot 2 \cdot 3 \cdot 5$
15. $2 \cdot 2 \cdot 2 \cdot 3 \cdot 5$
17. 1, 15; 3, 5
19. 1, 34; 2, 17
21. 1, 45; 3, 15; 5, 9
23. 1, 59
25. 1, 100; 2, 50; 4, 25; 5, 20; 10, 10
27. 1, 25; 5, 5
29. 1, 52; 2, 26; 4, 13
31. 1, 71
33. 2^4
35. $2 \cdot 3 \cdot 5$
37. $2^2 \cdot 3 \cdot 7$
39. $5 \cdot 13$
41. $2 \cdot 23$
43. $3^2 \cdot 11$
45. $3^2 \cdot 5 \cdot 7$
47. $2^2 \cdot 5 \cdot 7$
49. 180
51. 12,584
53. 25
55. 196
57. 4
59. 36
61. 441
63. 9
65. composite; *Sample answer:* 5 is a factor of the total.
67. yes; The rest of the even whole numbers have 2 as a factor.

69. 6
71. 26 yd

Section 1.4
Review & Refresh
1. 1, 20; 2, 10; 4, 5
3. 1, 56; 2, 28, 4, 14; 7, 8
5. sometimes
7. always

Concepts, Skills, & Problem Solving
9. 6
11. 12
13. 14
15. 13
17. 1
19. 4
21. 15
23. 9
25. 1
27. 15
29. 21
31. 1
33. Sample answer: 10, 15
35. Sample answer: 37, 74
37. 8 arrangements
39. yes; The common prime factors are 2^2 and 3.
41. 6
43. 1
45. 8
47. 12
49. Sample answer: 16, 32, and 48; Multiply 16 by 1, 2, and 3.
51. always; A prime number has no factors besides 1 and itself.
53. 24; 7 magnets, 2 robot figurines, 1 freeze-dried ice cream
55. **a.** The GCF of the three numbers is 1.
    b. 18; The GCF of 72, 54, and 36 is 18, leaving one banana left over.

Section 1.5
Review & Refresh
1. 6
3. 38
5. 216
7. (2, 4)
9. (4, 7)

Concepts, Skills, & Problem Solving
11. 21
13. 20
15. 6
17. 18
19. 72
21. 132
23. 84
25. 44
27. 108
29. 66
31. 350
33. 15
35. 42
37. 36

39. 126
41. 60 min
43. you: 7 mi; your friend: 6 mi
45. always; The LCM is the product of the prime factors.
47. never; The GCF is at most the lesser number, and the LCM is at least the greater number.

Chapter 2
Section 2.1
Review & Refresh
1. 40
3. 70
5. 12; Sample answer:

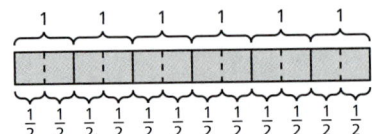

7. 12; Sample answer:

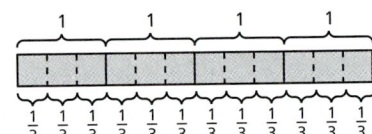

9. 10^3
11. D

Concepts, Skills, & Problem Solving
13. $\frac{1}{6}$
15. $\frac{2}{21}$
17. $\frac{1}{10}$
19. $\frac{8}{21}$
21. $\frac{1}{24}$
23. $4\frac{1}{6}$
25. $\frac{2}{5}$
27. $\frac{9}{49}$
29. $\frac{13}{21}$
31. $\frac{3}{10}$
33. $>; \frac{9}{10} < 1$
35. $=; \frac{7}{7} = 1$
37. 2
39. 2
41. 2
43. $1\frac{1}{2}$
45. $1\frac{3}{14}$
47. $36\frac{2}{3}$
49. $6\frac{4}{9}$
51. $11\frac{3}{8}$
53. no; $2\frac{1}{2} \times 7\frac{4}{5} = \frac{5}{2} \times \frac{39}{5} = 19\frac{1}{2}$

55. a. 7 ft² **b.** $10\frac{1}{3}$ ft²

57. $2\frac{1}{12}$ **59.** $\frac{27}{125}$

61. $\frac{121}{144}$ **63.** 4 mi

65. a. Sample answer:

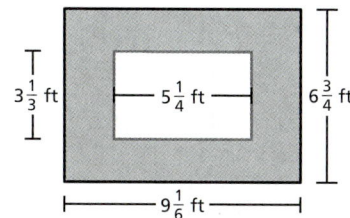

b. Sample answer: Subtract the area of the fountain from the total area of the garden; Use rectangles to find the area of each piece of the garden and add these areas.

c. $44\frac{3}{8}$ ft²; Sample answer: subtract; fewer calculations

67. Answer: $\frac{7}{24}$; Meaning: $\frac{7}{8}$ of $\frac{1}{3}$;

Sample model:

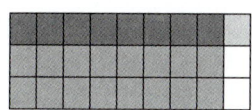

Sample application: The path around a park is $\frac{1}{3}$ mile long. You jog $\frac{7}{8}$ of the path. How far do you jog?

69. $9\frac{1}{2} \times 8\frac{3}{4} \times 7\frac{5}{6} = 651\frac{7}{48}$; $1\frac{4}{9} \times 2\frac{5}{8} \times 3\frac{6}{7} = 14\frac{5}{8}$;

Sample answer: Use the greatest (or least) digits for the whole numbers, then use the remaining digits to form the greatest (or least) fractional parts. Guess and test combinations of whole numbers and fractions to find the greatest (or least) product.

Section 2.2
Review & Refresh

1. $\frac{21}{40}$ **3.** $\frac{1}{6}$

5. B **7.** C

9. 14 ft² **11.** $\frac{15}{16}$ yd²

Concepts, Skills, & Problem Solving

13. 4 **15.** $\frac{1}{8}$

17. $\frac{5}{2}$ **19.** $\frac{2}{3}$

21. $\frac{1}{7}$ **23.** $1\frac{1}{2}$

25. 16 **27.** $\frac{1}{14}$

29. $\frac{1}{3}$ **31.** 3

33. $\frac{2}{27}$ **35.** $\frac{27}{28}$

37. $20\frac{1}{4}$

39. yes; The division is correct.

41. $\frac{3}{25}$

43. Sample answer: A pipe with a length of $\frac{5}{6}$ yard is cut into four equal pieces. How long is each piece? $\frac{5}{24}$ yd

45. Sample answer: You need $\frac{2}{3}$ pound of potting soil to repot 1 plant. How many plants can you repot with a 10-pound bag of potting soil? 15

47. $\frac{12}{5}$ **49.** $\frac{1}{8}$

51. =; When you divide a fraction by 1, the quotient is the fraction.

53. >; When you divide a fraction by a fraction less than 1, the quotient is greater than the fraction.

55. $\frac{1}{144}$ **57.** 4

59. $\frac{5}{6}$ **61.** $\frac{1}{10}$

63. when the simplified fraction has a 1 in the numerator; The reciprocal will have a 1 in the denominator.

65. a. 6 **b.** $17\frac{7}{9}$ **c.** $\frac{1}{4}$

67. a. 6 bowls; 10 plates

b. 2 plates; $\frac{1}{2}$ pt

c. 3 bowls and 6 plates; Sample answer: Used a table of values.

Section 2.3
Review & Refresh
1. $\frac{7}{8}$
3. $\frac{1}{12}$
5. 56
7. 540
9. 70 in.^{3}
11. B

Concepts, Skills, & Problem Solving
13. *Sample answer:* A pepper plant is $1\frac{1}{6}$ feet tall and a cherry tomato plant is $5\frac{5}{6}$ feet tall. How many times taller is the cherry tomato plant than the pepper plant? 5
15. 3
17. $9\frac{3}{4}$
19. $3\frac{18}{19}$
21. $\frac{9}{10}$
23. $12\frac{1}{2}$
25. $1\frac{1}{5}$
27. $\frac{2}{7}$
29. $1\frac{5}{18}$
31. no; $3\frac{1}{2} \div 1\frac{2}{3} = \frac{7}{2} \div \frac{5}{3} = 2\frac{1}{10}$
33. 3
35. 3
37. $6\frac{9}{26}$
39. $13\frac{1}{5}$
41. $3\frac{1}{3}$
43. $\frac{7}{108}$
45. no; There are 2 full groups of $1\frac{1}{6}$ plus one piece remaining, which represents $\frac{1}{7}$ of $1\frac{1}{6}$.
47. 4; *Sample answer:* $12 \div 2\frac{3}{4} = 4\frac{4}{11}$ and $8\frac{1}{2} \div 1\frac{1}{3} = 6\frac{3}{8}$, so you can make 4 full batches.

Section 2.4
Review & Refresh
1. $4\frac{1}{3}$
3. $2\frac{1}{12}$
5. 3
7. 1
9. 729
11. square
13. parallelogram

Concepts, Skills, & Problem Solving
15. 0.6457
17. 11.029
19. 22.899
21. 18.572
23. 40
25. 144
27. 29.937
29. 1.46
31. 4.366
33. 2.644
35. 5.611
37. 20.417
39. 117.2583
41. yes; The addition is correct.
43. 0.888 ton
45. 6.772
47. 10.343
49. 21.582
51. 16.5916
53. 33.6
55. 28.546 AU
57. 10.26 AU

Section 2.5
Review & Refresh
1. 6.14
3. 26.349
5. 26
7. 3
9. B
11. 2

Concepts, Skills, & Problem Solving
13. 3.15
15. 0.21
17. 33.6
19. 115.04
21. 21.45
23. 13.888
25. 2.4
27. 45.78
29. 610.5
31. 0.0342
33. yes; $9 \times 45 = 405$ and the factors have 4 decimal places.
35. 30.06 lb
37. 0.024
39. 0.000072
41. 0.03
43. 0.000012
45. 109.74
47. 3.886
49. 13.7104
51. 51.3156
53. $3.24
55. Carlton Centre: 731.44 ft; Burj Khalifa: 2715.84 ft; Q1: 1059.44 ft; Federation Tower: 1226.72 ft; One World Trade Center: 1774.48 ft; Gran Torre Santiago: 984 ft
57. 45.4
59. 4.355
61. 2.016
63. 150.183
65. 1.887
67. $14.03
69. Each number is 0.1 times the previous number; 0.0015, 0.00015, 0.000015
71. Each number is 1.5 times the previous number; 25.3125, 37.96875, 56.953125

73. Answers should include, but are not limited to:
   a. menu with main items, desserts, beverages, and prices
   b. guest check for 5 people showing items, prices, and subtotal
   c. tax and total with tax are shown
   d. amount rounded to nearest dollar, 20% tip, and total cost including tip are shown

Section 2.6
Review & Refresh
1. 30.32 **3.** 1.8004
5. 1, 26; 2, 13 **7.** 1, 50; 2, 25; 5, 10
9. B **11.** D

Concepts, Skills, & Problem Solving
13. 16,648 people **15.** 4162 people
17. 31 **19.** $12\frac{13}{24}$
21. 73 **23.** $53\frac{1}{118}$
25. 60 **27.** $47\frac{110}{173}$
29. 13
31. no; The answer should be 109.
33. 42 ft
35. a. 9
   b. from top shelf down: 22, 22, 19, 0, 0
37. Sample answer: $36,000 \div 900 = 40$

Section 2.7
Review & Refresh
1. 6 **3.** 27
5. B **7.** 15.267

Concepts, Skills, & Problem Solving
9. $0.54 \div 0.6 = 0.9$, $0.54 \div 0.9 = 0.6$; Rewrote $0.6 \times 0.9 = 0.54$.
11. 4.2 **13.** 0.5
15. 4.31 **17.** 6.2
19. 5.58 **21.** 0.15
23. yes; The long division is correct.
25. $29.95 **27.** 12
29. 52.1 **31.** 0.8
33. 11.7 **35.** 8.3

37. 0.23 **39.** 352.5
41. 7200 **43.** 40
45. 12.5 **47.** 180
49. 48 **51.** 9.60
53. 6.04
55. no; $0.32 \overline{)146.4} \longrightarrow 32 \overline{)14,640}$
57. 1.62 **59.** 10.12
61. 8.046 **63.** 7.1
65. 4.8 ft
67. =; Both quotients can be written as $666 \div 74$.
69. <; The second quotient has a lesser divisor.
71. a. Australia: 52.6625 sec
   United States: 52.9725 sec
   Canada: 53.2225 sec
   b. no; The team total would have been 210.89 seconds.
73. a. 9
   b. 3; Sample answer: Use the same method as part (a) to find the total number of weeks needed, then subtract the number of weeks in part (a).
75. a. 3
   b. 9
   c. no; The perimeter is measured in feet and the area is measured in square feet.
   d. perimeter: 2 times greater; area: 4 times greater

Chapter 3
Section 3.1
Review & Refresh
1. 7 **3.** 1.5
5. 64 **7.** 81
9. 540, 450; GCF: 90; LCM: 2700

Concepts, Skills, & Problem Solving
11. Add 2 parts of iced tea for every 1 part of lemonade added.
13. 2 to 5, or 2 : 5 **15.** 2 to 6, or 2 : 6
17. 12 : 16
19. Sample answer: 1 out of every 7 contestants wins a prize.
21. Sample answer: 2 days in a weekend per 5 weekdays

23. a. 4; The number of sunny days is 4 times the number of rainy days.
 b. 5 : 1
25. equivalent
27. not equivalent
29. not equivalent
31. equivalent
33. Sample answer: 9 : 3; $\frac{9}{3} = \frac{3}{1}$
35. Sample answer: 12 : 12; $\frac{12}{12} = \frac{6}{6}$
37. 18
39. 21
41. Sample answer: 2 cups of water and 4 cups of cornstarch, 3 cups of water and 6 cups of cornstarch; $\frac{1}{2} = \frac{2}{4} = \frac{3}{6}$
43. 21
45. yes
47. no; The ratio of boys to people in the competition is 3 : 5, and there is no ratio of a whole number to 9 that is equivalent to the ratio 3 : 5.
49. 67.5 in.; Sample answer: 2 : 3 : 4 is equivalent to 15 : 22.5 : 30, and 15 + 22.5 + 30 = 67.5.
51. 5 girls; Sample answer: $\frac{12}{10} = \frac{18}{15}$

Section 3.2
Review & Refresh
1. equivalent
3. not equivalent
5. $\frac{1}{2}$
7. $13\frac{7}{16}$

Concepts, Skills, & Problem Solving
9. 800 m
11. beginner trail: 400 m, expert trail: 1600 m
13. 6 h
15. 2
17. 9
19. 27
21. 12 blueberries, 4 strawberries
23. 4 blueberries, 8 strawberries
25. 36 blueberries, 12 strawberries
27. 288
29. 20 boys, 28 girls
31. $200

Section 3.3
Review & Refresh
1. 2
3. 15
5. 8
7. 5
9. 1
11. 28 yd

Concepts, Skills, & Problem Solving
13. 1440; Sample answer: 2 × 8 = 16 and 180 × 8 = 1440
15. 495; Sample answer: (2 ÷ 2) × 5.5 = 5.5 and (180 ÷ 2) × 5.5 = 495
17.

Burgers	3	6	9
Hot dogs	5	10	15

3 : 5, 6 : 10, 9 : 15

19.

Adults	2	1	3	18
Children	14	7	21	126

2 : 14, 1 : 7, 3 : 21, 18 : 126

21.

Plums	14	42	6	48
Grapes	7	21	3	24

14 : 7, 42 : 21, 6 : 3, 48 : 24

23. 16
25. $60
27. 1840
29. a. 392; no; Sample answer: 392 < 400
 b. $16,400; no; Sample answer: $16,400 < $16,800
31. yours; $\frac{2}{22} > \frac{3}{37}$
33. A = 12, B = 18
35. A = 33, B = 15
37. a. yes; Sample answer: Find a ratio equivalent to 161 : 28 that has 12 on the right.
 b. 21
39. 96
41. $270

Section 3.4
Review & Refresh
1.

Chickens	8	16	24
Eggs	6	12	18

8 : 6, 16 : 12, 24 : 18

3. seven and one tenth
5. thirteen and six tenths
7. Sample answer: 8 : 4, 2 : 1

Concepts, Skills, & Problem Solving
9.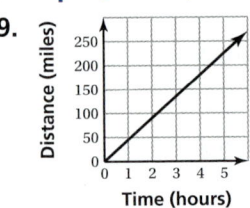

11.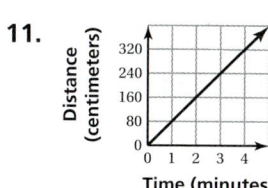

13.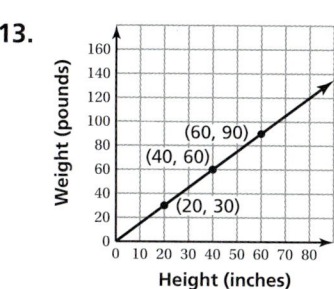

15.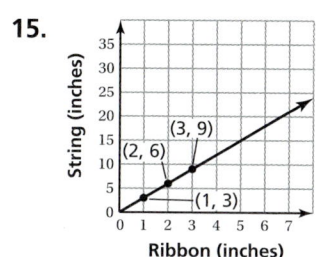

17.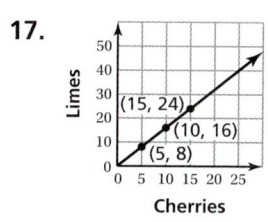

19. a.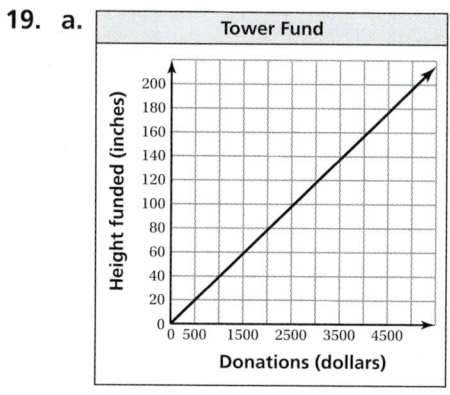

  b. $114.75

21. you; The graph that represents your earnings is steeper.

23.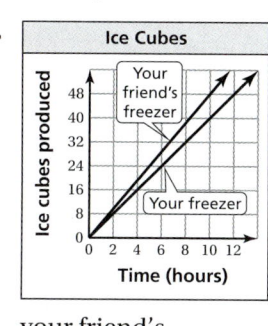

your friend's

25. a. 70

  b.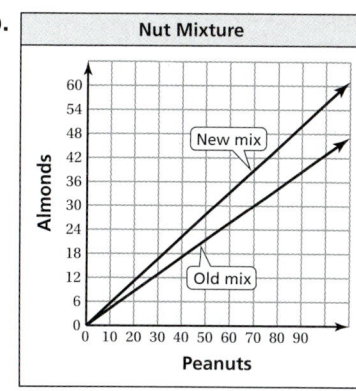

  The new mix has a greater ratio of almonds to peanuts.

  c. yes; The new mixture is more expensive to make.

Section 3.5
Review & Refresh

1.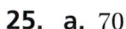

3.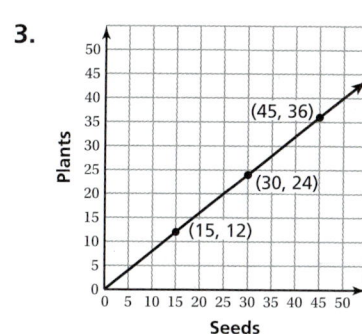

5. $\frac{2}{3}$ **7.** $1\frac{7}{12}$

9. 19.241   **11.** 7.919

Concepts, Skills, & Problem Solving

13. Answers will vary. Multiply the hand clap rate in Exploration 2(a) by 2.5.

15. Answers will vary. Multiply the hand clap rate in Exploration 2(a) by 11.25.

17. $20/guest   **19.** 6 necklaces/h

21. 19 students/class   **23.** 110 calories/serving

25. $2.5/oz   **27.** $\frac{1}{6}, \frac{17}{6}$

29. a. $20

  b. 9

31. equivalent
33. equivalent
35. your friend; 2.5 min
37. 17
39. 1.2 h

Section 3.6
Review & Refresh
1. 51 beats/min
3. $20/volunteer
5. C
7. 2 × 37
9. 2 × 2 × 3 × 11, or $2^2 × 3 × 11$
11. 18^4

Concepts, Skills, & Problem Solving
13. 1 gal; 1 gal = 4 qt, 2 L ≈ 2.1 qt, and 4 qt > 2.1 qt
15. 6
17. 2.5
19. 24
21. 5
23. 4
25. 45.92 or 46.67
27. 28.8 or 29.09
29. 191.01 or 192.82
31. 5.91 or 5.85
33. a.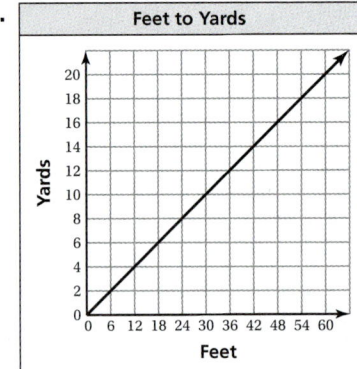
    b.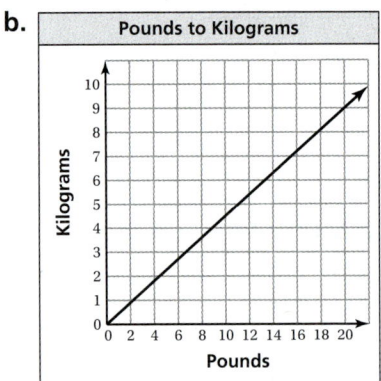
35. a. about 60.67 m or about 59.7 m
    b. about 8.04 km or about 7.91 km
37. 5.66 or 5.7
39. 8.07 or 8.06
41. 0.0175
43. a. 300 qt/h
    b. 4.75 L/min or 4.72 L/min

45. <
47. >
49. >
51. 111.6 or 111.80
53. 3.9 or 3.94
55. about 1.05 mi or about 1.06 mi
57. a. less than; Your speed is about 45.45 miles per hour.
    b. 1.32 min/mi
    c. yes
59. a. 1 cm = $\frac{1}{2.54}$ in.
    b. about 0.3937007874; Some conversion facts are rounded.

Chapter 4
Section 4.1
Review & Refresh
1. 3000
3. 1.5
5. 1.69 or 1.72
7. 10
9. 13
11. 52
13. 3 : 1
15. 4 : 1

Concepts, Skills, & Problem Solving
17.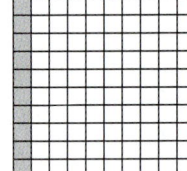
19.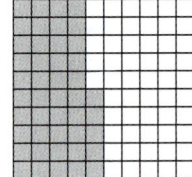
21. $\frac{9}{10}$
23. $\frac{7}{100}$
25. $\frac{79}{100}$
27. $1\frac{22}{25}$
29. $2\frac{6}{25}$
31. $\frac{1}{250}$
33. 10%
35. 55%
37. 8%
39. 1.2%
41. 185%
43. 300.5%
45. no; $\frac{56}{100} = 56\%$
47. 25%
49. $1\frac{1}{4}$; No, you have more than you need.
51. 81.25%
53. 82.5%

55. a. $\frac{7}{40}, \frac{17}{1000}, \frac{9}{3125}, \frac{3}{200}$

b. no; Alaska's area is more than 10 times as great as Florida's.

c. AK, TX, CA, MT, NM, AZ, NV, CO, OR, WY, MI, MN, UT, ID, KS, NE, SD, WA, ND, OK, MO, FL, WI, GA

57. Shade half of one of the squares.

Section 4.2
Review & Refresh
1. 14%   **3.** 0.4%

5.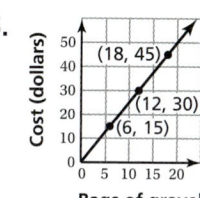

7. $\frac{2}{21}$ **9.** $4\frac{1}{5}$

Concepts, Skills, & Problem Solving
11. 51%, 0.51   **13.** 0.78
15. 0.185   **17.** 0.33
19. 0.4763   **21.** 1.66
23. 0.0006   **25.** 74%
27. 89%   **29.** 99%
31. 48.7%   **33.** 368%
35. 3.71%
37. no; $0.86 = 0.8\overline{6} = 86\%$
39. D   **41.** A
43. 34%   **45.** $\frac{9}{25}$, 0.36
47. $\frac{203}{1250}$, 0.1624
49. a. 14%
   b. $\frac{43}{50}$

Section 4.3
Review & Refresh
1. 0.12   **3.** 0.00046

5.

Cost (dollars)	2	3	13.5
Time (minutes)	50	75	337.5

2 : 50, 3 : 75, 13.5 : 337.5

7. 162.9   **9.** 5.7084
11. 17

Concepts, Skills, & Problem Solving
13. $\frac{3}{10}$, 40%, 0.65, 80%   **15.** $\frac{1}{8}$, 15%, 0.25, $\frac{27}{100}$
17. 20%   **19.** $\frac{13}{25}$
21. 76%   **23.** 0.12
25. 140%   **27.** 80%
29. 0.63, $\frac{13}{20}$, 68%   **31.** 0.15%, 0.015, $\frac{3}{20}$
33. $\frac{21}{50}, \frac{87}{200}$, 43.7%, 0.44
35. Russia, Brazil, United States, India
37. 21%, 0.211, $\frac{11}{50}, \frac{111}{500}$
39. D   **41.** C
43. a. dolphin, rabbit, lion, squirrel, tiger
   b. *Sample answer:* 33%
   c. first

Section 4.4
Review & Refresh
1. $\frac{1}{8}$, 0.33, 35%

3. $\frac{1}{5}$, 22%, $\frac{13}{50}$, 0.28, 0.41

5. $\frac{9}{2000}$ **7.** 12

9. 0.62   **11.** 120

Concepts, Skills, & Problem Solving
13. $30   **15.** $240
17. 4   **19.** 9
21. 3   **23.** 3
25. 36   **27.** 17.48
29. 8.36   **31.** 47.85
33. 39.6   **35.** 365
37. a. $3.15   b. $48.15

Selected Answers **A9**

39. 140   **41.** 84

43. 80   **45.** 25

47. 20

49. no; $5 \div 20\% = 5 \div \dfrac{1}{5} = 25$

51. 18   **53.** 125,000

55. =   **57.** >

59. 48

61. a. 432 in.²

    b. 37.5%; The width is 18 inches, and $\dfrac{18}{48} = 37.5\%$.

63. *Sample answer:* Because 30% = 2 × 15%, 30% of n = 2 × 12 = 24; Because 45% = 3 × 15%, 45% of n = 3 × 12 = 36.

65. yes; $\dfrac{x}{100} \cdot y = \dfrac{y}{100} \cdot x$

67. 97.2%

Chapter 5

Section 5.1
Review & Refresh

1. 25%   **3.** 45%

5. 22.5

7. $1\dfrac{7}{18}$ **9.** $2\dfrac{23}{27}$

Concepts, Skills, & Problem Solving

11. 20 − 12; $8   **13.** 20 × 6; $120

15. Terms: g, 12, $9g$; Coefficients: 1, 9; Constant: 12

17. Terms: $2m^2$, 15, $2p^2$; Coefficients: 2, 2; Constant: 15

19. Terms: $8x$, $\dfrac{x^2}{3}$; Coefficients: 8, $\dfrac{1}{3}$; Constant: none

21. a. Terms: 2ℓ, $2w$; Coefficients: 2, 2; Constant: none

    b. The coefficient 2 of ℓ represents that there are 2 lengths on the rectangle. The coefficient 2 of w represents that there are 2 widths on the rectangle.

23. g^5 **25.** $5.2y^3$

27. $2.1xz^4$ **29.** $(5d)^2$

31. 9   **33.** 11

35. 10   **37.** 6

39. 5   **41.** 9

43. 4   **45.** 24

47. $15; $105   **49.** 32; 16; 8

51. 23   **53.** $2\dfrac{5}{6}$

55. 22   **57.** 46

59. 24   **61.** 34 mm; 118 mm

63. *Sample answer:* 8 min at 250 ft/min; 9 min at $222\dfrac{2}{9}$ ft/min

65. 64 in.³

Section 5.2
Review & Refresh

1. Terms: $4f$, 8; Coefficient: 4; Constant: 8

3. Terms: $9h^2$, $\dfrac{8}{9}p$, 1; Coefficients: 9, $\dfrac{8}{9}$; Constant: 1

5. 2361.6 or 2400

7. $\dfrac{4}{5}$ **9.** $\dfrac{2}{15}$

Concepts, Skills, & Problem Solving

11. 10: you have $10; $5.25n$: total cost; 5.25: price per sandwich; n: number of sandwiches

13. 100: you have $100; $6.75n$: total cost; 6.75: price per sandwich; n: number of sandwiches

15. 3 • 12   **17.** 6 + 10

19. 15 + 17   **21.** 5 • d

23. $s - 6$ **25.** t^3

27. yes; The expression is correct.

29. a.

Days	1	2	3	4	5
Total Samples	15	30	45	60	75

    b. $15n$

31. *Sample answer:* the sum of n and 6; 6 more than a number n

33. *Sample answer:* a number b less than 15; 15 take away a number b

35. $\dfrac{y}{4} - 3$; 2   **37.** $8x + 6$; 46

39. a. $2 + 3g$

    b. $26

41. a. $140 - 15n$

    b. 20; There are $140 - 15(8) = 20$ people left after the eighth round.

43. $\dfrac{x}{4}$

Section 5.3
Review & Refresh
1. $10 + p$
3. $b \div 15$
5. $2^2 \times 3^2$
7. 3×7^2
9. 11.592
11. 13.641

Concepts, Skills, & Problem Solving
13.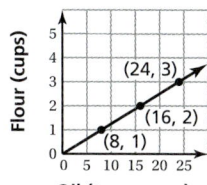
15. B
17. A
19. Comm. Prop. of Mult.
21. Assoc. Prop. of Mult.
23. Add. Prop. of Zero
25. no; The statement illustrates the Commutative Property of Addition.
27. $(14 + y) + 3 = (y + 14) + 3$ Comm. Prop. of Add.
    $= y + (14 + 3)$ Assoc. Prop. of Add.
    $= y + 17$ Add 14 and 3
29. $7(9w) = (7 \cdot 9)w$ Assoc. Prop. of Mult.
    $= 63w$ Multiply 7 and 9.
31. $(0 + a) + 8 = a + 8$ Add. Prop. of Zero
33. $(18.6 \cdot d) \cdot 1 = 18.6 \cdot (d \cdot 1)$ Assoc. Prop. of Mult.
    $= 18.6d$ Mult. Prop. of One
35. $(2.4 + 4n) + 9$
    $= (4n + 2.4) + 9$ Comm. Prop. of Add.
    $= 4n + (2.4 + 9)$ Assoc. Prop. of Add.
    $= 4n + 11.4$ Add 2.4 and 9.
37. $z \cdot 0 \cdot 12 = (z \cdot 0) \cdot 12$ Assoc. Prop. of Mult.
    $= 0 \cdot 12$ Mult. Prop. of Zero
    $= 0$ Mult. Prop. of Zero
39. a. the amount per box
    b. $120x$
41. $7 + (x + 5); x + 12$
43. $(7 \cdot 2) \cdot y$
45. $(17 + 6) + 2x$
47. $w \cdot 16$
49. a. Sample answer: From the expression, the 37 hats you sold cost $14 each and the x hats your friend sold cost $10 each.
    b. x is at most 51.

Section 5.4
Review & Refresh
1. $(s + 4) + 8 = s + (4 + 8)$ Assoc. Prop. of Add.
   $= s + 12$ Add 4 and 8.
3. $3(4n) = (3 \cdot 4)n$ Assoc. Prop. of Mult.
   $= 12n$ Multiply 3 and 4.
5. 12 boys, 20 girls
7. 14 boys, 8 girls
9. 123
11. $260\frac{8}{31}$

Concepts, Skills, & Problem Solving
13. $6s$; The sum of 4 parts and 2 parts is 6 parts.
15. $10b - 60$
17. $56 + 7y$
19. $18n + 9$
21. $90 - 54w$
23. $3 + \frac{1}{4}x$
25. $78 + 6z$
27. $25x - 25y$
29. $n + 2 + 3m$
31. A
33. B
35. $10(103 - x) = 1030 - 10x$
37. $29 + 8x$
39. $19y + 5$
41. $4n - 3$
43. 30
45. $1\frac{5}{6}y$
47. $7y$
49. no; $8x - 2x + 5x = 11x$
51. a. $42(10 - x) + 56(5 - y) = 700 - 42x - 56y$
    b. $322
53. $7(x + 3) + 8 \cdot x + 3 \cdot x + 8 - 9 = 2(9x + 10)$

Section 5.5
Review & Refresh
1. $2n + 16$
3. $7b - 21$
5. $5 + p$
7. $11d$
9. equivalent
11. C
13. A

Concepts, Skills, & Problem Solving
15. Sample answer: $4(3 + 4)$; Use a common factor as the height and find the lengths.
17. $7(1 + 2)$
19. $11(2 + 1)$
21. $12(5 - 3)$
23. $28(3 + 1)$
25. $19(1 + 5)$
27. $6(3 - 2)$
29. $14(7 - 5)$
31. $3(24 - 13)$
33. yes; yes; you can factor c out of each expression.
35. $2(x + 5)$
37. $13(2x - 1)$
39. $9(4x + 1)$
41. $2(9p + 13)$
43. $24(1 + 3n)$
45. $4(19d - 6)$
47. $2(9t + 19x)$
49. $5(2x - 5y)$

51. Sample answer: $8(x+2)$, $4(2x+4)$

53. B **55.** A

57. $6

59. The first solution calculates the total spent and the total earned, then subtracts. The second solution uses the Distributive Property first. *Sample answer:* second; There are fewer calculations.

Chapter 6

Section 6.1

Review & Refresh

1. $3(2+9)$ **3.** $6(7+4n)$

5. C   **7.** 3

9. 5   **11.** 34 cm

Concepts, Skills, & Problem Solving

13. $6.75x = 33.75$; 5   **15.** $y - 9 = 8$

17. $9b = 36$ **19.** $54 = t + 9$

21. $n - 9.5 = 27$

23. yes; "5 less than a number n" means "$n-5$" and "is 12" means "$=12$".

25. $90 = \frac{3}{4}d$

27. $30 = 4x$ **29.** 102.8

31. a. Answer should include, but is not limited to: The equation should have the form
total amount paid $=$ total price $+$ (total price $\times$ rate).

b. yes; Multiplying r by 100 gives the percent for the sales tax.

Section 6.2

Review & Refresh

1. $x + 9 = 15$ **3.** $d \cdot 7 = 63$

5. 2   **7.** 10

9. B   **11.** 25

13.

Snakes	2	8	24
Mice	5	20	60

$2:5$, $8:20$, $24:60$

Concepts, Skills, & Problem Solving

15. $t = 1$ **17.** yes

19. no   **21.** no

23. $y = 10$ **25.** $r = 22$

27. $k = 12$ **29.** $f = 46$

31. $j = 1\frac{1}{4}$ **33.** $m = 3.7$

35. no; Subtract 7 to get $x = 6$.

37. $366 = x - 30$; 396 m

39. 10,641 mi²   **41.** $w - 13 = 15$; $w = 28$

43. $9 = n - 7$; $n = 16$ **45.** $k + 11 = 29$; $k = 18$

47. $46 = 18 + d$; $d = 28$ **49.** $b = 11$

51. $m = 30$ **53.** $r = 47$

55. 16 ft   **57.** 27 ft

59. $a = 22$; $b = 31$; $c = 13$; The constant sum is 75, and $a + 53 = 75$, $b + 44 = 75$, and $c + 62 = 75$.

61. a. $5.25

b. no; It costs $9.75 to ride each ride once.

Section 6.3

Review & Refresh

1. $y = 11$ **3.** $p = \frac{5}{8}$

5. $\frac{2}{45}$ **7.** $\frac{7}{10}$

9. 0.36   **11.** 11.3353

Concepts, Skills, & Problem Solving

13. $x = 1$ **15.** $z = 7$

17. $t = 30$ **19.** $r = 32$

21. $z = 7$ **23.** $k = 6$

25. $w = 12.5$ **27.** $v = 45$

29. $m = \frac{3}{20}$ **31.** $k = 2\frac{2}{3}$

33. $b = 7.2$ **35.** $m = 6$

37. $3x = 9$

$3x \cdot \frac{1}{3} = 9 \cdot \frac{1}{3}$

$x = 3$

39. $20x = 1200$; 60   **41.** 11

43. 20

45. length: 20 in.; width: 5 in.

Section 6.4

Review & Refresh

1. $x = 9$ **3.** $x = 6$

5. $4\frac{3}{8}$ **7.** $\frac{8}{33}$

9. 256 in.²; 30,720 in.²   **11.** 5.5

13. 9.3 or 9.32   **15.** 3.2

17. 9.3

Concepts, Skills, & Problem Solving

19. $y = 175x$

21. yes   **23.** no

25. no   **27.** no

29. yes   **31.** yes

33. independent variable: w, dependent variable: A

35. independent variable: p, dependent variable: t

37. *Sample answer:* your test score

39. *Sample answer:* the amount of data you use

41. 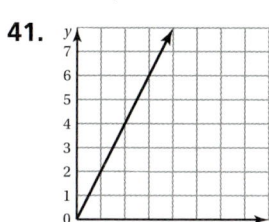   **43.**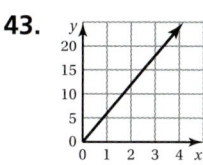

45. 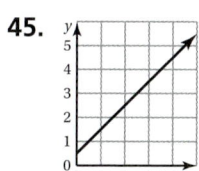   **47.**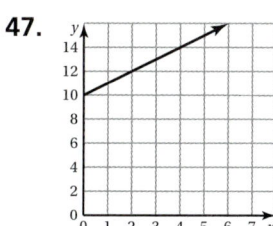

49. 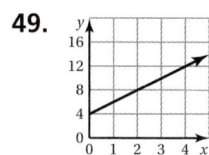   **51.**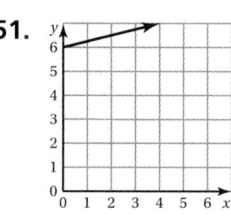

53. $c = 1.5t + 5$

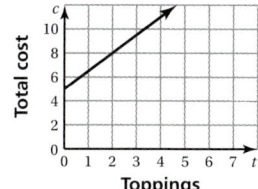

55. a. $r = 160 - x$

    b. independent variable: x, dependent variable: r

    c. 145

57. $d = \dfrac{2}{3}t$ **59.** $d = 66t$

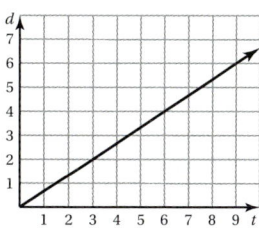   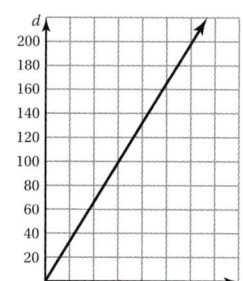

61. 11

63. 2

65. *Sample answer:* $y = 2x - 2$

67. 14 in./min

69. a. no; You can only buy whole numbers of tickets.

    b. $c = 10n$

Chapter 7

Section 7.1
Review & Refresh

1. 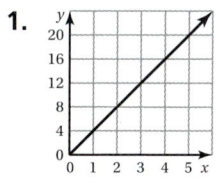   **3.**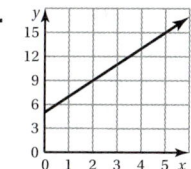

5.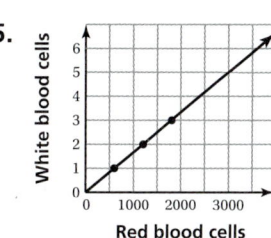

7. $2^2 \cdot 3 \cdot 5$ **9.** $2 \cdot 3^2 \cdot 7$

11. 2.891

Concepts, Skills, & Problem Solving

13. 12 units²   **15.** 24 units²

17. 840 mm²   **19.** 3750 cm²

21. 894 mi²   **23.** 6 in.²

25. 0.64

27. *Sample answer:*

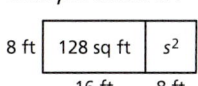

29. 9   **31.** 5

33. n^2bh or n^2A

Section 7.2
Review & Refresh
1. 72 in.²
3. 255 mi²
5. Comm. Prop. of Mult.
7. C

Concepts, Skills, & Problem Solving
9. 14 units²
11. 6 cm²
13. 1620 in.²
15. 1125 cm²
17. yes; $A = \frac{1}{2}bh$
19. about 50.2 in.²
21. $b = 4\frac{2}{3}$ ft
23. 120 ft²
25. 36 units²
27. 6
29. $A = \frac{1}{2}ab$, where a and b are the diagonal lengths; *Sample answer:* The formula is the same.

Section 7.3
Review & Refresh
1. 63 in.²
3. 25 ft²
5. trapezoid
7. C

Concepts, Skills, & Problem Solving
9. 16 units²
11. 25 cm²
13. 125 m²
15. 55 mi²
17. 28 in.²
19. 105 ft²
21. 16 ft²
23. 20 units²
25. 20 km
27. 4 mm
29. 648 ft²
31. *Sample answers:* $b_1 = 2$ ft, $b_2 = 3$ ft; $b_1 = 1.5$ ft, $b_2 = 3.5$ ft
33. 18 ft
35. a. $A = \frac{1}{2}ab$, where a and b are the diagonal lengths
    b. yes; *Sample answer:* The formulas are the same.

Section 7.4
Review & Refresh
1. 15 ft²
3. 108 m²
5. 75
7. 210
9. 40

Concepts, Skills, & Problem Solving
11. front:  side: top:

10

13. front: 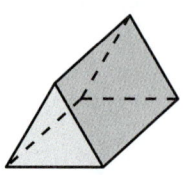 side: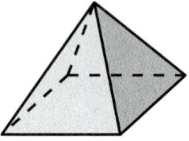
top: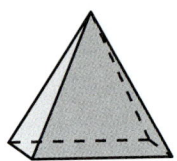

9

15. 10 faces, 24 edges, and 16 vertices

17.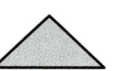
19. ...

21.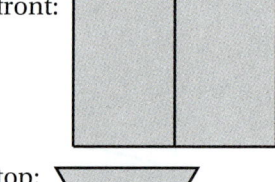

23. front: 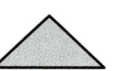 side:  top: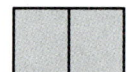

25. front: side:
top:

27. front: 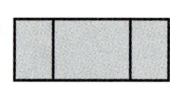 side: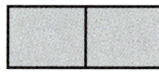
top:

29.

31. *Answer should include, but is not limited to:* an original drawing of a house; a description of any solids in the drawing

33. a.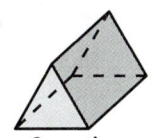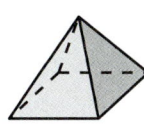

6 vertices, 9 edges   5 vertices, 8 edges

b. More than one solid can have the same number of faces, so knowing the number of edges and vertices can help you to draw the intended solid.

Section 7.5
Review & Refresh
1. front: side: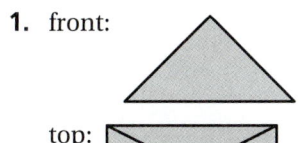
top:

3. front: side: top: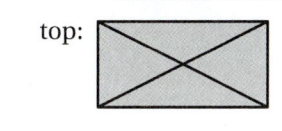

5. 22   **7.** 12
9. $x = 11$ **11.** $m = 65$
13. 17

Concepts, Skills, & Problem Solving
15. 94 units²; $2(5)(3) + (4)(3) + 2(4)(5) = 94$
17. 162 units²; $2(3)(7) + 2(6)(7) + 2(3)(6) = 162$
19. 198 cm²   **21.** 52 in.²
23. $48\frac{5}{6}$ mi²   **25.** 740 m²
27. 17.6 ft²   **29.** 324 m²
31. 136 ft²   **33.** $\frac{2}{3}$ ft²
35. 294 cm²   **37.** 83 ft²
39. box 2; It should cost less to make.
41. 2 qt

Section 7.6
Review & Refresh
1. 82 ft²   **3.** 384 in.²
5. B   **7.** A
9. 95%   **11.** 0.75%

Concepts, Skills, & Problem Solving
13. 133 units²; $7 \cdot 7 + 4 \cdot \left(\frac{1}{2}\right) \cdot 7 \cdot 6 = 133$

15. 119 in.²   **17.** 552 cm²
19. 195.6 in.²   **21.** 8.3 in.²
23. 81 micrometers²   **25.** 4
27. no; You can place the four triangles on top of the square and it covers the entire square. But when you lift up the triangles, they do not touch. So, they do not form a pyramid.

Section 7.7
Review & Refresh
1. 51 ft²   **3.** 533 yd²
5. $y - 9$; 3

Concepts, Skills, & Problem Solving
7. $\frac{1}{18}$ unit³   **9.** $\frac{3}{10}$ in.³
11. $\frac{8}{125}$ ft³   **13.** $3\frac{1}{8}$ cm³
15. $h = 20$ cm   **17.** $w = 16$ in.
19. a. Sample answer: 297 in.³
  b. no; The container only holds 196 cubic inches.
21. 1000; Multiply by the conversion factor $\frac{1 \text{ cm}^3}{1000 \text{ mm}^3}$; Multiply by the conversion factor $\frac{1000 \text{ mm}^3}{1 \text{ cm}^3}$.
23. 1152 cm³
25. Answers should include, but not limited to:
  a. a sketch of a tree house that has a surface area of at most 1400 square feet and a volume of at least 250 cubic feet
  b. For the dimensions to be reasonable, the tree house should be able to fit people and fit in a tree.

Chapter 8
Section 8.1
Review & Refresh
1. $\frac{3}{20}$ mm³   **3.** 8 ft³
5. $2(9z - 11)$ **7.** $3(14n - 9s)$

Concepts, Skills, & Problem Solving
9. Sample answer: losing 6 points on a test
11. Sample answer: diving 45 feet below the surface
13. -6 **15.** 600
17. 37,500   **19.** -56

21. 5 **23.** 83; −47

25. [number line with points at −8 and 8, range −8 to 8]

27. [number line with points at −9 and 9, range −12 to 12]

29. [number line with points at −20 and 20]

31. [number line with points at −18 and 18]

33. [number line with points at −150 and 150]

35. [number line with points at −400 and 400]

37. 5   **39.** −15

41. a. the average water level
   b. −4

43. no; The flag starts at 0, moves to 8, left to −4, and right to 9; The left team would need to pull the flag 19 feet. The right team would need to pull the flag 1 foot.

Section 8.2
Review & Refresh
1. 83   **3.** 75
5. B   **7.** $1\frac{1}{5}$
9. $\frac{1}{14}$

Concepts, Skills, & Problem Solving
11. *Sample answer:* test controls; occurs before rocket topping sequence complete
13. >   **15.** >
17. <   **19.** <
21. no; −3 < −1   **23.** −3, −1, 0, 2, 3
25. −4, −3, −2, 3, 4   **27.** −13, −3, 0, 4, 8
29. −16, −8, 1, 7, 12   **31.** −20, −10, −5, 15, 25
33. −9; −3
35. always; Positive integers are greater than negative integers.
37. a. Florida, Louisiana, Arkansas, Tennessee, California
   b. California, Louisiana, Florida, Arkansas, Tennessee
   c. sea level
39. no; In order for the median to be below 0°F, at least 6 of the temperatures must be below 0°F.

Section 8.3
Review & Refresh
1. <   **3.** >
5. D   **7.** 75
9. 70   **11.** 0.042
13. 25.69661

Concepts, Skills, & Problem Solving
15. [number line with point at $\frac{1}{4}$]

17. [number line with points at $-\frac{2}{3}$ and $\frac{2}{3}$]

19. [number line with points at −2.15 and 2.15]

21. [number line with points at −0.4 and 0.4]

23. [number line with points at $-2\frac{1}{4}$ and $2\frac{1}{4}$]

25. >   **27.** <
29. <   **31.** <
33. >
35. −5, −4.9, −4.35, −4.3, −4
37. $-1, -\frac{1}{2}, -\frac{1}{4}, \frac{1}{8}, \frac{3}{4}$
39. $-1, -\frac{3}{4}, -\frac{5}{8}, -\frac{1}{20}, 0$
41. Golfer B
43. Higher on: Thursday, Friday, Saturday; Lower on: Monday, Tuesday, Wednesday;
$-\frac{3}{25} > -\frac{7}{20} > -\frac{27}{50} > -\frac{13}{20}$,
$-\frac{13}{20} < -\frac{16}{25} < -\frac{53}{100} < -\frac{1}{3}$
45. 1, 2, and any integer less than −3

Section 8.4
Review & Refresh
1. −3.2, −1.8, −1.3, 0.6, 2.4
3. $\frac{1}{4}, \frac{1}{2}, \frac{2}{3}, \frac{3}{4}, 2$

5.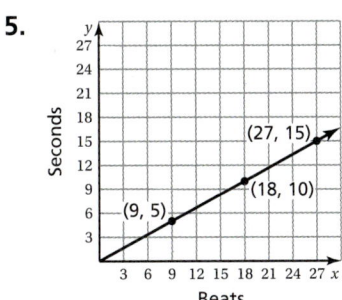

7. 13
9. 4

Concepts, Skills, & Problem Solving

11. dolphin; $|-22| > |-13|$
13. shark; $|-40| > |32|$
15. 23
17. 68
19. $\frac{1}{6}$
21. $\frac{5}{8}$
23. 1.026
25. 6.308
27. no; $|14| = 14$
29. =
31. <
33. >
35. <
37. 3; 1
39. $-4, -3, |-3|, |-4|, |5|$
41. $-20, -19, -18, |-18|, |-22|, |30|$
43. -6
45. a. $-50°C$
   b. neither can be negative
47. always; The absolute value is the positive distance from zero on a number line.
49. B; You owe $25, so debt = $25.
51. C; You owe less than $25, so debt < $25.
53. Sample answer: $x = -2, y = -3$

Section 8.5
Review & Refresh

1. 35
3. 4.7
5. C
7. (graph)

9. Associative Property of Multiplication
11. Multiplication Property of One

Concepts, Skills, & Problem Solving

13. reflection in the x-axis
15. reflection in the x-axis
17. $(-3, -2)$
19. $(1, 2)$
21. $(0, -4)$
23. $(-4, -4)$
25. $(4, -4)$

27–37.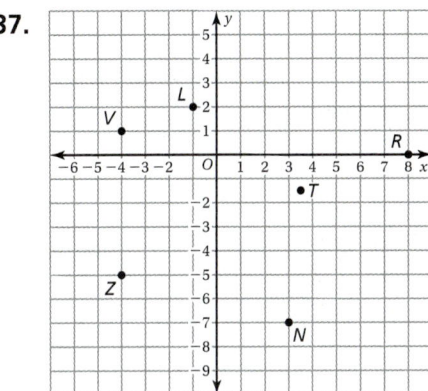

27. Quadrant II
29. Quadrant IV
31. x-axis
33. Quadrant IV
35. Quadrant II
37. Quadrant III
39. yes; The description is correct.
41. Flamingo Café
43. Sample answer: $(5, -1), (5, -2)$
45. a. $(3, -2)$ b. $(-3, 2)$
47. a. $(-5, 6)$ b. $(5, -6)$
49. a. $(-9, -3)$ b. $(9, 3)$
51. a. $(0, 1)$ b. $(0, -1)$
53. a. $(-3.5, -2)$ b. $(3.5, 2)$
55. a. $\left(-5\frac{1}{2}, -3\right)$ b. $\left(5\frac{1}{2}, 3\right)$
57. $(-4, -5)$
59. $(2, 2)$
61. $(8, -8)$
63. $(0, 2)$
65. $(-6.5, 10.5)$
67. $\left(-\frac{1}{3}, \frac{2}{3}\right)$
69. yes; Sample answer: The order of the reflections does not matter. You are still reflecting the points in both axes.
71. Quadrant III
73. Quadrant I, Quadrant IV, or the positive x-axis

75. origin

77. never; All points in Quadrant III have negative y-coordinates.

79.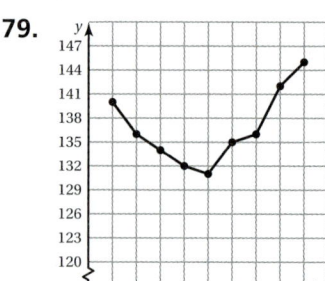

decreased from year 1 to year 5 and increased from year 5 to year 9

81. a.

Week	1	2	3	4	5	6	7	8	9
Miles	22	24	26	24	28	27	30	30	33

Week	10	11	12	13	14	15	16	17	18
Miles	35	38	40	40	40	36	33	24	14

b.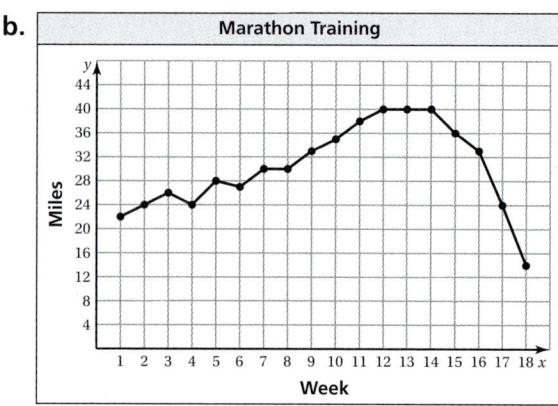

c. *Sample answer:* The number of miles per week increases so that people get used to running longer distances. Then towards the end of the program, the number of miles decreases so that people can recover and be fully prepared for the marathon.

Section 8.6
Review & Refresh

1. (1, 4) **3.** (−4, −1)
5. 0.62   **7.** 1.33
9. 8

Concepts, Skills, & Problem Solving

11.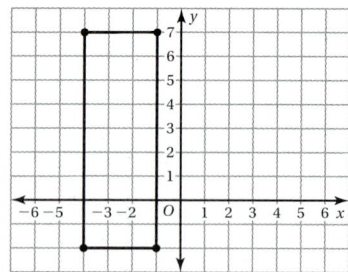

13.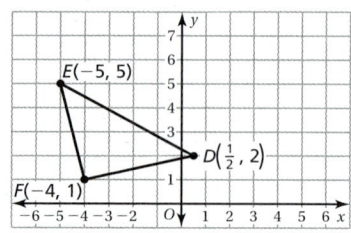

15.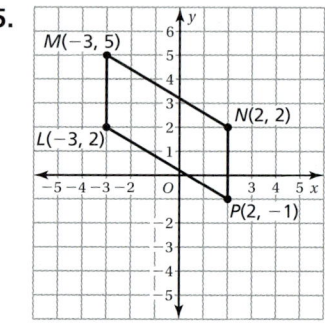

17.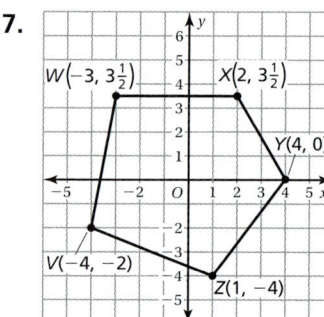

19. 5   **21.** 2
23. 15   **25.** 8.5
27. 8.75
29. 22 units; 30 units^2 **31.** 38 units; 88 units^2
33. *Sample answer:* (3, −3); (−5, −3)
35. 41 mi^2
37. *Sample answer:*

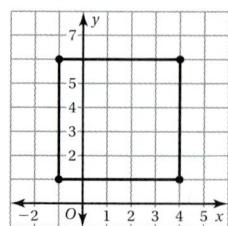

39. Sample answer:

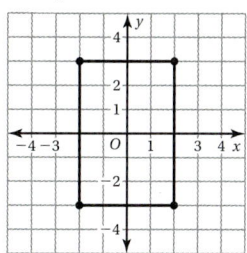

41. rectangle; *Sample answer:* Each base is 4 units and the heights are a units and b units.

Section 8.7
Review & Refresh
1. 4   **3.** 8
5. $x = 14$ **7.** $x = 9$
9. 12 in.²   **11.** 127.5 mi²

Concepts, Skills, & Problem Solving
13. 5 or any number greater than 5
15. 10 or any number less than 10
17. $a > 6$ **19.** $b \geq -3$
21. $\dfrac{x}{3} \leq 5$ **23.** no
25. no   **27.** yes
29. A   **31.** C
33.
35.
37.
39.
41.
43.
45. $x \leq 1$; A number x is at most 1.
47. $x > 0$; A number x is more than 0.
49. no; The graph shows $x < -1$, not $x > -1$.
51. a. $b \leq 3$;

b. $\ell \geq 18$;

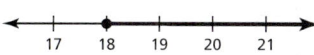

53. a. $s \leq 140$;

b. $s > 140$;

c. no; There are 300 milligrams of sodium in one serving, which is greater than 140.

55. sometimes; The only time this is not true is if $x = 5$.
57. $p + 50 \leq 425$

Section 8.8
Review & Refresh
1. no   **3.** yes
5. 56 mm²   **7.** $7\dfrac{1}{2}$ yd²
9. 12^4

Concepts, Skills, & Problem Solving
11. $2x + 1 < 9$; x is less than 4
13. $x < 9$;

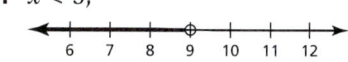

15. $5 \geq y$;

17. $6 > x$;

19. $n > 12$;

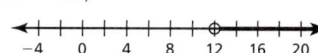

21. $c \geq 99$;

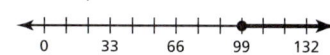

23. $3 < x$;

25. $\dfrac{1}{4} \leq n$;

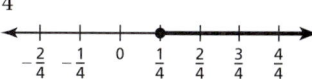

27. $v \leq 81$;

29. no; add 9 to both sides to get $37 \geq t$.
31. $p + 5 < 17$; $p < 12$
33. $8n < 72$; $n < 9$
35. $225 \geq \dfrac{3}{4}w$; $300 \geq w$
37. $x + 18.99 \leq 24$; $x \leq \$5.01$

Selected Answers   **A19**

39. when visiting more than 4 times in a year; $\$30 < \$7.50x$ when $4 < x$

41. *Sample answer:* the number of gallons of milk you can buy with \$20; the length of a park that has an area of at least 500 square feet

43. 25

45. $11 > s$;

47. $n < 27.6$;

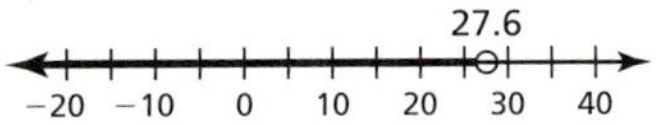

49. $b \leq 6$;

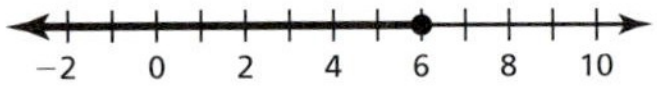

51. $x > 6.8$; $80x > 2 \cdot 272$ when $x > 6.8$.

53.

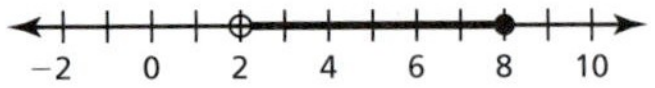

55.

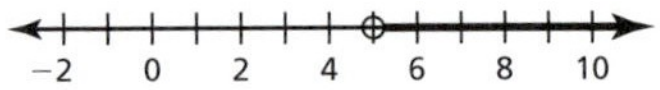

57.

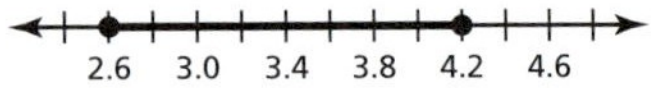

59. $400 \leq x \leq 500$

61. no; *Sample answer:* $a = 9$, $b = 8$, $x = 3$, $y = 2$; $a - x = 6$, $b - y = 6$; so, $a - x = b - y$.

63. no; *Sample answer:* $a = 6$, $b = 5$, $x = 10$, $y = 4$

$$\frac{a}{x} \overset{?}{>} \frac{y}{b}$$

$$\frac{6}{10} \overset{?}{>} \frac{4}{5}$$

$$0.6 \not> 0.8$$

Chapter 9

Section 9.1

Review & Refresh

1. $x > 24$

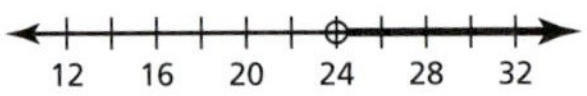

3. $9 > k$

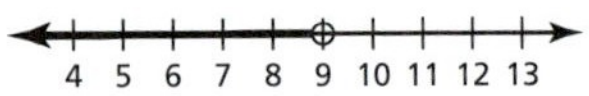

5. yes

7. yes

9. $0.2, 24\%, \frac{1}{4}, 0.32, \frac{7}{20}$

Concepts, Skills, & Problem Solving

11. 12; yes

13. *Sample answer:* 9th; no

15. yes; The answers will vary.

17. yes; The answers will vary.

19. **a.** 18; 18 players are on the team

b. *Sample answer:* Use a tape measure; inches

c. *Sample answer:* "What are the heights of players on an NBA championship team?"; The heights are spread out, but most of the heights (in inches) are in the mid-to-low 80s.

21.

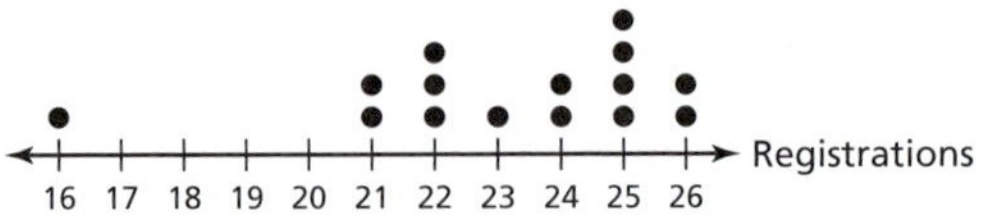

Most of the registrations are in a cluster from 21 to 26. The peak is 25. There is a gap between 16 and 21.

23. *Sample answer:* 45 mi/h; Most of the data cluster around 45 and 45 miles per hour is a common speed limit.

25. **a.** yes; It is a statistical question because you would anticipate variability in the hours spent on homework each night by students.

b. Most of the hours cluster around 2. The peak is 2. There is no gap.

c. Most students spend between 1 and 3 hours on homework during a school night.

27. *Sample answer:* rain gauge; inches

29. *Sample answer:* "How many letters are there in the English alphabet?"; "How many letters are there in a word?"; The number of letters in the English alphabet is fixed but different words will have different numbers of letters.

Section 9.2

Review & Refresh

1. yes; The answers will vary.

3. no; There is only one answer.

5. $\frac{21}{25}$

7. $3\frac{53}{100}$

9. 1.3

11. 4.2

Concepts, Skills, & Problem Solving

13. 12

15. 2

17. 103

19. a. yes; The answers will vary.
   b. 3.45 minutes

21. *Sample answer:* 20, 21, 21, 21, 21, 22, 20, 20.5, 20.5, 21.5, 21.5, 22

23. The Tigers' mean height is greater; no; neither data set has outliers.

25. a. $3.50; Divide $84 by 24 people.
   b. *Sample answer:* The class mean will not change since the new student's weekly allowance is the same as the mean.

Section 9.3
Review & Refresh
1. 4.7
3. 36.25
5. D
7. 600, 120; 5 : 100, 30 : 600, 6 : 120
9. 63.5 ft²

Concepts, Skills, & Problem Solving
11. 5.5
13. median = 7; mode = 3
15. median = 92.5; mode = 94
17. median = 17; mode = 12
19. no; The data were not ordered from least to greatest.
21. singing
23. mean = 7.61; median = 7.42; no mode
25. mean = 2.7; median = 2.6; mode = 2.2
27. With outlier: mean = 103, median = 85, mode = 85;
Without outlier: mean = 85, median = 85, mode = 85;
mean
29. With outlier: mean = 101, median = 102, mode = 110;
Without outlier: mean = 105.625, median = 106, mode = 110;
mean
31. mean = 35.875; median = 44; mode = 48;
Sample answer: The median is best because the mean is less than most of the data and the mode is the greatest value.
33. mean = 12; median = 8; mode = 2;
Sample answer: The median is best because the mean is greater than most of the data and the mode is the least value.

35. *Sample answer:* Both the median and mode are the best measures for the high temperatures; Both the mean and median are the best measures for the low temperatures; Median and mode were chosen for high temperatures because they are both close to most of the values. Mean and median were chosen for the low temperatures because there were two modes.

37. 10; you need to work 12 more hours to make the mean 10 hours but only 10 more hours to make the median and mode 10 hours.

39. a. no; The price is the mode, but it is the lowest price. Most fitness wristbands cost more.
   b. By advertising the lowest price, they are likely to draw more customers to the store.
   c. *Sample answer:* Knowing all the measures can help you know whether the store has many models in your price range.

Section 9.4
Review & Refresh
1. mean = 7.1; median = 7; mode = 4
3. mean = 16; median = 16; modes = 15, 17
5. >
7. >
9. 480 mm²
11. 96.6 ft²

Concepts, Skills, & Problem Solving
13. *Sample answer:*

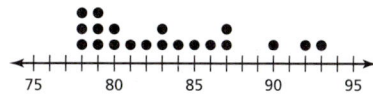

15. 33
17. 23
19. 7.3
21. 5
23. 7
25. 11

27. range = 122; The numbers of tornados in Alabama for several years vary by no more than 122; IQR = 39; The middle half of the numbers of tornados in Alabama vary by no more than 39; outlier = 145

29. yes; *Sample answer:* If all the data values in the set were the same number, the range and interquartile range would be the same.

31. a. range = 172 points; IQR = 42 points
   b. 193 points; range = 101 points; IQR = 34 points; range

33. a. Show A: mean = 20, median = 19.5, range = 13, IQR = 5;
Show B: mean = 21, median = 20.5, range = 23, IQR = 6;

The mean and the median ages for the shows, and the interquartile ranges of ages for the shows, are about the same. The range of the ages for Show A is less than the range for Show B, so the ages for Show B are more spread out.

b. Show A: The measures do not change by a large amount because 21 is towards the middle of the data set.
Show B: The mean, median, and IQR all change by small amounts but the range changes by a large amount because 36 is an outlier of the data set.

Section 9.5
Review & Refresh
1. range = 22; IQR = 13

3.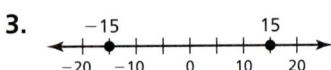

5.
```
  -16           16
◄──●──┼──┼──┼──●──►
-20 -10  0  10  20
```

7. faces: 7; edges: 15; vertices: 10

9. $14s = 49$

11. $\dfrac{36}{g} = 9$

Concepts, Skills, & Problem Solving
13. $7.20

15. 4; The values differ from the mean by an average of 4.

17. 4.4; The values differ from the mean by an average of 4.4.

19. 2; The values differ from the mean by an average of 2.

21. 1.45; The values differ from the mean by an average of 1.45.

23. range = 14, IQR = 8, MAD = 4; The prices vary by no more than $14, the middle half of the prices vary by no more than $8, and the admission prices differ from the mean price by an average of $4.

25. Your collection: mean = 1929, median = 1930, no mode, range = 54, IQR = 48, MAD = 23.75;
Your friend's collection: mean = 1929, median = 1929.5, no mode, range = 15, IQR = 6; MAD = 3.5;
Sample answer: The measures of center for the data sets are almost identical. But the measures of variation for your friend's coin collection are much less than the measures for your coin collection. This means that the years of the coins in your friend's collection are closer together than the years of the coins in your collection.

27. *Sample answer:* The MAD for gumballs is greater than the MAD for baseballs. In general, guesses for gumballs will *deviate* more because there are many more in the jar, making it harder to guess and producing a larger range of guesses.

29. *Sample answer:* MAD; The range only uses two data values from a set and is greatly affected by outliers. The interquartile range ignores outliers but also only uses two data values from a set. When calculating the mean absolute deviation of a data set, you use all of the values.

Chapter 10
Section 10.1
Review & Refresh
1. MAD = 2; The data values differ from the mean by an average of 2.

3. MAD = 9; The data values differ from the mean by an average of 9.

5. $5n + 40$

7. $28b + 42$

9. $p = 24$

11. $d = 12$

Concepts, Skills, & Problem Solving
13. *Sample answer:* How many times did 40 or more customers visit the store?

15. Books Read

Stem	Leaf
0	9
1	5 7 9
2	0 5 6 6 9
3	1 2
4	0

Key: 1 | 5 = 15 books

17. Test Scores

Stem	Leaf
6	2 5 8 9
7	
8	0 1 2 5 7 7 8
9	0 1 5 7

Key: 8 | 1 = 81%

19. Bikes Sold

Stem	Leaf
7	8 9
8	1 6
9	6 9 9
10	0 5 8
11	2 5

Key: 11 | 2 = 112 bikes

21. no; The stem of 3 should be included.

23. mean = 56.6 cases; median = 53 cases; modes = 41 cases, 43 cases, 63 cases; range = 56 cases; IQR = 20 cases

25. 97 cases; It increases the mean.

27. a. Sample answer:

Stem	Leaf
6	2 5 7 7 8 8 9
7	0 0 0 1 1 3 4 6 8 9
8	2 3 5 5 6 6 7 7 8 8 8 9 9

Key: 6 | 7 = 67°F

b. Sample answer: about 77°F

Section 10.2
Review & Refresh

1. Blog Posts

Stem	Leaf
0	4 4 5 6 7 8 9
1	1 2 2 5
2	0

Key: 2 | 3 = 23 blog posts

3. 45

5. 22.4

7. C

Concepts, Skills, & Problem Solving

9. Sample answer:

Interval	Tally	Total
20–29	I	1
30–39	III	3
40–49	IIII II	7
50–59	IIII	4

11.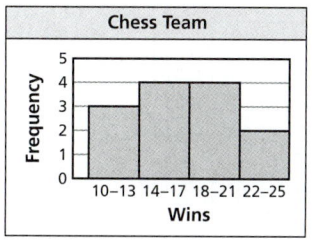

13.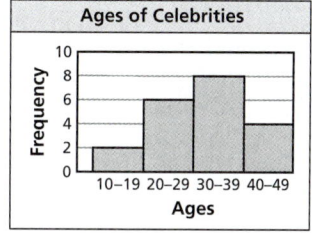

15.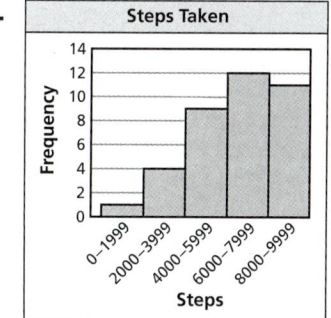

17. a. 4–5 magazines read

b. 20 students

c. 85%

19. a. no; The histogram shows that only one state fell in the interval of 40%–44.9%. This state did not necessarily have 40% of possible voters vote.

b. yes; 36 states are between 50% and 64.9%.

c. no; The 55%–59.9% interval has the highest frequency, but does not necessarily contain the mode of the data.

21. stem-and-leaf plot; You need to know the specific data values, the intervals in the histogram do not give enough information.

23. a.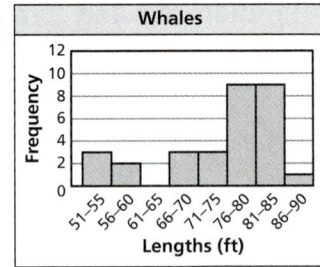

b.

c. *Sample answer:* The second histogram has four intervals and it does not have a gap as in the first histogram.

Section 10.3
Review & Refresh
1.

3.

5. 40 km/h

Concepts, Skills, & Problem Solving
7. flat

9. symmetric

11. skewed left

13. symmetric; The data on the left are a mirror image of the data on the right.

15. yes; *Sample answer:* When the stem-and-leaf plot has most of its values on the bottom, the distribution is skewed left because when the data are placed in a dot plot or histogram, most of the data values will be on the right. When the stem-and-leaf plot is a mirror image on the top and bottom for the numbers of values, the distribution is symmetric. When the stem-and-leaf plot has most of its values on the top, the distribution is skewed right because when the data is placed in a dot plot or histogram, most of the data values will be on the left.

17. a. skewed right; A: Mode, B: Median, C: Mean; *Sample answer:* A is the value with the highest frequency, about half of the data are to the left of B and about half are to the right of B

b. skewed left; A: Mean, B: Median, C: Mode; *Sample answer:* C is the value with the highest frequency, about half of the data are to the left of B and about half are to the right of B

Section 10.4
Review & Refresh
1. skewed left

3. median = 70; $Q_1 = 65.5$; $Q_3 = 75$; IQR = 9.5

5. $2\frac{1}{5}$ **7.** $2\frac{1}{8}$

Concepts, Skills, & Problem Solving
9. mean = 7; median = 7.5; The distribution is skewed left, so the median is the most appropriate measure of center.

11. median and IQR **13.** mean and MAD

15. Exercise 11: The price of jeans is centered around 42 dollars and the middle half of the prices vary by no more than 6 dollars; Exercise 12: The time spent volunteering centered around approximately 5.78 hours and the values differ from the average by about 1.47 hours.

17. The typical vertical jump height is about 30.45 inches. The data values differ from the mean by an average of about 2.31 inches.

19. *Sample answer:* Pile A; Pile A has less variability than Pile B.

21. a. median = 3 tokens

b. *Sample answer:* You will be most likely to pick a voucher for 1 token because that is the mode.

c. no; *Sample answer:* In part (a), The most appropriate measure to describe the center was found. In part (b), the number of tokens that appears most was found.

Section 10.5
Review & Refresh
1. mean and MAD
3. >
5. >
7. $6(2x - 3)$
9. $5(12g - 5h)$

Concepts, Skills, & Problem Solving
11. *Sample answer:* The top plot has a distribution that is skewed left while the bottom plot has a distribution that is skewed right. The top plot has less variability in its scores within its quartiles compared to the bottom plot. The bottom plot has a lower median and quartile values in comparison to the top plot.

13. (box-and-whisker plot: 5, 8, 9, 10.5, 12; Quiz score)

15. (box-and-whisker plot: 68, 84.5, 87.5, 94, 100; Score)

17. (box-and-whisker plot; Ski length (centimeters))

19. (box-and-whisker plot; Days camping)

    12 days camping

21. a. about 75%

    b. above 345 meters; The right whisker is longer.

    c. 84 m; The middle half of the data varies by no more than 84 meters.

23. skewed left; The left whisker is longer than the right whisker, and most of the data are on the right.

25. symmetric; The whiskers are about the same length, and the median is in the middle of the box.

27. a. School 1 is skewed left and School 2 is skewed right.

    b. School 2; The range and the IQR of School 2 are greater than those of School 1.

    c. class from School 1; At any time School 1 has more data on the left than School 2.

29.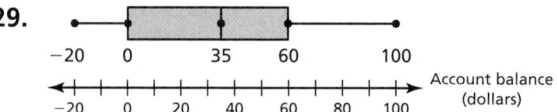
Account balance (dollars)

31. *Sample answer:* 10, 15, 20, 20, 25, 30, 30, 35, 40, 40, 45, 50

33. *Sample answer:*

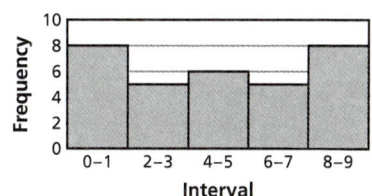

35. *Sample answer:* box-and-whisker plot; A stem-and-leaf plot of a large data set would be impractical. The size of the display would be massive and difficult to use. While a histogram can show the variability of a data set, it is not as accurate and obvious as a box-and-whisker plot due to intervals and bar size.

Chapter A
Section A.1
Review & Refresh
1. $6 : 4$
3. $4 : 10$
5. 4
7. 7

Concepts, Skills, & Problem Solving
9. $4; -6$; *Sample answer:* 4 is farther right, $|-6| > |4|$

11. $-\frac{4}{5}; -1.3$; *Sample answer:* $-\frac{4}{5}$ is farther right, $|-1.3| > \left|-\frac{4}{5}\right|$

13. 2
15. 10
17. $\frac{1}{3}$
19. $\frac{5}{9}$
21. 3.8
23. $\frac{15}{4}$
25. 18.26
27. $5\frac{1}{6}$
29. <
31. >
33. <
35. >

37. no; The absolute value of a number cannot be negative.

39. *Sample answer:* -4

41. a. airplane fuel   b. butter

43. $-7.2, -6.3, |5|, |-6.3|, 8$

45. $-\frac{1}{2}, \left|\frac{1}{4}\right|, \frac{5}{8}, \left|-\frac{3}{4}\right|, \left|-\frac{7}{8}\right|$

47. 3:00 P.M.; 12:00 P.M.

49. $n \leq 0$

51. false; The absolute value of zero is zero, which is neither positive nor negative.

Section A.2
Review & Refresh

1. < **3.** >

5. 6.406 **7.** $\dfrac{5}{8}$

9.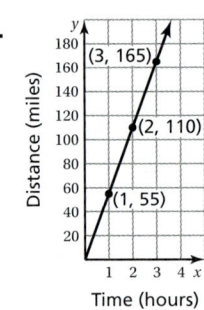

Concepts, Skills, & Problem Solving

11. integers with the same sign; -10; negative

13. $(-2) + 4$; 2 **15.** $-5 + 2$; -3

17. -10 **19.** 7

21. 0 **23.** 10

25. -4 **27.** -11

29. -4 **31.** -34

33. no; $-10 + (-10) = -20$

35. $48

37. Sample answer: $-26 + 1$; $-12 + (-13)$

39. Use the Associative Property to add 13 and -13 first. -8

41. Sample answer: Use the Commutative Property to switch the last two terms. -12

43. Sample answer: Use the Commutative Property to switch the last two terms. 11

45. -4 **47.** 21

49. -85

51. 7 units to the left of p

53. a distance of q units away from p

55. -3

57. yes; $650 + 530 + 52 + (-28) + (-75) = 1129$

59. Sample answer: You filled your water bottle with 12 ounces of water this morning and then drank 12 ounces.

61. $d = -10$ **63.** $m = -7$

65. a. $p = -q$; Sample answer: Subtract q from each side.

 b. $p < -q$; Sample answer: Subtract q from each side.

 c. $p > -q$; Sample answer: Subtract q from each side.

Section A.3
Review & Refresh

1. 15 **3.** -5

5. 31 **7.** 114

9. D

Concepts, Skills, & Problem Solving

11. Sample answer: $\dfrac{1}{3}, \dfrac{5}{3}$; $\dfrac{5}{3} + \left(-\dfrac{4}{3}\right) = \dfrac{1}{3}$

13. $-1\dfrac{4}{5}$ **15.** $-\dfrac{5}{14}$

17. $-2\dfrac{5}{6}$ **19.** -57.19

21. no; The sum is -3.95.

23. Sample answer: You earn $1.25 doing chores and buy a sandwich for $1.25. You have no money left.

25. $-$5.35

27. The sum will be positive when the addend with the greater absolute value is positive. The sum will be negative when the addend with the greater absolute value is negative. The sum will be zero when the numbers are opposites.

29. Sample answer: Use the Commutative Property to switch the last two terms; -6.21

31. Sample answer: Use the Commutative Property to switch the last two terms and the Associative Property to regroup; $4\dfrac{1}{10}$

33. Sample answer:

$-4.3 + \dfrac{4}{5} + 12$

$= -4.3 + 12 + \dfrac{4}{5}$ Comm. Prop. of Add.

$= 7.7 + \dfrac{4}{5}$ Add -4.3 and 12.

$= 7.7 + 0.8$ Write $\dfrac{4}{5}$ as a decimal.

$= 8.5$ Add.

35. 8 **37.** $\dfrac{1}{2}$

Section A.4
Review & Refresh
1. $\frac{1}{3}$
3. $-5\frac{1}{2}$
5. 19.923
7. $-1 + (-3)$

Concepts, Skills, & Problem Solving
9. 2
11. 4
13. $-2 + 5$; $-2 - (-5)$; 3
15. 13
17. -5
19. -10
21. 3
23. 17
25. 1
27. -22
29. 20
31. $-2 - 9$; *Sample answer:* The air temperature is 9°C colder, so subtract 9.
33. -14 m
35. *Sample answer:* Write the subtraction as addition. Then use the Commutative Property to switch the last two terms; -7
37. Use the Associative Property to add 8 and -8 first; -5
39. *Sample answer:* Use the Commutative Property to switch the first two terms and the Associative Property to regroup; -28
41. -5
43. -17
45. a. February b. 130°F
47. always; It's always positive because the first integer is always greater.
49. never; It's never positive because the first integer is never greater.
51. when a and b have the same sign and $|a| \geq |b|$ or $b = 0$

Section A.5
Review & Refresh
1. 4
3. -19
5. 64 ft³
7. $-2, 1, |3|, |-4|, 6$

Concepts, Skills, & Problem Solving
9. $\frac{1}{9}, \frac{3}{9}$; $\frac{1}{9}, \frac{2}{9}$
11. $1\frac{1}{2}$
13. -3.5
15. -2.6
17. $-18\frac{13}{24}$
19. $\frac{1}{18}$
21. 14.963

23. no; $\frac{3}{2} - \frac{9}{2} = \frac{3}{2} + \left(-\frac{9}{2}\right) = -3$

25. *Sample answer:* A judge deducts $\frac{5}{8}$ point from an athlete's score, then removes the deduction after watching a tape of the athlete from another angle.

27. no; $\frac{1}{12}$ oz

29. *Sample answer:* Use the Commutative Property to switch the first two numbers; $\frac{2}{3}$

31. *Sample answer:* Write the subtraction as addition. Then use the Commutative Property to switch the middle two numbers and the Associative Property to add the first two numbers and the last two numbers; 7

33. *Sample answer:* Write the subtraction as addition. Then use the Commutative Property to switch the last two numbers and the Associative Property to regroup; 3.8

35. 3.2
37. 10.6
39. 9.21
41. 5.556
43. $13\frac{11}{12}$
45. $-1\frac{7}{8}$ mi
47. 4.48
49. $7\frac{5}{9}$
51. $2\frac{7}{10}$
53. a. 4.03 in.
    b. The rainfall is 1.73 inches below the historical average for the year.
55. sometimes; It is positive only if the first fraction is greater.
57. 2; 8; 85

Chapter B

Section B.1
Review & Refresh
1. 5.1
3. $1\frac{3}{5}$
5. 8
7. 19.125

Concepts, Skills, & Problem Solving
9. -8
11. -20
13. -21
15. 12
17. 27
19. 12
21. 0
23. -30
25. 78
27. 121

29. $-320{,}000$
31. -36
33. 0
35. -59
37. 54
39. 12
41. -3
43. yes; The opposite of 10^2 is -100.
45. $1792, -7168$
47. about 45.83 min; *Sample answer:* Solve $22{,}000 + (-480t) = 0$.
49. -25

Section B.2
Review & Refresh
1. 80
3. 28
5. $0.24, \dfrac{1}{4}, 28\%$
7. $0.69, \dfrac{7}{10}, 71\%, 0.84, \dfrac{9}{10}$
9. $-3 + 3; -3 - (-3); 0$

Concepts, Skills, & Problem Solving
11. integers with different signs; -2; negative
13. -2
15. -5
17. -7
19. 3
21. -3
23. 0
25. -4
27. -13
29. no; The quotient should be positive.
31. 15
33. 4
35. -2
37. -2
39. -8
41. 65
43. 5
45. **a.** -2
  b. Score -6 or less in round 4.
47. *Sample answer:* $-20, -15, -10, -5, 0$; Start with -10, then pair -15 with -5 and -20 with 0. The sum of the integers must be $5(-10) = -50$.

Section B.3
Review & Refresh
1. -2
3. 6
5. 35.28
7. 74.16441
9.

Hours	2	8	$\dfrac{4}{3}$
Dollars Earned	18	72	12

$2 : 18; 8 : 72; \dfrac{4}{3} : 12$

Concepts, Skills, & Problem Solving
11. repeats; 7 is not a factor of a power of 10
13. repeats; 24 is not a factor of a power of 10
15. $0.\overline{09}$
17. $-0.\overline{7}$
19. $1.8\overline{3}$
21. $1.041\overline{6}$
23. $-2.9\overline{4}$
25. $8.6\overline{81}$
27. $-\dfrac{9}{10}$
29. $-\dfrac{129}{500}$
31. $-2\dfrac{8}{25}$
33. $6\dfrac{3}{250}$
35. **a.** 0.55 **b.** $\dfrac{11}{20}$
37. $<$
39. no; $9\dfrac{5}{6} = 9.8\overline{3}$
41. $-\dfrac{7}{3}, -\dfrac{3}{4}, 0.5, \dfrac{2}{3}, 1.2$
43. $-\dfrac{8}{5}, -1.4, -0.9, \dfrac{1}{4}, 0.6$
45. $-\dfrac{7}{2}, -2.8, -\dfrac{5}{4}, 1.3, \dfrac{4}{3}$
47. $-1\dfrac{91}{200}, -1\dfrac{5}{11}, -1.45, -\dfrac{7}{5}$
49. math quiz
51. **a.** when a is negative, and a is not a factor of a power of 10
  b. when a and b have the same sign, $a \neq 0 \neq b$, and a and b are factors of a power of 10

Section B.4
Review & Refresh
1. 0.3125
3. $6.\overline{72}$
5. 36 in.^{2}
7. 121 ft^2

Concepts, Skills, & Problem Solving
9. *Sample answer:* $0.7 \cdot 0.5; 0.35$
11. negative; The numbers have different signs.
13. $\dfrac{1}{3}$
15. $2\dfrac{1}{2}$
17. -0.012
19. 0.36
21. $-4\dfrac{17}{27}$
23. no; $-2.2 \times (-3.7) = 8.14$
25. $3\dfrac{5}{8}$ gal
27. $-\dfrac{3}{8}$
29. -172
31. $-1\dfrac{1}{14}$

33. $4951.45; The total length is $191\frac{11}{12}$ yards.

35. $1\frac{1}{5}$

37. $-2\frac{1}{10}$

39. $-5\frac{11}{24}$

41. *Sample answer:* $\frac{1}{2} \times \left(-\frac{5}{4}\right) \times \frac{-2}{5}$

Section B.5
Review & Refresh

1. -0.655

3. $\frac{1}{8}$

5. Terms: $14, z, 6f$; Coefficients: $1, 6$; Constant: 14

7. Terms: $42m, 18, 12c^2$; Coefficients: $42, 12$; Constant: 18

Concepts, Skills, & Problem Solving

9. *Sample answer:* $0.06 \div 0.6; 0.06 \div 0.1; 0.1; 0.6$

11. -6

13. $3.\overline{63}$

15. -2.45

17. $\frac{2}{5}$

19. no; $-\frac{2}{3} \div \frac{4}{5} = -\frac{2}{3} \times \frac{5}{4}$

21. 6

23. $-1\frac{47}{100}$

25. 1.3

27. $\frac{1}{9}$

29. a. -0.02 in.

  b. 0.03 in.; $\dfrac{-0.05 + 0.09 + (-0.04) + (-0.08) + 0.03}{5} = -0.01$

31. a. always; *Sample answer:* The number of decimal places in the product is the sum of the numbers of decimal places in the factors.

  b. sometimes; *Sample answer:* $0.5 \div 0.25 = 2$, $0.5 \div 0.3 = 1.\overline{6}$

Chapter C

Section C.1
Review & Refresh

1. $-\frac{1}{2}$

3. $-6\frac{1}{2}$

5. $15\%, 1450\%, 14.8, 15\frac{4}{5}$

Concepts, Skills, & Problem Solving

7. no; Expression 2 simplifies to $-5x$.

9. Terms: $t, 8, 3t$; Like terms: t and $3t$

11. Terms: $2n, -n, -4, 7n$; Like terms: $2n, -n$ and $7n$

13. Terms: $1.4y, 5, -4.2, -5y^2, z$; Like terms: 5 and -4.2

15. no; The terms are $3x, -5, -2x,$ and $9x$ and the like terms are $3x, -2x,$ and $9x$.

17. $11x + 2$

19. $4b - 5$

21. $-2.3v - 5$

23. $3 - \frac{1}{2}y$

25. $10.2x$; each hiker carries 10.2 pounds of equipment; 40.8 lbs

27. *Sample answer:* $15x^2 - 6x^2 + 2x^2 + 2y + y$;

$15x^2 - 6x^2 + 2x^2 + 2y + y$
$= [15x^2 + (-6x^2) + 2x^2] + (2y + y)$
$= 11x^2 + 3y$, and $8x^2 + 3x^2 + 3y = 11x^2 + 3y$

29. Find the difference of the two distances. Divide the difference by 60 minutes to determine the distance per minute. This difference of the distances traveled in miles per minute can be multiplied by the number of minutes to find the difference of the distances traveled.

31. a. 153 in.^{2};
Area $= 240 - 32x + x^2 = 240 - 32(3) + 3^2$
$= 240 - 96 + 9 = 153$ in.^{2}

  b. *Sample answer:* England

Section C.2
Review & Refresh

1. $15f$

3. $-11z - 3$

5. $\frac{2}{5}$

7. D

Concepts, Skills, & Problem Solving

9. $(2x + 7) - (2x - 4) = 11$

11. $2b + 9$

13. $6x - 18$

15. $-3\frac{3}{10}z - 11$

17. $\frac{3}{2}x + 2y - 3.5$

19. $-3g - 4$

21. $-7y + 20$

23. $-3\frac{1}{8}c + 15\frac{5}{8}$

25. $-6m - \frac{1}{4}n + 8\frac{1}{3}$

27. no; Your friend dropped the second set of parentheses instead of adding the opposite of the second expression.

29. no; If the variable terms are opposites, the sum is a numerical expression.

31. a. $(55w + 145)$ dollars;
$(10w + 120) + (45w + 25) = 55w + 145$

b. $145; $55

c. *Sample answer:* You would need to have a total of $650 in the accounts.
$$55w + 145 = 650$$
$$55w = 505$$
$$w = 9.\overline{18}$$
After 10 weeks, you can buy the new phone.

Section C.3
Review & Refresh
1. $6b - 1$
3. $-6z - 5$
5. -36
7. -56
9. $|3.5| < |-5.8|$

Concepts, Skills, & Problem Solving
11. $6.5(3x + 4) - 6.5(x + 2), 6.5(2x + 2)$;
$6.5(3x + 4) - 6.5(x + 2) = 19.5x + 26 - 6.5x - 13$
$= 13x + 13$;
$6.5(2x + 2) = 13x + 13$
13. $3a - 21$
15. $-15m + 20$
17. $13.5s + 27$
19. $-2p + 2\frac{2}{5}$
21. $8y + 16$
23. $6d + 49.8$
25. $1\frac{1}{2}y - \frac{3}{8}z$
27. $3\frac{1}{3}a + 15b$
29. no; $-3(4 - 5b + 7) = -3(11 - 5b)$
$= -3(11) - (-3)(5b)$
$= -33 + 15b$
31. $-20a - 28$
33. $-4\frac{1}{4}p + 25$
35. $\frac{2}{7}n - 1\frac{4}{7}$
37. $7200 + 57d$; For the 6 snow fans, it costs $7200 to buy them and $57 per day to operate them.
39. *Sample answer:*

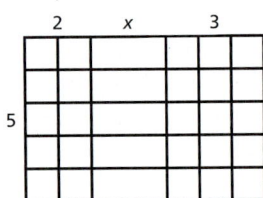

$5x + 25$

41. a. $8s + 16$; *Sample answer:*

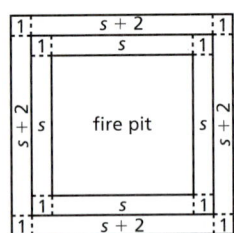

b. $4s + 8n - 4$; Row 1 is $4s + 4$ and row 2 is $4(s + 2) + 4 = 4s + 12$, so row n is $4(s + 2(n - 1)) + 4$.

Section C.4
Review & Refresh
1. $8k - 40$
3. $\frac{1}{2}g + 1\frac{1}{2}$
5. -10.3
7. 2
9. -8

Concepts, Skills, & Problem Solving
11. *Sample answer:* $\frac{2}{3}\left(\frac{5}{4}m + n\right)$; Factor out $\frac{2}{3}$ from each area. Because $\frac{2}{3}$ is the width of the smaller rectangles, the two lengths are $\frac{5}{4}m$ and n.
13. $16(2z - 3)$
15. $3(y - 8)$
17. $2(3 + 8k)$
19. $4(5z - 2)$
21. $4(9a + 4b)$
23. $3(4 + 3g - 10h)$
25. $\frac{1}{3}(b - 1)$
27. $2.2(x + 2)$
29. $0.8(w + 4.5)$
31. $4\left(h - \frac{3}{4}\right)$
33. $\frac{3}{8}\left(z + \frac{8}{3}\right)$
35. $\frac{5}{2}\left(k - \frac{4}{5}\right)$
37. yes; Your friend factored out the $\frac{2}{3}$ properly from the sum and correctly rewrote the expression.
39. $-6\left(-3z + \frac{5}{2}\right)$
41. $-\frac{1}{2}(x - 12)$
43. $-\frac{1}{4}(2x + 5y)$
45. *Sample answer:* Because the mat is a square, all sides are the same length. The perimeter is $12x - 32$, and to find the dimension of each side, divide each term by 4. The length of each side is $3x - 8$ and $4(3x - 8) = 12x - 32$.
47. $30 + 9.75p$; For each kit, the pen costs $30 and each package of plastic costs $9.75.

Chapter D

Section D.1
Review & Refresh
1. $p \geq 3$

3. $d \geq -7$

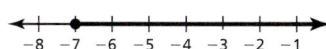

5. 0.84
7. D

Concepts, Skills, & Problem Solving
9. Sample answer:

x	$\frac{4}{5}$	8	2	$\frac{14}{5}$
y	$\frac{1}{2}$	5	$\frac{5}{4}$	$\frac{7}{4}$

yes; The values of the ratios are $\frac{8}{5}$.

11. $\frac{1}{2} : 3; \frac{1}{6}$; The amount of mint leaves is $\frac{1}{6}$ the amount of chopped watermelon.

13. $\frac{3}{4} : \frac{1}{2}; 1\frac{1}{2}$; The amount of lime juice is $1\frac{1}{2}$ the amount of mint leaves.

15. $80, \frac{1}{12}, \frac{3}{4}; 20 : \frac{1}{6}; 80 : \frac{2}{3}; 10 : \frac{1}{12}; 90 : \frac{3}{4}$

17. $\frac{1}{12}, 1\frac{1}{2}, \frac{1}{48}; \frac{1}{24} : \frac{1}{2}; \frac{1}{12} : 1; \frac{1}{8} : 1\frac{1}{2}; \frac{1}{48} : \frac{1}{4}$

19. yes; There are 2 boys for every 3 girls.
21. 300 m²
23. 108 pounds of mulch; 72 pounds of gravel

Section D.2
Review & Refresh
1. $\frac{3}{2}, \frac{4}{3}, \frac{4}{9}; \frac{3}{4} : \frac{1}{3}; \frac{3}{2} : \frac{2}{3}; 3 : \frac{4}{3}; 1 : \frac{4}{9}$
3. >
5. =

Concepts, Skills, & Problem Solving
7. $210°; \frac{360°}{h} \times \frac{7}{12} h = 210°$
9. 60 mi : 1 h
11. $\frac{1}{10}$ lb : 1 day
13. $2.40 : 1 lb
15. 54 words : 1 min
17. $2\frac{1}{2}$ oz : 1 pt
19. $75, 100, \frac{2}{3}; 25 : \frac{1}{3}; 50 : \frac{2}{3}; 75 : 1; 100 : \frac{4}{3}$
21. 2.2 million people per year
23. no; $75 \div 1.5 = 50$ lb per yr; $38.4 \div 0.75 = 51.2$ lb per yr
25. a. whole milk b. orange juice
27. a. Blue: more than 1500 gallons per minute
       Green: 1000–1499 gallons per minute
       Yellow: 500–999 gallons per minute
       Red: less than 500 gallons per minute
    b. Sample answer: If a firefighter pumps water out at too high a rate, the pipes in the ground could burst.

Section D.3
Review & Refresh
1. 6 in. : 1 yr
3. 7000 steps : 1 h
5. -13
7. -18
9. $x = 150$
11. $x = 75$

Concepts, Skills, & Problem Solving
13. 24 min
15. 48.6 min
17. yes
19. yes
21. no
23. no
25. no
27. yes
29. no; $\frac{8}{3} \neq \frac{9}{4}$
31. yes
33. no
35. yes
37. you: 1.1 beats per second, friend: 1.2 beats per second; No, the rates are not equivalent.
39. yes; The value of the ratio of height to base for both triangles is $\frac{4}{5}$.
41. a. no; Sample answer: Adding two pennies and two dimes to the coins will give a ratio of 5 pennies : 4 dimes. This ratio is not equivalent to 3 pennies : 2 dimes.
    b. yes; Sample answer: Adding two pennies and two dimes to the coins will give a ratio of 6 pennies : 6 dimes. This ratio is equivalent to 4 pennies : 4 dimes.
43. 30 min; $\frac{60}{2300} = \frac{x}{1150}, x = 30$

45. a. *Sample answer:* Machines at a factory that produce an output of a certain amount per unit of time.
    b. *Sample answer:* Running 10 laps during gym class. The time it takes to run each lap is rarely exactly the same.

47.

x	10	4	6	1
y	5	2	3	0.5

$\frac{10}{5} = \frac{4}{2} = \frac{6}{3} = \frac{1}{0.5}$

Section D.4
Review & Refresh
1. no

3 and 5.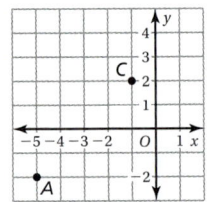

7. D

Concepts, Skills, & Problem Solving
9. 150 mi
11. 90 mi
13–27. Explanations will vary.
13. $y = 16$
15. $b = 20$
17. $v = 10.5$
19. $n = 15$
21. $y = 7\frac{1}{3}$
23. $d = 15$
25. $n = 10$
27. $k = 5.4$
29. yes; Both cross products give the equation $3x = 60$.
31. $\dfrac{n \text{ winners}}{85 \text{ entries}} = \dfrac{34 \text{ winners}}{170 \text{ entries}}$
33. $\dfrac{100 \text{ meters}}{x \text{ seconds}} = \dfrac{200 \text{ meters}}{22.4 \text{ seconds}}$
35. $x = 16$
37. $x = 1$
39. $x = 4$
41. 3 trombones
43. 90 adults
45. $15
47. 4 bags
49. about 3.72
51. about 40.5
53. $d = 12$
55. no; The relationship is not proportional. It should take more people less time to build the swing set.

57. $\dfrac{1}{5}$; $\dfrac{m}{k} = \dfrac{\frac{n}{2}}{\frac{5n}{2}} = \dfrac{n}{2} \cdot \dfrac{2}{5n} = \dfrac{1}{5}$

Section D.5
Review & Refresh
1. $x = 28$
3. $x = 1\frac{1}{3}$
5. -7
7. 11
9. $p \geq 12$

11. $w \geq -5$

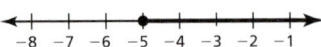

Concepts, Skills, & Problem Solving
13.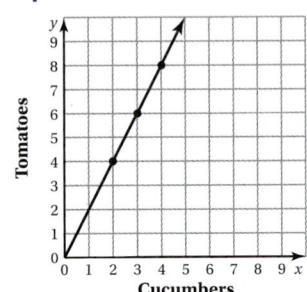

yes; The line passes through the origin.
15. no; The line does not pass through the origin.
17. yes; The line passes through the origin.
19. yes; The line passes through the origin.
21. (0, 0): You earn $0 for working 0 hours; (1, 15): You earn $15 for working 1 hour; (4, 60): You earn $60 for working 4 hours; $15 : 1 h
23. yes; $k = 1$; The equation can be written as $y = kx$.
25. no; The equation cannot be written as $y = kx$.
27. $k = 24$; $y = 24x$
29. $k = \dfrac{9}{8}$; $y = \dfrac{9}{8}x$
31. yes; $k = 13$; The cost of 1 ticket is $13; $182
33. about 76,000 mg

Section D.6
Review & Refresh
1. no; The line does not pass through the origin.
3. $13p$
5. $\dfrac{3}{20}b + 4$
7. $7 + z > 5$

Concepts, Skills, & Problem Solving

9. Sample answer: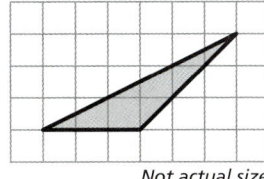
   Not actual size

11. 100 mi
13. 200 mi
15. 75 in.
17. 3.84 m
19. 17.5 mm
21. 1 cm : 10 mm; 1
23. a. 30 cm; 31.25 cm²   b. 9600 m; 3,200,000 m²

25.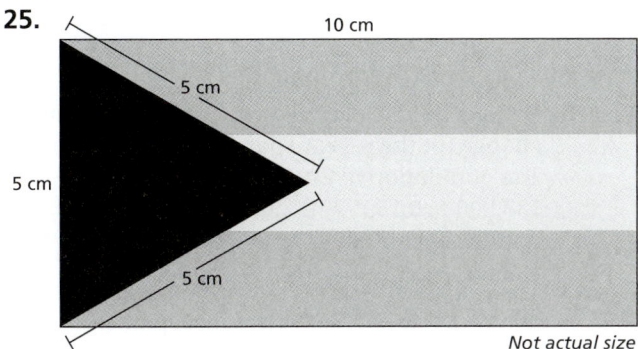
    Not actual size

27. The value of the ratio of the perimeters is the scale factor and the value of the ratio of the areas is the square of the scale factor. *Sample answer:* If 2 similar figures have a scale of $a:b$, then the ratio of their perimeters is $a:b$ and the ratio of their areas is $a^2:b^2$.

Chapter E

Section E.1
Review & Refresh
1. 4.5 in.
3. $10p - 12$
5. D

Concepts, Skills, & Problem Solving
7. $\frac{5}{6}$; Sample answer: convert both to decimals, $0.8\overline{3} > 0.82$

9. 63%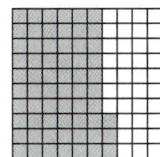
11. 60%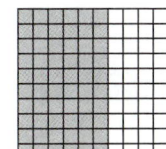
13. 0.55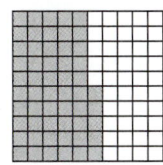

15. 355.4%

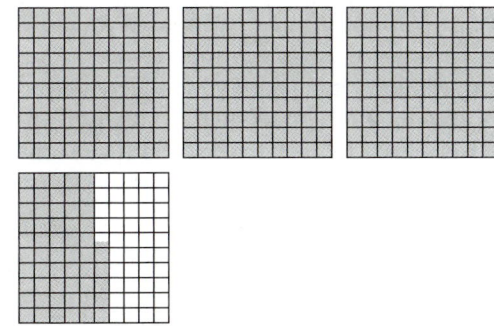

17. 4.1%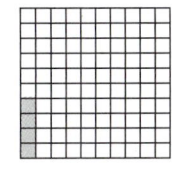
19. $0.49\overline{92}$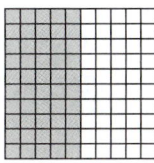

21. 0.29, 29%
23. 0.875, 87.5%
25. $0.\overline{7}, 77.\overline{7}\%$
27. 4.5, 450%
29. $2.8\overline{3}, 283.\overline{3}\%$
31. $0.00\overline{13}, 0.\overline{13}\%$
33. 0.66, 66.1%, $\frac{2}{3}$, 0.667
35. A
37. D
39. Grade 7, Grade 6, Grade 8
41. $1.\overline{1}$ times more shots
43. 14 or more students; $0.44 \times 30 = 13.2$. Because 13.2 is a decimal and over 44% of students read a book last month, round the value up to 14.

Section E.2
Review & Refresh
1. 0.42, 42%
3. $1.\overline{4}, 144.\overline{4}\%$
5. 3
7. -2.5

Concepts, Skills, & Problem Solving
9. 16
11. 50
13. 40%
15. $\frac{12}{25} = \frac{p}{100}; p = 48$
17. $\frac{9}{w} = \frac{25}{100}; w = 36$
19. $\frac{a}{124} = \frac{75}{100}; a = 93$
21. $\frac{a}{40} = \frac{0.4}{100}; a = 0.16$
23. yes; Your friend wrote and solved the correct percent proportion.
25. $6000
27. $\frac{14.2}{w} = \frac{35.5}{100}; w = 40$
29. $\frac{a}{\frac{7}{8}} = \frac{25}{100}; a = \frac{7}{32}$
31. $8.40

33. **a.** a scale along the vertical axis

    b. 6.25%; *Sample answer:* Although you do not know the actual number of votes, you can visualize each bar as a model with the horizontal lines breaking the data into equal parts. The sum of all the parts is 16. Raven has the least parts with 1, which is $100\% \div 16 = 6.25\%$.

    c. 31 votes

35. $5x$

Section E.3
Review & Refresh

1. $\dfrac{30}{100} = \dfrac{9}{w}$; $w = 30$

3. $\dfrac{p}{100} = \dfrac{20}{36}$; $p = 55.\overline{5}$

5. 14

7. $4\dfrac{1}{10}$

9. A

Concepts, Skills, & Problem Solving

11. 40%

13. $a = 0.2 \cdot 150$; 30

15. $35 = 0.35 \cdot w$; 100

17. $29 = p\% \cdot 20$; 145%

19. $51 = p\% \cdot 300$; 17%

21. yes; The percent was converted to a decimal and multiplied by the "whole".

23. $200

25. 8%

27. 70 years old

29. If the percent is less than 100%, the percent of a number is less than the number; 50% of 80 is 40; If the percent is equal to 100%, the percent of a number is equal to the number; 100% of 80 is 80; If the percent is greater than 100%, the percent of a number is greater than the number; 150% of 80 is 120.

31. false; If W is 25% of Z, then $Z : W$ is $100 : 25$, because Z represents the whole.

33. 92%

Section E.4
Review & Refresh

1. $a = 0.25 \cdot 64$; 16

3. $5 = 0.05 \cdot w$; 100

5. $-\dfrac{2}{7}$

7. $\dfrac{1}{3}$

Concepts, Skills, & Problem Solving

9. 20%

11. 6%

13. decrease; 66.7%

15. increase; 225%

17. decrease; 12.5%

19. decrease; 37.5%

21. 12.5% decrease

23. **a.** $11.\overline{1}\%$

    b. 4000 years old; *Sample answer:* The amount of the error and the original amount are the same, giving the same percent of error.

25. decrease; 25%

27. decrease; 70%

29. **a.** about 16.95% increase

    b. 161,391 people; *Sample answer:* The percent of change for the 6-year span is a 16.95% increase. The population in 2017 is 138,000, so $138,000 \times 0.1695 = 23,391$ increase in population for 2023. $138,000 + 23,391 = 161,391$ people

31. Box B is acceptable, Box A is unacceptable

33. less than; *Sample answer:* Let x represent the number. A 10% increase is equal to $x + 0.1x$, or $1.1x$. A 10% decrease of this new number is equal to $1.1x - 0.1(1.1x)$, or $0.99x$. Because $0.99x < x$, the result is less than the original number.

35. $8000

Section E.5
Review & Refresh

1. increase; 25%

3. decrease; $66.\overline{6}\%$

5. $-\dfrac{2}{21}$

7. $6\dfrac{3}{8}$

Concepts, Skills, & Problem Solving

9. 85% off $90;

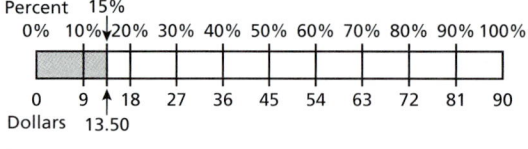

11. $35.70

13. $76.16

15. $53.33

17. $450

19. $172.40

21. 20%

23. $55

25. $175

27. "Multiply $42 by 0.9" and "Multiply $42 by 0.1, then subtract from $42." The first method is easier because it is only one step.

29. a. Store C   b. at least 11.82%

31. a. 25%   b. 6.5%   c. 20%

33. no; $31.08

Section E.6
Review & Refresh

1. $9.60

3. $x < -3$

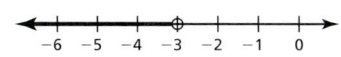

5. $w \leq -9$

Concepts, Skills, & Problem Solving

7. 8%; $315

9. a. $300   b. $1800

11. a. $292.50   b. $2092.50

13. a. $1722.24   b. $6922.24

15. 3%   17. 4%

19. 2 yr   21. 1.5 yr

23. $1440   25. 2 yr

27. $2720   29. $6700.80

31. $8500   33. 5.25%

35. 4 yr

37. 12.5 yr; Substitute $2000 for P, $2000 for I, 0.08 for r, and solve for t.

39. Year 1 = $520, Year 2 = $540.80, Year 3 = $562.43

English-Spanish Glossary

English | Spanish

A

absolute value *(p. 364, 504)* The distance between a number and 0 on a number line; The absolute value of a number a is written as $|a|$.

valor absoluto *(p. 364, 504)* La distancia entre un número y 0 en una recta numérica; El valor absoluto de un número a es escrito como $|a|$.

additive inverse *(p. 511)* The opposite of a number

inverso aditivo *(p. 511)* El opuesto de un número

algebraic expression *(p. 202)* An expression that contains numbers, operations, and one or more variables

expresión algebraica *(p. 202)* Una expresión que contiene números, operaciones, y uno o más variables

B

base (of a power) *(p. 4)* The base of a power is the repeated factor.

base (de una potencia) *(p. 4)* La base de una potencia es el factor repetido.

box-and-whisker plot *(p. 484)* A data display that shows the variability of a data set along a number line using the least value, the greatest value, and the quartiles of the data

diagrama de cajas y bigotes *(p. 484)* Una presentación de datos que muestra la variabilidad de un conjunto de datos a lo largo de una línea de números usando el valor menor, el valor mayor y los cuartiles de los datos

C

coefficient *(p. 202)* The numerical factor of a term that contains a variable

coeficiente *(p. 202)* El factor numérico de un término que contiene una variable

common factors *(p. 22)* Factors that are shared by two or more numbers

factores comunes *(p. 22)* Factores compartidos por dos o más números

common multiples *(p. 28)* Multiples that are shared by two or more numbers

comunes múltiplos *(p. 28)* Múltiplos compartidos por dos o más números

complex fraction *(p. 575)* A fraction that has at least one fraction in the numerator, the denominator, or both

fracción compleja *(p. 575)* Una fracción que tiene al menos una fracción en el numerador, el denominador, o ambos

composite figure *(p. 294)* A figure made up of triangles, squares, rectangles, and other two-dimensional figures

figura compuesta *(p. 294)* Una figura hecha de triángulos, cuadros, rectángulos, y otras figuras bidimensionales

constant *(p. 202)* A term without a variable

constante *(p. 202)* Un término que no tiene una variable

constant of proportionality *(p. 656)* The number k in the direct variation equation $y = kx$; the multiplicative relationship between two quantities

constante de proporcionalidad *(p. 656)* El número k en la ecuación de variación directa $y = kx$; la relación multiplicativa entre dos cantidades

conversion factor *(p. 143)* A rate that equals 1

coordinate plane *(p. 370)* A plane formed by the intersection of a horizontal number line and a vertical number line

cross products *(p. 641)* In the proportion $\frac{a}{b} = \frac{c}{d}$, the products $a \cdot d$ and $b \cdot c$ are called cross products.

factor de conversión *(p. 143)* Una tasa que es igual a 1

plano de coordenadas *(p. 370)* Un plano formado por la intersección de una recta numérica horizontal y una recta numérica vertical

productos cruzados *(p. 641)* En la proporción $\frac{a}{b} = \frac{c}{d}$, los productos $a \cdot d$ y $b \cdot c$ son llamados productos cruzados.

D

---

dependent variable *(p. 266)* The variable that represents the output values of a function

discount *(p. 704)* A decrease in the original price of an item

variable dependiente *(p. 266)* La variable que representa los resultados de una función

descuento *(p. 704)* Una disminución en el precio original de un artículo

---

E

---

edge *(p. 306)* A line segment where two faces of a polyhedron intersect

equation *(p. 246)* A mathematical sentence that uses an equal sign, =, to show that two expressions are equal

equation in two variables *(p. 266)* An equation that represents two quantities that change in relationship to one another

equivalent expressions *(p. 216)* Expressions with the same value

equivalent rates *(p. 136, 634)* Rates that have the same unit rate

equivalent ratios *(p. 109, 629)* Two ratios that describe the same relationship

evaluate *(p. 10)* To use the order of operations to find the value of a numerical expression

exponent *(p. 4)* The exponent of a power indicates the number of times the base is used as a factor.

arista *(p. 306)* Un segmento de recta donde dos caras de un poliedro se intersecan

ecuación *(p. 246)* Una oración matemática que usa un signo de igualdad, =, para mostrar que dos expresiones son iguales

ecuación en dos variables *(p. 266)* Una ecuación que representa dos cantidades que cambian en relación mutua

expresiones equivalentes *(p. 216)* Expresiones con el mismo valor

tasas equivalentes *(p. 136, 634)* Tasas que tienen la misma tasa de unidad

razones equivalentes *(p. 109, 629)* Dos razones que describen la misma relación

evaluar *(p. 10)* Usar el orden de operaciones para hallar el valor de una expresión numérica

exponente *(p. 4)* El exponente de una potencia indique cuantas veces el base sirve como un factor.

F

---

face *(p. 306)* A flat surface of a polyhedron

cara *(p. 306)* Una superficie plana de un poliedro

factor pair *(p. 16)* A set of two nonzero factors whose product results in a definite number

factor tree *(p. 16)* A diagram used to find the factors of a composite number

factoring an expression *(p. 228, 610)* Writing a numerical expression or algebraic expression as a product of factors

first quartile *(p. 434)* The median of the lower half of a data set; represented by Q_1

five-number summary *(p. 484)* The five numbers that make up a box-and-whisker plot

frequency *(p. 463)* The number of data values in an interval

frequency table *(p. 463)* A chart that groups data values into intervals

par de factor *(p. 16)* Un conjunto de dos factores distintos de cero cuyo producto resulta en un número definido

árbol de factores *(p. 16)* Un diagrama utilizado para hallar los factores de un número compuesto

factorizando una expresión *(p. 228, 610)* Escribiendo una expresión numérica o algebraica como un producto de factores

primer cuartil *(p. 434)* La mediana de la mitad inferior de un conjunto de datos; representado por Q_1

resumen de cinco números *(p. 484)* Los cinco números que componen un diagrama de cajas y bigotes

frecuencia *(p. 463)* El número de valores de datos en un intervalo

table de frecuencia *(p. 463)* Un gráfico que agrupa valores de datos en intervalos

---

G

graph of an inequality *(p. 386)* A graph that shows all the solutions of an inequality on a number line

greatest common factor (GCF) *(p. 22)* The greatest of the common factors shared by two or more numbers

gráfica de una desigualdad *(p. 386)* Una gráfica que muestra todas las soluciones de una desigualdad en una recta numérica

máximo factor común (MFC) *(p. 22)* El mayor de los factores comunes compartido por dos o más números

---

H

histogram *(p. 464)* A bar graph that shows the frequency of data values in intervals of the same size

histograma *(p. 464)* Un gráfico de barras que muestra la frecuencia de valores de datos en intervalos del mismo tamaño

---

I

independent variable *(p. 266)* The variable that represents the input values of a function

inequality *(p. 384)* A mathematical sentence that compares expressions; contains the symbols <, >, ≤, or ≥

integers *(p. 346, 503)* The set of whole numbers and their opposites

variable independiente *(p. 266)* La variable que representa los valores entradas de una función

desigualdad *(p. 384)* Una oración matemática que compara las expresiones; contiene los símbolos <, >, ≤, o ≥

números enteros *(p. 346, 503)* El conjunto de números naturales y sus opuestos

interest *(p. 710)* Money paid or earned for the use of money

interquartile range *(p. 434)* A measure of variation for a data set, which is the difference of the third quartile and the first quartile

inverse operations *(p. 253)* Operations that "undo" each other, such as addition and subtraction, or multiplication and division

interés *(p. 710)* Dinero pagado o ganado por el uso del dinero

rango de intercuartiles *(p. 434)* Una medida de variación para un conjunto de datos, el cual es la diferencia del tercer cuartil y el primer cuartil

operaciones inversas *(p. 253)* Operaciones que deshacer unos de otros, tales como suma y resta, o multiplicación y división

---

K

kite *(p. 298)* A quadrilateral that has two pairs of adjacent sides with the same length and opposite sides with different lengths

cometa *(p. 298)* Un cuadrilátero que tiene dos pares de lados adyacentes con la misma longitud y lados opuestos con longitudes diferentes

---

L

leaf *(p. 458)* Digit or digits on the right of a stem-and-leaf plot

least common multiple (LCM) *(p. 28)* The least of the common multiples shared by two or more numbers

like terms *(p. 223, 592)* Terms of an algebraic expression that have the same variables raised to the same exponents

linear expression *(p. 598)* An algebraic expression in which the exponent of each variable is 1

hoja *(p. 458)* Dígito o dígitos a la derecha de un digrama de tallo y hojas.

mínimo común múltiplo (MCM) *(p. 28)* El menor de los factores comunes compartido por dos o más números

términos semejantes *(p. 223, 592)* Términos de una expresión algebraica que tienen las mismas variables elevadas a los mismos exponentes

expresión lineal *(p. 598)* Una expresión algebraica en la que el exponente de cada variable es 1

M

markup *(p. 704)* The increase from what a store pays to the selling price

mean *(p. 420)* The sum of the data divided by the number of data values

mean absolute deviation (MAD) *(p. 440)* An average of how much data values differ from the mean

measure of center *(p. 426)* A measure that describes the typical value of a data set

measure of variation *(p. 434)* A measure that describes the spread, or distribution, of a data set

sobreprecio *(p. 704)* El aumento de lo que una tienda paga al precio de venta

media *(p. 420)* La suma de los datos dividido por el número de valores de datos

desviación media absoluta (MAD) *(p. 440)* Un promedio de cuántos valores de datos difieren de la media

medida de centro *(p. 426)* Una medida que describe el valor típico de un conjunto de datos

medida de variación *(p. 434)* Una medida que describe la extensión, o distribución, de un conjunto de datos

English-Spanish Glossary

median *(p. 426)* For a data set with an odd number of ordered values, the median is the middle value; For a data set with an even number of ordered values, the median is the mean of the two middle values.

metric system *(p. 142)* A decimal system of measurement, based on powers of 10, that contains units for length, capacity, and mass

mode *(p. 426)* The data value or values that occur most often; Data can have one mode, more than one mode, or no mode.

multiplicative inverses *(p. 54)* Two numbers whose product is 1

mediana *(p. 426)* Para un conjunto de datos con un número impar de valores ordenados, la mediana es el valor del medio; Para un conjunto de datos con un número par de valores ordenados, la mediana es la media de los dos valores del medio.

sistema métrico decimal *(p. 142)* Un sistema de medición, basado en potencias de 10, que incluye unidades para longitud, capacidad, y masa

moda *(p. 426)* El valor o valores de datos que ocurre(n) con más frecuencia; Datos pueden tener una moda, más que una moda, o ninguna moda.

inversos multiplicativos *(p. 54)* Dos números cuyo producto es 1

N

negative numbers *(p. 346)* Numbers that are less than 0

net *(p. 312)* A two-dimensional representation of a solid

numerical expression *(p. 10)* An expression that contains only numbers and operations

números negativos *(p. 346)* Números que son menos de 0

red *(p. 312)* Una representación bidimensional de un sólido

expresión numérica *(p. 10)* Una expresión que contiene solamente números y operaciones

---

O

opposites *(p. 346)* Two numbers that are the same distance from 0 on a number line, but on opposite sides of 0

order of operations *(p. 10)* The order in which to perform operations when evaluating expressions with more than one operation

origin *(p. 370)* The point, represented by the ordered pair (0, 0), where the horizontal and vertical number lines intersect in a coordinate plane

outlier *(p. 422)* A data value that is much greater than or much less than the other values in a data set.

opuestos *(p. 346)* Dos números que están a la misma distancia de 0 en una recta numérica, pero en lados opuestos de 0

orden de operaciones *(p. 10)* El orden en el que para realizar operaciones al evaluar expresiones con más de una operación

origen *(p. 370)* El punto, representado por el par ordenado (0, 0), donde las rectas numéricas horizontales y verticales se intersecan en un plano de coordenadas

valor atípico *(p. 422)* Una valor de datos que es mucho mayor o mucho menor que los otros valores en un conjunto de datos

English-Spanish Glossary

P

percent *(p. 164)* The value of a part-to-whole ratio where the whole is 100

percent of change *(p. 698)* The percent that a quantity changes from the original amount;

$$\text{percent of change} = \frac{\text{amount of change}}{\text{original amount}}$$

percent of decrease *(p. 698)* The percent of change when the original amount decreases;

$$\text{percent of decrease} = \frac{\text{original amount} - \text{new amount}}{\text{original amount}}$$

percent error *(p. 700)* The percent that an estimated amount differs from the actual amount;

$$\text{percent error} = \frac{\text{amount of error}}{\text{actual amount}}$$

percent of increase *(p. 698)* The percent of change when the original amount increases;

$$\text{percent of increase} = \frac{\text{new amount} - \text{original amount}}{\text{original amount}}$$

perfect square *(p. 5)* The square of a whole number

polygon *(p. 285)* A closed figure in a plane that is made up of three or more line segments that intersect only at their endpoints; for example, parallelograms and triangles

polyhedron *(p. 306)* A solid whose faces are all polygons

positive numbers *(p. 346)* Numbers that are greater than 0

power *(p. 4)* A product of repeated factors

prime factorization *(p. 16)* Writing a composite number as a product of its prime factors

principal *(p. 710)* An amount of money borrowed or deposited

prism *(p. 306)* A polyhedron that has two parallel, identical bases; The lateral faces are parallelograms.

porcentaje *(p. 164)* El valor de una razón de número de partes por un entero en donde el entero es 100

porcentaje de cambio *(p. 698)* El porcentaje que cambia una cantidad de la cantidad original

$$\text{porcentaje de cambio} = \frac{\text{cambio}}{\text{cantidad original}}$$

porcentaje de disminución *(p. 698)* El porcentaje de cambio cuando la cantidad original disminuye;

porcentaje de disminución

$$= \frac{\text{cantidad original} - \text{nueva cantidad}}{\text{cantidad original}}$$

error porcentual *(p. 700)* El porcentaje en que una cantidad estimada difiere de la cantidad real;

$$\text{error porcentual} = \frac{\text{error}}{\text{cantidad correcta}}$$

porcentaje de aumento *(p. 698)* El porcentaje de cambio cuando la cantidad original aumenta;

Porcentaje de aumento

$$= \frac{\text{nueva cantidad} - \text{cantidad original}}{\text{cantidad original}}$$

cuadrado perfecto *(p. 5)* El cuadrado de un número natural

polígono *(p. 285)* Una figura cerrada en un plano, hecha de tres o más segmentos de líneas que intersectan solamente à sus puntos finales; por ejemplo, paralelogramos y triángulos

poliedro *(p. 306)* Un sólído tridimensional cuyas caras son todas polígonos

números positivos *(p. 346)* Números que son mayores que 0

potencia *(p. 4)* Un producto de factores repetidos

factorización prima *(p. 16)* Escribir un número compuesto como el producto de sus factores primeros

principal *(p. 710)* Una cantidad de dinero prestada o depositada

prisma *(p. 306)* Un poliedro que tiene dos bases idénticas y paralelas; Las caras laterales son paralelogramos.

proportion *(p. 640)* An equation stating that the values of two ratios are equivalent

proportional *(p. 642)* Two quantities that form a proportion are proportional.

pyramid *(p. 306)* A polyhedron that has one base; The lateral faces are triangles.

proporción *(p. 640)* Una ecuación que se indica que los valores de dos razones son equivalentes

proporcional *(p. 642)* Dos cantidades que forman una proporción son proporcionales.

pirámide *(p. 306)* Un poliedro que tiene una base; Las caras laterales son triángulos.

Q

quadrants *(p. 370)* The four regions created by the intersection of the horizontal and vertical number lines in a coordinate plane

quartiles *(p. 434)* Values that divide a data set into four equal parts

cuadrantes *(p. 370)* Las cuatro regiones creadas por la intersección de las rectas numéricas horizontales y verticales en un plano de coordenadas

cuartiles *(p. 434)* Valores que dividen un conjunto de datos en cuatro partes iguales

R

range *(p. 434)* The difference of the greatest value and the least value of a data set

rate *(p. 136, 634)* A ratio of two quantities using different units

ratio *(p. 108, 628)* A comparison of two quantities; The ratio of a to b can be written as $a : b$.

ratio table *(p. 122, 629)* A table used to find and organize equivalent ratios

rational number *(p. 358, 503)* A number that can be written as $\frac{a}{b}$ where a and b are integers and $b \neq 0$

reciprocals *(p. 54)* Two numbers whose product is 1

repeating decimal *(p. 562)* A decimal that has a pattern that repeats

rango *(p. 434)* La diferencia del valor mayor y el valor menor de un conjunto de datos

tasa *(p. 136, 634)* Una razón de dos cantidades usando unidades diferentes

razón *(p. 108, 628)* Una comparación de dos cantidades; La razón de a a b puede escribirse como $a : b$.

tabla de razones *(p. 122, 629)* Una tabla usada para encontrar y organizar las razones equivalentes

número racional *(p. 358, 503)* Un número que puede ser escrito como $\frac{a}{b}$ donde a y b son enteros y $b \neq 0$

recíprocos *(p. 54)* Dos números cuyo producto es 1

decimal periódico *(p. 562)* Un decimal que tiene un patrón que se repite

S

scale *(p. 662)* A ratio that compares the measurements of a drawing or model with the actual measurements

scale drawing *(p. 662)* A proportional, two-dimensional drawing of an object

scale factor (of a scale drawing) *(p. 663)* The value of a scale when the units are the same

escala *(p. 662)* Una razón que compara las medidas de un dibujo o modelo a las medidas reales

dibujo a escala *(p. 662)* Un dibujo bidimensional y proporcional de un objeto

factor de escala (de un dibujo a escala) *(p. 663)* El valor de una escala cuando las unidades son las mismas

scale model *(p. 662)* A proportional, three-dimensional model of an object

simple interest *(p. 710)* Money paid or earned only on the principal

simplest form (of an algebraic expression) *(p. 592)* An algebraic expression is in simplest form when it has no like terms and no parentheses.

solid *(p. 306)* A three-dimensional figure that encloses a space

solution *(p. 252)* A value that makes an equation true

solution of an equation in two variables *(p. 266)* An ordered pair (x, y) that makes an equation true

solution of an inequality *(p. 385)* A value that makes an inequality true

solution set *(p. 385)* The set of all solutions of an inequality

statistical question *(p. 414)* A question for which a variety of answers is expected; The interest is in the distribution and tendency of those answers.

statistics *(p. 414)* The science of collecting, organizing, analyzing, and interpreting data

stem *(p. 458)* Digit or digits on the left of the stem-and-leaf plot

stem-and-leaf plot *(p. 458)* A data display that uses the digits of data values to organize a data set; Each data value is broken into a stem (digit or digits on the left) and a leaf (digit or digits on the right).

surface area *(p. 312)* The sum of the areas of all the faces of a polyhedron

modelo a escala *(p. 662)* Un modelo proporcional y tridimensional de un objeto

interés simple *(p. 710)* Dinero pagado o ganado solamente en el principal

mínima expresión (de una expresión algebraica) *(p. 592)* Una expresión algebraica está en su mínima expresión cuando tiene ningunos términos semejantes y ningunos paréntesis.

sólido *(p. 306)* Una figura tridimensional que encierra un espacio

solución *(p. 252)* Un valor que hace una ecuación verdadera

solución de una ecuación en dos variables *(p. 266)* Un par ordenado (x, y) que hace que una ecuación sea verdadera

solución de una desigualdad *(p. 385)* Un valor que hace una desigualdad verdadera

conjunto de solución *(p. 385)* El conjunto de todas las soluciones de una desigualdad

pregenta estadística *(p. 414)* Una pregunta para la cual se espera una variedad de repuestas; El interés es en la distribución y la tendencia de aquellas respuestas.

estadísticas *(p. 414)* La ciencia de recolectar, organizar, analizar e interpretar datos

tallo *(p. 458)* Dígito o dígitos a la izquierda de un diagrama de tallo y hojas

diagrama de tallo y hojas *(p. 458)* Un representación de datos que usa los dígitos de valores de datos para organizar un conjunto de datos; Cada valor de datos es roto en un tallo (dígito o dígitos a la izquierda) y una hoja (dígito o dígitos a la derecha).

área de la superficie *(p. 312)* La suma de las áreas de todas las caras de un poliedro

T

term *(p. 202)* A part of an algebraic expression; a number or variable by itself, or the product of numbers and variables

terminating decimal *(p. 562)* A decimal that ends

third quartile *(p. 434)* The median of the upper half of a data set; represented by Q_3

término *(p. 202)* Un component de una expresión algebraica; un número o variable por sí mismo, o el producto de números y variables

decimal finito *(p. 562)* Un decimal que termina

tercer cuartil *(p. 434)* La mediana de la mitad superior de un conjunto de datos; representado por Q_3

U

unit analysis *(p. 143)* A process used to decide which conversion factor will produce the appropriate units

unit rate *(p. 136, 634)* A rate that compares a quantity to one unit of another quantity

U.S. customary system *(p. 142)* A system of measurement that contains units for length, capacity, and weight

análisis de unidades *(p. 143)* Un proceso utilizado para decidir qué factor de conversión producirá las unidades apropiadas

tasa unitaria *(p. 136, 634)* Una tasa que compara una cantidad a una unidad de otra cantidad

sistema estadounidense *(p. 142)* Un sistema de medición que incluye unidades para longitud, capacidad, y peso

V

value of a ratio *(p. 109, 628)* The number $\frac{a}{b}$ associated with the ratio $a:b$

variable *(p. 202)* A symbol that represents one or more numbers

Venn diagram *(p. 21)* A diagram of overlapping circles used to show the relationships between two or more sets

vertex (of a solid) *(p. 306)* A point where three or more edges intersect

volume *(p. 325)* A measure of the amount of space that a three-dimensional figure occupies; Volume is measured in cubic units such as cubic feet (ft^3) or cubic meters (m^3).

valor de una razón *(p. 109, 628)* El número $\frac{a}{b}$ asociadas con la razón $a:b$

variable *(p. 202)* Un símbolo que representa a uno o más números

Diagrama de Venn *(p. 21)* Un diagrama de círculos solapados usado para mostrar las relaciones entre dos o más conjuntos

vértice (de un sólido) *(p. 306)* Un punto donde tres o más aristas se intersecan

volumen *(p. 325)* Una medida de la cantidad de espacio que una figura tridimensional ocupa; Volumen es medido en unidades cúbicas como pies cúbicos ($pies^3$) o metros cúbicos (m^3).

Index

A

Absolute value, 363–368
  comparing, 365–366
  definition of, 364, 504, 538
  finding, 364
  on number line, 504, 532
  subtracting rational numbers using, 529, 530, 531, 532
Addition
  Associative Property of, 215–218, 520
  Commutative Property of, 215–218, 520, 531, 593, 598, 599
    example and non-example chart for, 234
  of decimals, 67–72
  Distributive Property over (See Distributive Property)
  of integers, 509–516
  key words and phrases for, 210
  of linear expressions, 597–602
  on number line, 67
  in order of operations, 10–14
  properties of, 215–220
  of rational numbers, 517–522
  solving equations using, 251–258
  solving inequalities using, 392
  and subtraction, as inverse operations, 253
Addition Property of Equality, 253–255
Addition Property of Inequality, 392
Addition Property of Zero, 217
Additive Inverse Property, 511, 551, 597
Airplanes, paper, 463
Algebra tiles, 597
Algebraic expressions, 591–596, See also Linear expressions
  definition of, 202
  evaluating, 201–208
    with one variable, 203–205
    with two operations, 204
    with two variables, 203
  factoring, 227–232, 609–614
    definition of, 228
    using Distributive Property, 228–230
    using greatest common factor, 229–230
  identifying parts of, 202
  identifying terms and like terms in, 592
  in simplest form, 592
  simplifying, 221–226, 591, 592, 593, 604, 605
    using Distributive Property, 222–224
    using models, 221
  writing, 209–214, 603
    order of operations in, 210
    using exponents, 202
Annual interest rate, 710, 711
Answering questions. See Questions
Area(s). See also Surface areas
  of composite figures, finding, 294, 300
  converting measures of, 287
  definition of, 6
  of parallelograms, 285–290
    finding, 286–288
    formula for, 284, 285, 286
    four square for, 332
  of polygons
    definition of, 286
    formulas for, 284
  of rectangles, formula for, 284, 285, 294, 300
  of rhombus, formula for, 287
  of squares, finding, 6
  of trapezoids, formula for, 284, 299, 300
  of trapezoids and kites, 297–304
    finding, 298–301
    formula for, 297
  of triangles, 291–296
    finding, 292–294
    formula for, 284, 291, 292, 294
  units of (square units), 6, 286
Area models
  dividing decimals using, 87, 573
  multiplying decimals using, 73, 567
  multiplying fractions using, 44, 45
Associative Property of Addition, 215–218, 520
Associative Property of Multiplication, 215–217, 569
Axis (axes), of coordinate plane. See x-axis; y-axis

B

Backwards, working, 279
Balance, in simple interest problem, 710
Balance point, finding, 419
Bar graphs. See also Histograms
  double
    comparing means in, 421
    using, 81, 699
  interpreting, 456
Bar notation, 562, 680
Base, exponential expressions, definition of, 4
Base, of geometric figures
  of parallelograms
    in area formula, 285, 286
    definition of, 286
  of prisms, 306
  of pyramids, 306
  of trapezoids, in area formula, 297, 299
  of triangles
    in area formula, 291, 292
    definition of, 292
    finding, 293
Box-and-whisker plots, 483–490
  definition of, 484
  drawing, 483–484
  five-number summary in, 484
  interpreting, 485–487
  Problem-Solving Plan for, 491

C

Calculator
  finding decimals using, 561
  finding patterns using, 3
  order of operations on, 9
Capacity, converting measures of, 142, 144
Cent, meaning of, 162, 163
Center, measures of, 425–432. See also Mean; Median; Mode
  choosing appropriate, 477–482
  definition of, 426
  finding, 426–428
    after changes in values of data set, 428
  Problem-Solving Plan for, 445
  with and without outliers, 422, 427
Challenge. See Dig Deeper
Change, percents of, 697–702
Chapter Exploration, In every chapter. For example, see: 2, 44, 106, 162, 200, 244, 284, 344, 412, 456
Chapter Practice, In every chapter. For example, see: 7–8, 50–52, 112–114, 167–168, 206–208, 249–250, 289–290, 349–350, 417–418, 461–462

Chapter Review, *In every chapter. For example, see:* 34–37, 96–99, 150–155, 190–193, 234–238, 274–277, 332–337, 400–405, 446–449, 492–495
Charts
  definition and example, 150
  example and non-example, 234, 274
  place value, 67
Check Your Answer, *Throughout. For example, see:* 24, 511, 531, 636, 686, 704, 705, 706
Choices
  eliminating, 39
  problem solving before looking at, 157
  reading all, before answering, 407
Choose Tools, *Throughout. For example, see:* 29, 58, 134, 256, 263, 272, 490, 687, 689
Circle graphs, 466–467, 691
Clock, 633
Closed circle symbol, 386
Coefficients, in algebraic expressions
  definition of, 202
  identifying, 202
Common Error, *Throughout. For example, see:* T-7, T-50, T-126, T-169, 210, 218, T-295, T-345, 372, T-458
Common factors. *See also* Greatest common factor
  definition of, 22
  dividing out, 563, 569
  identifying, 21
Common multiples
  definition of, 28
  identifying, 27
  least, 27–32
Commutative Property of Addition, 215–218, 520, 531, 593, 598, 599
  example and non-example chart for, 234
Commutative Property of Multiplication, 215–216, 551, 569
Comparing
  absolute values, 365–366
  data displays, 466–467
  decimals
    with fractions and percents, 175–180
    with other decimals, 359
  fractions
    with decimals and percents, 175–180

    with other fractions, 358
  integers, 351–356
  means, 421–422
  mixed numbers, 358
  percents, with decimals and fractions, 175–180
  positions of objects, 363
Complex fractions, 575
Composite figures
  areas of, finding, 294, 300
  definition of, 294
Composite numbers
  characteristics of, 2
  definition of, 16
  prime factorization of, 15–20
Conjectures, analyzing, 27
Connecting Concepts, *In every chapter. For example, see:* 33, 95, 149, 189, 233, 273, 331, 399, 445, 491
Constant of proportionality, 656
Constants, in algebraic expressions
  definition of, 202
  identifying, 202
  as like terms, 223, 592
Conversion factors, 143–144
  converting measures using, 143–144
  definition of, 143
Converting measures, 141–148
  of area, 287
  estimating, 141
  Problem-Solving Plan for, 95
  within same system, 142
  between systems, 143
  using conversion factors, 143–144
  using ratio tables, 142–143
  using unit rates, 142–143
  of volume, 328
Coordinate plane, 369–376. *See also* Graph(s)
  axes of, 267, 369–373
  definition of, 370
  drawing polygons in, 377–382, 399
  finding distances between points in, 378–380
  negative numbers in, 369–373
  origin of, 370
  plotting ordered pairs on, 371–373
  quadrants of, 370
  ratio relationships in, 129–134
  reflecting a point in, 371–372
Counterexamples, using, 215
Critical Thinking, *Throughout. For example, see:* 20, 60, 114, 168, 229, 258, 290, 356, 390, 459

Cross products
  definition of, 641
  solving proportions using, 649
Cross Products Property, 641, 649, 687
Cubes
  definition of, 314
  finding surface areas of, 314
  finding volumes of, 326
  unit
    definition of, 325
    using, 325
Cubic units
  converting, 328
  definition of, 325
Cumulative Practice, *In every chapter. For example, see:* 39–41, 101–103, 157–159, 195–197, 239–241, 279–281, 339–341, 407–409, 451–453, 497–499
Customary system. *See* U.S. customary system

D

Data
  answering statistical questions using, 413–416, 429
  changes in values of, 428
  definition of, 412
  grouping, 433
  mean absolute deviation of, 439–444
    definition of, 440
    describing variation using, 478
    finding, 440–442
    interpreting, 440–442
  measures of center of (*See* Center, measures of)
  range of, 433–438
    definition of, 434
    finding, 434–436
Data displays
  in box-and-whisker plots, 483–490
  comparing, 466–467
  in dot plots, 415–416
  in histograms, 463–470, 472–474
  shapes of distributions in, 471–476
  in stem-and-leaf plots, 455, 457–462
Decimals, 679–684, *See also* Rational numbers
  adding, 67–72, 519
  in algebraic expressions, 591, 592, 609
  comparing and ordering, 175–180, 359

converting between fractions and, 561–566
dividing, 87–94
    by decimals, 89–91
    inserting zeros in, 90
    by rational numbers, 573, 574, 575
    using models, 87
    by whole numbers, 88
exploring, 561
models representing, 169–170
multiplying, 73–80
    by decimals, 75–77
    by rational numbers, 567, 568, 569
    using models, 73
    by whole numbers, 74
ordering, 679
in proportions, 640, 649
repeating, 562
in set of rational numbers, 358
subtracting, 67–72, 531, 532
terminating, 562, 563
writing as percents, 170–171, 680
writing fractions as, 171, 681
writing percents as, 170–172, 680, 716

Decomposition
  definition of, 294
  finding areas of composite figures using, 294
  finding areas of trapezoids and kites using, 298–301

Decrease, percents of, 698, 699
Deductive reasoning, definition of, 285
Definition and example chart, 150, 538
Denominators, analyzing, 561

Dependent variables
  definition of, 266
  in equations in two variables, 266–267

Deviation
  definition of, 439
  mean absolute (*See* Mean absolute deviation)

Diagrams
  interpreting, 135
  tape (*See* Tape diagrams)
  Venn (*See* Venn diagrams)

Diamonds, finding numbers of faces, edges, and vertices of, 308

Different signs
  adding integers with, 509, 510, 511, 512
  adding rational numbers with, 518, 519
  dividing integers with, 555, 556
  dividing rational numbers with, 573, 574, 575
  multiplying integers with, 549, 550, 551
  multiplying rational numbers with, 568
  subtracting integers with, 523, 524, 525
  subtracting rational numbers with, 530, 531, 532

Different Words, Same Question, *Throughout. For example, see:* 23, 63, 110, 184, 211, 247, 314, 365, 386

Differentiation, *See* Scaffolding Instruction

Dig Deeper, *Throughout. For example, see:* 6, 52, 111, 168, 205, 248, 288, 350, 418, 460

Dimensions, finding, 609. *See also* Missing dimensions

Discounts, 703–708
  definition of, 704
  finding, 704

Distance(s)
  finding actual, 662
  from mean, finding, 439
  on number line, 529, 532
  between points in coordinate plane, finding, 378–380

Distance formula, 269, 273

Distributions, shapes of, 471–476
  in box-and-whisker plots, 485–486
  choosing appropriate measures using, 477–480
  identifying and describing, 471–474, 486

Distributive Property, 221–226, 603–608
  combining like terms using, 223
  definition of, 222
  factoring algebraic expressions using, 228–230, 610, 611
  factoring numerical expressions using, 228
  for multiplying integers, 550
  simplifying algebraic expressions using, 222–224, 592, 593, 604, 605

Dividends. *See also* Division
  definition of, 82

Division
  of decimals, 87–94
    by decimals, 89–91
    inserting zeros in, 90
    using models, 87
    by whole numbers, 88
  of fractions, 53–60
    by fractions, 55
    summary triangle for, 96
    using models, 53–57
    by whole numbers, 56
  fractions as form of, 11
  of integers, 555–560
  key words and phrases for, 210
  of mixed numbers, 61–66
    using models, 61–62
  and multiplication, as inverse operations, 260, 393
  in order of operations, 10–14, 210
  of rational numbers, 573–578
  solving equations using, 259–264
  solving inequalities using, 393, 395
  solving problems using, 83
  of whole numbers, 81–86

Division Property of Equality, 261–262, 692, 693
Division Property of Inequality, 393, 395

Divisors. *See also* Division
  definition of, 82

Dot plots, 415–416
  constructing, 415–416, 433
  definition of, 415
  identifying shapes of distributions using, 472
  using, 415–416, 435

Dot symbol, 4
Dot-to-dot pictures, 344

Double bar graphs
  comparing means in, 421
  using, 81

Double number line
  converting measures using, 144
  creating, 121
  using, 124

Drawings, *See* Scale drawings

E

Edges, of solids
  definition of, 306
  finding number of, 306–308
  identifying, 305

Eliminating choices, as test-taking strategy, 39

ELL Support, *In every lesson. For example, see:* T-5, T-61, T-106, T-175, T-200, T-244, T-284, T-351, T-412, T-463

Equal sign (=), 246

Equality
  Addition Property of, 253–255
  Division Property of, 261–262, 692, 693

Multiplication Property of, 260, 641, 648, 649, 686
Subtraction Property of, 253–255
Equations, 245–250
   definition of, 246
   equivalent, 253, 261
   example and non-example chart for, 274
   expressions compared to, 246
   percent, 691–696, 704, 705
   proportion as, 640
   solutions to
      checking, 252–254
      definition of, 252
   solving
      using addition or subtraction, 251–258
      using models, 251, 259
      using multiplication or division, 259–264
      using tape diagrams, 251, 259
   writing
      in one variable, 245–250
      in two variables, 265–272
Equations in two variables, 265–272
   definition of, 266
   graphing, 267–269
   solutions to
      definition of, 266
      identifying, 266
   in tables, 267–269
   using, 266
   writing, 265–272
Equivalent equations, 253, 261
Equivalent expressions, 215–220
   definition of, 216
   identifying, 215
   writing, using properties of addition and multiplication, 216–217
Equivalent fractions, writing, with denominators of 100, 165
Equivalent rates, definition of, 136, 634
Equivalent ratios, 109–111
   definition of, 109, 185
   identifying, 110–111
   in proportions, 640, 641, 642, 649
   tables of, 121–128
   using, 135
Eratosthenes, Sieve of, 2
Error, percent, 700
Error Analysis. *See* You Be the Teacher
Estimating
   as test-taking strategy, 101
   unit conversions, 141

Evaluate, definition of, 10
Example and non-example chart, 234, 274, 616, 668
Experiments, performing, 463
Explain your reasoning, *Throughout. For example, see:* 26, 48, 128, 188, 232, 254, 330, 366, 431, 476
Exponent(s), 3–8
   definition of, 4
   in order of operations, 10–14
   writing algebraic expressions using, 202
   writing expressions using, 3–4
Exponential expressions, 3–8
   finding values of, 5–6
   using order of operations with, 10–14
   writing, 3–4
Expressions. *See also specific types of expressions*
   equations compared to, 246

F

Faces, of solids
   definition of, 306
   finding number of, 306–308
   identifying, 305
   lateral
      of prisms, 306
      of pyramids, 306
Factor(s)
   common (*See* Common factors)
   conversion, 143–144
      converting measures using, 143–144
      definition of, 143
   prime
      using, 21, 27
      writing numbers as products of, 15–20
   repeated, 3–4
Factoring expressions, 227–232, 609–614
   common, dividing out, 563, 569
   definition of, 228, 610
   Problem-Solving Plan for, 233
   using Distributive Property, 228–230
   using greatest common factor, 228–230
Factoring out, 228
Factorization, prime. *See* Prime factorization
Factor pairs
   definition of, 16

finding, 16–17
   using, 16–17
Factor trees, 15–20
   definition of, 16
   using, 15–20
Fair shares, finding, 419
Figures. *See* Composite figures; *specific figures*
First (lower) quartile, of data set, 434
Five-number summary, 484
Flat distribution, 472
Formative Assessment Tips, *Throughout. For example, see:* T-18, T-68, T-131, T-165, T-224, T-255, T-329, T-378, T-421, T-487
Four square, 190, 332
Fractions, 679–684, *See also* Rational numbers
   absolute value of, 504
   adding, 518, 519
   in algebraic expressions, 591–593, 604, 605, 609, 610
   complex, 575
   converting between decimals and, 561–566
   comparing, 175–180, 358
   dividing, 53–60, 73–75
      by fractions, 55
      summary triangle for, 96
      using models, 53–57
      by whole numbers, 56
   equivalent, with denominators of 100, 165
   factoring out, 610
   as form of division, 11
   on graphs, 657, 658
   improper, writing mixed numbers as, 47–48, 62
   models representing, 163
   multiplying, 45–52, 567–569
      using models, 44, 45–48
   on number line, 357
   ordering, 175–180, 358, 679
   in proportions, 640, 641, 642, 648, 649, 650
   in rates, 634, 635, 636
   in ratios, 628, 629, 630
   in set of rational numbers, 358
   simplifying, 46, 563, 569
   subtracting, 530, 531, 532
   writing as decimals, 171, 681
   writing as percents, 165–166, 171, 681
   writing percents as, 164–165
Frequency
   definition of, 463

in histograms, 464–467
Frequency tables, 463–466
    definition of, 463
    displaying data from, in histograms, 464–466, 473–474

G

GCF. See Greatest common factor
Graph(s)
    bar (See Bar graphs)
    circle, 466–467, 691
    dot-to-dot pictures on, 344
    of equations in two variables, 267–269
    fractions on, 657, 658
    identifying relationships in, 129
    of inequalities, 386–387
    of integers, on number line, 346–348, 351–354
    line, 373
    of proportional relationships, 655–660
    ratio relationships in, 129–134
    of rational numbers, on number line, 358–360
    time series, 373
Graphic organizers
    definition and example chart, 150, 446, 538
    example and non-example chart, 234, 274, 516, 668
    four square, 190, 332
    information frame, 34, 492, 580
    summary triangle, 96, 400, 716
Greater than or equal to symbol (≥), 384
Greater than symbol (>), 384. See also Comparing
Greatest common factor (GCF), 21–26
    definition of, 22
    factoring expressions using, 228–230, 610
    finding, 21–26
        using list of factors, 22
        using prime factorizations, 22
    finding two numbers given, 23
    Problem-Solving Plan for, 33
Grid paper, constructing solids using, 311, 319
Grouping data, 433
Grouping symbols, in order of operations, 10–14
Guessing, using intelligent, 451

H

Height
    of parallelograms, in area formula, 285, 286
    of trapezoids, in area formula, 297, 299
    of triangles
        in area formula, 291, 292
        finding, 293
Higher Order Thinking. See Dig Deeper
Histograms, 463–470
    definition of, 464
    identifying shapes of distributions using, 472–474
    information frame for, 492
    making, 464–467
    using, 465–467
Horizontal method
    for adding linear expressions, 598
    for subtracting linear expressions, 599
Hourly wages, 200

I

Improper fractions, writing mixed numbers as, 47–48, 62
Increase, percents of, 698
Independent variables
    definition of, 266
    in equations in two variables, 266–267
Inequality(ies), 383–390
    Addition Property of, 392
    definition of, 384
    Division Property of, 393, 395
    graphing, 386–387
    key phrases for, 384
    Multiplication Property of, 393–394
    solutions to
        checking, 385–387
        definition of, 385
        graphing, 386–387
        set of, 385
    solving, 391–397
        using addition or subtraction, 392
        using multiplication or division, 393–395
        using reciprocals, 394
        using tape diagrams, 391
    Subtraction Property of, 392
    writing, 384, 387
Inequality symbols, 384

Information frame, 34, 492, 580
Integers, 345–350
    adding, 509–516
    Additive Inverse Property of, 511, 551
    comparing, 351–356
    definition of, 346, 503
    dividing, 555–560
    graphing on number line, 346–348, 351–354
    multiplying, 549–554, 580
    negative, 346–348, 352–354
    on number line, 503
    opposite of, 346–347
    ordering, 351–356
    positive, 346–348, 352–354
    reasoning with, 353
    subtracting, 523–528
    in set of rational numbers, 358
    summary triangle for, 400
Integer counters, 509, 523, 549, 555, 597
Intelligent guessing, using, 451
Interest
    definition of, 710
    simple (See Simple interest)
Interquartile range (IQR), 434–438
    definition of, 434
    describing variation using, 478
    finding, 435–436
    interpreting, 435–436
Inverse operations
    addition and subtraction as, 253
    definition of, 253
    multiplication and division as, 260, 393
Inverses, multiplicative, definition of, 54
Invert, meaning of, 54
IQR. See Interquartile range

J

Justify your answer, Throughout. For example, see: 14, 95, 113, 141, 188, 262, 328, 398, 432, 474

K

Key, to stem-and-leaf plots, 458
Kites, areas of, 297–304
    finding, 298–301

L

Lateral faces
    of prisms, 306
    of pyramids, 306

Index A51

LCM. *See* Least common multiple
Leafs. *See also* Stem-and-leaf plots
   definition of, 458
Learning Target, *In every lesson. For example, see:* 3, 45, 107, 163, 201, 245, 285, 345, 413, 457
Least common multiple (LCM), 27–32
   definition of, 28
   finding, 28–32
      with three numbers, 29
      using lists of multiples, 28–30
      using prime factorizations, 28–30
Length, converting measures of, 142–145
Less than or equal to symbol (≤), 384
Less than symbol (<), 384. *See also* Comparing
Like terms
   in addition, 598
   combining, using Distributive Property, 223
   definition of, 223, 592
   example and non-example chart of, 616
   identifying, 592
   in simplifying expressions, 604, 605
   in subtraction, 599
Line graphs
   creating, 373
   using, 373
Linear expressions
   adding, 597–602
   definition of, 598
   factoring, 609–614
   simplifying, 604, 605
   subtracting, 597–602
   writing, 603
Logic, *Throughout. For example, see:* 20, 52, 113, 168, 250, 264, 290, 376, 390, 470
Lower quartile, of data set, 434

M

MAD. *See* Mean absolute deviation
Magic squares, 258
Markups, 703–708
Mass, converting measures of, 142
Mean(s), 419–424
   comparing, 421–422
   definition of, 420
   describing center using, 478
   distance from, finding, 439

finding, 419–420, 429
   after changes in values of data set, 428
   with and without outliers, 422, 427
Mean absolute deviation (MAD), 439–444
   definition of, 440
   describing variation using, 478
   finding, 440–442
   interpreting, 440–442
Meaning of a Word, 54, 162, 163, 346, 439, 471, 626
Measures. *See* Center, measures of; Conversion; Variation, measures of
Median, 425–432
   definition of, 425, 426
   describing center using, 478
   finding, 425–429
      after changes in values of data set, 428
      with and without outliers, 427
Mental Math, 516, 648
Metric system, 142–148
   converting measures in, 143–145
   definition of, 142
Missing dimensions, finding
   of rectangles, 227
   of rectangular prisms, 327
   of triangles, 293
Mixed numbers
   comparing, 358
   converting to decimals, 562
   dividing, 61–66
      using models, 61–62
   multiplying, 47–52
   in order of operations, 63
   writing as improper fractions, 47–48, 62
Mode, 425–432
   definition of, 425, 426
   finding, 426–429
      after changes in values of data set, 428
      with and without outliers, 427
Model(s)
   of decimals, 169–170
   dividing decimals using, 87
   dividing fractions using, 53–57
   dividing mixed numbers using, 61–62
   of fractions, 163
   multiplying decimals using, 73
   multiplying fractions using, 44, 45–48
   of percents, 163, 169, 181–185
   scale models, 662

simplifying algebraic expressions using, 221
solving equations using, 251, 259
Modeling Real Life, *In every lesson. For example, see:* 2, 44, 106, 166, 205, 248, 288, 348, 416, 460
Multiple Representations, *Throughout. For example, see:* 6, 47, 131, 164, 218, 247, 312, 373, 416, 466
Multiples. *See* Common multiples
Multiplication
   Associative Property of, 215–217, 569
   Commutative Property of, 215–216, 551, 569
   of decimals, 73–80
      by decimals, 75–77
      using models, 73
      by whole numbers, 74
   Distributive Property of (*See* Distributive Property)
   and division, as inverse operations, 260, 393
   of fractions, 45–52
      using models, 44, 45–48
   of integers, 549–554, 580
   key words and phrases for, 210
   of mixed numbers, 47–52
   in order of operations, 10–14
   properties of, 215–220
   of rational numbers, 567–572
   repeated, 551
   solving equations using, 259–264
   solving inequalities using, 393–394
   solving proportions using, 648
   symbols for, 4, 11
   by zero, 54
Multiplication Property of Equality, 260, 641, 648, 649, 686
Multiplication Property of Inequality, 393–394
Multiplication Property of One (1), 217, 551, 569, 592
Multiplication Property of Zero, 217
Multiplicative Inverse Property, 260, 569
Multiplicative inverses, definition of, 54

N

Negative numbers, 345–350
   absolute values of, 505
   in coordinate plane, 369–373
   definition of, 346

as differences, 524
factoring out, 611
on number line, 346–348, 352–354, 503
as products, 549, 550
as quotients, 555, 556
as sums, 509, 510, 511, 512, 518, 519
writing, 346
Negative sign (−), 346, 347
Nets
constructing solids using, 319
definition of, 312
finding surface areas using
of pyramids, 320–322
of rectangular prisms, 312, 315
of triangular prisms, 313
Nonlinear expressions, 598
Not equal to symbol (≠), 252
Not less than or equal to symbol, 385
Number(s). See specific types of numbers
Number line
absolute value on, 364–366, 504, 532
adding on, 67
box-and-whisker plots along, 484
in coordinate plane, 370
dot plots on, 415–416
double
converting measures using, 144
creating, 121
using, 124
finding differences on, 524, 529
finding distances on, 529, 532
finding products on, 549
finding sums on, 510, 517
fractions on, 357
integers on, 346–348, 351–354
multiplying fractions using, 45
ordering numbers using, 175–178
rational numbers on, 358–360
subtracting on, 67
using, 503
vertical
in coordinate plane, 370
rational numbers on, 358
temperatures on, 345, 354
Number Sense, *Throughout. For example, see:* 8, 52, 134, 165, 204, 264, 350, 421, 466
Numerical expressions
definition of, 10
evaluating, 10–14
factoring, 227–232
definition of, 228
using Distributive Property, 228

using greatest common factor, 228–229
order of operations in, 9–14
writing, 209–210

O

One (1), Multiplication Property of, 217, 551, 569, 592
Open circle symbol, 386
Open-Ended, *Throughout. For example, see:* 14, 51, 113, 165, 204, 247, 290, 349, 382, 389, 415
Operations. *See also specific operations*
order of (*See* Order of operations)
words and phrases for, 210
Opposite(s)
additive inverse as, 511
definition of, 346, 605
of integers, 346–347
on number line, 503
of rational numbers, 358
Order of operations, 9–14
on calculators, 9
comparing different, 9
definition of, 10
determining, 9
using, 10–14
with exponents, 10–14
with mixed numbers, 63
in writing algebraic expressions, 210
Ordered pairs
creating, from ratios, 130
definition of, 370
identifying, 370
negative numbers in, 369–373
plotting, 371–373
as solution to equations in two variables, 266
as vertices of polygons, 377–380
Ordering
decimals, 175–180, 359
fractions, 175–180, 358
integers, 351–356
percents, 175–180
rational numbers, 358–360
Origin, of coordinate plane, definition of, 370
Original price, finding, 705
Outliers
definition of, 422
finding measures of center with and without, 422, 427
identifying, using interquartile range, 436
removing, 427

P

Pairs. *See* Factor pairs; Ordered pairs
Paper airplanes, 463
Parallelograms
areas of, 285–290
finding, 286–288
formula for, 284, 285, 286
four square for, 332
base and height of, 285, 286
Parentheses
in multiplication, 11
in order of operations, 10–14
Part
in percent equation, 692
in percent proportion, 687
Patterns, 8, 80, 214, 554, 560
finding
with calculator, 3
in dividing by fractions, 53
Per, meaning of, 162
Percents, 163–168, 679–684
of change, 697–702
comparing and ordering, 175–180
of decrease, 698, 699
definition of, 163, 164
discounts and markups, 703–708
finding, 686, 692–693
finding whole from, 183–185
of increase, 698
models representing, 163, 169, 181–185
of number, finding, 182, 190
Problem-Solving Plan for, 189
simple interest, 709–714
solving problems involving, 181–188
symbol for (%), 164
writing as decimals, 170–172, 680
writing as fractions, 164–165
writing decimals as, 170–171, 680, 716
writing fractions as, 165–166, 681
Percent equations, 691–696, 704, 705
Percent error, 700
Percent models, 685
Percent proportion, 685–690
Perfect numbers, definition of, 20
Perfect squares
definition of, 5
finding factors of, 18
identifying, 5
Performance Task, *In every chapter. For example, see:* 1, 43, 105, 161, 199, 243, 283, 343, 411, 455
Place value chart, 67

Points, in coordinate plane
  finding distances between, 378–380
  reflecting, 371–372
Polygons. *See also specific types*
  areas of
    definition of, 286
    formulas for, 284
  definition of, 285
  drawing, in coordinate plane, 377–382, 399
  types of, 285
Polyhedrons, definition of, 306
Positions of objects, comparing, 363
Positive numbers, 345–350
  absolute value of, 505
  definition of, 346
  as differences, 524
  on number line, 346–348, 352–354, 503
  as products, 549, 550
  as quotients, 555, 556
  as sums, 509, 510, 511, 512
  writing, 346
Positive sign (+), 346
Powers, 3–8
  definition of, 4
  finding values of, 5
  information frame for, 34
  writing expressions as, 4
Practice Test, *In every chapter. For example, see:* 38, 100, 156, 194, 238, 278, 338, 406, 450
Precision, *Throughout. For example, see:* 52, 94, 180, 188, 211, 268, 321, 382, 424, 463
Price
  original, 705
  sale, 704
Prime factor(s)
  using, 21, 27
  writing numbers as product of, 15–20
Prime factorization, 15–20
  definition of, 16
  finding greatest common factor using, 22
  finding least common multiple using, 28–30
  using, 18
  Venn diagrams representing, 21, 27
  writing, 17
Prime numbers
  characteristics of, 2
  definition of, 16

Principal, definition of, 710
Prisms
  definition of, 306
  rectangular (*See* Rectangular prisms)
  surface areas of, 311–318
    finding, 312–315
  triangular
    definition of, 306, 313
    finding surface areas of, 313
Problem solving
  with fractions, using models, 45–47
  before looking at choices, 157
  with percents, 181–188
  with rates, 137–138
  strategies for, 33, 537
  using division, 83
  using tape diagrams, 45
  using unit rate for, 635
Problem Solving, *Throughout. For example, see:* 13, 51, 113, 186, 207, 250, 382, 390, 470
Problem-Solving Plan, *In every chapter. For example, see:* 12, 49, 125, 185, 205, 248, 301, 380, 416, 474
Problem-Solving Strategies, 33
Proportion(s)
  definition of, 640
  percent, 685–690
  ratios forming, 640, 641
  solving
    using mental math for, 648
    using multiplication for, 648
  solving ratio problems with, 647–654
  writing, 649, 650
Proportional relationships
  definition of, 642
  graphing, 655–660
  identifying, 639, 642, 656
Proportionality, constant of, 656
Protractor, 633
Pyramid(s)
  definition of, 306
  square
    definition of, 320
    finding surface areas of, 320
  surface areas of, 319–324
    finding, 319–322
  triangular
    definition of, 320
    drawing, 307
    finding surface areas of, 321
Pyramidions, 322

Q

Quadrants, of coordinate plane, definition of, 370
Quadrilaterals, drawing, in coordinate plane, 378
Quartiles, of data set
  definition of, 434
  first (lower), 434
  third (upper), 434
Questions
  answering easy first, 195, 339
    relaxing after, 239
  reading all choices before answering, 407
  reading before answering, 497
  statistical (*See* Statistical questions)
  using intelligent guessing for, 451
Quotients. *See also* Division
  definition of, 82

R

Range, of data set, 433–438
  definition of, 434
  finding, 434–436
  interquartile (*See* Interquartile range)
Rates, 135–140, 633–638, *See also* Unit rates
  definition of, 136, 634
  equivalent, definition of, 136, 190
  unit, 135–140
    converting measures using, 142–143
    definition of, 136
    finding, 136–138
    solving rate problems using, 137–138, 635
  writing, 633
Ratio(s), 107–114, 627–632
  converting measures using, 141–148
  definition and example chart for, 150
  definition of, 107, 108, 628
  equivalent, 109–111, 629, 630
    definition of, 109, 629
    identifying, 110–111
    identifying proportional relationships using, 640, 641, 642, 643
  solving proportions using, 649
  tables of, 121–128
  using, 135
  graphing, 129–134

    identifying, 627
    interpreting, 109
    phrases indicating, 108
    Problem-Solving Plan for, 149
    scale as, 662
    tape diagrams representing, 115–120
    in unit rates, 135–140
    using, 107
    value of, 109, 628
    writing, 107–111
  Ratio problems 647–654
  Ratio tables, 107, 121–128, 627–632
    completing, 122–125, 627, 629
    converting measures using, 142–143
    creating, 121
    definition of, 122
    equivalent ratios in, 629
    finding percent of number using, 182
    finding whole from percent using, 183
    graphing relationships in, 130–132
  Rational numbers, 357–362, 503–508
    absolute values of, 504
    adding, 517–522
    comparing and ordering, 358–360, 505
    converting between different forms of, 561–566
    definition of, 358, 503
    dividing, 573–578
    factoring out, 610
    graphing on number line, 358–360
    multiplying, 567–572
    on a number line, 503
    opposite of, 358
    subtracting, 529–536
  Reading, *Throughout. For example, see:* 108, 116, 248, 252, 347, 385, 386, 392, 434
  Real World. *See* Modeling Real Life
  Reasoning
    deductive, definition of, 285
    with integers, 353
  Reasonableness, checking for, 570, 688
  Reasoning, *Throughout. For example, see:* 11, 48, 114, 117, 207, 257, 287, 353, 376, 418, 457, 459
  Reciprocals
    definition of, 54, 574
    multiplication by, 574
    solving inequalities using, 394
    writing, 54

  Rectangles
    area of, formula for, 284, 285, 294, 300
    dimensions of, finding, 227
  Rectangular prisms
    definition of, 306, 312
    drawing, 307
    finding missing dimensions of, 327
    finding surface areas of, 312, 315
    identifying parts of, 305
    volumes of, 325–330
      finding, 326–328
      formula for, 326
      Problem-Solving Plan for, 331
  Remainders, in division, 82
  Remember, *Throughout. For example, see:* 4, 46, 130, 176, 260, 267, 286, 320, 393
  Repeated factors, 3–4
  Repeated multiplication, 551
  Repeated Reasoning, *Throughout. For example, see:* 8, 209, 350, 362, 522, 714
  Repeating decimals, 562
  Response to Intervention, *Throughout. For example, see:* T-0B, T-58, T-116, T-160B, T-219, T-273, T-289, T-342B, T-423, T-491
  Review & Refresh, *In every lesson. For example, see:* 7, 50, 112, 167, 206, 249, 289, 349, 417, 461
  Rhombus, areas of, formula for, 287
  Right triangles, areas of, formula for, 294
  Ruler, converting measures using, 141

S

  Sale price, finding, 704
  Same signs
    adding integers with, 509, 511
    adding rational numbers with, 519
    dividing integers with, 555, 556
    dividing rational numbers with, 573, 574, 575
    multiplying integers with, 549, 550, 551
    multiplying rational numbers with, 568
    subtracting integers with, 523, 524, 525
    subtracting rational numbers with, 530, 531, 532
  Scaffolding Instruction, *In every lesson. For example, see:* T-4, T-46, T-108, T-164, T-216, T-246, T-298, T-392, T-420, T-458

  Scale, definition of, 662
  Scale drawings, 661–666
    crating, 661
    definition of, 662
  Scale factor
    definition of, 663
    example and non-example chart for, 668
    finding, 663
  Scale models, definition of, 662
  Self-Assessment for Concepts & Skills, *In every lesson. For example, see:* 5, 48, 117, 165, 204, 247, 287, 347, 415, 459
  Self-Assessment for Problem Solving, *In every lesson. For example, see:* 6, 57, 111, 166, 205, 288, 348, 416, 460
  Shapes of distributions. *See* Distributions, shapes of
  Sieve of Eratosthenes, 2
  Signs. *See* Different signs; Same signs
  Simple interest, 709–714
    definition of, 710
    understanding, 709
  Simplest form, 592
  Simplification
    of algebraic expressions, 221–226, 591, 592, 593
      using Distributive Property, 222–224
      using models, 221
    of fractions, 46
  Skewed, definition of, 471
  Skewed distributions
    in box-and-whisker plots, 486
    choosing appropriate measures with, 478
    definition of, 472
    identifying, 472–473
  Solids, 305–310. *See also specific solids*
    constructing
      using grid paper, 311, 319
      using nets, 319
    definition of, 306
    drawing, 305, 307
    finding numbers of faces, edges, and vertices of, 306–308
    identifying parts of, 305
    nets representing, 312
    surface areas of (*See* Surface areas)
  Solutions
    to equations
      checking, 252–254
      definition of, 252
    to equations in two variables
      definition of, 266
      identifying, 266

to inequalities
    checking, 385–387
    definition of, 385
    graphing, 386–387
    set of, 385
Solution sets, for inequalities, 385
Solving directly, as test-taking strategy, 39
Square, area of, 6
Square, four, 190, 332
Square, magic, 258
Square pyramids
    definition of, 320
    finding surface areas of, 320
Square units
    converting, 287
    definition of, 6, 286
Squares, perfect
    definition of, 5
    finding factors of, 18
    identifying, 5
Statements, interpreting, 433
Statistical questions, 413–416
    definition and example chart for, 446
    definition of, 414
    identifying, 413–414
    using data to answer, 413–416, 429
Statistics. *See also* Data
    definition of, 414
STEAM Video, *In every chapter. For example, see:* 1, 43, 105, 161, 199, 243, 283, 343, 411, 455
Stem, definition of, 458
Stem-and-leaf plots, 457–462
    interpreting, 455, 459–460
    making, 457–460
Structure, *Throughout. For example, see:* 17, 69, 114, 184, 204, 254, 375, 381, 490
Subtraction
    and addition, as inverse operations, 253
    of decimals, 67–72
    of integers, 523–528
    key words and phrases for, 210, 248
    of linear expressions, 597–602
    on number line, 67
    in order of operations, 10–14, 210
    of rational numbers, 529–536
    solving equations using, 251–258
    solving inequalities using, 392
Subtraction Property of Equality, 253–255
Subtraction Property of Inequality, 392

Success Criteria, *In every lesson. For example, see:* 3, 45, 107, 163, 201, 245, 285, 345, 413, 457
Summary triangle, 96, 400, 716
Surface areas
    of cubes, finding, 314
    definition of, 312
    of prisms, 311–318
        finding, 312–315
    of pyramids, 319–324
        finding, 319–322
    units of (square units), 312
Symbols
    closed circle, 386
    equal sign ($=$), 246
    greater than ($>$), 384
    grouping, 10–14
    inequality, 384
    less than ($<$), 384
    multiplication, 4, 11
    negative sign ($-$), 346, 347
    not equal to ($\neq$), 252
    open circle, 386
    percent (%), 164
    positive sign ($+$), 346
Symmetric distribution
    in box-and-whisker plots, 486
    choosing appropriate measures with, 478
    definition of, 472

T

Tables
    equations in two variables in, 267–269
    frequency (*See* Frequency tables)
    ratio (*See* Ratio tables)
Tape diagrams, 115–120
    drawing, 116
    interpreting, 116
    multiplying fractions using, 45
    ratios represented with, 115–120
    solving equations using, 251, 259
    solving inequalities using, 391
Teaching Strategies, *Throughout. For example, see:* T-10, T-23, T-46, T-109, T-285, T-312, T-358, T-370, T-434, T-460
Temperatures
    below zero, 345
    reading and describing, 345, 354
Terminating decimals, 562, 563
Terms, in algebraic expressions
    definition of, 202
    identifying, 202

like
    combining, using Distributive Property, 223
    definition of, 223
Test-taking strategies, 39, 101, 157, 195, 239, 279, 339, 407, 451, 497
Thermometers, 345
Third (upper) quartile, of data set, 434
Three-dimensional figures. *See* Solids
Time, in simple interest problem, 711
Time series graphs, 373
Trapezoids, areas of, 297–304
    finding, 298–301
    formula for, 284, 297, 299, 300
Triangles
    areas of, 291–296
        finding, 292–294
        formula for, 284, 291, 292, 294
    base and height of, 291, 292
        finding missing, 293
    right, 294
    summary, 96, 400
Triangular prisms
    definition of, 306, 313
    finding surface areas of, 313
Triangular pyramids
    definition of, 320
    drawing, 307
    finding surface areas of, 321
Try It, *In every lesson. For example, see:* 4, 46, 110, 170, 202, 266, 321, 352, 414, 458
Two-dimensional figures. *See specific types*

U

Uniform distribution, 472
Unit analysis, definition of, 143
Unit cubes
    definition of, 325
    using, 325
Unit rates, 135–140, 633–638
    converting measures using, 142–143
    definition of, 136, 634
    finding, 136–138, 634, 635
    on graphs, 657
    solving rate problems using, 137–138, 635
Upper quartile, of data set, 434
U.S. customary system, 142–148
    converting measures in, 142–145
    definition of, 142
Using Tools, *Throughout. For example, see:* 7, 48, 167, 186, 289, 295, 302, 323, 361, 396

V

Value of ratios, definition of, 109, 628
Variability, of data sets, in box-and-whisker plots, 484
Variables
  case of letters for, 211
  definition of, 202
  dependent
    definition of, 266
    in equations in two variables, 266–267
  independent
    definition of, 266
    in equations in two variables, 266–267
  in like terms, 592
  in lowercase and uppercase, 594
  one, evaluating expressions with, 203–205
  two, evaluating expressions with, 203
Variation, measures of, 433–438
  choosing appropriate, 477–482
  definition of, 434
  finding, 434–436
Venn diagrams
  common multiples represented with, 27
  creating, 21
  definition of, 21
  prime factorization represented with, 21, 27
Vertex (vertices)
  of polygons, ordered pairs as, 377–380
  of solids
    definition of, 306
    finding number of, 306–308
    identifying, 305

Vertical method
  for adding linear expressions, 598
  for subtracting linear expressions, 599
Vertical number line
  in coordinate plane, 370
  rational numbers on, 358
  temperatures on, 345, 354
Volumes
  converting measures of, 328
  of cubes, finding, 326
  definition of, 325
  of rectangular prisms, 325–330
    finding, 326–328
    formula for, 326
    Problem-Solving Plan for, 331
  units of (cubic units), 325

W

Wages, hourly, 200
Weight, converting measures of, 142, 144
Which One Doesn't Belong?, *Throughout. For example, see:* 5, 56, 131, 165, 204, 261, 268, 307, 379, 473
Whole
  in percent equation, 693
  in percent proportion, 687
Whole numbers
  dividing decimals by, 88
  dividing fractions by, 56
  dividing whole numbers by, 81–86
  integers as, 503
  multiplying decimals by, 74
Words and phrases
  for equations, 246
  for inequalities, 384
  for operations, 210
Word sentences
  writing as equations, 246–248
  writing as inequalities, 384
Working backwards, 279
Writing, *Throughout. For example, see:* 11, 69, 124, 171, 217, 268, 287, 293, 347, 459

X

x-axis, 369–373
  definition of, 267, 370
  reflecting a point in, 371–372
x-coordinates, 370–373
  definition of, 370

Y

y-axis, 369–373
  definition of, 267, 370
  reflecting a point in, 371–372
y-coordinates, 370–373
  definition of, 370
You Be the Teacher, *Throughout. For example, see:* 13, 51, 113, 167, 206, 257, 290, 350, 430, 462

Z

Zero (0)
  Addition Property of, 217
  division by, 556
  inserting, in division of decimals, 90
  Multiplication Property of, 217
  multiplying by, 54
  on number line, 503, 504
  as sums, 509, 510, 511, 512
  temperatures below, 345

Credits

Front matter
viii Ryan McVay/DigitalVision/Getty Images; **ix** Blackzheep/ iStock/Getty Images Plus; **x** janulla/ iStock/Getty Images Plus; **xi** Talaj/iStock/Getty Images Plus; **xiii** scotto72/E+/Getty Images; **xiv** iZonda/ iStock/Getty Images Plus; **xv** Tal Inbar/Wikipedia; **xvi** NASA/Terry Virts; **xvii** damedeeso/iStock/Getty Images Plus; **xviii** EMFA16/iStock/Getty Images Plus; **xix** asiseeit/E+/Getty Images; **xx** eyfoto/iStock/Getty Images Plus; **xxi** goir/iStock/Getty Images Plus; **xxii** clintspencer/iStock/Getty Images Plus

Chapter 1
0 *top* zentilia/Shutterstock.com; *bottom* OnstOn/iStock/Getty Images Plus; **1** Ryan McVay/DigitalVision/Getty Images; **2** INTERFOTO/Alamy Stock Photo; **8** WestLight/iStock/Getty Images Plus; **9** manaemedia/iStock/Getty Images Plus; **14** frentusha/iStock/Getty Images Plus; **15** YuriyZhuravov/Shutterstock.com; **16** farbeffekte/Shutterstock.com; **18** *right* studioaraminta/iStock/Getty Images Plus; *left* carlosalvarez/iStock/Getty Images Plus; **20** chinaface/E+/Getty Images; **24** *top* jmatzick/Shutterstock.com; *bottom* Mike Flippo/Shutterstock.com, LauriPatterson/E+/Getty Images; **25** Glenda M. Powers/Shutterstock.com; **26** pialhovik/iStock/Getty Images Plus, cookelma/iStock/Getty Images Plus, CSA-Plastock/iStock/Getty Images Plus, tbd/E+/Getty Images; **29** monkeybusinessimages/iStock/Getty Images Plus; **30** *right* Constantne/iStock/Getty Images Plus; *left* Mlenny/E+/Getty Images; **32** yuyangc/Shutterstock.com; **33** OnstOn/iStock/Getty Images Plus; **35** Wendy Nero/Shutterstock.com; **36** Ioana Drutu/Hemera/Getty Images; **37** *right* hawk111/iStock/Getty Images Plus, IvonneW/iStock/Getty Images Plus; *left* Zimiri/iStock/Getty Images Plus; **38** AnthonyRosenberg/iStock/Getty Images Plus, VitalisG/iStock/Getty Images Plus

Chapter 2
42 *top* zentilia/Shutterstock.com; *bottom* OnstOn/iStock/Getty Images Plus; **43** Blackzheep/iStock/Getty Images Plus; **44** Firmafotografen/iStock/Getty Images Plus; **50** *left* fivespots/Shutterstock.com; *right* bluehand/Shutterstock.com; **57** *top* monkeybusinessimages/iStock/Getty Images Plus; *bottom* PARKJUNGHO/iStock/Getty Images Plus; **60** *left* ©iStockphoto.com/Michael Plumb; *right* g215/Shutterstock.com; **64** *top* ©iStockphoto.com/bonchan; *bottom* Dmytro Aksonov/E+/Getty Images; **66** *top right* AlexLMX/iStock/Getty Images Plus; *center left* damedeeso/iStock/Getty Images Plus, Kuzmik_A/iStock/Getty Images Plus; *bottom right* RomarioIen/iStock/Getty Images Plus; **69** bmcent1/iStock/Getty Images Plus; **70** Vacclav/Shutterstock.com; **72** SSSCCC/Shutterstock.com; **77** *top* bagi1998/iStock/Getty Images Plus; *bottom* andrejco/iStock/Getty Images Plus; **78** ©iStockphoto.com/suriyasilsaksom; **79** KENCKOphotography/Shutterstock.com; **81** *top* S.Dashkevych/Shutterstock.com; *bottom* auremar/Shutterstock.com; **83** avid_creative/E+/Getty Images; **84** *top* Nikada/iStock/Getty Images Plus; *bottom* jimfeng/iStock/Getty Images Plus; **86** urfinguss/iStock/Getty Images Plus; **93** dolgachov/iStock/Getty Images Plus; **94** loops7/iStock/Getty Images Plus; **95** *top right* Africa Studio/Shutterstock.com; *center left* swissmediavision/iStock/Getty Images Plus; *bottom left* OnstOn/iStock/Getty Images Plus; **97** *top* kali9/E+/Getty Images; *bottom* MikeyGen73/iStock/Getty Images Plus; **98** *right* elmvilla/iStock/Getty Images Plus; *left* 3DSculptor/iStock/Getty Images Plus; **100** Hstarr/iStock/Getty Images Plus

Chapter 3
104 *top* zentilia/Shutterstock.com; *bottom* OnstOn/iStock/Getty Images Plus; **105** janulla/iStock/Getty Images Plus; **107** *top right* fstop123/iStock/Getty Images Plus; *bottom left* Constantinos/Shutterstock.com; *bottom right* Adyna/DigitalVision Vectors/Getty Images; **108** Vladimir Wrangel/Shutterstock.com; **109** *right* MLB Photos/Contributor/Major League Baseball Platinum/Getty Images; *left* rusm/E+/Getty Images, Tazzy1/iStock/Getty Images Plus; **111** *right* anankkml/iStock/Getty Images Plus; *left* MrPants/iStock/Getty Images Plus; **112** *Exercise 14* Lightspring/Shutterstock.com; *Exercise 12* Constantinos/Shutterstock.com; **113** SpiffyJ/iStock/Getty Images Plus; **114** Route55/iStock/Getty Images Plus; **115** 4x6/iStock/Getty Images Plus; **116** stockcam/iStock/Getty Images Plus; **117** Terryfic3D/iStock/Getty Images Plus; **118** *top* OSTILL/iStock/Getty Images Plus; *bottom* wesvandinter/iStock/Getty Images Plus; **119** yulkapopkova/Vetta/Getty Images; **120** *top left* Lepas/Shutterstock.com; *center right* desert_fox99/iStock/Getty Images Plus; *bottom right* Mirko_Rosenau/iStock/Getty Images Plus, GlobalP/iStock/Getty Images Plus, thawats/iStock/Getty Images Plus, juliaart/iStock/Getty Images Plus; **121** 4kodiak/iStock/Getty Images Plus; **124** Petr Malyshev/Shutterstock.com; **125** *top* Mike Flippo/Shutterstock.com; *bottom* busypix/iStock/Getty Images Plus; **127** *right* GeorgeManga/DigitalVision Vectors/Getty Images; *left* Jacob Wackerhausen/E+/Getty Images; **128** Antagain/iStock/Getty Images Plus; cinoby/iStock/Getty Images Plus, GlobalP/iStock/Getty Images Plus; **129** baibaz/iStock/Getty Images Plus; **132** *top left* ElementalImaging/E+/Getty Images; *bottom right* skodonnell/E+/Getty Images; *bottom left* exopixel/iStock/Getty Images Plus, Givaga/iStock/Getty Images Plus; **134** R. Gino Santa Maria/Shutterstock.com; **135** *top* irmetov/DigitalVision Vectors/Getty Images; *bottom* vita khorzhevska/Shutterstock.com; **137** REUTERS/James Stirton/Handout (AUSTRALIA); **138** *top* Watcha/iStock/Getty Images Plus; *bottom* ©iStockphoto.com/Gord Horne; **140** *top* mvaligursky/GlobalP/iStock/Getty Images Plus; *bottom* FatCamera/iStock/Getty Images Plus; **145** *top* Tom Merton/OJO Images/Getty Images; *bottom* Doug James/Shutterstock.com; **147** Derek Wong/Mackinac Bridge at Sunset in 2008/CC-BY-3.0; **148** *left* DenisTangneyJr/iStock/Getty Images Plus; *right* ©iStockphoto.com/Paul Tessier; **149** *right* Jean Thompson; *center* Barcin/iStock/Getty Images Plus; *bottom* OnstOn/iStock/Getty Images Plus; **151** scanrail/iStock/Getty Images Plus; **152** Marcelo Horn/E+/Getty Images; **153** kentarus/E+/Getty Images; **154** ilbusca/E+/Getty Images; **155** bowdenimages/iStock/Getty Images Plus; **156** ©iStockphoto.com/Ermin Gutenberger; **158** Val_Iva/iStock/Getty Images Plus, blueringmedia/iStock/Getty Images Plus

Chapter 4
160 *top* zentilia/Shutterstock.com; *bottom* OnstOn/iStock/Getty Images Plus; **161** Talaj/iStock/Getty Images Plus; **162** frank600/iStock/Getty Images Plus; **166** Sauliakas/iStock/Getty Images Plus; **168** 4x6/iStock/Getty Images Plus; **172** vvvita/iStock/Getty Images Plus; **174** andyKRAKOVSKI/iStock/Getty Images Plus; **177** 4x6/iStock/Getty Images Plus; **180** ©iStockphoto.com/Eric Isselée; **185** offstocker/iStock/Getty Images Plus; **186** JackF/iStock/Getty Images Plus; **187** *left* LoopAll/iStock/Getty Images Plus; *right* m-gucci/iStock/Getty Images Plus; **188** EMPPhotography/E+/Getty Images; **189** *top* GeorgePeters/E+/Getty Images; *center* Coprid/iStock/Getty Images Plus; *bottom* OnstOn/iStock/Getty Images Plus; **192** pomarinus/E+/Getty Images; **193** BlackJack3D/iStock/Getty Images Plus; **194** fresher/Shutterstock.com

Chapter 5

198 *top* zentilia/Shutterstock.com; *bottom* OnstOn/iStock/Getty Images Plus; **201** *top right* Rawpixel/iStock/Getty Images Plus, Floortje/iStock/Getty Images Plus, esseffe/E+/Getty Images, kreinick/iStock/Getty Images Plus; *center left* pittawut/Shutterstock.com; *center right* Goran Bogicevic/Shutterstock.com; *bottom left* GlobalP/iStock/Getty Images Plus; **205** rgmeier/iStock/Getty Images Plus; **208** Mikalai_Manyshau/iStock/Getty Images Plus, cyano66/iStock/Getty Images Plus, fergregory/iStock/Getty Images Plus, MATJAZ SLANIC/E+/Getty Images, SmallArtFish/iStock/Getty Images Plus, hceliktas/iStock/Getty Images Plus; **212** *top* Kateryna Larina/Shutterstock.com; *bottom* DNY59/iStock/Getty Images Plus; **218** GibsonPictures/E+/Getty Images; **220** IftodeIulian/DigitalVision Vectors/Getty Images; **226** Kandfoto/iStock/Getty Images Plus; **230** *top* Inhabitant/Shutterstock.com; *bottom* Stockbyte/Stockbyte/Getty Images; **233** *top* mediaphotos/iStock/Getty Images Plus; *bottom* OnstOn/iStock/Getty Images Plus; **235** JohnnyGreig/iStock/Getty Images Plus; **237** *right* sidewaysdesign/iStock/Getty Images Plus; *left* liveslow/iStock/Getty Images Plus; **238** *right* Aptyp_koK/Shutterstock.com; *left* Anges van der Logt/Shutterstock.com

Chapter 6

242 *top* zentilia/Shutterstock.com; *bottom* OnstOn/iStock/Getty Images Plus; **243** scotto72/E+/Getty Images; **248** *top* Albert Russ/Shutterstock.com; *bottom* iwka/iStock/Getty Images Plus; **250** *top right* waymoreawesomer/iStock/Getty Images Plus, Zheka-Boss/iStock/Getty Images Plus, clovercity/iStock/Getty Images Plus; *left* ©iStockphoto.com/Kenneth C. Zirkel; **255** *top* alex-mit/iStock/Getty Images Plus; *bottom* ©iStockphoto.com/Jeremy Wee, ©iStockphoto.com/Jan Will; **257** TerryKelly/iStock/Getty Images Plus; **258** Raywoo/Shutterstock.com; **259** ©iStockphoto.com/Vladyslav Otsiatsia; **262** fototrav/iStock Unreleased/Getty Images Plus; **264** *top* monkeybusinessimages/iStock/Getty Images Plus; *bottom* ©iStockphoto.com/Eric Isselée; **266** SerrNovik/iStock/Getty Images Plus; **269** *top* macrovector/iStock/Getty Images Plus; *bottom* bnoragitt/iStock Unreleased/Getty Images Plus; **271** angelinast/iStock/Getty Images Plus; **272** *Exercise 57* d1sk/iStock/Getty Images Plus; *Exercise 59* irmetov/DigitalVision Vectors/Getty Images; **273** *top* imagedepotpro/Vetta/Getty Images; *bottom* OnstOn/iStock/Getty Images Plus; **275** Dmytro Aksonov/E+/Getty Images; **276** *top* Wesley Tolhurst/Shutterstock.com; *bottom right* Vadim Sadovski/Shutterstock.com; **277** *top left* Siraphol/iStock/Getty Images Plus; *center right* johnkellerman/iStock/Getty Images Plus; *bottom left* Yobro10/iStock/Getty Images Plus; **278** *top* ntzolov/E+/Getty Images; *bottom* ©iStockphoto.com/XiXinXing;

Chapter 7

282 *top* zentilia/Shutterstock.com; *bottom* OnstOn/iStock/Getty Images Plus; **283** *left* iZonda/iStock/Getty Images Plus; *right* Coprid/iStock/Getty Images Plus; **288** Roy Jankowski/Westend61/Getty Images; **296** Terrance Emerson/Shutterstock.com; **301** jsp/iStock/Getty Images Plus; **304** Kharidehal Abhirama Ashwin /Shutterstock.com; **308** *left* KanKankavee/iStock/Getty Images Plus; *right* gionnixxx/iStock/Getty Images Plus; **309** ©iStockphoto.com/Hedda Gjerpen; **310** *Exercise 23* ©iStockphoto.com/Rich Koele; *Exercise 27* design56/Shutterstock.com; *center* ©iStockphoto.com/rzdeb; **315** Diane Schuster/Fermi Gamma-ray Space Telescope/NASA; **317** Ivo Petkov/iStock/Getty Images Plus; **322** *top* DEA / A. DAGLI ORTI / Contributor; *bottom* Patryk Kosmider/Shutterstock.com; **324** *top right* Tupungato/Shutterstock.com; *center left* scanrail/iStock/Getty Images Plus; **328** ©iStockphoto.com/William Britten; **330** *Exercise 19 left* ©iStockphoto.com/Jill Chen; *Exercise 19 right* ©iStockphoto.com/LongHa2006; **331** OnstOn/iStock/Getty Images Plus; **333** *right* hrstkinkr/iStock/Getty Images Plus; *left* Beeldbewerking/iStock/Getty Images Plus; **334** ncognet0/iStock/Getty Images Plus; **337** Niki Crucillo/Shutterstock.com; **338** *top* ©iStockphoto.com/AlexMax; *center* PeterG/Shutterstock.com; *bottom* U.S. Geological Survey

Chapter 8

342 *top* zentilia/Shutterstock.com; *bottom* OnstOn/iStock/Getty Images Plus; **343** Tal Inbar/Wikipedia; **348** *top* ©iStockphoto.com/Egor Mopanko; *bottom* valdum/iStock/Getty Images Plus; **349** jennyt/Shuttertock.com; **350** catolla/iStock/Getty Images Plus; **351** *top* NASA/Kim Shiflett; *bottom* NASA; **356** Traveladventure/E+/Getty Images; **357** *a.* Astronaut Stephen S. Oswald/NASA; *b.* JSC/NASA; *c.* NASA; *d.* NASA; *e.* Astronaut Eileen M. Collins/NASA; *f.* NASA; **360** gregepperson/iStock/Getty Images Plus; **362** *top* ©iStockphoto.com/jclegg, ©iStockphoto.com/spxChrome, ©iStockphoto.com/Larua Eisenberg; *bottom* adventtr/E+/Getty Images; **366** *top* Shane W Thompson/Shutterstock.com; *bottom* wundervisuals/E+/Getty Images; **369** ©iStockphoto.com/ingmar wesemann; **373** IgorKirillov/iStock/Getty Images Plus; **374** *Exercise 13* rmnunes/iStock/Getty Images Plus; *Exercise 14* BNMK0819/iStock/Getty Images Plus; *Exercise 15* DieterMeyrl/E+/Getty Images; **380** CampPhoto/iStock/Getty Images Plus; **382** vovan13/ iStock/Getty Images Plus; **383** *top* ARSELA/E+/Getty Images; *center* Volosina/ iStock/Getty Images Plus; *bottom* viach80/ iStock/Getty Images Plus; **387** *top left* NASA/Johns Hopkins University Applied Physics Laboratory; *bottom left* cokacoka/iStock/Getty Images Plus; *bottom right* coddy/iStock/Getty Images Plus; **389** ©iStockphoto.com/George Peters; **390** Andrija1/iStock/Getty Images Plus; **395** mladn61/iStock/Getty Images Plus; **397** *top* dodo4466/DigitalVision Vectors/Getty Images; *center* JonathanLesage/iStock/Getty Images Plus; *bottom* Karin Hildebrand Lau/Shutterstock.com; **398** pialhovik/iStock/Getty Images Plus; **399** *top* ElementalImaging/iStock/Getty Images Plus; *bottom* OnstOn/iStock/Getty Images Plus; **404** kastanka/iStock/Getty Images Plus; **406** Harvepino/iStock/Getty Images Plus

Chapter 9

410 *top* zentilia/Shutterstock.com; *bottom* OnstOn/iStock/Getty Images Plus; **411** NASA/Terry Virts; **412** Jupiterimages/PHOTOS.com/Getty Images Plus; **413** *top* LeventeGyori/Shutterstock.com; *bottom* Farinosa/iStock/Getty Images Plus; **414** AnastasiaRasstrigina/iStock/Getty Images Plus; **416** Eric Isselée/iStock/Getty Images Plus; **418** bazilfoto/iStock/Getty Images Plus; **419** Rob Byron/Shutterstock.com; **420** *top* bowdenimages/iStock/Getty Images Plus; *bottom* ZargonDesign/iStock Unreleased/Getty Images Plus; **422** ©iStockphoto.com/Eric Isselée; **424** *top* OSTILL/iStock/Getty Images Plus; *bottom* Dash_med/iStock/Getty Images Plus; **425** Hein Nouwens/Shutterstock.com; **427** *right* andresr/E+/Getty Images; *left* only_fabrizio/iStock/Getty Images Plus; **429** USO/iStock/Getty Images Plus; **433** ChrisBoswell/iStock/Getty Images; **435** Charlie Hutton/Shutterstock.com; **438** *top* lvcandy/DigitalVision Vectors/Getty Images; *bottom* ©iStockphoto.com/Jason Lugo; **441** *left* Ganko/Shutterstock.com; *right* Mark Herreid/Shutterstock.com; **442** Ganko/Shutterstock.com; **444** *top* nikolay100//iStock Unreleased/Getty Images Plus; *center* tab62/Shutterstock.com; **445** *top* valentinrussanov/E+/Getty Images; *bottom* OnstOn/iStock/Getty Images Plus; **450** GlobalP/iStock/Getty Images Plus; **451** Six Dun/iStock/Getty Images

Chapter 10

454 *top* zentilia/Shutterstock.com; *bottom* OnstOn/iStock/Getty Images Plus; **455** damedeeso/iStock/Getty Images Plus; **457** Elzbieta Szpak/Shutterstock.com; **460** ©iStockphoto.com/Pekka Nikonen; **462** Ralwel/iStock/Getty Images Plus; **464** stockshoppe/Shutterstock.com; **465** *top* ©iStockphoto.com/susaro; *bottom* skynesher/E+/Getty Images; **467** Tomasz Trojanowski/Shutterstock.com; **470** KenCanning/iStock/Getty Images Plus; **471** fresher/Shutterstock.com, **474** mmaxer/Shutterstock.com; **476** ranplett/E+/Getty Images; **477** rasslava/iStock/Getty Images Plus; **478** Pingebat/iStock/Getty Images Plus; **480** Allevinatis/iStock/Getty Images Plus; **482** *top left* richcarey/iStock/Getty Images Plus; *center right* Boris Ryaposov/Shutterstock.com; *center left* aphrodite74/E+/Getty Images; **484** JackF/iStock/Getty Images Plus; **485** Sebastian Knight/Shutterstock.com; **487** mountainpix/Shutterstock.com; **489** *top* njpPhoto/iStock Unreleased/Getty Images Plus; *center* Ron_Thomas/E+/Getty Images; *bottom* GlobalP/iStock/Getty Images Plus; **490** *top* zhuda/Shutterstock.com; *bottom* LawrenceSawyer/E+/Getty Images; **491** *top* Rob Marmion/Shutterstock.com; *bottom* OnstOn/iStock/Getty Images Plus; **496** shapecharge/E+/Getty Images

Chapter A

500 *top* zentilia/Shutterstock.com; *bottom* OnstOn/iStock/Getty Images Plus;
501 EMFA16/iStock/Getty Images Plus; **503** Cisco Freeze/Shutterstock.com;
506 aldomurillo/E+/Getty Images; **507** JackF/iStock/Getty Images Plus;
508 theartist312/iStock/Getty Images Plus; **513** ekolara/iStock/Getty Images Plus; **520** ultrapro/Shutterstock.com; **522** *right* anna1311/iStock/Getty Images Plus; *left* Heide Hellebrand/Shutterstock.com; **526** HPphoto/iStock/Getty Images Plus; **535** *left* PhotoAlto/Alamy Stock Photo; *right* ©iStockphoto.com/RonTech2000; **537** *right* kbeis/DigitalVision Vectors/Getty Images; *left* OnstOn/iStock/Getty Images Plus; **541** GlobalP/iStock/Getty Images; **542** pat138241/iStock/Getty Images Plus

Chapter B

546 *top* zentilia/Shutterstock.com; *bottom* OnstOn/iStock/Getty Images Plus;
547 *left* asiseeit/E+/Getty Images; *right* AlonzoDesign/DigitalVision Vectors/Getty Images; **552** *top* ET-ARTWORKS/iStock/Getty Images Plus, Artulina1/iStock/Getty Images Plus; *bottom* sorbetto/DigitalVision Vectors/Getty Images; **553** HEPALMER/Vetta/Getty Images; **554** inthevisual/iStock/Getty Images Plus; **558** CLFProductions/Shutterstock.com; **560** *top* ahmetemre/iStock/Getty Images Plus; *bottom* LOVE_LIFE/E+/Getty Images;
561 aldomurillo/E+/Getty Images; **562** ET-ARTWORKS/iStock/Getty Images Plus; **564** *left* Tsuji/iStock/Getty Images Plus; *right* joecicak/E+/Getty Images;
565 TokenPhoto/E+/Getty Images; **566** GracedByTheLight/iStock/Getty Images Plus; **570** *top* Aiden-Franklin/Vetta/Getty Images; *bottom* OSTILL/iStock/Getty Images Plus; **576** Allies Interactive/Shutterstock.com, olegtoka/iStock/Getty Images Plus; **579** *top* ultrapro/iStock/Getty Images Plus; *bottom* OnstOn/iStock/Getty Images Plus; **582** Liem Bahneman/Shutterstock.com;
583 Liem Bahneman/Shutterstock.com; **584** *right* ©iStockphoto.com/susaro; *left* William Silver/Shutterstock.com; **586** Laborant/Shutterstock.com

Chapter C

588 *top* zentilia/Shutterstock.com; *bottom* OnstOn/iStock/Getty Images Plus; **589** *top left* eyfoto/iStock/Getty Images Plus; *center* Floortje/iStock/Getty Images Plus, Creativeye99/iStock/Getty Images Plus, Azure-Dragon/iStock/Getty Images Plus; **594** Horst Petzold/Shutterstock.com;
596 *Exercise 25 left* photo25th/Shutterstock.com; *Exercise 25 center* ©iStockphoto.com/Don Nichols; *Exercise 29* solarseven/iStock/Getty Images Plus; **600** OSTILL/iStock/Getty Images Plus; **601** Suzanne Tucker/Shutterstock.com; **602** GeorgeManga/DigitalVision Vectors/Getty Images;
608 *top* LUIGIALESI/iStock/Getty Images Plus; *bottom* macrovector/iStock/Getty Images Plus, PirinaI/iStock/Getty Images Plus;
612 *top* Andrew_Rybalko/iStock/Getty Images Plus; *bottom* Witthaya/iStock/Getty Images Plus; **614** *right* 4x6/iStock/Getty Images Plus; *left* Image provided courtesy of 3Doodler; **615** OnstOn/iStock/Getty Images Plus;
618 3quarks/iStock/Getty Images Plus; **619** *right* vicmicallef/iStock/Getty Images Plus; *left* artisteer/iStock/Getty Images Plus

Chapter D

624 *top* zentilia/Shutterstock.com; *bottom* OnstOn/iStock/Getty Images Plus;
625 goir/iStock/Getty Images Plus; **626** *left* Bombaert/iStock/Getty Images Plus; *right* Nerthuz/iStock/Getty Images Plus; **627** GaryAlvis/E+/Getty Images;
628 natalie-claude/iStock/Getty Images Plus; **629** mashuk/iStock/Getty Images Plus; **630** *top* Grafner/iStock/Getty Images Plus; *bottom* alxpin/E+/Getty Images; **631** Yasonya/iStock/Getty Images Plus, HandmadePictures/iStock/Getty Images Plus; **632** *left* ZUMA Press, Inc./Alamy Stock Photo; *right* Sergey Peterman/Shutterstock.com; **636** *top* vm/E+/Getty Images; *bottom* Neil Lockhart/Shutterstock.com; **638** donstock/iStock/Getty Images Plus;
639 andrearoad/iStock Unreleased/Getty Images Plus; **643** pepifoto/iStock/Getty Images Plus; **645** 4x6/iStock/Getty Images Plus; **646** *top* ssucsy/E+/Getty Images; *center* TokenPhoto/E+/Getty Images; *bottom* Ninell_Art/iStock/Getty Images Plus; **647** Mr. Klein /Shutterstock.com; **651** *top* ©AMNH/D. Finnin; *bottom* AlexeyFyodorov/iStock/Getty Images Plus; **653** NASA/Carla Thomas;
655 Creativ/iStock/Getty Images Plus; **658** coffeekai/iStock/Getty Images Plus;
663 *top* Andrew B Hall/Shutterstock.com; *bottom* labrlo/iStock/Getty Images Plus; **664** *top* Serg64/Shutterstock.com; *bottom* Emmanouil Filippou/E+/Getty Images; **667** *top* Daniel_M/iStock/Getty Images Plus; *bottom* OnstOn/iStock/Getty Images Plus; **670** catolla/iStock/Getty Images Plus

Chapter E

676 *top* zentilia/Shutterstock.com; *bottom* OnstOn/iStock/Getty Images Plus;
677 clintspencer/iStock/Getty Images Plus; **679** jangeltun/iStock/Getty Images Plus; **682** *top* IvonneW/iStock/Getty Images Plus; *bottom* NASA; **684** *left* McIninch/iStock/getty Images Plus; *right* Avesun/iStock/Getty Images Plus;
688 Courtesy Kristy Doyle; **689** President's Council on Fitness, Sports & Nutrition; **690** OSTILL/iStock/Getty Images Plus; **694** FatCamera/E+/Getty Images; **696** Photo Melon/Shutterstock.com; **698** aldomurillo/iStock/Getty Images Plus; **700** *top* Kseniia_Designer/iStock/Getty Images Plus, Creativ/iStock/Getty Images Plus; *bottom* leonello/iStock/Getty Images Plus;
701 bazilfoto/iStock/Getty Images Plus; **702** tbradford/iStock/Getty Images Plus; **704** final09/iStock/Getty Images Plus, -M-I-S-H-A-/iStock/Getty Images Plus; **705** *top* venakr/iStock/Getty Images Plus; *bottom* Fruit_Cocktail/iStock/Getty Images Plus; **706** tihomir_todorov/iStock/Getty Images Plus;
707 ©iStockphoto.com/Albert Smirnov; **708** ©iStockphoto.com/Lori Sparkia;
712 ©iStockphoto.com/anne de Haas; **714** *right* Creativ/iStock/Getty Images Plus; *left* Nerthuz/iStock/Getty Images Plus; **715** *top* kostsov/iStock/Getty Images Plus; *bottom* OnstOn/iStock/Getty Images Plus; **717** janrysavy/E+/Getty Images; **718** *top* ©iStockphoto.com/Susan Chiang;
bottom ©iStockphoto.com/ted johns; **719** Creativ/iStock/Getty Images Plus;
720 posteriori/iStock/Getty Images Plus

Cartoon illustrations: Tyler Stout
Design Elements: ©iStockphoto.com/Gizmo; Songquan Deng/Shutterstock.com; Juksy/iStock/Getty Images Plus

Mathematics Reference Sheet

Conversions

U.S. Customary
1 foot = 12 inches
1 yard = 3 feet
1 mile = 5280 feet
1 acre = 43,560 square feet
1 cup = 8 fluid ounces
1 pint = 2 cups
1 quart = 2 pints
1 gallon = 4 quarts
1 gallon = 231 cubic inches
1 pound = 16 ounces
1 ton = 2000 pounds
1 cubic foot ≈ 7.5 gallons

U.S. Customary to Metric
1 inch = 2.54 centimeters
1 foot ≈ 0.3 meter
1 mile ≈ 1.61 kilometers
1 quart ≈ 0.95 liter
1 gallon ≈ 3.79 liters
1 cup ≈ 237 milliliters
1 pound ≈ 0.45 kilogram
1 ounce ≈ 28.3 grams
1 gallon ≈ 3785 cubic centimeters

Time
1 minute = 60 seconds
1 hour = 60 minutes
1 hour = 3600 seconds
1 year = 52 weeks

Temperature
$C = \dfrac{5}{9}(F - 32)$

$F = \dfrac{9}{5}C + 32$

Metric
1 centimeter = 10 millimeters
1 meter = 100 centimeters
1 kilometer = 1000 meters
1 liter = 1000 milliliters
1 kiloliter = 1000 liters
1 milliliter = 1 cubic centimeter
1 liter = 1000 cubic centimeters
1 cubic millimeter = 0.001 milliliter
1 gram = 1000 milligrams
1 kilogram = 1000 grams

Metric to U.S. Customary
1 centimeter ≈ 0.39 inch
1 meter ≈ 3.28 feet
1 kilometer ≈ 0.62 mile
1 liter ≈ 1.06 quarts
1 liter ≈ 0.26 gallon
1 kilogram ≈ 2.2 pounds
1 gram ≈ 0.035 ounce
1 cubic meter ≈ 264 gallons

Number Properties

Commutative Properties of Addition and Multiplication
$a + b = b + a$
$a \cdot b = b \cdot a$

Associative Properties of Addition and Multiplication
$(a + b) + c = a + (b + c)$
$(a \cdot b) \cdot c = a \cdot (b \cdot c)$

Addition Property of Zero
$a + 0 = a$

Multiplication Properties of Zero and One
$a \cdot 0 = 0$
$a \cdot 1 = a$

Multiplicative Inverse Property
$n \cdot \dfrac{1}{n} = \dfrac{1}{n} \cdot n = 1, n \neq 0$

Distributive Property:
$a(b + c) = ab + ac$
$a(b - c) = ab - ac$

Properties of Equality

Addition Property of Equality
If $a = b$, then $a + c = b + c$.

Subtraction Property of Equality
If $a = b$, then $a - c = b - c$.

Multiplication Property of Equality
If $a = b$, then $a \cdot c = b \cdot c$.

Division Property of Equality
If $a = b$, then $a \div c = b \div c$, $c \neq 0$.

Properties of Inequality

Addition Property of Inequality
If $a > b$, then $a + c > b + c$.

Subtraction Property of Inequality
If $a > b$, then $a - c > b - c$.

Multiplication Property of Inequality
If $a > b$ and c is positive, then $a \cdot c > b \cdot c$.

Division Property of Inequality
If $a > b$ and c is positive, then $a \div c > b \div c$.

Perimeter and Area

Square	Rectangle	Parallelogram	Triangle	Trapezoid
$P = 4s$ $A = s^2$	$P = 2\ell + 2w$ $A = \ell w$	$A = bh$	$A = \dfrac{1}{2}bh$	$A = \dfrac{1}{2}h(b_1 + b_2)$

Surface Area

Prism

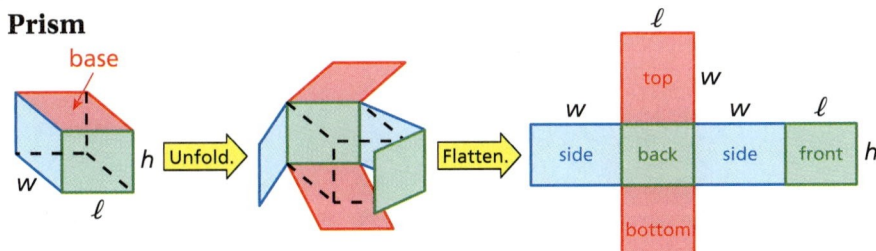

$S =$ areas of bases $+$ areas of lateral faces

Pyramid

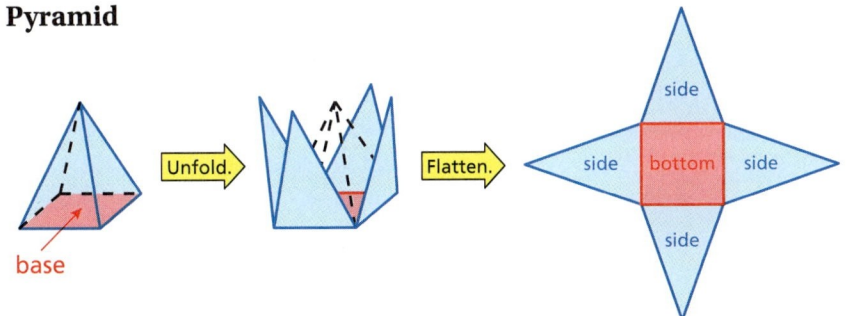

$S =$ area of base $+$ areas of lateral faces

Volume of a Rectangular Prism

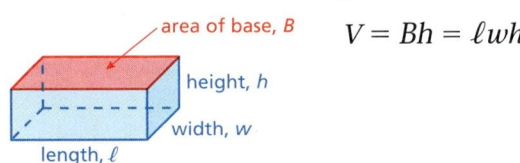

$V = Bh = \ell w h$

Simple Interest

Simple interest formula
$I = Prt$

The Coordinate Plane

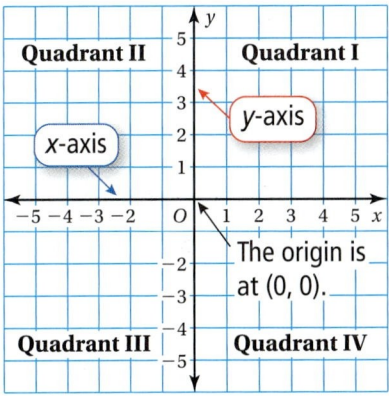

The origin is at (0, 0).